AA004038

70th Highway Geology Symposium (HGS 2019)

Better Highways Through Applied Geology

Portland, Oregon, USA
21-24 October 2019

ISBN: 978-1-7138-0875-6

Printed from e-media with permission by:

Curran Associates, Inc.
57 Morehouse Lane
Red Hook, NY 12571

Some format issues inherent in the e-media version may also appear in this print version.

Copyright© (2019) by Highway Geology Symposium (HGS)
All rights reserved.

Printed by Curran Associates, Inc. (2020)

For permission requests, please contact Highway Geology Symposium (HGS) at the address below.

Highway Geology Symposium (HGS)
C/O John Pilipchuk
1589 Mail Service Center
Raleigh, NC, 27699-1589

Phone: (919) 707-6851
Fax: (919) 250-4237

jpilipchuk@ncdot.gov

Additional copies of this publication are available from:

Curran Associates, Inc.
57 Morehouse Lane
Red Hook, NY 12571 USA
Phone: 845-758-0400
Fax: 845-758-2633
Email: curran@proceedings.com
Web: www.proceedings.com

70TH HIGHWAY GEOLOGY SYMPOSIUM

"Better Highways Through Applied Geology"

October 21-24, 2019

Local Organizing Committee Members

Tim Shevlin
Darren Beckstrand
Krystle Pelham
John Pilipchuk
Jamie Schick
Doug Anderson
Scott Burns

Field Trip Planning Committee

Doug Anderson
Stephen Hay
Evan Garich
Marc Fish

Technical Review Committee

John Pilipchuk
Jeff Dean
Peter Ingraham
Chris Ruppen
Richard Lane
Krystle Pelham

Grateful Acknowledgments

We would like to thank the following people who helped make made this Symposium possible.

Simon Boone
Curren Mohney
Scott Burns
Field Trip Speakers
ODOT Maintenance Districts: Troutdale and The Dalles
Oregon State Parks and Recreation
HGS Sponsors and Exhibitors
HGS Steering Committee
The Beering Committee
Moderator Volunteers
Student Volunteers
Delaney Meeting and Event Management
Port of Cascade Locks

On the Cover: Downtown Portland with Mt. Hood on the horizon.

Left: The White Stag sign, also known as the "Portland Oregon" sign, is a lighted neon-and-incandescent-bulb sign located atop the White Stag Building, at 70 NW Couch Street (Wikipedia, 2019)

70th ANNUAL HIGHWAY GEOLOGY SYMPOSIUM
Table of Contents

70TH HIGHWAY GEOLOGY SYMPOSIUM

"Better Highways Through Applied Geology"

October 21-24, 2019

Portland Marriott Downtown Waterfront Hotel
Portland, Oregon

2019 Proceedings

BETTER HIGHWAYS THROUGH APPLIED GEOLOGY
2019 HIGHWAY GEOLOGY SYMPOSIUM
Oregon highways connecting coasts, mountains, and deserts
70th HGS - PORTLAND, OREGON

Hosted By: The Oregon Department of Transportation

70th ANNUAL HIGHWAY GEOLOGY SYMPOSIUM
Hotel Floor Plan

MAIN LOBBY

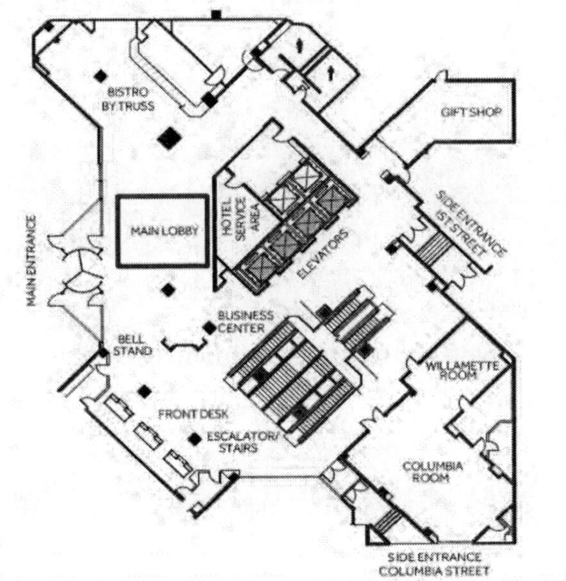

LOWER LEVEL 1

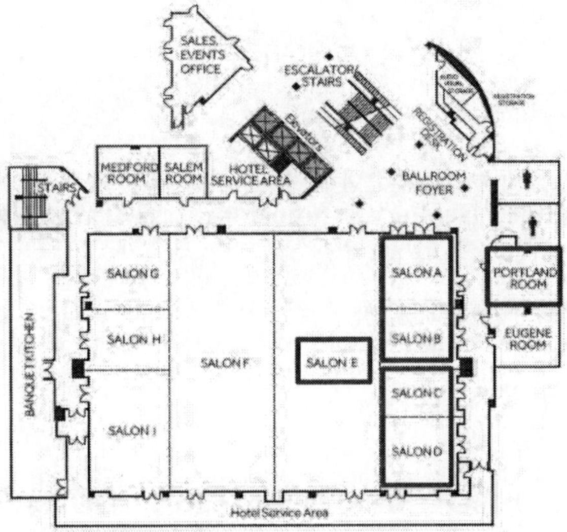

2ND FLOOR

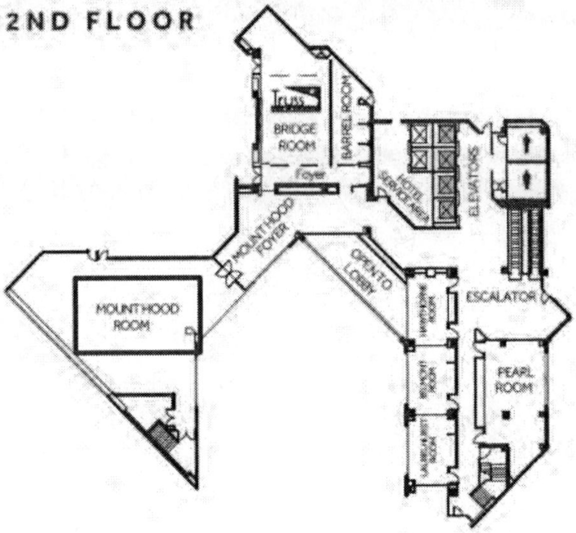

3RD FLOOR

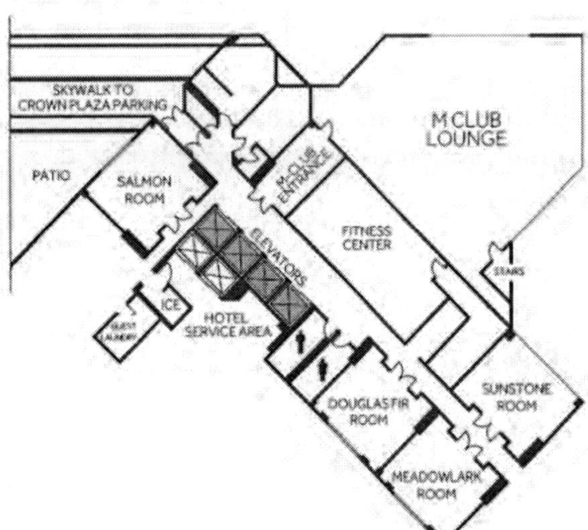

PORTLAND MARRIOTT DOWNTOWN WATERFRONT • 1401 SW Naito Parkway • Portland, OR 97201 • Phone 503.226.7600, Sales Fax 503.226.1209 • portlandmarriott.com

At a Glance Schedule of Events

70th Highway Geology Symposium
Portland, Oregon
October 21-24, 2019

Monday, October 21

11:00 AM – 5:00 PM
Highway Geology Symposium Registration Open
Location : Portland Room

1:00 PM – 5:00 PM
Transportation Research Board Technical Session: "Looking Ahead at Informed Decision-Making for Engineering Geologists"
Location: Salon C - D

5:00 PM – 6:15 PM
HGS National Steering Committee Meeting
Location: Salon A - B

6:00 PM – 8:30 PM
Highway Geology Symposium Exhibitor Area Open
Location: Salon E

6:30 PM – 8:30 PM

Ice Breaker Social - Sponsored by Landslide Technology
Location: Salon E (Exhibitor Area)

Tuesday, October 22

6:30 AM – 5:00 PM
Highway Geology Symposium Registration Open
Location : Portland Room

6:30 AM – 8:30 AM
Continental Breakfast
Location: Salon E (Exhibitor Area)

8:00 AM – 8:30 AM
Welcome and Opening Remarks
Tim Shevlin, HGS Organizing Committee Chair
Keynote Speaker
Curran Mohney, C.E.G., Engineering Geology Program Leader, Oregon DOT
Location: Salon A - D

Tuesday, October 22 (continued)

8:00 AM – 5:00 PM
Highway Geology Symposium Exhibitor Area Open
Location: Salon E

8:30 AM – 2:00 PM
HGS Guest Field Trip "Tour of Portland Landmarks"
Guest Field trip Lunch - Sponsored by Apex Rockfall Mitigation LLC
Pick-up Location: Hotel Front Lobby

Technical Session 1: Young Author Presentations

Location: Salon A - D
Chris Ruppen, GeoStabilization International, Moderator

8:30 AM – 8:50 AM
Young Author Presentation: Geographic Information Systems (GIS) and Landslide Stabilization
Author: Andrew Ferguson

8:50 AM – 9:10 AM
Young Author Presentation: Geohazard Mitigation Alternative Foundation Backfill for Emergency Construction
Author: Joseph McElhany

9:10 AM – 9:30 AM
Young Author Presentation: Rock Mechanics and its Effects on Spillway Modification Design
Authors: Coralie Wilhite, K. Pattermann, J. Hilmar, and Vanessa Bateman

9:30 AM – 9:50 AM
Young Author Presentation: Case History and Remediation of a Troublesome Rock Cut in Georgia, Vermont
Authors: Ethan Thomas, P.C. Ingraham, J.R. Smerekanicz, and T.D. Eliassen

9:50 AM – 10:20 AM
Morning Coffee Break
Location: Salon E

Technical Session 2 : Young Author Presentations

Location: Salon A - D
Krystle Pelham, NHDOT, Moderator

10:20 AM – 10:40 AM
Young Author Presentation: The Assessment and Remediation of Wabasha St. Rock Fall
Authors: Anya Brose, Lee Peterson, Ryan Peterson

Tuesday, October 22 (continued)

10:40 AM – 11:00 AM
Young Author Presentation: Glaciolacustrine Earthflow Slides on US95
Author: Collin McCormick

11:00 AM – 11:20 AM
Young Author Presentation: Flexible Ring Nets as a Solution for Debris Flow Protection in Environmentally Sensitive Areas
Authors: Saleh Feidi, Mallory Jones, Bill Kane

11:20 AM – 11:40 AM
Young Author Presentation: Verification of Tabulated Design Grout-Ground Bond Strength
Author: Brian J. Forsthoff

11:40 AM – 12:00 PM
Young Author Presentation: Replacing Deteriorating Retaining Walls Along the Million Dollar Highway
Authors: Brett Arpin and Todd Schlittenhart

12:00 PM – 1:30 PM
Lunch - Sponsored by Jensen Drilling
Location: Mount Hood Room

Technical Session 3 : Young Author and Gorge Presentations

Location: Salon A - D
Stephen Hay, Oregon DOT, Moderator

1:30 PM – 1:50 PM
Young Author Presentation: Implementation and Application of Geotechnical Asset Management in Colorado
Author: Nicole Oester

1:50 PM – 2:10 PM
Young Author Presentation: Preliminary Rockfall Evaluation for the Historic Columbia River Highway State Trail, Segment E
Author: Noah Kimmes

2:10PM – 2:30 PM
Oneonta Tunnel – Restoration, Fire, Restoration
Author: George Freitag

2:30PM – 3:00 PM
Field Trip Preview & Lessons learned about slope stability and erosion after the forest fire in the Columbia Gorge, Oregon of 1991: implications for the fire of 2017 & Field Trip Preview
Dr. Scott Burns

Tuesday, October 22 (continued)

3:00 PM – 3:30 PM
Afternoon Break
Location: Salon E (Exhibitor Area)

Technical Session 4

Location: Salon A – B
Evan Garich, WFL-HD, Moderator

3:30 PM – 3:50 PM
The Characteristic Friction Angle, Its Determination and Use
Author: G. Norris

3:50 PM – 4:10 PM
Road re-opening after landslides the contribution of Remote Sensing
Authors: A Brunetti, P. Mazzanti, S. Moretto, A. Rocca, S. Romeo

4:10 PM - 4:30 PM
Poisson's Ratio Assessed from Ultrasonic versus Load Test
Authors: S. Elfass, E. Saint-Pierre, R. Watters. And G. Norris

4:30 PM – 4:50 PM
Settlement Monitoring of a Trial Embankment in Philadelphia Determining Site Specific Parameters for Large Embankment Construction
Author: Sarah McInnes

4:50 PM – 5:10 PM
Precision Presplitting – Changes to Design Methodology Based on Young's Modulus
Authors: Anthony Konya, Dr. Calvin J. Konya

Technical Session 5

Location: Salon C – D
Nicole Oester, CDOT, Moderator

3:30 PM – 3:50 PM
Blasting 2M+ Yards in Gneiss for a New Phoenix, AZ Freeway
Author: Robert Cummings

3:50 PM – 4:10 PM
Case Study of a Failed Tied-Back Retaining Wall in A Colluvium Slope Under Landslide Conditions
Author: Craig S. Lee

4:10 PM - 4:30 PM
Helicopter Sluicing for Rockfall Risk Mitigation in Response to the 2016 Kaikōura Earthquake
Authors: Rori Green

4:30 PM – 4:50 PM
Triggering Mechanisms Of The Landslide And Rockfall Events Of The Historic February 2019 Rainfall Event In The Tennessee Valley
Authors: David Freistaedter and Michael Laney

4:50 PM – 5:10 PM
Rockfall Hazard Assessment for the I-90 Snoqualmie Pass Corridor Snowbridges Project, Washington State
Authors: Kyle Obermiller, Reda Mikhail, David Findley

Free evening to explore and dine in Portland

Wednesday, October 23

6:00 AM – 7:00 AM
To-Go Continental Breakfast - Sponsored by Cascade Drilling
Location: Meet in Hotel Front Lobby

7:00 AM – 7:30 AM
Load buses for Field Trip
Pick-up Location: Meet in Hotel Front Lobby

7:30 AM – 5:30 PM
Highway Geology Symposium Field Trip
Lunch - Sponsored by Geobrugg North America
Beverages - Sponsored by Golder Associates
Snacks - Sponsored by IDS Georadar
(NO GLASS ALLOWED INSIDE BUSES)

6:30 PM – 7:30 PM
Highway Geology Symposium Social Hour
Sponsored by Access Limited Construction
Location: Salon E
Everyone Welcome!

7:30 PM – 8:30 PM
Location: Salon A -D
Highway Geology Symposium Banquet Dinner
<u>Ticketed Event</u>

8:30 PM – 9:30 PM
Highway Geology Symposium Banquet
Keynote Speaker
Catastrophic Missoula Floods, by Dr. Scott Burns
Young Author Awards Sponsored by Ameritech Slope Constructors
Everyone Welcome!

Thursday, October 24

6:30 AM – 9:00 AM
Continental Breakfast - Sponsored by GeoStabilization International
Location: Salon E

8:00 AM – 10:30 AM
Highway Geology Symposium Exhibitor Area Open
Exhibitors can break down after morning coffee break

Thursday, October 24 (continued)

Technical Session 6

Location: Salon A – B

Benjamin George, Landslide Technology, Moderator

8:00 AM – 8:20 AM
Estimating Rockfall Kinetic Energy as a Function of Rock Mass
Author: John Duffy

8:20 AM – 8:40 AM
An Innovative Solution for Debris Flow Barriers: Better Performance with Less Maintenance
Authors: Chiara Morstabilini, Luca Gobbin, Marco Luigi Deana, Daniele Lepore

8:40 AM - 9:00 AM
Advances in Rockfall: Protection: A Preliminary Design Tool for Attenuators Estimating Rockfall Kinetic Energy as a Function of Rock Mass
Authors: Helene Hofmann and Tim Shevlin

9:00 AM – 9:20 AM
Application of Rockfall Simulation to Risk Analysis
Author: Timothy J. Pfeiffer

9:20 AM – 9:40 AM
Assessment of Unstable Rock Columns at the Tieton Royal Columns, SR-12, Oak Creek Wildlife Area, Naches, WA
Author: William C. B. Gates

9:40 AM – 10:00 AM
A Review of Scaling as Rock Slope Remediation Method
Authors: Gabrielle Mellies, John Nichols Maureen Matthew

Technical Session 7

Location: Salon C – D

Palo Giscombe, Oregon DOT, Moderator

8:00 AM – 8:20 AM
The Meandering Mundo Mud Pot, or How Salton Sea Tectonics Affects International Trade
Author: Dean G. Francuch and Carolina Zamora

8:20 AM – 8:40 AM
Mitigation of Chronic Bluff Retreat with an Engineered Riprap Revetment BNSF Bellingham Subdivision, Whatcom County, Washington
Authors: Matthew Grizzell, Stephens

8:40 AM - 9:00 AM
TDOT's Response to Landslides due to February 2019 Floods
Author: Robert Jowers

9:00 AM – 9:20 AM
Large-Scale Earthquake-Induced Landslide Repair Following New Zealand's Kaikoura Earthquakes
Author: Colby Barrett

9:20 AM – 9:40 AM
OR213 MP 10 Spangler Hill Emergency Slide Mitigation & Monitoring
Author: Max Gummer

9:40 AM – 10:00 AM
Disseminating Geological and Geotechnical Information to End-users
Authors: Marc Fish, Tracy Trople, and Jim Struthers

Thursday, October 24 (continued)

10:00 AM – 10:30 AM
Morning Coffee Break
Location: Salon E

Technical Session 8
Location: Salon A – B
Mike Marshall, GRI, Moderator

10:30 AM – 10:50 AM
Investigation, Design and Construction of Rock Slopes and Foundations for the New Genesee Arch Bridge Letchworth State Park, New York
Authors: Jay R. Smerekanicz, Mark F. McNeilly, Jeffrey D. Lloyd

10:50 AM – 11:10 AM
Pfeiffer Canyon Bridge Failure in Context of Risk
Author: Gresham D. Eckrich

11:10 AM - 11:30 AM
Rockfall and Risk: A Perspective from Managing Risks for Dams and Levees
Author: Vanessa C. Bateman

11:30 AM - 11:50 AM
Evidence for the Value of Risk-Based Life-Cycle Management for Geohazards and Geotechnical Assets
Author: Mark Vessely

11:50 AM – 12:10 PM
Extended Q&A for session presenters

Technical Session 9
Location: Salon C – D
Marc Fish, Washington Department of Transportation, Moderator

10:30 AM – 10:40 AM
Route T Landslide Study and Repair, Ray County, MO
Authors: John Szturo and Wayne Duryee

10:50 AM – 11:10 AM
The Emergency Repair for Federally Owned Roads Program
Author: James Arthurs

11:10 AM - 11:30 AM
US 60 Pinto Creek Bridge Foundation Optimization with Emphasis On Cost And Constructability In A Challenging Geologic Environment
Authors: Daniel N. Fréchette and Patrice P. Brun

11:30 AM - 11:50 AM
Slide Ridge Culvert Replacement
Author: Robert E. Kimmerling

11:50 AM – 12:10 PM
Ground Anchor Testing – Matching Test Elements to Ensure Critical Anchor Attributes are Verified, or "Why We Don't Push Ropes"
Author: Martin Woodard, David Scarpato, David Wood and Peter Ingraham

12:10 PM - 12:30 PM
Closing Remarks and Adjournment
Location: Salon C - D

ENGINEERING GEOLOGY COMMITTEE

AFP00

2019 TRB Midyear Meeting at the 70th Highway Geology Symposium (HGS), Portland, Oregon

Date: Monday, October 21, 2019, 1:00 PM – 5:00 PM

Location: Salon C - D

Session Theme: Looking Ahead at Informed Decision-Making for Engineering Geologists

Time	Topic	Discussion Lead/Presenter
1:00 pm - 1:10 pm	Introduction	Ty Ortiz *Colorado DOT*
1:10 pm - 1:35 pm	Laser scanning of highway aggregates – Results of a transportation pooled fund study	Warren Chesner *Chesner Engineering, P.C.*
1:35 pm – 2:00 pm	Glaringly obvious things we need to consider with data management, but all too often do not	Vanessa Bateman *US Army Corps of Engineers*
2:00 pm – 2:25 pm	Developing a corridor health index to allocate budgets – Chiniak highway project – Kodiak, Alaska	Benjamin George & Trevor Strait *Landslide Technology and HDL Engineering*
2:25 pm – 2:40 pm	BREAK	
2:40 pm – 3:05 pm	Numerical solutions for problematic photogrammetrically derived digital elevation models and the consequence for slope stability analysis	Justin Lindeman, *Cal Engineering & Geology*
3:05 pm – 3:30 pm	Slope deformation and change detection using ground-based LIDAR and remote-monitored instrumentation, OSU project on remote sensing and modeling of coastal bluff and rock slope deterioration	Curran Mahoney *Oregon DOT*
3:30 pm – 4:10 pm	Simple decisions from complex data: How we can leverage emerging tools to manage rock slope risk with better information	Dave Gauthier *BGC Engineering* Jean Hutchinson *Queen's University*
4:10 pm – 5:00 pm	Open Discussion	All

Keynote Speaker: Curran Mohney

"Welcome to Portland and Engineering Geology"

Curran Mohney, RG, CEG
Engineering Geology Program Leader, Oregon DOT

Curran is presently the Engineering Geology Program Leader for the Oregon Department of Transportation. The Engineering Geology Program at ODOT encompasses site characterization, subsurface exploration, slopes and embankments, geologic hazards, groundwater, geotechnical instrumentation, and planning and research activities. In this role, he has also implemented elements of Geotechnical Asset Management including the Unstable Slopes (Landslide/Rockfall) program for ODOT.

Curran is a Registered Geologist and Certified Engineering Geologist in Oregon with over 25 years of experience in Oregon and the Western States. He has been the Engineering Geology Program Leader since 2004. Prior to this, he has been a Staff and Project-level geologist for Consulting firms and the Mining Industry as well as for ODOT. He is a graduate of the Geology program at Portland State University. During his professional career, he has been involved in the investigation, design, and mitigation of literally hundreds of landslides and rockfalls.

Banquet Speaker:

"Cataclysms on the Columbia: The Great Missoula Floods"

Scott Burns, RG, LG, CEG, PhD

Scott just completed his 49th year of teaching at the university level, with past positions in Switzerland, New Zealand, Washington, Colorado and Louisiana before coming to Portland State University 29 years ago. He has a BS and MS from Stanford University and his PhD from the University of Colorado. He is an engineering geologist and environmental geologist who also studies soils. His areas of expertise are landslides, radon gas, heavy metals in soils, Missoula Floods, and terroir of wine. He has over 100 publications including two books and has had 48 MS and PhD students complete degrees under him. He has been Chair of three different geology departments and also has been an Associate Dean. He has been president of AEG, chair of the engineering geology division of GSA, and also president of IAEG (first American president in its 54 year history). He was chair of the HGS the last time there was a meeting in Portland in the early 1990's. burnss@pdx.edu

One of the greatest set of geological events to ever have occurred in North America was given the name, the Missoula Floods. The talk will focus on the incredible story of discovery and development of the idea of the floods by J Harlen Bretz and then will discuss the effect of the floods on the development of the landscape of 16,000 square miles of the Pacific Northwest. The floods occurred between 15,000 and 18,000 years ago.

Highway Geology Symposium History, Organization, And Function

Inaugural Meeting

Established to foster a better understanding and closer cooperation between geologists and civil engineers in the highway industry, the Highway Geology Symposium (HGS) was organized and held its first meeting on March 14, 1950, in Richmond Virginia. Attending the inaugural meeting were representatives from state highway departments (as referred to at that time) from Georgia, South Carolina, North Carolina, Virginia, Kentucky, West Virginia, Maryland, and Pennsylvania. In addition, a number of federal agencies and universities were represented. A total of nine technical papers were presented. W.T. Parrott, an engineering geologist with the Virginia Department of Highways, chaired the first meeting. It was Mr. Parrott who originated the Highway Geology Symposium.

It was at the 1956 meeting that future HGS leader, A.C. Dodson, began his active role in participating in the Symposium. Mr. Dodson was the Chief Geologist for the North Carolina State Highway and Public Works Commission, which sponsored the 7th HGS meeting.

East and West

Since the initial meeting, 70 consecutive annual meetings have been held in 35 different states. Between 1950 and 1962, the meetings were east of the Mississippi River, with Virginia, West Virginia, Ohio, Maryland, North Carolina, Pennsylvania, Georgia, Florida, and Tennessee serving as host state.

In 1962, the symposium moved west for the first time to Phoenix, Arizona where the 13th annual HGS meeting was held. Since then it has alternated, for the most part, back and forth from the east to the west. The Annual Symposium has moved to different location as follows:

Organization

Unlike most groups and organizations that meet on a regular basis, the Highway Geology Symposium has no central headquarters, no annual dues and no formal membership requirements. The governing body of the Symposium is a steering committee composed of approximately 20 to 25 engineering geologist and geotechnical engineers from state and federal agencies, colleges and universities, as well as private service companies and consulting firms throughout the country. Steering committee members are elected for three-year terms, with their elections and re-elections being determined principally by their interests and participation in and contribution to the Symposium. The officers include a chairman, vice chairman, secretary, and treasurer. all of whom are elected for a two-year term. Officers, except for the treasurer, may only succeed themselves for one additional term.

A number subcommittees conduct the affairs of the organization. The lack of rigid requirements, routing and relatively relaxed overall functioning of the organization is what attracts many participants.

Meeting sites are chosen two to four years in advance and are selected by the Steering Committee following presentations made by representatives of potential host states. These presentations are usually made at the steering committee meeting, which is held during the Annual Symposium. Upon selection, the state representative becomes the state chairman and a member pro-tem of the Steering Committee.

The symposia are generally scheduled for two and one-half days, with a day-and-a-half for technical papers plus a full day for the field trip. The Symposium usually begins on Wednesday morning. The field trip is usually Thursday, followed by the annual banquet that evening. The final technical session generally ends by noon on Friday. In recent years this schedule has been modified to better accommodate climate conditions and tourism benefits.

The Field Trip

The field trip is the focus of the meeting. In most cases, the trips cover approximately 150 to 200 miles, provide for six to eight scheduled stops, and require about eight hours. Occasionally, cultural stops are scheduled around geological and geotechnical points of interests. To cite a few examples: in Wyoming (1973), the group viewed landslides in the Big Horn Mountains; Florida's trip (1976) included a tour of Cape Canaveral and the NASA space installation; the Idaho and South Dakota trips dealt principally with mining activities; North Carolina provided stops at a quarry site, a dam construction site, and a nuclear generation site; in Maryland, the group visited the Chesapeake Bay hydraulic model and the Goddard Space Center. The Oregon trip included visits to the Columbia River Gorge and Mount Hood; the Central mine region was visited in Texas; and the Tennessee meeting in 1981 provided stops at several repaired landslide in Appalachia regions of East Tennessee.

In Utah (1988) the field trip visited sites in Provo Canyon and stopped at the famous Thistle Landslide, while in New Mexico, in 1990, the emphasis was on rockfall treatments in the Rio Grande River canyon and included a stop at the Brugg Wire Rope headquarters in Santa Fe. Mount St, Helens was visited by the field trip in 1994 when the meeting was in Portland, Oregon, while in 1995 the West Virginia meeting took us to the New River Gorge Bridge that has a deck elevation of 876 feet above the water.

In Cody, Wyoming the 1996 field trip visited the Chief Joseph Scenic Highway and the Beartooth Uplift in northwest Wyoming. In 1997 the meeting in Tennessee visited the newly constructed future I-26 highway in the Blue Ridge of East Tennessee. The Arizona meeting in 1998 visited the Oak Creek Canyon near Sedona and a mining ghost town at Jerrome, Arizona. The Virginia meeting in 1999 visited the "Smart Road" Project that was under construction. This was a joint research project of the Virginia Department of Transportation and Virginia Tech University. The Seattle Washington meeting in 2000 visited the Mount Rainier area. A stop during the Maryland meeting in 2001 was the Sideling Hill road cut for I-68 which displayed a tightly folded syncline in the Allegheny Mountains.

The California field trip in 2002 provided a field demonstration of the effectiveness of rock netting against rock falls along the Pacific Coast Highway. The Kansas City meeting in 2004 visited the Hunt Subtropolis which is said to be the "world's largest underground business complex". It was created through the mining of limestone by way of the room and pillar method. The Rocky Point Quarry provided an opportunity to search for fossils at the North Carolina meeting in 2005. The group also visited the US-17 Wilmington Bypass Bridge which was under construction. Among the stops at the Pennsylvania meeting were the Hickory Run Boulder Field, the No.9 Mine and Wash Shanty Museum, and the Lehigh Tunnel.

The New Mexico field trip in 2008 included stops at a soil nailed wall along US-285/84 north of Santa Fe and a road cut through the Bandelier Tuff on highway 502 near Los Alamos where rockfall mesh was used to protect against rockfalls. The New York field trip in 2009 included the Niagara Falls Gorge and the Devil's Hole Trail. The Oklahoma field trip in 2010 toured the complex geology of the Arbuckle Mountains in the southern part of the state along with stops at Tucker's Tower and Turner Falls.

In the bluegrass state of Kentucky, the 2011 HGS field trip included stops at Camp Nelson which is the site of the oldest exposed rocks in Kentucky near the Lexington and Kentucky River Fault Zones. Additional stops at the Darby Dan Farm and the Woodford Reserve Distillery illustrated how the local geology has played such a large part in the success of breeding prized Thoroughbred horses and made Kentucky the "Birthplace of Bourbon".

In Redding, California, the 2012 field trip included stops at the Whiskeytown Lake, which is one in a series of lakes that provide water and power to northern California. Additional stops included Rocky Point, a roadway construction site containing Naturally Occurring Asbestos (NOA), and Oregon Mountain where the geology and high rainfall amounts have caused Hwy 299 to experience local and global instabilities since first constructed in 1920.

The 2013 field trip of New Hampshire highlighted the topography and geologic remnants left by the Pleistocene glaciation that fully retreated approximately 12,000 years ago.

The field trip included stops at various overlooks of glacially-carved valleys and ranges; the Old Man of the Mountain Memorial Plaza, which is a tribute to the famous cantilevered rock mass in the Franconia Notch that collapsed on May 3, 2003; the lacustrine deposits and features of the Glacial Lake Ammonoosuc; views of the Presidential Range; bridges damaged during Tropical Storm Irene in August 2011; and the Willey Slide, located in the Crawford Notch where all members of the Willey family were buried by a landslide in 1826.

The 2014 field trip presented a breathtaking tour of the geology and history of southeast Wyoming, ascending from the high plains surrounding Laramie at 7000 feet to the Medicine Bow Mountains along the Snowy Range Scenic Byway. Visible along the way were a Precambrian shear zone, and glacial deposits and features. From the glacially carved Mirror Lake and the Snowy Range Ski Area, the path wound east to the Laramie Mountains and the Vedauwoo Recreational Area, a popular rock climbing and hiking area before returning to Laramie.

In Sturbridge, MA, the 2015 field trip focused on the Connecticut Valley, a Mesozoic rift basin that signaled the breakup of Pangea, and the Berkshires, which represents the collision and amalgamation of an island arc system with the North American Laurentian margin.

The field trip in 2016 was an urban setting along the western edge of Colorado Springs and around Manitou Springs. Stops included the Pikeview Quarry, Garden of the Gods Visitor Center, and several other locations where rockfall and debris flow mitigation, post-flooding highway embankment repair, and a nonconformity in the rock records that spans 1.3 billion years were observed.

The 2017 field trip provided an opportunity to view the geology of northern Georgia. Stops included the Bellwood Quarry, which, at one time was run by the City of Atlanta and served as a prison labor camp. It will eventually serve as a 2.4 billion-gallon water storage facility for the City of Atlanta upon completion of a tunnel to connect the quarry to two water treatment plans and three pump stations. Additional stops included the Buzzi Unicem Cement Plant to get a close-up view of the Clairmont Melange, The Cooper Furnace near the Allatoona Dam, and the New Riverside Ochre-Emerson Barite mine.

The 2018 field trip in Portland Maine provided a good overview of the geology of coastal Maine. Field trip stops included a stop at the Sherman Salt Marsh near Newcastle which was recently restored to its natural state after the dam that carried US Highway 1 washed out during a 2005 storm. Additional stops included the site of the 1996 landslide near Rockland Harbor that consumed several homes and the rock slope remediation project at the Penobscot Narrows Bridge near Prospect Maine. A lobster lunch along the shore of Penobscot Bay was one of several highlights of the field trip.

Technical Sessions and Speakers

The Highway Geology Symposium technical sessions most commonly include case histories and state-of- the- art papers with highly theoretical papers the exception. The papers presented at the technical sessions are published in the annual proceedings. All of the HGS Proceedings, from the first 1950 Symposium to present, were digitized and are available at the HGS website. Banquet speakers are also a highlight and have been varied through the years.

A Medallion Award was initiated in 1970 to honor those persons who have made significant contributions to the Highway Geology Symposium. The selection was and is currently made from the members of the national steering committee of the HGS.

Member Recognition

Emeritus Members: A number of past members of the national steering committee have been granted Emeritus status. These individuals, usually retired, resigned from the HGS Steering Committee, or are deceased, have made significant contributions to the Highway Geology Symposium. A total of 42 persons have been granted Emeritus status. Ten are now deceased.

Dedications: Several Proceedings volumes have been dedicated to past HGS Steering Committee members who have passed away. The 36th HGS Proceedings were dedicated to David L. Royster (1931 - 1985, Tennessee) at the Clarksville, Indiana Meeting in 1985. In 1991 the Proceedings of the 42nd HGS held in Albany, New York were dedicated to Burrell S. Whitlow (1929 - 1990, Virginia). The 64th HGS Proceedings were dedicated to Earl Wright (1931 – 2012) at the North Conway, New Hampshire meeting. The 65th proceedings were dedicated to Nicholas Priznar (1952 – 2014) at the Laramie, Wyoming meeting. The 67th HGS held at Colorado Springs, Colorado dedicated the proceedings to Vern McGuffy (1934 – 2016). The proceedings for the 68th HGS held in Marietta, Georgia were dedicated to Richard (Dick) Cross (1944 – 2016). The proceedings for the 69th HGS are dedicated to Dave Bingham (1932-2018) and Joe Gutierrez (1926-2018).

Young Author Award: The Highway Geology Symposium has always encouraged participation of Young Professionals, realizing that Young Professionals are the future of the Organization. This participation was taken formal in 2014, with the formation of an annual National Young Author Competition, where Young Authors have the opportunity to prepare papers and present their work. To participate, Young Author's must be up to 35 years old or younger, the principal author of the paper and the sole presenter of the paper at the Symposium. Papers are reviewed and judged based on Technical Presentation of the Paper (including Geology), Originality of the Work, Applicability of the Work to Others and Paper Layout. One Young Author is selected each year to receive the coveted Young Author Award, with presentation of the award conducted at the annual Symposium banquet.

List of Highway Geology Symposium Meetings

No.	Year	HGS Location	No.	Year	HGS Location
1st	1950	Richmond, VA	2nd	1951	Richmond, VA
3rd	1952	Lexington, VA	4th	1953	Charleston, WV
5th	1954	Columbus, OH	6th	1955	Baltimore, MD
7th	1956	Raleigh, NC	8th	1957	State College, PA
9th	1958	Charlottesville, VA	10th	1959	Atlanta, GA
11th	1960	Tallahassee, FL	12th	1961	Knoxville, TN
13th	1962	Phoenix, AZ	14th	1963	College Station, TX
15th	1964	Rolla, MO	16th	1965	Lexington, KY
17th	1966	Ames, IA	18th	1967	Lafayette, IN
19th	1968	Morgantown, WV	20th	1969	Urbana, IL
21st	1970	Lawrence, KS	22nd	1971	Norman, OK
23rd	1972	Old Point Comfort, VA	24th	1973	Sheridan, WY
25th	1974	Raleigh, NC	26th	1975	Coeur d'Alene, ID
27th	1976	Orlando, FL	28th	1977	Rapid City, SD
29th	1978	Annapolis, MD	30th	1979	Portland, OR
31st	1980	Austin, TX	32nd	1981	Gatlinburg, TN
33rd	1982	Vail, CO	34th	1983	Stone Mountain, GA
35th	1984	San Jose, CA	36th	1985	Clarksville, TN
37th	1986	Helena, MT	38th	1987	Pittsburg, PA
39th	1988	Park City, UT	40th	1989	Birmingham, AL
41st	1990	Albuquerque, NM	41st	1991	Albany, NY
43rd	1992	Fayetteville AR	44rd	1993	Tampa, FL
45th	1994	Portland, OR	46th	1995	Charleston, WV
47th	1996	Cody, WY	48th	1997	Knoxville, TN
49th	1998	Prescott, AZ	50th	1999	Roanoke, VA
51st	2000	Seattle, WA	52nd	2001	Cumberland, MD
53rd	2002	San Luis Obispo, CA	54th	2003	Burlington, VT
55th	2004	Kansas City, MO	56th	2005	Wilmington, NC
57th	2006	Breckinridge, CO	58th	2007	Pocono Manor, PA
59th	2008	Santa Fe, NM	60th	2009	Buffalo, NY
61st	2010	Oklahoma City, OK	62nd	2011	Lexington, KY
63rd	2012	Redding, CA	64th	2013	North Conway, NH
65th	2014	Laramie, WY	66th	2015	Sturbridge, MA
67th	2016	Colorado Springs	68th	2017	Marietta, GA
69th	2018	Portland, ME	70th	2019	Portland OR

Emeritus Members of the Steering Committee

Emeritus Status is granted by the Steering Committee

R.F. Baker
John Baldwin
David Bingham
Vernon Bump
Virgil E. Burgat
Robert G. Charboneau
Hugh Chase
Jim Coffin
Dick Cross
A.C. Dodson
Walter F. Fredricksen
Brandy Gilmore
Russell Glass
Robert Goddard
Joseph Gutierrez
Mike Hager
Rich Humphries
Charles T. Janik
John Lemish
Bill Lovell
A. David Martin
Henry Mathis
William McCasland
George S. Meadors, Jr.
David Mitchell
Harry Moore
W.T. Parrot
Paul H. Price
Nicholas Priznar
David L. Royster
Bill Sherman
Willard L. Sitz
Mitchell Smith
Jim Stroud
Steve Sweeney
Sam Thornton
Berke Thompson
Mike Vierling
Burrell Whitlow
W.A. "Bill" Wisner
Earl Wright
Ed J. Zeigler

Medallion Award Recipients

The Medallion Award was instituted in 1969 to recognize individuals who have made significant contributions to the Highway Geology Symposium over many years. The award is a 3.5" medallion mounted on a walnut shield and appropriately inscribed. The Medallion Award is presented during the banquet at the annual symposium.

Hugh Chase	1970	Earl Wright	1997
Tom Parrott	1970	Russell Glass	1998
Paul Price	1970	Harry Ludowise	2000
K.B. Woods	1971	Sam Thornton	2000
R.J. Edmondson	1972	Bob Henthorne	2004
C.S. Mullin	1974	Mike Hager	2005
A.C. Dodson	1975	Joseph A. Fischer	2007
Burrell Whitlow	1978	Ken Ashton	2008
Bill Sherman	1980	A. David Martin	2008
Virgil Burgat	1981	Michael Vierling	2009
Henry Mathis	1982	Dick Cross	2009
David Royster	1982	John F. Szturo	2010
Terry West	1983	Christopher Ruppen	2012
Dave Bingham	1984	Jeff Dean	2012
Vernon Bump	1986	Eric Rorem	2014
C.W. "Bill" Lovell	1989	John Pilipchuk	2015
Joseph A. Gutierrez	1990	Peter Ingraham	2016
Willard McCasland	1990	Richard Lane	2017
W.A. "Bill" Wisner	1991	Steve Sweeney	2018
David Mitchell	1993	John Duffy	2018
Harry Moore	1996	Krystle Pelham	2018

Young Author Award Winners

2014 Simon Boone, "Performance of Flexible Debris Flow Barriers in a Narrow Canyon"

2015 Cory Rinehart, "High Quality H20: Utilizing Horizontal Drains for Landslide Stabilization"

2016 Todd Hansen, "Geologic Exploration for Ground Classification: Widening of the I-70 Veterans Memorial Tunnels"

2017 James Arthurs, "Construction of Transportation Infrastructure in Weathered Volcanic Ash Soils"

2018 Brian Felber, "Geotechnical Challenges for Bridge Foundations & Roadway Embankment Design in Peats and Deep Glacial Lake Deposits"

HGS National Steering Committee Officers

Ken Ashton CHAIRMAN
West VA Geological Survey
1 Mont Chateau Road
Morgantown, WV 26508
Phone: (304) 594-2331
Cell: (304) 216-3025
Fax: (304) 594-2575
Email: ashton@geosrv.wvnet.edu

Krystle Pelham VICE-CHAIRMAN
New Hampshire Dept. of Transportation
PO Box 483
Concord, NH 03302
Phone: (603) 271-1657
Email: Krystle.Pelham@dot.nh.gov

Bill Webster SECRETARY
CalTrans
5900 Folsom Blvd.
Sacramento, CA 95819
Phone: (916) 662-1183
Fax: (916) 227-1082
Email: bill_webster@dot.ca.gov

John Pilipchuk TREASURER
NCDOT Geotechnical Engineering Unit
1020 Birch Ridge Drive
Raleigh, NC 27699-1589
Phone: (919) 707-6850
Fax: (919) 250-4237
Email: jpilipchuk@ncdot.gov

HGS National Steering Committee Members

Vanessa Bateman
Headquarters, USACE
441 G Street NW
Washington, DC 20314-1000
Phone: 202-761-7423
Email: Vanessa.c.bateman@usace.army.mil

Jeff Dean
Terracon
4701 North Stiles Avenue
Oklahoma City, OK 73015
Phone:405 445-3280
Email: jeff.dean@terracon.com

John D. Duffy
Caltrans (Retired)
128 Baker Ave.
Shell Beach, CA 93449
Phone: (805) 440-9062
Email: JohnDuffy@charter.net

Tom Eliassen
VT AOT (Retired)
15 Cliff Street, Apt. 2
Montpelier, VT 05602
Phone: (802) 498-4993
Email tomeli@myfairpoint.net

Mark Falk
Wyoming DOT
5300 Bishop Blvd.
Cheyenne, WY 82001
Phone: (307) 777-4205
Cell (307) 631-5015
Email: mark.falk@wyo.gov

Kyle Halverson
Chief Geologist
Kansas Department of Transportation
Bureau of Structures and Geotechnical Services
700 SW Harrison St.
Topeka, KS 66603
Office: 785-291-3860
Cell: 785-845-4332
Email: kyle.halverson@ks.gov

HGS National Steering Committee Members (continued)

Bob Henthorne
Mid-States Materials
1800 Brickyard Road
Topeka, KS 66618
Phone: (785) 640-2477
Email: bhenthorne@midstatesmaterials.com

Peter Ingraham
Golder Associates Inc.
670 North Commercial Street, Suite 103
Manchester, NH 03101-1146
Phone: (603) 668-0880
Fax: (603) 668-1199
Email peter_ingraham@golder.com

Jody Kuhne
North Carolina DOT
P. O. Box 3279
US Highway 74
Asheville, NC 28802
Phone: (828) 298-3874
Email: jkuhne@ncdot.gov

Richard Lane
NHDOT (Retired)
213 Pembroke Hill Rd.
Pembroke, NH 03275
Phone: (603) 485-3202
Email: lanetrisbr@hotmail.com

Sarah McInnes
PA DOT
District 6-0
7000 Geerdes Blvd.
King of Prussia, PA 19406
Phone: (610) 205-6544
FAX: (610) 205-6599
Email: smcinnes@pa.gov

Victoria Porto
PA DOT (Retired)
10 Pine Lake Drive
Carlisle, PA 17015
Phone: (717) 805-5941
Email: vamporto@aol.com

Erik Rorem
Geobrugg North America, LLC
20483 Whistle Punk Rd. 97702
Bend, OR 97702
Phone: 1 505 690 7144
Email: erik.rorem@geobrugg.com

Christopher A. Ruppen YOUNG AUTHOR COMMITTEE
Geostabilization International
3808 Sunflower Road
New Brighton, PA 15066
Phone: (724) 272-7532
Email: chris.ruppen@gsi.us

Stephen Senior
Ontario Min of Trans. (Retired)
11 Dewbourne Ave.
Richmond Hill, ON L4B 3G7 Canada
Phone: (416) 235-3734
Fax: (416) 235-4101
Email: sa.senior@rogers.com

Deana Sneyd
Petrologic Solutions, Inc.
3997 Oak Hill Road
Douglasville, GA 30135
Phone: (678) 313-4147
Email: dsneyd@gmail.com

HGS National Steering Committee Members (continued)

John F. Szturo
HNTB Corporation
715 Kirk Drive
Kansas City, MO 64105
Phone: (816) 527-2275 (Direct Line)
Cell: (913) 530-2579
Fax: (816) 472-5013
Email: jszturo@hntb.com

Terry West
Earth and Atmospheric Science Dept.
Purdue University
West Lafayette, IN 47907-1297 Phone:
(765) 494-3296
Fax: (765)496-1210
Email: trwest@purdue.edu

Richard Wilson
S&ME, Inc.
2020 Liberty Road, Suite 105
Lexington, KY 40505
Phone: (859) 293-5518
Cell: (502) 682-1203
Email: rwilson@smeinc.com

Highway Geology Symposium
Past, Present, and Future Symposium Contact List

2013	New Hampshire	Krystle Pelham	603-271-1657	Krystle.Pelham@dot.state.nh.us
2014	Wyoming	Jim Coffin	307-777-4205	Jim.coffin@wyo.gov
2015	Massachusetts	Peter Ingraham	603-688-0880	peter_ingraham@golder.com
2016	Colorado	Ty Ortiz	303-921-2634	Ty.ortiz@state.co.us
2017	Georgia	Deana Sneyd	678-313-4147	Dsneyd61@gmail.com
2018	Maine	Krystle Pelham	603-271-1657	Krystle.Pelham@dot.state.nh.us
2019	Oregon	Tim Shevlin	503-423-7258	tim.shevlin@geobrugg.com
2020	North Carolina	John Pilipchuk	919-707-6851	jpilipchuk@ncdot.gov
		Jody Kuhne	828-298-3874	jkune@ncdot.gov

70th ANNUAL HIGHWAY GEOLOGY SYMPOSIUM
Sponsors

The following companies have graciously contributed toward the sponsorship of the Symposium. The HGS relies on sponsor contributions for refreshment breaks, field trip lunches and other activities. We gratefully appreciate the contributions made by these sponsors.

Platinum Sponsors

Geobrugg North America, LLC.
22 Centro Algodones
Algodones, New Mexico 87001
P: (505) 771 4080
Fax: (505) 771 4081
www.geobrugg.com

Geobrugg supplies Geohazard Mitigation and Safety Systems from nets and meshes of high-tensile steel wire. Many years of experience and a global network with branches and partners in over 50 countries ensures fast, thorough, and cost-effective solutions for specific customer requirements. We are partners, consultants, developers, and project managers for our customers.

Jensen Drilling Company
1775 Henderson Ave.
Eugene, Oregon 97403
P: (800) 762-7435
www.jensendrilling.com

Jensen Drilling Company provides a wide variety of drilling expertise, ranging from small projects to multi-million-dollar construction projects throughout the United States. Technical capabilities include, air and mud rotary drilling, large diameter flooded reverse circulation wells, anchor and tiebacks, drilling and grouting, air percussion drilling from 3-inch to 24-inch diameter, horizontal and vertical drain drilling.

www.geobrugg.com/attenuator

GEOBRUGG
BRUGG
Safety is our nature

Geobrugg North America, LLC | 22 Centro Algodones | Algodones, NM 87001 USA |
T +1 505 771 4080 | www.geobrugg.com

Jensen Drilling Company provides a wide variety of drilling expertise, ranging from small projects to multi-million-dollar construction projects throughout the United States. Technical capabilities include, air and mud rotary drilling, large diameter flooded reverse circulation wells, anchor and tiebacks, drilling and grouting, air percussion drilling from 3-inch to 24-inch diameter, horizontal and vertical drain drilling.

History:

Jensen Drilling Company has over 53 years of experience in the drilling industry. Our company was founded on our excellence in horizontal drain drilling and since then our progressive business philosophy has moved us to where we are now a drilling company that offers a complete service to the drilling industry. *"Our goal is to provide excellent service by being attentive and responsive to the needs of our clients" – Jerry Jensen*

Jensen Drilling Company
1775 Henderson Ave.
Eugene. OR 97403
(800) 762-7435
www.jensendrilling.com

Contacts:
Jerry Jensen, President – jerryj@jensendrilling.com
Scott Weaver, Vice President – scottw@jensendrilling.com
Chris Humphries, Sec/Tres – chris@jensendrilling.com
Casey Cooper, Estimator – casey@jensendrilling.com

70th ANNUAL HIGHWAY GEOLOGY SYMPOSIUM

Gold Sponsors

Ameritech Slope Constructors INC
Geotechnical Contractors

Ameritech Slope Constructors, Inc
PO Box 2702
Asheville, North Carolina
28802
P: 828-633-6352
www.ameritech.pro

Ameritech Slope Constructors, Inc. is a specialty contracting company specializing in Civil/Geotechnical Construction projects, including rock and soil slope stabilization, rock scaling, rock bolting, high strength steel mesh drapes and barriers as well as dry mix shotcrete. We also drill and break large boulders and overhanding ledges using nonexplosive rock removal methods and mechanical rock splitters.

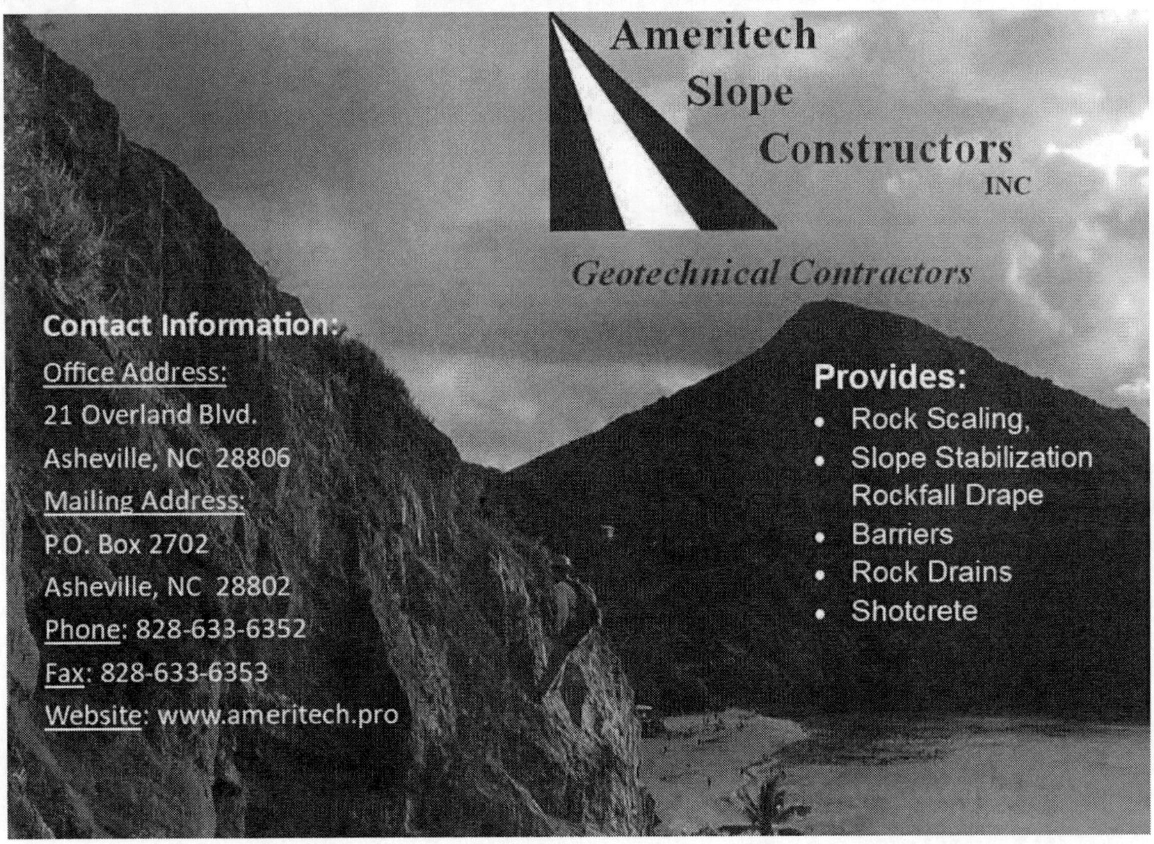

70th ANNUAL HIGHWAY GEOLOGY SYMPOSIUM

Silver Sponsors

ACCESS LIMITED
CONSTRUCTION

Access Limited Construction
1102 Pike Lane
Oceano, California 93445
P: (805)-592-2230

www.accesslimitedconstruction.com

Access Limited Construction is a Specialty Contractor located in San Luis Obispo, California. An industry leader, we provide rockfall mitigation, slope stabilization, and difficult drilling services for transportation, energy, mining and private sector clients. With our fleet of Spyder Excavators, we can access steep terrain and hard to reach projects throughout the United States from the East Coast to Hawaii.

LANDSLIDE
TECHNOLOGY
A DIVISION OF CORNFORTH CONSULTANTS

Landslide Technology
10250 SW Greenburg Road, Suite 111
Portland, Oregon 97223
P: (503) 452-1200
www.landslidetechnology.com

Landslide Technology, a division of Cornforth Consultants, is a geotechnical engineering firm based in Portland, Oregon. Currently in its 36th year of operation, we were established to focus specifically on transportation projects including landslides, rock slope stability, emergency response to sudden failures, and other challenging geotechnical and geologic assignments.

70th ANNUAL HIGHWAY GEOLOGY SYMPOSIUM

Bronze Sponsors

Acker Drill Company
PO Box 830
Scranton, Pennsylvania 18501
P: (501) 479-2009
www.ackerdrill.com

Acker Drill Company was founded in 1916 and has been exporting its products worldwide since 1960. Today, we are recognized as a world leader in the manufacture of drill rigs and tooling for the geotechnical, environmental, mineral exploration, and civil engineering industries.

AMS, Inc.
105 Harrison Street
American Falls, Idaho 83211
P: (208) 226-2017
www.ams-samplers.com

AMS, Inc is a world leading manufacture of soil sampling equipment from hand augers, bucket augers, soil probes and core/split core samplers for all types of soils testing from environmental, groundwater, geotechnical, geophysical, site investigations. AMS, Inc is also the sole mfg of the PowerProbe and PowerCore Drilling equipment that is capable of surface and down-hole asphalt and concrete coring and geotechnical (SPT) sampling for sub-surface investigations for new / used roadway repairs.

70th ANNUAL HIGHWAY GEOLOGY SYMPOSIUM

Bronze Sponsors (continued)

Apex Rockfall Mitigation, LLC
3229 Springfield Road
Grand Junction, Colorado 81503
P: (970) 314-7302
www.apexrockfall.com

Apex Rockfall Mitigation is a leader in the rockfall and geo-hazard stabilization industry. Apex prides itself in providing innovate solutions across the county to safe guard slopes of all natures, specializing in limited access and rope access work.

Cascade Drilling, LP
13600 SE Ambler Road
Clackamas, Oregon 97015
P: (503) 572-3090
www.cascade-env.com

Cascade is a field services contractor that partners with our clients to provide seamless geotechnical and environmental solutions from concept to completion. We provide the, industry's most comprehensive in-house suite of field services to support your geotechnical and environmental projects no matter how routine of complex.

70th ANNUAL HIGHWAY GEOLOGY SYMPOSIUM

Bronze Sponsors (continued)

Durham Geo Slope Indicator
12123 Harbour Reach Drive, Suite 106
Mukilteo, Washington 98275
P: (425) 493-6249
www.durhamgeo.com

Durham Geo Slope Indicator (DGSI) is a designer and manufacturer of geotechnical instrumentation, materials testing equipment and environmental pumps. Our expertise is based of decades of experience. Our instrumentation is integrated into some of the World's most important assets and construction projects.

GeoStabilization International
543 31 Road
Grand Junction, Colorado 81504
P: (855)-579-0536
www.geostabilization.com

GeoStabilization International® focuses on bringing new technologies to the geohazard repair industry to reduce project time, cost, and minimize environmental impact. Through many years of training, experience, and this founding philosophy, GeoStabilization engineers and constructors now stand as the most qualified and most experienced in the industry.

70th ANNUAL HIGHWAY GEOLOGY SYMPOSIUM

Bronze Sponsors (continued)

GOLDER

Golder Associates
670 North Commercial Street, Suite 103
Manchester, New Hampshire 03101-1146
Phone: (603) 668-0880
Fax: (603) 668-1199
www.golder.com

Golder is respected across the globe for providing consulting, design and construction services in our specialist areas of earth and environment. Our highly skilled engineers, scientists, project managers and other technical specialists are committed to helping clients achieve project success on projects around the globe.

IDS GeoRadar NORTH AMERICA

IDS GeoRadar
14828 w. 6th Avenue, Unit 12-B
Golden, Colorado 80401
P: (303)-726-6024
www.idsgeoradar.com

IDS GeoRadar manufactures, sells, and services radar instruments. IBIS mine and geohazard monitors, and GPR tools, lead the sector in features, affordability and utility.

70th ANNUAL HIGHWAY GEOLOGY SYMPOSIUM

Bronze Sponsors (continued)

Rocscience, Inc
54 Saint Patrick Street
Toronto, Ontario M5T 1V1
P: (416) 698-8217
www.rocscience.com

Rocscience is a world leader in developing 2D & 3D software for civil, mining, and geotechnical engineers. For over 20 years, we've used leading-edge research to build geotechnical tools used by over 7,000 engineers around the world for slope stability, excavation design, and geotechnical analysis.

Williams Form Engineering Corp
7601 N Columbia Boulevard
Portland, Oregon 97203
P: (503) 285-4548
www.williamsform.com

Williams Form Engineering Corporation has been offering Ground Anchors, Concrete Anchors, Post Tensioning Systems, and Concrete Forming Hardware to the construction industry for over 95 years.

Call Us - (805) 592-2230
1102 Pike Lane, Oceano, California 93445
info@accesslimitedconstruction.com
ACCESS LIMITED
CONSTRUCTION
SPECIALIZED IN:
ROCKFALL MITIGATION
SLOPE STABILIZATION
DIFFICULT DRILLING
www.AccessLimitedConstruction.com

LANDSLIDE
TECHNOLOGY
A DIVISION OF CORNFORTH CONSULTANTS
10250 S.W. Greenburg Road, Suite 111
Portland, Oregon 97223
Phone 503-452-1200
www.landslidetechnology.com
Landslides
Rockfall/Rock Slopes
Geotechnical Asset Management
Emergency Response
Instrumentation
SPRAT Rope Access Team

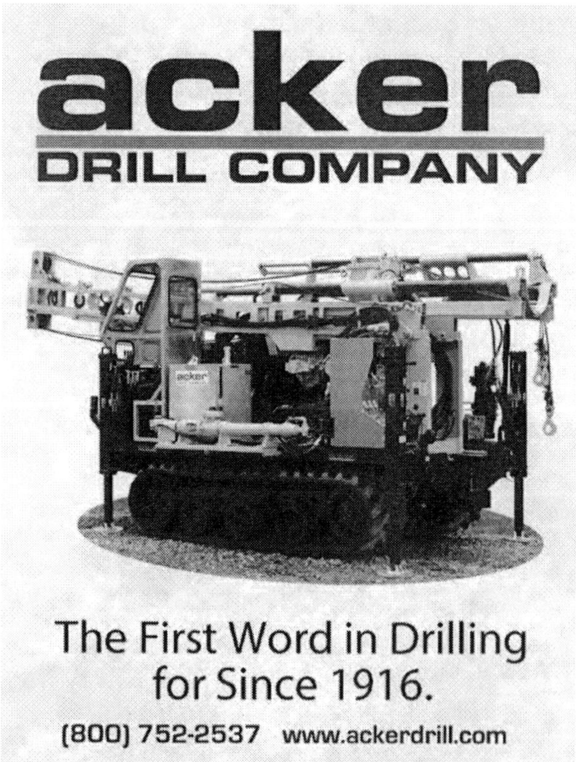
acker
DRILL COMPANY
acker
The First Word in Drilling for Since 1916.
(800) 752-2537 www.ackerdrill.com

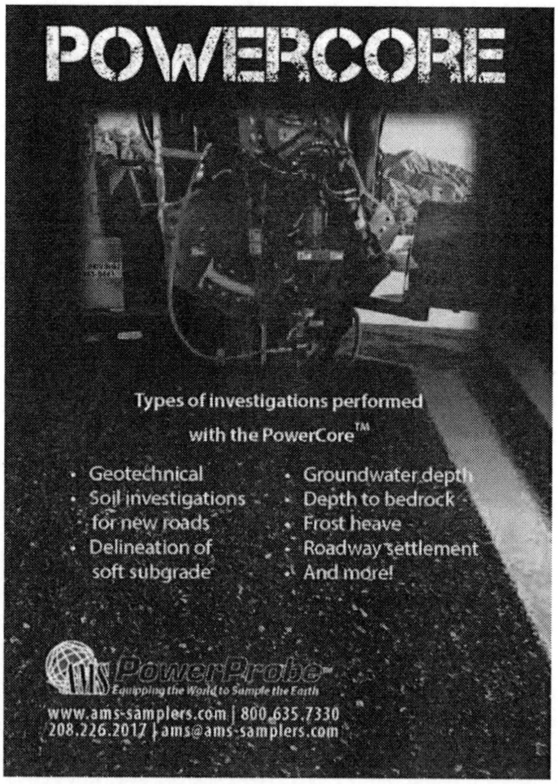
POWERCORE
Types of investigations performed with the PowerCore™
Geotechnical
Soil investigations for new roads
Delineation of soft subgrade
Groundwater depth
Depth to bedrock
Frost heave
Roadway settlement
And more!
AMS PowerProbe
Equipping the World to Sample the Earth
www.ams-samplers.com | 800.635.7330
208.226.2017 | ams@ams-samplers.com

"Specializing in Limited Access Slope Stabilization"
APEX
Rockfall Mitigation LLC
Providing a wide varitey of solutions ranging from- Scaling, rockfall netting, drapes, rockfall barriers, debris flow, rock anchors, soil anchors, shotcrete cable lashing, trim blasting, rock dowels and much more.

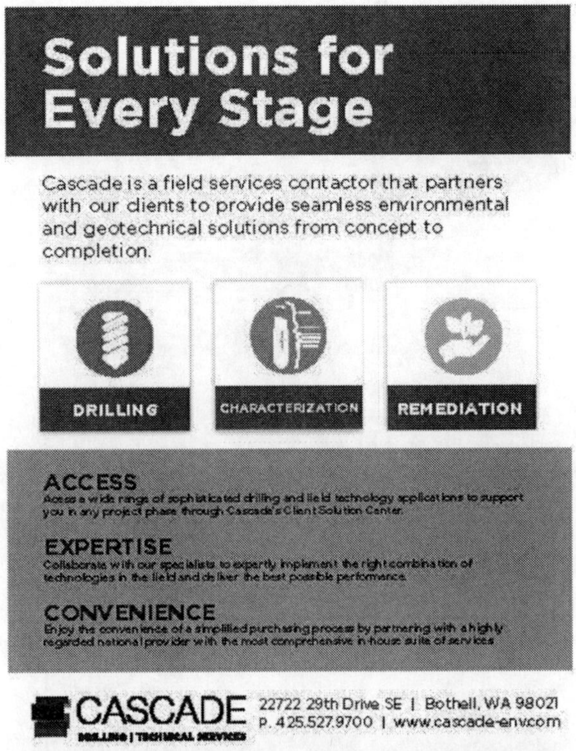
Solutions for Every Stage
Cascade is a field services contactor that partners with our clients to provide seamless environmental and geotechnical solutions from concept to completion.
DRILLING
CHARACTERIZATION
REMEDIATION
ACCESS
Access a wide range of sophisticated drilling and field technology applications to support you in any project phase through Cascade's Client Solution Center.
EXPERTISE
Collaborate with our specialists to expertly implement the right combination of technologies in the field and deliver the best possible performance.
CONVENIENCE
Enjoy the convenience of a simplified purchasing process by partnering with a highly regarded national provider with the most comprehensive in-house suite of services.
CASCADE
DRILLING | TECHNICAL SERVICES
22722 29th Drive SE | Bothell, WA 98021
P. 425.527.9700 | www.cascade-env.com

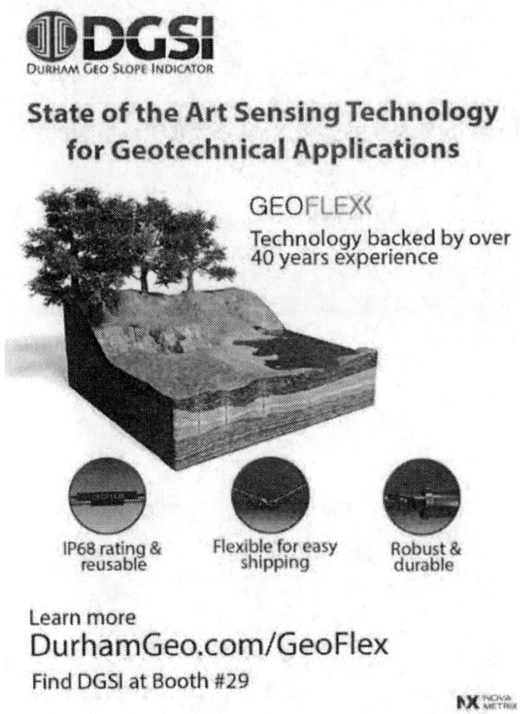
DGSI
DURHAM GEO SLOPE INDICATOR
State of the Art Sensing Technology for Geotechnical Applications
GEOFLEX
Technology backed by over 40 years experience
IP68 rating & reusable
Flexible for easy shipping
Robust & durable
Learn more
DurhamGeo.com/GeoFlex
Find DGSI at Booth #29

GeoHazard Mitigation Experts
› Grouting & Foundation Repairs
› Landslide Remediation
› Rockfall Mitigation
...and more!
24 Hour Emergency Response
GSI
Service across the United States & Canada
855-579-0536
GeoStabilization.com

Rockfall Mitigation
Ground Engineering
Environmental Services
Shoreline and River Studies
Soil and Rock Slope Stability
Environmental Health and Safety
Hazard Assessment Rugged Terrain Design
Pavement Design and Management Remediation
We thrive
on challenges
golder.com
GOLDER

IDS
GeoRadar
Critical Slope Monitoring
IDSNA Geomatics Services Group
Contact our team for Critical Slope Monitoring
and Risk Reduction Monitoring deployments.
IDSNA Inc.
14828 W. 6th Ave., Ste. 12-B
Golden, CO., 80401, United States
Tel: +1 303-232-3047 Ext. 121
EM: idsna.geo@idsgeoradar.com
IDS
GeoRadar
NORTH AMERICA

Slide2
RocFall
Dips
SWedge
TOOLS FOR
SLOPE STABILITY
Try the full version of any Rocscience software for free: rocscience.com
rocscience
Geotechnical tools,
inspired by you.

WILLIAMS

70th ANNUAL HIGHWAY GEOLOGY SYMPOSIUM
Exhibitor Display Locations

Exhibit Directory
Alphabetical by Company Name

Access Limited Construction	7
Acker Drill	15
Ameritech Slope Constructors, Inc	2
AMS INC	13
Apex Rockfall Mitigation LLC	17
Atlas Pipe Piles	22
BGC Engineering	6
Cascade Drilling	12
Central Mine Equipment Company	26
Durham Geo Slope Indicator	29
Geobrugg North America, LLC	18
Geokon	11
GeoStabilization International LLC	4
Gilson Company, Inc.	23
Golder Associates Inc	9
GRI	28
HI-TECH Rockfall Construction Inc.	8
IDS GeoRadar	14
Jensen Drilling Company	1
Juniper Unmanned	10
KANE GeoTech	19
Landslide Technology	21
Maccaferri Inc.	20
NHAZCA S.r.l.	30
Olson Engineering, Inc.	27
Pacific Blasting	25
Rocscience, Inc	24
Shannon & Wilson	3
Trumer Schutzbauten	5
Williams Form Engineering Corp.	16

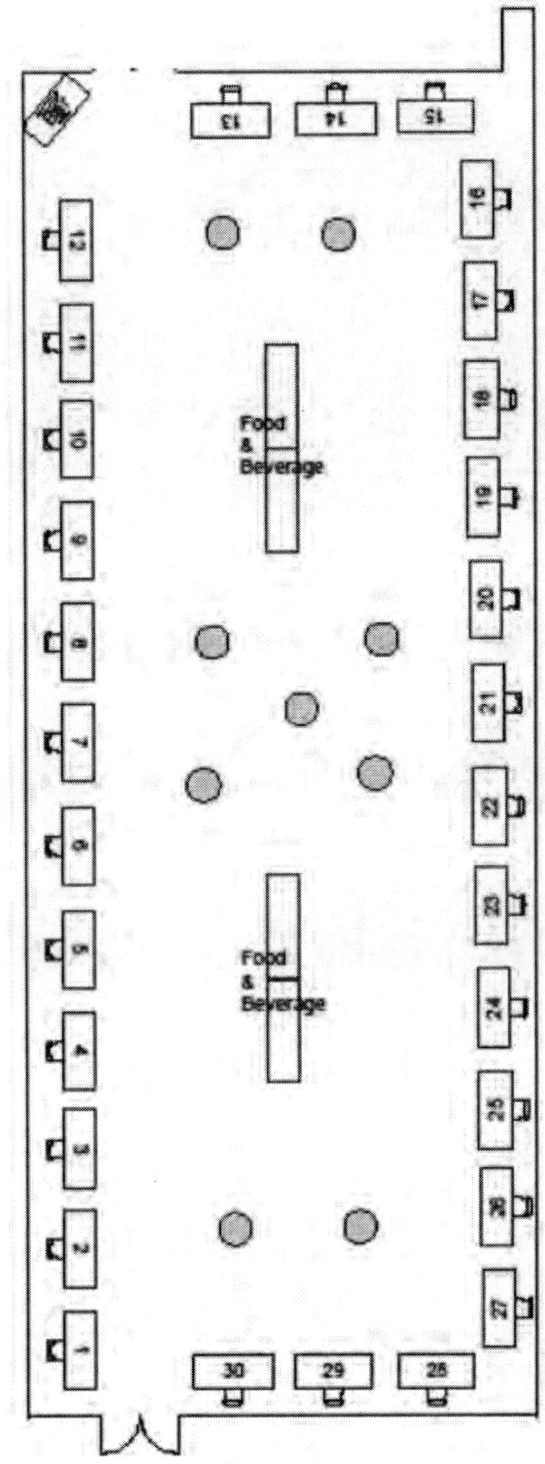

70th ANNUAL HIGHWAY GEOLOGY SYMPOSIUM
Exhibitors

Thank you to all participating exhibitors. The exhibit booths are in the Salon E.

Access Limited Construction
1102 Pike Lane
Oceano, California 93445
P: (805)-592-2230
www.accesslimitedconstruction.com

Acker Drill Company
PO Box 830
Scranton, Pennsylvania 18501
P: (501) 479-2009
www.ackerdrill.com

Ameritech Slope Constructors, Inc.
PO Box 2702
Asheville NC 28802
P: (828) 633-6352
Fax: (828) 398-2041
www.ameritech.pro

AMS, Inc.
105 Harrison Street
American Falls, Idaho 83211
P: (208) 226-2017
www.ams-samplers.com

Apex Rockfall Mitigation, LLC
3229 Springfield Road
Grand Junction, Colorado 81503
P: (970) 314-7302
www.apexrockfall.com

70th ANNUAL HIGHWAY GEOLOGY SYMPOSIUM
Exhibitors

Atlas Pipe Piles
1855 East 122nd Street
Chicago, Illinois 60633
P: (312)-275-1608
www.atlaspipepiles.com

BGC Engineering, Inc.
Suite 211
701 - 12th Street
Golden, Colorado
USA, 80401
Tel: (720)-598-5982
www.bgcengineering.ca

Cascade Drilling, LP
13600 SE Ambler Road
Clackamas, Oregon 97015
P: (503) 572-3090
www.cascade-env.com

Central Mine Equipment Company
4215 Rider Trail North
Earth City, Missouri 63045
P: (800) 325-8827
www.cmeco.com

Durham Geo Slope Indicator
12123 Harbour Reach Drive, Suite 106
Mukilteo, Washington 98275
P: (425) 493-6249
www.durhamgeo.com

70th ANNUAL HIGHWAY GEOLOGY SYMPOSIUM
Exhibitors

GEOBRUGG®
BRUGG
Safety is our nature

Geobrugg North America, LLC.
22 Centro Algodones
Algodones, New Mexico 87001
P: (505) 771 4080
Fax: (505) 771 4081
www.geobrugg.com

GEOKON®

Geokon
48 Spencer Street
Lebanon, New Hampshire 03766
P: (603) 448-1562
www.geokon.com

GSI
GeoStabilization International®

GeoStabilization International
543 31 Road
Grand Junction, Colorado 81504
P: (855)-579-0536
www.geostabilization.com

GILSON® COMPANY, INC.

Gilson Company, Inc.
PO Box 200
7975 N Central Drive
Lewis Center, Ohio 43035
P: (800)-444-1508
www.globalgilson.com

GOLDER

Golder Associates
670 North Commercial Street, Suite 103
Manchester, New Hampshire 03101-1146
Phone: (603) 668-0880
Fax: (603) 668-1199
www.golder.com

70th ANNUAL HIGHWAY GEOLOGY SYMPOSIUM
Exhibitors

GRI
9750 SW Nimbus
Beaverton, Oregon 97008
P: (503) 641-3478
www.gri.com

Hi-Tech Rockfall Construction, Inc.
PO Box 674
Forest Grove, Oregon 97116
P: (503) 357-6508
Fax: (503) 357-7323
www.hitechrockfall.com

IDS GeoRadar
14828 w. 6th Avenue, Unit 12-B
Golden, Colorado 80401
P: (303)-726-6024
www.idsgeoradar.com

Jensen Drilling Company
1775 Henderson Ave.
Eugene, Oregon 97403
P: (800) 762-7435
www.jensendrilling.com

Juniper Unmanned
15000 W 6th Ave, Suite 330
Golden, Colorado 80401
P: 720-440-9960
www.JuniperUnmanned.com

70th ANNUAL HIGHWAY GEOLOGY SYMPOSIUM
Exhibitors

KANE GeoTech, Inc.
7400 Shoreline Drive, Suite 6
Stockton, California 95219
P: 209-472-1822
www.kanegeotech.com

Landslide Technology
10250 SW Greenburg Road, Suite 111
Portland, Oregon 97223
P: (503) 452-1200
www.landslidetechnology.com

Maccaferri Inc
9210 Corporate Blvd., Suite 220
Rockville, Maryland 20850
P: (240)-818-4116
www.maccaferri.com/us

NHAZCA S.r.l
Via Vittorio Bachelet 12
Rome, Italy 00185
P: +39-0695065820
www.nhazca.it

Olson Engineering, Inc.
12401 W. 49th Avenue
Wheat Ridge, Colorado 80033
P: 303-423-1212
www.olsonengineering.com

70th ANNUAL HIGHWAY GEOLOGY SYMPOSIUM

Exhibitors

Pacific Blasting
3183 Norland Avenue
Burnaby, British Columbia V5B 3A9
P: 604-291-1255
www.pacificblasting.com

Rocscience, Inc
54 St. Patrick Street
Toronto, Ontario M5T 1V1
P: 416-698-8217
www.rocscience.com

Shannon & Wilson
3990 Collins Way, Suite 100
Lake Oswego, Oregon 97035
P: (503) 210-4750
www.shannonwilson.com

Trumer Schutzbauten
720-999 West Broadway
Vancouver, British Columbia V5Z1K5
P: (604) 732-0325
www.trumer.cc

Williams Form Engineering Corp.
8165 Graphic Drive
Belmont, Michigan 49306
P: 616-866-0815
www.williamsform.com

Contributing Level Sponsor

McMillen Jacobs Associates
1011 Western Avenue, Suite 706
Seattle, Washington 98104
P: (206) 588-8193
www.mcmjac.com

70th ANNUAL HIGHWAY GEOLOGY SYMPOSIUM
October 21-24, 2019

Abstracts & Notes

Tuesday, October 22, 2019

Tuesday, October 22, 2019 (continued)

Thursday, October 24, 2019

Thursday, October 24, 2019 (continued)

10:00 AM – 10:30 AM
Morning Coffee Break
Location: Salon E

Geographic Information Systems (GIS) and Landslide Stabilization

Andrew Ferguson, PE
GeoStabilization International® (GSI®)
844 21 1/2 Rd
Grand Junction, CO 81505
(855)-579-0536
andrew@gsi.us

Jeff Hamlin
GeoStabilization International® (GSI®)
844 21 1/2 Rd
Grand Junction, CO 81505
(855)-579-0536
jeff.hamlin@gsi.us

Prepared for the 70th Highway Geology Symposium, October 2019

Acknowledgements

The author(s) would like to thank the individuals/entities for their contributions in the work described.

Derrick Hayes, PE – GeoStabilization International®
Ian Gates – GeoStabilization International®

Disclaimer

Statements and views presented in this paper are strictly those of the author(s), and do not necessarily reflect positions held by their affiliations, the Highway Geology Symposium (HGS), or others acknowledged above. The mention of trade names for commercial products does not imply the approval or endorsement by HGS.

Copyright Notice

Copyright © 2019 Highway Geology Symposium (HGS)
All Rights Reserved. Printed in the United States of America. No part of this publication may be reproduced or copied in any form or by any means – graphic, electronic, or mechanical, including photocopying, taping, or information storage and retrieval systems – without prior written permission of the HGS. This excludes the original author(s).

ABSTRACT

Understanding the geology and typical engineering properties of subsurface soil conditions is essential to identify possible failure mechanisms, perform site reconnaissance, and perform preliminary design analysis on landslides. States, local municipalities, and other organizations provide free access to Geographic Information Systems (GIS) including geologic and topographic maps, satellite imagery, LiDAR imagery, well logs, and geologic surveys. This paper will briefly discuss some of the free online resources available, how they can be applied to geohazard professional practice and their importance to successful landslide mitigation design.

Access to free online GIS tools has dramatically increased over recent years and shall grow with use as geospatial technologies continue to develop. The readily available information can be used to create multiple cross-sections at various locations, inspect the site along with the road platform, identify possible failure mechanisms and/or essential features, understand subsurface conditions, and create preliminary slope stability models. The application of these tools into engineering practice can provide a greater understanding of the existing geologic conditions and aid in designing the appropriate repair for landslides.

Case histories will be provided from completed projects to demonstrate the successful application of these tools in landslide mitigation. Projects benefiting from these applications include a landslide stabilization on Highway 36 in California, a deep-seated emergency landslide located on Highway 101 in Oregon, and preliminary design analysis on a landslide on Highway 138W in Oregon.

INTRODUCTION

Geohazard Professionals (i.e., engineers and geologists) strive to develop and install innovative solutions that protect people and infrastructure from the dangers of geohazards, such as landslides. The geology and engineering properties of subsurface soil conditions must be understood to implement these solutions effectively. As technology improves, the ability to understand site conditions also improves. As a result, geohazard professionals can now gather more information about a project location than ever before.

Technologies such as online GIS, available for free to the public, have created a platform where geohazard professionals can obtain information that previously required comprehensive subsurface investigations. This paper highlights several free online GIS resources and how to implement these technologies in practice successfully. The case studies provided in this paper are from completed projects showing the successful application of the free available GIS resources. The objective of this paper is to explain why geohazard professionals should incorporate GIS tools into their practice.

FREE AVAILABLE INFORMATION

Online GIS

Geographic Information Systems (GIS) is geographic data that is stored, managed, and analyzed in geospatial technologies where the information can be used for various applications. The information gathered can be visualized and assessed by the user for quick access to geographic data. Federal, state, and private entities collect and maintain online GIS databases where the information is easily accessible. This paper will outline various GIS tools available, where they can be found, and how the information applies to landslide stabilization.

Geologic Maps

Geologic maps provide information about subsurface conditions necessary for designing geohazard repair solutions. Geologic maps are available online from several sources, including the United States Geological Survey (USGS) and state or county GIS databases (Websites, 1). Valuable information required for proper design can be obtained or inferred from geologic maps, including structural orientation, geologic history, and rock/soil type. This information helps the designer make critical assumptions about important engineering material parameters in place of conventional geotechnical investigation techniques. Geologic data from these sources can often be incorporated into existing maps to help provide a more thorough understanding of site conditions. These maps can help identify potential concerns that may impact design, such as larger landslide complexes or adverse structural or bedding orientation.

Other Maps

Other readily available resources that are commonly used by geohazard professionals include topographic maps and landslide location maps. Topographic maps are helpful during the initial assessment phase and can be useful in determining slope severity and creating cross-sections to be used in modeling software. These maps can also be used to visually identify moderate to larger landslides or landslide complexes that may affect the project area. Landslide location maps, where available, provide information about the susceptibility of an area to sliding and can be useful determining the location and frequency of movement on historically active slides. All of these maps are typically available in digital format for free online.

LiDAR Imagery

Light Detection and Ranging (LiDAR) surveying is another useful tool in landslide mitigation that has become more readily available in recent years. LiDAR surveying collects data by using light pulsed lasers to measure ranges to the earth, which generate precise, three-dimensional geographic information of surface characteristics (Website, 2). LiDAR data is collected and provided through agreements with a variety of organizations including national, state, and private companies. State, federal, and private websites share information when possible to provide users with a large amount of data. Online resources offer interfaces where users can access and download pre-computed raster images, Google Earth KMZ files, and point cloud data (Website, 3).

LiDAR, when available, provides the ability to analyze the topography of a landslide in more detail than satellite imagery or topographic maps alone. Critical features of landslides, including head scarps, toe bulging, and limits of the failure are often easily identifiable. Topographic features, including ravines, rivers, and slope geometries are also more easily recognizable. Point cloud data can be downloaded and processed to generate cross-sections in areas of interest. When information derived from onsite investigation is limited or unavailable, cross-sections created from LiDAR datasets can provide a slope geometry to perform a back analysis. Figure 1 displays different perspectives of processed LiDAR data.

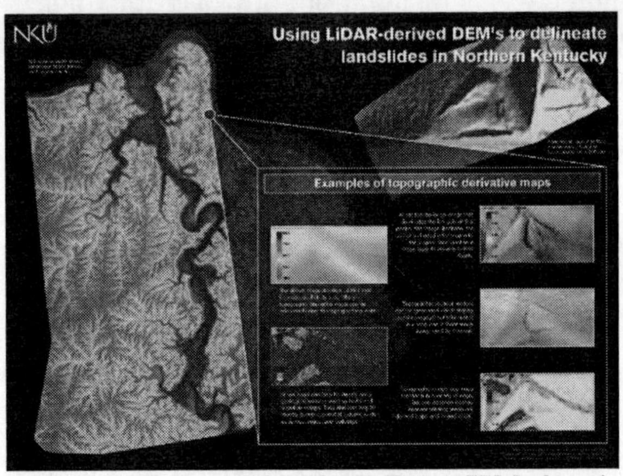

Figure 1: Processed LiDAR Data (Website, *4*)

Satellite Imagery

Free satellite imagery data from agencies around the world can be used for reference and informational purposes by merely performing an online search. Programs such as Google Earth allow users to take measurements, gather elevation data, view the site from the road platform, and even view images taken at various dates over time. The programs include roadway information, city and county information, and global positioning coordinates for reference. Overlay files such as images, LiDAR, township & range, soil surveys, and many other records can be placed into these programs for visualization.

The information obtained from these free available resources is valuable in understanding the geology and critical features of landslides. When street views are available, the user can gain a deeper understanding by virtually visiting the site to get a 360-degree view. Geologic features including slope geometries, rock outcrops, and natural drainage paths can be identified above, below, and within the landslide. Key features of the landslide including head scarps, limits of the failure, cut/fill interfaces, pistol butted trees, and bulging can often be identified. Assumptions on subsurface profiles can even be made based on approximated measurements obtained from satellite imagery.

Well Logs

Well logs are used in landslide analysis, especially when subsurface investigation and/or testing has not been performed. Most states require wells to be drilled by a licensed driller/operator to protect groundwater resources from contamination. Drilling reports, also referred to as well logs, are typically required for documentation as part of the permitting process. State water resource departments maintain the well logs and the information can be found on GIS data hubs online. Subsurface investigations involving drilling performed by transportation departments or other entities can also be a resource for locating well logs.

Well logs typically include information such as date and location of drilling, drilling conditions, observed subsurface material types, static water levels. The driller/operator records the observed conditions during the well's installation; however, the logs may not have been reviewed by an engineer or geologist. Therefore, judgment should be used when reviewing well logs and incorporating data into the design solution. Additional information such as geologic maps and/or web soil surveys may also be needed to verify drilling conditions. Figure 2 displays well logs that can be obtained from online GIS databases.

Figure 2: GIS Well Log Map (Website, *5*)

Web Soil Surveys

Web soil surveys provide a comprehensive database of soil maps and soil data throughout the United States. General soil information gathered by the USDA Natural Conservation Service (NRCS) is maintained in real-time on the interactive Web Soil Survey application for online access (Website, 6). Currently, there are data and soil maps for more than 95 percent of the counties nationwide. The NRCS also provides access to historical soil survey reports in pdf format archived by state for reference. Programs such as the SoilWeb app by UC Davis incorporate web soil survey data provided by NRCS for use in interactive web browsers, Google Earth, and smartphones (Website, 7).

The web soil surveys are useful in landslide analysis because they give insight into the near-surface soil conditions as determined by the USDA. Typical surveys will provide engineering and chemical properties for soils up to approximately six feet deep. Engineering properties from the surveys include USCS and AASHTO soil classification, USDA texture, hydrologic properties, sieve analysis, liquid limits, plasticity index, and soil PH. This information can be used to estimate important soil strength parameters, create soil profiles, identify the potential presence of water, and estimate the corrosion potential of the soils. Figure 3 displays soil data from the web soil survey application.

Figure 3: Web Soil Survey Data (Website, *6*)

APPLICATION

Once the available tools have been identified for use in a landslide repair, the next step is to apply them to geohazard mitigation practice. Free resources such as online interactive GIS programs and other free design tools provide designers with site-specific information necessary to perform an analysis. In summary, the free available information online can be used to create cross-sections, make visual site observations, identify possible failure mechanisms, understand subsurface conditions, and make detailed measurements. This information can then be used to create a back analysis and evaluate repair options.

Back Analysis

The purpose of a back analysis is to create a reasonable model of the site conditions at the time of failure. A slope stability analysis is performed with the measured critical failure surface, having a factor of safety equal to one (Textbooks, 1.). The engineering properties of the soil/rock are estimated based on an iterative process of adjusting material parameters until the failure surface is accurately re-created using various soil/rock mechanics parameters. In the absence of Geotechnical reports, the free available information can meet or exceed the minimum requirements to create a back analysis. The free, accessible data can also be used along with a geotechnical report to supplement information not provided.

Preliminary design

Once a representative back analysis is created, the project then moves into the design phase of the landslide repair. During the design phase, the designer evaluates repair options based on cost and constructability. The designer uses the back analysis to create a repair model used to determine the driving and resisting forces required to stabilize the landslide. Structural elements and supporting calculations will be derived from the resting forces.

CASE HISTORIES

This paper contains three case studies showing how successful application of GIS tools was used in landslide mitigation. The projects were designed based on limited geotechnical subsurface information available at the time of design; therefore, online GIS was instrumental in evaluating these sites. The case studies will outline the design methodology used to obtain supporting information, determine failure mechanisms, and ultimately provide design-build repair options. Lastly, the case studies will demonstrate how the application of these tools was used for successful project outcomes.

Case Study 1:

Mad River, California

Background

This landslide was in Trinity County near Mad River, California. In the area of interest, a V-shaped head scarped extended from the outboard slope face to the inboard ditch approximately 105 LF measured along the guardrail. Vertical displacement on the road platform had been a reoccurring issue requiring yearly maintenance to keep the road safe for the traveling public. Traditional solutions had been evaluated but were not economically feasible. Figure 4 displays the pre-construction conditions of the area of interest.

Figure 4: Existing Head Scarp Near Mad River, California

Information obtained from site visits indicated the road platform was originally constructed approximately 80 feet slope distance below the current roadway elevation. The new road platform appeared to be built entirely of fill material with no clear delineations of the native soil contact. Rock outcrops were observed upslope with increased exposure along the western side of the road. Drainage appeared to be concentrated where the head scarp extended into the inboard ditch. Inclinometers and piezometers were also observed throughout the slope face; however, none of the data collected from these devices could be obtained.

Design Methodology

A back analysis was created to evaluate possible failure mechanisms and estimate the engineering properties of the rock and soil. The site visit provided crucial information for field measurements and general assumptions. However, no subsurface information was available to aid in the analysis. Supporting resources used to create a representative back analysis were geologic maps, Google Earth, and web soil surveys (Website, 3 and 6).

Web soil surveys were used to estimate engineering parameters of native soils in the area. The survey indicated the native soils were coarse-grained and contained fine-grained sediments with variable plasticity. The USGS hydrologic soil group conditions classified the native soils as a group C for slow infiltration rates (Website, 6). Poor drainage of the native soils was assumed to be one of the driving factors in the landslide. The general soil classification provided a starting point for typical engineering properties of the native soils.

Geologic maps were used to refine bedrock parameters used in the back analysis. These maps indicated that bedrocks in the project area were of Cretaceous and Jurassic age and were composed of massive graywacke, minor amounts of platy shale, thin-bedded chert, and mildly metamorphosed sedimentary and volcanics of the Franciscan Formation. Several north-northwest striking, large scale, thrust and strike-slip faults were identified adjacent to the region, suggesting bedrock in the area was highly-fractured. It is interesting to note that directly above the road platform at this site, a massive, highly-fractured outcrop is juxtaposed against highly-weathered marine sedimentary rocks of seemingly different origin. Based on geologic maps of the area, one of the nearby larger-scale strike-slip faults project directly into the project area and could help explain the juxtaposition of these different rock types. These geologic maps were also used as a reference to infer native soil properties derived from this parent material. The geologic maps helped estimate the typical engineering properties of the rock used in the back analysis. They also provided an understanding of the standard bedding and jointing of the rock in the area which was used to estimate reasonable bond capacities for anchor elements. Figure 5 shows the geologic map of the case study near Mad River, CA.

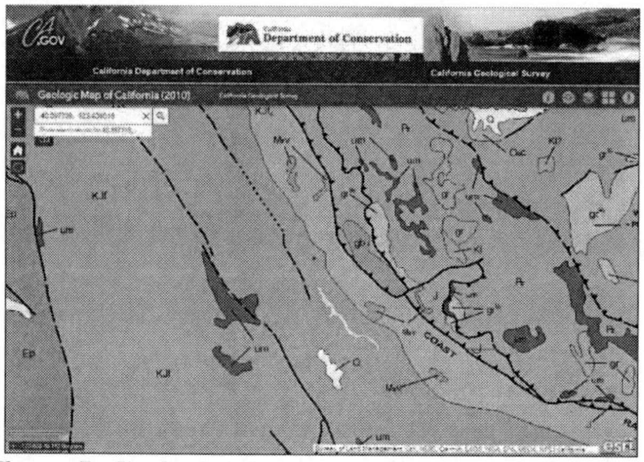

Figure 5: Geologic Map of Case Study 1 (Website, *8*)

Google Earth was used to provide additional information used in analyzing the site (Website, 3). Satellite imagery provided important information including the head scarp location, slope conditions outside of the landslide limits, preferential drainage pathways, and historical imagery dating back to 1993. Street-view images were used to "revisit" the site as questions arose during the design phase. These images revealed the highly-fractured nature of the bedrock

Web soil surveys were used to understand the native soil conditions in the area. The survey indicated the surficial soils were of coarse-grained with fines varying from low plastic to plastic. The hydrologic soil group conditions classified the native soils as a group C for slow infiltration rates (Website, 6). Poor drainage of the native soils was assumed to be one of the driving factors in the landslide. The general soil classification provided a starting point for typical engineering properties of the native soils.

The back analysis was evaluated using a non-circular failure surface in Slide by Rocscience (Figure 6; Website, 9). The critical failure surface was modeled as a deep-seated failure extending from the inboard ditch and extending down to the location of the previous road platform. The slide was assumed to be confined within the fill along native contact. Sensitivity analyses were performed until the critical failure matched conditions observed in the field. The model created in Slide by Rocscience can be seen below in the Figure.

The repair consisted of a 20-foot-high soil nail wall with up to 50-foot embedment hollow bar injection anchors and reinforced shotcrete per FHWA GEC 7 design guidance (Federal Government Reports, 1). Horizontal drains were installed with up to 20-foot embedment for the length of the repair, and in areas groundwater was encountered during drilling. The anchors were conservatively designed with allowable bond capacities non-dependent on bond values for a homogenous soil for full embedment. However, all the anchors were installed with at least 10 feet embedment into bedrock. Design assumptions, including subsurface profiles and anchor bond capacities, were vetted during construction and field testing. The model created in Slide by Rocscience for the repair can be seen in Figure 6.

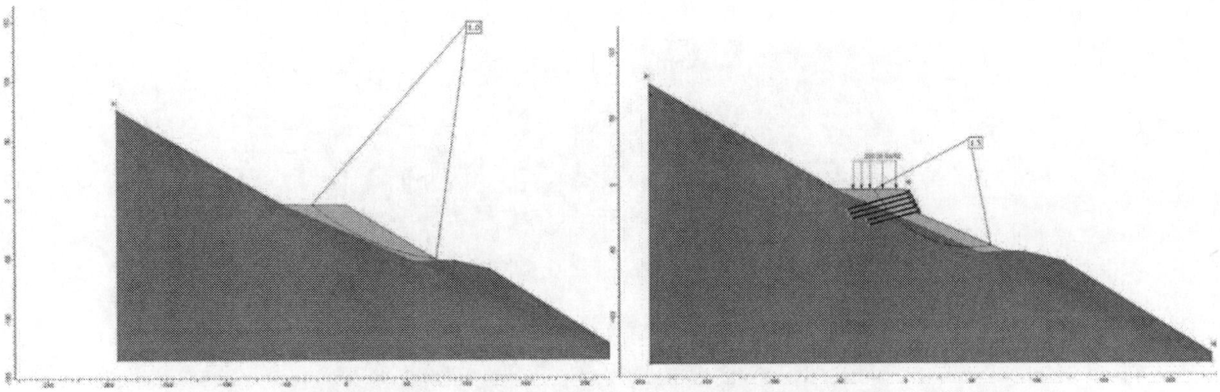

Figure 6: Back Analysis and Repair Analysis for Case Study 1 (Website, 9)

Project Outcome

The desired outcome of the project was to stabilize the landslide quickly with an economically feasible solution. The project duration lasted approximately six weeks, and the highway remained open during construction with limited impact on traffic. While important slide information collected from previously installed inclinometers and piezometers were never obtained, data derived from free online resources and field reconnaissance helped provide crucial subsurface information necessary for the design. The final product of the project can be viewed in Figure 7.

Figure 7: Final Repair for Case Study 1

Case Study 2: Hwy 101, Oregon

Background

Case study two was located along Hwy 101 south of Coos Bay, Oregon. A massive winter storm passed through the area resulting in approximately 14 inches of rainfall within three days. Multiple landslides were triggered due to the large concentration of moisture, and this landslide of interest sat on the flanks of a larger landslide complex. Figure 8 displays the pre-construction conditions of the area of interest.

Figure 8: Site Photo of Highway 101

Visual observation from the initial reconnaissance visit revealed two scarps along the trace of the slide; the shorter scarp, which had up to approximately 8 inches of offset, was over 230 LF measured along the guardrail and extended to the outboard fog line. The longer scarp, with approximately 3 inches of offset, was over 300 LF and impacted all three travel lanes (and paved shoulder) while extending over 45 feet inboard to the fog line.

Design Methodology

A detailed site investigation to obtain geotechnical information was not able to be performed due to the emergency response time needed to stabilize the landslide. Continued rainfall during the initial site visit made conditions too dangerous to access the slope. Additionally, the dense vegetation along the slope face prevented a clear view or reasonable access to the area. Although access and site information was limited, an analysis was able to be performed successfully as a direct result of the free available GIS resources. Supporting resources used to analyze the landslide were well logs, satellite imagery, geologic maps, soil surveys, and LiDAR imagery.

Well logs provided by the Oregon Well Resources Department were used to gain an understanding of typical subsurface conditions in the area (Website, 10). There were no well logs available in the immediate vicinity of the project. However, nearby well logs were used to approximate soil thickness and to estimate depths to bedrock. Based on the available information, the native soil was estimated to be approximately 15 to 20 feet thick overlying highly weathered bedrock. General soil observations, bedrock conditions, and water level elevations were documented from the well log, so reasonable assumptions could be made in the back analysis.

Satellite imagery from Bing Maps and Google Earth were used to assess the conditions of the surrounding area and identify key features in the landslide (Website, 3 and 11). These resources were used to estimate the location of the fill and native material interface below the road platform based on slope geometries and apparent upslope conditions. Historic street view images indicated the larger head scarp had been paved over in recent years. They also indicated a drainage channel located approximately 250 LF downslope incised the hillside oblique to the movement of the slide mass. The channel appeared to undermine the landslide resulting in a loss of resisting forces contributing to the slope instability.

Geologic maps and web soil surveys were used to understand the geology in the area, so engineering properties of the subsurface materials could be estimated. Geologic maps provided by the Oregon Department of Geology and Mineral Industries indicated the bedrock in the region was composed of severely folded and deformed marine sedimentary and mafic volcanic rocks of Cretaceous age (Website, 12). The overlying soils derived from these deeply weathered parent rocks were inferred to be weak with moderate cohesion and moderately lower phi angle values. A USGS web soil survey indicated the overlying soils consisted of a low to high plasticity clay with silt with slow infiltration rates (Website, 6). Soil chemical properties were also evaluated, indicating the soil PH ranged between 5 and 6 in the upper soils. Figure 9 shows a geologic map of the project area procured from the Oregon Department of Geology and Mineral Industries Website (DOGAMI).

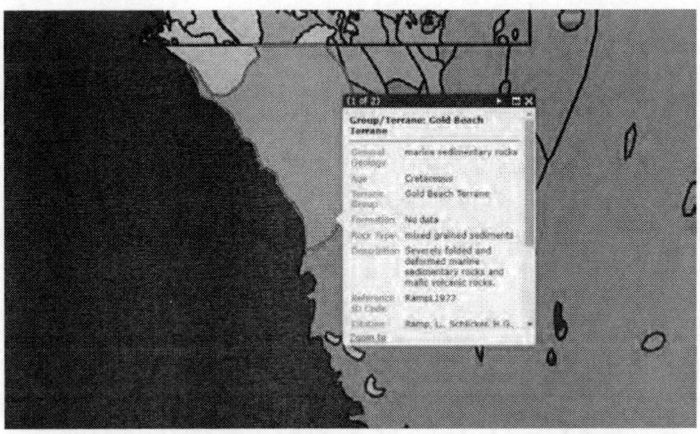

Figure 10: Geologic Map of Case Study 2 (Website, *12*)

Due to the emergency nature of the landslide and the unsafe slope conditions, detailed measurements of the slope were not available. However, LiDAR imagery was available and used to create representative cross-sections at multiple locations. Open Topography resources were used to download LiDAR data from the Oregon Department of Geology and Mineral Industries (Website, 12). The data was then processed using free geospatial analysis software to export cross-sectional profiles of the slide mass. The profiles were analyzed, and initial observations identifying the erosional channel at the toe of the slope were supported.

An understanding of the subsurface and geologic conditions was solidified by successfully incorporating free online GIS information into the site research. The available information made it possible to create a representative back analysis to estimate engineering soil and rock properties at the time of failure. Sensitivity analyses were performed to evaluate the effects of varying water and layer profile depths until the critical failure surfaces matched conditions observed in the field. The representative back analysis showed the two failures were deep-seated and likely failed through the weathered bedrock layer. The model created in Slide by Rocscience for the repair can be seen in Figure 10.

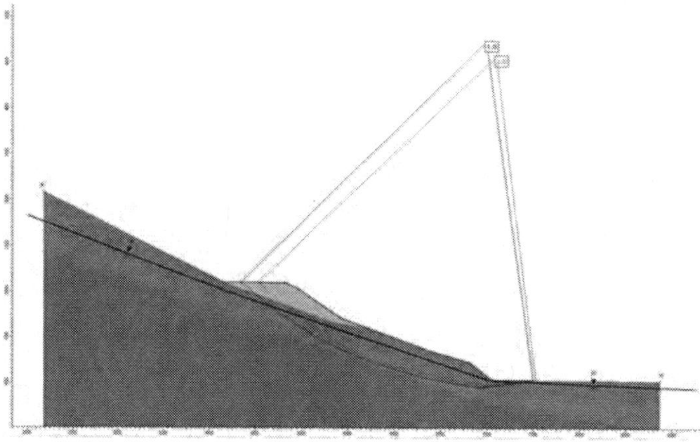

Figure 10: Back Analysis of Highway 101 (Website, *9*)

The repair consisted of a soil nail wall varying in height up to 19 feet tall at the critical section. Hollow bar soil nails were installed in varying lengths up to 60 feet with a reinforced shotcrete facing. Subsurface conditions and anchor lengths required injection anchors so grout encapsulation would be maintained during installation. Horizontal drains were installed along the base of the soil nail wall for anticipated groundwater and connected to toe manifold for positive drainage. Design assumptions were vetted during construction based on real time drilling feedback and anchor testing. The model created in Slide by Rocscience for the repair can be seen in Figure 11.

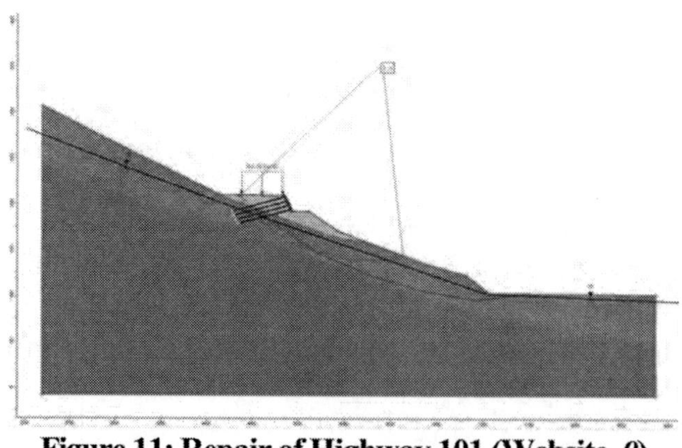

Figure 11: Repair of Highway 101 (Website, *9*)

Project Outcome

The objective of the design was to stabilize a sizeable deep-seated emergency landslide as fast as possible with limited information. Geotechnical investigations were not feasible because the emergency landslide occurred on a vital stretch of highway that needed to be re-opened to traffic. As a result, access to free innovative GIS tools was a crucial factor in the successful stabilization of the landslide. Work was performed in lift construction from the platform where possible and a construction bench below as required. The project was designed and priced in less than three days, and construction took approximately eight weeks to complete. Figure 12 shows the final repair of this project.

Figure 12: Final Repair of Highway 101

Case Study 3: Hwy 138W, Oregon

Background

Case study three was located on Hwy 138W in Douglas County south of Elkton, OR. As seen below in Figure 13, a large head scarp extending a couple of feet beyond the inboard fog line was observed for approximately 195 LF as measured along the outboard fog line. The head scarp was clearly delineated as it approached the bottom of the slope with signs of toe bulging. Another smaller head scarp appeared to extend along the outboard shoulder of the road surface. Figure 13 displays the pre-construction conditions of the area of interest.

Figure 13: Site Photo of Highway 138W

Visual observations taken from the site visit indicated there was a rock buttress along the base of the landslide extending north beyond the limits of the observed failure. The buttress appeared to be providing adequate resisting forces north of the slide mass, but additional support would be required for the southern section showing distress. A road was cut into the side of the slope for buttress installation, and there were multiple construction benches observed along the slope face. Drainage was identified as a possible problem on the inboard side of the roadway. Dense vegetation made it hard to identify additional key features.

Design Methodology

Accurate field measurements were difficult to obtain because of the dense vegetation above and below the road platform. The dense vegetation also prevented a clear view of the area upslope of the landslide. The benches constructed along the slope face made it difficult to determine the slope geometry without limiting assumptions. Supporting resources to effectively analyze the landslide were satellite imagery, geologic maps, web soil surveys, well logs, and LiDAR imagery. A back analysis was then created with the additional information obtained.

Information obtained from satellite imagery using Google Earth was used to analyze the topography of the area, including the geometry immediate above and below the landslide (Website, 3). Using Google Earth, the bottom of the slope could be easily identified with gross measurements indicating horizontal and vertical distances from the road platform to the apparent toe of the slide. Above the road platform there appeared to be an irrigated field and a ravine where the drainage appeared to be concentrated to the critical section of the landslide. Lastly, the geometry of the slopes surrounding the slide mass indicated this section of road was constructed mainly of fill materials.

Geologic maps provided by the Oregon Geologic Survey and USGS were then consulted to get a general understanding of the local geology of the area (Website, 1). Bedrock, if encountered, would consist of sedimentary rocks, including sandstone, siltstone, and mudstone. Web soil surveys were then used to get a general understanding of the typical surface soil conditions in the area. The results of the web soil survey indicated the native soils consisted of low to high plastic silts and silty clays low to moderate infiltration rates when thoroughly wet. Figure 14 below displays how the geologic map was procured for the case study south of Elkton, OR.

Figure 14: Geologic Map of Case Study 3

Once a good understanding of the geology and surface soils was achieved, the next step was to determine the stratigraphy of the area. Two well logs were found in the nearby vicinity using the Oregon Water Resources Well Report Query (Website, 10). The first well log was in the irrigated field upslope from the project location and the second well log was a subsurface investigation performed by the Oregon Department of Transportation (ODOT) located on Hwy 138W MP 4.7 (approximately 0.10 miles away). The well logs provided detailed observations made during installation, including drilling conditions, material layer thicknesses, and static water levels.

LiDAR imagery provided by the Oregon Department of Geology and Mineral Industries and Open Topography was also used to analyze the site (Website, 12). The LiDAR supported observations made from the satellite imagery indicating the runoff was concentrated at the landslide, and the road platform was constructed with a significant amount of built-up fill material. Key features were clearly identifiable, including the shape/limits of the slide mass, the benches built into the slope face, and bulges. Lastly, the downloadable LiDAR data was used to generate multiple cross-sections, including a cross-section at the critical section.

From the freely available information, supplemented with the site visit, gave enough detailed information about the landslide to prepare a representative back analysis of the failure. The critical failure surface was modeled with a deep-seated non-circular failure extending through the native and fill materials. Soil profiles and groundwater depths were estimated based on the boring log information provided by ODOT. Sensitivity analyses were performed to evaluate the effects of varying depths of native soil, rock, and groundwater. The final analysis showed failure surfaces at approximately the same locations as measured in the field. The model created in Slide by Rocscience for the back analysis can be seen in Figure 15 below.

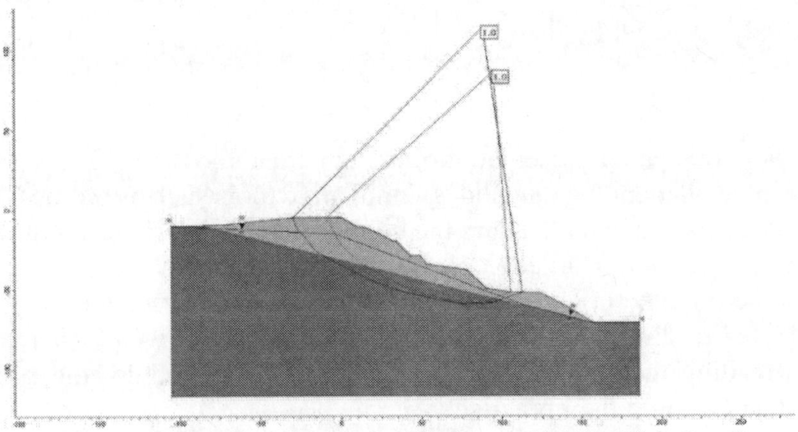

Figure 15: Back Analysis for Highway 138W (Website, *9*)

The proposed repair consisted of a soil nail wall in varying height up to 15-feet in the critical section. An array of hollow bar injection anchors varying in lengths up to 60-foot embedment and reinforced shotcrete was designed in the critical section. The wall heights and embedment lengths were optimized based on scarp measurements and cross-sections taken at various locations throughout the slide mass. Hollow bar injection anchors were selected and designed based on anticipated drilling conditions for a sandy fill material and silty clay w/sand native material identified in the free available information. The model created in Slide by Rocscience for the repair can be seen in Figure 16 below.

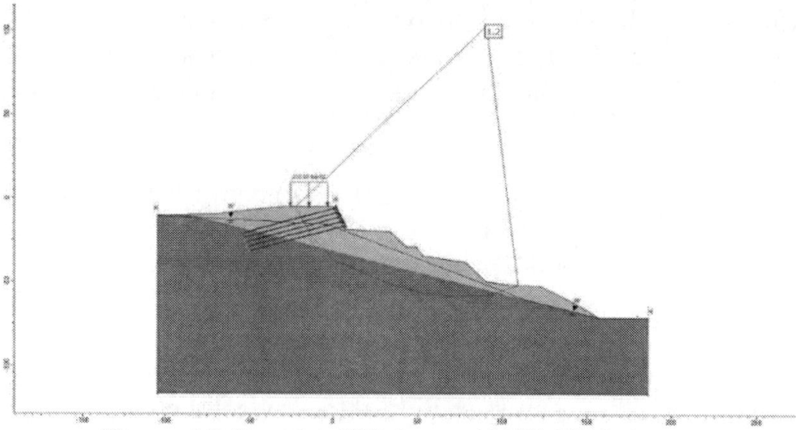

Figure 16: Repair of Highway 138W (Website, *9*)

Project Outcome

The object of this design was to stabilize the sizeable deep-seated landslide with limited impact to the traveling public. The proposed design consisted of a 1.2 factor of safety, which was considered adequate for a landslide of this size and magnitude. The projected duration of the project would last approximately five to six weeks. The work could be performed from the existing road platform or a construction bench in a top-down construction. Design assumptions would be vetted during installation by a quality assurance program involving the documentation of drilling conditions and proof testing on production anchors.

CONCLUSION

In conclusion, this paper presented the reader with some of the free available information to help engineers solve challenging landslides. Innovative tools such as online GIS were identified so Geohazard Professionals can understand the geologic and subsurface conditions of a site when geotechnical reports are not available. Next, the paper showed how these tools could be incorporated into the engineering practice of creating back analyses and evaluating repairs for landslides. Lastly, case studies demonstrated how these tools were used to stabilize actual landslides. Upon reading this paper, Geohazard Professionals should understand the power of incorporating Online GIS into their practice.

REFERENCES:

- Federal Government Manuals

1. Lazarte, C.A., Robinson, H., Gomez, J.E., Baxter, A., Cadden, A., Berg, R. *Soil Nail Walls Reference Manual*. Report FHWA-NHI-14-007, U.S. Department of Transportation, 2015.

- Textbooks

1. Turner, A.K. and Schuster, R.L. *Landslides Investigation and Mitigation.* Transportation Research Board, Special Report 247, 1996, pp. 363-365.

- Website

1. United States Geological Survey. *Maps*. https://www.usgs.gov/products/maps/overview. Accessed Jul 20, 2019.

2. National Ocean Service. *What is LiDAR?* https://oceanservice.noaa.gov/facts/lidar.html. Accessed Jul. 29, 2019

3. Google. *Google Earth*. https://earth.google.com/web/. Accessed Jul. 20, 2019.

4. Amundsen, J. *LiDAR and Landslides*. http://www.trishock.com/academic/lidar.shtml. Accessed Jul. 30, 2019.

5. Idaho Department of Water Resources. Wells and Areas of Drilling Concern Map. https://idwr.maps.arcgis.com/apps/webappviewer/index.html?id=c9c65a636adb478fbc63057bc267d741. Accessed Jul. 30, 2019.

6. National Resource Conservation Service. *Soil Surveys*. https://www.nrcs.usda.gov/wps/portal/nrcs/soilsurvey/soils/survey/state/. Accessed Jul. 20, 2019.

7. UC Davis California Resource Soil Lab. *SoilWeb Apps*. https://casoilresource.lawr.ucdavis.edu/soilweb-apps/. Accessed Jul. 20, 2019.

8. California Department of Conservation. Geologic Map of California (2010). http://maps.conservation.ca.gov/cgs/gmc/. Accessed Jul. 31, 2019.

9. Rocscience Inc. Slide. https://www.rocscience.com/. Accessed Jul. 20, 2019.

10. Oregon Water Resources Department. *Well Report Query*. https://apps.wrd.state.or.us/apps/gw/well_log/. Accessed Jul. 20, 2019.

11. Bing. *Bing Maps*. https://www.bing.com/maps. Accessed Jul. 20, 2019.

12. Oregon Department of Geology and Mineral Industries. *Interactive Maps*. https://www.oregongeology.org/gis/index.htm. Accessed Jul. 20, 2019.

Geohazard Mitigation Alternative Foundation Backfill For Emergency Construction

Joseph McElhany, EIT

KANE GeoTech, Inc.

7400 Shoreline Drive, Suite 6, Stockton, California 95219

1441 Kapiolani Boulevard, Suite 1115, Honolulu, Hawaii, 96814

(209) 472-1822

joey.mcelhany@kanegeotech.com

William Kane, Ph.D., PG, PE

KANE GeoTech, Inc.

7400 Shoreline Drive, Suite 6, Stockton, California 95219

1441 Kapiolani Boulevard, Suite 1115, Honolulu, Hawaii 96814

(209) 472-1822

Prepared for the 70th Highway Geology Symposium, October 2019

Acknowledgements

The author(s) would like to thank the individuals/entities for their contributions in the work described:

Brian Forsthoff, MSCE

Mallory Jones, GIT

William Kane, Ph.D., PG, PE

Access Limited Construction, Inc.

Disclaimer

Statements and views presented in this paper are strictly those of the author(s), and do not necessarily reflect positions held by their affiliations, the Highway Geology Symposium (HGS), or others acknowledged above. The mention of trade names for commercial products does not imply the approval or endorsement by HGS.

Copyright Notice

Copyright © 2019 Highway Geology Symposium (HGS)

All Rights Reserved. Printed in the United States of America. No part of this publication may be reproduced or copied in any form or by any means – graphic, electronic, or mechanical, including photocopying, taping, or information storage and retrieval systems – without prior written permission of the HGS. This excludes the original author(s).

ABSTRACT

The design of geohazard mitigation protection barriers for debris flow, landslides, and rockfall are governed by lateral loads. After impact, the loads are distributed through the system with post foundations transferring the loads to the ground. Cast-In-Drill-Hole (CIDH) foundations can be designed to resist the lateral loading by calculating the depth required using the properties of the in-situ surrounding soil. However, geohazard mitigation systems are often installed in remote locations where large drilling equipment may not be appropriate due to limited access or inclined terrain. Alternatively, ground improvement techniques for in-situ soils can be applied to support foundations. Ground improvement is a technique that modifies the existing soil properties by improving strength. Two types of ground improvement backfill are Controlled Density Fill (CDF) and Soilcrete. Both types of backfill improve soil lateral support resulting in reduced CIDH foundation depth. For emergency construction projects, the use of these ground improvement options potentially eliminates the need for large drilling equipment and decreases overall construction duration and costs. Three case studies will be presented where Soilcrete and CDF were used to improve foundation support. This resulted in meeting emergency construction time constraints and design requirements for the foundations.

INTRODUCTION

Since the 1950s different techniques have been used around the world to help improve existing ground conditions for foundation and embankment applications in geotechnical engineering (1). These include embankment support and widening, bridge abutment support, bridge pier support, retaining wall foundations, and deep pile foundations. The idea behind ground improvement is to use native soil material blended with cementitious material. This technique helps increase soil stability and strength while also reducing the soil settlement and permeability. Different ground improvement techniques have been developed for a range of applications of geotechnical engineering. These techniques include jet grouting, dynamic compaction, and deep soil mixing method (DMM). This paper focuses on the dry soil mixing method (Soilcrete) and the wet mixing method (Controlled Density Fill) for backfill support.

Soilcrete Background

Since 2003, Soilcrete has been used by the Federal Highway Administration (FHWA) for deep soil mixing and has developed the FHWA Design Manual for Deep Mixing for Embankment and Foundation Support (1). The report provides deep soil mixing design and construction guidelines for embankment and foundation support. Soilcrete is a mix of native soil material with a cementitious base, admixtures, and binders that affect the overall flowability and or strength of the Soilcrete. The native soil for mixing is often loose, unconsolidated soils that are not stiff or include large aggregate size cobbles or boulders, Figure 1. Silts and sands react better with cement mixing rather than clayey or organic soils which require an increase in binder materials to achieve similar strengths (1). The cement component of the mix is typically a readily available Portland type II cement. The angular aggregate for the mix is ranged between ½-in to ¾-in; this is to provide a well-cemented mix that has low void possibilities that would be evident using native sized boulders or cobbles. In addition, angular aggregate helps increase the frictional shear strength component of the mix more than a rounded cobble size rock.

The phase relationship of the soilcrete strata is critical to developing a competent Soilcrete mixture. The water to binder (w:b) ratio is critical for developing strength workability of the Soilcrete. Deep soil mixing projects w:b ratios vary based on the minimum strength required, workability, and or flowability. Guidelines usually suggest w:b ratios ranging from less than 1.0 for shallow foundations to greater than 1.5 for deep foundations (1). Values used in the case studies presented used a w:b ratio of 0.45 for a shallow foundation 4-ft deep. Admixtures can be fluidifiers, dispersants, and or retarding agents are also used in Soilcrete mixtures help with flowability, water volume reducer, or substitution for air entrainment within a mix (1). Binders used in mix designs differ from admixtures as they subsist of chemically reactive materials such as cement and fly ash.

Figure 1 - Soilcrete Mixture on-site in Camarillo, California.

Controlled Density Fill Background

Controlled Density Fill (CDF) is a flowable slurry cementitious mix. The slurry type is classified by the American Standards for Material Testing (ASTM) as a Controlled Low Strength Material (CLSM) (2). The main differences between the CLSM and the Soilcrete types are the CLSM is a low strength material compared to Soilcrete. CLSM is a self-compacting and self-leveling flowable material. Soilcrete is vibrated and inserted into lifts, whereas CLSM is not. CLSM is considered as a substitution in lieu of compacted backfill soil, Figure 2. CLSM mix components vary slightly than soilcrete. CLSM main base is the cementitious material typically Portland cement type II / V and binders also called pozzolans. The most used pozzolans within a CLSM mix is fly ash for void filling. Fine aggregate is added to the mix but includes smaller size sand particles to provide flowability and low strength. Due to the high flowability of the CLSM mix, admixtures are used for air entrainment and pumping aid use in hose applications. The water to cement (w:c) ratio is critical for developing strength workability of the CLSM; published w:c ratios range from greater than 0.50 to less than 1.0 for low strength flowable materials (2). For the purpose of this paper, only the strength characteristics of the CLSM backfill and Soilcrete will be compared and explained for design applications of shallow foundations.

Figure 2 – Control Density Fill Mixture in Felton, California.

GEOHAZARD MITIGATION REINFORCED CONCRETE FOUNDATIONS

Rockfall and debris flow mitigation systems have been developed by geohazard mitigation manufacturers to contain and dissipate kinetic rockfall and debris flows with high-strength steel wire mesh and or ring nets. Nets and meshes cover the mitigated area and distribute the loading impact to support wire ropes. The heights are supported using steel flanged column posts, which are subjected to loading from the rockfall and debris flow impact. Manufacturers have field-tested and developed loading inflicted upon posts for foundation design. Due to the lateral loading scenarios of the rockfall and debris flow events, post foundation design for geohazard mitigation systems are governed by substantial lateral loads, that results in shear load and moment components.

For structures subjected to large lateral loads and moments, deep foundations have been developed using different design criteria. These include the American Institute of State Highway Officials (AASHTO) and the Federal Highway Administration (FHWA) to resist overturn, reduce lateral displacement, and distribute axial loading to competent subsurface geologic layers. The bending of laterally loaded foundations reduces with depth and embedded in more competent geologic layers. As overburden depth increases the lateral passive pressure resistance to the movement of the foundation under loading decreases. For geohazard mitigation systems, the magnitude of the lateral loading is applied near the foundation surface. Different manufacturers have developed post base plates that incorporate options for inclined upslope anchor micropiles to reduce the applied shear and rotation of the foundation (3). The inclined anchors are designed to be encased within the foundation to improve resistance to foundation rotation. Anchors for the post base plate attachment are also used as vertical micropiles that extend below the foundation. Vertical anchors help reduce rotation of the foundation but do not effectively resist shear. Upslope inclined micropiles are a reliable solution to reduce laterally loaded deep piles, but inefficient in unfavorable loose soil strata and bedrock anchors.

TYPICAL FOUNDATION TYPE ISSUES

Soils near the foundation surface often do not provide adequate lateral passive pressure resistance because of inadequate strength from loose or not properly compacted, low consolidation, and or liquefiable soils. Piles are commonly designed with increased depth or diameter to improve passive resistance and bending curvature. Pile foundations often require permanent steel casing to improve the structural pile capacity. Steel casing can reduce the pile skin friction between the geologic strata and concrete.

Cast-In-Drill-Hole (CIDH) Pile Foundations

Deep foundations require a substantial amount of steel and concrete material for construction. Depending on the pile depth, additional steel rebar for concrete confinement, bar alignment, and flexural bending may be required. The time required for deep foundations to reach design drilling depths increases overall materials and labor costs. For deep foundations, large drill rigs with augers are often attached to large construction equipment such as drill rig semi-trucks and excavators. Geohazard mitigation systems are commonly placed in limited site access areas including sloped surfaces, debris channels, and secluded locations. These areas make large standard drilling rigs accessibility limited or impossible due to machinery safety constraints, Figure 3.

Figure 3 - Typical Deep Foundation Pile.

Tieback Micropile Anchored Foundations

Tieback micropiles are efficient when bedrock is shallow for anchorage, Figure 4. Tieback micropile anchors are often deep in soil strata due to low achievable anchor grout/ground bond strength. Theoretical design anchor grout/ground bond strengths are used in the design and tend to be conservative for soil conditions. Increase in the drill hole diameter correlates to an increase in the skin friction but is more efficient in anchor length reduction. Locating the micropile tieback anchors from previous observations tend to be difficult depending on drill equipment used. Inclined drill rigs can provide an accurate drill hole where drills mounted to wagons or hand drills can be difficult. The inclination of the tie-back micropile is critical to the design for shear and moment resistance.

Figure 4 – Post Foundation Upslope Micropile Tieback Anchor.

SOILCRETE AND CDF LATERAL SUPPORT ALTERNATIVES

An advantage of Soilcrete and CLSM backfill is that it increases the surrounding lateral support of the foundation. The lateral pressure resistance is increased by the soil compressive strength correlating to cohesion. The FHWA has conservative design guidelines relating low strength backfill support to compacted soil cohesion. Soilcrete and CLSM also help to prevent erosion, liquefaction, or saturated soils around the foundation. Often, debris flow system foundations are installed within debris channels. Erosion of compacted soils from flowing water within debris channels can cause loss of lateral pressure support.

Using Soilcrete and CLSM increases the stiffness of the topsoil, resulting in reduced foundation depth and diameter. Depending on project size, Soilcrete and CLSM can reduce the overall material cost and construction. Construction drilling and concrete installation due to less foundation volume and steel reinforcement. Using native soil materials can also be used for Soilcrete. This method was done in Camarillo, California with the native colluvial soil sieved to provide a well-graded soil with uniform aggregate. Using native soil can reduce imported fill and additional CLSM volume costs. Soilcrete and CLSM do not require the use of large excavation equipment for foundation construction. Smaller and less expensive equipment such as backhoes and walking mobile excavators can be used.

Potential Issues

Occasionally foundation excavations can be constrained due to property limits, underground utilities, and terrain topography. Environmental constraints and large soil excavations can result in permits requirements. Vertical excavations below 5-ft in unstable soils can often lead to temporary shoring.

Lateral Support Design Factors

The focus of CLSM and Soilcrete in this report is the design strength factors. Per FHWA the strength for Soilcrete and CLSM can be determined using a materials unconfined compression strength. The CLSM and Soilcrete compression strengths are both influenced by multiple mixture materials including; binder mixing ratios, additive types, curing temperature, and curing duration. Studies show that Soilcrete strength increases with improved mixing efficiency, curing time, organic soil, and water content (1). Therefore, a decrease in the water to base (w:b) ratio of a mixture increases the unconfined compressive strength. This theory is true for similar concrete mixtures. Higher water to cement (w:c) ratios results in a higher slump and high flowability with strength reduction, Figure 5.

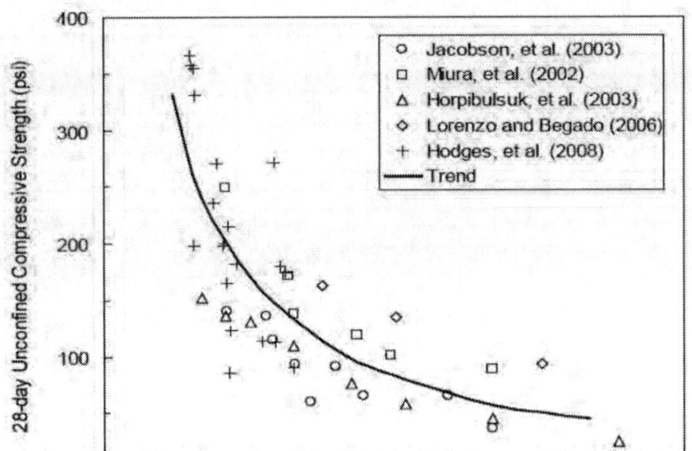

Figure 5 - FHWA Figure 29. Compressive Strength Sample Comparison.

Soilcrete and CLSM Design Strengths

Laboratory-prepared samples to determine unconfined compression strengths were tested by the FHWA for wet and dry mixing methods. For the dry mixing (Soilcrete) method the unconfined compression strength ranged from 2-psi to 400-psi and the wet mixing method 20-psi to 4,000-psi curing for 28-days (1). Typical values for Soilcrete unconfined compressive strength range between 75-psi to 150-psi (1). CLSM samples were tested at 28-days curing resulting in unconfined compressive strengths ranging from 56-psi to 1,000-psi (2). The mean of the samples was between 300-psi to 500-psi (2). Values range for CLSM unconfined compressive strength from 50-psi to 100-psi for most applications (4). The National Ready Mixed Concrete Association (NRMCA) guideline specification for CLSM specifies that backfill is typically less than 300-psi at 28-days for most applications with a maximum of 1,200-psi (5). The rationale of low strength is to provide a similar alternative to compacted backfill. CLSM at low strengths of less than 150-psi is excavatable by hand tolls and conventional smaller excavators. CLSM at higher strengths exceeding 150-psi are considered non-excavatable (5). Compacted soils can achieve cohesive strengths up to 2,100-psf (15-psi) (6).

INPUT : Camarillo Springs, CA

Testing Data - Break Results				
Specimen Sample No.	Curing Age (Days)	Maximum Compression Load Applied (LBS)	Maximum Compression Strength at Failure (PSI)	Maximum & Minimum Values
1	2	8,810	700	Minimum
2	2	9,000	720	
3	2	9,220	730	Maximum

INPUT : Felton, CA

Testing Data - Break Results				
Specimen Sample No.	Curing Age (Days)	Maximum Compression Load Applied (LBS)	Maximum Compression Strength at Failure (PSI)	Maximum & Minimum Values
1	21	1,750	140	Minimum
2	28	2,340	190	Maximum
3	28	2,120	170	

INPUT : Riverside, CA

Testing Data - Break Results				
Specimen Sample No.	Curing Age (Days)	Maximum Compression Load Applied (LBS)	Maximum Compression Strength at Failure (PSI)	Maximum & Minimum Values
1	22	7,720	270	
2	22	7,890	280	Maximum
3	22	7,360	260	Minimum

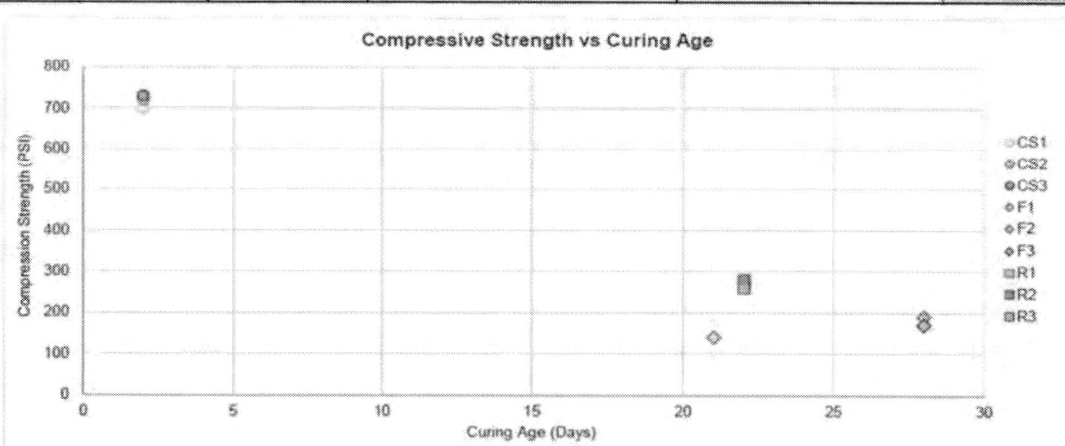

Figure 6 - KANE GeoTech Backfill Support Testing Results.

Additional CLSM and Soilcrete samples have been tested specifically for geohazard mitigation post foundations. The tests were to determine unconfined compression strengths to compare with published results. The compression strength data for backfill support is presented in Figure 6. Three Soilcrete samples provided 700-psi unconfined compressive strength in 2-days of curing. Six CLSM samples provided unconfined compression strengths with a maximum of 280-psi and a minimum of 140-psi, Figure 6. The results have a similar comparison in compression strength with referenced published results.

Soilcrete and CLSM Design Guidelines

The Soilcrete and CLSM strength design criteria have not been accepted as a worldwide standard of practice due to the differences in the strength envelope development (1). It is recommended by the FHWA that a reasonable but conservative strength envelope be used for design. The total stress should include a negligible internal friction angle of Ø = 0-deg and tensile strength, Figure 7. The standard of practice in the United States does not rely on the tensile strength of the mixed ground for design (1). The conservative approach incorporates a total stress internal friction angle of Ø = 0-deg and a cohesion (c) intercept of c = 1/2(qu) where qu is the unconfined compression strength (1). For design, the foundation backfill support should include a minimum factor of safety (FOS). The standard of practice for foundation FOS includes a minimum of 3.0. The FOS provides resistance against the variability in the mixture strength and the shearing capacity through the backfill support zone. Using this design method provides a minimum unconfined compression strength required for the backfill support.

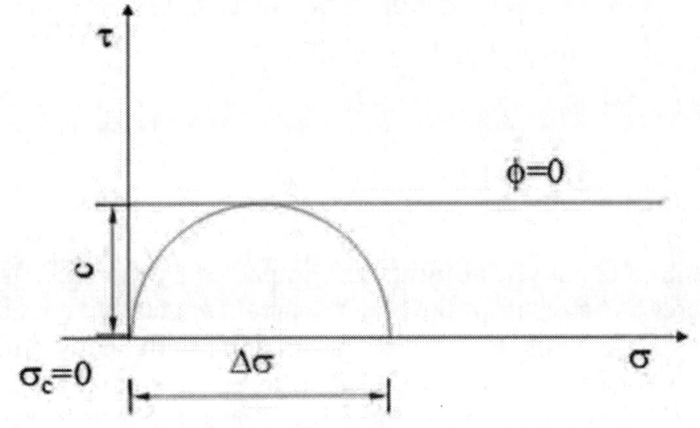

Figure 7 - Typical Mohr Circle for Unconfined Compression.

Often geohazard mitigation system post foundations are located within loose non-stiff soils such as colluvial and alluvial soil deposits. These soil deposits include matrices of silty sands, pebbles, and cobble size rock. Locations of certain geohazard mitigation projects can determine if geotechnical subsurface and material testing can be performed. Secluded and difficult access areas for drilling equipment can result in limited geotechnical information. Conservative designs are often used for geologic subsurface materials where limited geotechnical information is achieved. Typical strength properties of compacted soils have been tested and provided for design use based on the soil classification, cohesion, and internal friction angle. These soil classifications range from well-graded clean gravel sand mixtures with no cohesion and high friction to inorganic clay and silts with higher cohesive strength from compaction and lower friction. The parameters for compacted soils aid in the foundation backfill design. The information from Hunt Geotechnical Engineering Investigation Manual is also referenced from the Naval Facilities Engineering Command (NAVFAC) DM 7 Manual (6).

Soilcrete and CLSM Design Implementation

Soil parameters for compacted silty sand, poorly graded sand-silts, and sand-silt clays have an internal friction angle of 33-deg to 34-deg with an estimated compacted cohesion of 1,050-psf (6). Deep pile foundations are often designed using the typical parameters with foundation design software for laterally loaded piles. To achieve the required minimum compression strength of the foundation, backfill a FOS of 3.0 is added to the 1,050-psf compacted cohesion to increase the required minimum cohesive strength to 3,168-psf. The FOS also ensures that the CLSM or Soilcrete will not fall below the required minimum that would be used for the similar compacted soil. Using the FHWA design criteria, the minimum unconfined compression strength (total stress state) can be determined. A minimum of 44-psi (6,336-psf) compressive strength is estimated to be adequate backfill support compared to a compacted silty sand type soil.

The rationale for the backfill estimated maximum compression strength at 1,000-psi is due to the lateral backfill interaction with the foundation (1). Typically, low strength concrete or slurry do not require structural steel to increase tension and shear strength. Low strength concrete has a limited young's modulus than reinforced concrete. The increase in strength, rigidity, and flexure resistance of the backfill around the foundation, limit the foundation's ability to interact similarly to the soil. The soil around the foundation allows normally designed flexure and displacement. The CLSM and Soilcrete at low strength allow the post foundation to react normally like a compacted or in-situ soil under lateral load.

CASE STUDIES

Camarillo Springs, California

In May 2013, Ventura County, California was impacted by the Springs Fire that scorched approximately 24,000-acres. As a result of the fire, the area's vegetative coverage and soil characteristics were drastically changed. These changes, along with heavy rainfall, caused a residential area at the base of Conejo Mountain, Camarillo Springs, to experience two major debris flow events (7). Following the major debris flows, the threat of El Niño predicted for the upcoming rainy season, an innovative rapid response mitigation design was put into action to be constructed prior to foreseen debris flow events. (7). To mitigate any additional debris flow hazards, KANE GeoTech designed five debris flow mitigation systems. Three of the five systems spanned along existing debris flow channels where post foundations were required.

To shorten the project construction timing, the post foundations were designed using Soilcrete backfill. The backfill reduced the foundation construction time in-place of deep piles requiring large on-site drilling rigs. By using the soilcrete backfill design approach, the same drilling equipment to install the mitigation system anchors were used to excavate the shallow foundation footprints, Figure 8. Accessibility for large drilling equipment was limited due to steep terrain at the foundation locations. The debris flow mitigation systems designed were Geobrugg UX180-H6 and Geobrugg SL-150 type systems. The Geobrugg UX180-H6 system post foundations were designed to resist the lateral loading of 95.4-kips and an axial compressive load of 67.4-kips (8). The Geobrugg SL-150 system post foundations were designed to resist lateral loading of 63.6-kips and an axial compressive load of 78.6-kips (8).

Figure 8 - Shallow Foundation Using Soilcrete Method.

The foundation location subsurface soils were colluvial with matrices of silty sands and cobbles (9). Compacted soil cohesive strength value of 1,050-psf was used for backfill support design to resist lateral deflection and reduce overall flexure of the foundation (6). The minimum design unconfined compressive strength required was 44-psi (6,336-psf). Due to high variability in Soilcrete design mixtures, a minimum of 300-psi unconfined compressive strength Soilcrete mixture was required. The soilcrete volume was relatively small at 3-CY per foundation. Therefore, the Soilcrete compaction lifts and effort during placement reduced overall variability in the Soilcrete strength. The Soilcrete was placed in 1-ft leveled lifts around the foundation. The Soilcrete was mixed on-site with samples collected for testing to verify minimum strength. The soilcrete contained Calportland cement Type II/V with a maximum aggregate size of 1/2-in. The water to cement (w:c) ratio was 0.45 with 4-sacks of cement per mix, Figure 9. The unconfined compression strength values for all three samples were well above the minimum required strength of 300-psi. The tested values were all between 700-psi to the 730-psi range (10).

Figure 9 - Contractor Mixing Soilcrete Backfill.

Construction was completed three weeks ahead of the projected schedule (7). The day following the construction completion, Camarillo, California experienced the season's heaviest rainfall. The storm continued for three days and as a result, the newly constructed barriers were impacted by debris flows (7). The higher elevation debris flow barriers (without post foundations) were heavily impacted while the lower barriers were not impacted.

Figure 10 – Completed Shallow Foundation with Soilcrete Backfill.

Felton, California

A shallow landslide failure occurred above an abandoned railroad in Felton, California. (11). The owner of the property uses the existing roadway to access their facilities. The shallow landslide material slid onto the roadway requiring debris removal and remediation to provide access along the roadway and prevent residential properties downslope of the roadway from the potential risk of debris impact. (12). Provided geotechnical subsurface soil classification below the roadway composed of silty fine sand, clayey sands, and loose-fill (11). The project site access roadway was relatively narrow and steep terrain. Due to the narrow roadway, the backfill support alternative was selected for the design. Due to the weather conditions, the excavated soil became

Figure 11 - Contractor Pumping CDF Backfill Around Foundation.

difficult to use for Soilcrete mixture. CLSM was used in lieu of the Soilcrete. The CLSM was pumped through long hoses from slurry vehicles on wider sections of the road, Figure 11. The debris flow mitigation system design was a Geobrugg SL-150 type system. The Geobrugg SL-150 system post foundations were designed to resist lateral loading of 66.7-kips and an axial compressive load of 82.3-kips (13).

Due to the narrow and sloped area below the roadway, the available passive pressure resistance was relatively low resulting in an increased strength required for the backfill, Figure 12. For design compacted soil cohesion of 10,800-psf (6) was used for backfill support to resist lateral deflection and reduce bending of the foundation. The minimum unconfined compressive strength required for the design was 150-psi (21,600-psf) (15). The CLSM volume was relatively small at 3-CY per foundation. Since the CLSM is a flowable and self-leveling and compacting backfill, no vibration or compaction lifts were performed during installation. The CLSM was pumped on-site with samples collected for testing to verify minimum compression strength. The CLSM contained ASTM C-150 cement with fine sand aggregate. The CLSM mixture slump was 6-in with 4-sacks of cement per mix (14). The unconfined compression strength values for all three samples were between 140-psi to 190-psi with the average at 167-psi exceeding the required 150-psi for design (15). Due to the cold and wet climate and long pumping distances, the flowability of the CLSM was high resulting in lower range CLSM values.

Figure 12 - Completed CDF Backfill Around Foundation.

Riverside County, California

East of Highgrove, California in Riverside County a residential property development has been under construction. KANE GeoTech provided design for a rockfall mitigation system at the northern end of the property and two debris flow mitigation barriers at the southern end of the property development. One of the debris flow barrier systems required post foundations for a Geobrugg SL-150 type system. The Geobrugg SL-150 system was located at the slope toe inside a channelized potential rockfall and debris flow chute. The nearest residential properties were within 300-ft of the chute opening downslope on fill housing pads. Due to the locations of the post foundations, no subsurface investigation was performed. The United States Department of Agriculture (USDA) soils maps provided limited information on subsurface geologic material. USDA classified the site as alluvial deposits of sandy silt (16). The post foundations were designed for Cast-in-drill-hole (CIDH) pile conditions.

Figure 13 - Contractor Pumping CDF Backfill Around Foundation.

During construction, the installation Contractor provided a request for information (RFI) for a change order on the post foundation design. The change was to substitute an augered CIDH foundation with an over-excavation with backfill support. The Geobrugg SL-150 system post foundations were designed to resist the lateral loading of 63.6-kips and an axial compressive load of 78.6-kips (17).

The backfill support design incorporated compacted soil with the cohesion of 1,050-psf (6) to resist lateral deflection and reduce bending of the foundation. The minimum design unconfined compressive strength required was 44-psi. For the CLSM, a minimum of 100-psi unconfined compressive strength was required. The CLSM volume was relatively large at 10-CY per foundation. Since the CLSM is a flowable and self-leveling and compacting backfill, no vibration or compaction lifts were performed. The CLSM was pumped on-site with samples collected for testing to verify minimum compressive strength, Figure 13. The CLSM mix contained Portland type II / V cement with fine aggregate. The CLSM mixture slump was 5-in with 2.3-sacks of cement per mix (18). The unconfined compression strength values for all three samples were between 260-psi to 280-psi with the average of 270-psi (19) exceeding the minimum 100-psi design requirement. Due to the warmer climate and admixtures, the CLSM compressive strength was improved with a reduction in water content, Figure 14.

Figure 14 - Completed CDF Backfill Around Foundation.

CONCLUSION

Post foundations for geohazard mitigation using backfill support provide an efficient alternative to tieback anchored and deep pile foundations. The tested Soilcrete and CLSM mix designs used in the case studies have been tested and provide similar results to the referenced documents. The backfill support alternative has resulted in a foundation design that requires less construction time, reduced labor, equipment, and materials. The reduction in overall construction time provides an efficient solution for areas in need of emergency rapid response construction using geohazard mitigation after post-wildfire events.

REFERENCES:

1. U.S. Department of Transportation Federal Highway Administration (FHWA). *Deep Mixing for Embankment and Foundation Support*. Publication No. FHWA-HRT-13-046. October 2013.
2. Alizadeh, V. *Design Application of Controlled Low Strength Materials as a Structural Fill.* Department of Civil Engineering and Mechanics, University of Wisconsin-Milwaukee. December 5, 2013.
3. Geobrugg AG. *SL-150 Post Foundation Anchor Forces, Product Manual SPIDER® SL-150 Shallow Landslide Barrier*. CH-8590 Romanshorn, Switzerland. Edition 01-N-FO/14, 2017 08-31.
4. American Society for Testing and Materials (ASTM). *Standard Test Method for Preparation and Testing of Controlled Low Strength Material (CLSM) Test Cylinders*. ASTM Designation D4832-02.
5. National Ready Mixed Concrete Association (NRMCA). *Guide Specification for Controlled Low Strength Materials (CLSM).*
6. Hunt, Roy E. *Geotechnical Engineering Investigation Manual*. New York: McGraw-Hill, Print 1984.
7. Jones, M. *Rapid Response to a Post Fire Debris Flow Event* Highway Geology Symposium (HGS), Colorado Springs, Colorado, 2016.
8. KANE GeoTech, Inc. *Camarillo Springs Debris Flow Investigation, Calculation Report Camarillo, California*. September 22, 2015.
9. Geodynamics, Inc. *Geotechnical Reconnaissance of Debris Flow Risk, Camarillo Springs, Camarillo, California*. April 14, 2015.
10. NV5 West, Inc. *Report of Unconfined Compression Test (ASTM D2166)*. 12/2/2015.
11. Harwood, Craig. *Geologic Investigation of Landslide on Railroad Right-of-way, Santa Cruz County, California*. 1/16/2018
12. Friar Associates, Inc. *Geotechnical Investigation, Railroad Landslide, Santa Cruz County, California*. 1/30/2018.
13. KANE GeoTech, Inc. *Zayante Canyon Landslide Mitigation, Engineering Design Calculation Report, Felton, California*. May 25, 2018.
14. Las Animas Concrete. *CDF Mix Design*, 2018.
15. Wallace-Kuhl & Associates. *Compression Test Report*. 3/7/2019.
16. United States Department of Agriculture (USDA). *NRCS Web Soil Survey Maps*.
17. KANE GeoTech, Inc. *SMR Riverside County Rockfall Mitigation, Engineering Design Calculation Report, Riverside County, California*. August 17, 2018.
18. Robertson's Ready-mix Concrete. *CDF Mix Design*. 6/18/2019.
19. Wallace-Kuhl & Associates. *Compression Test Report*. 7/22/2019.

Rock Mechanics and its Effects on Spillway Modification Design

Coralie Wilhite, PG, SPRAT I
US Army Corps of Engineers
1325 J Street
Sacramento, CA 95814
(916) 557-7864
Coralie.P.Wilhite@usace.army.mil

Kenneth Pattermann, PE, GE
US Army Corps of Engineers
1325 J Street
Sacramento, CA 95814
(916) 557-6980
Kenneth.R.Pattermann@usace.army.mil

Jerilynn Hilmar
US Army Corps of Engineers
1325 J Street
Sacramento, CA 95814
(916) 557-7231
Jerilynn.Hilmar@usace.army.mil

Vanessa Bateman, PG, PE
US Army Corps of Engineers
441 G St. NW #3168
Washington DC 20314
(202) 761-7423
Vanessa.C.Bateman@usace.army.mil

Prepared for the 70th Highway Geology Symposium, October, 2019

Acknowledgements

Love life and sparkle while you are at it!

Disclaimer

Statements and views presented in this paper are strictly those of the author(s), and do not necessarily reflect positions held by their affiliations, the Highway Geology Symposium (HGS), or others acknowledged above. The mention of trade names for commercial products does not imply the approval or endorsement by HGS.

Copyright Notice

Copyright © 2019 Highway Geology Symposium (HGS)

All Rights Reserved. Printed in the United States of America. No part of this publication may be reproduced or copied in any form or by any means – graphic, electronic, or mechanical, including photocopying, taping, or information storage and retrieval systems – without prior written permission of the HGS. This excludes the original author(s).

ABSTRACT

Success Lake is a U.S. Army Corps of Engineers dam and reservoir located about 6 miles upstream of the city of Porterville, CA on the Tule River and provides flood damage risk reduction, agricultural water supply, and recreation. An engineering study was completed in 1999 that proposed to raise the spillway 10 feet and lengthen it across the channel from 200 to 365 feet to provide additional flood risk reduction and irrigation water supply. Following that, a baseline risk assessment was completed that found the main risk drivers to be extreme loading events and confirmed that the proposed raise and widening would further reduce the flood related risks. General site geology consists of moderately fractured diorite, with large bodies of quartzite and metavolcanics with known areas of block failure and sliding. The existing spillway is unlined with a 3 foot-wide x 200 foot-long reinforced concrete sill that is embedded into rock. Modifications will include widening the spillway to the right by blasting and excavating, re-assessing the side slope rock stability, and investigating a landslide that occurred in the left abutment during original construction. New structural features include a 10 foot high reinforced concrete ogee weir, a 100 foot wide concrete apron with a downstream concrete headcut cutoff wall, reinforced concrete sidewalls anchored into rock, and relocation of a public access road. Additional geologic site investigations include: geologic mapping, seismic refraction survey (p-wave), structural geologic data gathering, subsurface drilling, downhole optical and acoustic televiewer, and lab testing. This data was processed to analyze all potential cut slopes and to design stable slopes based on block stability analyses for, sliding, wedge failure, and toppling. The extensive geologic investigations and kinematic analyses have decreased design and construction risk by reducing the number of onsite unknowns and allowing the design team to proceed forward with increased confidence in slope geometries and less risk of significant differing site conditions during construction.

INTRODUCTION

Success Lake is a U.S. Army Corps of Engineers dam and reservoir located about 6 miles east of the city of Porterville, California (**Figure 1**). Completed in 1961, Success Dam sits on the Tule River and provides flood damage risk reduction, agricultural water supply, and recreation to the local community and central valley. Draining an area of 630 square miles the Tule originates as three forks, that join to form Lake Success. The main project features include a zoned earthen embankment dam, an auxiliary dike (Frazier Dike) 3 ½ miles northeast of the main dam, and an emergency spillway.

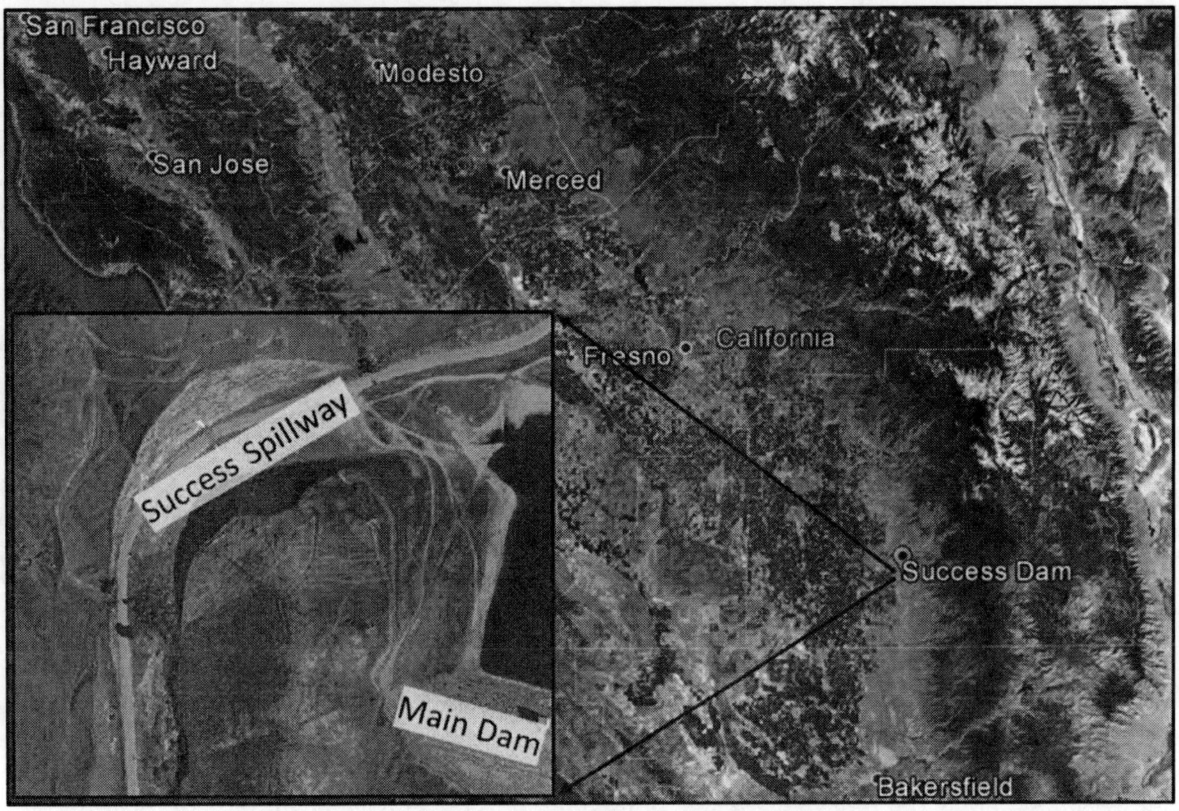

Figure 1: Site Location

An engineering study was completed in 1999 that proposed to raise the spillway 10 feet and lengthen it from 200 to 365 feet to provide additional flood risk reduction and irrigation water supply. A subsequent baseline risk assessment found the main risk drivers to be extreme loading events and confirmed that the proposed raise and widening would further reduce the flood related risks.

This proposed spillway modification was approved and work began in the Fall of 2018. Work within the spillway will be completed in two phases and this paper will focus on design of the Phase I (which includes widening of the spillway and relocation of the public access road)

slope angles for the right abutment. The work presented here will inform design as well as reduce uncertainty in the geologic model and associated project risk during design and construction.

BACKGROUND

Success spillway was constructed as the emergency outlet structure for Success Dam. It is located about 700 feet northwest of the right abutment of the main dam, with a large nob of existing rock between the structures (**Figure 2**). The spillway was blasted and excavated through colluvium and the underlying bedrock.

Figure 2: Spillway features – looking downstream (west)

The existing spillway is unlined with a reinforced concrete (RC) control sill at elevation 652.5 feet (NGVD 1929) located at station 10+00 that measures 3 feet wide at the top and 8 feet wide at the bottom, and is 6 feet deep. The existing sill is 200 feet long (across the channel). Reinforced concrete key blocks were also embedded into the abutments at the sill location measuring 4.5 feet thick and ranges from 8 to 12 feet wide. Number 8 steel reinforcing bars were drilled and embedded 8 feet into rock on 5 foot centers and terminated with an S-style hook into the RC. The new proposed RC ogee weir will be at elevation 662.5 feet (10 feet higher), approximately 365 feet long with RC sidewalls extending approximately 50 feet upstream and 100 to 200 feet downstream. There will also be a RC apron extending approximately 100 feet downstream of the weir with a RC headcut cutoff wall approximately 30 feet deep at the end of the apron.

A rock and soil slide occurred during original construction of the spillway. It is located on the left abutment between stations 7+50 and 9+50 and appears to be joint controlled (**Figure 2**). The spillway expansion will be to the right, however, the mechanics of this slide were considered during the investigations and analysis of the right spillway abutment to see if similar sliding could occur.

Other small scale rock falls, wedge failures, slides, and topples have occurred since construction, but in general the right spillway abutment appears to be relatively stable at the current slope of 1:1 to date.

Historical Flow Through The Spillway

The spillway has only operated once since initial construction. This event occurred in December of 1966 and had a maximum flow of 8,300 cfs through the spillway. Erosion within the excavated spillway was minimal upstream of the existing sill and downstream for about 450 feet. After that point, erosion began to cut into the spillway floor with a max depth of about 15 feet from original grade and headcutting moving upstream (**Figure 3**).

Figure 3: 1966 spillway flow

Regional and Site Specific Geology

This project sits in the foothills of the Sierra Nevada Mountains and exhibits some of the key features of the geologic processes that are common in this region. The earliest regional deposits were a thick series of sedimentary and volcanic rocks associated with shallow intrusions deposited in the late Paleozoic to early Mesozoic. The early Nevadan Orogeny produced volcanic deposits. All of the above rocks were intensely folded and metamorphosed during a period of compression and folding. Late Nevadan Orogenic events produced massive igneous intrusions in multiple stages: the early stage being hornblende gabbro and diorite sills, dikes, and stocks; later stages being granodiorite and quartz monzonite batholiths; and end stages being small bodies of granite, aplite, and pegmatite dikes and quartz veins cutting all previously implaced rocks. The intrusion of those bodies produced contact metamorphism that was superimposed on earlier phases of metamorphism. The Nevadan Orogeny was followed by a long period of erosion, removing immense thicknesses of rock to expose the granitic core of the Sierra Nevada. Finally, uplift during the Pliocene and Pleistocene as well as subsequent erosion established the current drainage pattern, fluvial deposits, and stream paths.

This has resulted in significant geological complexity at the dam due to the deposits, compression, intrusions, uplift, and erosion that formed the regional geology. The project foundation report describes multiple different rock types on site such as quartzite and quartz-mica schist, meta-basalt, grabbro, micro-gabbro (diabase), diorite, and granite. However, in geologic logs, the most common rock types encountered are granite, quartzite, and meta-basalt.

Additionally, some borings (e.g., 1B-07-33, 1F-19-08 & 1F-19-11) exhibit relatively, thick highly weathered, sections that appear to be structurally controlled causing variability in the weathering grades across the site and at depth. Furthermore, differential weathering has resulted in sporadic and unpredictable occurrences of moderately/slightly/unweathered core stones floating in a groundmass of highly weathered rock.

Reverse and normal faulting on small orders of magnitude have occurred in the bedrock at the dam site, but no evidence of faulting is observable in the alluvium. Fault, lineation and joint orientations on site are complex and vary somewhat depending on the rock type and depth in which they are encountered. Shear zones were also mapped onsite when no direction of movement could be determined for a structure.

FIELD INVESTIGATIONS

Numerous field investigations have taken place onsite before and after spillway construction. All historical investigations were reviewed and through project development and planning it was determined that additional field investigations were needed. To date, the 2019 field efforts have included: structural geologic mapping, rock mass classifications, sub surface explorations (rock coring), seismic refraction lines, and laboratory testing. All investigations related to the right spillway abutment are summarized below and shown in **(Figure 4)**.

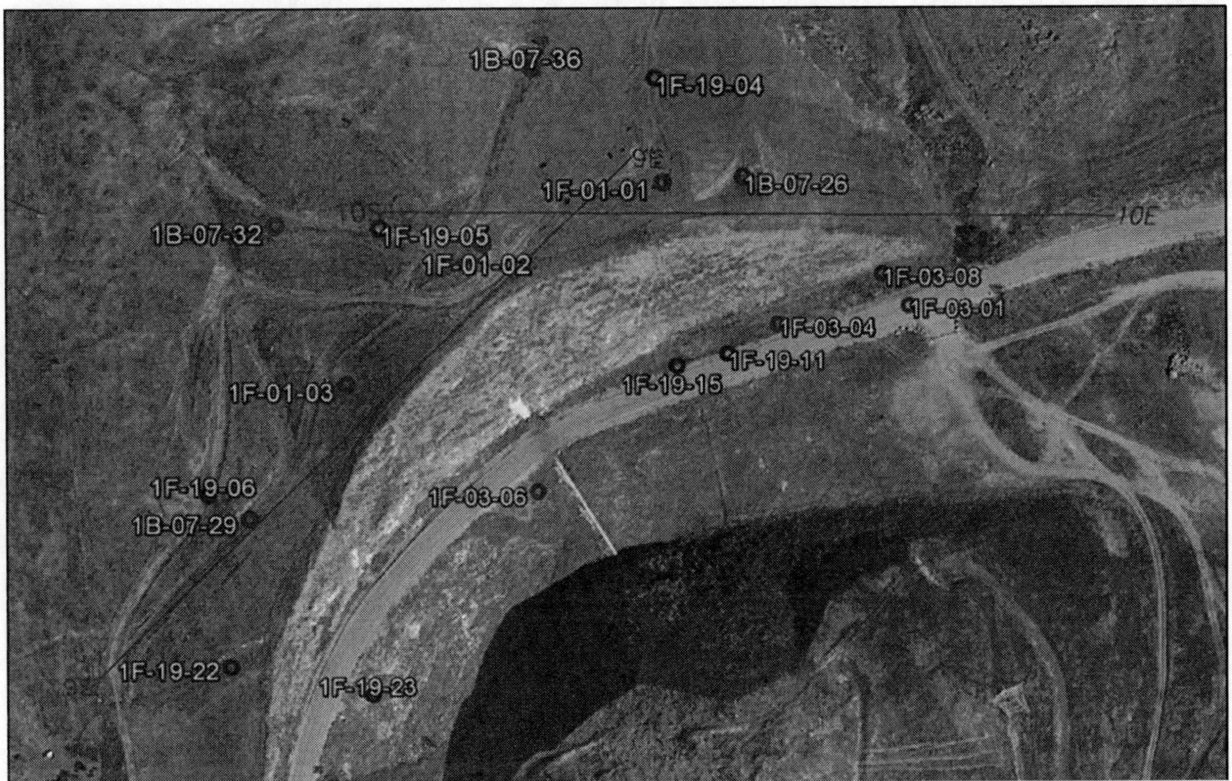

Figure 4: Map of borings and seismic lines for the right spillway abutment

Pre-2019 Field Investigations

Numerous field investigation programs have been performed for the Success Project. Some of which are directly relevant to the right abutment expansion of the spillway. Due to inconsistent funding sources, drilling for the spillway enlargement was conducted over several years, including, 2001, 2003, and 2007. The 2001 drilling consisted of 8 rock core borings (HQ size) in the invert and right abutment with depths ranging from 20 to 119 feet. Some of these depths were too shallow considering the potential for spillway erosion. The 2003 drilling consisted of 10 rock core borings (HQ size) focused mostly in the invert for a proposed new reinforced concrete ogee weir at station 5+50. It was found from the 2003 borings that the foundation conditions for the proposed ogee weir were not satisfactory. For example, a few borings consisted of decomposed granite for depths up to 25 feet, which would have to be removed during construction and replaced with mass concrete. This unsatisfactory foundation was also verified during the 2019 seismic lines and prompted the team to move the proposed ogee weir location further downstream where the foundation rock is of much better quality. The 2003 exploration program also included rock bolt bond zone verification testing and unconfined compressive strength testing. The 2007 drilling had two purposes: 1) to identify potential borrow sources for a major dam rehabilitation (which was later found not to be required through risk analyses) and 2) for widening the spillway. The 2007 drilling consisted of 39 rock core borings with depths ranging from 40 to 140 feet. Most of the 2007 borings were drilled outside of the spillway area.

2019 Right Abutment Geologic and Joint Mapping

Based on field observations, it was found that the current right abutment cut face is generally composed of granitic rocks, with pockets and veins of quartzite and meta-basalt. It was also found that in general a relatively thick (approximately 25 to 60 foot) zone of decomposed to highly weathered rock exists in the current cut face with moderately weathered to better rock below. This rock extends along most of the spillway slope from the toe and approximately 20 to 40 feet in height, with the highest quality material around the existing key. Two zones of highly weathered rock extend down to the spillway floor approximately 80 and 170 feet upstream of the existing sill. Both zones are approximately 20 feet wide and appear to be composed of meta-basalt, which has weathered more rapidly than the surrounding granitic rock. Additionally, these two zones are intensely fractured and some surfaces show striations indicating motion, but since the spillway floor is covered and no evidence of the two features exist in the left abutment, their trace is unknown across the spillway site. Additionally, three Rock Mass Classifications were performed in moderately and slightly weathered rock and provide a general rating of "Good" rock with high intact rock strength, joint spacing that varied from about 3 inches to approximately 7 feet, and RQD values ranging from 57%-72%.

2019 Right Abutment Sub-Surface Explorations and Televiewer Logging

Subsurface investigations included drilling 7 rock core borings located on and at the toe of the right abutment (**Figure 4**). Optical and acoustic televiewer logs were performed in 5 of these borings (1F-19-04, -05, -06, -15, & -22) and provided data for hundreds of structural features. This was combined with the surface mapping to create the data set of 1133 measurements used in these kinematic analyses and slope modeling. There is limited data for the site piezometric surface, but televiewer data showed that the borings located high on the right abutment had ground water elevations of approximately 621, 669, and 680 feet (for comparison, the existing spillway invert is at elevation 651 feet. Additionally, it is important to note that the site experienced higher than average rain fall this year and the borings were drilled toward the end of the rainy season, therefore, these measurements are assumed to represent relatively high groundwater elevations for the right abutment. Samples from the 2019 borings were also sent out for analysis of unconfined compressive strength (UCS) and direct shear (angle of internal friction – ϕ). Results of these tests were used in the analyses performed and are summarized in the Material Properties Used in Analysis Section below.

Direct Shear Testing

Direct shear testing was conducted according to ASTM D5607. Samples were selected from the current 2019 investigations with shallow natural open fractures that were typically covered with iron (Fe) staining from groundwater, rain infiltration, and weathering processes. All samples were tested on existing fractures in the rock, rather than on sawn surfaces. The testing is being conducted in several phases to assess the results from the first phase before selecting rock cores for the subsequent phases. The first phase included testing samples in all types of weathering grades including decomposed, highly, moderately, slightly, and unweathered. During collection and testing, condition and weathering of the fracture was considered and logged, including the level of Fe staining, angle of the joint, types of surface minerals, and degree of asperities (or joint roughness coefficients). The cores were tested at stress levels 1 to 2 times the magnitude of the insitu stress at the fracture depth. Interestingly, the moderately weathered surface gave a higher friction angle than the slightly weathered surface, which is likely due to samples surface condition and fracturing. Laboratory specification sheets were made for the lab as shown in **Figure 5**. Residual shear stress was used as the failure criterion with the results for each weathering grade shown in **Figure 6**. A second phase of testing will be performed during the next subsurface investigation program that will take place this summer.

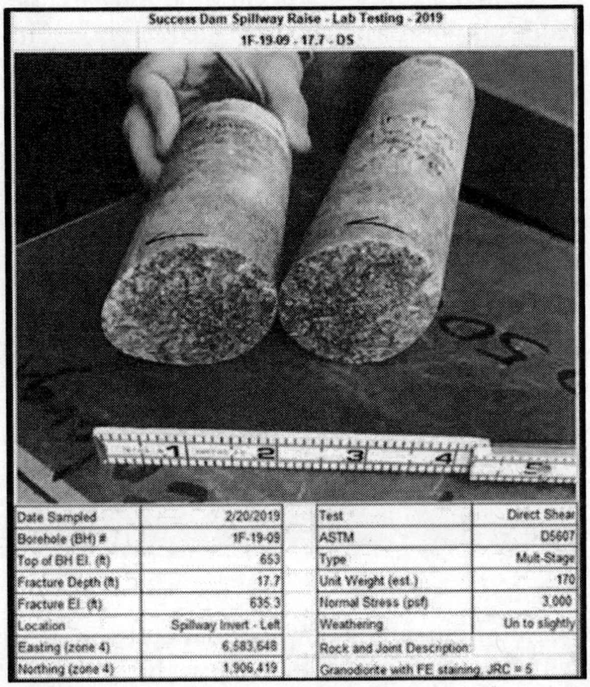
Success Dam Spillway Raise - Lab Testing - 2019

1F-19-09 - 17.7 - DS

Date Sampled	2/20/2019	Test	Direct Shear
Borehole (BH) #	1F-19-09	ASTM	D5607
Top of BH El. (ft)	653	Type	Mult-Stage
Fracture Depth (ft)	17.7	Unit Weight (est.)	170
Fracture El. (ft)	635.3	Normal Stress (psf)	3,000
Location	Spillway Invert - Left	Weathering	Un to slightly
Easting (zone 4)	6,583,648	Rock and Joint Description:	
Northing (zone 4)	1,906,419	Granodiorite with FE staining, JRC = 5	

Figure 5: Direct shear lab specification test sheet - rock core

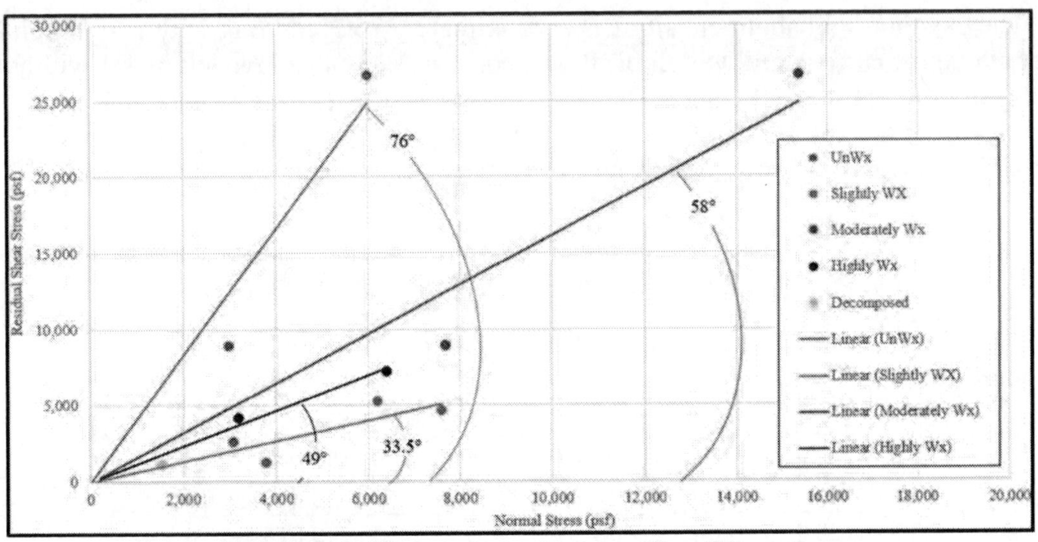

Figure 6: Direct shear test results - Phase 1

2019 Right Abutment Seismic

Thirteen seismic refraction lines were run onsite in early 2019. Lines 9 and 10 run along the top of the right spillway abutment and were analyzed with respect to the known geology, boring logs, geophysical data, and field mapping. This data was then imported into a 3-dimensional environment and compared to nearby boring logs to produce a better understanding of the geologic conditions between borings. It was determined that a value of 3000 fps represented the approximate change in rock quality from highly weathered to moderately weathered or better. This values is approximate since the known, rapidly varying, apparently joint controlled weathering profiles provide complexity and uncertainty in picking changes in weathering surfaces. Also, differential weathering on site is known to result in sporadic occurrences of slightly and unweathered core stones floating in a groundmasses of highly weathered and decomposed rock. Even with this uncertainty, these geophysical analyses provide a better understanding of rock quality across the project site and were used to inform modeling.

Summary of Recent Investigations

Combining all historical and recent field investigations, a site geologic model was created. This shows that geologic materials vary from igneous to metamorphic and that rock quality generally increases with depth. It also shows that the quartzite typically remained slightly to unweather no matter its location, but that the granitic and metavolcanic rocks tend to decrease in weathering and fracturing with depth.

DATA ANALYSIS

Analysis of all historical, field, and laboratory data presented above was used to characterize the geology and material parameters of the right spillway slope. The slope was then broken down into sections and combined with the field data to determine slope design grades for the Phase I construction by running kinematic and slope stability analyses.

The existing right abutment slope is approximately 850 feet long with a high point about 134 feet above the current spillway floor. It was constructed with a layback of 1:1 without benching, slope support, or drainage. The slope curves as it extends downstream (approximately E-W on the upstream end and approximately N-S on the downstream end). Because of the changing slope orientation, kinematic analyses were run for 4 different slopes (**Table 1**).

Table 1: Analyzed slope orientations

Slopes Analyzed - Right Spillway Abutment				
Slope	Strike	Dip	Max Height	Approximate Location (Centerline Stationing)
A	080	170	115	5+25 to 8+00
B	065	155	135	8+00 to 10+50
C	050	140	105	10+50 to 13+50
D	035	125	60	13+50 to 14+25

A number of design considerations were used for these analyses: 1) The existing 1:1 slopes have experienced small volume block failures and have formed a small wedge of soil and rock along the toe of the slope (note, this wedge is only since the 1966 flow event discussed above); 2) Slope design does not need to prevent all failures, but must ensure that the OGEE weir structure, between station 7+00 and 10+00 (which will be constructed during Phase II), will not be damaged and will remain in place during spill events; 3) Failure of slopes upstream and downstream of the OGEE weir are acceptable as long as the volume is small enough that it does not block or impeded flow; and 4) After new grading, any remaining potential slope failures must be protected to ensure site safety during construction and along the new roadway bench.

Based on these design considerations, initial kinematic analyses were run in RocScience, Dips7.0. Initial analyses started with the existing slope parameters to determine if the performance since construction is indicative of the type, nature, and extent of failures that could occur or if larger scale failures are kinematically possible. Sensitivity analyses were then run by varying slope grades, material properties, and water tables. The results of these analyses were then modeled in RocScience Swedge6.0 and RocTopple1.0 to determine factors of safety (FS) and potential failure volumes. A global slope stability limit equilibrium analysis was also modeled in GeoStudio Slope/W to assess FS, especially through highly weathered to decomposed materials.

Material Properties Used in Analyses

Historical and current laboratory tests were used to determine the material properties for each weathering grade on site. As described above, geology across the site is variable and precise locations of each rock type in the slope was difficult to determine. Therefore, average material properties for the right abutment slopes were used for analysis and are summarized in **Table 2**.

Table 2: Material properties

Weathering Grade	Average Unconfined Compressive Strengths (psi)	Unit Weight (lbs/ft^3)	Angle of Internal Friction (Phase 1 Lab Results) (ϕ)
Unweathered	29,000	165	76
Slightly Weathered	20,500	155	33*
Moderately Weathered	8,500	150	58
Highly Weathered	1,500	145	49

*Used as lower bound for slightly to highly weathered material for Phase I analyses

Most samples followed a general trend of strengthening in both unconfined compressive strength and angle of internal friction (ϕ) as the weathering grades improved. However, the friction angles for the slightly weathered samples were comparatively low and may have been due to defects in the core including healed micro-fractures, reduced asperity strength, material infilling, or sheared surfaces. These unusual results will be examined further and supplemented with additional tests in Phase 2. Until that time, the ϕ from slightly weathered rock will be used as a basis for the analyses as it may represent the lower end of joint strengths on site.

Stereographic Kinematic Analysis

All structural joint measurements from surface geologic mapping and downhole televiewer for the right spillway abutment were combined to create a representative stereographic plot of the geologic structure. The resultant stereoplot **(Figure 7)** shows that the geologic structures are scattered for the low angle jointing, but that a few prominent, higher angle, joints are persistent throughout the site. Using this stereonet, 7 key joint sets were selected as representative for the right spillway abutment **(Table 3)**. These joint sets were then used to create a simplified stereoplot, which was used to run the kinematic analyses for rock wedge, planar, and topple failures.

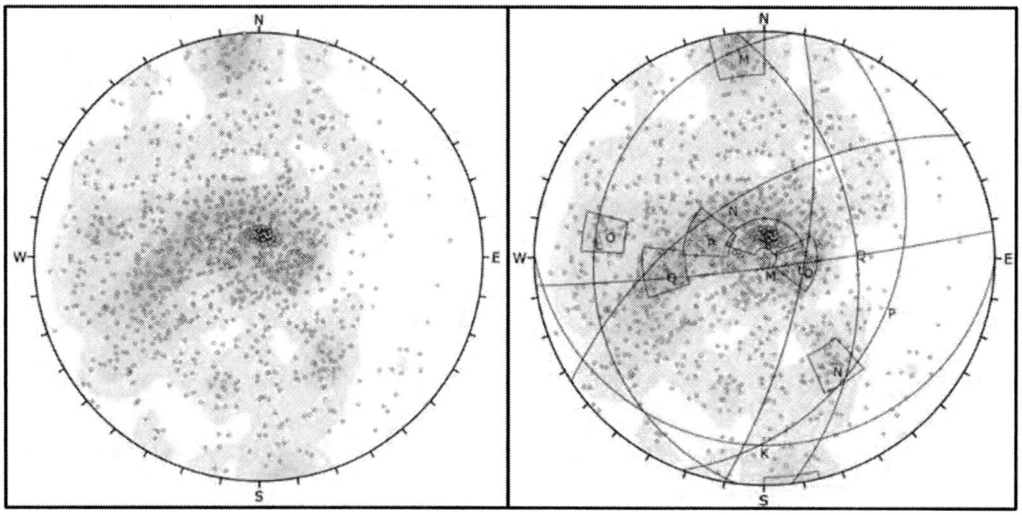

Figure 7: Right spillway wall pole plot and pole plot with seven representative joint sets. Fisher distribution, equal angle, lower

Table 3: Representative joint sets and simplified stereoplot

Set Name	Strike (right)	Dip	Dip Direction
Set K	091	11	181
Set L	187	17	277
Set M	083	85	173
Set N	237	58	327
Set O	010	70	100
Set P	021	30	111
Set Q	353	46	083

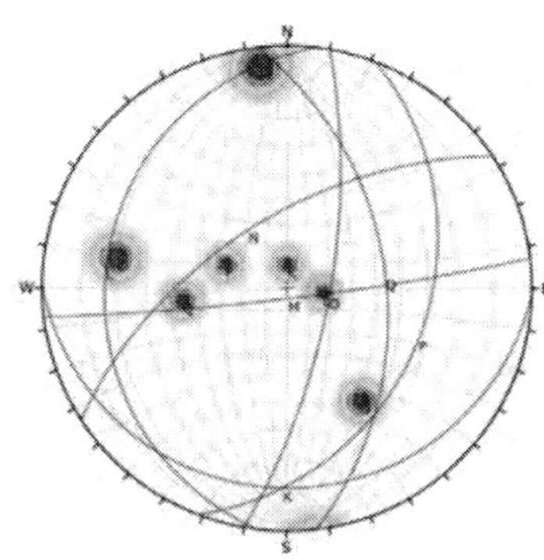

Kinematic analyses were run for each of the four slope orientations listed in **Table 1**. Initial analyses used the existing 1:1 slopes and ϕ=33 degrees. This initial analysis does not account for the variation in joint orientations as seen on the stereonet or friction angles found in testing. Therefore, secondary sensitivity analyses were run by varying ϕ to determine the strength at which the slopes may become unstable. This is also necessary to check based on the performance of the existing 1:1 slope, where evidence of small scale slides have been observed in the field.

The results of the initial and sensitivity analyses are shown in **Table 4** and are summarized as follows: 1) All slope orientations have potential for topple failure even in shallow (less than 2:1) slopes; 2) The existing 1:1 slopes with the initial ϕ of 33 degrees do not show potential for wedge or planar failure; 3) As the ϕ decreases, the potential for wedge and planar failure increases; and 4) As the slope orientation changes from near east-west to near north-south, the prominent joint orientations become less favorable and potential failure modes increase.

Table 4: Kinematic analysis results

Location	Details	Slope	Friction Angle	Planar	Wedge Analysis	Flexural Toppling	Direct Toppling	Notes
Slope A Max height 115 ft	Dip Direction = 170 (SE)	1:1 (45°)	33	None	None	None	2 (K,P)	K & P can topple at any slope angle - Likely small blocks (max ~5x5ft)
	Sensitivity Analysis Dip Direction = 170 (SE) Strike 080	1:1 (45°)	25	None	1 (P&Q)	None	2 (K,P)	Sensitivity run for 1:1 slopes -Wedge failure potential at a phi of 25degrees
Slope B Max height 135ft	Dip Direction = 155 (SE) Strike (80-15) = 065	1:1 (45°)	33	None	None	None	2 (K,P)	K & P can topple at any slope angle - Likely small blocks (max ~5x5ft) Variability on P could create planar failures if phi is 30 degrees or less
	Sensitivity Analysis Dip Direction = 155 (SE) Strike (80-15) = 065	1:1 (45°)	30/25	1(P)	1 (P&Q)	None	2 (K,P)	Sensitivity run for 1:1 slopes -Planar failure potential at 30 degrees (P) -Wedge failure potential at a phi of 25degrees
Slope C Max height 105ft	Dip Direction = 140 (SE) Strike (65-15) = 050	1:1 (45°)	33	None	None	None	2 (K,P)	K & P can topple at any slope angle - Likely small blocks (max ~5x5ft) Variability on P could create planar and wedge failures if phi is 30 degrees or less
	Sensitivity Analysis Dip Direction = 140 (SE) Strike (65-15) = 050	1:1 (45°)	30/30/25	1(P)	1 (P&M) 1(P&Q)	None	2 (K,P)	Sensitivity run for 1:1 slopes -Planar failure potential at 30 degrees (P) -Wedge failure potential at a phi of 30 degrees (P&M) & 25 degrees (P&Q)
Slope D Max height 60ft	Dip Direction = 125 (SE) Strike (50-15) = 035	1:1 (45°)	33	None	None	None	2 (K,P)	K & P can topple at any slope angle - Likely small blocks (max ~5x5ft) Variability on P could create planar and wedge failures if phi is 30 degrees or less
	Sensitivity Analysis Dip Direction = 125 (SE) Strike (50-15) = 035	1:1 (45°)	30/30/26	1(P)	1 (P&M) 1(P&Q)	None	2 (K,P)	Sensitivity run for 1:1 slopes -Planar failure potential at 30 degrees (P) -Wedge failure potential at a phi of 30 degrees (P&M) and 26 degrees (P&Q)

Figure 8 shows representative examples of the kinematic analyses for wedge, planar, and topple failures on Slope C. This slope has potential for all types of failures depending on varying ϕ values. Topples at occur at ϕ=33 degrees, planar and wedge at ϕ=30 degrees, and a secondary wedge at ϕ=25 degrees. As **Table 4** shows, this sensitivity analysis was performed for each of the 4 slope orientations. The results were then modeled for slope stability in SWedge and RocTopple (next section) to determine FS and potential block size of each failure type.

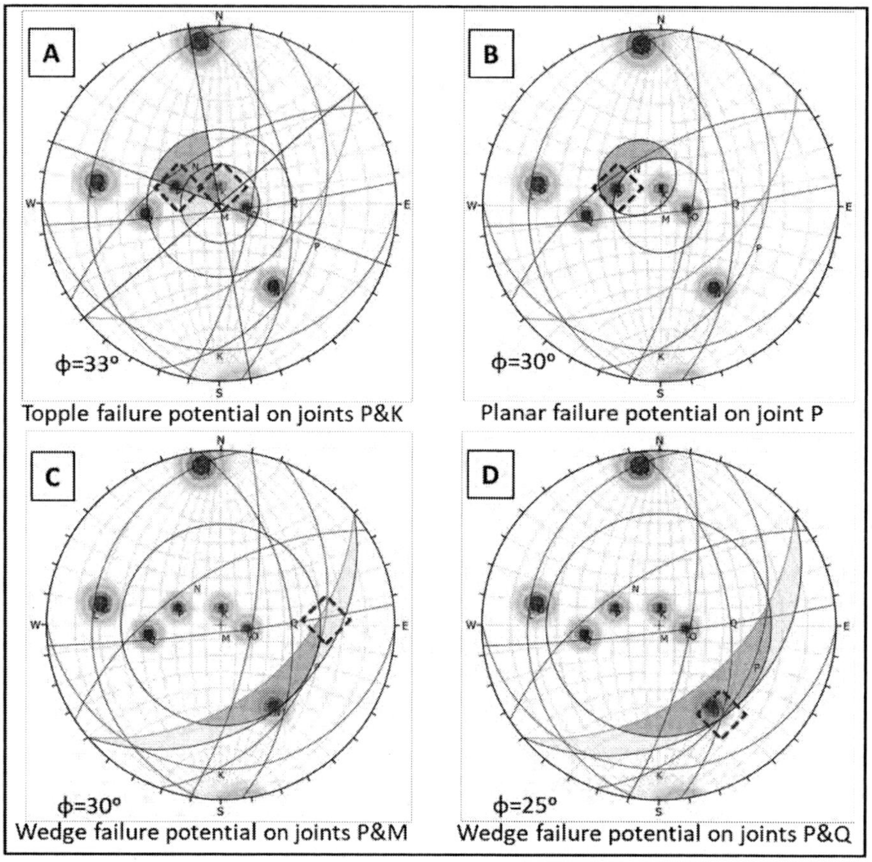

Figure 8: Representative kinematic analyses for Slope C

Block Failure Analysis

Based on the results of the stereographic kinematic analysis, modeling was performed to determine likelihood of failure, size, and factor of safety (FS). RocTopple was used to get a general look at the toppling potential and size. SWedge was used to model wedge and planar failures. All RocScience analyses used the material parameters stated above, an initial slope of 1:1, and a water pressure based on 50% filled joints.

Findings:

- Stereographic analysis showed that all four slope orientations have potential for toppling failures. Potential for topple failures remains even to impractically shallow slope angles which cannot be constructed onsite due to level of effort and cost. Modeling joints P and

K in RocTopple with respect to the other joint sets showed that toppling on P is most likely with O as the back plane (FS=1.06) and that toppling on K is unlikely due to the low base angle. Based on these analyses, historical performance, and the rock mass ratings (which showed that individual block sizes should not exceed approximately 7 x 7 feet) for the site, it was determined that topple failures with large enough volumes to block the spillway or damage the ogee weir are highly unlikely and low consequence. Therefore, any potential topple blocks identified during construction should either be scaled out, spot anchored, or held in place by mesh netting and that the risk associated with any remaining topple failure potential is acceptable based on the design parameters.

- Stereographic analysis showed that all four slope orientations have a potential for wedge failures on joint sets P and Q at ϕ=25 degrees, but when these values are input into SWedge for the 4 slope orientations the results showed, "no wedge is formed". Since the initial stereonet showed variation in joint orientations, P and Q were varied in SWedge and found that a wedge could be forced by varying Q by more than 10% of its representative value. These wedges formed with FS greater than one, very small volumes, and breaks that extended long distances into the slope face. Therefore, a wedge forming on joint sets P and Q was considered unlikely and low risk.

- Stereographic analysis also showed that slopes C and D have a potential for wedge failures on joint sets P and M at ϕ=30 degrees. Modeling in SWedge showed that wedges formed by P and M increase in volume and decrease in FS in the downstream direction. The largest and least stable wedge found during modeling was in Slope D with a resultant FS = 0.92 and a volume of 120,033 ft^3 (**Figure 9 - B**). The slope was then varied to determine if acceptable FS and volumes could be obtained. It was found that the FS would remain consistent at 0.92, but the volume would decrease as the slope grades shallowed. At a slope of 1.2H:1V the volume decreased to 32,104 ft^3 which is still too large, but at a slope of 1.4H:1V, the volume decreases to 617 ft^3, and finally the wedge potential would disappear at a slope of 1.5H:1V. Similarly, Slope C initially produces a wedge with a FS = 0.91 and a volume of 10,107 ft^3 and at a slope of 1.2H:1V the wedge disappears. This shows that slopes C and D are stable at a grade of 1.5H:1V, but that steeper slopes would be reasonable for design in some areas since the potential failure volumes are small enough that they could either be scaled, stabilized by anchoring, or allowed to fail. This is especially true for Slope D, which is at the far downstream end of the spillway, only about 60 feet tall, and over 400 feet from the proposed OGEE weir.

- Stereographic analysis showed that slopes B, C, and D have potential for planar failures on joint set P. These potential planar failures were also modeled in SWedge by adding a 3rd plane to the analysis (Basal Plane) (this is also supported by the wedge analyses since P is the base plane of all identified wedge failures). Joint set P was input as the basal plane, then all other joint sets were paired and input to find potential failures. The lowest FS was found in slope D, when M & N were input to produce a FS=0.46 and a volume of 2,910 ft^3. When a slope of 1.2H:1V was modeled for M & N, the FS increased to 0.71 and the volume also increases to 4,894 ft^3. The largest volume with a FS less than one was found when M & Q were input and resulted in a FS=0.91 and a volume of 44,968 ft^3 (**Figure 9 – B**). When a slope of 1.2H:1V was modeled for M & Q, the FS increased to

1.01 and the volume decreased to 24,311 ft^3. For a slope of 1.5H:1V the failure potential on M & N goes away and FS for Q & M increases to 2.1. Therefore, 1) Slope C is near stable at a slope of 1.2H:1V with volumes that could be scaled, anchored, or allowed to fail and 2) Slope D becomes stable at a slope of 1.5H:1V as the FS increases over 2 and volume decreases to 423 ft^3 and at slopes nearing 1.5H:1V the volumes could be scaled, anchored, or allowed to fail.

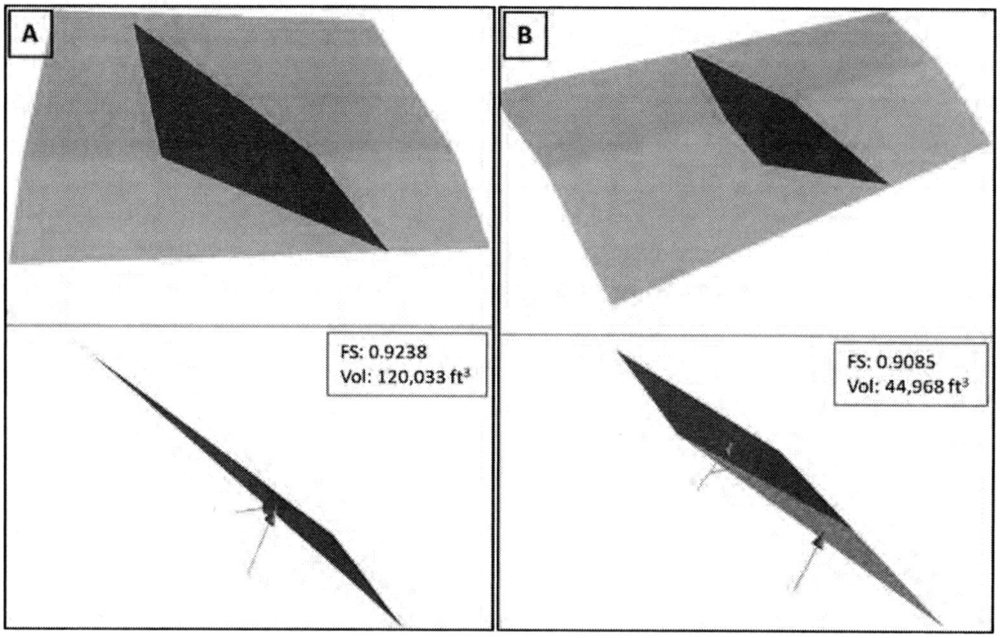

Figure 9: A) Potential wedge failure and B) Potential planar failure, on slope D at ϕ=30 degrees and a 1:1 slope

Limit Equilibrium Slope Stability Analysis

Even though the site is mainly composed of highly weathered to unweathered rock with a mantle of decomposed rock and colluvium, a limit equilibrium (LE) analysis was also used to provide factor of safety guidance on global slope stability analysis using the shear strength data from the rock joint testing, and various levels of piezometric lines. This analysis was used to supplement the kinematic and block failure analysis and several different models were developed based on varying sequences of rock weathering profiles based on borehole stratigraphy at different locations along the spillway right abutment. Three piezometric lines were used per model that represent normal, unusual, and extreme conditions based on estimated combinations of reservoir seepage and rainfall infiltration. The design slope stability factors of safety proposed for the normal, unusual, and extreme piezometric conditions were 1.4, 1.2, and 1.0 respectively. Based on the kinematic and block analysis, the right abutment was modeled in GeoStudio with a lower slope of 1.2H:1V, a 40-wide bench road (which is located just above the PMF water profile elevation), and a 1.5H:1V slope above the bench as shown in **Figure 10**. The results of the slope stability analysis for one of the models with a normal piezometric condition and a resulting FS of 1.43 is shown in **Figure 11**. Modeling slopes steeper than shown in **Figure 11** produces lower FS and therefore the LE model supports the findings of the block analyses.

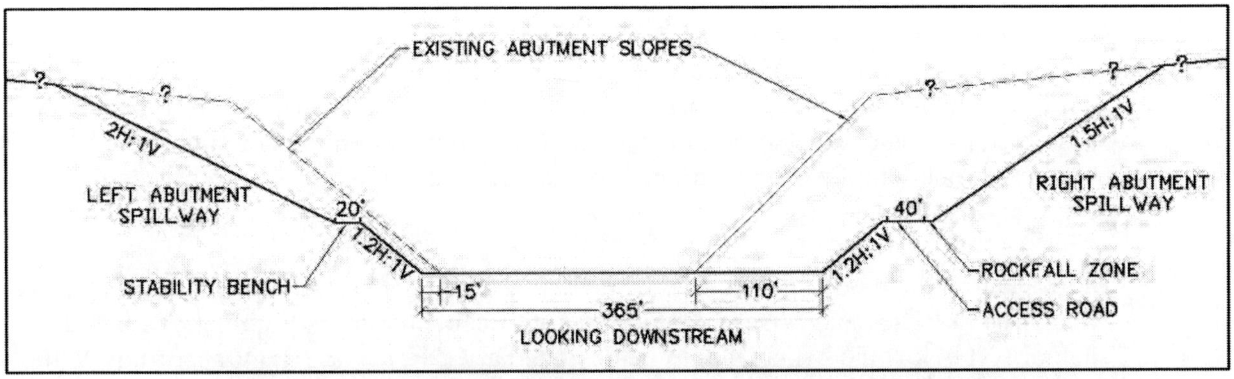

Figure 10. Proposed design slopes of spillway at ogee weir

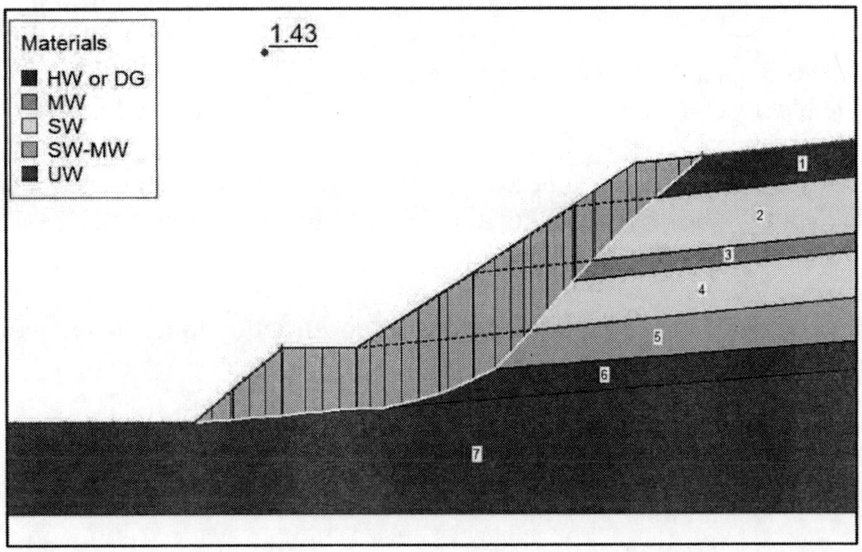

Figure 11. Right abutment (LE) slope stability - normal piezometric condition

CONCLUSIONS

Site characterization, kinematic analysis, and modeling show that the right spillway abutment varies in stability based on the slope orientation, structural data, and rock properties. Findings from the kinematic analysis were modeled in RocScience and the whole generalized slope was modeled in GeoStudio to provide a cross check of the results. The results from these analyses generally support one another in their findings and show that there is a feasible path forward for the right abutment spillway expansion design. They also provide the design team with confidence for a design path forward that will decrease risk during both construction and long term operation.

Based on the information and analyses presented here, the right abutment will be designed with the following parameters:

- All slopes in colluvium, alluvium, soil, decomposed rock, or similar and greater than five (5) feet in thickness shall be designed at 2H:1V.
- Slopes above the proposed road bench should be designed at 1.5H:1V for moderately weathered or worse rock (some of these areas will include sections of slightly weathered rock, but the extents are not considered large enough to warrant steeper slopes at this time).
- Slopes below the road bench should be designed at 1.2H:1V when in moderately weathered or better rock. Areas containing significant amounts of highly weathered rock or soil below the bench should be designed at 1.5H:1V or 2H:1V according to the first two bullets.
- Scaling shall be performed to remove all small scale potentially unstable blocks.
- Spot anchoring shall be installed as excavation progresses to stabilize any potentially unstable blocks that may still exist in the graded and scaled slopes.
- All slopes in moderately weathered or better rock and taller than 15 feet shall be covered with double twist hexagonal mesh for site safety from small scale rock topples, falls, and slides.
- Finally, all slopes shall be monitored during construction to field verify these parameters and ensure localized reinforcement is applied as needed.

REFERENCES:

- Federal Government Reports
  1. U.S. Army Corps of Engineers, Sacramento District. *Foundation Report, Success Dam Project, Tule River, California, Volume I.* USACE, 1961.
  2. U.S. Army Corps of Engineers, Sacramento District. *Tule River Basin, Lake Success Enlargement, Spillway Exploration and Rock Anchor Testing: Geotechnical Report – Preliminary Draft.* USACE, 2003

- Testing Organizations
  1. ASTM Standard D5607, 2016, "*Performing Laboratory Direct Shear Strength Tests of Rock Specimens Under Constant Normal Force*," ASTM International, West Conshohocken, PA, 2016

Case History and Remediation of a Troublesome Rock Cut in Georgia, Vermont

Ethan J. Thomas
Vermont Agency of Transportation
2178 Airport Road
Berlin, Vermont. 05641
(802) 595-6752
Ethan.Thomas@vermont.gov

Jay R. Smerekanicz, P.G.
Golder Associates Inc.
670 North Commercial St., Ste. 103
Manchester, NH 03101
(603) 668-0880
Jay_Smerekanicz@golder.com

Peter C. Ingraham, P.E.
Golder Associates Inc.
670 North Commercial St., Ste. 103
Manchester, NH 03101
(603) 668-0880
Peter_Ingraham@golder.com

Thomas D. Eliassen
Peters Construction Consultants Inc.
15 Cliff Street, Apt. 2
Montpelier, VT 05602
(802) 498-4993
Eliassentde@outlook.com

Prepared for the 70th Highway Geology Symposium, October, 2019

Acknowledgements

For example: The author(s) would like to thank the individuals/entities for their contributions in the work described:

Ameritech Rock Slope Constructors

Chris Benda, P.E. – Golder Associates Inc

Bailey L. Theriault, P.G. – Golder Associates Inc

Christopher Eddy – Vermont Agency of Transportation

Disclaimer

Statements and views presented in this paper are strictly those of the author(s), and do not necessarily reflect positions held by their affiliations, the Highway Geology Symposium (HGS), or others acknowledged above. The mention of trade names for commercial products does not imply the approval or endorsement by HGS.

Copyright Notice

Copyright © 2019 Highway Geology Symposium (HGS)

All Rights Reserved. Printed in the United States of America. No part of this publication may be reproduced or copied in any form or by any means – graphic, electronic, or mechanical, including photocopying, taping, or information storage and retrieval systems – without prior written permission of the HGS. This excludes the original author(s).

ABSTRACT

In the early 1970's shortly after construction, two significant rockfall events occurred at a rock cut located on I-89 NB at Mile Marker 107.50 in the Town of Georgia, Vermont. The rockfalls blocked both travel lanes of the northbound barrel of the Interstate. Smaller chronic rockfall events continued into the late 1990's. In 2007, the rock cut was ranked within Vermont's Rockfall Hazard Rating System as the 9th highest hazard rock cut.

The rock cut was constructed within the Lower Cambrian Dunham Formation consisting of steeply dipping curved beds of dolomite forming the east limb of a syncline with its axis nearly parallel to the roadway. Remnant bedding dips moderately to steeply to the west with at least three major joint sets parallel to cleavage planes, forming planar and wedge failure features. Significant rockfall events have originated from these areas.

In 2014, the rock cut was programmed for remediation. Due to the complex geology of the site, and poor blasting practices in the original construction, remediation options were utilized consisting of trim blasting, hand scaling, rock dowel and shear key installation, rock drains, and constructing shotcrete buttresses.

The condition of the rock cut presented challenges during construction including trim blasting to address back-break from original construction, additional rock reinforcement to mitigate poorer than anticipated deteriorated rock conditions, and construction of a substantial shotcrete buttress to address a remaining overhanging rock mass.

INTRODUCTION

Vermont's Interstate system was constructed in the early 1960s through the 1980s. Due to the mountainous nature of Vermont, the development of the Interstate system necessitated the construction of numerous rock cuts throughout the state. Most of these rock cuts were constructed using presplit blasting techniques and utilized a standard rock slope cut angle of 4V:1H (76°) without consideration of geologic structural control. As these rock cuts age, many become unstable and produce rockfalls that have potential to reach the roadway and impede traffic flow.

This paper presents a case study of a rock cut which produced significant rockfall events shortly after construction, the Vermont Agency of Transportations' (VTrans) response to managing rockfalls at this location, and subsequent rock slope remediation techniques used at the rock cut in question.

ROCK SLOPE INFORMATION AND ROCKFALL HISTORY

The subject rock cut is located on Interstate 89 (I-89) Northbound at Mile Marker 107.50 in the northwestern section of the state in the Town of Georgia, as shown on the location map in **Figure 1** below. The rock cut is 2,352 feet long, lies on the east side (right side) of the northbound lanes of I-89 between project Stations (Sta) 5699+78 south to Sta 5676+26, and ranges between 20 to 60 feet high. **Figure 2** shows a view of the rock cut looking south-east showing the general conditions of the rock cut prior to the remedial construction discussed later in this paper.

In 1972 the first significant rockfall occurred at the rock cut between Sta 5687+00 to Sta 5687+50. This rockfall overwhelmed the catchment ditch and spilled into the roadway blocking the majority of I-89 Northbound. During clean up, traffic was forced to skirt close to the median rock cut.

Figure 3 and **Figure 4** below show the unstable rock mass exposed after the rockfall. At the time, VTrans Chief Geologist Frank Lanza recommended that the unstable rock mass be removed utilizing controlled blasting techniques using a maximum drill hole diameter of 3" on a 1.5 foot to 2 foot spacing to produce a stable, near vertical wall.

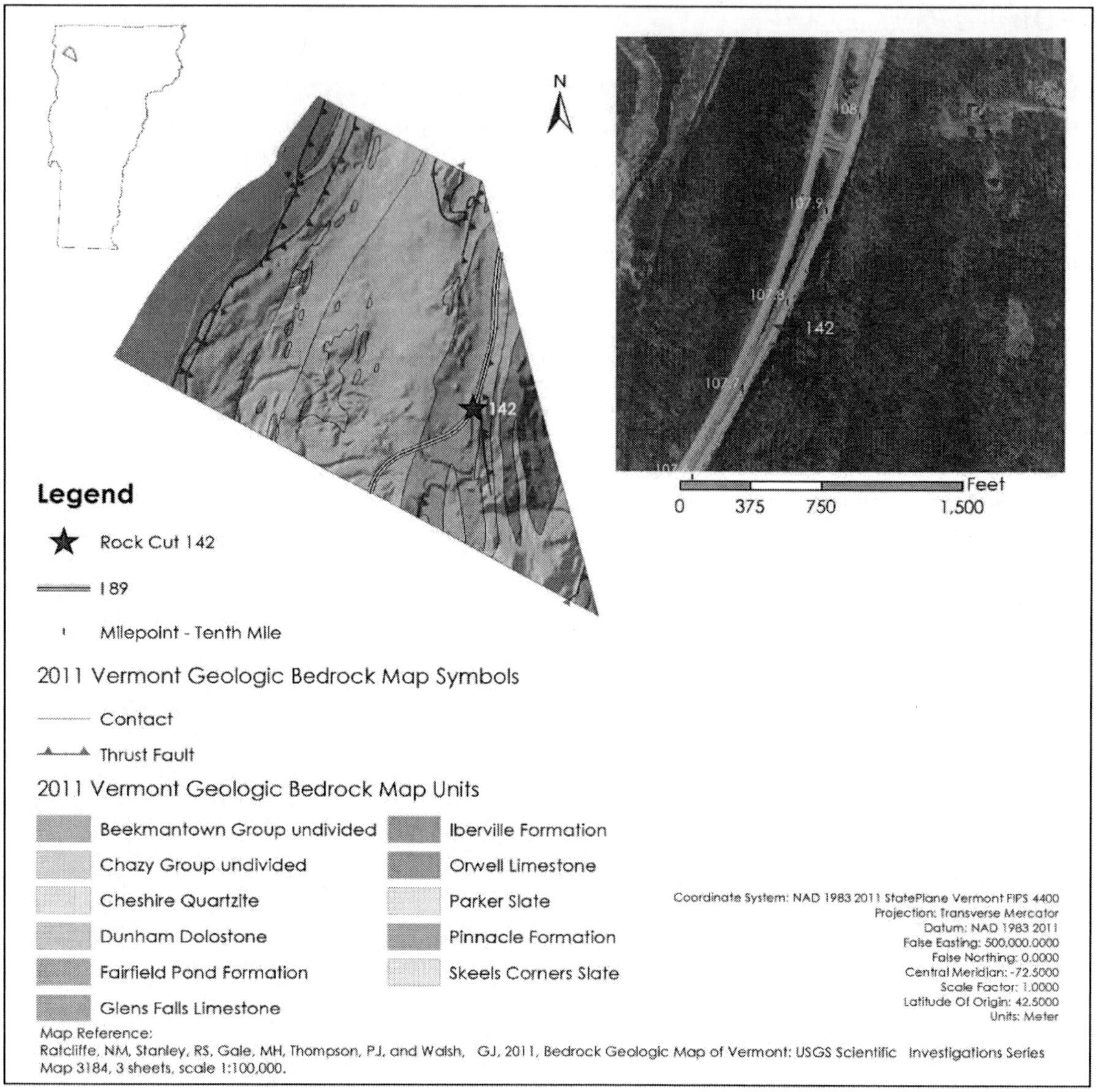

Figure 1 – Location and geologic map of project area. Red star denotes the location of the rock cut, lying within the Dunham Dolostone.

Figure 2 – View of rock cut looking to the south east. Note the extensive vegetation along the crest and the variable slope height.

Figure 3 – Unstable rock mass exposed after rockfall in 1972 near project Sta 5687+00 to Sta 5687+50. View is to the north north-east. Red dots denote joint sets parallel to overhanging rock face.

Figure 4 – View of unstable rock mass looking east north-east near project Sta 5687+00 to Sta 5687+50. Red dots denote west dipping tabular rock masses exposed after rockfall.

These remediation options were not performed, and on October 11, 1978 the rock mass failed again producing the second significant rockfall event at this rock cut. The fallen rock overwhelmed the catchment ditch and spilled into the roadway, blocking traffic as shown in **Figure 5** below. The rockfall resulted in naturally removing the unstable rock mass identified in 1972.

At some point between the 1978 rockfall and the 1990s, trim blasting was performed to remove unstable overhangs present at Sta 5691+50 to Sta 5692+00 and Sta 5686+00 respectively.

As the rock cut continued to age, smaller, chronic rockfall events continued into the late 1990s leading to frequent cleanout of the ditch line by maintenance personnel. An additional substantial rockfall occurred in February 1999.

In 2007 VTrans adopted and implemented a Rockfall Hazard Rating System (RHRS) based on the FHWA/Oregon DOT RHRS. The subject rock cut was designated as Rock Cut #142 within the RHRS inventory and was ranked as an "A" rock cut, where rockfalls will happen and rocks will reach the roadway. "A" ranked rock cuts are further differentiated from each other by developing a Rockfall Hazard Rating Score (RHR Score). This numerical value represents the hazard associated with a rockfall event at the particular rock cut. and rock cuts with high RHR Scores represent significant hazards. The subject rock cut received an RHR Score of 558 making it the 9th highest hazard rock cut in the inventory.

No. 284 ●●● #1 GEORGIA Wednesday, October 11, 1978

Free Press Photo by STU PERRY

Rock Slide

State Highway Department personnel Bernie Camley of Georgia and Wilbur Richards of St. Albans try to figure out how to clear away tons of rock that fell Monday night on Interstate 89, one mile from Exit 18 in Georgia. District Highway Engineer Sanford Brigham said the rocks — which do not block traffic completely — will have to be dynamited. He said a similar slide occurred in the same place six years ago.

Figure 5 – Newspaper clipping showing the rockfall event of 1978.

REGIONAL GEOLOGY

The bedrock composing the rock cut is mapped as the Lower Cambrian Dunham Dolostone Formation which consists of buff and pink mottled and massive dolostone, or light gray, pinkish gray weathering and massive to poorly bedded dolostone. The site is located within the overturned eastern limb of the Hinesburg-St. Albans synclinorium.

ROCK SLOPE DESIGN

Due to continued deterioration, increase in frequency of rockfall cleanup in the catchment ditch and the high RHR Score, the rock cut was programmed for remediation in 2014 with construction scheduled in the 2018 construction season.

Rock slope design was based on the results of review of rockfall history, field visits to photograph and record current rock slope conditions, identify areas of instability, to collect discontinuity data and assess properties of discontinuities.

GENERAL ROCK SLOPE CONDITIONS

Figures 6 through **8** show the general conditions of the rock cut by project stationing prior to construction activities. Of particular note are the following:

- Sta 5692+00 to Sta 5692+50 (see **Figure 6**): Steeply dipping curved beds of the east limb of a syncline with its axis nearly parallel to the roadway.
- Sta 5687+00 to Sta 5687+50 (see **Figure 7**): Vegetated face consisting of moderately to steeply dipping beds similar to the area exposed in Sta 5692+00 to Sta 5692+50. This is the source area for the rockfall events in 1972 and 1978. A large overhanging rock mass was exposed approximately 40 feet up on this cut after vegetation was removed during construction as discussed later in this paper.
- Sta 5685+00 to Sta 5685+50(see **Figure 8**): Recent wedge failure, possibly associated with karst solutioning. This section also contains a large rock block with dilated discontinuities, indicating movement has occurred.

Figure 6 – Rock cut conditions between Sta 5692+00 to Sta 5692+50. Note curved beds of the eastern limb of a syncline (left), and presplit blasting mitigation following the 1978 rockfall event (right), which occurred down station.

Figure 7 – View of rock cut conditions between Sta 5687+50 to Sta 5687+00. This is the area that produced rockfall in 1972 and 1978. View of rock cut conditions is obscured by vegetation.

Figure 8 – Large overhanging block of dolostone and recent wedge failure along solutioned joints near Sta 5685+33.

Groundwater evidenced by ice buildup was present in a few locations on the rock slope, and VTrans has observed groundwater emanating from bedding plane surfaces and karst features in several areas.

KINEMATIC ANALYSIS

VTrans used rock mass discontinuity information to conduct preliminary kinematic analyses and support rock slope mitigation design--including a possible recut of the slope. Fifty-five discontinuities were measured along the base of the rock cut, where remnant bedding dips moderately to steeply to the west with at least three major joint sets parallel to cleavage planes, forming planar and wedge failure features.

Stereonets created using Rocscience's Dips V5.0 program, presenting the kinematic analyses are presented in **Figure 9**. The discontinuities form three distinct sets (above the 4% contour), consisting of bedding, joints and cleavage. The kinematic analysis indicates that at the original cut slope configuration, i.e. 4V:1H, the slope is susceptible to planar sliding failure and wedge failure (**Figure 9A**). This conclusion is evidenced by the past rock slope failures and observed conditions. The kinematic analysis indicated that the rock slope failure mechanisms can be reduced by reducing the slope cut angle to as low as 1V:1H (**Figure 9B**). The curvilinear nature of some of the features associated with the syncline made use of the kinematic analysis highly variable based on location on the slope, therefore, kinematic analysis was used only to assess gross trends of structural orientations.

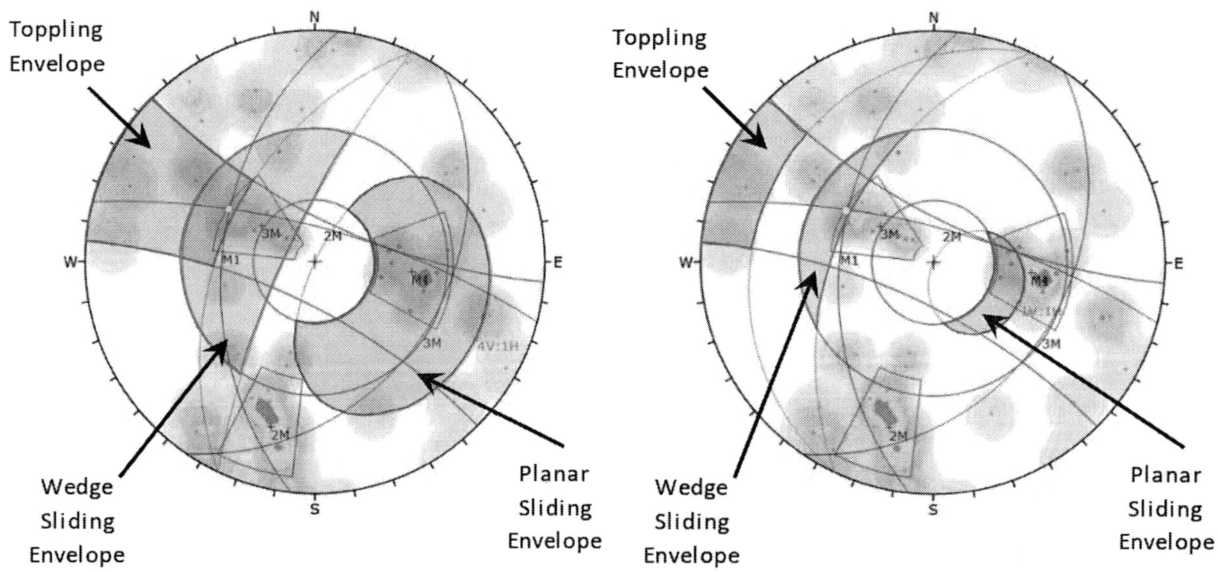

Figure 9A and 9B - Stereonets of the pre-mitigation slope conditions (Figure 9A, left), and of the mitigation design conditions (Figure 9B, right)

ROCK SLOPE REMEDIATION OPTIONS

Due to the complex geology of the site, the poor conditions of the rock cut due to blasting damage, the presence of large, unstable overhangs and the rockfall history of the site, two viable remediation options were considered. The first was cutting the entire rock cut back to a kinematically stable 1V:1H slope angle. The second option was to apply a wide range of location-specific remediation options consisting of hand scaling, rock dowel installation, rock drain installation, shotcrete buttresses, trim blasting to remove unstable overhangs, and installation of shear keys. Due to the high traffic counts of this area, the difficulty and cost of constructing a cross over for blasting work, it was decided that the rock slope stabilization measures of the second option would be implemented. Estimated quantities for the above-mentioned rock slope remediation items are shown below in **Table 1**.

Table 1 - Estimated Rock Slope Remediation Quantities for Rock Cut #142

Rock Slope Remediation Item	Estimated Quantity
Trim Blasting	5,000 Cubic Yards
Hand Scaling	150 Hours
Rock Dowel Installation	600 Linear Feet
Shear Key Installation	100 Linear Feet
Drain Holes in Rock	40 Linear Feet
Shotcrete	20 Cubic Yards
Shotcrete Nails	60 Linear Feet

The as bid cost of the project was $1,048,203.

CONSTRUCTION

Rock slope remediation construction began in the summer of 2018 starting with the setup of traffic control. One lane of the northbound barrel and the shoulder adjacent to the rock cut were closed during construction to allow the Contractor adequate room to store equipment, scale and trim the rock cut, and remove the scaled rock.

Clearing and grubbing of the crest, rock cut face, and ditch line was then performed. Once the vegetation was removed it became readily apparent that the rock slope conditions at the source area for rockfall in 1972 and 1978 were worse than assumed in the mitigation design. A fractured overhanging rock mass was exposed near the upper part of the slope, as shown in **Figures 10** and **11** below. The overhang was 6 feet to 8 feet deep, by 12 feet tall, by 32 feet wide. Field adaptation of the design for this overhang is discussed below.

Figure 10 – Overall view of rock cut at Sta 5687+50 to 5687+00 after vegetation has been removed, showing the extent of the overhang.

Figure 11 – Close up view of unstable rock mass exposed after vegetation removal.

Trim blasting was performed at four designated trim blast areas using five blasting events, and was followed by extensive hand and machine scaling. **Figure 8** above shows a typical unstable rock mass condition that was designated for remediation through trim blasting, and is identified as Trim Blast Area 1. Blasting in Trim Blast Areas 2, 3, and 4 was difficult due to venting of blast gases and explosive energy out of open back joints/back break in seams present in these areas. After the three blasts performed in Trim Blast Area 2, the Contractor opted to remove the remaining rock by machine scaling and hand scaling. Hand scalers were not readily available on-site during blasting activities and the blasters were forced to scale rock near the crest utilizing their drill rig after Trim Blast 2. This method was time consuming, risky for the blaster, and resulted in damage to the drill rig leading to a delay in drilling and blasting operations.

Hand scaling was conducted by a team of rock slope specialty contractors using rope access techniques. Hand scaling activities began at the southern section of the rock cut and progressed northward. Hand scaling proved difficult in areas where massive dolostone blocks were present. Although these blocks had open joints large enough to accommodate air bags, many of the blocks were keyed into the rock mass leading to difficult and time-consuming hand scaling in these areas. The trim blast areas were thoroughly scaled, and loose, blast damaged rock was eventually removed. Stable, intact dolostone blocks were left in many of the trim blasting areas and were deemed stable. This produced a rough look in the trim blast areas.

Rock dowel installation was performed by the rock slope specialty contractor. In some areas, hand scaling had removed blocks that had originally been planned to be stabilized with rock dowels. Additional rock dowel reinforcement was deemed necessary in the area of historical rockfalls as well as the fold limb exposed at Sta 5692+00 to Sta 5692+50.

Rock dowels were #11 Grade 80 galvanized steel and were fully grouted. The estimated linear footage of rock dowels in the mitigation design was 600 linear feet, but due to the severity of encountered conditions, 879 feet of dowels were used in the project.

Shear keys were originally intended to be installed to support relatively small, partially overhanging blocks of dolostone. Steel for shear keys consisted of #11 Grade 80 galvanized steel, encased in a block of concrete. The estimated linear footage of shear keys was 100 linear feet. As work progressed, some areas designated for shear key installation were thoroughly scaled and the need for shear keys in these areas was reduced. In some areas it was determined that rock dowels would be needed instead, and shear keys were eliminated in these areas. Twelve (12) shear keys were installed between at Sta 5685+80 and Sta 5687+30 for a total of 50.5 linear feet of steel for shear keys.

Dry mix steel fiber-reinforced shotcrete was utilized on this project. Originally shotcrete was intended to be used as dental shotcrete to fill small voids located in the presplit face at Sta 5692+00 (see **Figure 12** below). With the 8-foot-deep void exposed under the fractured rock mass at Sta 5687+00 (mentioned earlier), it was determined that the most appropriate remediation option was to install rock dowels within the fractured rock mass and create a shotcrete buttress along the base of the overhang. Removal of the overhanging rock mass would possibly encroach on the right-of-way, leading to significant project delays. It was estimated that 35 cubic yards of shotcrete, supported by a series of short rock nails, would be needed to address this area.

Drains were also used in the mitigation design, consisting of open drill holes inclined upward into the rock cut below known seeping areas and karstic features. The drains were designed to reduce groundwater pressures at the face, thus reducing ice buildup in winter that could cause rockfalls. Most of the drains produced water immediately following drilling.

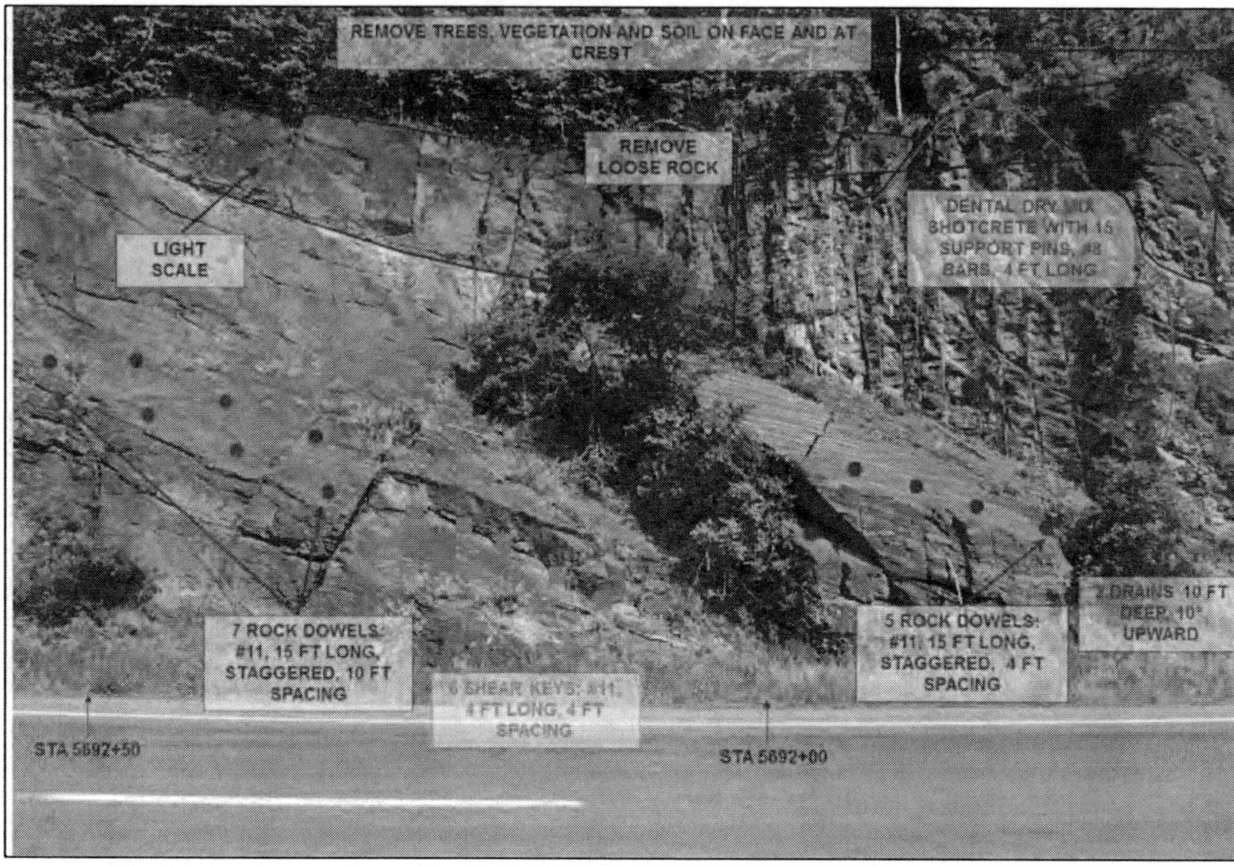

Figure 12 – Annotated photo showing planned remediation options at Sta 5692+00 to Sta 5692+50. Note the area highlighted in blue designating dental shotcrete.

Figure 13 below shows the condition of the rock cut at Sta 5687+00 to Sta 5687+50 after shotcrete, rock dowel installation, and shear key installation.

Figure 13 – Annotated photo of Sta 5692+00 to Sta 5692+50 showing rock slope reinforcement installed. Yellow text and dots represent rock dowel locations, green text and dots represents rock nail locations, red text and dots represent shear key locations.

CONCLUSIONS

Remediation of Rock Cut #142, which historically has produced significant rockfall events, was completed in the 2018 construction season. Due to vegetation growth that obscured rock slope conditions, complex structural geology, back break from poor blasting practices during original construction, and difficult scaling conditions, overruns occurred on the project. The total cost of the project was $1,466,000 which is $417,797 above the as-bid estimate.

After remediation the rock cut was reevaluated and downgraded from an "A" ranked rock cut to a "B" ranked rock cut. A "B" ranked rock cut is defined as a rock cut having possible rockfall potential and/or rockfalls that may reach the roadway. The reasoning for this rating was the overhangs remaining in the trim blast areas. Although this area was thoroughly scaled, it is expected that these areas will degrade through ice jacking each winter and may produce rockfalls in the next 10 to 20 years that will likely be contained by the improved ditch design.

Rock slope remediation projects differ vastly from conventional civil transportation infrastructure projects. As in almost all rock slope mitigation construction projects, the final product did not match that intended by the design. However, flexibility of the design to accommodate encountered conditions following clearing and grubbing, trim blasting and hand scaling allowed the project to be completed within a reasonable construction schedule, albeit above the original estimated cost by about 40%. This project is another example supporting that all rock slope projects should include contingencies to address unknown conditions that will be encountered once slope work begins.

REFERNCES

Rocscience Inc. 1998, Dips Version 5.0 - Graphical and Statistical Analysis of Orientation Data. www.rocscience.com, Toronto, Ontario, Canada.

The Assessment and Remediation of the Wabasha Street Rockfall

Anya Brose
Itasca Consulting Group
111 Third Avenue South
Suite 450
Minneapolis, MN 55401
612-371-4711
abrose@itascacg.com

Lee Petersen
Itasca Consulting Group
111 Third Avenue South
Suite 450
Minneapolis, MN 55401
612-371-4711
lpetersen@itascacg.com

Ryan Peterson
Itasca Consulting Group
111 Third Avenue South
Suite 450
Minneapolis, MN 55401
612-371-4711
rpeterson@itascacg.com

Prepared for the 70th Highway Geology Symposium, October, 2019

Acknowledgements

The authors would like to thank the City of Saint Paul Public Works for their contributions in the work described.

Brent Christensen- City of Saint Paul Public Works
Glenn Pagel- City of Saint Paul Public Works
Saint Paul Fire Department

Disclaimer

Statements and views presented in this paper are strictly those of the author(s), and do not necessarily reflect positions held by their affiliations, the Highway Geology Symposium (HGS), or others acknowledged above. The mention of trade names for commercial products does not imply the approval or endorsement by HGS.

Copyright Notice

Copyright © 2019 Highway Geology Symposium (HGS)

All Rights Reserved. Printed in the United States of America. No part of this publication may be reproduced or copied in any form or by any means – graphic, electronic, or mechanical, including photocopying, taping, or information storage and retrieval systems – without prior written permission of the HGS. This excludes the original author(s).

ABSTRACT

In April 2018, a limestone block weighing approximately 384,000 lbs. toppled off a 100-ft tall bluff along Wabasha Street in Saint Paul, Minnesota. The block broke into slabs along bedding planes, slid down the slope, overtopped a concrete wall, and blocked the west sidewalk and both lanes of Wabasha Street. The block was comprised of Platteville limestone, which overlies the Glenwood shale and St. Peter sandstone at the bluff where the fall occurred. Itasca Consulting Group (Itasca) was retained by the City of Saint Paul (City) to investigate the bluff stability after the fall and to propose remedial actions.

Site data collected in a field investigation allowed Itasca to determine the primary causes of the rockfall and the stability of the remaining limestone blocks and sandstone. Various remediation options were developed and presented to the City along with a list of factors to be considered by the City (such as location of unstable rock, land ownership, risk, cost, schedule, etc.). The final remediation plan was based on the following City guidelines:

1. No removal of materials from private property
2. No installation of support materials on private property
3. No exposure of workers to the unstable bluff.

These factors led to a gabion wall rockfall barrier installed between the toe of the slope and Wabasha Street.

INTRODUCTION

A rock slide on April 30, 2018 blocked Wabasha Street between Plato Boulevard and Cesar Chavez Street in Saint Paul, Minnesota. A Platteville limestone block weighing approximately 384,000 lbs. toppled off the bluff, broke into pieces, slid down the slope, overtopped a low concrete wall, and blocked the west sidewalk and lanes of Wabasha Street (Figure 1). After the limestone block fell, the affected section of Wabasha Street was closed until an investigation could be performed. Upon examining the bluff after the rockfall, another potentially unstable limestone block was identified adjacent to the block that had fallen.

Itasca was retained by the City to perform a site investigation to determine the cause of the April 2018 rockfall and to develop remediation options to protect Wabasha Street and the adjacent sidewalk from future rockfalls. Then, the City determined remediation guidelines. Since the bluff is owned by various private entities, the City requested that no installation or removal of material occur on private property and that there be no exposure of workers to the unstable bluff face. Following these guidelines, Itasca proposed a remediation plan that involved cleaning the bluff face from rockfall debris and vegetation, excavating a catchment behind the existing concrete wall, and building a gabion wall adjacent to the concrete wall.

Figure 1 – Aerial view of April 2018 rockfall.

Site Geology

There were five major rock units present along the bluff where the rockfall occurred. These geologic units are shown in Figure 2. Understanding the nature of the rock units played an important role in determining the cause of the rockfall and in engineering an appropriate solution for protecting the road from future rockfall events. Consequently, these units will be described in detail in the sections below.

Figure 2 – Geologic units present at the rockfall site.

St. Peter Sandstone

The base of the bluff is composed of St. Peter Sandstone, which is a quartz sandstone containing greater than 98% silica. At the rockfall location, there is approximately 60 feet of exposure of St. Peter Sandstone, with a total unit thickness of approximately 150 feet. The St. Peter Sandstone is weak to very weak, with a UCS of approximately 0-1000 psi (CNA, 2005). The St. Peter Sandstone is not water bearing at this elevation.

The St. Peter Sandstone was actively raveling along the bluff (Figure 3). In addition, trees and shrubs grew from many of the open joints along the bluff.

Figure 3 – Raveling St. Peter sandstone with tree roots growing in the joints.

Glenwood Shale

The Glenwood shale lies under the Platteville limestone and is about 5-ft thick. The Glenwood shale is composed of soft and hard, green to blue shale layers of varying thickness. This unit is also water bearing at some locations on site. The Glenwood shale does not have sharp contacts with either the overlying Platteville limestone or the underlying St. Peter sandstone—the contacts are transitional to both formations.

The Glenwood shale layer appeared to be heavily eroded. In some places along the bluff, the shale layer was eroded to loose shale fragments, which a rock hammer could easily penetrate (Figure 4). Additionally, due to the water-bearing nature of this layer, it was common to see roots and vegetation growing from the shale.

Figure 4 – Example of a heavily eroded section of the Glenwood shale.

Platteville Limestone

At the project site, the Platteville limestone consists of two layers: the upper, weathered limestone and the lower, less weathered limestone. The weathered limestone consists of the upper three members of the formation: Carimona, Magnolia, and Hidden Falls. The weathered Platteville is about 10 ft thick at the site. This unit is characterized by relatively strong rock pieces cut by weakened joints and seams.

The lower, less weathered limestone consists primarily of the Mifflin and Pecatonica members. The Mifflin and Pecatonica members of the Platteville limestone were defined by thin, wavy shale layers. The Mifflin member appeared to be more massive at this site, only cut by persistent subvertical joints, whereas the Pecatonica member was defined by blockier, more variable jointing.

Soil

Soil is the top layer at most locations on the site and is a mixture of natural soils, fill and top soil. The soil is intermixed with the prior slope stabilization and limestone slabs. The soil is also partially filling some of the large subvertical joints between limestone blocks.

ROCKFALL DESCRIPTION

Rock Block and Slope Description

The rock block that fell in April 2018 was in the lower, less weathered Platteville limestone, primarily the Mifflin member, plus some overlying weathered members. The block was defined by four subvertical joints and the dimensions of the block were 14 ft by 17 ft by 11 ft.

Once the block dislodged and hit the slope surface, the weaker shaley layers caused the block to break into slabs. The high strength of the lower Platteville caused some of the sheets to remain intact even after the block fell (Figure 5). Because the slabs were thin and wide, sliding rather than tumbling or rolling was the principal rockfall mechanism. This is significant when determining the impact and extent of future rockfalls.

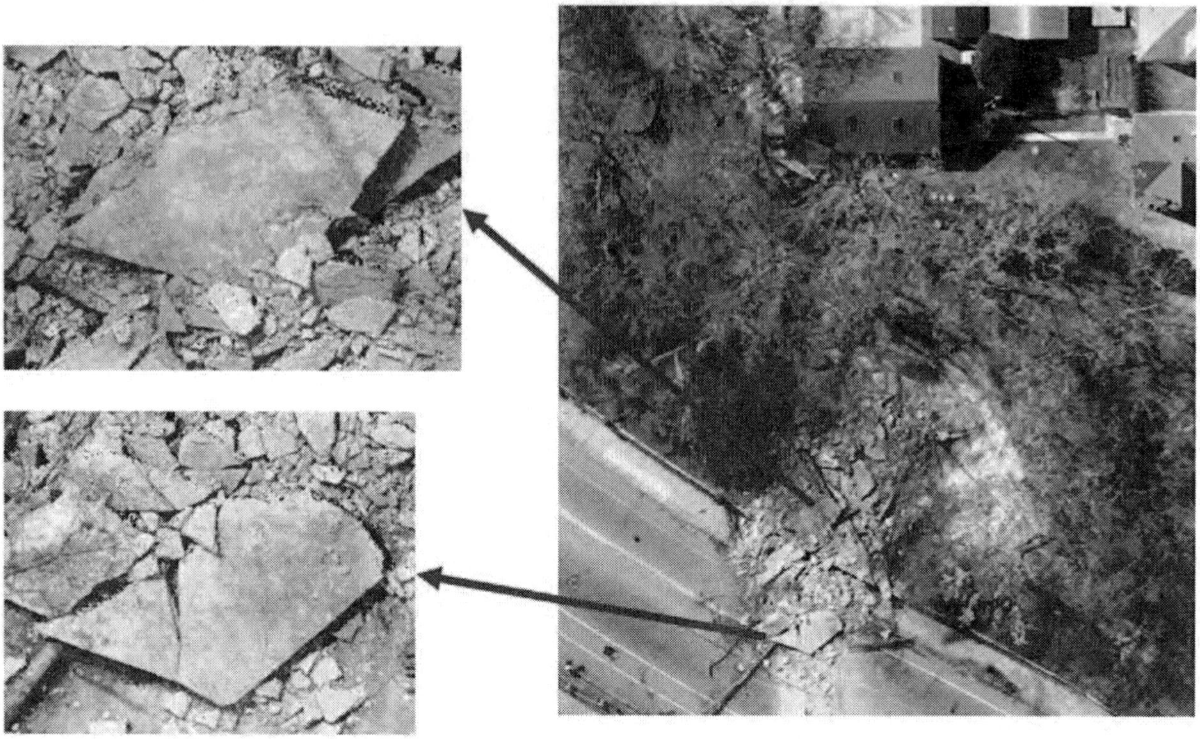

Figure 5 – Examples of intact limestone sheets.

Some of the upper, weathered limestone fell with the rock block, but a section of weathered Platteville remained, creating an overhang above the location where the rock block existed (Figure 6). After the rockfall, the upper, weathered limestone also continued to support a few feet of soil and vegetation.

Figure 6 – Image showing the upper, weather limestone overhang.

SITE INVESTIGATION

To understand the geology of the site, to determine the cause of the rockfall, and to identify areas requiring remedial action, a site investigation was performed in May and June 2018. The site investigation included photogrammetry and surveying, site mapping, and remote monitoring.

Photogrammetry, Scanning, and Surveying

To capture the bluff geometry, photogrammetry data was collected using a drone. Later, ground-based scanning provided additional coverage. Both techniques produced a point cloud that was used as a base to place and organize site information. Figure 7 shows a horizonal section through the limestone unit of the bluff. The arrows correlate the image of the bluff face to related locations in the photogrammetry. The red block indicates the approximate geometry of the block that fell.

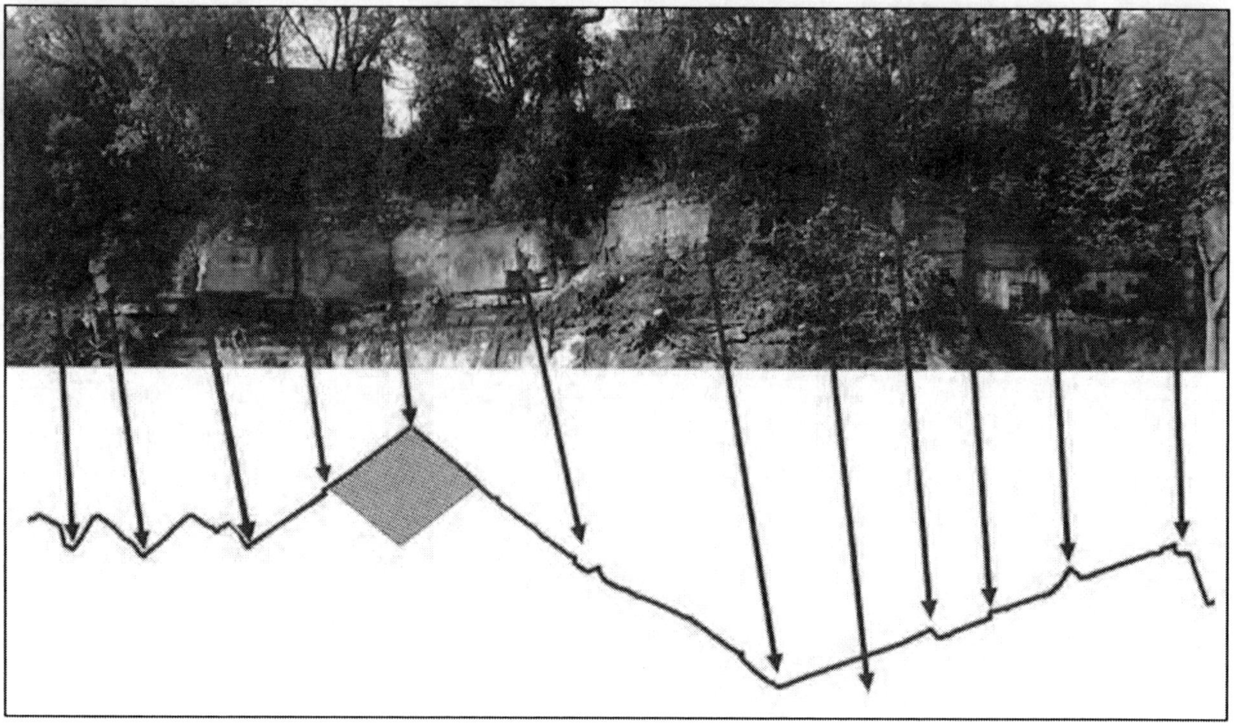

Figure 7 – Horizontal slice of the photogrammetry aligned with an image of the bluff.

In addition to photogrammetry and scanning, significant features across the bluff were surveyed to incorporate into the site model.

Site Mapping

To understand the nature and geometry of the rock units along the bluff, the site was extensively mapped, and the collected data was incorporated into the digital model of the site. A 90-ft boom lift was used to access the lower regions of the bluff. From the lift, most areas could be mapped and measured. To map the upper section of the bluff, which was inaccessible from the lift, the St. Paul Fire Department assisted in rope access. Joint location, characteristics, and orientation were recorded. These joints were then numbered and incorporated into the digital model. Examples of limestone joints and sandstone joints mapped on site are shown in Figure 8 and Figure 9.

Figure 8 – Example of mapped limestone joints.

Figure 9 – Example of mapped sandstone joints.

Remote Monitoring

To monitor potential movement of the rocks along the bluff while the site investigation was underway (and later during design and construction), a total of 51 points were monitored, with points in limestone, shales, and sandstone. Measurements were taken approximately 15 times per day. All monitored points were reflectorless—no prisms were installed. Remote monitoring data was reviewed daily to detect any potential movement. Reviewed data was compiled into PowerPoint presentations for quality assurance and future reference.

SITE INVESTIGATION FINDINGS

Slope geometry and changes with time

The pre-rockfall slope had two large bowl-shaped recesses that are laterally concave and characterized by near-vertical sandstone faces in the upper slope, transitioning to a flatter slope near street level. Prominent sandstone noses, which are vertically and laterally convex, are present on either side of the bowl-shaped recesses (see Figure 10). The slope was excavated to a uniform convex shape in 1929—the bowls have since formed by decades of erosion. The amount of erosion was analyzed using historic plans provided by the City. Plans of the 1929 Wabasha Street construction were compared to the present-day photogrammetry to visualize the significant amount of sandstone erosion that had occurred over the past 89 years. A cross-section taken in the south bowl is presented in Figure 11, showing the profile of the as-built cliff geometry (green) and the present geometry (black). The figure shows that some sections of the sandstone slope have eroded of the order of 50 ft in about 90 years. This rate of retreat is expected to continue, further undercutting the limestone.

Figure 10 – Photograph of the site highlighting the north and south bowls, and the north and south noses.

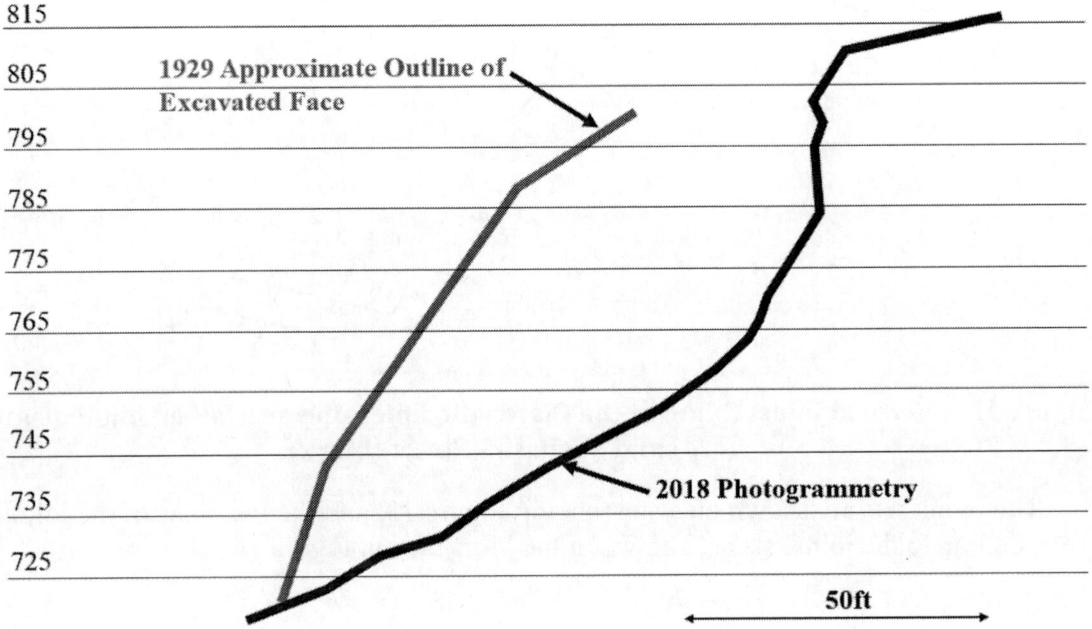

Figure 11 – Past and current slope profiles in the south bowl.

Bluff Discontinuities

A plan view of the location and orientation of the jointing in the limestone is shown in Figure 12. Note that the extent of the joints is unknown, so the length of the blue lines cannot be interpreted as fact. The vertical jointing in the limestone is cut by horizontal bedding planes.

The limestone joints were vertical to subvertical. There were three major joint sets in the limestone, striking at 320, 250, and 280 degrees. Many of the vertical joints were open and rough. One joint was measured to be open at least 17 ft deep. This led to the belief that the joints may extend deep into the bluff face.

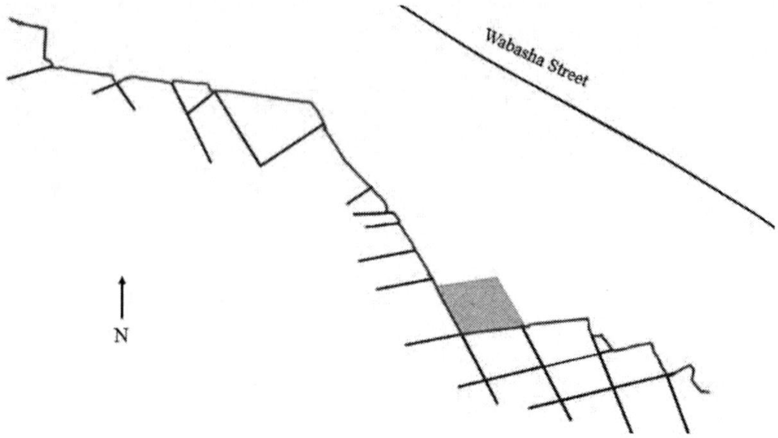

Figure 12 – Vertical joints (blue) in the Platteville limestone overlaying photogrammetry (red) and Wabasha Street (black).

The joints sets are shown on a stereonet in Figure 13, and the locations of the joints corresponding to the joint sets are shown on the bluff face in Figure 14.

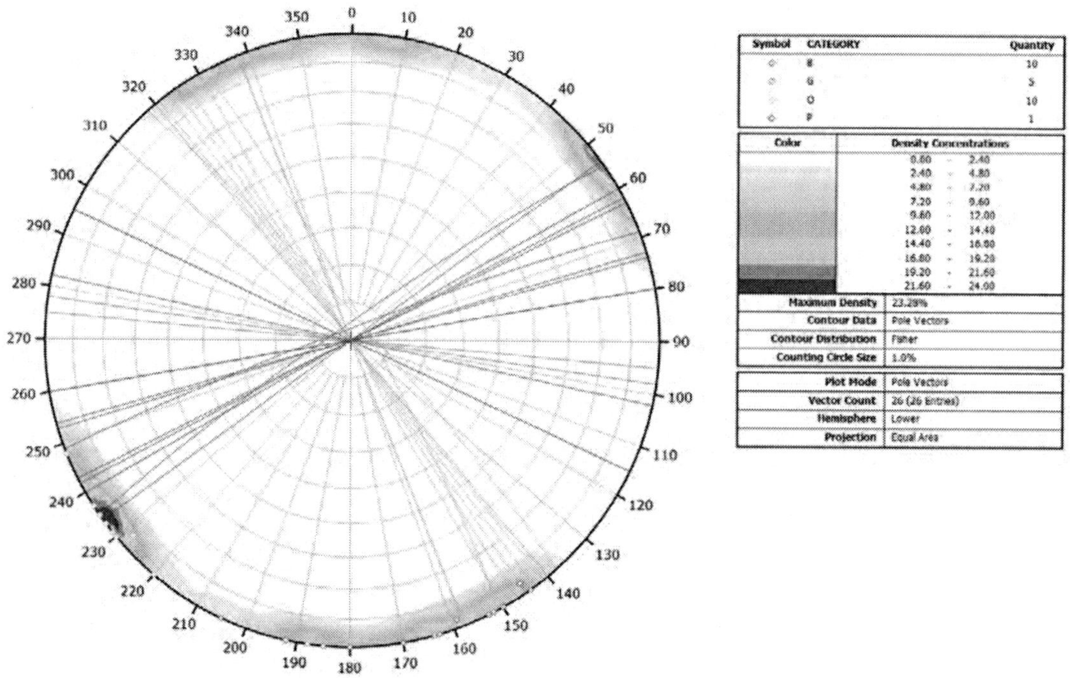

Figure 13 – Stereonet showing joint sets in the limestone.

Figure 14 – Joint locations in the limestone, with colors corresponding to the joint sets in Figure 13.

One of the major northwest striking joints doglegs to the west, then doglegs again back to approximately the same orientation. This joint is highlighted in Figure 15. Due to this variation in joint orientation, it was not certain which joints could be projected straight back from the bluff face, making it challenging to approximate the size of limestone blocks.

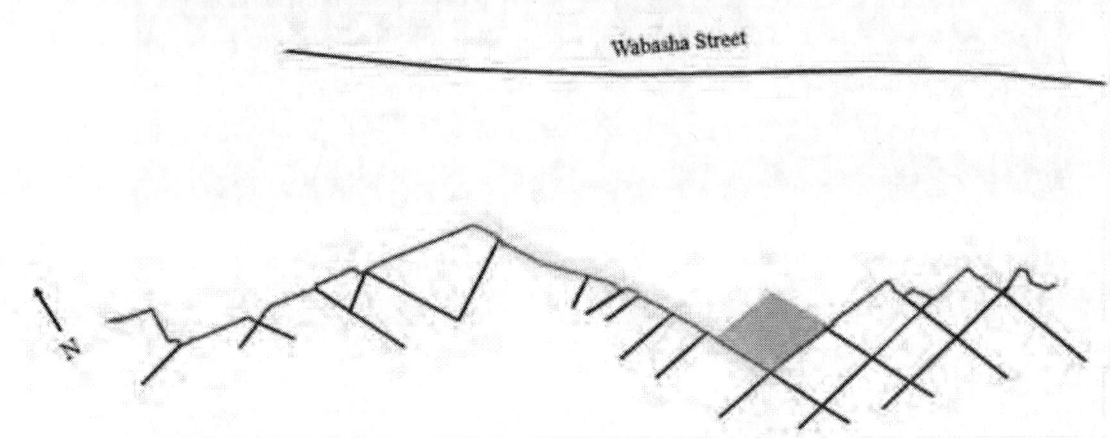

Figure 15 – Location of joint that doglegs to the west (highlighted).

The sandstone is more jointed and broken than typically seen in the Twin Cities area. The sandstone and the shale appeared to be actively raveling underneath the limestone, especially in locations where the shale was water bearing.

Trees, shrubs, and vines were present across the bluff. Trees were observed to have both positive and negative effects: the roots penetrate rock, loosen, and deteriorate, but also hold rock and soil in place.

ROCKFALL CONTRIBUTING CAUSES

Past rockfalls at the site were the result of four contributing causes. Future rockfalls will result from the same causes, which are:

- Surface water either pools on the surface and percolates into the soil and rock or runs off over the cliff face.

- Groundwater seeps out of the Glenwood shale. The St. Peter sandstone and Glenwood shale at the site has significantly eroded in the past 89 years, on the order of 50 ft.
- Both the surface and groundwater cause freeze-thaw action and the rapid erosion experienced. The erosion in the shale and sandstone leads to undercutting the Platteville limestone.
- The vertical joints in the Platteville limestone form discrete rock blocks, which are subject to undercutting and toppling. An example of an undercut limestone block is shown in Figure 16.

Figure 16 – An undercut limestone block located along the Wabasha Street bluff.

SITE REMEDIATION

Based on the site investigation findings, four site features were identified as potentially needing remedial measures. Referring to Figure 17, the identified site features include the following:

1. Weathered Platteville limestone: The weathered limestone could fall off the bluff and into the street.
2. Lower, hard Platteville limestone blocks: The limestone blocks could topple out of the bluff and into the street.
3. Glenwood shale at the limestone-shale-sandstone interface: The interface could erode underneath existing limestone blocks, accelerating the process of blocks toppling off the bluff.
4. Soil and sandstone: Soil or sandstone masses could slide off the bluff and reach the street.

Figure 17 – Identified remedial areas, numbered.

FACTORS INFLUENCING CHOSEN REMEDIATION OPTION

Itasca proposed multiple remediation options to the City. The solutions included different methods to either stabilize or remove the potentially unstable materials. Itasca also proposed factors to be considered when choosing the final solution. Selected factors are discussed in the sections below.

Private Property

A challenging aspect regarding the remediation involves the land ownership along the bluff. The bluff is owned by multiple private entities, one being a homeowner. Figure 18 shows the land ownership along the bluff. The limestone block that fell in April 2018 and most of the overhanging, loose limestone blocks are on homeowner property. For this reason, supporting or removing the limestone blocks would be a greater liability risk to the City, since they would be constructing on private property.

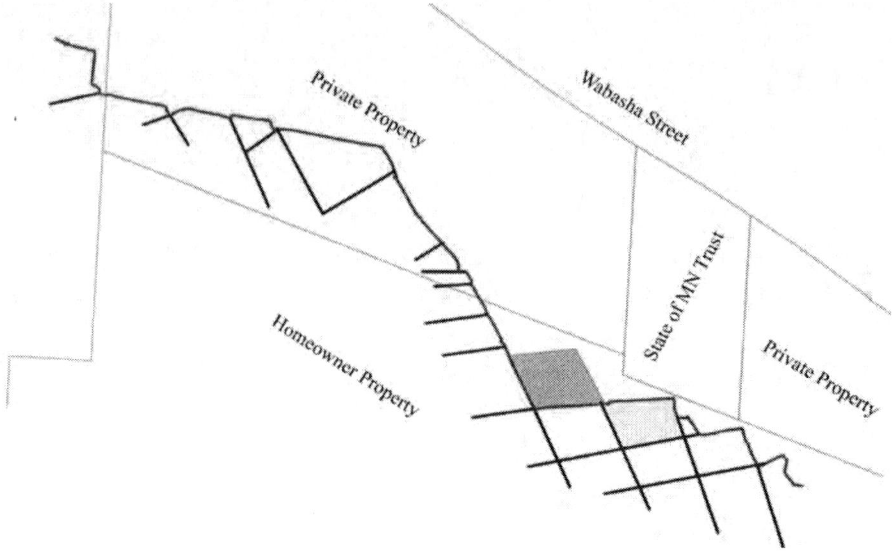

Figure 18 – Land designations along the Wabasha Street bluff.

Schedule

Wabasha Street is a main corridor leading into downtown Saint Paul. Because of the frequency of use, there was public pressure to find and execute a solution efficiently so that the street as well as the sidewalk may reopen for commuters.

Aesthetics

The bluff along Wabasha street provides natural beauty for this neighborhood. Several remediation options may result in changing the appearance of the bluff (i.e., bolting, netting, walls). Consequently, the visual aesthetics of the final solution were considered.

Public Safety and Construction Risk

The overhanging limestone blocks, the weathered limestone, and the soil mass on top and along the slope all create a public safety risk as well as a risk to construction workers for this project. There is a public sidewalk that runs along the bluff side of Wabasha Street that the city desired to keep open, meaning that the public will remain relatively close to the bluff.

Likewise, construction workers would be close to potential hazards, including the steep sandstone slopes and the overhanging limestone blocks. Remediation options that either limit workers' exposure to the unstabilized bluff or establish necessary safety measures during construction were considered.

CITY OF SAINT PAUL REMEDIATION GUIDELINES

The City of Saint Paul determined the guidelines for which Itasca should propose their final recommendation. The guidelines include:

- No removal of materials from private property.

- No installation of support materials on private property.
- No exposure of workers to the unstable bluff.

Following these guidelines, Itasca proposed the following remediation plan:

- Cleanup site debris and vegetation.
- Excavate catchment behind existing concrete wall.
- Construct gabion wall at the base of the bluff, adjacent to the existing concrete wall.

Since the wall was to be built on city property, it would involve no interference with private property. To address the safety risk, cleanup and construction of the gabion wall was to be performed from a safe distance from the bluff face, putting no workers at risk. The gabion wall would be built from local Prairie Du Chien limestone, helping camouflage the wall and maintain the natural beauty of the area.

Gabion Wall Design

Choosing the height required to provide the necessary storage volume was complex. (Because the rockfall mode was sliding of rock slabs, not rolling and bouncing, only storage volume was important.) Three design measures were applied:

- Assessment of how the April 2018 rockfall spread, in order to estimate the length of wall that would stop a future rockfall.
- Conduct a *3DEC* back analysis of the April 2018 rockfall, and apply the calibrated model to a future rockfall.
- Estimate the rate at which future slope raveling would fill the storage volume. This consideration also established the frequency that the area behind the gabion wall must be cleaned.

The gabion wall was designed to be 12 ft tall, 9 ft wide, and 260 ft long. The wall cross-section is shown in Figure 19. Most of the wall volume was constructed using 4- to 8-inch Prairie Du Chien limestone. The rock facing the street was constructed using beige stone to match the bluff. Wire mesh coating used in the gabion baskets consisted of zinc overcoated with PVC to extend the life of the gabion wall. A geotextile filter material was placed at the base of the gabion baskets.

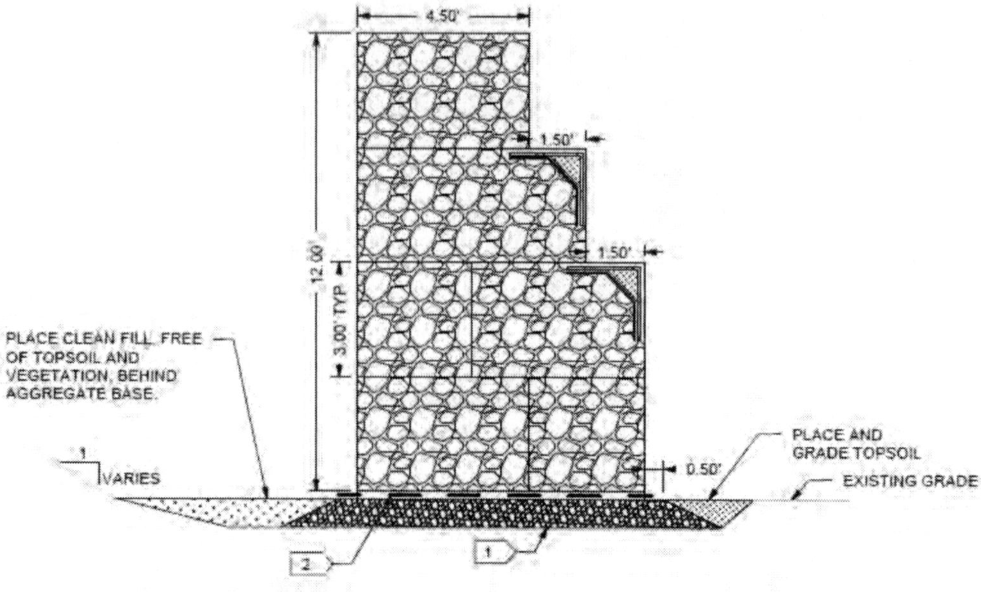

Figure 19 – Wabasha Street gabion wall section.

CONCLUSIONS

After a rockfall in April 2018, Itasca performed a site investigation to determine the cause of the rockfall and to mitigate future public safety risks. From observation data, photogrammetry, and historical information, Itasca provided the City of Saint Paul with multiple remediation options. The chosen option involved cleaning the debris from the April 2018 event, excavating a catchment, and building a gabion wall to prevent any future rockfalls from harming the public. This option was ideal because it did not interfere with private property and it kept construction workers at a safe distance from the bluff face.

REFERENCES:

- Report
    1. CNA Consulting Engineers (2005). City of Saint Paul, Assessment of Saint Peter Sandstone Mines- Wabasha Street to Lilydale.

Glaciolacustrine Earthflow Slides on US 95
Naples, ID

Collin McCormick, P.E.
Landslide Technology
10250 SW Greenburg Rd. Suite 111
Portland, OR 97223
503-452-1200
collinm@landslidetechnology.com

Prepared for the 70th Highway Geology Symposium, October, 2019

ABSTRACT

Failure of a cut slope along US 95 just south of Bonners Ferry, Idaho impacted traffic along a major highway linking commerce between the US and Canada in March 2017. Initial field reconnaissance revealed groundwater seeps emanating from thin sand layers within glaciolacustrine silt deposits. Piping occurred in the seepage areas, resulting in flow slides and slumping of destabilized portions of the cut slope. Liquefied slide debris traveled down the cut slope over long runout distances, flowed into the ditch, and then onto the highway and across to the opposite side. This highly-fluid earthflow was difficult to contain. Concrete barriers placed along the edge of pavement were pushed by the leading edge of the advancing earthflow. Retrogression of the head scarp caused additional slumping and subsequent lobes of earthflow debris posed a continuing hazard to traffic.

Temporary mitigation measures were developed to address the immediate earthflow hazards. A temporary debris retention barrier was rapidly constructed using readily available materials to contain liquefied slide debris.

Supplemental temporary mitigation measures were necessary to maintain highway safety through winter. In late summer 2017, the slope was laid back in the headscarp area to improve local stability and a network of French drains were installed in the slide debris perched on the cut slope to reduce water runoff and groundwater pressures.

Acknowledgements

The author would like to thank the individuals/entities for their contributions in the work described:

Charlie While, P.G. – Idaho Transportation Department

Jeff Drager, P.E. – Idaho Transportation Department

George Machan, P.E. – Landslide Technology

Benjamin George, P.E. – Landslide Technology

Darren Beckstrand, C.E.G. – Landslide Technology

Disclaimer

Statements and views presented in this paper are strictly those of the author(s), and do not necessarily reflect positions held by their affiliations, the Highway Geology Symposium (HGS), or others acknowledged above. The mention of trade names for commercial products does not imply the approval or endorsement by HGS.

Copyright Notice

Copyright © 2019 Highway Geology Symposium (HGS)

All Rights Reserved. Printed in the United States of America. No part of this publication may be reproduced or copied in any form or by any means – graphic, electronic, or mechanical, including photocopying, taping, or information storage and retrieval systems – without prior written permission of the HGS. This excludes

INTRODUCTION

US Route 95 through the northern Idaho Panhandle serves as a major transportation corridor connecting commerce between the US and Canada. The location map is shown in Figure 1. Repeated slumping and earthflow slides occurred in the right cut slope at US 95 Milepost (MP) 498 near Naples, Idaho the spring of 2017. Earthflows had long runout distances causing liquefied slide debris to displace concrete barrier rails and enter the travel lanes. Multiple road closures were required to clean up the roadway. Hazardous landslide conditions persisted as retrogression of the head scarp recharged potential for earthflow slide debris to liquefy and impact the roadway.

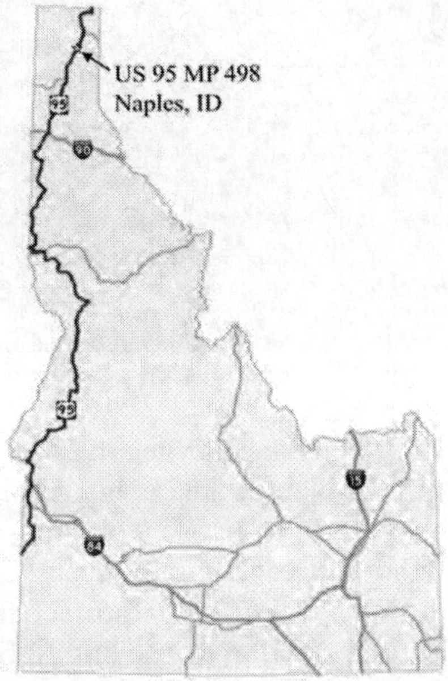

Figure 1. US Route 95 MP 498 Location Map

The Idaho Transportation Department retained Landslide Technology under an emergency contract to assist the Department with initial interpretation of the slide and development of mitigation options. Emergency mitigation measures were developed to address the immediate hazards to traffic. A temporary debris retention barrier was rapidly constructed using readily available materials to contain liquefied slide debris. Supplemental temporary mitigation measures were necessary to maintain highway safety through the following winter until a more permanent mitigation could be developed. In late summer 2017, the slope was laid back in the headscarp area to improve local stability and a network of French drains were installed in the slide debris perched on the cut slope to control water runoff and reduce groundwater pressures.

SITE CONDITIONS AND GEOLOGY

The right cut slope at MP 498 is approximately 80 feet tall by 540 feet long and was originally excavated at a 1.5H:1V slope angle. US 95 is a two-lane, undivided highway at this location. Maintenance crews reported the site has a long history of surficial erosion and sloughing during heavy precipitation and snow melt events requiring frequent ditch cleaning. Erosion and sloughing was also reported in the cut slope adjacent to the railroad and below the highway.

Historic aerial imagery available since 1992 indicates vegetation has been mostly unable to establish itself on the slope as shown in the oblique aerial imagery from 2014 shown in Figure 2.

Figure 2. Google Earth oblique aerial view of US95 MP498 cut slope in 2014.

Slope materials consist of sensitive, glaciolacustrine silt deposits, susceptible to flow slides when saturated. The regional geology of the Naples area is heavily influenced by past glaciation. Pleistocene glaciation advanced south from British Columbia and carved out the Purcell Trench all the way to Coeur d'Alene, Idaho. Retreat of more recent Holocene glaciation deposited the glaciolacustrine materials that form the slopes in the project area. The geologic map of the Naples are is shown in Figure 3 with the MP 498 site highlighted in red.

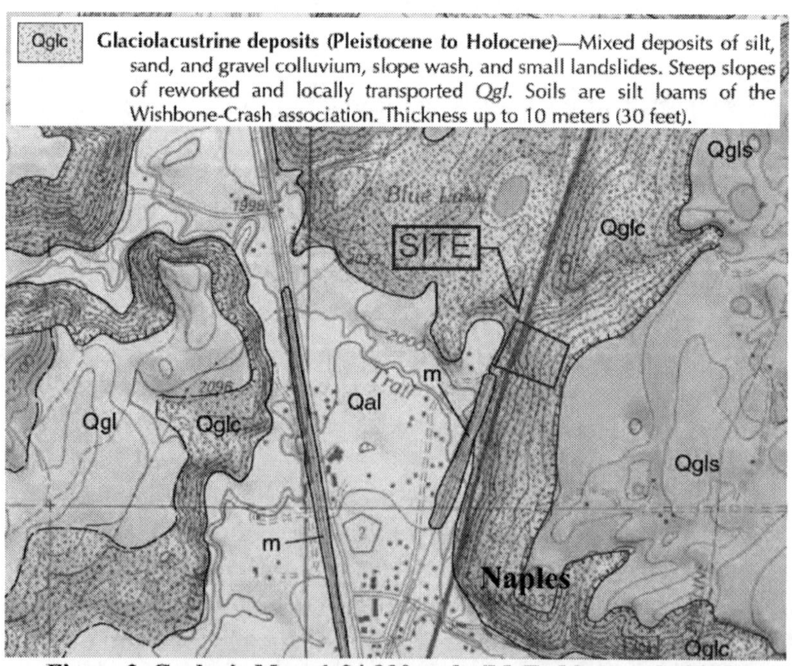

Figure 3. Geologic Map, 1:24,000 scale (McFadden et al., 2009)

LANDSLIDE CONDITIONS

In the spring of 2016, a slump formed a head scarp at top of the northern end of the cut slope. Photos 1 and 2 show the slump condition in August 2016. The site remained in this condition until winter.

Photo 1 and 2: Condition of northern slump and headscarp in August 2016

The Idaho Panhandle saw near-record precipitation in the winter and spring of 2016/2017. Historic climate records from the Bonners Ferry weather station, 10 miles north of MP 498, indicate that on average approximately 20 inches of precipitation falls from October through May. In the 2016/2017 season over 30 inches of precipitation fell in this time period and snow depths peaked later than usual at 37 inches in mid-February. Increasing temperatures culminating in a rain-on-snow event decreased the snow pack to zero by March 13, 2017. Precipitation and snow depth from the 2016/2017 winter-spring season are shown in Figure 4.

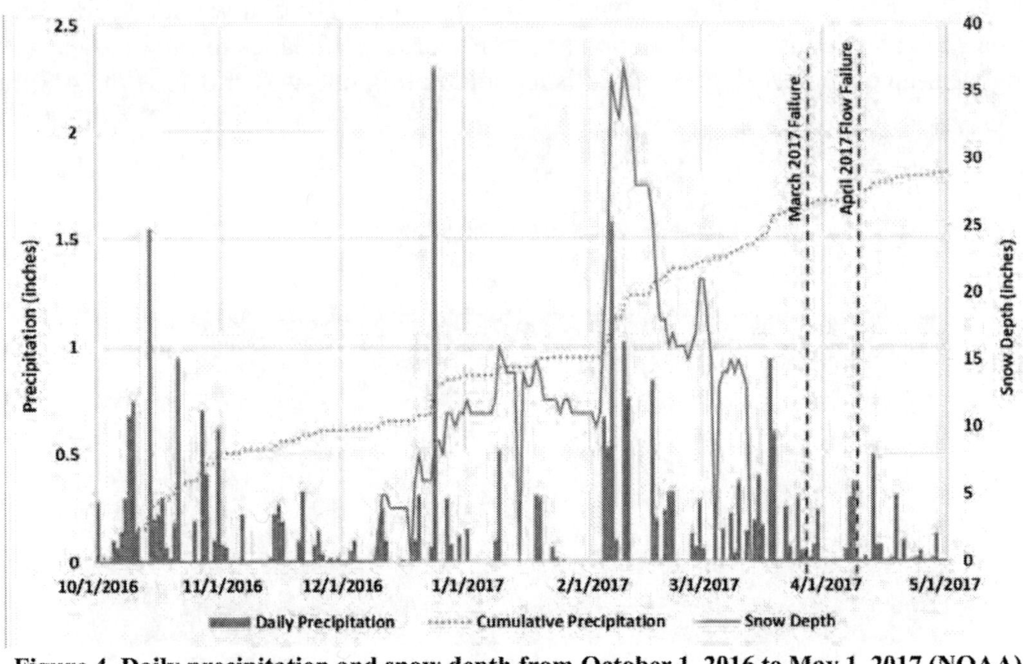

Figure 4. Daily precipitation and snow depth from October 1, 2016 to May 1, 2017 (NOAA)

Retrogressive slumping and earthflow slides occurred in the cut slope between March 21 and March 29 after the rapid melt off. Oversteepened headscarps calved and slumped, increasing the amount of unstable material in the slide mass. Continued calving and slumping of the headscarp caused trees to tilt and fall, creating an additional hazard to traffic. The ongoing accumulation of soft, saturated slide debris caused the toe of the slide mass to progressively bulge out of the cut slope and run down towards the highway. Liquefied slide debris had long runout distances, which made material containment difficult. Slide debris and earthflows reached and covered the northbound lane of the highway impacting traffic. ITD closed the northbound (NB) lane of US 95 and traffic control was established to maintain alternating one-way traffic. Concrete barriers were installed to contain slide debris and protect traffic; however, subsequent flow slide events pushed the barriers west towards traffic as shown in Photos 3 and 4. A secondary line of barriers was installed to improve slide debris containment.

Photo 3 and 4. Concrete barrier rails displaced after slide debris impact

On April 7th just after 2:00 pm, a rapid earthflow slide developed in the southern portion of the cut slope. ITD promptly closed the roadway before the accelerating flow slide broke through the concrete barriers and deposited slide debris and trees across both lanes of the highway (Photo 5). Following cleanup of the debris, ITD established alternating one-way traffic with 24/7 flaggers.

Photo 5. Liquefied slide debris crossing US 95 on April 7, 2019

FIELD RECONAISSANCE

Landslide Technology performed geologic and engineering reconnaissance of the site to assist ITD with initial interpretation of the slide conditions and development of emergency mitigation. Soils observed in the cut slope consisted of layered glaciolacustrine deposits of clayey silt to sandy silt. The upper 10 feet of material exposed in the headscarp was stiff to very stiff silt and fine sandy silt. This stiff upper layer was underlain by soft to very soft clayey silt. In April 2017, groundwater seeps were observed 5 to 10 feet below the top of the headscarp emanating from thin interbedded sand layers. Sag ponds had formed in the area behind down-dropped blocks of the stiff surface layer and perched slide debris. Lower cut slope materials were covered by slide debris from the upper slope. However, the lower slope materials appeared to be stiffer and similar in composition to the fine sandy silt exposed in the upper head scarp.

ITD observations indicated the slide had consistently bulged out of the slope about one third to halfway up the cut slope. Pavement distress was not observed in the roadway which seemed to confirm that a deep-seated slide was not active under the roadway. Potential paleo-landslide features were observed on the natural hill slope above the cut including muted or eroded scarps and small ridges. However, the slope above the slide had been previously logged by the property owner in 2011 and natural slope features may have been disturbed by logging activity.

Evidence of piping erosion was observed in seepage areas along the cut slope below the highway and above the railroad to the west. The piping erosion formed overhanging blocks of more cohesive glaciolacustrine interbeds that eventually would calve off. A similar mechanism likely occurred in the cut slope above the highway contributing to slumping of destabilized portions of the cut slope and formation of flow slides.

EMERGENCY MITIGATION

Landslide Technology worked with ITD engineering and maintenance personnel to develop an action plan to accomplish the following objectives for emergency response and reestablishment of normal roadway operations:

1) Maintain safe conditions for public, Department personnel, and contractors.

2) Shift from manned to signalized traffic control when hazardous conditions are contained

3) Clean roadside debris and reestablish capacity of the roadside ditch.

4) Open northbound lanes when landslide activity has stopped (or slowed to a creep).

The first action identified to meet the aforementioned objectives was to construct a more robust temporary containment barrier that would retain slide debris and allow ITD to eventually resume two-way traffic. A simple gravity barrier system was developed to increase containment capacity and resistance against sliding from rapid flow slide impact. The roadside ditch was cleared of slide debris and an improved containment barrier was constructed using readily available materials from a nearby ITD maintenance yard. A row of concrete ecology blocks were stacked, two high, approximately 4 feet inboard of the roadside concrete barrier rails. The space between the ecology blocks and concrete barrier rail was backfilled with 12 inches of sand backfill material. The top of the sand backfill was then lined with geotextile separation fabric and the remaining space between the concrete barrier and ecology blocks was backfilled with native soil from the site for additional weight to resist sliding. The sand layer allowed the saturated native backfill soils to drain and stiffen over time. Barrier construction is shown in Photos 6 and 7.

Photo 6 and 7. Temporary debris containment barrier construction

A series of survey points was established along the slide allowing ITD maintenance personnel to collect regular measurements and monitor slope deformations. By late-May of 2017, slumping and flow sliding had ceased, groundwater levels had decreased appreciably, and the surface of the slope began to dry. Anticipating reactivation of slide movement and further retrogression of the head scarp the following winter, ITD obtained a contractor under emergency contract in September 2017 to remove slide debris and lay back the upper slope. The contractor also installed French drains after regrading the cut slope to intercept and direct surface water and shallow groundwater away from the slope. A 10-foot wide bench was constructed from the roadway elevation on the north and south sides of the slope up to approximately 50 feet vertical distance above the roadway to allow drill rig access for geotechnical subsurface explorations. The slope below the bench was graded to 1.25H:1V and the slope above the bench to approximately 2.5H:1V.

Photo 8 and 9. Regraded slope conditions in September 2017

During winter and spring of 2017/2018, additional slumping occurred in the upper slope. However, the temporary debris containment successfully prevented slide debris from entering the roadway. ITD was able to maintain two-way traffic through the slide area throughout the season.

PERMANENT MITIGATION

A geotechnical subsurface exploration program was completed at the site in October 2017. Slope inclinometers and vibrating wire piezometers were installed to monitor slope deformations and groundwater levels over the 2017/2018 winter.

A permanent mitigation design was developed in 2018 to remove slide debris and prevent continued upslope retrogression of the slide. A rock inlay was designed to replace disturbed glaciolacustrine soils and slide debris with free-draining, high strength rockfill material. The additional confinement applied by the rock inlay will increase local stability at the cut face and decrease potential for piping erosion processes from high gradients and seepage forces during major precipitation and snow melt events. A schematic of the rock inlay design is shown in Figure 5. A contract has been awarded for the rock inlay and construction is underway with expected completion in late summer 2019.

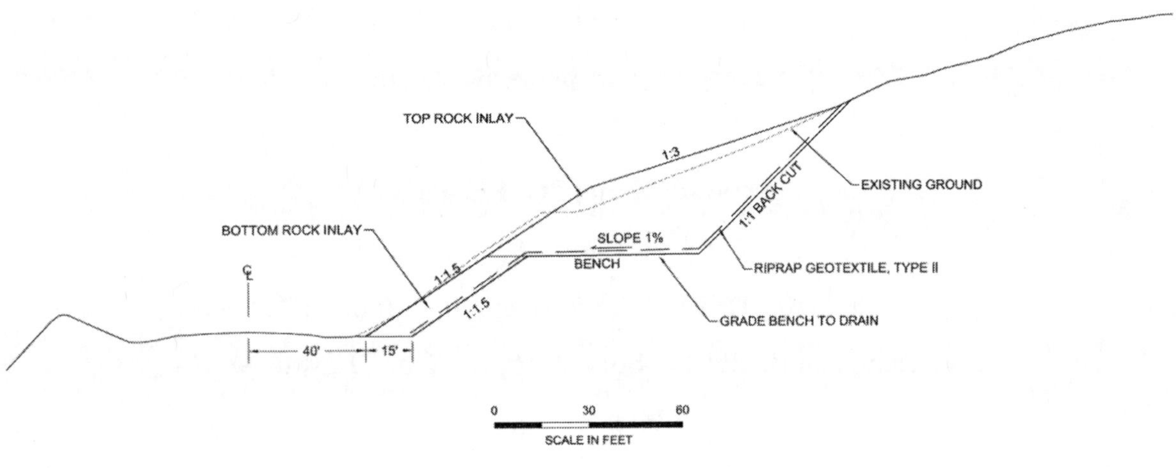

Figure 5. Rock Inlay Schematic

REFERENCES

National Centers for Environmental Information. *Global Historical Climatology Network – Daily*. National Oceanic and Atmospheric Administration (NOAA). ncei.noaa.gov/cdo-web/search. Accessed July 2, 2019.

McFaddan, M.D.; Breckenridge, R.M.; Burmester, R.F.; Lewis, R.S., 2009, Geologic Map of the Naples Quadrangle, Boundary and Bonner County, Idaho: Idaho Geologic Survey Digital Web Map 109, scale 1:24,000

Flexible Ring Nets as a Solution for Debris Flow Protection in Environmentally Sensitive Areas

Saleh Feidi

Geobrugg North America, LLC

22 Centro Algodones, Algodones, NM 87001

(909) 706-8964

Mallory A. Jones, GIT

KANE GeoTech, Inc.

7400 Shoreline Drive, Suite 6

Stockton, California 95219

(209) 472-1822

mallory.jones@kanegeotech.com

William Kane, PG, PE, PhD

KANE GeoTech, Inc.

7400 Shoreline Drive, Suite 6, Stockton, California 95219

1441 Kapiolani Boulevard, Suite 1115, Honolulu, Hawaii, 96814

(209) 472-1822

Prepared for Submission to the 70th Highway Geology Symposium

Portland, Oregon

ACKNOWLEDGEMENTS

The author(s) would like to thank the individuals/entities for their contributions in the work described:

Les Firestein

Geobrugg North America

Access Limited Construction, Inc.

Disclaimer

Statements and views presented in this paper are strictly those of the author(s), and do not necessarily reflect positions held by their affiliations, the Highway Geology Symposium (HGS), or others acknowledged above. The mention of trade names for commercial products does not imply the approval or endorsement by HGS.

Copyright

Copyright © 2019 Highway Geology Symposium (HGS)

All Rights Reserved. Printed in the United States of America. No part of this publication may be reproduced or copied in any form or by any means – graphic, electronic, or mechanical, including photocopying, taping, or information storage and retrieval systems – without prior written permission of the HGS. This excludes the original author(s).

ABSTRACT

Debris flow events generally initiate in remote, undeveloped areas with extreme topographic relief and water channelization. These areas are also the natural habitat and water resource to a large population of wildlife and are often adjacent to densely populated areas. For environmental reasons, large, permanent debris basins and ridged structures such as Sabo dams are not an option for debris flow mitigation. Nevertheless, options are available for mitigation that can have very little impact to the environment. One of these is the installation of flexible debris flow nets within the debris channels.

Flexible debris nets are designed to allow the natural flow of the stream and smaller sediment to continue the unabated down-stream. They also retain material during large events. The nets are installed by anchoring wire rope anchors into the sides of the channels and hanging flexible ring nets from support ropes attached to the wire rope anchors.

Standard nets are designed according to the width of the channel. Typically, a width exceeding approximately 50-feet requires the installation of two posts with foundations to decrease sagging of the top support rope.

Following an extremely large debris flow event in January 2018, a system of flexible debris nets was designed for five canyons above the town of Montecito, California. Environmental constraints required that no disturbance of the channel was allowed. Due to the large channel widths, and the fact that foundations could not be constructed for posts, wide, flexible debris flow nets, "Super VX" were designed. The Super VX debris nets have a high basal opening to allow the flow of water and sediment to continue down-stream. Fish and wildlife pass beneath the nets while offering protection against the threat of another large debris flow event.

Environmental constraints also included cataloging vegetation, and wildlife which also impacted the design and construction of the nets. In this situation, only flexible debris nets could be constructed.

INTRODUCTION

Every year debris flows cause substantial damage and loss of life. Torrential rains can trigger large masses of vegetation, soil, and rock to flow catastrophically from mountain valleys and canyons out into inhabited areas. For the most part these events can be relatively predictable as far as location is concerned but timing is often unknowable. As a result, permanent structures have been installed in strategic locations. These structures require significant planning, engineering, permitting, and construction operations.

Debris basins are widely used throughout the world. They consist of a large, excavated basin constructed to fill with sediment and then overflow. Once the event is over, the basin is cleaned out. Sometimes basins can be used to mine for construction aggregate once the event is over.

Another common approach to debris flow mitigation is the construction of dams to catch and hold the debris while allowing water to exit. These are commonly known as Sabo dams. They were originally developed in Japan and are widely used there.

In recent years, flexible debris flow barriers composed of ring nets and wire ropes have become increasingly popular. These barriers are designed to be installed within a stream channel and use the upstream channel for storage while water is allowed to drain through the barrier. The flexible barriers have the advantage of not requiring large tracts of land to be disrupted and can be designed and constructed rapidly.

MITIGATION MEASURES

Debris Basins

Excavating debris basins is a common mitigation approach. They are generally constructed in the depositional area of a debris flow torrent or channel. Debris basins consists of an earth dam or other barrier constructed across a drainageway or other suitable location for collecting sediment. The dam is provided with properly designed spillways to dispose of excess runoff water at safe velocities that will not damage the dam or other improvements (1). Basins are often large, flat areas at the mouth of debris channels where high velocity flows able to carry large sediment can lose velocity, Figure 1. As the flows lose velocity, they are no longer able to carry large, heavy sediment decreasing the particle size and sediment load of the flow. Debris basins are designed to retain significant amounts of material and if overtopped, are designed to keep flows channelized and minimizing flow energy.

Debris basins are designed and required to be maintained and cleaned overtime and basically become ineffective if retained debris material is not excavated and removed. In unburned watersheds, debris basins are cleaned out once they are 25 percent full. The number of years it takes to reach that level varies. In burned watersheds, where the potential for mudflows is higher, debris basins are cleaned out once they are five percent full. A watershed that has had more than 20 percent of its area burned within the previous five years is considered a burned watershed. For some debris basins in burned watersheds, this may lead to multiple cleanouts within a year, (2). Debris basins are designed using estimated volume of past events; however, their design is often constrained by available space. This is especially true for areas that are heavily populated. Debris basins have recently been designed to be more environmentally sensitive by including designs for fish passage.

Figure 1 - Debris basin in Montecito, California

Sabo Dams

A Sabo dam is a structure that is constructed in a stream or riverbed that consists of vertical bars, a metal grate, of concrete bollards. During normal stream flow, water flows normally through, under, or around the obstruction. During a debris event, boulders, trees and other objects are caught by the obstructions while the water is allowed to flow past. Friction and arching behind the captured debris lead to the capture of additional material. The buildup of material behind the dam also causes the stream to lose energy, thus turning the Sabo dam into a check dam.

Although they can be built on many scales, Sabo dams tend to be large and require the clearing and regrading of large construction sites. There is plentiful research on the construction of Sabo dams, with respect to types and locations of obstructions. However, Sabo dams are rarely constructed from a standard methodology and generally custom creations based on the ideas of a particular designer. Figures 2 and 3 show Sabo dams under construction in Hiroshima and Fukuoka Prefectures in Japan.

Figure 2 - Sabo Dam under construction in Hiroshima, Japan

Figure 3 - Sabo Dam construction in Fukuoka Prefecture, Japan

Flexible Debris Flow Barriers

To date, the majority developmental work on debris flow nets has been conducted by the Swiss government in conjunction with the Swiss company Geobrugg, AG, Romanshorn, Switzerland. Other manufacturers also provide debris nets but have not produced a substantial body of test data or literature.

European countries pioneered the development of rockfall protection barriers in the mid-20th Century. Rockfall barriers were the logical extension of snownets already installed in mountainous regions. Early snownets were composed of wire rope nets which held snowfall until spring melting thereby preventing avalanche formation. Post-thaw inspection of the nets showed them to have caught boulders which had fallen from above. This led to research and development of rockfall barriers on a full scale by Brugg Cable (now Geobrugg), Maccaferri, and other wire and wire rope manufacturers.

The California Department of Transportation (Caltrans), an early American adopter of rockfall barriers, observed that the barriers were effective in stopping small debris flows. This led to increased interest and research in the use of flexible nets to stop debris events.

Existing methods for determining debris flow protection were meant for large watersheds and large-scale structures such as basins and check dams (3). Early research on debris nets, including the use of anti-submarine ring nets, was conducted by the Caltrans and the United States Geological Survey (USGS). They installed a flexible ring net at the base a small flume (4). Other researchers were also conducting research on flexible nets in Japan, Europe and other countries.

Conventional debris flow net design is based on field observations and full-scale testing in controlled situations. Other publications addressed the design of debris flow protection systems (5) (6) (7) (8).

After catastrophic debris flows in Switzerland in 2005, the Swiss government partnered with Geobrugg to conduct a major research program to determine if the nets could be used as lightweight, low-cost, environmentally sound replacements for concrete check dams and debris basins. The goal was to develop a standardized approach to debris flow mitigation using flexible high-strength steel ring nets (9).

The basic debris flow protection system consists of a custom ring net engineered to resist the velocities and dynamic and static pressures unique to debris flows. Support ropes are installed into channel banks and transfer debris impact and pressure loads from ring nets to the ground. Excessive energy is absorbed by net braking elements in the support ropes. In addition, the system net rings allow the passage of water and fine sediment beneath and through the net, Figure 4.

Figure 4 - Flexible Debris Flow Barrier constructed in the Nambé Pueblo in New Mexico allows natural steam to flow and fish to pass below

The principle behind debris nets is to catch debris flows close to the source, usually in mountain canyons, stop the massive flow, and then, if desired, allow the material to be placed back in the channel to allow natural process to return to normal sediment transport conditions.

Flexible debris nets have been installed in hundreds of locations around the world to protect people and infrastructure in a low-impact, environmentally sound way.

Flexible debris barriers have shown themselves to be especially valuable in areas of the American West where a wildfire/torrential rain/debris flow cycle is often the case. The barriers can be installed relatively quickly with little environmental disruption. However, they still are held to many of the constraints the larger, and more disruptive, debris flow protection measures are subjected to.

ENVIRONMENTAL CONSTRAINTS

Due to the environmental constraints common in the United States, it is extremely difficult to construct new structures in the mountainous, undeveloped areas where most debris flows initiate. These areas are often on federal, or state land. However, some cases, these areas are privately owned.

While environmental protection regulations are developed to protect wildlife and the environment, they generally are not event or site-specific. This poses difficulties when debris flows occur, and emergency responses are necessary to protect the lives of people and their communities. At times, regulatory environmental assessments, studies, and inventories are required to occur prior to mitigation construction, rather than a rapid response to protect from future events.

California Environmental Quality Act

The California Environmental Quality Act (CEQA) is a statue that requires state and local agencies to identify significant environmental impacts of their actions and to avoid or mitigate those impacts, if feasible, (10). Activities or "projects" performed by a state or local agency, or private projects that require state or local agency approval, must comply with CEQA. A CEQA study is required to receive discretionary approval from a government agency when there may be either a direct physical change or a reasonably foreseeable indirect change in the environment. This applies to the majority and of development and construction in California, especially in mountainous, undeveloped areas where most debris flows initiate.

There are, however, exemptions to environmental constraints for emergency purposes. CEQA guidelines state that specific actions may be necessary to prevent or mitigate an emergency. However, this does not include long-term projects designed to mitigate or prevent a situation that has a low probability of occurrence in the short-term. It is general knowledge that debris flows have a high probability to impact the same area when rainfall is rates exceed a certain threshold. Therefore, in most cases, mitigation measures should be considered exempt from CEQA requirements, but that is not always the case.

In some cases, environmentally sensitive circumstances and areas and such as the presence of endangered species and USGS mapped "blue line" streams may call for additional environmental impact reports (EIR) to assess the potential for negative consequences should the project move forward. These EIRs can have a major influence on the progression of projects, especially in these environmentally sensitive areas.

Even in this case, regulatory agencies may still require in-depth EIRs depending on the environmental sensitivity of the area. Depending on the state and locale, the list of agencies required to approve a project can be quite long. These can include the US Forest Service, the US

Army Corps of Engineers, state Fish and Wildlife agencies, state Parks agencies, as well as local engineering, building, and permitting agencies. In addition, public hearings often bring out environmentalist who can bring up additional issues that could slow or stop net installation. Because it is so difficult to construct mitigation, people and communities are often left unprotected from known debris flow hazards while a project undergoes numerous reviews and revisions.

Environmental Aspects of Mitigation

Debris Basins

Debris basins have been a common debris flow mitigation measure used throughout the United States in the past. These basins have also proved to be affective in minimizing or containing debris flows. Due to the need for a large, flat area, new basins are very difficult to construct in populated areas impacted by debris flows. They require a large, deep excavation producing massive amounts of waste material that is required to be hauled offsite creating an even longer construction duration, Figure 6. These basins also have a negative impact of fish by creating a barrier making it difficult for fish to continue spawning upstream if not designed especially for fish passage. These special designs are new technology that require additional engineering.

Figure 5 - Large Debris Basin in Southern California (11)

Sabo Dams

Sabo dams are used throughout the world for debris flow mitigation. These large, mostly concrete structures are constructed in areas where debris flows occur to retain material to lessen the velocity and acceleration of debris flows, Figure 5. These dams have proved to be quite affective in mitigating large scale debris flows. However, these dams have a large impact environmentally in the areas they are constructed. A large footprint, large amount of concrete and other materials, and long construction duration make these structures extremely impractical for construction in environmentally conscious areas especially in the United States. These large concrete wall structures also create an impassable barrier preventing fish and wildlife form proceeding upstream.

Figure 6 - Large footprint of concrete Sabo Dam on the Jeneberang River, Indonesia (12).

Flexible Debris Flow Nets

Flexible debris flow barriers are a relatively new concept that have recently been implemented worldwide for debris flow protection. They provide a high level of protection with very little environmental impact. Their high strength steel wire ring-nets are anchored into the sides of the debris channels creating a very small overall footprint. Flexible barriers can also be designed to allow fish and wildlife passage below or through them. The flexible debris barriers can be installed rapidly and in remote areas giving them a lot of flexibility for location making them ideal for emergency temporary or permanent protection.

In many locations of the United States, debris flow hazards are often inevitable following a wildfire. In environmentally sensitive areas that only require protection while revegetation establishes, flexible debris flow barriers can be installed for protection and quickly removed following revegetation.

RESPONSES TO DEBRIS FLOW HAZARDS

Japan

Japan is known for having thousands of designated debris flow hazard residential areas. The number has significantly increased over the years due to development of areas near mountainous areas on top of flat, debris fans. In Japan, intense heavy rainfall sometimes observed to last for several days often causing large-scale debris flow events, Figure 7. These storms can bring an average of 15-in to 20-in in a day with intense bursts of rainfall of rainfall often reaching 5-inches of rainfall in an hour. Debris flows resulting from the prolonged rainfall events in Japan are devastating and have caused thousands of deaths in recent time.

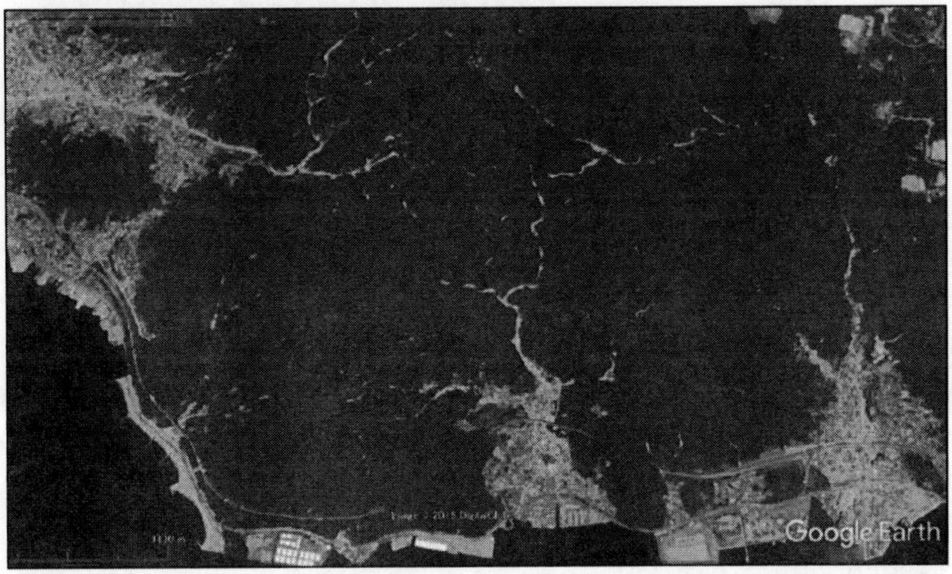

Figure 7 - Aerial image of Hiroshima, Japan after 2018 debris flows

Fukuoka, located in western Japan experienced extremely heavy rainfall events July 5, 2017. Rainfall intensities reached over 4-inches/hour in some areas resulting in debris flows and slope failures. A total of 41 people was killed throughout the area. Road closures also isolated villages which caused difficulty in providing emergency services to residents of the impacted areas.

Areas near Hiroshima, also in western Japan, also recently experienced large debris flow events. July 6 and 7, 2018, nearly 17-in of intense rainfall within a twenty-four-hour period. Rainfall continued totaling over 26-in of rain causing debris flows and slope failures throughout the region. A total of 109 people was killed and 5 missing as a result of the storm. A total of 87 deaths were caused by debris flows or slope failures during this event. Many deaths were caused by ineffective evacuations. Only 22.1% of the residents who lived in the hazard areas evacuated (13).

Temporary Emergency versus Permanent Mitigation in Japan

To prevent secondary disasters in areas already impacted by debris flow, Japan often uses flexible debris flow barriers as a part of emergency actions immediately following debris flow

Figure 8 - Installation of Flexible Barrier for emergency protection in Hiroshima Japan

Events, Figures 8 and 9. These barriers are quickly erected and provide immediate protection. After constructing the flexible barriers, large, concrete Sabo dams were constructed as permanent protection structures in both Fukuoka and Hiroshima. These permanent structures provide a high level of protection, however large areas and a lot of construction materials are necessary for construction.

Montecito, California

The Santa Ynez Mountains above Montecito, California was scorched by the Thomas Fire that burned during December 2017 into January 2018. Montecito was then impacted by extremely large debris flows on January 9, 2018 resulting in twenty-three deaths, and billions of dollars in damaged property and infrastructure. Debris flows initiated in five major canyons, four of which contained debris basins at the mouth of the canyons. These basins were quickly overwhelmed and overtopped during the flow event. A fifth canyon did not have a debris basin resulting in the immediate plugging and breaching of a culvert causing major damages. Due to the extremely environmentally sensitive area, lack of available space, lack of available funding, and the necessity for a rapid response following the first initial flow, flexible debris barriers were believed to be the only viable option to provide the community protection from future debris flows.

A local non-profit group fund-raised to finance the engineering, design, and construction with private money. However, environmental constraints including blue line stream concerns, the presence of critical habitat for steelhead salmon, and the reports of other endangered species, posed difficult hurdles to overcome prior to moving forward with construction of the emergency protection.

A number of compromises had to be made to obtain agreement form the numerous groups and agencies involved. For example, to comply with environmental concerns of constructing post foundations in wide channels within the running stream, more substantial barriers without posts were designed for channels which were too wide for a standard narrow channel VX barrier. These "Super VX" barriers were specifically located in areas that did not require tree removal, did not intersect with existing hiking trails, and consisted of a large opening at the bottom to allow wildlife to pass beneath.

Following compliance with all environmental requirements and regulations, a "first round" of debris flow barriers were able to be designed and constructed. Originally sixteen barriers were included in this first round of construction, however, due to the additional, in-depth and costly environmental requirements and permitting fees, a large portion of the funding was spent before construction proceeded. One requirement was the necessity of provided a maintenance and removal fund of $0.30 on every project. This reduced the number of potential protective nets by almost 1/3. As a result, only four debris barriers were constructed. The potential for constructing a series of debris flow barriers to creating larger debris volume storage and energy dissipation providing protection to the community below was significantly decreased due to the financial and time-consuming environmental requirements.

CONCLUSION

Although used throughout the world, due to the environmental constraints seen in the United States common debris flow mitigation practices such as excavating debris basins and constructing Sabo dams are not practical options for mitigation. Flexible debris flow barriers can not only be rapidly engineered and constructed, they are also able to be modified for environmental purposes, have a very small footprint, and can be easily removed from environmentally sensitive areas if protection is no longer necessary. For this reason, flexible debris flow barriers are a practical, time and cost-effective option for debris flow protection in the United States and around the world.

Governmental agencies should be encouraged to pre-approve the construction of debris nets when faced with extreme situations, especially post-wildfire. They can also incorporate the nets into a post-fire emergency response. Doing so will prevent a catastrophe before one occurs.

REFERENCES

1. University of Missouri Extension. Design Criteria for Debris Basin Design. https://extension2.missouri.edu/g1528 Accessed July 24, 2019.
2. Department of Public Works, Los Angeles County. "How Do Dams and Debris Basins Help Los Angeles?" https://dpw.lacounty.gov/lacfcd/sediment/dcon/366.pdf Accessed August 1, 2019.
3. Bradley, J. B., Richards, D. L., Bahner, C. D. Debris Control Structures – Evaluation and Countermeasures. Hydraulic Engineering Circular 9, 3rd Edition, Federal Highway Administration, FHWA-IF-04-016 HEC-9. 2005.
4. De Natale, J. S. et al. (1996). "Response of the Geobrugg Cable Net System to Debris Flow Loading." Report, California Polytechnic State University, San Luis Obispo, California.
5. Mitzuyama, et al. Development of Debris Flow. 1992.
6. Rickenmann, D. (1999). "Empirical Relationships for Debris Flows." Natural Hazards, 19 (1), 47-77.
7. Rickenmann, D. (2001). "Estimation of Debris Flow Impact on Flexible Wire Rope Barriers." Birmensdorf, interner Bericht, unver ffentlicht. (In German)
8. PWRI (1988). "Technical Standard for Measures Against Debris Flows (Draft)." Ministry of Construction, Japan.
9. Wendeler, C., Gröner, E., Denk, M. (2017). "Zehn Jahre Erfahrung mit flexiblen Murgangbarrieren." ("Ten years of experience with flexible debris barriers.") Austrian Journal of Engineers and Architects, v. 162, p. 209-214. (In German)
10. California Environmental Quality Act. California Natural Resource Agency. http://resources.ca.gov/ceqa/more/faq.html Accessed July 25, 2019.
11. Department of Public Works, Los Angeles County. "Sediment Management. Debris Basins." https://dpw.lacounty.gov/lacfcd/sediment/debrisbasins.aspx Accessed August 1, 2019.
12. Japan International Cooperation Agency. "The Current Situation of the Jeneberang River." https://www.jica.go.jp/project/english/indonesia/0800040/news/general/110218_01.html Accessed August 1, 2019.
13. Flood List. "Japan – Floods and Landslides Leave 80 Dead, 28 Missing." http://floodlist.com/asia/japan-floods-july-2018 Accessed August 1, 2019.

Verification of Tabulated Design Grout-Ground Bond Strength

Brian Forsthoff, MSCE
KANE GeoTech Inc.
7400 Shoreline Drive, Stockton, CA 95219
(209) 472-1822
Brian.forsthoff@kanegeotech.com

Prepared for the 70th Highway Geology Symposium, October, 2019

Acknowledgements

The author would like to thank the individuals/entities for their contributions in the work described:

Dr. William Kane – KANE GeoTech Inc.
Access Limited Construction
APEX Rockfall
HI-TECH Rockfall Construction

Disclaimer

Statements and views presented in this paper are strictly those of the author(s), and do not necessarily reflect positions held by their affiliations, the Highway Geology Symposium (HGS), or others acknowledged above. The mention of trade names for commercial products does not imply the approval or endorsement by HGS.

Copyright Notice

Copyright © 2019 Highway Geology Symposium (HGS)

All Rights Reserved. Printed in the United States of America. No part of this publication may be reproduced or copied in any form or by any means – graphic, electronic, or mechanical, including photocopying, taping, or information storage and retrieval systems – without prior written permission of the HGS. This excludes the original author(s).

ABSTRACT

Slope stabilization and rockfall protection systems often utilize ground anchorage consisting of steel wire rope or steel threaded rebar grouted in-place inside boreholes. The anchors resist tensile forces, keeping the system stable. The primary mechanism for anchor pullout resistance is friction between the grout and surrounding ground. The maximum resistance mobilized before the anchor begins to pull out of the ground (failure) is often called the "bond strength". Bond strength can vary depending on a multitude of factors such as subsurface conditions and construction methodology.

Current ground anchor design recommendations come from government agencies such as the Federal Highway Administration (FHWA) and committees such as the Post Tensioning Institute (PTI). The methodologies these sources provide are admittedly conservative. In addition, engineers must cautiously assume loss of life can be a consequence of system failure. The combination of these aspects often results in a generally overly conservative approach often with unnecessarily deep anchors. This increases the complexity of construction requirements and can potentially render the project unfeasible.

Verification anchors during the design phase of the project can give more accurate in-situ bond strengths and result in an overall cost savings despite an initial mobilization cost. The in-situ bond strengths provide a more accurate representation of the actual situational components than the values provided from tabulated data sources. Drilling depths can be reduced leading to potentially substantial cost savings. The purpose of this paper is to compare tabulated data with actual in-situ properties based on test results and emphasize the importance of predesign anchor testing. Sacrificial anchor test results show that ultimate grout-ground bond stress can be much higher than recommended preliminary design values. Currently there are no requirements for mandatory predesign testing.

INTRODUCTION

Geohazard engineering is a relatively new and developing field. As the need to expand roads and infrastructure rises, so does the demand for stable land and protection from hazards such as rockfall and shallow landslides. Landslide and rockfall mitigation systems are becoming more common as they become more critical to site expansion. Efforts to convert space into safe, usable land can be highly profitable, especially in populated areas where real estate is at a premium. As the engineering and construction of these systems become more frequent, the data concerning their performance becomes both more available and increasingly reliable. However, despite research over the last 50 years, some of the design standards upon which engineers rely remain unchanged. The factors that affect anchor performance are well documented, but varying ground conditions make reliable predictions difficult.

Engineering of rockfall and slope stabilization systems involves the use of ground anchorage. Components typically consist of steel wire rope or steel threaded rebar "threadbar" grouted inside boreholes. This anchorage acts to transfer the impact or structural loads to the surrounding ground. The Federal Highway Administration (FHWA) and Post-tensioning Institute (PTI) are the two main agencies that provide the standards for ground anchorage performance upon which engineers generally rely.

The FHWA defines ground anchors as "cement grouted, prestressed tendons that are installed in soil or rock" (*8*), Figure 1. Tendons are placed in boreholes ranging from 2.5 in. (63 mm) up to 12 in. (305 mm) in diameter. Anchor lengths can vary from 5 ft. (1.5 m) to over 50 ft. (15 m) long. They are used extensively as common practice in landslide stabilization, rock bolts, and excavation support. Their potential cost savings versus other methods, as well as their ability to reduce land required for stabilization, have made permanent ground anchors popular.

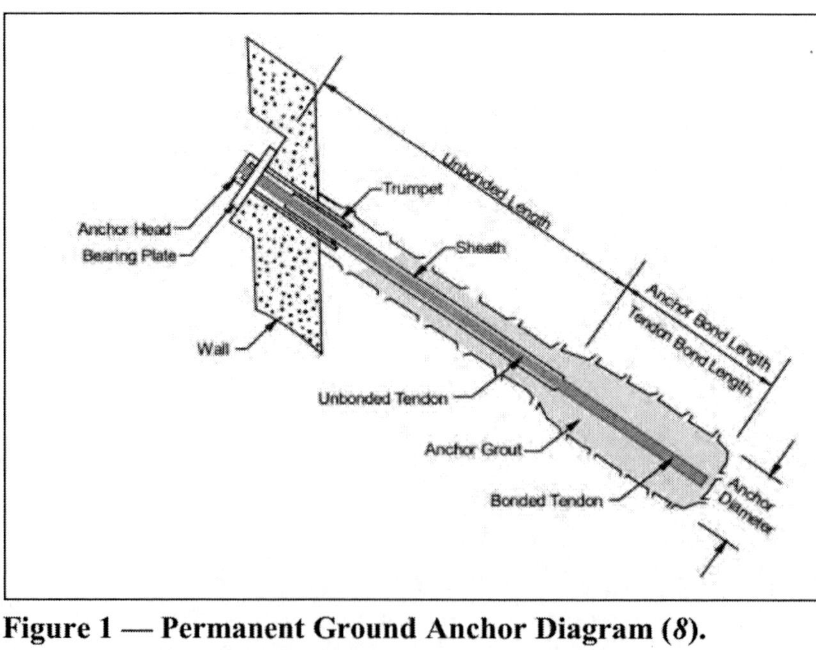

Figure 1 — Permanent Ground Anchor Diagram (*8*).

Use of permanent ground anchors in the United States was rare until the late 1970s. Prior to that they were commonly used for reinforcement of temporary retaining structures. The application of ground anchors for permanent purposes was often met with hesitancy by other state officials due to lack of design and installation procedures, knowledge of long-term effects such as creep or corrosion, exact application guidance (location and quantity), and quality assurance (*6*).

At first, only a few states allowed ground anchors as an alternate design to conventional concrete retaining walls, Figures 2 and 3. The alternate design provided significant cost savings which encouraged the FHWA to do further research. They conducted an instrumentation program to increase understanding on how the systems functioned and analyzed their performance over time. It was not until the 1980s that permanent ground anchors were widely accepted (*6*). As their popularity grew, so did the necessity for a standard set of guidelines for engineers and contractors to consider.

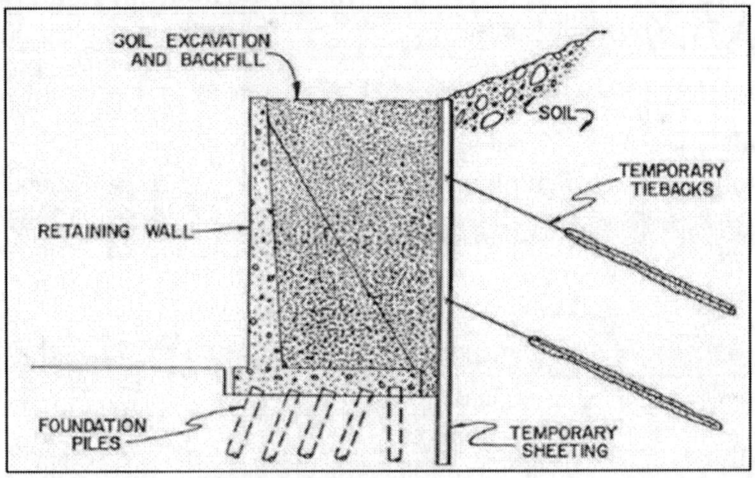

Figure 2 — Conventional Retaining System Design Prior to Permanent Ground Anchors (*8*).

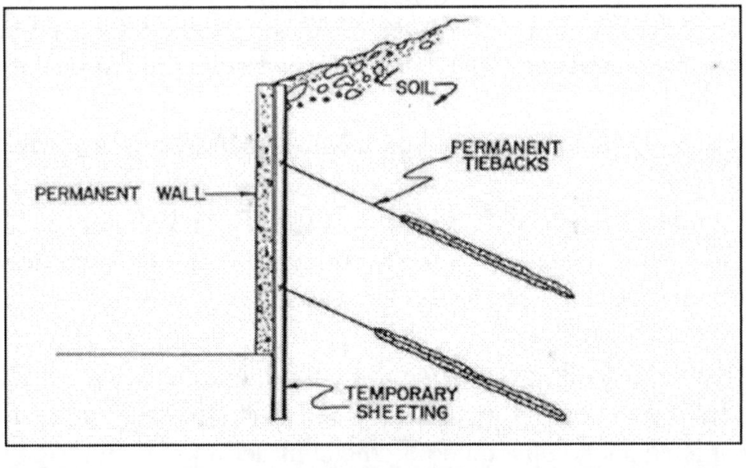

Figure 3 — Alternate Permanent Ground Anchors (*8*).

In 1974, the Prestressed Concrete Institute (PCI) released "Tentative Recommendations for Prestressed Rock and Soil Anchors" to provide engineers and contractors guidelines for installing and designing ground anchorage. The referenced document included "typical values" for grout-ground bond strength. Since then, except for adding a few category types, there have been minimal changes to the original values, Table 1. In 1976, members from the PCI formed a new committee the Post-Tensioning Institute (PTI) and have periodically updated its original 1974 design recommendations (1980, 1896, 1996, 2004, 2014). New design guidelines and specifications are more detailed (e.g. corrosion protection, acceptance criteria, etc…) and provide more guidance on bond stress influences (e.g. borehole roughness, rock joint influences, rock mineral lubrication, etc…). In 1999, the Federal Highway Administration (FHWA) published Geotechnical Circular No.4 which provided federal guidelines for the design and installation of permanent ground anchors in connection with structural foundations, soil and rock instabilities, earth retaining systems, and for ground modification techniques.

Table 1 — Typical Range of Average Ultimate Rock-Grout Bond Stresses

Rock Type by Publication		Average Ultimate Bond Stress
PCI 1974 (*2*)	PTI 2014 (*3*)	Rock-Grout (PSI)
Granite & Basalt	Granite & Basalt	250 - 450
Dolomite Limestone	Dolomite Limestone	200 - 300
Soft Limestone	Soft Limestone	150 - 220 (PCI 1974) 150 - 200 (PTI 2014)
Slates & Hard Shales	Slates & Hard Shales	120 - 200
Soft Shales	Soft Shales	30 - 120
Sandstone	Sandstone	120 - 250
Concrete	Concrete	200 - 400
Not Included	Weathered Sandstone	100 - 120
Not Included	Chalk	30 - 155
Not Included	Weathered Marl	25 - 35

FHWA currently references the PTI values which are extensively used in the preliminary design of ground anchors. A simplified typical design process for ground anchors is as follows:

1. Site reconnaissance – site investigation and/or subsurface exploration (i.e. Geotechnical borelogs).
2. Preliminary grout-ground values are determined using PTI/FHWA values.
3. Preliminary required bond zone lengths are determined from anticipated design loads.
4. Assumed bond values are verified or reconsidered based on test results.

Regarding rockfall and shallow landslide mitigation, current design practice is that pre-construction anchor testing is not common and low bond strengths are usually assumed in the initial design. The completed anchors installed by the contractor are then tested to verify performance. This process lacks opportunity to potentially reduce anchor bond lengths. Costly, inefficient anchor lengths are possibly determined, and often the engineer will not test the anchor to failure to better approximate in-situ bond values. This is likely due to time constraints or tendon structural limitations (i.e. the tendon will rupture due to increasing loads.) Engineers often rely on the PTI bond values since approximate in-situ values and are not usually obtained during the preconstruction phase of the project.

However, the PTI's average bond stress assumptions can lead to conservative bond lengths (*1*). This can result in unnecessarily deep anchors that consequentially increase initial cost and time estimates (*5*). For larger construction projects (i.e. public entities) this may be assumed with little consequence due to large budgets. However, in the case of smaller, private entities (i.e. homeowners), the extra drilling can prove expensive, rendering the project unfeasible. The engineer usually determines a minimum bond and un-bonded length that the contractor must achieve. The contractor must proceed with the engineer's conservative design or have an in-house engineer calculate alternate bond lengths further delaying the project. Understanding how anchors interact and reviewing current design practices will help with suggesting a solution to this problem.

GROUND ANCHOR INTERACTION

Ground anchor functionality relies on its ability to transfer the load from the structure to the surrounding ground prior to pulling out (i.e. grout-ground failure). A general process of load transfer is listed below:

1. Loading is applied to the tendon.
2. Loading is transferred from the tendon to the grout surrounding the tendon.
3. The grout annulus resists pullout/uplift force by shear resistance between the grout and the surrounding ground.

Advancements in construction techniques such as pressure grouting, post-grouting, and under-reaming allows the grout to infiltrate voids or expand the grout pack in soil. The fracturing or expansion of grout increases shear resistance, thus increasing pullout resistance, Figure 4. However, these practices are typically performed by specialty contractors and can be costly compared to conventional techniques. Conservatively, simple construction techniques are often assumed (i.e. gravity-grouted anchors) and lower grout-ground bond values are used for preliminary design.

When the external forces exceed the shear resistance generated between the grout and ground, the bonding between the grout and surrounding ground deteriorates, Figure 5. This causes the grout column to move and insufficient residual resistance is generated throughout the bonding zone. This results in the anchor pulling out the ground.

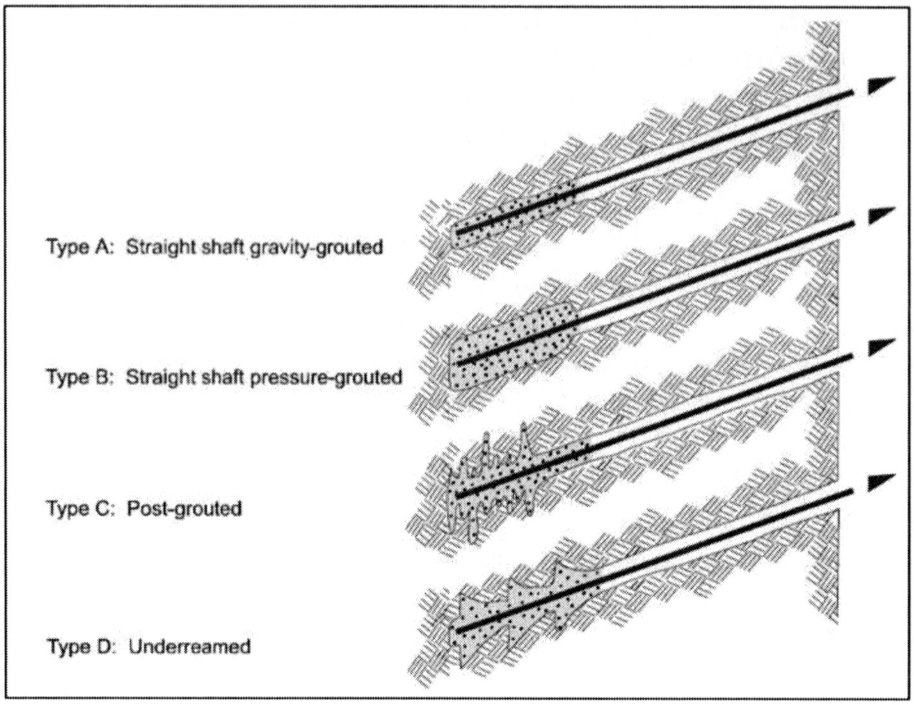

Figure 4 — (a) Gravity-Grouted Anchor. (b) Pressure-Grouted Anchor. (c) Post-Grouted Anchor. (d) Anchor Grout Via Underreaming. (*8*).

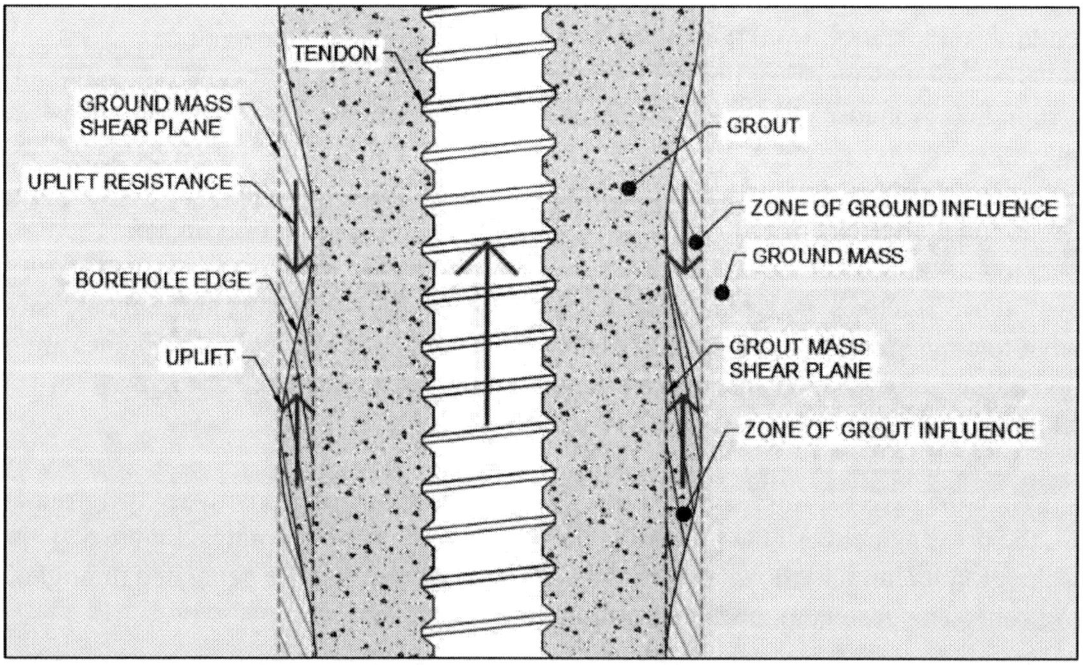

Figure 5 — Shear Plane Develops Due to Grout-Ground Interlocking When the Borehole is Non-Uniform (Rough).

BOND LENGTH DESIGN

Since mobilized shear strength is the primary mechanism for resistance, relative movement must occur between the grout-ground interface to develop friction. Typically, when the anchor is loaded in tension, loading is transferred to the top of the grout column and propagates downward toward the bottom of the bond zone, Figure 6. If enough resistance cannot be developed by the time the load transfers to the bottom of the grout column, then anchor pullout (failure) is likely.

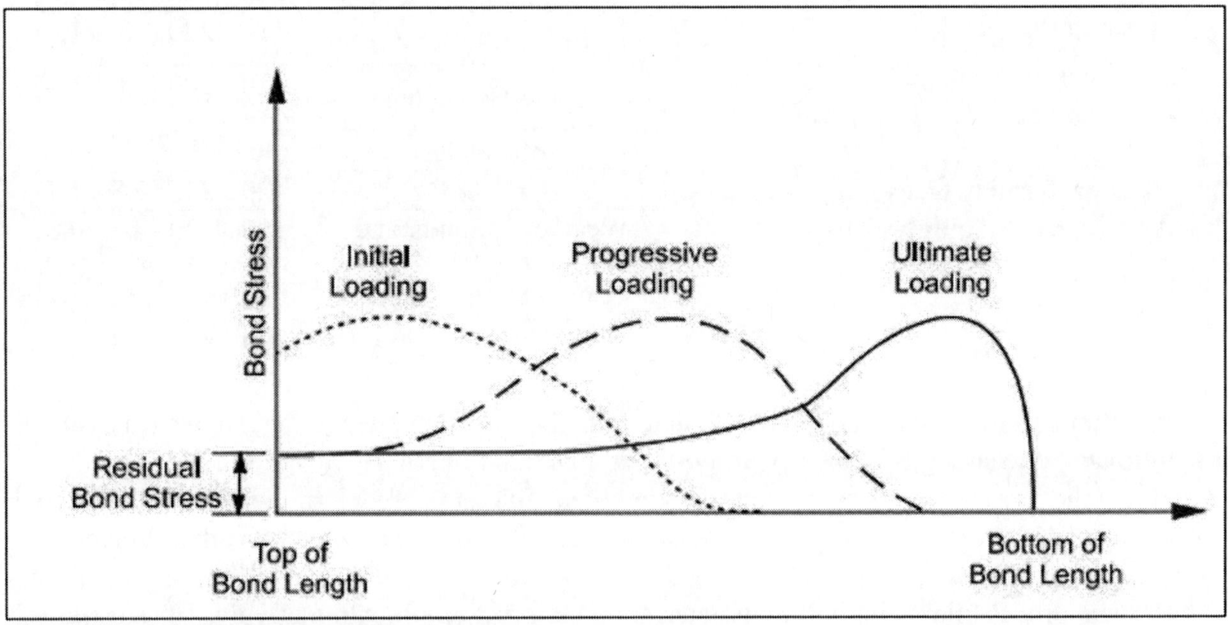

Figure 6 — Loading Sequence of a Ground Anchor (*8*).

The FHWA (1999) formula for calculating ultimate pull-out resistance is:

DL = (BL)(UTL)/SF

Where:
DL = Design Load
BL = Bond Length
UTL = Ultimate Transfer Load (Table 2)
SF = Safety Factor
(Minimum = 2.0))

Table 2 — Presumptive Ultimate Values of Load Transfer. (*8*).

Rock Type	Estimated Ultimate Transfer Load Kips/ft (kN/m)
Granite or Basalt	50 (730)
Dolomitic Limestone	40 (580)
Soft Limestone	30 (440)
Sandstone	30 (440)
Slates and Hard Shales	25 (360)
Soft Shales	10 (150)

The PTI (2014) formula for calculating required bond length to resist pullout is:

Lb = (DL)(FS)/(π)(d)(t)
Where:
Lb = Bond Length
DL= Design Load
π = 3.14
d = Diameter of the Drill Hole
t = average ultimate bond strength along interface between grout and ground (Table 3)
FS = Factor of safety on average ultimate bond strength (Minimum = 2.0)

Table 3 — Typical Range of Average Ultimate Rock-Grout Bond Stresses (*3*).

Rock Type	Average Ultimate Rock-Grout Bond Stress PSI (MPa)
Granite & Basalt	250 - 450 (1.7 - 3.1)
Dolomite Limestone	200 - 300 (1.4 - 2.1)
Soft Limestone	150 - 200 (1.0 - 1.4)
Slates & Hard Shales	120 - 200 (0.8 - 1.4)
Soft Shales	30 - 120 (0.2 - 0.8)
Sandstone	120 - 250 (0.8 - 1.7)
Weathered Sandstone	100 - 120 (0.7 - 0.8)
Chalk	30 - 155 (0.2 - 1.1)
Weathered Marl	25 - 35 (0.15 - 0.25)
Concrete	200 - 400 (1.4 - 2.8)

Equations from the PTI and FHWA to determine bond length are similar. They both assume the ultimate bond strength is linearly proportional to anchor depth. However, the FHWA simplifies the process and assumes that the borehole diameter is predetermined and between four to six inches (102 mm to 152 mm). One advantage of using the PTI's method is that it gives the engineer flexibility to use various borehole diameters without requiring back-calculations. Also, the PTI's approach allows flexibility in determining bond strength. Alternatively, 10% of the minimum compressive strength of the rock (up to 600 psi or 4.1 MPa) can be used as the assumed nominal grout-rock bond strength (*3*). Littlejohn and Bruce (*1*) state that the above methods for calculating bond lengths assume the following:

1. Transfer of loading is uniform throughout the entire anchor
2. Rock or soil matrix is homogenous and isotropic
3. Failure takes place by sliding along grout-ground interface (i.e. smooth borehole, cohesive soils) or by shearing across the ground-grout interface in whichever medium is weaker (i.e. rough borehole, rock and granular soils)
4. There are no inherent planes of weakness or discontinuities along which failure can be induced
5. There is no local debonding at the grout-ground interface

The general critique of this approach is the following:

1. There is evidence that suggests loading is not uniform along the grouted anchor (4, *12*).
2. No consideration is given to the type of cement used in the grout which can significantly alter shearing resistance (*11*).

3. No allowance for bedding planes, faults, voids, or joints within the rock strata (*13*)
4. No guidance on what to do in the case of soil/rock conditions. How does loading transfer from rock to soil or vice-versa? It is unclear if it is appropriate to assume a weighted average of the bond strength for soil and rock combinations or to conservatively ignore contributions from soil-grout bonding.
5. The FHWA approach of using the “Ultimate Transfer Load” predetermines anchor borehole diameter and requires additional calculations if design borehole diameters do not match.
6. Presumptive values can lead to unnecessarily long bond lengths (*1, 5, 8*)

The FHWA states that using this approach may lead to calculated bond lengths "significantly greater" than what is needed to resist loading (*8*). Other researchers similarly state that using the average bond stress value can lead to "extraordinary and wastefully long bond zones" (*5*). Original bond stress values the PCI developed in 1970's recommended anchor pullout factor of safety of 1.5 to 2.5, Table 4. Currently, The PTI requires that a minimum safety factor of 2.0 be used when designing bond lengths. However, the ultimate bond stresses for particular rocks (e.g. shales) have large design ranges. For example, the recommended ultimate grout-ground bond values for shale ranges from 30 to 120 psi (0.2 to 0.8 MPa). If a conservative ultimate ground-grout bond strength value of 30 psi (0.2 MPa) is assumed for design, but the in-situ bond was determined to be 120 psi (0.8 MPa), then the anchor will be four times deeper than it needed to be. Also accounting for the minimum safety factor of two specified for design gives a nominal grout-ground bond stress of 15 psi (0.1 MPa). If in-situ bond capacity was actually 120 psi (0.8 MPa), then the theoretical safety factor against pullout would be eight, meaning, the anchor was installed eight times deeper than it needed to be. When this effect is applied to tens or hundreds of anchors the cost due to this conservative design increases significantly. The following scenarios emphasize this point:

<u>Scenario One:</u>
Calculate the bond length of a threaded-bar anchor with a 3 in. diameter borehole in soft shale (assume 30 psi bond stress) with a design load of 50 kips in tension.

<u>Following PTI, 2014:</u>

BL = (P)(SF)/((π)(d)(t))
P = 50 kips (50,000 lbf), SF = 2, π = 3.14, d = 3 in, t = 30 psi
BL = (50,000 lbf)(2)/((π)(3 in)(30 psi)) = 354 in = **<u>30 ft</u>**.

<u>Scenario Two:</u>
Perform the same calculation except assume an ultimate bond stress of 120 psi.
BL = (50,000 lbf)(2)/((π)(3 in)(120 psi)) = 99 in = **<u>8 ft</u>**

Table 4 — Tabulated Rock-Grout Bond Values Which Have Been Recommended For Design (*1*)

Rock Type (Country of Study)	Recommended Ultimate Bond PSI (MPa)	Recommended Factor of Safety
Igneous		
Medium Hard Basalt (India)	830 (5.73)	3 - 4
Weathered Granite (Japan)	215 - 360 (1.5 - 2.5)	Not Provided
Basalt (Britain)	560 (3.86)	2.8 - 3.2
Granite (Britain)	700 (4.83)	3.1 - 3.5
Serpentine (Britain)	225 (1.55)	2.6 - 3.5
Granite & Basalt (USA)	250 – 450 (1.72 - 3.10)	1.5 - 2.5
Metamorphic		
Manhattan Schist (USA)	2.8	4
Slate & Hard Shale (USA)	0.83 - 1.38	1.5 - 2.5
Calcareous Sediments		
Limestone (Switzerland)	405 (2.83)	2.8
Tertiary Limestone (Britain)	400 (2.76)	2.9 - 3.3
Soft Limestone (USA)	150 - 220 (1.03 - 1.52)	1.5 - 2.5
Dolomitic Limestone (USA)	200 - 300 (1.38 - 2.07)	1.5 - 2.5
Arenaceous Sediments		
Hard Coarse-Grained Sandstones (Canada)	620 (4.29)	1.75
Weathered Sandstone (New Zealand)	100 - 120 (0.69 - 0.85)	3
Well-Cemented Mudstone (New Zealand)	100 (0.69)	2.0 - 2.5
Bunter Sandstone (Britain)	175 (1.2)	3
Bunter Sandstone (Britain) (UCS > 290 psi)	260 (1.8)	3
Hard Fine Sandstone (Britain)	325 (2.24)	2.7 - 3.3
Sandstone (USA)	120 - 250 (0.83 - 1.73)	1.5 - 2.5
Argillaceous Sediments		
Weak Shale (Canada)	50 (0.35)	Not Provided
Soft Sandstone & Shale (Britain)	54 (0.37)	2.7 - 3.7
Soft Shale (USA)	30 - 120 (0.21 - 0.83)	1.5 - 2.5

Scenarios One and Two highlight the potential disparity in anchor design. The resulting difference between Scenario One and Scenario Two is 22 ft. per anchor. Assuming an installation cost of $100 to $125 per linear foot of drilling, the cost of extra drilling would range from $2,200 - $2,750 per anchor. As an example, even a small shallow landslide stabilization system can easily have 40 to over 100 anchors. Assuming 40 anchors installed at a minimum of $100 per foot of drilling would cost the client an additional $88,000.

Some states (e.g. Hawaii and California) require homeowners to stabilize unstable land before they can sell or reside within the house. The extra cost in the conservative design could place a burden on the landowner. A case could be made that poor engineering practices were employed. Conservative designs are safe for the client and engineer. The client receives a quality end-product while the engineer's liability is greatly reduced. However, it is also the engineer's responsibility to produce a cost-effective design for the client.

Bond stresses can vary greatly in the same kind of material. Barley (1988) documented the installation of 10,000 ground anchors installed in Europe. Figure 7 shows that of the 151 anchors that pulled-out (failure) in sandstone only nine anchors (6%) were within the presumptive range of the PTI. This highlights the following:

1. Grout-ground bond can be highly variable. Expected rock type is insufficient for estimating bond lengths.
2. Preproduction or predesign anchor testing can be highly beneficial. Accurate bond lengths can be determined which allows the contractor to reduce the anticipated drilling requirements. This consequentially reduces construction time and total materials required.

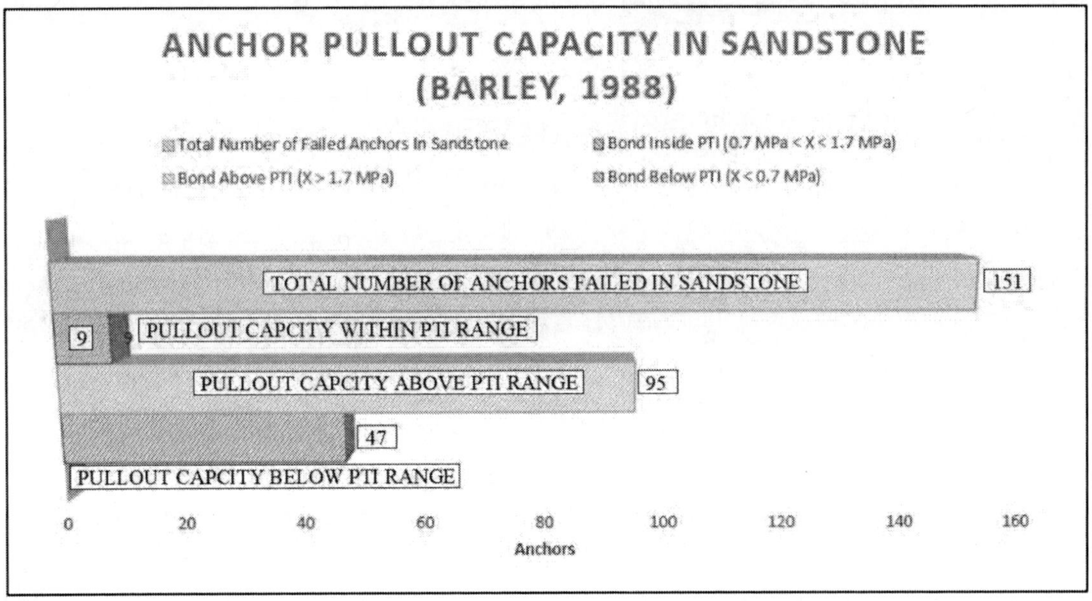

Figure 7 — Comparison of the Anchor's Ultimate Grout-Ground Bond Strength that Failed in Sandstone Compared to the PTI's Presumptive Range of Ultimate Bond Stresses (151 Total Anchors) (*10*).

To better estimate expected bond stresses in soil, attempts to approximate bonding based on Standard Penetrating Test (SPT) values have been done, Table 5. One aspect with this approach is that it requires SPT soil testing "$(N1)_{60}$" values. On occasion, the subsurface strata and strength properties are available as the result of exploration drilling for previous site assignments (e.g. wells and foundations). SPT values could be used to assist the engineer to better estimate bond lengths at no additional cost to the client. However, if the information is not readily available (e.g. rockfall hazard along road), it would be not be cost efficient to specifically test for soil/rock properties to approximate the bond stresses.

Installing a test anchor is potentially more time and cost efficient and gives a better approximation of in-situ bond stresses. Predesign and preproduction testing are two common forms of anchor load test that assist in approximating the bonding properties of expected ground conditions.

Table 5 — Presumptive Ultimate Values of Soil Nail Pullout Resistance per Unit Length (7)

Soil Type	Relative Density/ Consistency	SPT $(N1)_{60}$ Range	Ultimate Pullout Resistance per Unit Length Kip/ft (kN/m)
Sand and Gravel	Loose	4 - 10	10 (146)
Sand and Gravel	Medium Dense	11 - 30	15 (219)
Sand and Gravel	Dense	31 - 50	20 (292)
Sand	Loose	4 - 10	7 (102)
Sand	Medium Dense	11 - 30	10 (146)
Sand	Dense	31- 50	13 (190)
Sand and Silt	Loose	4 - 10	5 (73)
Sand and Silt	Medium Dense	11 - 30	7 (102)
Sand and Silt	Dense	31 - 50	9 (131)
Silt-clay mixture (LP	Stiff	10 - 20	2 (29)
Silt-clay mixture (LP)	Hard	21 - 40	4 (58)
Note: Values are for borehole with diameters between 4 in to 6 in.			

PREDESIGN ANCHOR TESTING

Predesign anchor testing can take place during the investigation phase and up until the construction phase begins. Evaluation of grout-ground bond stresses and creep potential can be determined, allowing for a more efficient anchor design. Typically, one to three anchors are installed and tested to failure, but additional anchors may be required depending on the size and number of grounds conditions encountered. The contractor must mobilize to the site solely for this purpose. The contractor performs the drilling, installation, and returns a few days later after the grout has cured enough to perform the load test. However, since construction preparations are not often ready to begin, the predesign anchor testing adds time, number of site visits, and therefore additional costs to the project.

Cost savings can be maximized if the timing of testing is after the initial design but before materials have been ordered by the contractor. This is most efficiently accomplished with a design-build contract. Prior to ordering materials, the contractor can drill, become familiar with ground conditions, and evaluate potential difficulties. This allows the contractor to acquire additional drilling equipment (if needed) and potentially reduce total construction time. Then the pullout capacity of the anchors can be determined, and bond lengths can be calculated more accurately. As a result, lower material quantities than originally estimated are ordered, and construction is often completed faster than anticipated. The reduction of drilling and materials provides significant cost savings, even considering the required pre-construction site visits.

Predesign anchor testing can be effective with design-bid-build contract. The predesign anchor testing can also be completed by a third-party driller, contracted only for work regarding the predesign anchors. A lump sum contract can be put in place for predesign anchor testing before the job is sent out for bids. Once test data is acquired, necessary adjustments can be made to the anchor design, which contributes to minimizing mid-construction design changes. The design is sent to various installation contractors so they can submit their bids with the improved anchor. Consequentially, the contractor can bid the job with less contingency and more accurate construction parameters allow the contractors to bid more competitively, likely lowering the cost for the client. In bridge foundation construction, Alabama Department of Transportation recently implemented a pre-bid load test program utilizing piles and drilled shafts which successfully allowed bidding contractors to better assess risk and reduce contingencies associated with the project. The pre-bid load test cost 0.2% of the total estimated overall cost of the project (*9*).

PREPRODUCTION ANCHOR TESTING

Preproduction testing occurs after the contractor has mobilized on site to build the final design, but before construction begins. This allows the contractor to verify anticipated anchor pull-out capacity and construction methodology. This type of testing may appear favorable to a contractor as the timing allows for condensed site visits. Other construction preparations can take place while preproduction testing is in progress. However, since preproduction anchor testing provides this information after a project design is in place, there is less flexibility regarding design revisions. Though anchor improvements can be achieved after testing, contracts, materials, and funds are less flexible. Project resources have already been allocated and therefore are less available or exhausted. Table 6 summarizes the comparison between testing types. Predesign anchor testing has the greater potential to reduce anchor depths. However, the initial cost increase associated with multiple site visits should be considered. If a very small quantity of anchors is being installed (i.e. less than 10), then the cost reduction due to reduced anchor depths may not offset the mobilization costs required for multiple site visits. In this case, preproduction anchor testing may be more beneficial if a flexible contract can be agreed on during the bid phase of the project. The lower initial cost associated with preproduction anchor testing may favorably compare with the cost savings with shorter anchor lengths.

Table 6 — Predesign and Preproduction Anchor Testing Comparison

Anchor Test Type	Predesign	Preproduction
Phase Implemented	Investigation to Design	Construction
Design Anchor Tested?	No	Yes
No. of Mobilizations Required of Contractor	2 minimum	1
Design Flexibility	High	Low
Initial Cost	Higher	Lower
Potential Savings	Higher	Lower

ANCHOR TEST DATA

A summary of collected data from various anchor tests is presented on Table 7 and Table 8. Anchor tests were performed in accordance to PTI or FHWA testing standards. While sacrificial testing, if loads were approaching the strength limits of the tendon, testing operations ceased as the safety of testing personnel was prioritized over obtaining more data. Ground conditions encountered include sandstone, shale, limestone, and granular soil conditions. The quality of rock ranged from moderately weathered to highly weathered rock.

In Table 7, as-built bonding stresses ranged from 23 psi to 99 psi. “As-built” is defined as the maximum bonding determined from production testing. This is not to be confused with the ultimate bond stress as failure was not induced in any of the anchors tested. Due to the tested anchors being critical elements after testing, determining ultimate capacities were not feasible. However, this data does highlight the variance when it comes to determining a design grout-ground bond strength. All anchors were gravity-grouted into place. Most of the production anchor tests yielded working bond stresses that were on the lower end of the bond strength ranges given by the PTI. This is likely attributed to the anchors that were conservatively designed initially and the contractor voluntarily installed the anchors deeper than required or used a larger diameter borehole to be confident in passing anchor test criteria (a common occurrence). This further increased the conservative nature of anchor construction.

In Table 8, the sacrificial anchors in limestone recorded bond strengths nearly double than the PTI working values. Ultimate bond stresses were not achieved due to concerns of the anchor rupturing. The anchors grouted in soil were much stronger than the recommended values. One of the soil anchors could not be pulled to failure as the testing equipment reached ultimate capacity and higher loading could not be achieved. However, this does support the point that preconstruction anchor testing can provide significantly stronger in-situ bond strengths than relying on tabulated sources.

Based on the experiences encountered during testing, the following items should be considered when sacrificial testing:

- Anchors installed in rock (even in shallow depths) could require high loads to induce failure. Upsizing the anchor tendon may be required to avoid rupture during testing.
- Ensure testing equipment can greatly exceed expected pullout loads

Table 7 — Summary of <u>Production</u> Anchor Testing Data (KANE GeoTech Inc.)

Job Location	Ground Conditions	Borehole Diameter (inches)	Bond Length (feet)	Max. Test Load (kips)	*Min. As-Built Grout-Ground Bond Strength (psi)	**PTI Working Bond Stress Range (psi)
Daly City, California	Soft Shale	3	19	49.0	22.9	15-60
Tiburon, California	Soft Shale	6	10	75.0	33.2	15-60
Felton, California	Sandstone	6	5	112.0	99.1	60-125
Harrisburg, Pennsylvania	Sandstone/Hard Shale	4	8	87.9	72.9	60-125
Harrisburg, Pennsylvania	Sandstone/Hard Shale	3	8	65.0	71.9	60-125
Harrisburg, Pennsylvania	Sandstone/Hard Shale	3.5	4	45.1	85.5	60-125
*Zero Anchor Failures Recorded						
**Assumed SF = 2						
Anchor Installation - Gravity Grouted Anchors.						

Table 8 — Summary of <u>Sacrificial Verification</u> Anchor Testing Data (KANE GeoTech Inc.)

Job Location	Ground Conditions	Borehole Diameter (inches)	Bond Length (feet)	Max. Test Load (kips)	Min. Field Tested Grout-Ground Bond Strength Achieved (psi)	**PTI Working Bond Stress Range (psi)
*Nazareth, Pennsylvania	Limestone	3.5	3	70.0	176	75-100
Mill Valley, California	Silty Gravel	4	11	93.0	56	5-10
*Riverside, California	Sandy Gravel	4	12	100	55	5-10
*Anchor did not fail. Testing stopped due to equipment limitations.						
**Assumed SF = 2						
Anchor Installation - Gravity Grouted Anchors.						

Engineers can design with greater confidence is if they are provided with more data. With the internet it is now possible to better document construction data and anchor test results. A national database with the following information would greatly help designers predict more accurately grout-ground bond stresses for future projects:

- Location
- Rock/soil type (e.g. weathered Sandstone, silty sand)
- Standard penetration test (SPT) blow counts or rock quality designation (RQD) percentages.
- Type of anchor (gravity grouted, pressure grouted, underreamed, etc...)
- Drilling Equipment
- Bond length
- Borehole diameter
- Ultimate bond stress

This would better equip designers to provide a more efficient preliminary design that would reduce anticipated construction time and costs. Anchors should always be tested due to the high variability of bond strengths and to verify the installation contractor's construction practices.

CONCLUSION

The design practice of ground anchors has moved from groundbreaking to complacency. In the infancy of permanent ground anchor construction in the United States there were many uncertainties. Nearly half a century later the same general design practices are still implemented despite a better understanding of ground anchor force interactions and a large volume of successful constructed anchors globally. Part of that success is due to the conservative approach engineers have adopted during that time. While safe, the cost of this tactic is excessively over-designed anchors that is costing the client resources. One-way engineers can design with greater certainty is to provide more data and establish a national database.

Regardless, the ability to better predict bond stresses does not relinquish the engineer or the contractor to verify anchor capacity. As empirical data shows, bond stresses can vary greatly thus testing requirements should remain as an integral part of the overall design. Predesign testing under the right circumstances can provide significant cost and time savings.

REFERENCES:

- Books
  1. Littlejohn, S. G., & Bruce, D. *Rock Anchors - State of the Art.* England: Foundation Publications LTD. 1977.
  2. PCI. *Tentative Recommendations for Prestressed Rock and Soil Anchors*. Precast Concrete Institute, 1974.
  3. PTI. *Recommendations for Prestressed Rock and Soil Anchors.* Post-Tensioning Institute (PTI), 1980, 1986, 1996, 2004, 2014.
  4. Xanthakos, P. P. *Ground Anchors and Anchored Structures.* New York: John Wiley & Sons, Inc. 1991.

- Conference Proceedings
  5. Bruce, D. *The Stabilization of Concrete Dams by Post-Tensioned Rock Anchors: The State of American Practice.* International Conference, Institution of Civil Engineers. (pp. 508-521). GeoSystems L.P. London, 1997.

- Federal Government Reports
  6. Cheney, R. S. *DP-68 - Permanent Ground Anchors - Volume 1, Final Report.* Report FWHA-DP-90-068-003. FHWA, U.S. Department of Transportation, 1990.
  7. Lazarte, C., Robinson, H., Gómez, J., Baxter, A., Cadden, A., & Berg, R. *Geotechnical Engineering Circular No. 7 - Soil Nail Walls - Reference Manual.* Report FHWA-NHI-14-007. FHWA, U.S. Department of Transportation. 2015.
  8. Sabatini, P., Pass, D., & Bachus, R. C. Geotechnical Engineering Circular No. 4 - Ground Anchors and Anchored Systems. Report FHWA-IF099-015. FHWA, U.S. Department of Transportation, 1999.

- Magazine
  9. Sternberg, S. and Thompson, R., Reducing Risk Through Prebid Load Testing – Case Study of the Mobile River Bridge and Bayway Project. *Deep Foundations* Deep Foundation Institute. July/August 2019 Issue. pp. 14-18. 2019.

- Journal
  10. Barley, A. D. (1988). *Ten Thousand Anchorages in Rock.* Ground Engineering, January 1988, pp. 20-39.
  11. Brown, E. (2014). *Rock engineering design of post-tensioned anchors for dams – A review.* Journal of Rock Mechanics and Geotechnical Engineering, vol 7, issue 1, pp 1-13.
  12. Ostermayer, H., & Scheele, F. Research on Ground Anchors in Non-Cohesive Soils. *Revue Francaise de Geotechnique, vol 3*, pp. 92-97, 1978.
  13. T. Dados, Anastasios. (1984). *Design of Anchors in Horizontally Jointed Rocks.* Journal of Geotechnical Engineering, vol 110, Issue 11.

Replacing Deteriorating Retaining Walls Along the Million Dollar Highway

Brett Arpin, P.E.
Yeh and Associates, Inc.
2000 Clay Street, Suite 200
Denver, CO 80211
(303) 781-9590
barpin@yeh-eng.com

Todd Schlittenhart, P.E.
Yeh and Associates, Inc.
2000 Clay Street, Suite 200
Denver, CO 80211
(303) 781-9590
tschlittenhart@yeh-eng.com

Prepared for the 70th Highway Geology Symposium, October, 2019

Acknowledgements

The author(s) would like to thank the individuals/entities for their contributions in the work described:

David Valentinelli, P.E. – Colorado Department of Transportation
Tom Allen, P.E. – Yeh and Associates, Inc.
Ben Arndt, P.E. – RJ Engineering

Disclaimer

Statements and views presented in this paper are strictly those of the author(s), and do not necessarily reflect positions held by their affiliations, the Highway Geology Symposium (HGS), or others acknowledged above. The mention of trade names for commercial products does not imply the approval or endorsement by HGS.

Copyright Notice

Copyright © 2019 Highway Geology Symposium (HGS)

All Rights Reserved. Printed in the United States of America. No part of this publication may be reproduced or copied in any form or by any means – graphic, electronic, or mechanical, including photocopying, taping, or information storage and retrieval systems – without prior written permission of the HGS. This excludes the original author(s).

ABSTRACT

US 550 traverses across steep canyon walls and over high mountain passes through the San Juan Mountains of southwest Colorado. The highway section between Ouray and Silverton, Colorado is commonly referred to as the Million Dollar Highway. Three retaining walls in various states of deterioration were leading to pavement damage and loss of roadway width along the already narrow road. South of Silverton, two additional steel crib walls were deteriorating and leading to roadway damage.

Replacing each retaining wall presented unique design, construction, environmental, and aesthetic challenges. Because of the historic and scenic nature of the highway, the facing for each retaining wall replacement had to look similar to the retaining wall it was replacing. This was accomplished by designing mechanically stabilized earth (MSE) walls for each site with facings that matched the previous wall. Four facing types were designed and constructed: timber, stone, galvanized wire mesh, and ungalvanized wire mesh. Using the MSE wall system as the base solution provided the flexibility to match facing to the aesthetic requirements while keeping a single wall type that could be constructed by numerous specialty contractors.

A minimum of one lane of the two lane roadway had to remain open during construction requiring shoring at three of the wall locations. Permanent soil nail walls were designed and constructed to maintain traffic and to reduce the required MSE wall reinforcement length. The five replacement retaining walls were successfully constructed along with two additional walls to accommodate additional roadway width.

INTRODUCTION

United States Highway 550 (US 550) is a primary route traveling north and south through southwestern Colorado. The highway was built through difficult mountain terrain including high mountain passes and along steep canyon walls. Red Mountain Pass, at an elevation of 11,075 feet, is at the apex of the highway between Ouray and Silverton. The highway was originally constructed in the 1880s and operated as a toll way until it was rebuilt in the 1920s. The highway is commonly referred to as the Million Dollar Highway however the origin of the name is still debated. Towns like Silverton and Ouray rely on US 550 to bring goods, services, and tourism to the area. Due to the importance of the highway corridor, the Colorado Department of Transportation (CDOT) proactively identified numerous walls that were in various states of deterioration resulting in roadway damage. A total of five retaining walls were replaced and two additional walls were constructed along the highway to improve the corridor. The locations of the retaining walls are shown in Figure 1.

Figure 1 – Retaining Wall Locations

DETERIORATING WALL CONDITIONS

Timber Face Crib Wall

This timber faced crib wall along US 550 consisted of approximately 12x4 rough sawn timbers (Figure 2). Vertical timbers holding horizontal lagging timbers were anchored behind the wall face with steel cables. The wall face had a notable curve in plan view and appeared to have settled near the center of the wall. The wall was not embedded leading to loss of backfill from below the timber facing. Surface water drainage from the roadway had resulted in erosion of the fill material on top of the wall. The inlet of a corrugated steel pipe was noted in the roadside ditch on the opposite side of the highway; however, the pipe could not be located daylighting on the downhill (wall) side of the roadway.

Figure 2 – Deteriorating Timber Faced Crib Wall

Stone-Faced Gravity Wall

This stone-faced gravity wall was also constructed directly below the highway. Sections of the wall had collapsed likely due to large rocks being pushed over the top of the wall after falling from the rock cut on the uphill side of the highway. The roadway shoulder was eroded away in the collapsed wall sections (Figure 3). A section of the stone-faced wall that had not collapsed is shown in Figure 4.

Figure 3 – Collapsed Section of Stone-Faced Gravity Wall (note timber faced crib wall in the background)

Figure 4 – Stone-Faced Gravity Wall Section Still Standing

Steel Crib Wall and I-beam Wall near Red Mountain Pass

This deteriorating wall consisted of two different wall facings: steel cribs and I-beams (Figure 5). Numerous elements of the crib wall were damaged, and some were missing. Backfill was raveling out between the steel elements. Additionally, the slope above the wall was over-steepened and experiencing erosion which was undercutting the roadway pavement. The steel crib wall and I-beam wall were separated by a rock outcrop with timber lagging above the outcrop and spanning between the two walls. As with the crib wall, backfill was raveling out of the I-beam wall through the gaps between beams. Steel bars were drilled into the rock at the base of the wall to provide support.

Figure 5 – Steel Crib Wall and I-beam Wall

Just north of the steel crib wall and I-beam wall, the highway pavement was cracked in several locations due to the over-steepened and eroding slope below the roadway (Figure 6). A culvert pipe daylighting on the slope below the roadway was contributing to the slope erosion and instability. A thickened pavement section was present along the edge of the road where pavement overlays were used to reestablish the roadway edge lost to erosion and instability.

Steel Crib Walls near Molas Pass

South of Silverton on the ascent of Molas Pass two additional steel cribs walls were in various stages of deterioration. One crib wall had numbers damaged steel elements and raveling of backfill material between the elements (Figure 7).

Figure 6 – Pavement Damage and Culvert North of the Steel Crib and I-beam Walls

Figure 7 – Steel Crib Wall with Damaged Elements and Backfill Raveling

A second steel crib wall in the same area also had damage to numerous facing elements (Figure 8). This wall was experiencing movement which resulted in a crack within the pavement behind the wall. Both crib walls had large concrete blocks anchoring the guardrail likely where the slope above the wall had eroded and undermined the guardrail and pavement.

Figure 8 – Crib Wall Experiencing Movement

RETAINING WALL DESIGN AND CONSTRUCTION

Geotechnical investigations were completed at each of the walls to characterize the subsurface conditions. Soil thickness and bedrock depths were highly variable both at varying distances behind and along each of the existing walls. Several wall types were evaluated at each site to identify the preferred solution. Mechanically stabilized earth (MSE) walls were selected because of the ability to tolerate differential movement between walls founded on rock and soil. MSE walls also provided the most flexibility for modifications to the site conditions found during construction. Additionally, because the highway is a scenic byway and is largely located within Forest Service property, aesthetics of the replacement walls was important. Wire basket faced MSE walls were selected as they provided the most options for applying varying facing types to the walls. Where bare steel was required for the facing to provide a rusted appearance, larger wire diameters were specified for wire baskets to provide for a longer design life.

Timber Faced MSE Wall

The geotechnical investigation for this wall encountered soils overlying quartzite with shale partings of the Uncompahgre Formation (*1*) at the approximate elevation of the bottom of the existing wall. In order to maintain a single lane of traffic during construction shoring was required to construct the necessary reinforcement length for the wall. A permanent soil nail wall was designed to both act as shoring during construction and to reduce the required MSE wall reinforcement length. The truncated base MSE wall (Figure 9) was constructed with a timber facing to match the facing of the previous wall (Figure 10). A new culvert was placed through the wall facing providing a connection to the inlet on the cutslope side of the highway. Construction modifications for this wall, and other walls on the project, included placing the wire MSE baskets to match the grade of the roadway rather than horizontally. This modification eliminated the need to cut the top of each basket to match the road grade.

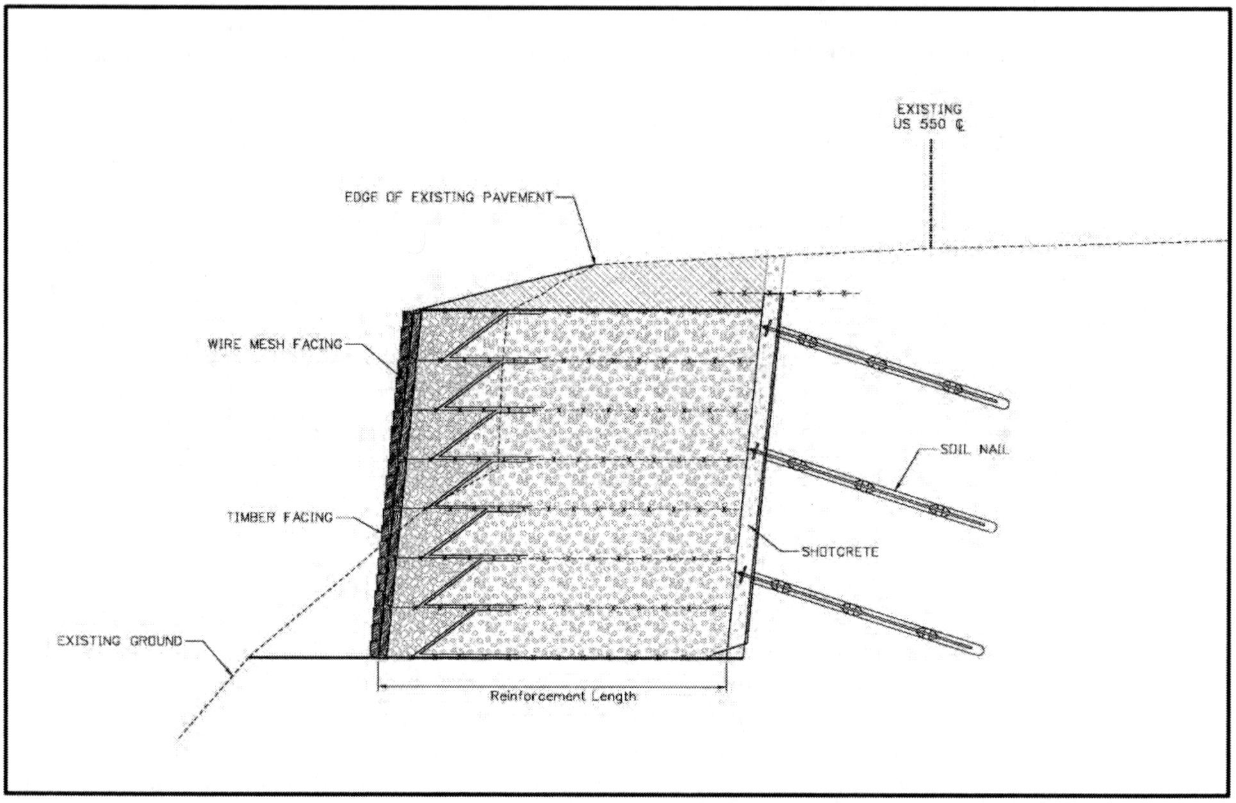

Figure 9 – Typical Section of Timber Faced MSE Wall and Soil Nail Wall

Figure 10 – Completed Timber Faced MSE Wall

Stone-Faced MSE Wall

Quartzite, also of the Uncompahgre Formation, was encountered at a shallow depth behind the stone-faced gravity wall. The previous wall appeared to have a relatively shallow depth which was likely stable because of the shallow bedrock and limited lateral loading. Because of aesthetics and the apparent stability of the stone-faced gravity wall, only sections of the wall that had collapsed were replaced (Figure 11). Stone facing rock was locally sourced to match the surrounding rock as much as practical. Using the wire mesh MSE wall baskets allowed for the bottom basket to be offset from baskets above to create footing to support the rock facing. Bedrock was excavated behind the MSE wall where required to achieve the minimum reinforcement length necessary for stability. Construction modifications including changing the structural connection between the MSE baskets and stone facing from using continues welded wire fabric to using discrete ties to save time and money.

Figure 11 – Stone Face MSE Wall

Wire Mesh Faced MSE Walls near Red Mountain Pass

The geotechnical investigation near the steel crib wall and I-beam wall encountered volcanic tuffs and rhyolite of the Burns Formation (*1*) at an elevation near the base of the existing walls. Because of the depth of soil and limited roadway width, shoring was necessary to construct the wall while maintaining a single lane of traffic during construction. A permanent soil nail wall was designed to provide shoring and to reduce lateral loading on the MSE wall.

North of this wall, two additional MSE walls were constructed to provide additional roadway width and repair pavement damage due to the over-steepened and eroding slopes. Shoring was not required for these walls due to shallow bedrock and shorter wall heights. These walls were constructed with ungalvanized wire baskets exposed to resemble the previous rusted steel walls (Figure 12). Stone was used to backfill the outer face of the baskets to provide a more aesthetically pleasing look compared to other wire mesh facing treatments.

In order to widen the roadway and create rockfall catchment, the existing rock cut slope adjacent to the highway and MSE walls was excavated. The rock removed from the excavation was crushed and used as wall backfill for the MSE walls on Molas Pass discussed in the next section.

Figure 12 – Wire Mesh Faced MSE Wall near Red Mountain Pass

Wire Mesh Faced MSE Walls near Molas Pass

The geotechnical investigation at one of the steel crib walls encountered a few feet of highly weathered bedrock above slightly weathered bedrock located at an elevation near the base of the existing wall. The highly weathered bedrock was also visible in the rock cut adjacent to the highway and wall and appeared to be a mineralized shear zone. Bedrock in the area is mapped as granitic intrusives associated with the Sultan Mountain stock (*2*). Because of the depth of soil, highly weathered bedrock, and limited roadway width, shoring was again necessary to construct the wall while maintaining a single lane of traffic during construction. A permanent soil nail wall was designed to provide shoring and to reduce lateral loading on the MSE wall. To improve bearing capacity and global stability of the wall founded on the highly weathered shear zone, micropiles with a concrete cap were designed and constructed to support the wall.

The geotechnical investigation for the second wall in this area encountered relatively shallow bedrock. The bedrock was excavated to achieve the necessary reinforcement length without the need for shoring. These walls were constructed with galvanized wire baskets to resemble the previous steel walls (Figure 13). Stone was again used to backfill the outer face of the baskets. Rock excavated from the cut near the wire mesh MSE walls near Red Mountain

Pass was used as backfill behind the facing stone. The backfill rock was tested for corrosivity as the area is known to be highly mineralized. The results of the corrosion testing showed that corrosivity met AASHTO specifications for nonaggressive backfill but numerous heavy metals were present in the rock. The steel wire of the galvanized wire mesh baskets was upsized to account for the higher potential of corrosion due to the heavy metals in the backfill material.

During excavation of the rock supporting one of the MSE walls, planar discontinuities with adverse orientation were discovered below the proposed wall location. The potentially unstable rockmass was stabilized using post-tensioned rock bolts.

Figure 13 – Wire Mesh Faced MSE Wall near Molas Pass

CONCLUSION

The five deteriorating retaining walls were replaced and two additional walls were constructed as part of two construction projects in 2016 and 2017 totaling approximately 11,000 square feet of wall facing at a total construction cost of approximately $6.6 million. Wire basket MSE walls were successfully used and adapted to the specific conditions at each site within the challenging mountain environment. The wire basket facing allowed flexibility in varying the facing types to meet the aesthetic requirements of CDOT, the Forest Service, and other stakeholders. Construction modifications to the MSE wall designs resulted in an improved and more aesthetically pleasing finished product.

REFERENCES

1. Burbank, W.S. and Luedke, R.G. *Geology of the Ironton Quadrangle, Colorado*, Map GQ-291, United States Geological Survey, Washington, D.C., 1964.
2. Yager, D.B. and Bove, D.J. *Geologic Map of Part of the Upper Animas River Watershed and Vicinity, Silverton, Colorado*, Miscellaneous Field Studies Map MF-2377, United States Geological Survey, Washington, D.C., 2002.

Implementation and Application of Geotechnical Asset Management in Colorado

Nicole Oester
Colorado Department of Transportation
4670 Holly St. Unit A
Denver, CO 80216
303-398-6603
nicole.oester@state.co.us

Robert Group
Colorado Department of Transportation
4670 Holly St. Unit A
Denver, CO 80216
303-398-6589
robert.group@state.co.us

Ty Ortiz
Colorado Department of Transportation
4670 Holly St.
Denver, CO 80216
303-398-6601
ty.ortiz@state.co.us

Prepared for the 70th Highway Geology Symposium, October, 2019

Acknowledgements

Disclaimer

Statements and views presented in this paper are strictly those of the author(s), and do not necessarily reflect positions held by their affiliations, the Highway Geology Symposium (HGS), or others acknowledged above. The mention of trade names for commercial products does not imply the approval or endorsement by HGS.

Copyright

Copyright © 2019 Highway Geology Symposium (HGS)

All Rights Reserved. Printed in the United States of America. No part of this publication may be reproduced or copied in any form or by any means – graphic, electronic, or mechanical, including photocopying, taping, or information storage and retrieval systems – without prior written permission of the HGS. This excludes the original author(s).

ABSTRACT

The Colorado Department of Transportation's (CDOT) Geoahzard Program uses a risk-based approach to manage the impact of geohazards on Colorado's transportation network. With the implementation of the Geohazards Management Plan (GMP) in 2015, CDOT started to take a more comprehensive approach to reduce the risk of geohazards through corridor-wide mitigation projects. The GMP utilizes event data to estimate the monetary risk of each tenth of a mile segment of roadway. Currently there are 52 geohazard corridors containing 3,437 segments with 2,873 reported geohazard events.

Data quality and availability is often a challenge, as CDOT relies on maintenance crews and state patrol accident information as the main source of event data. This can result in incomplete, and sometimes inconsistent reporting. The first 4 years of implementation have been forward-looking and focused on collecting more comprehensive event data for the GMP. The intent of this paper is to discuss the value gained by applying asset management principles, describe the challenges of implementation, and the future goals for CDOT's GMP.

By looking at specific corridors that have events being reported and corridors that have been mitigated, CDOT can begin to look at the effectiveness of the GMP. Also, geohazard corridors that have been mitigated can be evaluated to determine if mitigation benefit can be correlated with a decrease in event reporting. With the framework of the original GMP in place, modifications can be made to better align asset management strategies with field conditions and include factors outside of the current inputs.

GEOHAZARD ASSET MANAGEMENT IN COLORADO

In 2014 the Colorado Department of Transportation (CDOT) Geohazards Program transitioned to using an asset management plan to allocate mitigation funding in order to address geohazards affecting the state's transportation system. The Geohazard Management Plan (GMP) is a living document that is guiding corridor selection for mitigation efforts through a risk based approach. Geohazards are categorized as:

- Rockfall
- Rockslide
- Landslide
- Debris Flow
- Sinkhole
- Embankment Failure

By gathering information detailing how events effect the roadway, the Geohazards Program is able to inventory and rank corridors based on risk.

Risk Based System

Segments of roadway are given a grade (A-F) based on information from geohazard events that are reported. Currently, events are used as an indication of condition. For example, if an area has multiple events reported per year the probability of that area experiencing another event is assumed to be higher, increasing the likelihood factor for that segment. Each event has an associated safety, mobility, and maintenance risk. Safety threat is based on accidents, injuries, and/or recorded fatalities associated with a geohazard event. Often, information about safety threats is relayed first-hand from maintenance personnel. Mobility threat takes into account partial and full closures as a result of a geohazard event. This information is gathered from accounts of maintenance responders, as well as a database from traffic operations that tracks reported closures. Maintenance threat is based on estimates of the cost incurred by internal and external maintenance crews to respond to the geohazard event. These costs are separated into a range at this stage of implementation. (1)

Risk Cost

Risk cost is calculated by assigning a dollar amount to each level of safety, mobility, and maintenance threat. These costs are then added together for a risk cost for each segment of roadway. By calculating a total risk cost for each segment, CDOT is able to quantitatively compare segments of roadway to one another. For example, a segment of roadway with a recorded rockfall event in the cut-slope and an embankment failure on the fill slope will have a risk cost that is additive of each of those events.

Segments and Corridors

Risk costs for each segment of roadway are added together for each geohazard corridor. Currently, CDOT has 3,437 tenth of a mile segments of roadway affected by geohazards. While these segments only total 350 miles of Colorado's 23,000 lane miles of transportation network, they can have significant impacts on the traveling public. Risk cost separated into the limits seen in Figure 1 to give each segment a letter grade. One benefit of using segments is that it allows for some discrepancies in event mile marker reporting. Another is that segments allow for multiple hazards to affect one single segment. High risk corridors are selected for feasibility studies which is the first step in designing mitigation to effectively reduce risk and increase slope performance.

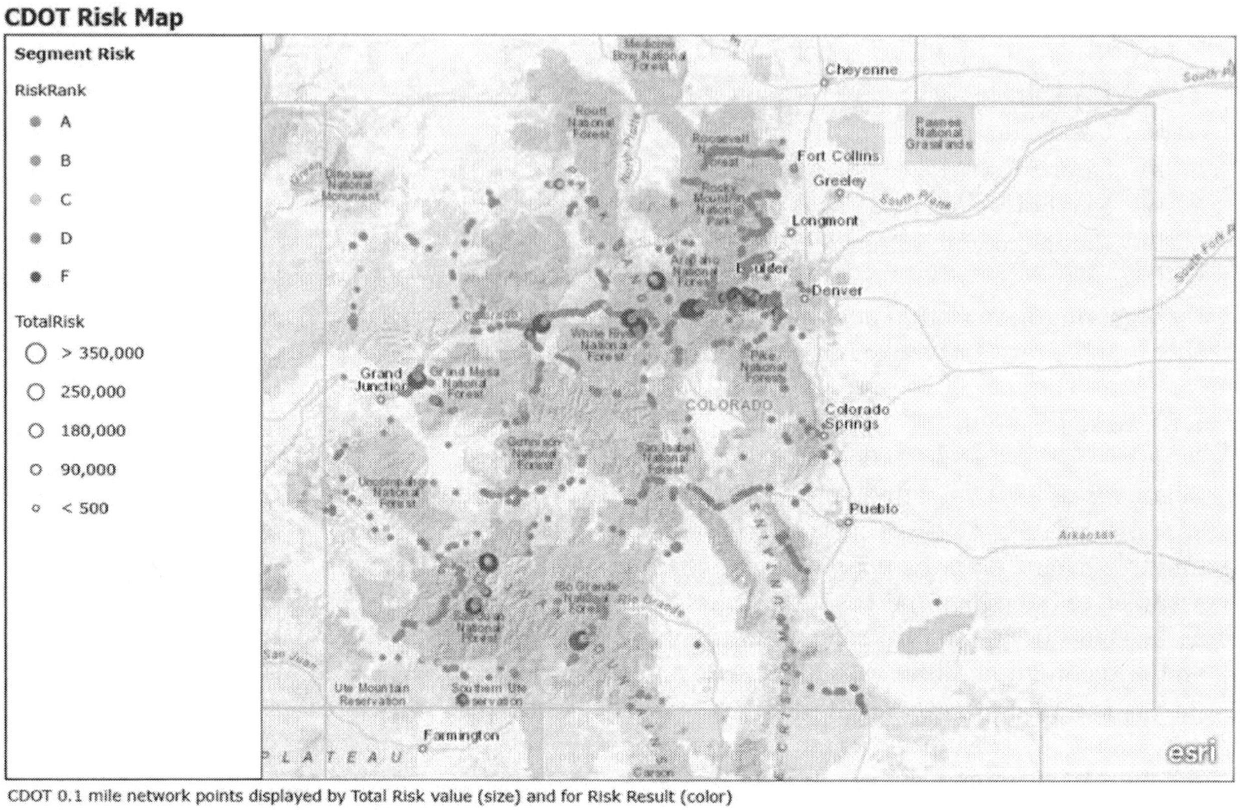

Figure 1 – CDOT Risk Map (segment total risk cost) created by Shannon and Wilson 2018

Corridor Feasibility Study

CDOT's GMP is intended to guide high level decision making, and is not meant to dictate specific sites for mitigation. Corridor feasibly studies allow for geologic conditions and other factors to be taken into consideration. In these feasibility studies, sites within the corridor are evaluated in order to select areas to mitigate such that the risk grade for the corridor can improved. In this stage expert judgment and available corridor event data is used to make decisions on site specific mitigation techniques with safety, mobility, and maintenance metrics in mind. Currently, CDOT does not incorporate feasibility study findings into the GMP, but future work will allow for integrating in depth corridor studies to show a risk grade which is more in line with true field conditions. Figure 2 displays a flow chart which outlines the current and aspirational inputs into the GMP.

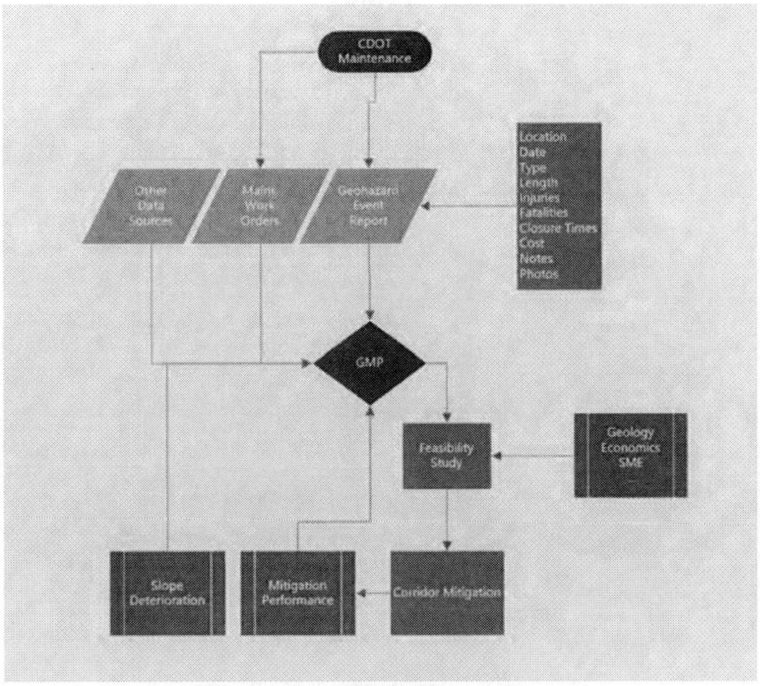

Figure 2 – GMP Flow Chart

CORRIDOR EXAMPLE

Since the GMP implementation in 2014, 5 corridor feasibility studies were completed. Since then, two corridor-wide mitigation projects were completed. In these projects, geohazard sites were selected for mitigation based on both Risk Cost and geological investigation. Because both projects were completed less than a year ago, it is difficult to quantify the benefit within the GMP framework. Last year, a feasibility study was completed for State Highway 133. This specific corridor has consistent maintenance reporting of events, as well as a multitude of geohazards affecting the roadway.

State Highway 133

State Highway 133 is an approximately 70-mile long corridor of two-lane highway located in western Colorado. The corridor stretches between Delta, Gunnison, and Pitkin counties. The geology is variable on either side of the mountain pass which cuts it in half. The rock on the South side of McClure pass consists of interbedded sandstone and shale which is susceptible to failures on both the cut and fill sides of the roadway. The North side of the pass consists tall cuts of variable sedimentary geology and steep intruded igneous slopes extending hundreds of feet above the roadway. Geohazards extend along the length of the corridor, and traffic impacts due to geohazards are not uncommon. While the Annual Daily Traffic (ADT) for the area is relatively low compared to other Colorado mountain corridors, the highway is the main north-south artery for locals and detour routes add approximately 140 miles to the commute. Traffic counts range from 3700 ADT near the town of Carbondale to as low as 1000 ADT near Paonia Reservoir.

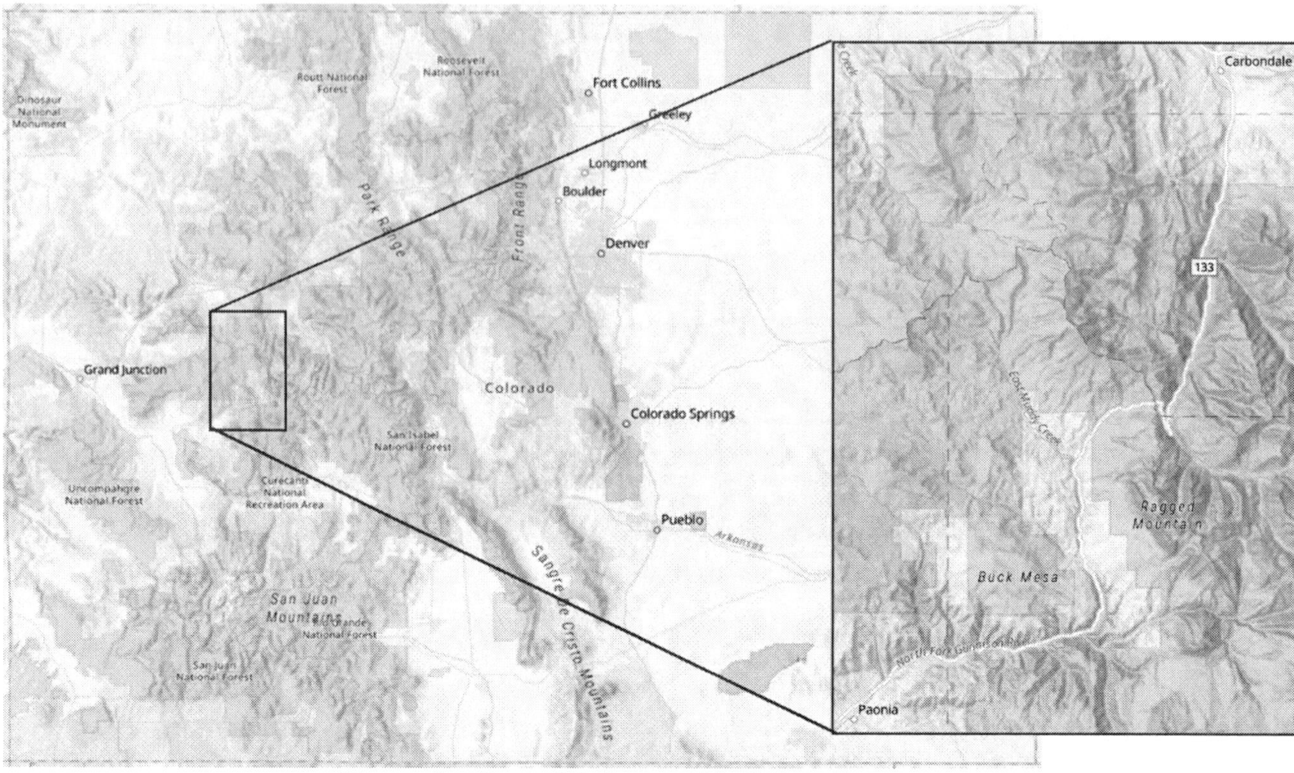

Figure 3 – State Highway 133 Colorado

Reported Events

Reported events range in size from small basketball size rocks that are seen during routine rock runs to large scale mobility disruptions as shown in Figure 4. In these cases, specialty contractors are utilized in order to mitigate the remaing hazard on the slope, and assist with downsizing and cleaning the material. In these cases, estimates for the work are included in the maintenance cost for that specific event.

Figure 4 – Large Rockslide on SH 133 near MM 29 Occurred on October 21, 2016

Small embankment failures are a common occurrence, and are especially active during the spring snow melt season. Often these regular maintenance costs are not accounted for unless there is direct communication with maintenance patrols to specifically request cost data for these fixes. Debris flows as seen in Figure 5 occur following heavy rains which come through the high country each summer. These debris flows often have hours of mobility impact and require the use of maintenance time and equipment to clean.

Figure 5 – Debris Flow on SH 133 near MM 53.6 Occurred on July 16, 2019

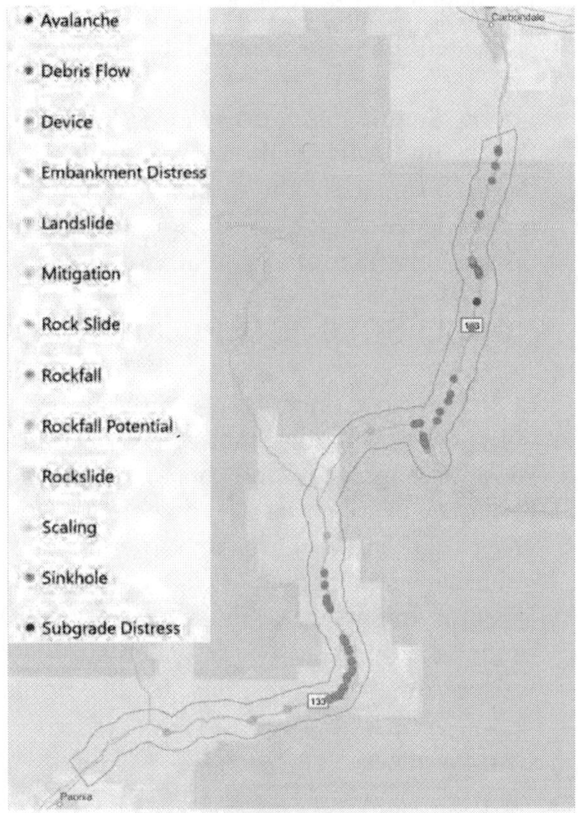

Figure 6 – Reported Geohazard Events on SH 133

Work Orders

2019 work order data indicates that maintenance has spent 2,582 working hours responding to geohazards between 1/1/2019 and 6/18/2019. Work order categories pertaining to geohazards include "rock runs" where maintenance crews drive the route to clear any rocks from the roadway, and "slope repairs". Hours also include traffic control provided by maintenance to respond to geohazard event. These categories are broad but they are another source of information to more accurately represent the maintenance risk cost of geohazards. These work orders equate to almost 40 hours spent responding to geohazards per mile of corridor in the first half of 2019. As noted in Economic Impacts from Geologic Hazard Events on Colorado Department of Transportation Right-of-Way, a large majority of documented work orders in Colorado are from "rock runs", but even so much of the work conducted by maintenance to address geohazards is undocumented especially if they are associated with small more routine events (2). Even in areas such as 133 where event reports and work orders are tracked relatively accurately, there is a large cost that is not being accounted for as a result of this non-recorded work.

Feasibility Study

The feasibility study for this corridor was conducted by Golder and Associates between 2017 and 2019. Using event data, on-site geologic investigations, and economic evaluations, sites were selected for mitigation to increase the performance of the corridor. Mitigation options were presented for those sites and the corridor is now in the design phases (3).

Future Evaluation of Corridor Effectiveness

Following the installation this the mitigation measures in the corridor, CDOT can begin to track the performance of the slopes by looking for a marked decrease in event reporting and maintenance work orders. Currently, CDOT is still in the process of developing a way to track mitigation effectiveness in order to include in the GMP risk calculations.

CHALLENGES OF IMPLEMENTATION/LIMITATIONS

In 2014, CDOT implemented GAM in an attempt to more effectively manage the risk associated with geohazards adjacent to CDOT's transportation network. As mention in the Geotechnical Asset Management for Transportation Agencies, Volume 2: Implementation Manual, a complete inventory is not necessary to begin an asset management system, but as an inventory begins to grow over the years, it is important to adjust the system to ensure that data is adding value to your management plan (4). A very important piece of GAM implementation is to continue inventorying assets, or in the case of CDOT, events in order to refine decision making processes. In the last four years of implementation, lessons learned and refinements surrounding data sources, emergency response, and geological considerations have been realized.

Data Sources

Currently, data collection is focused heavily on accounts of maintenance responders. These accounts come in various forms, and completeness is variable. Maintenance personnel have been an integral part of data collection for the initial phases of CDOT's GMP. These data are being collected in various forms ranging from quick phone calls and emails to detailed event report sheets. Information requested is:

- Location
- Date Occurred
- Type of Event
- Length of Hazard
- Injuries or Fatalities
- Closure Time
- Approximate Cost
- Notes
- Photos

In addition, efforts to collect information from CDOT traffic incident records and Colorado State Patrol reports are being made. While relevant geologic information is often missing from these records, they are underutilized sources which assist in the programs attempt to collect comprehensive data from around the state. Often these records only include route, general MM, and occasionally closure times, but using these in conjunction with other sources such as maintenance reports is often the first step in quality assurance.

Data Collection Efforts

Historical records contained 1379 geohazard events. Since 2014, 1494 geohazard events have been recorded. This is in large part because of outreach efforts directed at maintenance crews across the state. From 2014 to 2017 reporting was up over three times with 123 reports in 2014 and 370 reports in 2017. There was a slight drop of reporting in 2018 most likely due to cyber security issues that affected the state's network. So far in 2019, over 300 event reports have been documented. One factor in reporting is the perception of a reportable event by maintenance staff. In order to work towards consistency in reporting from various maintenance patrols, the Geohazards program conducts annual information sessions to encourage reporting and request feedback on ways to make reporting straightforward. Outreach from CDOT's Geohazard program in conjunction with maintenance level work being done in the regions helps ensure that reporting continues.

Emergency Response and Mitigation

Within the current GMP there is not a systematic way to account for work being done on an emergency basis. Often these costs are captured, but there are not factors to indicate that mitigation work was completed and that the sites are no longer high risk.

Geological Considerations

Some geological conditions are not currently being captured through the use of event occurrence as a surrogate for condition. For example, occasionally slopes are in better condition following a rockslide event. In essence, the geohazard has mitigated itself by failing, and the condition is better. To account for such cases, a factor could be applied to the event count following a detailed field investigation. This factor would take into account geologic conditions following the event such as block size, slope height, and ditch effectiveness to reduce the calculated risk for that segment of roadway. Currently, CDOT is addressing this though corridor feasibility studies, as mentioned above.

FORWARD-LOOKING GMP

With the GAM framework set, modifications can be made to better align asset management strategies with field conditions and include factors outside of the GAM inputs. By including variable data sources at different levels, asset management decisions can be made using both condition and risk based techniques.

Variable Data Sources

In the first years of implementation collecting a large quantity of data was important in order to make sure that risk costs were accurately represented within the GAM calculations. Now that CDOT has collected close to 5 years of event data, the program can begin to refine those risk costs and look to modify the GMP in order to better represent the risk cost of the segment and include geologic conditions in the likelihood.

Maintenance Work Orders

Maintenance work orders are currently an underutilized data source in CDOT's geohazard asset management practice. These data are being collected for maintenance budgeting purposes, but currently our framework is not set up to use them as a direct input. By averaging the time spent by maintenance in a single corridor, maintenance costs can be more accurately accounted for.

News and Social Media

Often, news articles and social media act as an initial report of geohazards affecting Colorado's transportation system. While this information should be reviewed for accuracy, often there are photos which offer valuable data points. Voumard et al. discusses in depth how Switzeralnd collected geohazards event data from local news articles sorted by Google Alerts (5). Utilizing search engine technology to sort through Colorado geohazard related articles could offer another significant data source, or at the very least give the Geohazards Program a tool for quality assurance to compare to other data sources such as work orders. Voumard et al. points out that some of this data can be inadvertently skewed by the media by including more information on events which have injuries or fatalities, and those which impact larger municipalities.

Inclusion of Mitigation and Deterioration Factors

As mentioned in the 133 example, CDOT does not currently have a way to include mitigation performance in the GMP risk calculations. Along those lines, regular deterioration of slopes is not directly input into the calculations, either. While event data should capture deterioration of slopes, the inconsistency in reporting may not be accurately capturing the true rate of declining performance of Colorado's aging slope assets. The GMP is an adaptable plan meant to be updated as data and research becomes available.

CONCLUSION

The implementation of the GMP has given CDOT a systematic way to prioritize mitigation efforts. The corridor approach to mitigation is intended to offer the most benefit in terms of safety, mobility, and maintenance. Following 4 years of data gathering efforts, it is apparent that collection must go beyond first hand maintenance accounts. While the information offered in those reports is invaluable and is a great starting point for the GMP, it would be of great benefit to reach beyond this to capture more comprehensive information to accurately represent event occurrence around the state. A system that could handle variable levels of information would be beneficial for states trying to implement a GAM system. A comprehensive GAM system would be able to take in data sources ranging from maintenance work orders, to detailed field investigations to better reflect the true risk of the geohazard.

REFERENCES:

1. Colorado Department of Transportation FY2018 Geohazards Management Plan, 2017

2. Vessely, M., Richrath, S., Weldemicael, E., *Economic Impacts from Geologic Hazard Events on Colorado Department of Transportation Right-of-Way*. Paper ID 17-02718, Transportation Research Board, 2017 Annual Meeting Compendium of Papers, 2017.

3. Golder and Associates, Colorado State Highway 133 MP 0.0 to 63 Geologic Hazard Assessment and Mitigation Feasibility Study, 2019

4. National Academies of Sciences, Engineering, and Medicine 2019. Geotechnical Asset Management for Transportation Agencies, Volume 2: Implementation Manual. Washington, DC: The National Academies Press. https://doi.org/10.17226/25364.

5. J. Voumard et al.: *Natural hazard events affecting transportation networks in Switzerland from 2012 to 2016*, Nat. Hazards Earth Syst. Sci., 18, 2093–2109, 2018

Preliminary Rockfall Evaluation for the Historic Columbia River Highway State Trail, Segment E

Noah Kimmes, E.I., G.I.T.
Landslide Technology
10250 SW Greenburg Rd. Suite 111
Portland, OR 97223
503-452-1200
noahk@landslidetechnology.com

Prepared for the 70th Highway Geology Symposium, October, 2019

ABSTRACT

The Historic Columbia River Highway (HCRH) is America's first scenic highway and was constructed between 1913 and 1922. During the 1940's and 50's many sections of the HCRH were removed and/or abandoned due to construction of a route closer to the elevation of the Columbia River, which would eventually become I-84. Since 1987, the Oregon Department of Transportation has been working with stakeholders to restore the HCRH as a bike and pedestrian trail. Segment E includes a 2.7-mile section of the trail approximately 50 miles east of Portland, OR that is currently under design and awaiting additional funding to construct. Segment E is subdivided into seven rockfall reaches, where the trail alignment runs along the base of slopes with slope between 80 and 280 feet. All slopes are composed of flows of the Columbia River Basalt (CRB) Group, exposed by the Missoula Floods of the late Pleistocene and have a history of frequent rockfall activity.

Preliminary rockfall evaluations for the seven slopes in Segment E included surface reconnaissance, on-slope (rope-accessed) reconnaissance and geologic mapping, kinematic analysis, rockfall simulations, and development of preliminary rockfall risk reduction measures. Preliminary rockfall risk reduction measures were required to meet specific rock retainment criteria for the trail and I-84 while limiting impacts to the aesthetic qualities of the Columbia River Gorge. This paper discusses the preliminary rockfall evaluations for all seven slopes, but focuses on the design challenges encountered at two specific slopes within Segment E.

Acknowledgements

The author would like to thank the individuals for their contributions in the work described:

Gerry Heslin, P.E. – Landslide Technology

Brent Black, C.E.G. – Landslide Technology

Darren Beckstrand, C.E.G. – Landslide Technology

Disclaimer

Statements and views presented in this paper are strictly those of the author(s), and do not necessarily reflect positions held by their affiliations, the Highway Geology Symposium (HGS), or others acknowledged above. The mention of trade names for commercial products does not imply the approval or endorsement by HGS.

Copyright Notice

Copyright © 2019 Highway Geology Symposium (HGS)

All Rights Reserved. Printed in the United States of America. No part of this publication may be reproduced or copied in any form or by any means – graphic, electronic, or mechanical, including photocopying, taping, or information storage and retrieval systems – without prior written permission of the HGS. This excludes

INTRODUCTION

The Historic Columbia River Highway (HCRH) is America's first scenic highway, constructed between 1913 and 1922. Construction of a roadway which would eventually become Interstate 84 during the late 1940s to early 1950s lead to many sections of the HCRH being removed and/or abandoned. The Columbia River Gorge National Scenic Area Act of 1986 directed the State of Oregon to connect the abandoned sections of the HCRH as a pedestrian and cyclist trail. Since 1987, the Oregon Department of Transportation (ODOT) has been working with stakeholders including the HCRH Advisory Committee, Oregon Parks and Recreation Department (OPRD), the State Historic Preservation Office, and Travel Oregon to restore the HCRH as a trail. The HCRH State Trail Plan is focused on 11 miles of the trail between Exit 51 on I-84 (Wyeth) and Ruthton County Park. This 11-mile portion of the trail is divided into eight segments (Segment A through Segment H).

Segment E of the HCRH trail is located along a 2.5 mile stretch of I-84 in northern Hood River County, Oregon. The alignment includes a 2.7-mile section of the trail that starts at Viento Trailhead and ends at Mitchell Point. This portion of the trail has been subdivided into seven rockfall reaches, as shown in Figure 1. The length, maximum height, and approximate station ranges for the seven rockfall reaches are provided in Table 1. The Federal Highway Administration Western Federal Lands Highway Division (WFLHD) retained David Evans and Associates, Inc. (DEA) to provide a preliminary design for Segment E. DEA retained Cornforth Consultants, Inc, and it's division, Landslide Technology (LT), to assist with geotechnical aspects of the preliminary trail design, including evaluation of the rockfall hazards and design of rockfall risk reduction measures along this segment.

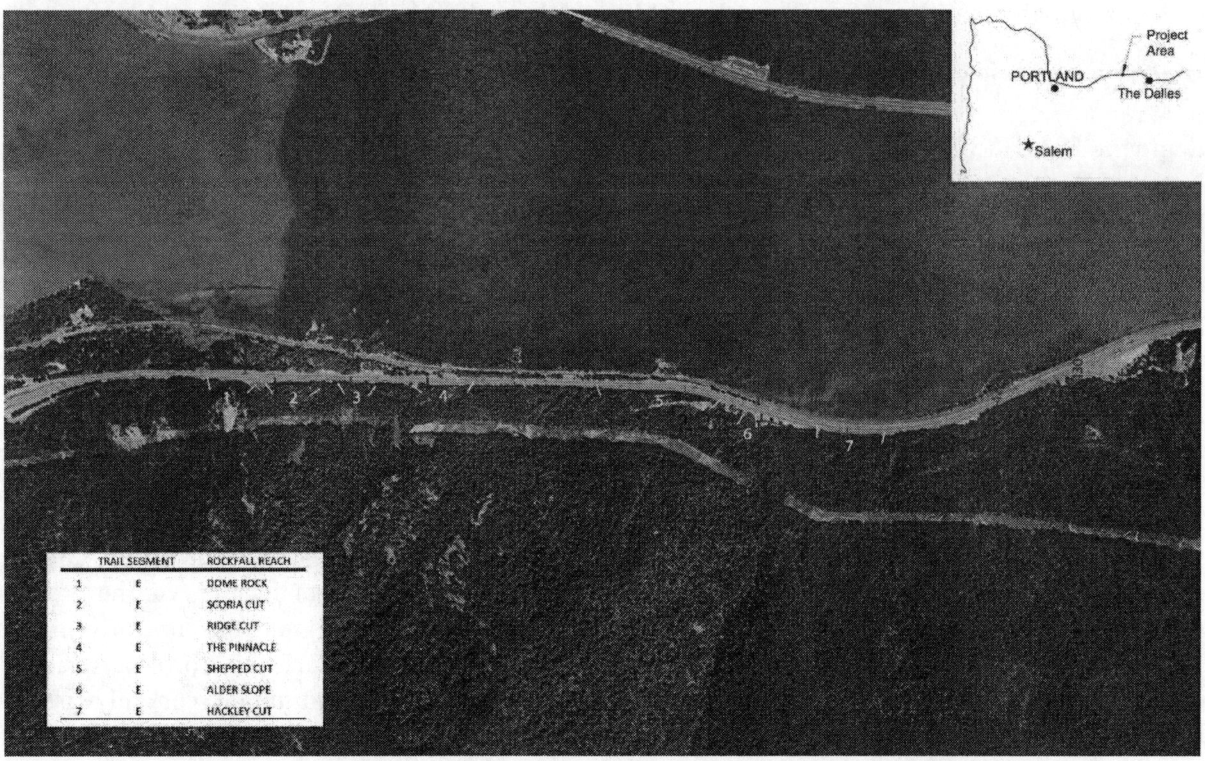

Figure 1: Location and vicinity map of Segment E of the HCRH State Trail

Table 1: Rockfall reach information

Rockfall Reach	Approx. Station Range	Length (feet)	Max. Slope Height(feet)
A - Dome Rock	320+50 to 325+75	525	335
B - Scoria Cut	330+00 to 334+50	450	120
C - Ridge Cut	338+00 to 344+00	600	110
D - The Pinnacle	348+50 to 355+50	700	90
E - Stepped Cut	370+50 to 385+00	1450	280
F - Alder Slope	386+25 to 388+25	200	80
G - Hackly Cut	399+50 to 405+50	600	200

Preliminary rockfall evaluations for the seven slopes in Segment E included surface reconnaissance, on-slope (rope-accessed) reconnaissance, and geologic mapping, kinematic analysis, rockfall simulations, and development of preliminary rockfall risk reduction measures. Preliminary rockfall risk reduction measures were required to meet specific rock retainment criteria for the trail and I-84 while limiting impacts to the aesthetic qualities of the Columbia River Gorge. This paper discusses the approaches used in the preliminary rockfall evaluations and design development for all seven slopes, then presents two case studies focusing on the design challenges encountered at two slopes: The Pinnacle and Stepped Cut (Rockfall Reaches D and E).

REGIONAL AND SITE GEOLOGY

Segment E of the Historic Columbia River Highway Trail is located in the Columbia Gorge National Scenic Area approximately 50 miles east of Portland, Oregon. The Columbia River Gorge was formed as the Columbia River cut through both the Cascade Volcanics and the underlying volcanic plateaus formed by the Columbia River Basalt (CRB) Group. A series of late Pleistocene floods (Missoula Floods) caused erosion and downcutting of the gorge leaving behind the steep basalt cliffs present throughout the gorge.

The CRB Group dominates the regional geology. The basalt flows present in the CRB Group are from the Miocene Epoch (approximately 16.7 to 5.5 million years ago) with the majority of the flows occurring during an approximate 1.1-million-year period (16.7 to 15.6 Ma). Estimated volumes of the basalt flows are up to 1,000 km^3. The area covered by the flows is approximately 200,000 km2 in Oregon, Washington and Idaho. The flows are thought to have originated from what is now the Yellowstone Hotspot, previously located in what is now northeastern Oregon, eastern Washington, and western Idaho. North-northwest trending linear fissures on the order of tens to hundreds of kilometers long released low viscosity basalt lava multiple times during the Miocene Epoch. The molten rock flowed westward from the fissures to form a series of overlapping basalt flows. Uplift, faulting and downcutting from the Columbia River and Missoula Floods have formed the outcrops, cliffs, and natural rock benches observed throughout the gorge present day.

At the western end of Segment E of the HCRH Trail, the Grande Ronde Formation of the CRB is present including the Grande Ronde Upper (Normal and Reverse Polarity Flow) members, as well as the Grande Ronde Lower member. The upper flows are exposed on the upper cliffs and overlay the lower flows typically near and below the elevation of I-84. Moving to the east end of Segment E, exposures of the Frenchman Springs member of the Wanapum Formation and the Pomona member of the Saddle Mountain Formation are present. The flows on the east end of Segment E are overlain by The Dalles Formation which consists of upper Miocene volcaniclastic and sedimentary deposits of breccia, tuff, conglomerate, and sandstone to mudstone. The basalt flows on the east end of Segment E are found at the top of the exposed cliffs to below highway elevation, with the Dalles Formation found on the upper plateau (*1*).

DESIGN CRITERIA

The proposed HCRH Trail parallels the south side of I-84 through Segment E. At six of the seven rockfall reaches, the proposed trail prism will be located in the existing rockfall catchment area (roadside ditch) along I-84. One objective for preliminary risk reduction options is to maintain the current level of rockfall protection along I-84 after the trail is constructed and to reduce the hazards from rockfall along the proposed HCRH Trail.

The proposed risk reduction measures must also follow the design objectives and considerations given in the I-84 Corridor Strategy Guidelines for the Columbia River Gorge National Scenic Area (*2*). These guidelines state that rockfall risk reduction measures should:

- Protect the traveling public
- Choose an alignment that precludes the need for mitigation
- Apply treatment away from the roadway edge where possible to minimize visual impacts
- Minimize structural solutions in the immediate foreground
- Minimize alteration of existing slopes
- Create slopes and structures that visually blend with the natural surrounding terrain

These guidelines for rockfall risk reduction measures place an emphasis on aesthetics. To the greatest extent possible, risk reduction measures with low visual impacts were utilized in the design. In particular, on-slope items such as draped mesh and mid-slope attenuators were used only where no other feasible options were available. Based on feedback from the trail stakeholders, risk reduction measures located at the base of the slope were given preference over on-slope methods.

After discussions with WFLHD and ODOT, it was determined that rockfall risk reduction designs capable of precluding 99% of impacting rockfalls and 90% of rolling rockfalls from reaching the trail and I-84 were desired (i.e. 90% retention of rolling rocks). Retention of rock is measured as the percentage of simulated rocks that do not pass the edge of the travel-way for the trail and the edge of pavement for I-84. Per discussions with ODOT and WFLHD, deflection of flexible rockfall barriers into the trail is an acceptable performance; however, no deflection is to be allowed onto the travel lanes of I-84.

ROCKFALL EVALUATION AND PRELIMINARY DESIGN DEVELOPMENT

Development of preliminary rockfall reduction measures for Segment E of the HCRH State Trail included the following:

- Review of existing site and historical information
- Meet with ODOT maintenance personnel to discuss historical rockfall events and maintenance efforts
- Complete site reconnaissance
- Analyze rock mass discontinuity data and perform kinematic analyses
- Model rockfall events to evaluate existing containment and preliminary rockfall risk reduction options
- Develop preliminary rockfall risk reduction options and cost estimates for each rockfall reach
- Evaluate proposed options for effectiveness, durability, constructability, and aesthetics

Site reconnaissance, discontinuities and kinematic analyses, and rockfall modeling are discussed in further detail in the following sections.

Site Reconnaissance

Several site visits were performed by engineering geologists and geotechnical engineers from LT in the spring and summer of 2018. The rock slopes adjacent to the proposed trail alignment were observed to document the geology, rock mass characteristics, and existing rockfall evidence. Existing fallout areas (i.e. roadside ditches) and rockfall protection measures were evaluated. To aid in the development of rockfall models, the size and dimensions of rockfall debris and in-place rock blocks, slope properties, and cross-section data were recorded. Potentially unstable rock blocks were located and measured to characterize potential failure mechanisms.

Rope access techniques were employed at select rock slope reaches to obtain access to upper, steeper slope sections to assess rock mass conditions, rockfall potential, and the constructability of risk reduction measures. Rope access techniques utilized included top-down access (i.e. rappelling off of anchors at the top of the slope to access points of interest on the slope) and ground-up access. Ground-up access was utilized at Rockfall Reach E, Stepped Cut, to access a prominent mid-slope bench on the slope. Due to the relatively large height of the slope (280 feet above I-84) and the potential for loose rocks to be knocked down the slope towards the highway, top-down access was considered to be hazardous. Ground-up access involved aid climbing from the ditch on bolts installed in the cut slope to the mid-slope bench (Figure 2).

Structural Mapping and Kinematic Analysis

The presence, orientation and condition of discontinuities in a rock slope have a major influence on slope stability and rockfall potential. Several techniques were employed at the seven rock slopes to define the geologic character, including structural mapping, joint set analysis, and kinematic analysis.

Figure 2: Aid climbing to access the mid-slope bench at Stepped Cut

Structural mapping was conducted on the seven slope sections along Segment E using LiDAR point clouds with the software program Cloud Compare® (version 2.8.1), and specifically the plugin "Compass". Compass is a structural geology toolbox for analysis of virtual outcrop point clouds. Within the Compass plugin, the Plane Tool and Trace Tool were used to extract structural measurements from the LiDAR point clouds. The Plane Tool allows the orientations of fully exposed planar structures to be measured while the Trace Tool allows the orientation of geologic structure without fully exposed planar surfaces to be estimated based on the apparent plane's intersection with the point cloud. A single LiDAR point cloud was produced and provided to LT for each slope section using terrestrial-based LiDAR collected by WFLHD. The resulting clouds produced 426 structural measurements across the seven slope sections. The structural measurements obtained from LiDAR data were supplemented with hand-collected measurements. Figure 3 shows a screen capture from CloudCompare® with structural measurements displayed as red planes overlaid on the gray, point cloud. The slope shown in this figure is Stepped Cut (Rockfall Reach E). Figure 4 shows the location of Figure 3 on a UAV-obtained aerial photograph of Stepped Cut.

Figure 3: Structural measurements obtained from CloudCompare® on the terrestrial LiDAR obtained point cloud of Stepped Cut. Structural measurements are displayed as red planes on the underlying, gray, point cloud.

Figure 4: Area of structural measurements on Stepped Cut shown on Figure 3

Structural data were plotted on equal-area lower-half, equatorial stereonets using the computer program DIPS (version 6.0) to estimate joint sets. Kinematic analyses, including analyses for planar, wedge, and toppling potential, were performed with DIPS. These analyses provide an indication of the types of failures that may be possible, but do not give a Factor-of-Safety (FS) for failures nor do they take observed slope performance into consideration. The slope orientations used in the analyses were estimated by fitting a plane to the overall orientation of the slope by using the Compass plugin in Cloud Compare. For slopes with variable orientations, multiple kinematic analyses were performed using different slope orientations.

Rockfall Modeling

Rockfall modeling was performed using the computer program RocFall (version 5.0). The cross sections used for modeling were developed from topographic maps based on aerial LiDAR data. Cross sections were selected along anticipated rockfall paths that intersect the trail alignment and I-84. Slope height, steepness, presence of launch features, and evidence of rockfall activity were considered when selecting analysis locations. Potential bounce heights and rockfall energies were utilized to evaluate the likelihood of rockfall reaching the trail and I-84 and to determine the location, height, and required capacity of preliminary rockfall risk reduction measures.

The potential shape and size of modeled rocks were based on field observations and reported historical rockfall information. The RocFall program utilizes a lumped mass model that analyzes individual rocks as points. Rockfall source zones were selected based on observed slope conditions. Rockfall source zones were typically distributed along bedrock outcrops in the modeled cross sections. The source zones included the cut face, the slopes immediately above the crest of the slopes, and isolated rock outcrops above the slope crest.

Material parameters were selected based on observed slope materials and site conditions. Surface materials assigned to the modeled cross sections include bedrock outcrops, talus cover, soil with vegetation, gravel ditch material, rockfall debris, trail fill slopes, and asphalt. Slope coefficients were assigned based upon site observations, published values and experience modeling similar geologic materials. The models were calibrated by varying input parameters so that simulated rockfall trajectories were consistent with reported and/or observed rockfall events and the information provided by ODOT maintenance personnel.

The analyses for each cross section included 1,000 simulated rockfall events for each slope configuration and rockfall size modeled. Initial models were performed to calibrate the analysis to existing conditions (prior to trail construction) so that calculated rockfall trajectories and rollout distances were consistent with observations and rockfall history. The rockfall sections were then modified to include the proposed trail geometry. Analysis points (collectors) were placed at the edge of the trail and at the edge of pavement for I-84 in the models to collect rockfall model data. The number and energy of rocks impacting and/or rolling onto the trail and I-84 were recorded. Additional rockfall models including preliminary rockfall risk reduction measures were performed to evaluate the effectiveness of the risk reduction measures.

CASE STUDY 1: ROCKFALL REACH D, THE PINNACLE

Rockfall Reach D, known as the Pinnacle, is a west- to north-facing rock cut, approximately 700 feet long, that ranges between 60 feet and 90 feet high, shown in Figure 5. The trail alignment follows the historic highway grade as it approaches the western outcrops, traveling parallel to talus slopes. The rock slope is vertical to overhanging at its western extent. At this location, a large pinnacle of rock is separated from the main rock mass by a large crack up to three feet wide. This pinnacle of rock is approximately 32 feet tall by 32 feet wide along its base, and 90 feet above the trail alignment at its top. The portion of the rock slope running parallel to I-84 is cut between 1/4H:1V to 1/5H:1V and tapers downward in height towards the east from approximately 90 feet to 60 feet.

Figure 5: UAV obtained aerial photograph of Rockfall Reach D, The Pinnacle.

The rock mass at Rockfall Reach D is comprised of basalt that appears relatively sound and tight. It is slightly weathered to fresh and medium hard (R3) to hard (R4). Weaker, more weathered zones were observed in the slope, usually associated with flow contacts. A flow contact is present on the west end of the slope section near the base of the slope. At the top of the contact, differential weathering has occurred and has undermined the base of the overlying basalt by up to three feet. This is common occurrence along flow contacts within the basalt flows throughout Segment E. In general, the rock mass is highly to moderately jointed (joint spacing between three inches and two feet). Structural mapping identified one primary and one secondary joint set with dip and dip directions of 89°/232° and 40°/354°, respectively. Kinematic analyses indicate that wedge failure is the most prevalent failure mode at the slope.

Rockfall Hazards

The western portion of the trail alignment to the west of the pinnacle appears to produce a relatively low volume of rockfall. Rockfall debris in this area that was estimated to be relatively recent is typically less than one foot in size. Rope access techniques were utilized to further observe the pinnacle of rock at the western extent of the rock slope (Figure 6). Although it appears that the bounding joint set on the east side of the pinnacle has eroded away to the point that a three-foot gap is present, the base of the column is comprised of relatively sound and tight rock. A basal or potential release joint was observed at the base of the rock column. This joint dips towards the northwest at an angle between 38 and 44 degrees. The joint is relatively tight (not open or inflated) and the surface is irregular. Given the favorable characteristics of the joint and the rock mass properties, it is likely that a relatively high internal shear strength condition exists. The basal joint does not appear to daylight on the west side of the column, indicating that it is currently not a though-going joint. Given the above and the fact that there is no historical evidence of recent large-scale failures along the rock slopes in Segment E, the likelihood of a large-scale block, or global failure of the pinnacle is relatively low.

Figure 6: Photo of the pinnacle of rock at the western extent of Rockfall Reach D. Rope access techniques were utilized to obtain structural measurements of the pinnacle.

A relatively low to moderate amount of rockfall debris, predominately less than one-foot in size, was observed in the ditch along the portion of the slope that parallels I-84. According to ODOT maintenance, the ditch was cleaned three to four years ago and little to no rock reaches I-84 near this slope. It appears that the frequency of rockfall at this slope is relatively low to medium. The source of rockfall at this slope appears to be the tall, near-vertical portions of the cut slope. The crest along the slope is relatively flat and upslope sources were not observed.

Rockfall Modeling Results

Six cross sections were modeled to simulate rockfall at Rockfall Reach D. Based upon rockfall history and site observations, 0.5- and 2-foot rocks were modeled. Models were performed under existing conditions and with the proposed trail alignment with no rockfall risk reduction measures. Rockfall retainment, energies, and bounce heights were evaluated at the edge of pavement of I-84 for the existing condition models. Existing condition models show that all six cross sections meet or exceed the required design criteria of 99% impacting and 90% rolling rock retainment for I-84. This is in agreement with ODOT maintenance records.

Models with the proposed trail alignment taken into account were evaluated at the inboard edge of the trail alignment and at the edge of pavement of I-84. The proposed trail alignment models show that all but five of the six sections fail to meet the 90% rolling rock retainment criteria at the inboard edge of the trail, and four of the sections also fail to meet the 99% impacting rock retainment criteria at the inboard edge of the trail. These models do show that I-84 continues to meet the required design criteria at all cross-section locations.

Preliminary Rockfall Risk Reduction Design

Rockfall modeling showed that when the trail is added, I-84 continues to meet the rockfall retainment criteria while the trail does not meet the retainment criteria at five of the six locations evaluated. Conceptual rockfall risk reduction measures were considered for rockfall reach D to meet the design retention criteria at the edge of the trail while continuing to meet the design retention criteria at I-84. These conceptual measures included scaling, gabion baskets, an inboard flexible rockfall barrier (FRB), rockfall attenuator, draped rockfall protection wire mesh, and enlarging and improving the fallout area. From these, two preliminary design options were developed and are discussed below.

Option 1: Draped Rockfall Protection Wire Mesh with Gabion Baskets

This option includes the installation if high-tensile strength draped rockfall protection wire mesh along the rock cut slope above the trail alignment, and gabion baskets along two portions of the slope. Two separate areas of draped rockfall protection wire mesh are recommended, with approximate limits shown on Figure 7. To minimize aesthetic impacts to the slope, it is anticipated that the draped mesh would consist of high-tensile strength mesh due to its ability to better contour to the slope when compared to traditional twisted wire mesh. Contouring the mesh to the slope will consist of installing short anchors strategically throughout the rock slope to hold the mesh closer to the rock surface where it would otherwise drape freely. Contouring the mesh will require scaling prior to installation of the anchors to reduce the rockfall risk for workers on the slope. To further minimize the aesthetic impacts to the slope, the draped mesh is expected to be treated with a weathering agent to allow it blend with the natural slope.

Gabion baskets are included with this option at the locations shown in Figure 7. Gabion baskets are located in areas where rockfall modeling showed that bounce heights are low enough that draped mesh is not required, but where a short barrier at the inboard edge of the trail is still required to meet the rock retainment criteria.

Rockfall modeling with the prosed rockfall risk reduction measures outlined in Option 1 shows that the rockfall retainment criteria are met at the inboard edge of the trail and at I-84 for all locations evaluated.

Option 2: Enlarge and Improve Fallout Area

This option includes making a rock cut to increase the size of the rockfall catchment area between the slope and the proposed trail alignment. This option would use controlled blasting techniques to construct a ¼H :1V to ½H:1V cut slope to create an additional 20 to 25 feet of fallout between the slope and the proposed trail alignment. Depending upon conditions encountered during the rock excavation, this option may require the use of rock dowels or rock bolts to reinforce select sections of the newly excavated rock slope. It is also recommended that any mid-slope offset benches created by blasting in lifts be limited to 2-feet or less in width or cut at 1H:1V to limit rocks from bouncing off of the bench and into the trail or the interstate. This option would have substantially higher costs and traffic disruptions than option 1.

Figure 7: Rockfall Reach D, The Pinnacle, risk reduction options

WFLHD has elected to move forward with the rockfall risk reduction measures presented in Option 1, with the substitution of a six-foot tall, low-deflection, FRB for all gabion baskets. An FRB was selected over gabion baskets to minimize visual obstructions between the trail and the rock slope.

CASE STUDY 2: ROCKFALL REACH E, STEPPED CUT

Rockfall Reach E, known as Stepped Cut, is a north-facing rock cut, approximately 1,450 feet long with a maximum height of 280 feet, shown in Figure 8. Stepped Cut is the largest rock slope in Segment E. Concrete barriers are currently installed along I-84. It appears the lower portion of the natural slope was cut during construction of I-84. The rock cut slope is relatively short (less than 30 feet high) in the westernmost 200 feet of Stepped Cut; however, the natural slope above the cut extends to 280 feet above the trail. Further towards the east, the height of the rock cut gradually increases to 130 feet. At this location, it appears that the top of the cut follows a natural bench within the rock mass located at elevation 250 feet. This bench gradually dips to the east over a distance of approximately 440 feet until it terminates at elevation 180 feet. At this point, it appears the cut transitioned higher on the slope to approximately elevation 290 feet (approximately 170 feet above the trail), where the top of the cut dips again to the east to an elevation of 250 feet. This section of Stepped Cut is relatively steep with an average inclination of 65 degrees.

Figure 8: UAV obtained aerial photograph of Rockfall Reach E, Stepped Cut.

The rock mass at Stepped Cut is comprised of several basalt flows. It is likely that the natural bench described above and the top of the rock outcrop are associated with contacts between basalt flows. The primary rock type observed in the rock cut is basalt that is moderately to slightly weathered and medium hard (R3) to hard (R4). Weaker, more weathered zones were observed in the slope, usually associated with flow contacts. In general, the rock mass is highly to moderately jointed (joint spacing between three inches and three feet). Within the center of the larger flows, the basalt is more massive with joint spacing between three and 10 feet. The wide joint spacing forms larger blocks and columns present above the midslope bench and just below the crest of the slope. Structural mapping identified two primary and one secondary joint set with dip and dip directions of 70°/324°, 89°/228°, and 38°/149°, respectively. Kinematic analyses indicate that wedge, planar, and toppling failure potentials are all statistically unlikely along the slope.

Rockfall Hazards

The trail is to be built between the base of the slope and the eastbound lanes of I-84. The current ditch is approximately 15 feet wide. There is a slight swale in the ditch along the toe of the rock slope. This ditch configuration helps to diminish the rotational energy of fallen rocks and reduce the rollout distance. The amount of rockfall observed in the ditch of the western-most 300 feet of this rockfall reach was relatively low to moderate and typically less than one-foot in size. The source of rockfall in this area appears to be the relatively short rock cut. The slope above the cut is heavily vegetated and does not appear to be a major source of rockfall.

The catchment area of the remainder of the slope contained significant rockfall debris between six-inches and three feet in size. The concrete barriers throughout this portion of the slope displayed signs of being struck by rockfall. Smaller (less than six-inches in size) pieces of rockfall debris were observed in the eastbound travel lane of I-84. According to ODOT maintenance personnel, the ditch was cleaned three to four years ago. Maintenance also stated that the concrete barriers were installed due to a history of rockfall debris reaching the eastbound travel lane of I-84. Since the barriers have been installed, the amount of rockfall debris reaching I-84 has been significantly reduced, but smaller rocks (less than six-inches in size) occasionally do reach the eastbound travel lane. Overall, the majority of Stepped Cut appears to be highly active.

The midslope bench in this area was accessed to evaluate constructability of mid-slope risk reduction measures and to further observe the conditions of the slope above the bench. The slope above the bench appears to be a separate flow. Along the bench, the base of the flow is highly jointed and brecciated. This material is weaker and more prone to differential weathering than the rock in the center of the flow. Rock in the center of the flow also has a wider joint spacing than the base of the flow. This results in the formation of larger rock blocks and columns that are subsequently being undermined by differential erosion. Over an approximately 300-foot-long section above the eastern extent of the midslope bench numerous large rock columns are losing basal support as shown in Figure 9. Although there is no historical evidence of large-scale failures at this site, given continued differential erosion and time, these areas and appear to be a potential large-scale failure concern.

Rockfall Modeling Results

Nine cross sections were modeled to simulate rockfall at Stepped Cut. Based upon rockfall history and site observations, 0.5- and 2-foot rocks were modeled. Models were performed under existing conditions and with the proposed trail alignment with no rockfall risk reduction measures. Rockfall retainment, energies, and bounce heights were evaluated at the edge of pavement of I-84 for the existing condition models. Existing condition models show that the concrete barriers are sufficient in meeting the 90% rolling rock retainment criteria for all sections, but are insufficient in meeting the 99% impacting rock retainment criteria (i.e.: more than 1% of falling rocks fly over the concrete barriers and impact the interstate) at six of the nine cross sections.

Figure 9: View looking south at undermined, massive columns above the midslope bench of Stepped Cut.

Models with the proposed trail alignment taken into account were evaluated at the inboard edge of the trail alignment and at the edge of pavement of I-84. The proposed trail alignment models show that all sections fail to meet the 99% impacting rock retainment criteria, and seven of the nine sections also fail to meet the 90% rolling rock retainment criteria.

Preliminary Rockfall Risk Reduction Design

Conceptual rockfall risk reduction measures were considered for Stepped Cut to meet the design rockfall retainment criteria at the edge of the trail and at the edge of pavement of I-84. Conceptual rockfall risk reduction measures that were considered included: scaling, rock bolting, rockfall sheds, rockfall canopies, FRBs, midslope rockfall attenuator, draped rockfall protection wire mesh, and enlarging and improving the fallout area. From these, two preliminary design options were developed and are discussed below.

Option 1: Draped Rockfall Protection Wire Mesh and Midslope Rockfall Attenuator with Rock Bolting

This option includes the installation of a midslope rockfall attenuator along the western and middle portions of the slope and high-tensile strength draped rockfall protection wire mesh along the eastern portion of the cut slope. This option also includes scaling and rock bolting to stabilize large columns above the bench. One area of draped mesh will be utilized along the eastern portion of the slope, as shown in Figure 10. A midslope attenuator with 12-foot tall posts will run along the prominent midslope bench in the middle portion of the slope, and then will step down across discontinuous rock outcroppings to the top of the short cut slope on the western portion of the rockfall reach. A portion of overlap between the attenuator and draped mesh is present near the area of the eastern portion of the midslope bench where rockfall modeling indicates that the attenuator alone is insufficient mitigation to meet the rockfall retainment criteria for the trail and I-84. The draped mesh is expected to be contoured and stained for the same reasons outlined in Option 1 for Rockfall Reach D (The Pinnacle). The attenuator panels will not be contoured to the slope. Pinning the attenuator panels to the slope results in poor self-cleaning of the system and decreased attenuation of rockfall energies.

Rockfall modeling with the prosed rockfall risk reduction measures outline in Option 1 shows that the rockfall retainment criteria are met at the inboard edge of the trail and at I-84 for all locations evaluated.

Option 2: Enlarge and Improve Fallout Area

This option includes making a rock cut to increase the size of the rockfall catchment area between the slope and the proposed trail alignment. This option would use controlled blasting techniques to construct a ¼H :1V to ½H:1V cut slope to create an additional 20 to 25 feet of fallout between the slope and the proposed trail alignment. Depending upon conditions encountered during the rock excavation, this option may require the use of rock dowels or rock bolts to reinforce select sections of the newly excavated rock slope. It is also recommended that any mid-slope offset benches created by blasting in lifts be limited to 2-feet or less in width or cut at 1H:1V to limit rocks from bouncing off of the bench and into the trail or the interstate. This option would have substantially higher costs and traffic disruptions than option 1.

WFLHD has elected to move forward with the rockfall risk reduction measures presented in Option 1.

Figure 10: Rockfall Reach E, Stepped Cut, Mitigation Options

REFERENCES

(1) McClaughry, J.D., Wiley, T.J., Conrey, R.M., Jones, C.B., & Lite, K.E. Jr. *Digital Geologic Map of the Hood River Valley, Hood River and Wasco Counties, Oregon*. Open File Report O-12-03, State of Oregon, Oregon Department of Geology and Mineral Industries, 2012.

(2) Oregon Department of Transportation. *I-84 Corridor Strategy: A Vision and Design Guidelines for Interstate 84 in the Columbia River Gorge National Scenic Area*. Oregon Department of Transportation, November, 2005.

Oneonta Tunnel – Restoration, Fire, Restoration

George Freitag, CEG
GRI
9750 SW Nimbus Avenue
Beaverton, OR, 97008
(503) 641-3478
gfreitag@gri.com

Michael Marshall, CEG
GRI
9750 SW Nimbus Avenue
Beaverton, OR, 97008
(503) 641-3478
mmarshall@gri.com

Matthew Shanahan, PE, GE
GRI
1101 Broadway, Suite 215
Vancouver, WA, 98660
(360) 213-1690
mshanahan@gri.com

Michael Zimmerman, PE, GE, CEG
Oregon Department of Transportation, Region 1
123 NW Flanders Street, Portland, OR 97209-4012
(503) 731-3095
michael.j.zimmerman@odot.state.or.us

Prepared for the 70th Highway Geology Symposium, October, 2019

Acknowledgement

The authors wish to thank Stan Kelsay, PE, for his mentorship regarding tunnel design and rock mechanics, and for co-founding GRI with Dave Driscoll, PE.

Disclaimer

Statements and views presented in this paper are strictly those of the author(s), and do not necessarily reflect positions held by their affiliations, the Highway Geology Symposium (HGS), or others acknowledged above. The mention of trade names for commercial products does not imply the approval or endorsement by HGS.

Copyright Notice

Copyright © 2019 Highway Geology Symposium (HGS)

All Rights Reserved. Printed in the United States of America. No part of this publication may be reproduced or copied in any form or by any means – graphic, electronic, or mechanical, including photocopying, taping, or information storage and retrieval systems – without prior written permission of the HGS. This excludes the original author(s)

ABSTRACT

Oneonta Tunnel was constructed in 1914 on the Historic Columbia River Highway (HCRH) in the Columbia River Gorge about 35 miles east of Portland, Oregon. The tunnel is 115-ft long and extends through a promontory composed of Columbia River Basalt (Miocene). The promontory is about 250 ft tall and is composed of five separate basalt flows with varied geologic and volcanic structures, including large- and small-diameter columnar joints, flow banding, flow-top vesicles, and variable weathering. Construction of US 30 and Interstate 84 removed many sections of the HCRH. In 1948 the tunnel was closed to all vehicular traffic due to a realignment of US 30. Between 2006 and 2009, the tunnel was restored for pedestrians and cyclists. The restoration design included removal of the backfill used to close the tunnel, installation of about 250 rock dowels in the crown, installation of mine straps in areas of jointed/weak basalt, installation of drainage panels and shotcrete liner, and a non-structural wood liner. A series of steel sets with concrete lagging were installed in an area of particularly weak and weathered basalt and thin roof cover. In September 2017, the entire wood liner of the tunnel burned during the Eagle Creek Fire and portions of the structural tunnel liner system were damaged. As part of a second planned tunnel restoration, the post-fire performance of the tunnel support system was evaluated in 2019 using a program of dowel tension testing, shotcrete coring and strength testing, and structural mapping. The 2009 shotcrete system was effective in resisting damage to the rock dowels from the wood liner fire.

INTRODUCTION

Oneonta Tunnel was constructed in 1914 as part of the Historic Columbia River Highway (HCRH) that connected Portland to Hood River, Oregon. The HCRH is located in the Columbia River Gorge (Gorge), a scenic location along the Oregon-Washington border adjacent to the Columbia River.

Figure 1 – Oneonta Tunnel Location about 30 miles east of Portland, Oregon (star)

The Oregon side of the Gorge is characterized by steep uplands mostly composed of volcanic rocks. Additionally, the section of HCRH near Oneonta Tunnel has one of the largest concentrations of high waterfalls in North America. The upland through which the tunnel was constructed is informally referred to as the *Oneonta Promontory* (OP).

The tunnel was closed to all vehicular traffic in 1948. Construction of US 30 in the 1950s and Interstate I-84 in the 1960s removed many sections of the HCRH and left many upland sections disconnected. Since passage of the 1986 Columbia River Gorge National Scenic Area Act, the Oregon Department of Transportation (ODOT) has worked to reconnect the abandoned sections of the HCRH as a pedestrian and cyclist trail.

The Eagle Creek Fire was a 50,000-acre destructive wildfire in the Gorge that started on September 2, 2017. The fire burned for three months before being declared completely contained. Because of fire damage, a six mile section of HCRH between Bridal Veil and Ainsworth State Park that included many popular scenic destinations such as Multnomah Falls and Oneonta Tunnel, remained closed until November 2018.

The purpose of this paper is twofold; first, to describe the geologic setting of the tunnel and outline how geologic variability affected restoration design; and second, to describe the original restoration effort in 2006-2009 and a current evaluation of the damage to the tunnel shotcrete and rock dowel system because of the 2017 Eagle Creek Fire.

GEOLOGIC SETTING

The Oneonta Tunnel is located in the Gorge portion of the Cascades Range physiographic province of Oregon. Rock in the walls of the Gorge include units of the Columbia River Basalt Group (CRBG), a series of Miocene (about 17 to 6 Ma) basalt flows that erupted from fissures and vents in northeastern Oregon, eastern Washington, and western Idaho. The Columbia River eroded through the basalt flows as the rock was tectonically uplifted. A series of Pleistocene catastrophic floods eroded the Gorge walls and left behind steep basalt cliffs. The local topography at the site is characterized by steep mountain slopes that predominantly descend to the north.

Figure 2 – Oneonta Promontory Showing Basalt Flows (1-5) and Tunnel

Oneonta Tunnel was constructed through the OP basalt outcrop at the mouth of Oneonta Creek where the Oneonta Creek Gorge meets the Columbia River Gorge. The HCRH crosses a bridge at the mouth of the Oneonta Gorge about 100 ft west of the tunnel. The roadway is situated at about elevation 45 ft and is about 15 ft above the bed of Oneonta Creek. Near vertical to overhanging basalt rock cliffs extend up to about elevation 300 ft south of the portals, and north of the portals along the northern terminus of the OP prior to reaching the HCRH.

The OP upland is composed of five separate flows of the Grande Ronde Member basalt (15 Ma) of the CRBG, designated from oldest (Flow 1) to youngest (Flow 5). The flow layer boundaries show an overall inclination to the west. Flows 2, 3, and 4 vary in thickness from about 45 to 90 ft. The bottom of Flow 1 is not exposed. The total thickness of the flows exposed at the OP is about 250 ft.

The flows exhibit varied geologic and volcanic structures, including large- and small-diameter columnar (vertical) cooling joints, flow banding, vesicles, and variable weathering. A set of near-vertical fractures with iron oxidation, weathering of mafic minerals, and secondary clay seams cross the west side of the tunnel. Following heavy precipitation these fractures transmit water to the tunnel crown.

ORIGINAL CONSTRUCTION

Oneonta Tunnel measures 115 ft in length, 20 ft in width, with a nominal vertical clearance of 19 ft. The tunnel is horseshoe-shaped in cross section. The tunnel was constructed using boring and mining techniques. Areas of weak rock were present in the center of the tunnel and resulted in local crown heights on the order of 25 to 30 ft. A timber liner was constructed to protect tunnel users from falling rocks inside the tunnel.

ODOT records show the S. P. White Company completed the tunnel during 1914 at a cost of \$6,684.88. This included \$4,140.00 for cutting the tunnel, \$523.05 for excavating the portals, and \$1,723.29 for 61,546 board feet of timbering. The tunnel did not open until after construction of the adjacent Oneonta Creek bridge in late 1914.

Figure 3 – Oneonta Tunnel East Portal, circa 1920.
Note railroad is located on current alignment of HCRH.

2006-2009 RESTORATION

Between 2006 and 2009, ODOT contracted for restoration of the tunnel. Restoration included installation of about 250, 10 ft long resin-grouted rock dowels in the tunnel, installation of mine straps in areas of jointed/weak rock, installation of strip drains, and construction of structural shotcrete liner and a non-structural wood liner. To address an area of particularly weak and weathered rock and thin roof cover associated with the Flow 1/Flow 2 contact, mine strap and a series of steel sets were installed near the East Portal. The tunnel was reopened in 2009 as a pedestrian-only feature.

Figure 4 – South Wall of Tunnel Showing Contact (Red Dots) Between Flow 1 and Flow 2. Green bar = 10 ft.

Figure 5 – Installation of Rock Dowels in Tunnel Crown

Figure 6 – Restored Tunnel with Wood Liner. View of East Portal.

2017 FIRE

In September 2017, the entire wood liner of the tunnel burned during the Eagle Creek Fire.

Figure 7 – Tunnel Wood Liner Burning.
Note fire-related rockfall debris in roadway.

2019 RESTORATION SHOTCRETE AND ROCK DOWEL EVALUATION

Shotcrete

As part of a second planned tunnel restoration, the post-fire performance of the tunnel support system was evaluated in February 2019 using a program of dowel tension testing, shotcrete coring and strength testing, and structural mapping. The shotcrete liner consisted of two layers: an inner layer with contact on the rock face of the tunnel, and an outer, rougher top coat or "cover" layer. The outer cover layer appeared to be locally damaged by spalling. The inner layer did not appear to be damaged by spalling. Two small (less than 1 sq ft in size) localized areas of the outer and inner layer appeared to be damaged by the fire, resulting in exposed sections of the strip drain. Some areas of the shotcrete were wet from tunnel seepage. Over the course of very cold weather in February 2019, icicles were noted due to seepage from the tunnel crown. Minor accumulations of ash, presumably from the burned timber liner, were locally observed on the shotcrete. Six cores of the shotcrete liner were drilled and collected. In general, the shotcrete cores were observed to be in good condition. The test results ranged from 4,374 to 6,070 psi with a corrected average of 5,080 psi. Each of the shotcrete cores tested close to or in excess of the tunnel design minimum compressive 28-day strength of 4,500 psi.

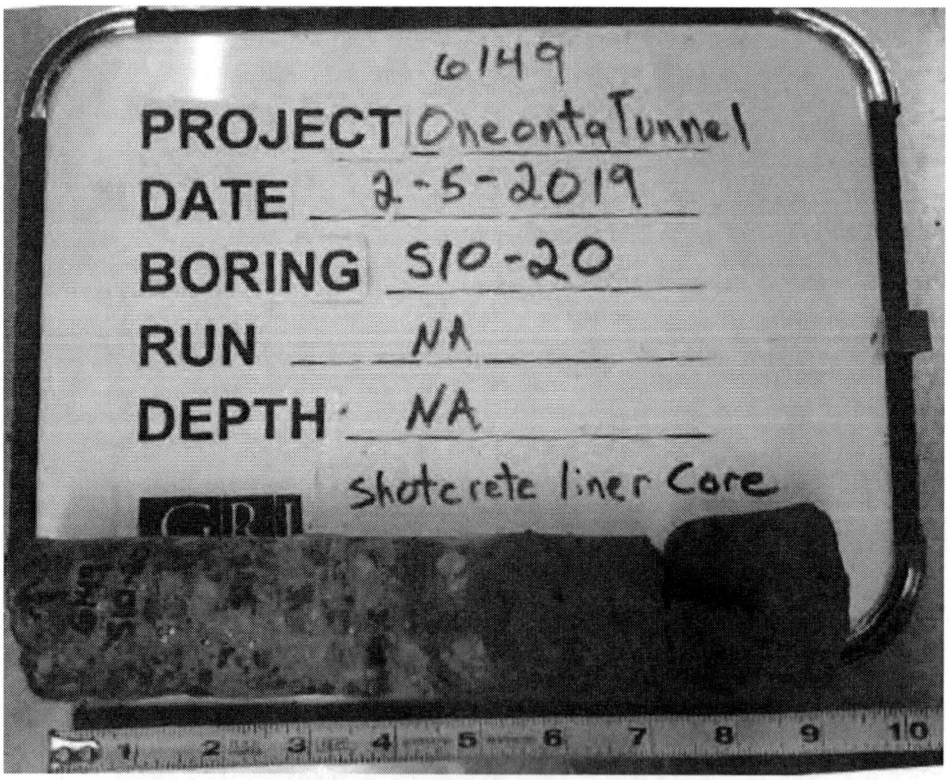

Figure 8 – Shotcrete core S10-20.
Note excellent bond of shotcrete (gray, left) to tunnel wall rock (brown, right).

Rock Dowels

In February 2019 seven rock dowels from the tunnel were exposed and tension-tested by a local specialty contractor. GRI personnel observed and documented the dowel evaluation and testing work. The dowels were tension-tested under a 30-kip load, which was the load applied to several test dowels during construction. Six of the dowels were located in the tunnel beneath shotcrete. One dowel (S110-20) was located outside the tunnel proper in the south wall of the East Portal and outside of an area of shotcrete cover.

Each dowel system (nuts, threaded bar, and backing plates) was observed for obvious evidence of fire-related damage. Where possible, the anchor resin in the tunnel wall rock space between the bar and the original drill hole was observed for fire-related damage.

Five of the seven dowels passed the 30-kip tension load test. One of the failed dowels (S10-10) was installed in the west portion of the tunnel in an area with numerous clayey seams and fractures noted during original construction. The other failed dowel (S110-20) was not covered by shotcrete and was noticeably damaged by the fire. Two other dowels also not covered with shotcrete were observed near failed dowel S110-20 and appeared damaged by the fire.

The remaining five dowels passed the load test and did not appear to have noticeable fire damage.

Figure 9– Dowel S50-15. Tension test = Pass.
Note green epoxy coating in good condition with no noticeable fire damage.

CONCLUSIONS

- Columbia River Basalt flow layering and secondary fracture structure in the Oneonta Promontory geology required design efforts to address geologic variability during the 2009 tunnel restoration. Flow boundaries, vertical fractures, and differential weathering required a system of steel sets, mine straps, rock dowels and shotcrete to achieve design goals.

- The 2009 restoration inner shotcrete system appeared relatively undamaged by the 2017 Eagle Creek fire. Final design for shotcrete repair is in progress at the time of this writing as part of the second tunnel restoration.

- The 2009 system of rock dowels in the tunnel appeared relatively undamaged by the 2017 Eagle Creek fire. As part of the planned second restoration, new rock dowels are planned in the area of the East Portal near dowel S110-20 which appeared to fail because of exposure to fire and high temperature.

- The shotcrete was effective in resisting fire-related damage to the rock dowel epoxy coating and resin grout.

The Characteristic Friction Angle, Its Determination and Use

Gary Norris
Emeritus Professor
University of Nevada Reno
1664 N. Virginia St, MS258
Reno, NV 89557
(775) 331-5582
norris@unr.edu

Prepared for the 70th Highway Geology Symposium, October, 2019

Disclaimer

Statements and views presented in this paper are strictly those of the author(s), and do not necessarily reflect positions held by their affiliations, the Highway Geology Symposium (HGS), or others acknowledged above. The mention of trade names for commercial products does not imply the approval or endorsement by HGS.

Copyright Notice

Copyright © 2019 Highway Geology Symposium (HGS)

All Rights Reserved. Printed in the United States of America. No part of this publication may be reproduced or copied in any form or by any means – graphic, electronic, or mechanical, including photocopying, taping, or information storage and retrieval systems – without prior written permission of the HGS. This excludes the original author(s).

ABSTRACT

The characteristic or phase transformation friction angle in soil mechanics is equally important as the constant volume, critical state friction angle in assessing stress-strain and volume change behavior of soil. It is the friction angle at which the soil reaches its maximum compressive volume change before beginning its dilatant turn-around. It occurs early on, prior to the peak friction angle, not post peak as with the constant volume friction angle. Hence it is of great use in establishing pre-peak stress strain behavior. With knowledge of it and the peak friction angle, drained and undrained stress strain behavior can be established. It supplies the basis for establishing the drained volume change response as well as the variation in Poisson's ratio with strain or stress level. The characteristic or phase change friction angle has been assessed from seven sands from four different natural environments exhibiting a wide range in particle size and natural variation in shape and surface roughness. Such characteristic friction angle data is characterized as a function of density state and confining pressure as well as its relation to peak friction angle. Its use in establishing volume change and Poisson's ratio with stain or stress level is demonstrated.

INTRODUCTION

In the drained triaxial test at constant confining pressure, the deviator stress (σ_d) is recorded versus the axial strain (ε_1) of the sample. The deviator stress normalized by the peak or failing value (σ_{df}) is the stress level (SL). However, the change in stress with increasing strain could as easily be characterized by the mobilized friction angle (ϕ_m), i.e.

$$\phi_m = \sin^{-1}[\,SL \cdot A / (SL \cdot A + 2)\,] \qquad SL = \sigma_d / \sigma_{df} \tag{1}$$

where

$$A = 2\sin\phi / (1 - \sin\phi) \qquad \phi = \text{peak friction angle} \tag{2}$$

Concurrent with the stress-strain response, the volumetric strain ($\varepsilon_v = \Delta V / V_o$) of the sample is also recorded. Such response is portrayed in Fig. 1. Of note are the characteristic or phase transformation friction angle (ϕ_c or ϕ_{PT}) and the constant volume or critical state friction angle (ϕ_{CV}). As shown, the characteristic friction angle occurs when the volumetric strain reaches its maximum compressive value. By contrast, the constant volume friction angle corresponds to the volume change flattening out and becoming constant post peak strength.

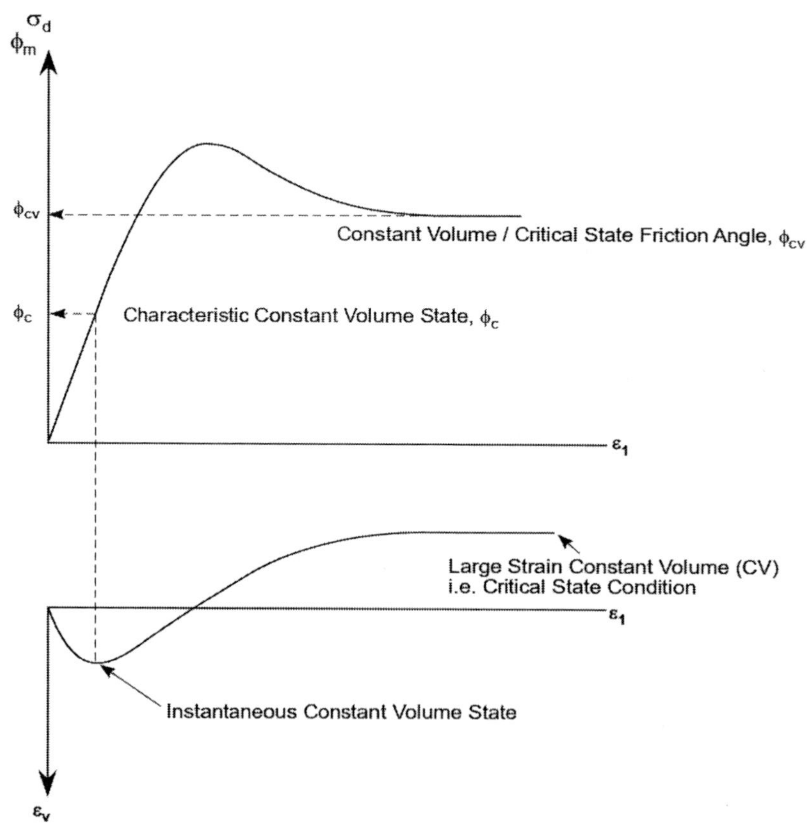

Figure 1 - Assessment of Characteristic (ϕ_c) and Constant Volume (ϕ_{cv}) Friction Angles of Sand (after Lade and Ibsen 1997)

Figure 2 shows how the characteristic friction angle is envisioned and how it changes with density as expressed in terms of porosity (n = volume of voids / total volume) or void ratio [e = volume of voids / volume of solids, which is equal to n / (1- n)]. The characteristic friction angle is that part of the peak friction angle (ϕ) less the volumetric dilatant component of strength ($\Delta\phi_f$). At the loosest state (the critical void ratio), the peak, the characteristic and the constant volume friction angles become one and the same. See Fig. 3.

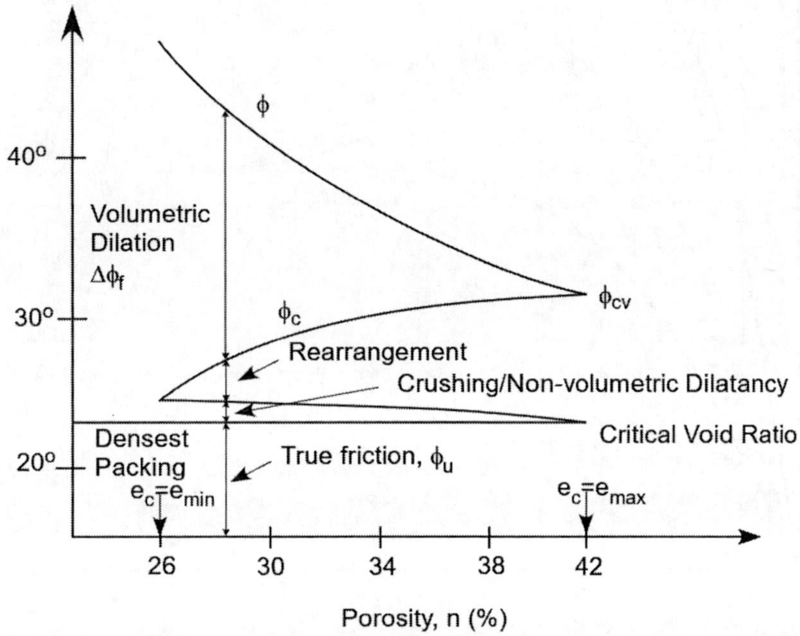

Figure 2 - Components of Peak Friction Angle (Modified from Mitchell 1993, after Rowe 1962)

As shown in Fig. 4, the characteristic friction angle becomes an integral factor in linking volume change behavior to developing stress as expressed in terms of the mobilized friction angle (ϕ_m) as given by the expression

$$d\varepsilon_v/d\varepsilon_1 = 1 - \phi_m/\phi_c \tag{3}$$

The integration of Eq. 3 leads to the volumetric strain

$$\varepsilon_v = \int (d\varepsilon_v / d\varepsilon_1) d\varepsilon_1$$

or

$$\varepsilon_v = \varepsilon_1 - \frac{1}{\phi_c} \int \phi_m d\varepsilon_1 \tag{4}$$

The volumetric strain, ε_v, at any ε_1 is the offset from a 1:1 line of ε_v versus ε_1 equal to the area under the ϕ_m versus ε_1 curve, divided by ϕ_c, up to the value of ε_1 in question.

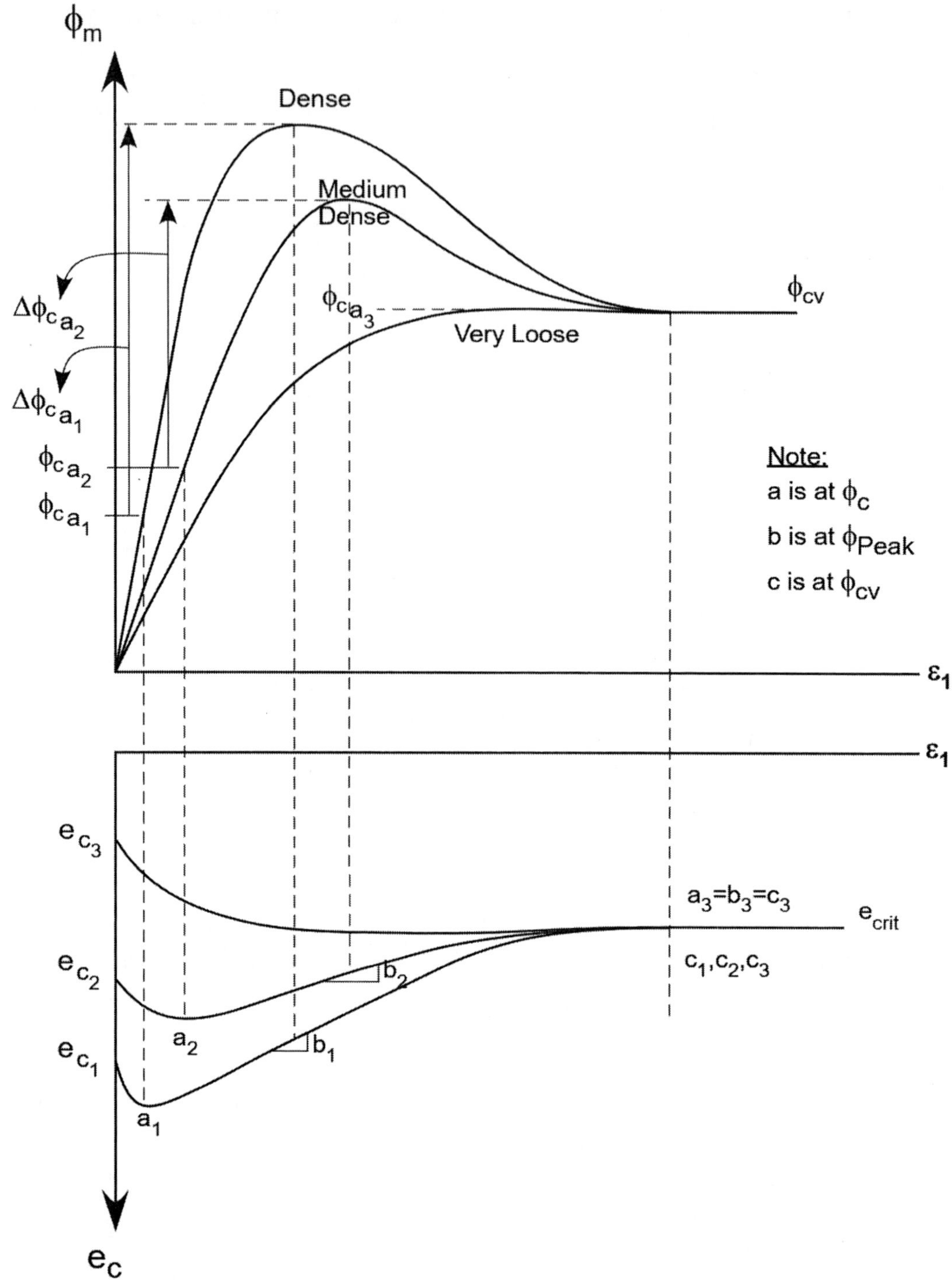

Figure 3 - Convergence of ϕ_c and ϕ to ϕ_{cv} at Very Loose Conditions (i.e. ϕ_{cv} as a special case of ϕ_c)

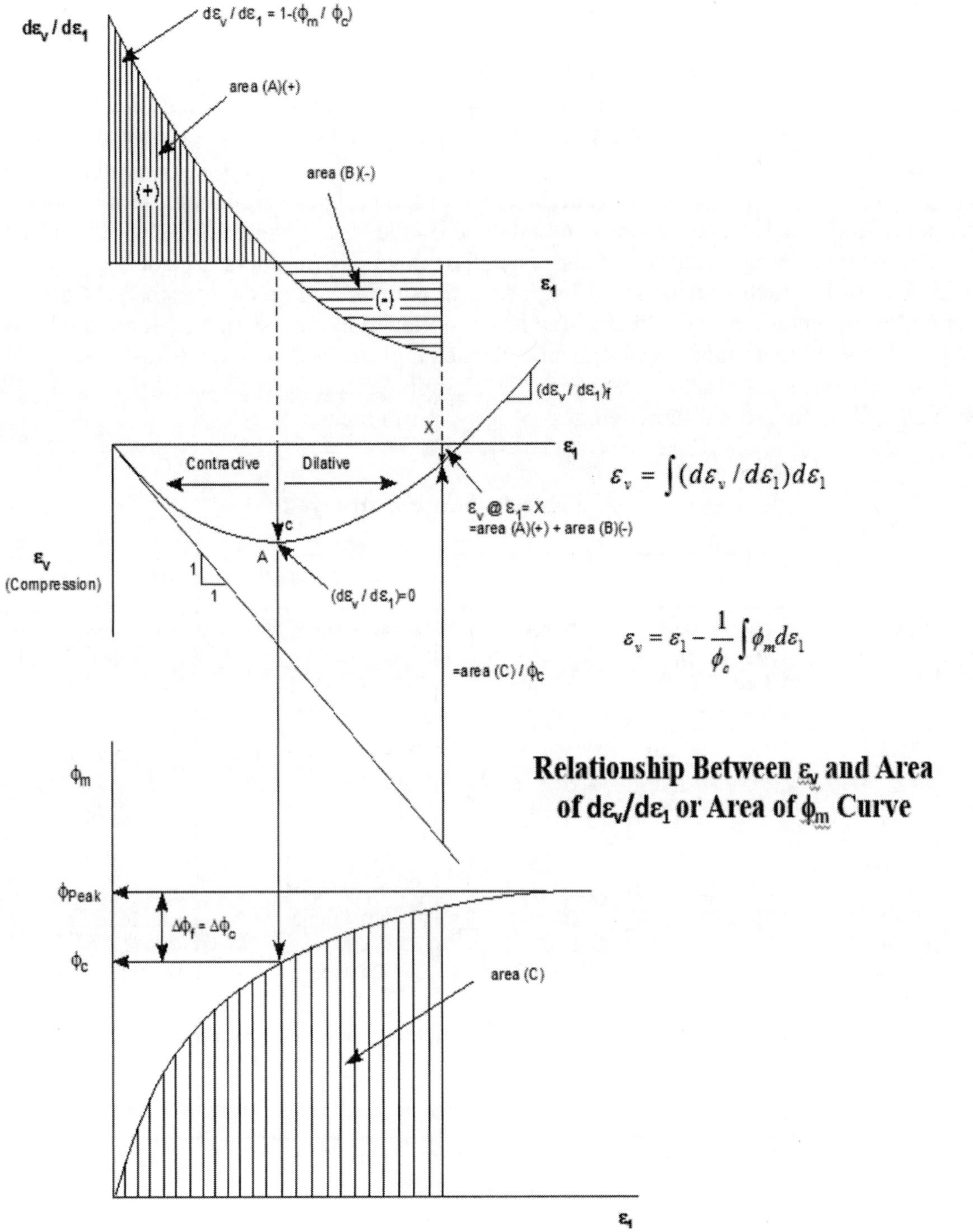

Figure 4 - Important Relationship Linking Stress-Strain and Volume Change Behavior: $d\varepsilon_v/d\varepsilon_1 = 1 - \phi_m/\phi_c$

MATERIALS EMPLOYED FOR THE EVALUATION OF THE CHARACTERISTIC FRICTION ANGLE

To evaluate the characteristic friction angle, data from 1:1 height to diameter ratio "frictionless cap and base" tests (Norris, 1981) taken from Norris (1977) were used. Table 1 provides information on the source of the natural sands tested, while Table 2 gives the values of the sphericity (ψ) roundness (ρ), surface roughness (sr), e_{max} and e_{min} (Norris, 1980) of the possible four different size fractions sieved from the original sands. The designation 7-8 indicates material passing the U.S. No. 7 sieve and retained on the No. 8, while 18-20, 45-50 and 120-140 have the same meanings relative to their respective sieve size numbers. The uniformity coefficient (C_u) of each fraction is 1.18 and each fraction is 2.83 size larger than the next smaller fraction. Furthermore, only the quartz fraction as obtained by heavy media separation was employed, and all materials were ultrasonic cleaned a prescribed time to clean particle surfaces of non-quartz coating. Mitchell, 1993 provides scanning electron microscope photographs of the particle surfaces from the four different geologic environments.

Table 1 - Sands Employed

Environment	Subtype	Locality[+]	Property Manager	Abbreviation
Beach	High Energy Wave Low Energy Wave	Felton Monterey Bay	PCA[++] (Felton Plant) PCA (Lapis Plant)	F for Felton M for Monterey
Eolian	Coastal Dune Desert Dune	Monterey Bay Kelso	PCA (Prattco Plant) Mr. Roy Parker	P for Prattco K for Kelso
River Channel /Primary		Ione Bear River	Owens-Illinois Joe Chevreaux	I for Ione BR for Bear River
Diagenetic		Le Seur, Minn	Gopher Silica	G for Gopher

[+] All localities are in California except for the diagenetic source.
[++] PCA is an abbreviation for Pacific Cement and Aggregate.

In all, eighteen size fractions were tested at three different confining pressures, 0.425, 1.20 and 3.40 kg/cm^2 (the same ratio between pressures as between size fractions). The record of the stress-strain-volume change response of the 144 tests undertaken as well as the details of the evaluation of shape and surface roughness parameters are given elsewhere (Norris, 1977).

Table 2 – Characteristics of Sands Tested

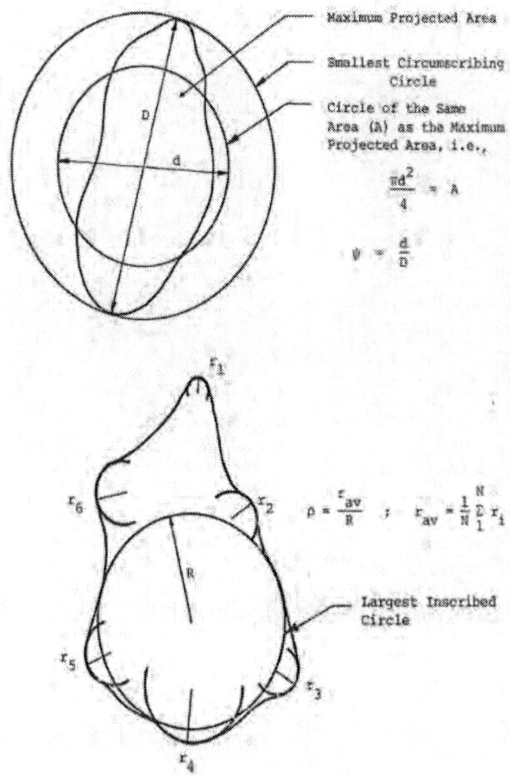

Material	ψ	ρ	sr*	e_{min}	e_{max}
7-8 M	0.835	0.612	0.03	0.489	0.720
F	0.831	0.496	0.03	0.540	0.820
BR	0.810	0.451	0.03	0.544	0.843
18-20 G	0.853	0.665	0.03	0.471	0.723
M	0.827	0.462	0.03	0.539	0.811
F	0.820	0.344	0.02	0.563	0.884
BR	0.806	0.287	0	0.588	0.888
40-50 G	0.845	0.636	0.03	0.486	0.733
K	0.821	0.454	0.04	0.566	0.805
P	0.796	0.379	0.04	0.596	0.892
F	0.806	0.289	0	0.654	1.002
I	0.800	0.291	0	0.705	1.049
120-140 G	0.826	0.559	0.01	0.539	0.799
K	0.775	0.461	0.04	0.638	0.894
P	0.762	0.332	0.04	0.680	1.012
I	0.780	0.313	0	0.733	1.104
BR	0.766	0.251	0	0.814	1.260
F	0.755	0.236	0	0.856	1.360

* See Norris et. al (2011) for definition of sr

EVALUATION

It was decided that the appropriate way to evaluate the characteristic friction angle was to employ Eqs. 3 and 4 to find that value of friction angle that yielded the best fit of the entire volume change curve of each test given the collected stress-strain and volume change data. Figure 5 is the Excel sheet from one test that indicates the numerical integration carried out to find the best fit value of ϕ_c (or ϕ_{PT}). The value of ϕ_{PT} (indicated in yellow) was varied until the best match of the calculated and recorded volume change curves was obtained. Figure 6 shows the type match for a test of loose sand and Fig. 7, that of the match for a test of a dense sand. Curves for plus and minus one degree from the best fit curve are shown superposed.

	A	B	C	D	E	F	G	H	I	J	K	L	M	N	O	P	Q
1																	
2					σ_3	1.2											
3					σ_{df}	3.62											
4					ϕ	37.0											
5					A	3.02											
6					ϕ_{PT}	29.0			0.821725								
7					ε_{50}(%)	0.821725											
8									from $\phi_{m,area}$	$(\varepsilon_{v,i+1}-\varepsilon_{v,i-1})$							Calc
9	%	%	%	ksc			(°)		%	$(\varepsilon_{1,i+1}-\varepsilon_{1,i-1})$	from col. J	from col. K		Recorded	Calculated		%
10	ε_1	ε_V	ε_V(-)	σ_d	SL	σ_d/σ_3	ϕ_m	$\phi_{m,area}$	ε_V	$d\varepsilon_V/d\varepsilon_1$	ϕ_m	σ_d		ν_{sec}	ν_{sec}	λ	ε_1
11	0.06029	0.06026	-0.06026	0.52063	0.144	0.434	10.268	0.310	0.050	1.000	0.014	0.001		0.000249	0.088521	3.19	0.063139
12	0.18087	0.13056	-0.13056	0.92064	0.254	0.767	16.096	1.899	0.115	0.814	5.382	0.248		0.139078	0.181026	3.19	0.168173
13	0.33159	0.28121	-0.28121	1.22085	0.337	1.017	19.705	4.597	0.173	0.666	9.676	0.485		0.075967	0.239026	3.19	0.303278
14	0.48232	0.33143	-0.33143	1.44621	0.400	1.205	22.087	7.747	0.215	0.250	21.754	1.413		0.156421	0.276916	3.19	0.452516
15	0.65314	0.36155	-0.36155	1.63051	0.450	1.359	23.862	11.671	0.251	0.088	26.443	1.927		0.223222	0.30809	3.19	0.616156
16	0.82395	0.36155	-0.36155	1.77387	0.490	1.478	25.150	15.857	0.277	0.019	28.442	2.183		0.2806	0.331813	3.19	0.776329
17	1.17564	0.3716	-0.3716	2.01273	0.556	1.677	27.137	25.051	0.312	0.000	29.000	2.258		0.341958	0.367393	3.163577	1.134359
18	1.54743	0.36155	-0.36155	2.19622	0.607	1.830	28.544	35.402	0.327	-0.039	30.129	2.419		0.383177	0.39445	3.101668	1.523448
19	1.94936	0.34147	-0.34147	2.34442	0.648	1.954	29.613	47.090	0.326	-0.076	31.201	2.580		0.412415	0.416492	3.02495	1.940768
20	2.34124	0.3013	-0.3013	2.45833	0.679	2.049	30.398	58.848	0.312	-0.110	32.181	2.735		0.435654	0.433372	2.949756	2.345146
21	2.77331	0.25108	-0.25108	2.56333	0.708	2.136	31.095	72.133	0.286	-0.126	32.652	2.812		0.454733	0.448443	2.867952	2.800563
22	3.61736	0.1406	-0.1406	2.73153	0.755	2.276	32.161	98.829	0.209	-0.142	33.116	2.890		0.480566	0.471045	2.711936	3.749239
23	4.47146	0.01004	-0.01004	2.85684	0.789	2.381	32.919	126.621	0.105	-0.176	34.108	3.064		0.498877	0.488235	2.575714	4.693904
24	6.06913	-0.29125	0.29125	3.08193	0.851	2.568	34.208	180.244	-0.146	-0.191	34.550	3.144		0.523994	0.512044	2.288159	7.177731
25	6.8328	-0.4419	0.4419	3.15754	0.872	2.631	34.622	206.526	-0.289	-0.182	34.287	3.096		0.532337	0.521132	2.17921	8.343083
26	7.6668	-0.5825	0.5825	3.23519	0.894	2.696	35.037	235.573	-0.456	-0.191	34.547	3.144		0.537988	0.529766	2.060854	9.787299
27	8.46061	-0.75324	0.75324	3.29291	0.910	2.744	35.340	263.506	-0.626	-0.204	34.905	3.210		0.544515	0.536984	1.968629	11.06359
28	9.29461	-0.91393	0.91393	3.3297	0.920	2.775	35.530	293.059	-0.811	-0.191	34.547	3.144		0.549165	0.543621	1.907957	11.98612
29	10.08842	-1.06458	1.06458	3.3669	0.930	2.806	35.721	321.339	-0.992	-0.197	34.721	3.176		0.552762	0.549176	1.845112	13.01947
30	10.82195	-1.21523	1.21523	3.41092	0.942	2.842	35.943	347.623	-1.165	-0.195	34.646	3.162		0.556147	0.553827	1.768803	14.39313
31	11.63585	-1.36587	1.36587	3.4504	0.953	2.875	36.141	376.957	-1.363	-0.189	34.469	3.129		0.558692	0.558555	1.698572	15.78725
32	12.41961	-1.51652	1.51652	3.47807	0.961	2.898	36.278	405.337	-1.558	-0.187	34.435	3.123		0.561053	0.562704	1.64834	16.87012
33	13.24357	-1.66717	1.66717	3.50859	0.969	2.924	36.428	435.290	-1.766	-0.151	33.365	2.933		0.562943	0.56669	1.591969	18.18017
34	14.08762	-1.7676	1.7676	3.54428	0.979	2.954	36.602	466.110	-1.985					0.562736	0.570457	1.524765	19.88829
35	17.5			3.62	1.000	3.017	36.965								0.5	1.3776	24.2959
36																	

Figure 5 - Excel Sheet of Numerical Integration of Test 77 Data

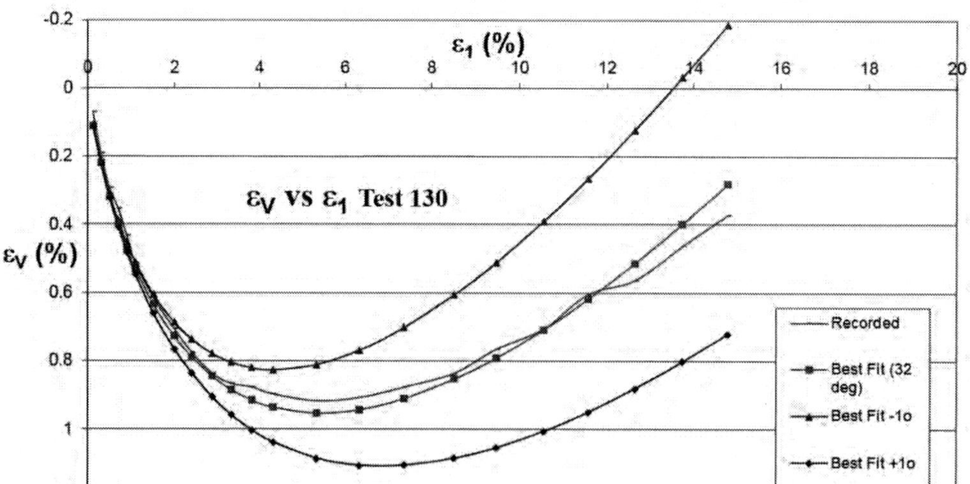

Figure 6 – Determination of Best Fit Value of ϕ_c of Loose Sand

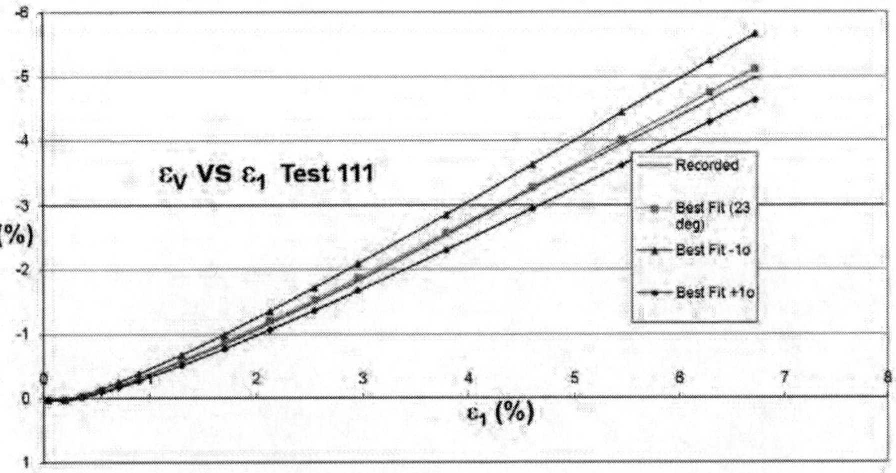

Figure 7 – Determination of Best Fit Value of ϕ_c of Dense Sand

RESULTS

If ϕ_m in Eq. 3 is set equal to the peak friction angle (ϕ), then the corresponding $\varepsilon_v/d\varepsilon_1$ at failure becomes the negative of the difference $\Delta\phi_f = \phi - \phi_c$ divided by ϕ_c. Figure 8 provides a comparison of the recorded vs predicted dilatant slope $(d\varepsilon_v/d\varepsilon_1)_f$ at failure from those tests at the confining pressure of 1.20 kg/cm^2. Similar results were obtained at the two other confining pressures.

The variation of the characteristic friction angle with density as expressed in terms of the void ratio of consolidation (e_c) relative to e_{min} is shown in Fig. 9. A simpler but more memorable relationship in terms of the porosity of consolidation (n_c) can be found in Fig. 10 for the 1.20 kg/cm^2 confining pressure. There is only a slight difference (in the fourth decimal place, i.e. 0.600x) for the other two confining pressures.

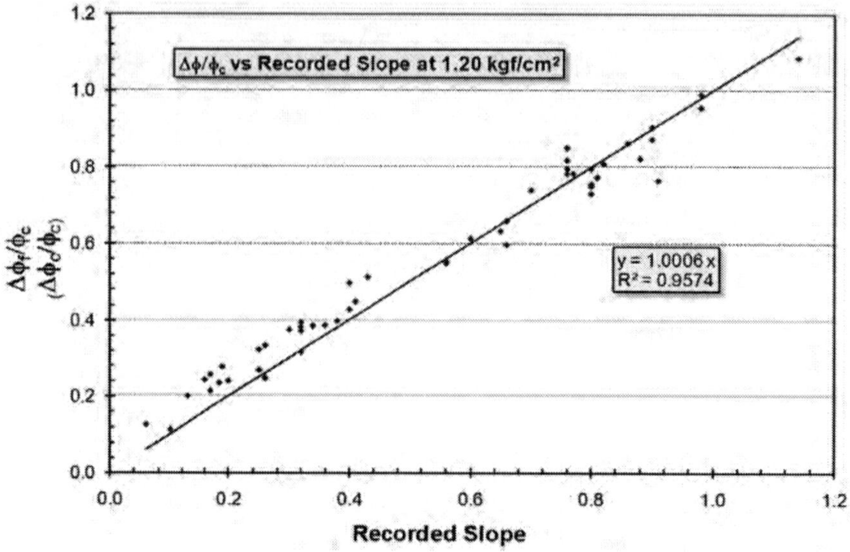

Figure 8 – Assessed (i.e.$\Delta\phi_f/\phi_c$) versus Recorded $(-d\varepsilon_v/d\varepsilon_1)_f$ Slope at Peak Resistance

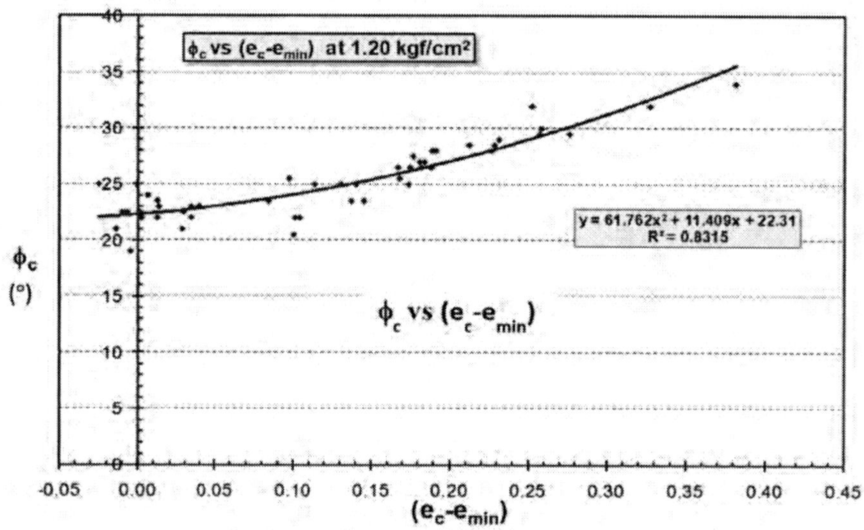

Figure 9 – ϕ_c as a Function of Density versus Recorded $(-d\varepsilon_v/d\varepsilon_1)_f$ Slope at Peak Resistance

Realize, however, Figs. 9 and 10 are for (virtually) the monosize sand fractions ($C_u = 1.18$) tested.

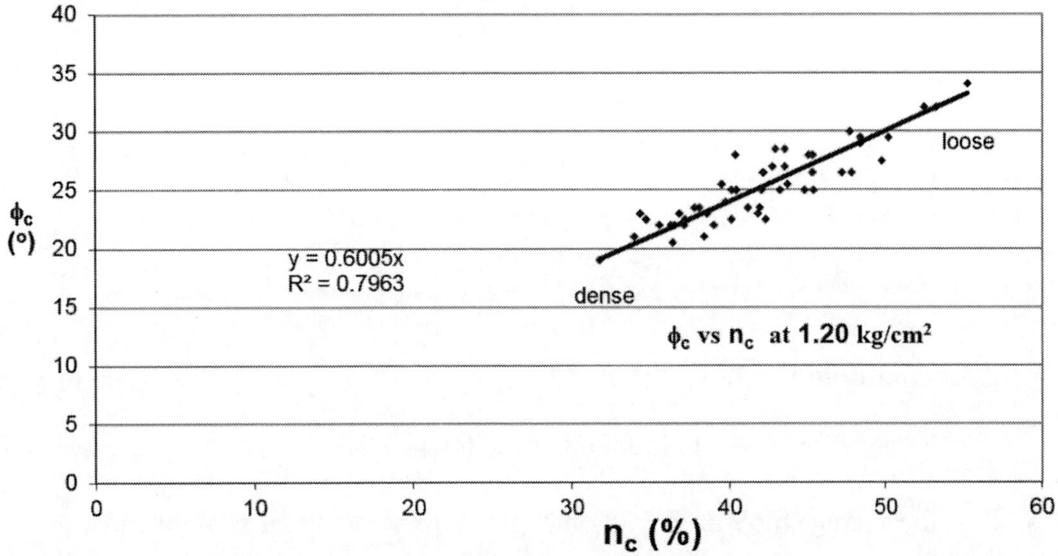

Figure 10 – A More Memorable Relationship

Figure 11 is a plot of the ratio of the peak to the characteristic friction angle versus density for tests at 1.20 kg/cm^2. The availability of this relationship suggests that knowing the peak friction angle, one might evaluate the characteristic friction angle knowing the density state, and then use the characteristic friction angle to predict the volume change behavior even for those tests where volume change was not recorded. From the drained response, the undrained response is then achievable (Ashour and Norris 1998).

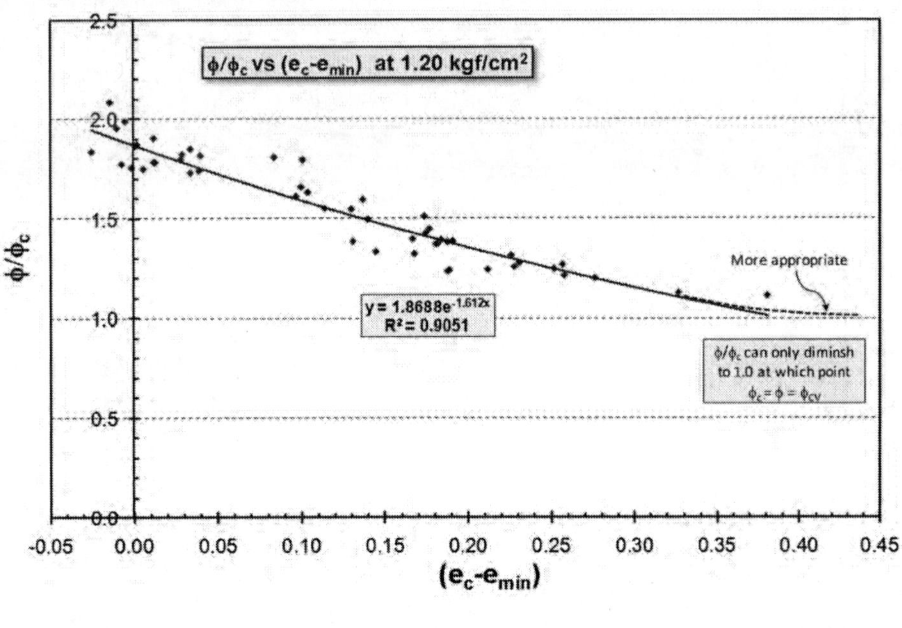

Figure 11 – Normalized Peak Frication Angle versus Density

The secant Poisson's ratios from the triaxial compression test (where $\varepsilon_v = \varepsilon_1 + 2\varepsilon_3$) is related to the characteristic friction as follows:

$$\nu_{sec} = -\frac{\varepsilon_3}{\varepsilon_1} = \frac{1}{2}\frac{(\varepsilon_1 - \varepsilon_v)}{\varepsilon_1} = \frac{1}{2}\left(1 - \frac{\varepsilon_v}{\varepsilon_1}\right) \quad (5)$$

Substituting for $\varepsilon_v/\varepsilon_1$ from Eq. 4,

$$\nu_{sec} = ½ \frac{1}{\varphi_c}(\int \varphi_m d\varepsilon_1)/\varepsilon_1 = ½ (\phi_{m,ave} / \phi_c) \quad (6)$$

where $\phi_{m,ave}$ is the average mobilized friction angle up to the value of ε_1 of interest.

In a similar manner, since $d\varepsilon_v = d\varepsilon_1 + 2d\varepsilon_3$

$$\nu_{tan} = ½(\phi_m / \phi_c) \quad (7)$$

Figure 5 shows the evaluation of ν_{sec} from Eq. 6 versus the recorded value [= ½ $(\varepsilon_1 - \varepsilon_v)/\varepsilon_1$] in the Excel sheet for Test 77. Figure 12 shows such variation as compared to the recorded variation with strain for Test 139.

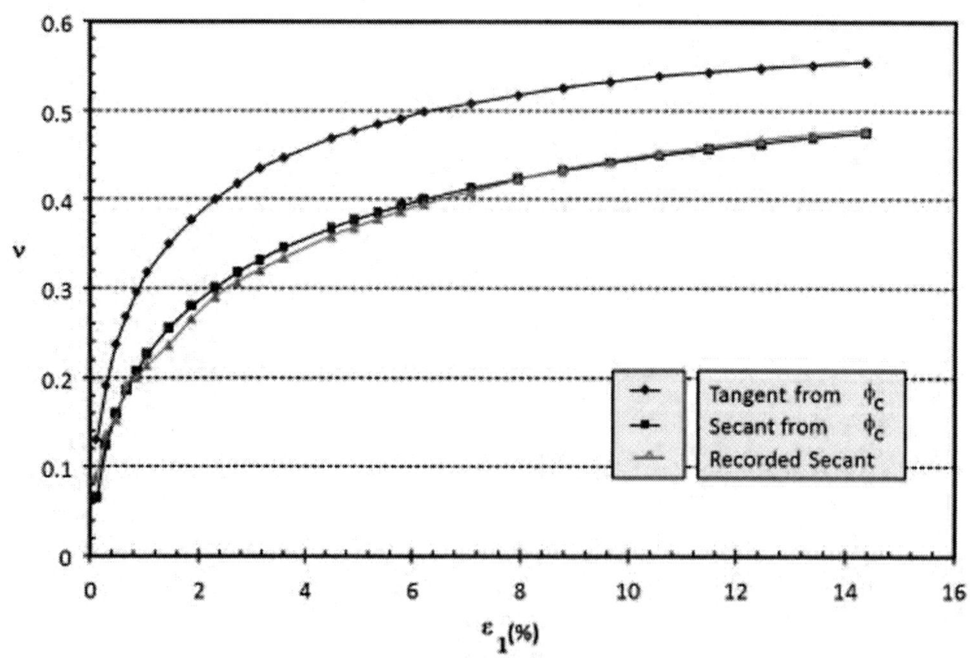

Figure 12 – Variation in Poisson's Rations (Tangent and Secant) with Strain, Test 139

SUMMARY AND CONCLUSIONS

The previous presentation has provided a relationship (Eqs. 3 and 4) linking the mobilized friction angle to volume change behavior via the characteristic (or phase change) friction angle as demonstrated relative to drained 'frictionless cap and base" triaxial tests on sands from different geologic environments yielding a variety of size fractions with a range in particle shape and surface roughness. The ability to establish the entire volume change curve from knowledge of the characteristic friction angle was shown. An approximate relationship for establishing the characteristic friction angle as a function of porosity (within reasonable bounds of error, independent of particle shape, surface roughness and confining pressure) was provided (Figs. 9 and 10). Likewise, the ratio of peak to characteristic friction angle with change in density was also shown (Fig. 11). This latter relationship provides the possibility of establishing the characteristic friction angle from the peak strength for the triaxial test where volume change was not recorded, e.g. on a partially saturated soil. Such value of characteristic friction angle might then be compared with that based on knowledge of porosity (Fig. 10) for confirmation. If there is agreement, the value obtained might then be used to establish the entire drained volume change curve, and from drained response, the undrained behavior is possible (Ashour and Norris 1998). While the aforementioned relationships are for virtually monosize fractions of sand (C_u = 1.18), a means for cataloging additional data for a variety of particle size distributions has been established.

As a second but still important issue, the relationship between the characteristic friction angle and Poisson's ratio was shown (Eq. 5 and Fig. 12). Knowledge of the variation of Poisson's ratio with stress level (or mobilized friction angle) allows assessment of shear strain and shear modulus from axial/normal strain and Young's modulus.

REFERENCES

Ashour, M. and Norris, G.M. (1998). "Formulation of Undrained Behavior of Saturated Sand from Drained Rebounded Response," Geotechnical Special Publications. Geotechnical Earthquake Engineering and Soil Dynamics III, Seattle, Washington, ASCE, Vol.1, No.75, pp. 361-372.

Lade, P.V. and Ibsen L.B. (1997). "A Study of the Phase Transformation and the Characteristic Lines of Sand Behavior," International Symposium on Deformation and Progressive Failure in Geomechanics, Asoaka and Adachi, Eds., Pergamon Press, Nagoya, Japan, pp 353-358.

Mitchell, J.K. (1993). "Fundamentals of Soil Behavior," Second Edition, John Wiley & Sons, Inc., N.Y., 1993, pp. 61-64, 372.

Norris, G. (1977). "The Drained Shear Strength of Uniform Quartz Sand as Related to Particle Size and Natural Variation in Particle Shape and Surface Roughness," Ph.D. Thesis, University of California, Berkeley.

Norris, G. (1980). "Shape and Surface Roughness Effects on Maximum and Minimum Void Ratios of Sand," Proceedings of the 18th Annual Engineering Geology and Soils Engineering Symposium, Boise, Idaho, pp. 187-197.

Norris, G. (1981). "Effect of End Membrane Thickness on the Strength of 'Frictionless' Cap and Base Tests," Laboratory Shear Strength of Soil, ASTM STP 740, Young and Townsend, Eds., Sept. 1981, pp 303-314.

Norris, G., Elfass, S. and Yang H-J. (2011). "Internal Friction as a Function of Surface Roughness," Proceedings of the 43rd Engineering Geology and Geotechnical Symposium, Las Vegas, Nevada, pp. 412-420.

Rowe, P.W. (1962). "Theoretical Meaning and Values of Deformation Parameters for Soil," The Strain Behavior of Soils, London: Foulis, pp. 143-194.

Blasting 2M+ Yards in Gneiss for a New Phoenix, AZ Freeway

Robert Cummings, P.E.
President, Saguaro GeoServices, Inc.
P.O. Box 44154
Tucson, AZ 85733-4154
(520)-321-4644
rcummings@saguarogeo.com

Prepared for the 70th Highway Geology Symposium, October, 2019

Acknowledgements

The author would like to thank the below-listed entities for their contributions in the work described:

The Arizona Department of Transportation
HDR, Inc.
Connect202 Partners

Disclaimer

Statements and views presented in this paper are strictly those of the author(s), and do not necessarily reflect positions held by their affiliations, the Highway Geology Symposium (HGS), or others acknowledged above. The mention of trade names for commercial products does not imply the approval or endorsement by HGS.

Copyright Notice

Copyright © 2019 Highway Geology Symposium (HGS)

All Rights Reserved. Printed in the United States of America. No part of this publication may be reproduced or copied in any form or by any means – graphic, electronic, or mechanical, including photocopying, taping, or information storage and retrieval systems – without prior written permission of the HGS. This excludes the original author(s).

ABSTRACT

The new South Mountain Freeway, part of the Arizona State Route 202 Loop, is much needed to alleviate traffic congestion along Interstate 10 through central Phoenix. A Design-Build-Maintain contract was procured through the P3 (Public-Private Partnership) framework, from Connect 202 Partners, a joint venture of Fluor Enterprises, Granite Construction, and Ames Construction with WSP/Parsons Brinckerhoff leading the design. The Arizona Department of Transportation, supported by its General Engineering Consultant (HDR, Inc.), has oversight, inspection and management responsibility for the $1.3B contract.

Saguaro GeoServices, Inc. (SGS) is the GEC team member responsible for review and oversight of the blasting. Along the southern (Pecos Segment) and Center Segment of the alignment, over 2.5 million cubic yards of Proterozoic granitic gneiss have been blasted and excavated, much of it near suburban houses, a Post Office, overhead power lines administered by the Salt River Project and Western Area Power Administration, buried water transmission mains and a pump station, sanitary and storm sewers, roadways in active use by traffic, and a Native American culturally significant site. Some 687 pre-blast surveys were performed at private residences and a casino.

Cuts ranged up to 144 ft in height. Lift heights ranged up to more than 40 ft but were most commonly 20 ft, and blast hole diameters ranged from 2.5 in. to 5.5 in. Adequate vibration and air overpressure control were paramount. Ground vibration and air overpressure (AO) were controlled by developing and updating the coefficients in predictive equations of the form

$$V \text{ (or AO)} = K^*(SD)^b$$

where V = predicted ground peak particle velocity (ppv, inches per second (ips)), AO = air overpressure (psi, converted to dBL), K = coefficient related to confinement, SD = scaled distance (physical distance in feet divided by the square root or cube root of the charge weight in lb per 8 ms delay period, $ft/lb^{1/2}$ or $ft/lb^{1/3}$ for ground vibration or AO, respectively, and b = coefficient representing the slope of the data points obtained when plotting the logarithm of the scaled distance against the logarithm of ppv measured by seismographs for each shot. For ground vibrations in construction blasting, the value of K is commonly close to 240 and b is near -1.62, but these coefficients were much different on this project, and were frequently updated based on collected data.

Conventional ppv criteria for surface structures are not applicable to buried pipelines. In order to address the impacts to buried pipelines, vertical attenuation relationships were developed by analyzing the data from buried seismographs. Fly rock control was attained by special blast hole stemming and, when crossing beneath overhead power lines, earthen pre- and post-covering were used. The entire project was successfully blasted using nonelectric methods of initiation.

INTRODUCTION

In southern Phoenix, AZ, Connect202 Partners ("C202P") recently completed major rock blasting and excavation operations on the Loop 202 South Mountain Freeway project, after producing over 2.5 million cubic yards of metamorphic rock in a setting complicated by dense, nearby housing development, public traffic that both crosses and parallels the new roadway, buried pipelines and a pump station for sewer and for potable water serving over 250,000 people, nearby overhead high-tension power lines, and Native American cultural resources. This paper will outline how careful implementation of conventional blasting techniques along with some novel instrumentation and monitoring techniques were successful in accomplishing this work, within the alternative delivery contracting framework known as Public-Private Partnership ("P3").

The South Mountain Freeway is Arizona's largest highway construction project ever, at a total cost of $1.77 billion (of which $917 million is design and construction). The completed facility will extend the Loop 202 system by adding 22 miles of new freeway around the southern side of Phoenix, connecting to Interstate 10 at each end (Figure 1).

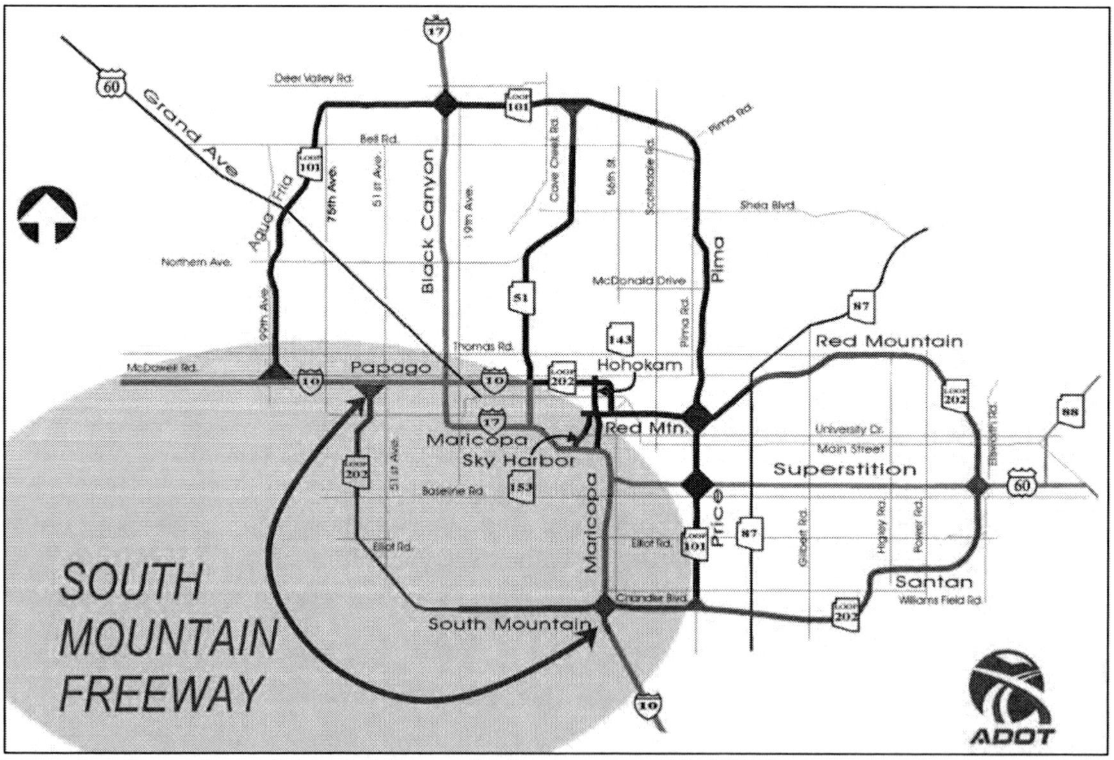

Figure 1 -- Location of South Mountain Freeway -- SR 202 Loop -- on the South Side of Phoenix, AZ

This new freeway will offer an alternative for I-10 through traffic, and bring much needed relief to the existing I-10 and I-17 system in central Phoenix. To expedite construction and save money over the project lifetime, the Arizona Department of Transportation (ADOT) elected to apply Arizona's new P3 statute to procure a Design-Build-Maintain (DBM) contract.

It is estimated that this method of procurement will shave 3 years and $100 million from the project.

C202P will design and build the facility, and then maintain it for a period of 30 years. C202P is a joint venture of Fluor Enterprises, Granite Construction, and Ames Construction. Ames Construction performed all the rock blasting and rock excavation. Along with Fluor, DBi Services will perform the maintenance during the operational period. The construction and operations team is supported by WSP/Parsons Brinckerhoff as the lead designer, with AZTEC Engineering Group, Stanley Consultants, and AMEC Foster Wheeler ("AFW", now Wood Group) in major engineering support roles.

On the Owner side, ADOT is assisted by HDR, Inc. as General Engineering Consultant (GEC). The GEC team is comprised of HDR core personnel and a number of specialty subconsultants, one of which is Saguaro GeoServices, Inc. (SGS) of Tucson, AZ. SGS joined the GEC in 2015 with responsibility for reviewing proposed and actual rock blasting practices relative to the Technical Provisions and the design documents. This includes review of various geotechnical submittals involving rock slopes, review of all blasting plans and blasting reports, evaluation of instrumentation and monitoring plans and reports relating to blasting, and carrying out regular field reviews during drilling and blasting activities and rock slope finishing.

Freeway Configuration

This project has been decades in the making and like most large highway projects, is not without controversy. The freeway concept dates to 1983, when it was introduced as the "Southwest Loop Highway". Environmental work and design concepts were approved in 1988, but in 1996 the project was shelved due to funding shortages. The pre-design process, including a full EIS, was resumed in 2001, and in 2004 Maricopa County voters approved a multimodal transportation system that provided a funding source. The Final EIS was released in 2014 and the Record of Decision was issued in 2015, clearing the way for the final procurement. Legal challenges were filed by various groups including the Gila River Indian Community (GRIC). These were eventually concluded after the 9th US Circuit Court of Appeals issued a ruling in 2017 that allowed the full project to move forward.

As now configured, the finished freeway will have 3 general-purpose lanes plus 1 HOV lane, in each direction. To reduce the footprint and therefore right-of-way, traffic is separated by a median barrier. At the crossroads, the design uses both conventional diamond interchanges and half-diverging diamond interchanges; the mainline passes over some crossroads and under others.

The roadway alignment was divided into four reaches, or Segments, according to the character of road and its surroundings. Rock excavation has only been required in the Pecos and Center Segments (Figure 2, next page).

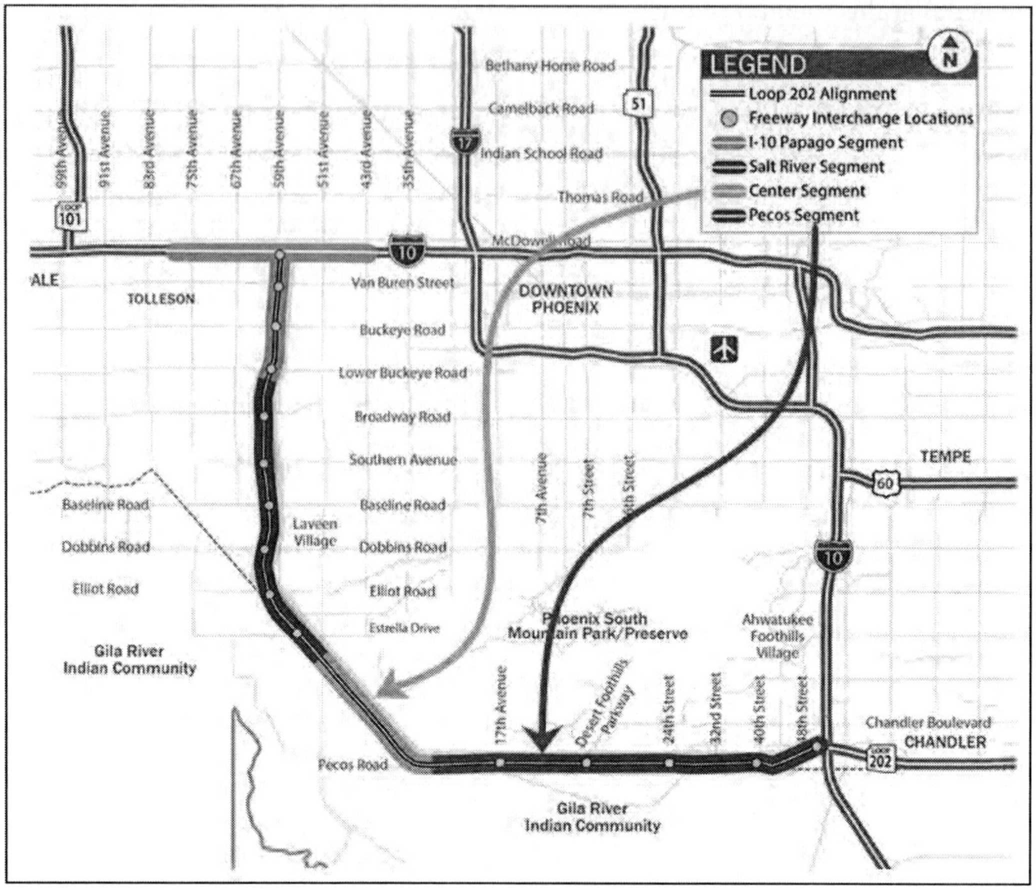

Figure 2 -- Location of Pecos and Center Segments

The Pecos Segment follows Pecos Road in the east-west direction, and is characterized by dense suburban development adjoining the right-of-way to the north, mostly undeveloped GRIC lands immediately to the south, overhead and buried utilities along and crossing the roadway, several significant crossroads and interchanges, and 5 rock cuts of significant height and length, as close as 165 ft from existing residences and much closer to utilities. Groundbreaking on the Pecos Segment was one of the earliest construction activities.

The Center Segment passes through a more rural setting. Only one housing development, currently under construction, is in proximity to the corridor. The Center Segment has several nearby Native American Traditional Cultural Properties (TCPs) however, a nearby casino, and scattered dwellings of the GRIC in its northern reaches. The main features of the Center segment are two major rock cuts, where the highway crosses through two southwest-trending ridges extending from the South Mountains to the valley floor. It is because of the TCPs and these rock cuts that Center Segment construction was delayed by lawsuits even after work on the endpoint segments had begun. One of the main objections from GRIC involved the rock cuts; GRIC asserted that the South Mountains are sacred lands and must remain whole. Ultimately ADOT and GRIC agreed that rock removed from those cuts must be used in such a way that it would remain physically connected to its source areas. The earthwork implications of this were solved by the design team. Center Segment rock has been used for rock fills extending out from the original cuts, and is currently being crushed as mineral aggregate and base course for pavement. Special studies and measures were put in place to protect TCPs, notably a site bearing petroglyphs within 650 ft of the south rock cut.

Geologic and Topographic Setting

The project lies in the Basin and Range Physiographic Province of Arizona. Extensional tectonics in Late Tertiary time resulted in a series of elongated uplifted mountain ranges separated by down dropped basins. The South Mountains represent such a block, although they are more east-west aligned.

The climate in the project area is very arid, averaging about 8 in. of rainfall per year. Most precipitation falls in the form of slow-moving winter storms or intense, short-lived monsoon-driven summer downpours that can be accompanied by dust storms and high-velocity wind outflows (microbursts). Temperatures range from average daily highs of near 106°F in June, July and August, to near 66°F in December and January. Vegetation is lower Sonoran desert scrub, characterized by cacti, creosote, low forbs, and sparse grasses.

The terrain is rolling to steep, becoming more rugged as one leaves the generally flat alluvial pediment and approaches the hillier uplands. Geologic materials exposed at the surface vary from thick alluvial deposits away from the mountain front to thin rocky soils and colluvium in the pediment areas, to rock outcrops with intermittent accumulations of thin soils in the uplands.

The alignment along the Pecos Segment runs east to west immediately south of the main block of the South Mountains. The Pecos Segment rock cuts are in a part of the southern foothills area that represents a pediment of the South Mountains. As noted above, within the Center Segment, the alignment turns to the north and crosses over two steep ridges. Further to the northwest, the alignment skirts the west flank of Alta Ridge, where a small amount of rock excavation is also required.

The geologic units exposed in the South Mountain bedrock block include an older (Precambrian) metamorphic core complex within which granitic gneiss, schistose interbeds, and irregularly distributed bodies of alaskite comprise the majority of the rock excavated. Reynolds (1985) has interpreted the South Mountain deformational history as follows:

- Precambrian deformation and development of crystalloblastic foliation;
- Middle Tertiary dilation associated with plutonism;
- Middle Tertiary mylonitization developing low angle foliation;
- Middle Tertiary fracturing, brecciation and detachment faulting;
- Middle Tertiary arching and antiform development; and
- Late Tertiary Basin and Range faulting and deformation.

The Precambrian deformation affecting the rock exposed in the project area resulted in folding, faulting, and development of a strongly foliated rock mass in a general NE-SW direction, roughly paralleling the direction of the two ridges in the Center segment. Mid-Tertiary plutonism resulted in bodies of granodiorite, granite, and alaskite intruding the Precambrian rocks. These Tertiary plutons and the Precambrian rocks were later intruded by northwest striking dikes, occurring mostly in the central portion of the mountain, which cut roughly 90 degrees to the NE-SW elongation of the South Mountain block. Mylonitization and low angle foliation of the rock fabric reportedly occurred in two more episodes after pluton emplacement and again after detachment faulting and arching (Reynolds, 1985). The western portions of the project contain almost exclusively gneiss; granitic occurrences noted in some early Center Segment geologic investigations are probably differentiates within the gneiss.

These multiple deformation events imposed a complex combination of faulting, fracturing, folding and mylonitization on the rock and resulted in a highly variable rock mass condition over relatively short distances (Figure 3).

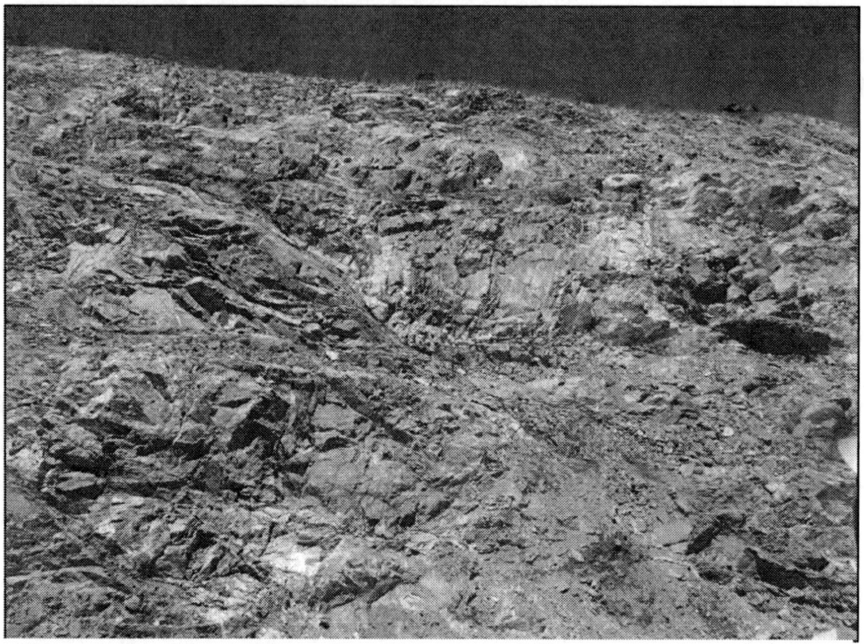

Figure 3 -- Wavy Complex Shears and Fractures Overlain on Foliation Patterns

Geotechnical Investigation

C202P carried out geotechnical investigations to supplement data collected decades before by Sergent, Hauskins, and Beckwith in support of the initial alignment studies. AFW used truck-mounted core drilling, seismic geophysics (both 24-channel conventional refraction, and passive refraction microtremor (ReMi) techniques), and surface mapping to assess rock mass condition and discontinuity patterns. Rock core was logged by hand. Rock fabric information was collected from surface exposures in both Segments. On the Pecos Segment, there were enough cut slope exposures for kinematic analysis. On the Center Segment, surface exposures of native outcrop were relatively sparse, so those measurements were supplemented by extensive discontinuity data from optical televiewer logging.

Discontinuity data were presented in stereonet form. The variability in rock mass fracturing complicated the interpretation of fracture trends, and correlation of trends from one place to another. Foliation and foliation shear zones were especially hard to pin down from rock core data. Stereonets usually displayed 3 or 4 discontinuity clusters, dominated by moderately-dipping but variable foliation, and including cross cutting joint sets and local shears. Because of foliation variability those trends varied widely. Clusters were not consistent from area to area, so for analysis the data were commonly aggregated through an entire cut.

Geophysical profiles showed an increase in seismic velocity downward from the surface, corresponding to a few feet of weathered or colluvial material having V_p (compression wave velocity) less than 2,000 ft/sec, underlain by a 5-15-ft-thick zone of weathered but in-place bedrock with V_p most commonly in the range of 4,500 to 6,500 ft/sec. Generally, V_p encountered below a depth of 15 ft approached the maximums of 9,500 to over 12,000 ft/sec.

Unconfined compressive strengths ranged from around 4,000 psi to over 20,000 psi. Global rotational shear mechanisms did not govern in stability analyses. For the large cuts C202P/WSP recommended maximum slope angles of 1:1 (H:V), chiefly on the basis of wedge and plane shear kinematics. In one location on the Center Segment, a ¾:1 slope was recommended near the crest to allow sufficient offset from a high-voltage transmission tower. Some interchange slopes were flatter due to the interchange design. Extensive rock fall modeling was performed to specify catchment widths and grades. Concrete protective barrier was included as part of the rock fall catchment.

A good example of the benefit of the P3 approach was the geotechnical evaluation of the Center segment. Construction on adjoining Segments was occurring, but for the Center Segment, scheduling of the geotechnical investigation had to await the resolution of the pertinent legal battles, after which all the access road building, core drilling and televiewer logging, geophysical surveys, laboratory testing, and data interpretation had to be completed before review and acceptance by the GEC and ADOT. Slope recommendations had been formulated, but the final geotechnical report was still in preparation when excavation had to begin.

Through intensive coordination among ADOT, the GEC, WSP, and C202P's management and construction engineering staff, it was decided to begin excavation according to the slope recommendations, at C202P's risk, subject to progressive geotechnical observations to evaluate the suitability of the slope recommendations. A program of regular slope inspections was developed involving C202P geotechnical and operations personnel as well as GEC representatives. A maximum lift height of 20 ft was recommended to allow the team to identify and remediate any adverse geotechnical conditions as they were exposed. As it turned out, some local slope modifications and improvements were required, but the overall slope configuration did not have to be changed.

PECOS SEGMENT

The Pecos segment contains 6 significant cut slopes, 5 of which required blasting (Cut 1 was ripped to grade). Figure 4 shows the distribution of the cuts.

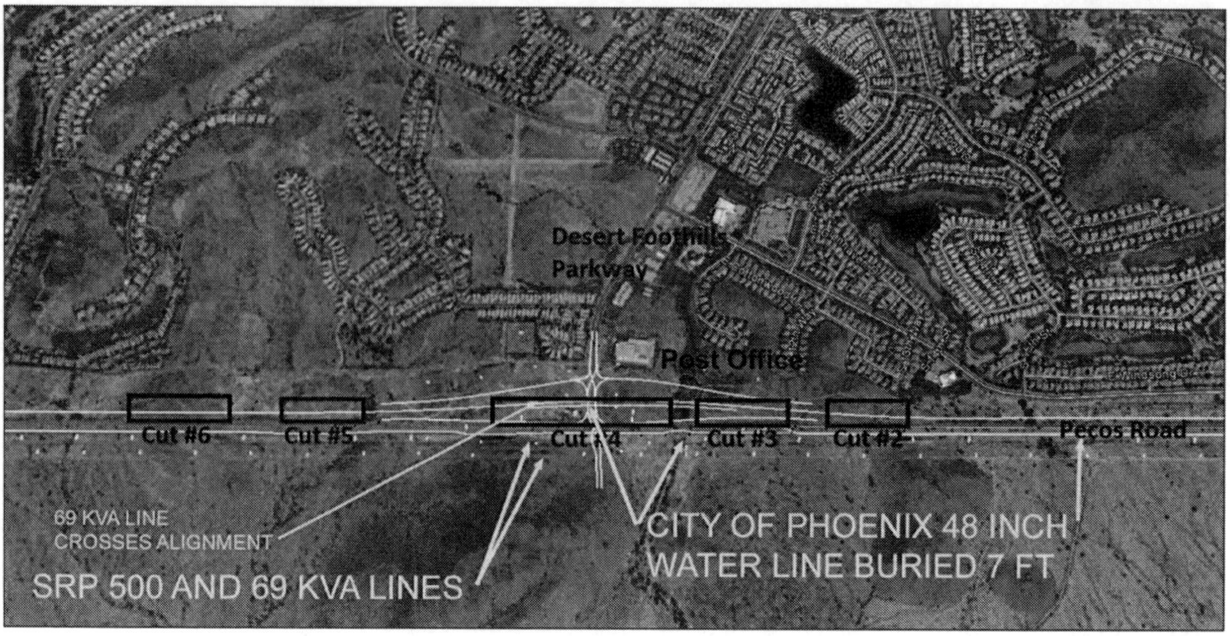

Figure 4 -- Pecos Segment Rock Cuts Requiring Blasting

Figure 4 shows that excavation was on the north side of existing Pecos Road, close to the existing subdivision developments. A major interchange is within the limits of Cut 4 at Desert Foothills Parkway. The mainline will pass beneath Desert Foothills Parkway in a long depressed section. Through most of the depressed roadway section, Cut 4 sides were ramp cross-slopes designed at 3H:1V and 4H:1V. The geometries of the rock cuts in Figure 4 are summarized in Table 1 below.

Table 1 – Pecos Segment Cut Slopes Requiring Blasting

Cut	Station Limits	Length, ft	Maximum Cut Height, ft	Cut Slope Angle(s) (H:V)
2	2256+00 – 2265+00	900	80	1:1
3	2268+00 - 2274+50	650	80	1:1
4	2277+50 – 2300+00	2,250	50	1:1, 3:1, 4:1
5	2308+00 – 2311+00	300	35	1:1
6	2317+00 – 2326+00	900	55	1:1

Pecos Blasting Constraints

As is apparent from Figure 4, rock excavation for the Pecos Road segment had to contend with nearby residential development, the Post Office at the intersection of Pecos and Desert Foothills Parkway, buried pipelines, a pump station, sequencing of excavation and detours on Desert Foothills Parkway, traffic control on roadways adjacent to the blasting areas, and overhead high-voltage power lines. Houses were as close as 160 ft from rock cut limits and less than 200 ft from the limits of large production blasts (Figure 5).

Figure 5 -- Location of Pecos Segment Mainline, Houses, and Post Office. Looking West from Cut 3 toward Cut 4.

Adjacent to the project, Pecos Road, a 4-lane facility, carries local traffic, some of which turns and uses Desert Foothills Parkway, a 4-lane collector street with signalized intersections. Traffic control in the form of blasting road closures, and protection of these roadways from fly rock, were also priorities.

The City of Phoenix Fire Department is responsible for administering blasting regulations that apply to residential and commercial structures. The City of Phoenix Water Services Department was concerned with the water and sewer systems and pump station. These agencies and their consultants reviewed each blasting plan for acceptability before implementation, and were present in the field for each blast when in the vicinity of significant infrastructure.

Houses also represented a major constraint to peak particle velocity and air overpressure produced by blasting. Masonry walls surround the residential developments and the Post Office. These presented the same vibration constraints as occupied structures.

Pipelines (domestic water and sewer) are buried just north of the cut limits and along Desert Foothills Parkway. The pump station is just west of Desert Foothills Parkway. The most critical pipeline was a 48-inch diameter water line, which will be discussed separately below.

Power lines pass directly over some blasts within Cut 4 and the main overhead power lines parallel Pecos road throughout the blast areas. These power lines (some of which are visible in the left side of Figure 5) are operated by Salt River Project (SRP), a large central

Arizona utility also responsible for hydropower generation from central Arizona dams and delivery of irrigation water. SRP's power lines crossing over the Pecos alignment are 69 kVa and those paralleling the roadway are 500 kVa. SRP imposed its own blasting requirements, reviewed blasting plans for approval, and had representatives present in the field for every blast near its facilities.

Table 2 gives relevant blasting vibration and air overpressure constraints imposed on the project. These were applicable to both the Pecos and Center Segments.

Table 2 – Blasting Constraints

Feature	Ground Vibration, ips	Air Overpressure, dBL	Agency
Residential and commercial buildings	1.0	129	City of Phoenix FD
Buried Pipelines (general) and pump station	2.0 for <20 Hz, 4.0 for >20 Hz	N/A	City of Phoenix
48-in.-diameter waterline	0.25 (revised to 0.39)	N/A	City of Phoenix
Overhead Power (monopole and lattice tower)	See Figure 6	133	Salt River Project
Buried Utilities	2.0 for <20 Hz, 4.0 for >20 Hz	N/A	Salt River Project

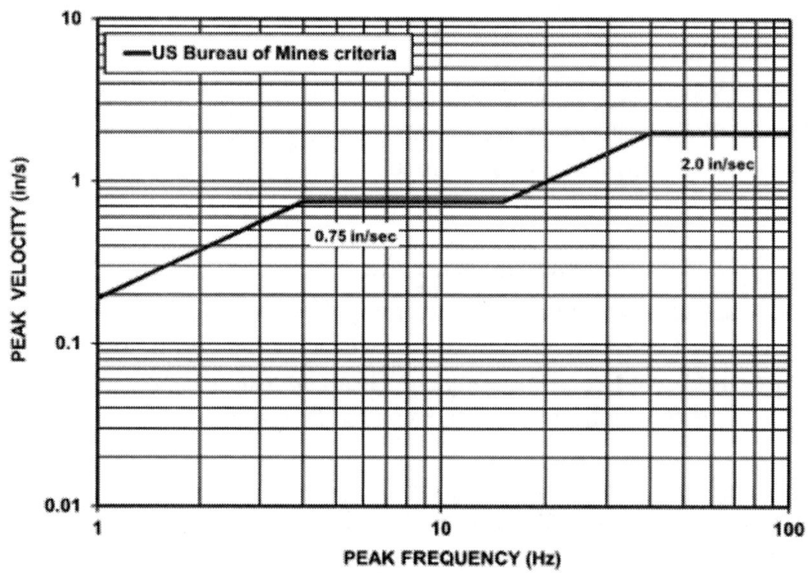

Figure 6 -- USBM Criteria for Above-Ground SRP facilities

These kinds of constraints are common when blasting in urban areas and for this project they combined to complicate the blast designs. Because different portions of blasts were different distances away from features representing different constraints, blast designs were done in great detail and had to consider which of the several surrounding features would control the blast loading and timing configuration and at what location within the blast. It was crucial that approved blast designs be followed in the field.

In addition to the above, fly rock control was critical in all cases. SRP imposed a requirement to cover any blast within 50 ft laterally of the outer conductor of any overhead power line, and reserved the right to extend the cover requirement based on blast configuration or prior fly rock occurrences. The City of Phoenix Fire Department also requires covering blasts whenever fly rock could pose a hazard but does not specify a distance. The project Technical Provisions required fly rock prevention as discussed later in this paper.

CENTER SEGMENT

The Center Segment contains 2 significant cut slopes (Figure 7), and a third much shallower cut slope to the northwest ("Alta Ridge" area, not shown on Figure 7). The North and South cuts were removed almost entirely by blasting; about half the height of the Alta Ridge cut slope and most of its ends were removed by ripping.

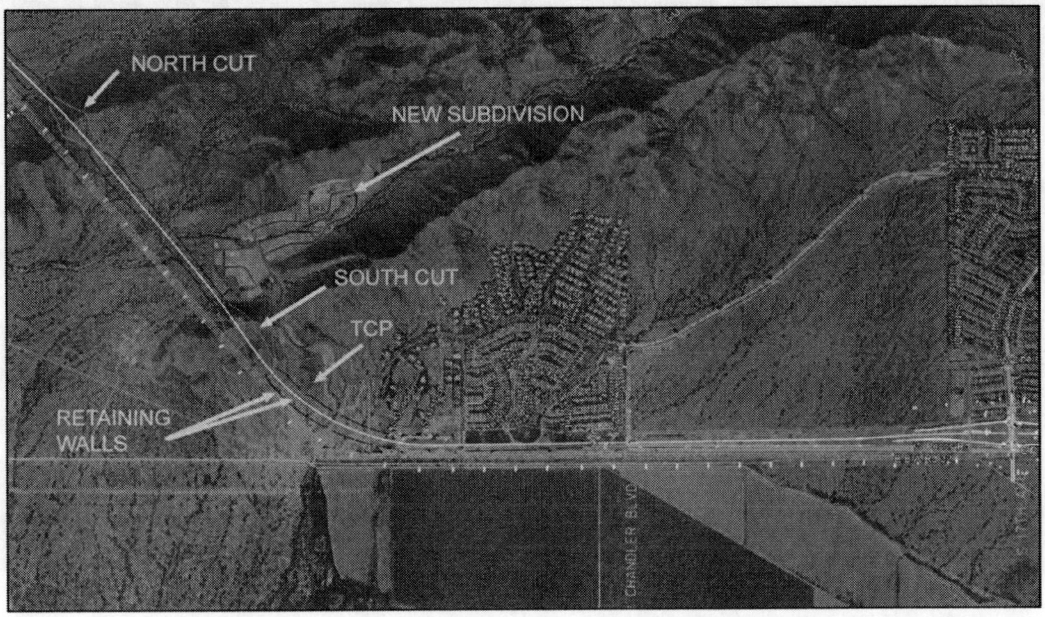

Figure 7 -- Center Segment Major Rock Cuts

Figure 7 also shows other notable surface features. The TCP noted is a petroglyph site that will be discussed in the next subsection. A summary of the cut slope geometries is given in Table 3.

Table 3 – Center Segment Cut Slopes Requiring Blasting

Cut	Station Limits	Length, ft	Maximum Cut Height, ft	Cut Slope Angle(s) (H:V)
South	2507+20 - 2520+40	1,320	140	1:1 RT, 1:1 and ¾:1 LT
North	2268+00 - 2274+50	1,160	170	1:1 both sides
Alta Ridge	2966+00 – 2971+70	570	24	½:1

Center Segment Blasting Constraints

As is apparent from Figure 7, rock excavation for the Center Segment was in an area of more rural character. Of significant concern were the nearby transmission towers. SRP Tower 84 was at the summit of the South Cut, a little less than 100 ft from the west cut slope crest. There was a similar tower near the crest of the North Cut that was a little further away. The subdivision, the petroglyph site, and the 48-in. diameter water line merited consideration but were of subordinate impact. Figure 8 shows this arrangement in more detail.

In general, the particle velocity and air overpressure limitations for the features around the Center segment were the same as for the Pecos Segment (Table 2).

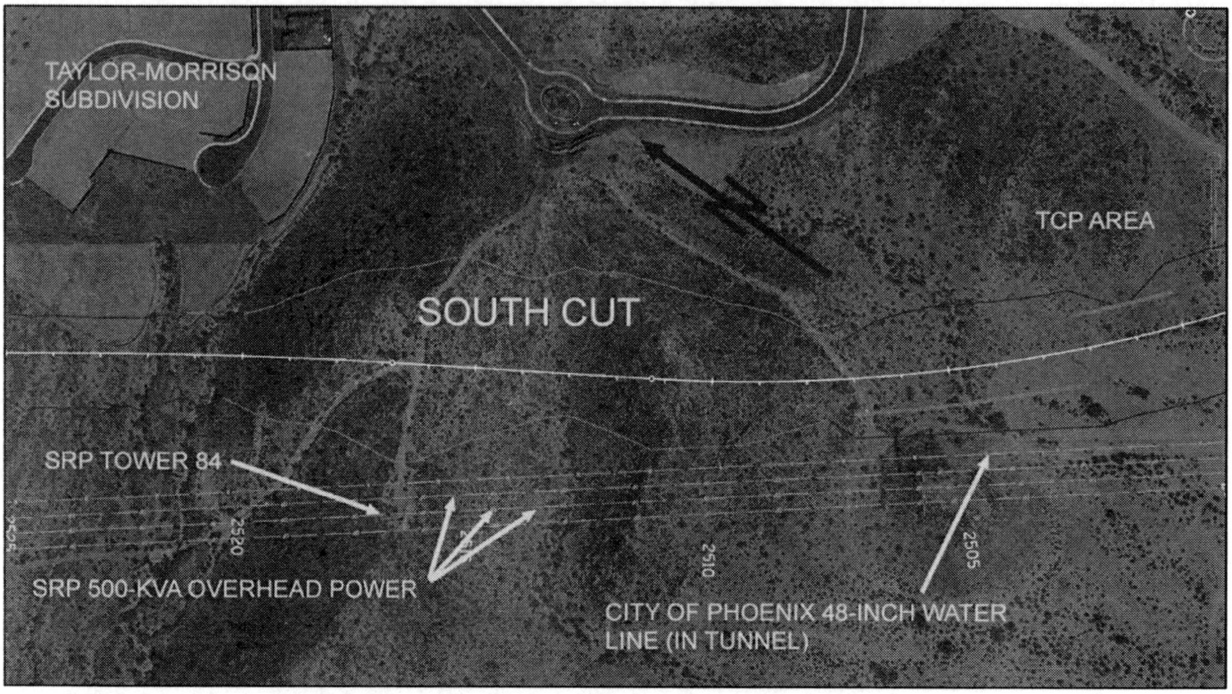

Figure 8 -- South Cut, Center Segment Features

Petroglyph Site

The petroglyph site (TCP area in Figures 7 and 8) is of great importance to the GRIC, and prior to construction an assessment of its stability was required. The site is around 650 ft from the closest point of the South Cut, and further away from areas of the South Cut where significant blast energy would occur. This distance was too far for a credible threat of direct fly rock impact. A field inspection carried out by SGS in 2016 showed that many of the petroglyphs had been produced on thin exfoliation slabs. There was concern for disturbance of those. SGS performed a waveform analysis for likely blasting excitation and found that the differential strains across the rock faces containing the exfoliation slabs would be exceedingly small, so there was considered to be near-zero likelihood of blast-induced loosening or fracturing of the slabs. The most plausible source of disturbance was found to be a single undercut boulder, itself unmarked by petroglyphs, but which if dislodged could result in disturbance of adjoining, marked boulders. The unmarked boulder was supported partly on a narrow ledge of rock to one side (Figure 9) and partly by native ground at its opposite rear corner.

Figure 9 -- Design-Basis Rock Block at the Petroglyph Site

SGS then performed a pseudostatic analysis to see if potential blast-related acceleration could crush or dislodge this support point. The contact area was measured and the crushing strength of the supporting rock was assumed to be around 4,500 psi. It was conservatively assumed that, during blasting excitation, lateral translational movements could reduce side friction against the adjoining blocks to zero, allowing freer vertical movement of the rock block. Assuming that blasting could induce vertical transient acceleration of the rock and thereby increase the dead weight force imposed on the supporting rock ledge, it was recommended that the blasting accelerations should be limited to less than 0.13g. It was found that the allowable charge weight per delay associated with this acceleration would be far greater than the allowable charge weight to meet the ppv limits at the transmission towers, which were far closer to the blasts.

In 2018, blasting was slated to begin in the Center Segment. By that time a consultant had been engaged specifically to oversee the preservation of the various TCP occurrences on behalf of the Tribe, and that consultant placed a semipermanent blasting seismograph at the petroglyph site. After a pre-blasting baseline monitoring period, the ground motions were monitored relative to a triggering value of 0.1 ips. In addition, a baseline LIDAR scan was obtained, and differential point cloud analyses were conducted periodically during blasting in the South Cut. Monitoring continued through January of 2019, and although ppv levels in excess of 0.1 ips were experienced on one occasion, no differential movements of the petroglyph blocks were ever recorded.

48-in. Diameter Water Line

The 48-inch diameter water line passes through the two ridges in a tunnel of circular cross section. The water line passes within about 200 ft below and to the west side of the South Cut (Figure 10, next page). According to the as-builts (dated 2001), the tunnel was supported around the crown with rock bolts, the pipeline was positioned within it on "timber sleepers" (wood blocking), and the space outside the tunnel was backfilled with low-density cellular concrete.

Before the Center Segment blasting commenced, there had been months of blasting near this pipeline on the Pecos segment. Prior to Center Segment excavation, the City had inspected the pipeline within this Segment and it appeared to be in good condition. Nonetheless, because this section of the pipeline was to remain in service, and because it is not easily repairable, special consideration was warranted. A vertical attenuation seismograph array was placed at the

South Cut margin prior to test blasting, to collect data that could aid in assessing the influence of blasting should any degradation in the pipeline occur during construction or in the future.

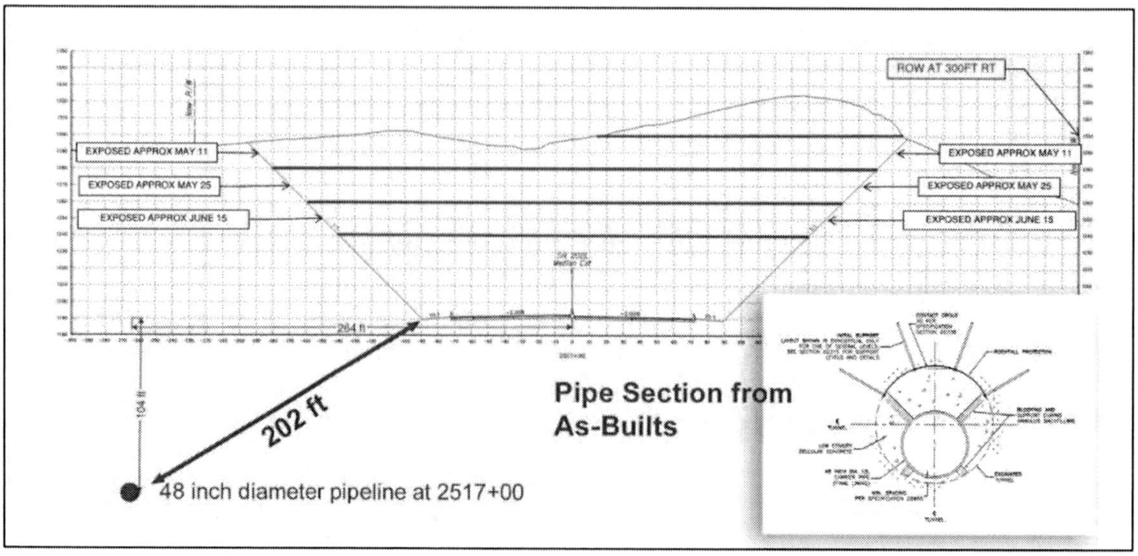

Figure 10 -- South Cut Typical Section with Pipeline Location and Cross Section (Inset)

ROCK EXCAVATION AND BLASTING APPROACH

Technical Provisions (TPs)

On this project, Technical Provisions (TPs) play the role that Standard Specifications and Special Provisions do on conventionally-procured design-bid-build projects. The TPs for blasting were tailored to this project and bore many similarities to Special Provisions on conventionally procured contracts, but the key differences had to do with avoiding prescriptive approaches and not resorting to reliance on convention. In a DBM contract, the Developer (in this case C202P) is encouraged to exert maximum flexibility and ability to innovate. In so doing, the Developer may weigh the risk and cost benefits of various approaches, some of which may not be conventional. For blasting, notable TP requirements included the following:

- Designation and Owner approval of a Rock Engineer/Blasting Professional (Arizona registrant with at least 10 years of relevant experience), Blasting Supervisors (minimum 10 years experience), and Blasters in Charge (7 years experience)
- Prevent or remove deleterious drill hole traces and machine scars (defined as scars more than 3 ft long, or traces/scars that total more than 25% of the length of drill holes used to form the faces, or machine scars traceable for more than 12 ft).
- Prevent fly rock when within 300 ft of any transmission line. Prevention measures include covering blasts with mats or soil. (Based on blast performance, the Developer's stemming practices were accepted in lieu of covering, except where SRP imposed more stringent requirements.)
- A Blast Monitoring Plan detailing the sensors to be used and their deployment, the method of analyzing data, the qualifications of technicians, and means of

adjusting blasting practices based on results. Also required were the methods of documenting the control of fly rock, and plans for coordination with the public.

- Seismic monitoring (ground vibrations and AO) whenever blasting occurs within 500 ft of any building, box culvert, retaining wall, bridge, pipeline, utility pole, or transmission tower.
- A Blasting Information Report which details blasting personnel, blasting approaches, blasting products, drilling and loading equipment, and operational information connected with blasting such as patterns, hole diameters and angles, lift heights, and so on.
- A Test Blasting program for each cut location where blasting is proposed. (Test blasts occurred in Pecos Segment Cuts 4, 5, and 6; and the South Ridge cut on Segment B)
- Pre-blast surveys to a distance of 2,500 ft. In all, 687 surveys were completed.
- Provisions of detailed blasting plans for each shot including pattern details, scaled distances, seismograph locations, and predicted ground vibrations and AO for each.
- Blasting Reports for each shot including notation of drilling issues, loading changes, cutoffs, and evaluations of seismograph measurements.
- Videos of each blast from two different locations.

Although not part of the TPs for blasting, the TPs elsewhere defined the requirements for a detailed public information program.

Blasting Patterns and Procedures

All blasting on the project was carried out with ANFO as the primary blasting agent. Bulk blasting agent was delivered in 50-lb bags on the Pecos Segment because of the smaller patterns. On the Center Segment, bulk ANFO trucks were chiefly used. Occasionally waterproof, packaged blasting agent was required. (This was more common on the Center Segment than on the Pecos Segment, possibly because of the amount of water that had been applied to remove loose material from the larger slopes there.) Primers varied; in smaller blast holes high-velocity emulsion was used, and in larger blast holes, cast boosters. The contractor began with delivered product and transitioned to a magazine located within the site perimeter as soon as permits for magazines and an ANFO silo could be obtained.

Electronic initiation was considered, but the decision was made to stay with conventional nonelectric initiation as a matter of Contractor preference. Detonators were typically 25/500 combination "Handi-dets" (25 ms surface delay, 500 ms downhole delay). Surface delays were typically 17 ms and 42 ms NTDs. Timing patterns were center pull whenever possible – a very common approach was to zigzag 42 ms surface delays up the center of the pattern and thereby control the charge weight per delay. Where the shot was wide and the proximity of sensitive features required eliminating redundant firing, additional time was added to the center pull; in some cases row timing was as slow as 134 ms. Normal production timing was such that the blast progressed parallel to centerline, and lifts were removed in slices across the cut area. Side control was generally excellent; in order to reduce the potential for high spots between shots using center pull timing, "looked" side relievers were occasionally used.

Blast hole diameters ranged from 2.5 in. for a few sinking cuts and bridge foundation shots, to 5.5 in. for the larger, deeper blasts on the Center Segment. Slopes were step-drilled in all cases, using reduced burden and spacing, reduced hole diameter, and staying at or 2 ft above slope grade, to reduce slope damage. On the Pecos Segment, holes on the slopes were typically 3 in. diameter, and on the Center Segment, 3.5 in. and 4 in. Patterns varied widely, typically falling within the ranges shown in Table 4. Production hole spacings were typically equal to production hole burdens. Where the pattern extended onto the slopes, the spacing of the more lightly loaded slope holes was reduced but the burden remained the same so that detonator lead lines could be kept straight down the row. For the Center Segment, the 15 ft patterns used for the 5.5 inch production holes (Table 4) would not work for the 4 in.-diameter slope holes, so an additional slope hole row was drilled from the toe outward, cutting the burden in half.

TABLE 4 – Blast Design Summary

	Hole Depths, ft			Production Hole Burden, ft			Powder factor, lb/cy		
Area	Avg	Min	Max	Avg	Min	Max	Avg	Min	Max
Pecos	20	7	29	8	2	11	1	0.7	4.0
Center	31	18	47	13	7	15	1	0.7	1.2

Subdrill was typically 30% of the burden except in areas where the cut graded out to daylight and it was necessary to overdrill to gain sufficient blast hole depth for confinement. Excessive subdrill was avoided as it wastes money, increases vibrations, and thickens the disturbed rock zone beneath structures and pavements. Stemming thickness typically equaled or was slightly less than the burden. Imported stemming comprised of crushed fine rock controlled stemming ejection, especially important on the Pecos Segment. In a few cases where there was no relief, sinking cut methods were used, employing an open center hole and closer-spaced, lightly-loaded burn holes arranged around the open hole. One instance of this produced the maximum powder factor of 4.0 lb/cy shown in Table 4, which was a confined sinking shot employing 2.5 in. diameter holes and small patterns to reduce subdrill. A more representative maximum powder factor for Pecos was close to 1.3 lb/cy.

Ground Vibration and Air Overpressure Control

Ground vibration (V) and air overpressure (AO) were controlled by developing and updating the coefficients in predictive attenuation equations of the form

$$V \text{ (or AO)} = K*(SD)^b$$

where V = predicted ground peak particle velocity (ppv, inches per second (ips)), AO = air overpressure (psi, converted to dBL), K = coefficient related to blast confinement, SD = scaled distance (physical distance in feet divided by the (i) square root or (ii) cube root of the charge weight in lb per 8 ms delay period, (i) $ft/lb^{1/2}$ or (ii) $ft/lb^{1/3}$ for (i) ground vibration or (ii) AO, respectively; and b = coefficient representing the slope of the data points obtained when plotting the logarithms of the scaled distances against the logarithms of the seismograph measurements for each shot. For ground vibrations in construction blasting, the value of K is commonly near 240 and b is near -1.62, but these attenuation coefficients were much different on this project, and were frequently updated.

Except for Test Blasts where seismograph arrays were employed, seismographs were deployed as required by the TPs, generally at the closest house, pipeline, and transmission towers. Generally, 4 to 6 seismographs were required for each blast on the Pecos Segment and 2 to 3 on the Center Segment. Wherever possible, sensors were buried. For the Center Segment, semipermanent remote stations with solar panels were needed at the tower foundations above the tall cuts. Vandalism of these semipermanent stations did occur on a few occasions.

Attenuation Equations

The initial attenuation coefficients were developed using linear arrays of multiple seismographs in the Test Blast phase. For the first Test Blast, 5 seismographs were aligned past the end of the blast at various distances. Because wave transmissions in the foliated rock were expected to be anisotropic, another array of 5 machines was aligned perpendicular to the roadway in the direction of nearby houses. Other compliance machines were placed at the nearest transmission poles and other features. The Test Blast data were extensive enough to develop attenuation parameters and perform a regression analysis for the mean, 95%, and upper bound envelopes, for both ground vibration and AO. The upper bound coefficients were initially used for blast design. Later, values representing the 95th percentile were calculated.

As ongoing production data were obtained, the attenuation coefficients were updated and new coefficients were obtained for each new cut. The attenuation coefficients that evolved were markedly different from the generalized values given above. For example, by the time the last few shots in Cut 4 were performed, the equation for ground vibration used $K = 11.2$ (50th percentile) and $K = 18$ (upper bound); $b = -0.911$ for both. The AO equation used $K = 164.5$ and $b = -0.08$. This difference in coefficients illustrates that wherever possible, vibration control for blasting should develop and use site-specific values.

Initially, ppv predictions were conservative, because the Developer's procedure was to use the maximum charge weight per delay and shortest distance in all scaled distance calculations. This is convenient, and common practice in bench blasting when blast hole depths do not vary much, but problems began to emerge when predicting vibrations for features close to grade-out points (where the cut daylights, or near the crests of slopes). Because of the terrain it was not uncommon on this project to have blast hole depths ranging from 7 ft to over 20 ft within the same shot. Assigning the shot's maximum charge weight to a closer (lower physical distance) but shorter hole will result in a calculated scaled distance that is too low. Using this scaled distance when comparing to actual field measurements will give the false presumption that the measured value was achieved at a low scaled distance. The measured ppv or AO theoretically comes from the blast hole(s) representing the minimum scaled distance. This may be from the closest hole, even if it is shallow, but that hole will not have the scaled distance calculated using the maximum charge weight. The resulting regression analysis will tend to under-predict the blasting effect at the probability level involved.

Early in the project, blast designs could be based on the upper bound values for scaled distances calculated through this practice. Later, however, as the work approached Desert Foothills Parkway and houses and pipelines were very close, this became unworkable for many shot designs. Blast designs needed to use scaled distances calculated using the charge weight that would present the minimum scaled distance with respect to the feature in question. For many blasts, multiple combinations of charge weight and physical distance had to be checked.

Vertical Attenuation Studies

It was known at the outset that there were many pipelines that could be impacted by blasting on this project and that ground deformations at the buried pipeline depth would need to be evaluated. Because it is impractical to routinely measure ppv on or near a buried structure that is subject to changing blast locations, and because it is known that that blasting-related ppv at depth will generally be lower than the blasting-induced surface ground motions at the same locations, C202P agreed to measure the vertical blast vibration attenuation at typical pipeline burial depths.

It is known that surface waves are responsible for most of the ppv measured in blasting, but that these are effective for only 1-2 wavelengths below the surface. Deeper buried features are impacted by body waves, to a relative degree that depends on the depth of burial.

C202P's vibration and monitoring consultant Aimone-Martin Associates (AMA) carried out vertical ground motion attenuation measurements. The first Test Blast on the Pecos Segment produced the first vertical attenuation data. This blast was in Cut 6, well away from the pipeline, on September 11, 2017. A seismograph sensor was placed and backfilled at the bottom of a 4-ft-deep hole drilled at a distance that would be similar to the smallest blast-pipeline separation distances to be encountered later. A paired surface sensor was placed at the same location. Similar tests were performed on 4 more occasions between September 11 and September 26, 2017. The peak particle velocity amplitudes at 4 ft averaged 50.7% of the surface amplitudes, for all components of motion, consistent with other studies (Aimone-Martin Associates, 2017). Additional vertical attenuation tests were conducted. In all, 39 tests were performed on the Pecos Segment, reflecting scaled distances ranging from 24 ft/lb$^{1/2}$ to 117 ft/lb$^{1/2}$. The ratio evolved, but at 7 ft depth the overall average ppv was about 70% of the surface ppv.

For the Center Segment test blasting, AMA installed a vertical attenuation array. The site of the array was a dozer pioneer road on the flanks of the hill just outside the west cut slope, at an elevation well below the starting elevation of blasting. The array consisted of a 5 inch diameter hole outfitted with sensors at the surface and at depths of 10, 20, 30, and 40 ft. To control the orientation of the sensors, they were taped to PVC pipe, one pipe for the sensors at 10 and 20 ft, and the other pipe for the sensors at 30 and 40 ft. The hole was backfilled with sand, the intent being to recover the sensors at the conclusion of testing by blowing out the sand. Data were collected for 14 blasts in the South Cut, representing scaled distances ranging from 7 ft/lb$^{1/2}$ to 40 ft/lb$^{1/2}$. Vertical attenuation was indicated as follows: at 10 ft depth the ppv was 56% of surface ppv, 41% at 20 ft, 37% at 30 ft, and 34% at 40 ft, indicating a progressive drop in surface wave influence with depth (Meins and Aimone-Martin, 2019).

The 48-inch Water Line

The chief constraint in many blast designs proved to be the 48-inch-diameter reinforced concrete potable water line operated by the City of Phoenix (see Table 2 and Figure 4). This pipe ran south down Desert Foothills Parkway and then turned east, following Pecos Road. The pipeline was scheduled for relocation, but the freeway construction schedule required that it remain in service for the most of the duration of rock excavation.

In the vicinity of the Pecos Road blasting, the water line was typically at a depth of 7 ft or less, and therefore subject to excitation by surface as well as body waves from blasting.

After blasting had begun in the fall of 2017, the City of Phoenix disclosed that its nondestructive testing had indicated breakage of hoop prestressing strands in a 24-ft-long section of pipe elsewhere on the project. The extent to which this condition might exist near the blast areas was not exactly known. The City noted that concrete in PCC pipe should be considered to have low, if not zero, tensile strength in the absence of effective reinforcement, and that PCC pipe failures with severe consequences were known to have occurred elsewhere. Because the pipe was the sole water source for Ahwatukee, the City imposed a very conservative ppv limit of 0.25 ips at the pipeline.

This presented challenges when designing blasts at the physical distances required. C202P's consultant cited studies of buried pipelines to show that pipe strains induced by elastic ground waves tend to be very low, but the City questioned the relevance of these studies, some of which were on welded steel pipelines. C202P's consultant presented an estimate of pipeline longitudinal, circumferential, and bending strains, for assumed blast-induced elastic particle displacements under a range of surface ppv, using an assumed ground motion frequency of 36 Hz, and the measured vertical attenuation factor at pipeline depth. For a surface ppv of 0.417 ips, giving rise to 0.25 ips at the -7 ft depth of burial, the peak particle displacements can be calculated to be around 0.001 in. However, the pipeline strain analysis was subject to unresolved questions about the magnitude of pipeline internal pressures (which were known to fluctuate), pipeline stiffness and in situ concrete strength properties, and the degree of coupling of the pipe to the substrate (soil-structure interaction).

As a result of further discussions and vertical attenuation studies, the City relaxed the maximum allowable ppv at the pipeline to 0.39 ips. Vertical attenuation studies at that location indicated that the vibrations at pipeline level (depth of 7 ft) would not be more than about 0.6 times the surface ppv. Therefore, the surface ppv criterion for blast design was revised to a more workable upper bound of 0.65 ips (0.39/0.6).

Fly Rock Control

Fly rock control is always a critical matter when blasting in urban areas. Special stemming and extra care when loading near vertical faces or fissures from previous blasting were the first line of defense against fly rock.

SRP required covering shots when blasting within 50 ft laterally of its power lines, and this situation occurred where the 69 kVa lines crossed the alignment at Cut #4. Earthen cover was used rather than blasting mats, not just because of the number of mats that would be required, but because there was insufficient vertical clearance below the power lines to safely handle and place mats. The first blasts used post-covering, which is more reliable than pre-covering (and then drilling blast holes through the cover material) because it provides a more continuous blanket and precludes stemming ejection. However post-covering with the required 3 ft of earthen cover proved unworkable for area blasting because the low-headroom equipment required could not reach the interior of the shot after it was tied in. For post-covering, the blast had to be tied in and covered in stages using equipment lightweight enough that it would not disturb the buried lead lines. With the approval of SRP, subsequent blasts were pre-covered with 3 ft of earthen material, and conventional stemming practices were used. This was proved sufficient (Figure 11).

Figure 11 -- Effect of Pre-Covering within 50 ft of 69 kVa Overhead Power Lines (Left Half of Shot) and Uncovered Blast Holes Outside 50-ft Limit (Right Half)

CLOSURE

Blasting for the SR 202 South Mountain Freeway was completed in July 2019, and is an example of how a team approach can surmount significant technical challenges while using relatively traditional blasting techniques. The keys to success were experienced personnel, thorough planning, flexibility and skill to adapt to changing requirements, and good communication. The DBM contract structure permitted the Developer to evaluate and allocate risk according to the conditions as they were revealed during excavation. The document flow and lines of communication that were established early in the project allowed all stakeholders a thorough understanding of the issues and the ability to focus knowledge to solve problems in an appropriate and timely matter.

REFERENCES:

Aimone-Martin Associates, Inc. *Vertical Attenuation Report : Update.* Internal Memorandum to P. Ekstrom, M. Trembly, and B. Kite, C202P, September 26, 2018, 3 p.

Meins, B, and Aimone-Martin, C., *Vertical Attenuation of Blast Vibrations.* Proceedings of the 45th Annual Conference on Explosives and Blasting Technique, Nashville, TN: International Society of Explosives Engineers, 2019, 10 p.

Reynolds, Stephen, J.,. *Geology of the South Mountains, Central Arizona.* Arizona Bureau of Geology and Mineral Technology, Geological Survey Branch Bulletin 195, 1985.

Road Re-Opening After Landslides: The Contribution of Remote Sensing

Alessandro Brunetti [1], Paolo Mazzanti [1,2], Serena Moretto [1], Alfredo Rocca [1], Saverio Romeo[3], Chiara Varone[1]

(1) NHAZCA S.r.l., spin-off of "Sapienza" University of Rome, Via Vittorio Bachelet 12, Rome, 00185, Italy

(2) Sapienza University of Rome, Department of Earth Sciences, Rome, Italy;

(3) CERI, Centro di Ricerca di Previsione, Prevenzione e Controllo dei Rischi Geologici, Rome, 00185, Italy

ABSTRACT

This paper highlights the potential and the contribution of remote sensing techniques to support the reopening strategies of transportation corridors after landslide events.
The PRIMO (PRompt Intervention MOnitoring) protocol designed and implemented by NHAZCA (Natural HAZards Control and Assessment) will be introduced, focusing on its application for the safe reopening of an Italian Regional Route (SP45, Subiaco-Jenne-Vallepietra) that was involved in a landslide on November 26, 2018.
Specifically, for the rapid and safe reopening of the Regional road, based on site monitoring of movement and monitoring of ongoing safe conditions during roadway operation, the following activities were carried out:

1) multi-temporal Terrestrial Laser Scanning and historical Satellite InSAR analysis (knowledge monitoring);
2) Risk Assessment;
3) Continuous Terrestrial InSAR monitoring for alert purposes (emergency monitoring);
4) Residual Risk Assessment;
5) Ground-based PhotoMonitoring™ to detect potential displacements and localized rock-falls able to generate risk conditions for the infrastructure (control monitoring).

The monitoring and consultancy activities carried out by NHAZCA supported the decision making for the characterization of the landslide phenomenon, allowing reopening of the Regional route a few weeks after the landslide event and ensuring the traffic safety conditions thanks to the continuous monitoring of the landslide process.

Introduction

Landslides are among the most frequent geohazard phenomena, causing a great impact on human activities (Froude & Petley 2018). In particular, landslide events greatly affect the costs for transportation infrastructure, causing direct (e.g. by physical damage) and indirect socio-economic losses (such as: travel delays transportation efficiency decreases) (Romeo et al. 2017, Klose et al. 2014). A single landslide along a national road in a developed country may cause losses up to several million US dollars (Mazzanti 2017). For example, losses on the order of 200 million US$ in 2009 have been estimated six months of road closures after rockslide occurrence on I-40 (North Carolina and Tennessee) (US DOT, 2013). A landslide occurred in 2007 over the A83 in UK, caused a loss of £80 k per day, totaling £1.2 million over a 15-day closure (Postance et al. 2017). Consequently, road re-opening strategies play a relevant role in containing the economic impact of landslides.

In this context, monitoring activities can be fundamental in supporting the safe re-opening of a road after a landslide event.

Different commercial remote sensing technologies have proven to be effective tools for different applications related to geohazards and transportation network management (Mazzanti 2017, Moretto et al. 2018). In very high-risk and emergency conditions, such as few hours/days after a landslide event, remote sensing techniques can be an advantageous monitoring solution, offering the opportunity to monitor the process/area of interest from a remote position.

The development of new processing algorithms and the enhancement of computing capabilities have contributed to the evolution of traditional remote sensing technologies (e.g., Laser Scanning) and to the spread of innovative techniques such as: Terrestrial and Satellite SAR Interferometry and PhotoMonitoring^{TM}.

Nowadays remote sensing techniques find different applications in civil engineering and engineering geology fields, providing contribution during planning, construction and operational phases and management activities (Antonielli et al. 2018, Brunetti & Mazzanti 2015, Mazzanti 2012). In addition, the tendency toward real-time and multi-parametric monitoring is leading toward response to new goals. In this paper the PRIMO (PRompt Intervention Monitoring) protocol, a methodological approach developed and adopted by NHAZCA for the quick on-site intervention in emergency conditions is described, in relation to its first application for the reopening strategies of a Regional Italian route (SP45 Subiaco-Jenne-Vallepietra, Central Italy).

REMOTE SENSING TECHNIQUES

The main remote sensing techniques applied worldwide are ascribable to radar-based technologies such as satellite and terrestrial SAR (Synthetic Aperture Radar) interferometry and techniques able to provide detailed point-based information like Terrestrial Laser Scanning and cutting-edge optical-base solutions like PhotoMonitoring^{TM}. In this chapter, the technologies used in the framework of the Subiaco-Jenne-Vallepietra road re-opening actions are described, including:

Satellite SAR Interferometry, Terrestrial SAR Interferometry, Terrestrial Laser Scanning, and PhotoMonitoring™.

Satellite SAR interferometry is a powerful tool to monitor wide areas or local site, ground or structures deformation with high accuracy, in all-time, with all-weather conditions (Ferretti et al., 2001). SAR images consist of a matrix of resolution cells containing two types of information: the amplitude and the phase. The phase difference between two SAR images collected in different times (Differential SAR interferometry – DInSAR) allows one to investigate the superficial deformations of targeted areas or objects (i.e. displacement occurred between the two acquisitions) along the satellite Line of Sight (LOS). Moreover, by analysing several satellite SAR images (Advanced-DInSAR analysis) acquired with specific characters over the same area, it is possible to retrieve the average displacement of measurement points with millimetre accuracy and the time series of displacement. Using archive SAR data collected by different Spatial Agencies, it is possible to perform historical A-DInSAR analyses since in the early 1990s. An example of the achievable results is reported in Figure 1, showing the average displacement rate of the measurement points and a time series of displacement in correspondence of a landslide process interfering with a highway.

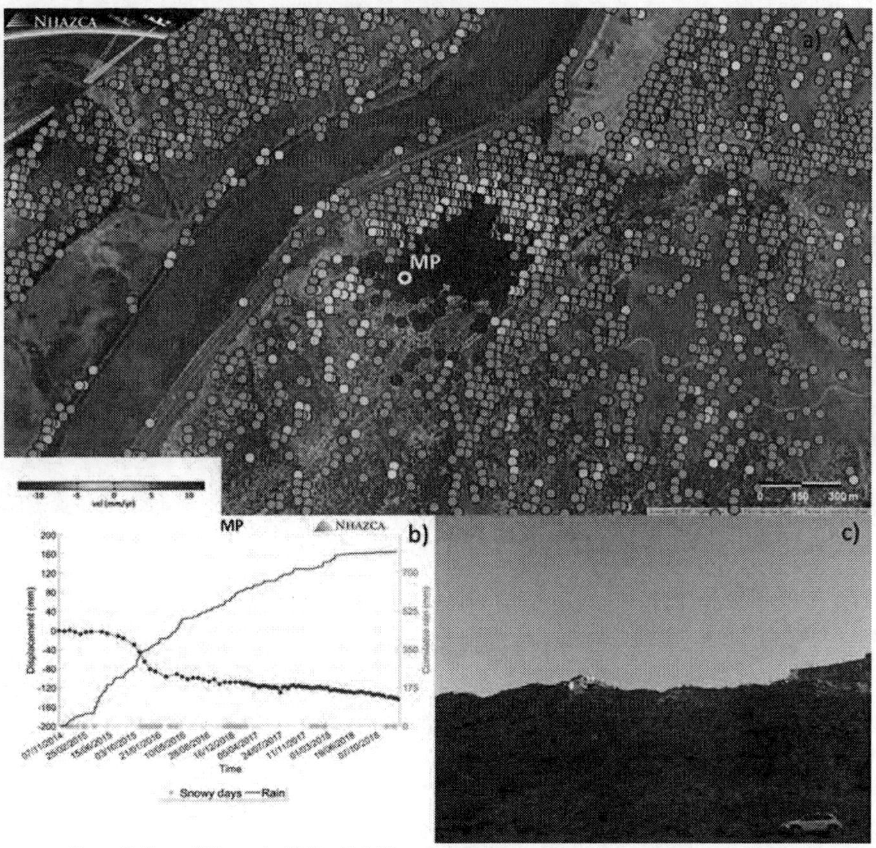

Figure 1-Example of Satellite A-DInSAR technique application to slope interfering with transportation network prone to instability phenomena. The main outputs are a) Displacement rate in mm/year; b) time series of displacement; c) view of the monitored slope.

Terrestrial SAR Interferometry (TInSAR) is a technical based on the analysis of the phase delay between SAR images collected at different times by moving the sensors along a linear structure and by combining the backscattered signals using focusing algorithms (Bozzano et al., 2011). An active radar sensor that emits microwaves and receives the return of scattering objects allows to measure the displacement over time. A sub-millimeter accuracy measurement can be obtained in optimal monitoring conditions for a site (very high signal to noise ratio values). The main output is a 2D displacement map of the scenario along the instrumental LOS.

An application of TInSAR technology to monitor a viaduct in Sicily (South Italy) exposed to slope instability phenomena is shown in Figure 2. During 22 months of monitoring, localized discontinuous displacements have been detected corresponding to the SP24 roadway and in the central sector of the landslide body, while the viaduct appeared to be stable (Moretto et al., 2018). The cumulative displacement along the LOS towards the sensor direction (Figure 2) was quantified to allow assessment of the deformational behavior of the landslide.

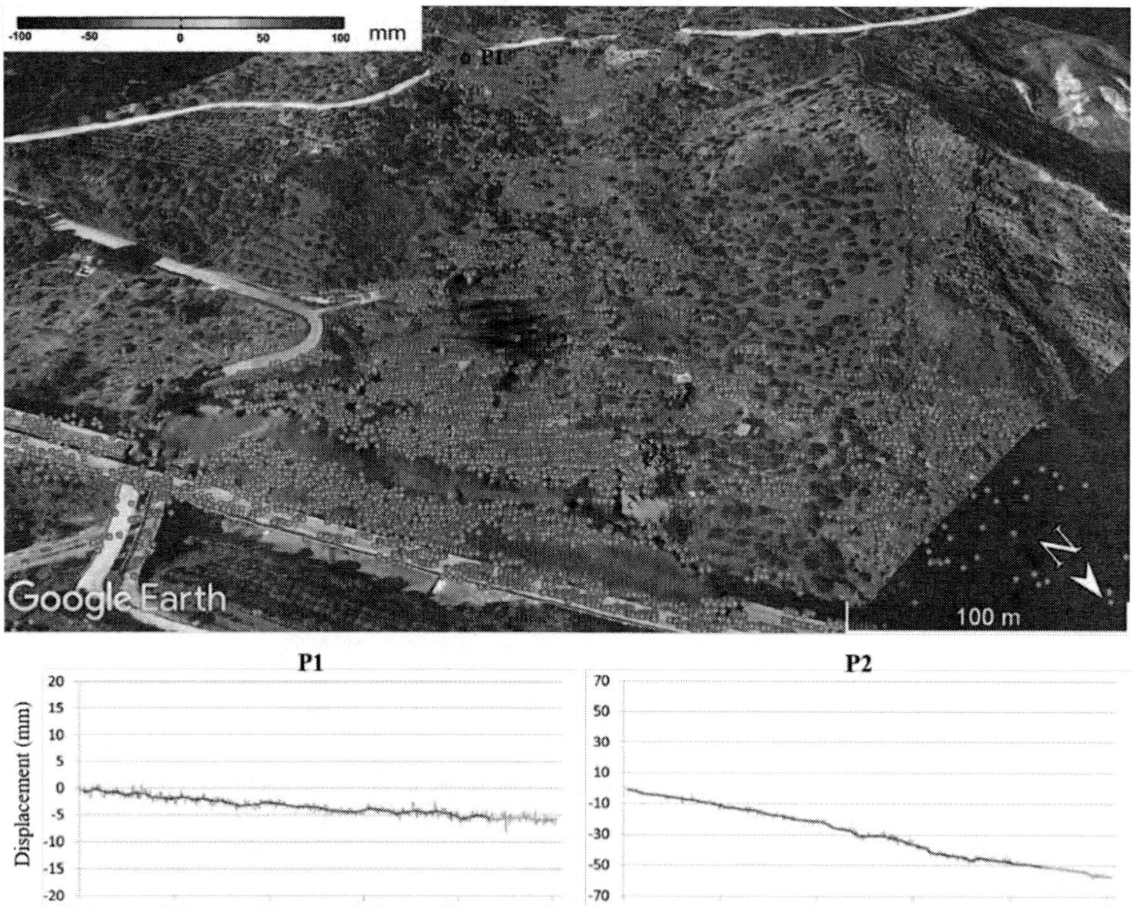

Figure 2 - Example of Terrestrial InSAR technique application to infrastructures exposed to slope instability hazards. Multi-temporal deformation maps (top) and time series of displacement (bottom) of the measurement points P1 and P2 located on the instable slope nearby the Imera viaduct in South Italy (modified from Moretto et al. 2018).

In addition to SAR interferometry, techniques capable of providing detailed point-based information are gaining importance. Among these, Terrestrial Laser Scanning (TLS) is one of the most applied. This technique is a ground-based active imaging method consisting in a device that emits a laser pulse and records its backscattered radiation, measuring the distance between the sensor and the reflecting target (Milenković et al., 2015; Mazzanti et al., 2018). The distance measurements are developed by considering the time-of-flight information of emitted and received laser signal by the sensor and permit retrieval of 3D high-resolution digital models of both natural and anthropogenic objects. Terrestrial Laser Scanning from different positions allows the 3D modelling of complex objects and/or areas of interest. Through the integration with a HR digital camera, it is possible to obtain three-dimensional models in true colors, suitable for better data visualization and interpretation (Figure 3). In addition, by comparing multitemporal TLS surveys, it is possible to quantify volumetric variations of monitored objects (i.e. landslides, rock scarps, etc.). An example of 3D modelling and the volume chances estimation due to rock falls is shown in Figure 3.

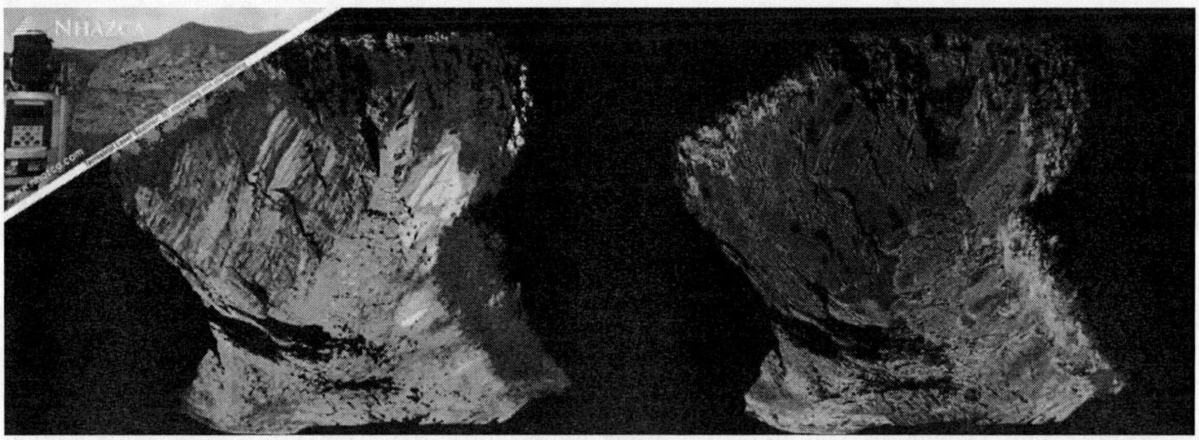

Figure 3 – Example of 3D high resolution model in true colors (left) and volume changes estimation (right) from TLS. Blue colors identify the volumetric losses while red colors the accumulation zones.

The increasing of optical and multispectral sensors, from low-cost and low-resolution to very high resolution (VHR) cameras, facilitates collection of a large amount of data suitable for geotechnical and structural monitoring. The PhotoMonitoring™ solution, with Change Detection (CD), Digital Image Correlation (DIC) techniques, represents an innovative remote sensing technology to identify and quantify changes of objects and/or areas from optical and multispectral data. The correlation of co-registered images collected at different time intervals allows the user to identify, describe and quantify any changes of the area of interest and to measure the "full-field" deformation on the surface of investigated area thanks to CD and DIC approaches respectively (Caporossi et al., 2018; Sutton et al., 2009; Lava et al., 2009). This technique has been successfully applied to monitor slopes prone to instabilities (Figure 4) in terms of shallow landslides as well as rockfalls.

Figure 4- Example of PhotoMonitoring^{TM} application for slope stability. a) Displacement map (in term of intensity and direction) of a slope nearby a linear infrastructure; b) Change detection analysis of a slope prone to rock falls (red circles).

PRIMO Protocol

Emergency management represents a crucial function in decision-making processes as it requires reliable analysis, planning, and assignment of available resources to respond to emergencies. In this context, the role of engineering geologists is becoming more and more prominent with regard to emergencies induced by geohazards.

The increasing demand of rapid intervention following geohazard events has brought us to develop an action strategy: the PRIMO (PRompt Intervention MOnitoring) protocol.

The PRIMO protocol can be described as an emergency response framework of actions based on the use of Remote Sensing techniques for quick intervention over roads, bridges, viaducts, dams etc, affected by extreme events or safety issues.

The PRIMO workflow is summarized in Figure 5, representing the main possible actions to be implemented after an extreme event.

The PRIMO protocol is based on the scope of monitoring, which is fundamental in geotechnical monitoring campaigns (Mazzanti 2017). According to Mazzanti (2017), three main purposes of monitoring are recognized as (Table 1):

1. knowledge monitoring, aimed to characterize the area/process under investigation supporting the identification and delimitation of potential risks;
2. control monitoring, aimed to quantitatively follow the evolution of well-known problems and phenomena;
3. emergency monitoring aimed at providing alerts when a well-known risk become unacceptable.

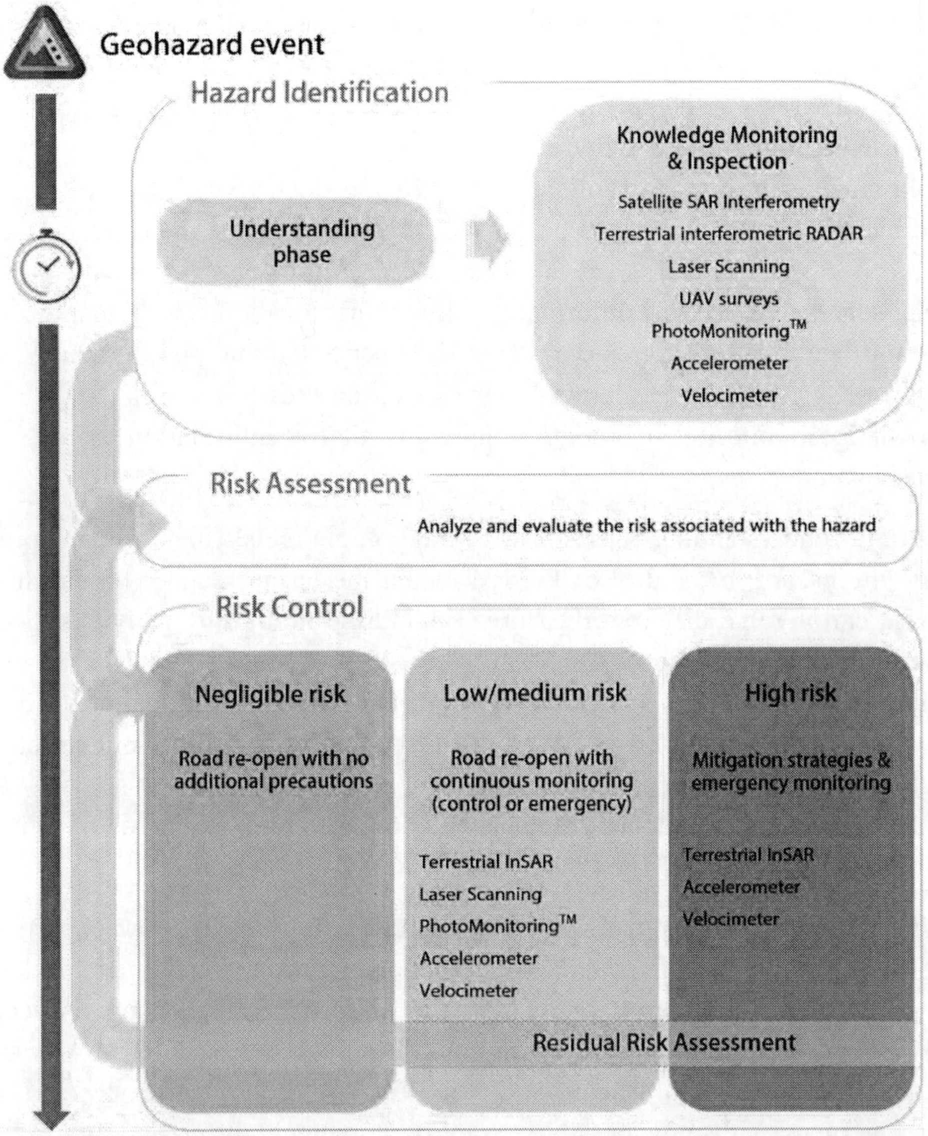

Figure 5. The PRIMO protocol workflow in the framework of risk assessment

Table 1. Monitoring purposes, common applications and instruments (from Mazzanti 2017)

	Common applications	Common instruments
Knowledge monitoring	Design phase Standard maintenance Screening after paroxysmal events (earthquakes, floods, etc.)	LiDAR, satellite SAR interferometry, terrestrial interferometric radar, GNSS, photogrammetry, observation well, piezometer, inclinometer, TDR, earth pressure cell, accelerometer, velocimeter, seismometer
Control monitoring	Construction phase in medium risk areas Advanced maintenance (critical segments) Verification of high risk area	LiDAR, satellite sar interferometry, terrestrial interferometric radar, GNSS, photogrammetry, total station, optical levelling, digital image correlation, observation well, piezometer, inclinometer, TDR, extensometer, earth pressure cell, stress-meter, load cell, strain gauge, fibre optic, pendulum, deflectometer, convergence gauge, surface and probe tiltmeter, liquid level gauge, crack gauge, accelerometer, velocimeter, seismometer
Emergency monitoring	Construction phase in high risk areas Early warning systems for operation in high risk areas	LiDAR, terrestrial interferometric radar, GNSS, total station, piezometer, inclinometer, extensometer, strain gauge, fibre optic, pendulum, surface and probe tiltmeter, liquid level gauge, crack gauge, TDR, convergence gauge, accelerometer, velocimeter, seismometer

After an extreme unexpected event leading to an emergency road closure (e.g. geohazard event, structural critical condition, etc.), a first cognitive phase must be carried out to characterize the phenomenon and assessing the potential risks (understanding phase in Figure 5). To this aim, the actions that can be implemented include:

a) bibliographic study and examination of existing documentation that can include current and historical data, theoretical analysis, informed opinions, the concerns of stakeholders, etc.
b) field surveys, to collect information about the location. Both manual (observational method) and instrumental surveys (UAV, Laser Scanning, photogrammetry, etc.) can be performed in this phase according to the specific needs;
c) knowledge monitoring, in order to acquire quantitative information about the process under investigation.

The results of the understanding phase allow evaluation of the risk, identifying appropriate actions to eliminate the hazard, or control of the risk when the hazard cannot be eliminated. The risk control phase can involve different measures, including monitoring activities, according to the level of the risk recognized (Figure 5):

a) negligible risk level - the road can be re-opened without additional measures;
b) low to medium risk level - the road can be re-open after or without structural mitigation measures;
c) high risk level – direct actions are needed in order to re-open the road, including structural mitigation strategies.

The ongoing monitoring of the hazards identified under low to high risk conditions can play a relevant role in shortening the time until a road can be reopened to traffic, and specialty monitoring technologies able to provide alerts (often automatic) in case the risk become unacceptable.
Taking advantage of the capabilities of new monitoring equipment, in terms of data acquisition, data processing and both temporal and spatial resolution, for each specific purpose (i.e. knowledge, control and emergency) it is now possible to rapidly implement different monitoring solutions.
In consideration of the site-specific conditions and monitoring purpose (control or emergency), different monitoring techniques are suitable under different risk categories, such as: Terrestrial SAR interferometry, PhotoMonitoring™, multi-temporal Laser Scanning, accelerometer, velocimeter, etc.
In the following section, the actions undertaken based on the methodological approach reported in Figure 5 to rapidly reopen a Regional Italian road after a landslide event are described.

Case study: the Italian SP45 road re-opening after a landslide event

On the night of 26th November 2018, a landslide occurred in the Municipality of Subiaco, in the vicinity of the Monastery of S. Scolastica, one of the oldest monasteries in Italy (Figure 6). Although the landslide involved only few hundred cubic meters not causing injuries and losses, it immediately generated a serious concern among citizens and the media. This was because the landslide interrupted the road connecting some villages and with the Monastery itself, which is travelled by several thousand visitors per year.

Figure 6. Overviews of the 26th November 2018 landslide (sources: http://www.confinelive.it and http://www.cinquequotidiano.it).

For the safe reopening of the SP45, according to the phased reopening strategy and PRIMO protocol (Hazard Identification, Risk Assessment and Risk Control) and data from the focused monitoring activity (knowledge, control and emergency), the following activities were carried out:

1) multi-temporal Terrestrial Laser Scanning and historical Satellite InSAR analysis (knowledge monitoring);
2) Risk Assessment;
3) Continuous Terrestrial InSAR monitoring for alert purposes (emergency monitoring);
4) Residual Risk Assessment;
5) Ground-based PhotoMonitoring^{TM} to detect potential displacements and localized rock-falls able to generate risk conditions for the infrastructure (control monitoring).

Following the first on-site inspections by technical experts to complete a Hazard Identification assessment, multi-temporal surveys using a long-range Terrestrial Laser Scanner (TLS) were performed. This stage of monitoring (knowledge monitoring) was aimed at identifying the unstable area and detecting possible changes/movements in progress in the slope. In this regard, in less than one month (19 December 2018 - 07 January 2019) ten TLS surveys were carried out from a fixed position (Figure 7a). Thanks to this first stage, some significant evidence was detected:

- no evidence of displacement was found for the area of the Monastery (green colors Figure 7c);
- displacements on the order of few centimeters were detected in limited portions of the slope, specifically on the right and left flanks of the landslide (red colors Figure 7c);
- erosive processes were concentrated in the central portion of the slope (blue colors Figure 7c).

Furthermore, a historical analysis of ground deformation was also performed using the DInSAR technique applied to COSMO-SkyMed images (Italian Space Agency - ASI). Such large-scale

satellite analysis did not identify critical issues or anomalies in the study area. Within the Monastery, adverse displacement - in terms of ground deformation - was once again ruled out.
Although the measurements from knowledge monitoring did not identify alarming trends, an active dynamic of the slope was detected that could represent a serious threat for travelers on the roadway. Considering the significant traffic on the road for locals and tourists, the potential risk exposure was rated high. Because the risk was not negligible and in order to ensure an appropriate level of safety, the road was reopened a few days after the first inspection with the following security measures:

- Closure of the road overnight;
- Reduction of the traffic speed;
- Emergency monitoring by continuous Terrestrial SAR Interferometry (24/7) providing alerts if movements exceed selected thresholds;
- Road closures based on regional weather alerts.

The main results from the Terrestrial SAR Interferometry, confirmed the evidence shown by the multi-temporal analysis using TLS (Figure 8). This technique permitted continuous monitoring of the slope in all weather and lightening conditions, thus representing a suitable technique for the real-time monitoring and early warning.
After 2 months of TInSAR surveys, the assessment of the residual risk yielded a medium level risk. With the reduced risk, emergency monitoring was halted and a less expensive control monitoring program was put in place.
A PhotoMonitoring™ campaign was carried out for several months using a high-resolution DSLR (Digital Single-Lens Reflex) camera to ensure maximum visibility during the entire period of operation. The images were acquired every 10 minutes and processed using Change Detection and Digital Image Correlation algorithms. The monitoring allowed identification of instability processes on the slope and quantification to millimeter accuracy.
Once again, the analysis showed localized displacements and changes corresponding to the left and right side of the landslide (Figure 9). In the central portion of the slope, during the last period of monitoring, even where affected by widespread vegetation growth, no appreciable movements were detected.

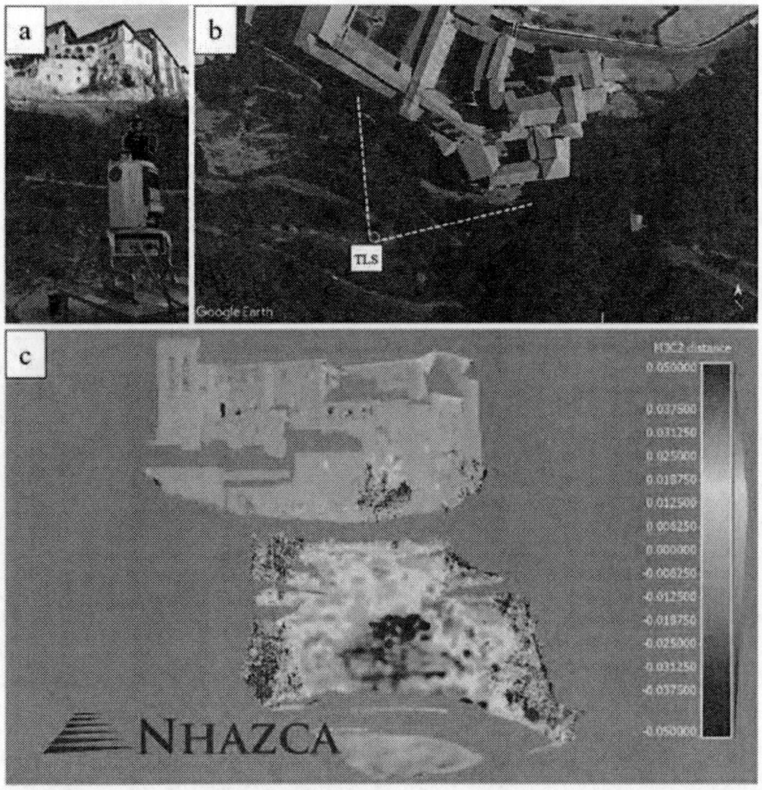

Figure 7. a) TLS placed on a fixed platform; b) map of the study area, the red circle indicates the TLS location; c) cumulated displacement by computing distances between two TLS point clouds (12/19/2018 - 01/07/2019).

Figure 8. Radar map with cumulative displacement (January-March) overlaid on RGB colored point cloud acquired by TLS. Negative values correspond to movements toward the radar sensor (distance shortening along the Line of Sight); positive values correspond to movements away from the sensor (distance increasing along the Line of Sight).

Figure 9. Example of results obtained with PhotoMonitoring^{TM} techniques during 1 day observation period. Slope changes are reported in green (ascribable to localized rock falls and surficial movements), while anthropic changes in pink. 2D displacement field have been obtained using proprietary software tools developed by NHAZCA implementing DIC technique. In the bottom left corner, the PhotoMonitoring^{TM} station.

CONCLUSIONS

Emergency management represents a crucial function in decision-making processes as it requires reliable analysis, planning, and assignment of available resources to respond to the emergencies. In this context, the role of engineering geologists is becoming more and more prominent, with particular regard to emergencies induced by geohazards.

The increasing demand of rapid intervention in emergency conditions, led us to develop a standard approach: thee PRIMO protocol, an emergency response protocol developed based on the use of remote sensing techniques as a quick answer to support asset management under emergency conditions.

The monitoring and consultancy activities carried out by NHAZCA supported the decision making for the characterization of the landslide phenomenon, allowing reopening of SP45 route (Subiaco-Jenne-Vallepietra) after the landslide event, and ensured traffic safety conditions thanks to the continuous monitoring of the landslide process.

The application of PRIMO protocol to safely re-open the SP45 showed the potential for standardized procedures and remote sensing techniques to support authorities engaged in assets management under emergency conditions.

REFERENCES

Antonielli B., Caporossi P., Mazzanti P., Moretto S., Rocca A., 2018. InSAR & PhotoMonitoring™ for dams and reservoir slopes health & safety monitoring. VINGT-SIXIEME CONGRES DES GRANDS BARRAGES Autriche, juillet 2018. DOI 10.3217/978-3-85125-620-8-227

Bozzano F., Cipriani I., Mazzanti P., Prestininzi A., 2011. Displacement patterns of a landslide affected by human activities: insights from ground-based InSAR monitoring. *Nat Hazards*, doi: 10.1007/s11069-011-9840-6.

Brunetti A, Mazzanti P. 2015. Monitoring an unstable road embankment for public safety purposes by terrestrial SAR interferometry. In: Proceedings of the 9th international symposium on field measurements in geomechanics (Sydney 9–11 September 2015) p 769–780

Ferretti, A., Prati, C. and Rocca, F. Permanent scatterers in SAR interferometry. *IEEE Trans. Geosc. and Remote Sens*, 39(1), 2001, 8-20

Froude, M. J. and Petley, D. N.: Global fatal landslide occurrence from 2004 to 2016, Nat. Hazards Earth Syst. Sci., 18, 2161-2181, https://doi.org/10.5194/nhess-18-2161-2018, 2018.

Klose, M., Damm, B., Terhorst, B. (2014): Landslide cost modeling for transportation infrastructures: a methodological approach. Landslides. DOI: 10.1007/s10346 -014-0481-1.

Lava P., Cooreman S., Coppieters S., De Strycker M., Debruyne D., 2009. Assessment of measuring errors in DIC using deformation fields generated by plastic FEA. *Opt. Las. Eng.* 47, 2009, 747-753.

Mazzanti P., Schilirò L., Martino S., Antonielli B., Brizi E., Brunetti A., Margottini C., Scarascia Mugnozza G. 2018. The Contribution of Terrestrial Laser Scanning to the Analysis of Cliff Slope Stability in Sugano (Central Italy). *Remote Sensing*, 10, 1475, 2018. doi:10.3390/rs10091475

Mazzanti P. 2017. Toward transportation asset management: what is the role of geotechnical monitoring? J Civil Struct Health Monit.

Mazzanti P. 2012. Remote monitoring of deformation. An overview of the seven methods described in previous GINs. Geotech News 30(4):24

Milenković, M.; Pfeifer, N.; Glira, P. Applying Terrestrial Laser Scanning for Soil Surface Roughness Assessment. *Remote Sensing*. 2015, 7, 2007–2045, doi:10.3390/rs70202007

Moretto S., Bozzano F., Brunetti A., Della Seta M., Majetta S., Mazzanti P., Rocca A. Valiante M, 2018. The 2015 Scillato Landslide (Sicily, Italy): deformational behavior inferred from Satellite & Terrestrial SAR Interferometry. 10th International Symposium on Field Measurements in Geomechanics July, 17 - 20, 2018, Rio de Janeiro, Brazil.

Postance B., Hillier J., Dijkstra T., Dixon N. 2017. Extending natural hazard impacts: an assessment of landslide disruptions on a national road transportation network. Environ. Res. Lett. 12 (2017) 014010

Romeo S., Di Matteo L., Melelli L., Cencetti C., Dragoni W., Fredduzzi A. Seismic-induced rockfalls and landslide dam following the October 30, 2016 earthquake in Central Italy. *Landslides,* 14(4), 2017, 1457-1465, doi:10.1007/s10346-017-0841-8.

Sutton M.A., Orteu J.J., Schreier H.W. 2009. Shape, motion and deformation measurements. Basic concepts, theory and applications, Springer Science, chap. V.

CASE STUDY OF A FAILED TIED-BACK RETAINING WALL IN A COLLUVIUM SLOPE UNDER LANDSLIDE CONDITIONS

Craig S. Lee, P.E.
S&ME Inc.
2020 Liberty Road, Suite 105
Lexington, Kentucky 40505
(859)-293-5518
clee@smeinc.com

Prepared for the 70th Highway Geology Symposium, October, 2019

Acknowledgements

The author(s) would like to thank the individuals/entities for their contributions in the work described:

Andrew Fiehler, P.E. – S&ME Lexington KY
Curtis Chandler – S&ME Lexington, KY

Disclaimer

Statements and views presented in this paper are strictly those of the author(s), and do not necessarily reflect positions held by their affiliations, the Highway Geology Symposium (HGS), or others acknowledged above. The mention of trade names for commercial products does not imply the approval or endorsement by HGS.

Copyright Notice

Copyright © 2019 Highway Geology Symposium (HGS)

All Rights Reserved. Printed in the United States of America. No part of this publication may be reproduced or copied in any form or by any means – graphic, electronic, or mechanical, including photocopying, taping, or information storage and retrieval systems – without prior written permission of the HGS. This excludes the original author(s).

ABSTRACT

THIS PAPER SEEKS TO DISCUSS HOW LAPSES IN PLANNING, CONTRACTING, UNDERSTANDING THE GEOLOGY AND LACK OF ADEQUATE CONSTRUCTION INSPECTION CREATED A SIGNIFICANT FAILURE THAT REQUIRED A MULTI-MILLION DOLLAR REPAIR. A POWER GENERATING COMPANY IN NORTHERN KENTUCKY WAS EXPANDING THEIR OPERATIONS TO ADDRESS THE NEW COAL COMBUSTION RESIDUAL (CCR) RULES ON FLY ASH AND GYPSUM HANDLING. THE PROJECT INCLUDED A NEW GYPSUM STORAGE BUILDING TO HANDLE GYPSUM GENERATED FROM THE SCRUBBERS. THE GRADING FOR THE NEW GYPSUM STORAGE BUILDING REQUIRED A CUT INTO AN ADJACENT HILLSIDE AND CONSTRUCTION OF AN 850 FOOT LONG TIEDBACK RETAINING WALL RANGING FROM FOUR TO 20 FEET TALL. SHORTLY AFTER INSTALLATION OF THE RETAINING WALL, AND DURING CONSTRUCTION OF THE FOUNDATIONS FOR THE NEW BUILDING, THE DBC NOTICED THE SOUTH WALL GRADE BEAM DEFLECTED LATERALLY TO THE NORTH AND RAISED FIVE INCHES IN ELEVATION SINCE INITIALLY PLACED. SUBSEQUENT EXPLORATION AND MONITORING OF THE RETAINING WALL AND THE SLOPE INDICATED A FAILURE SURFACE THAT WAS MUCH DEEPER AND LONGER THAN ANTICIPATED IN THE DESIGN GEOTECHNICAL REPORT AND EXTENDED UNDER THE TIEDBACK WALL. PORTIONS OF THE TIEDBACK WALL EVENTUALLY MOVED 25 INCHES TO THE NORTH YET REMAINED RELATIVELY PLUMB.

THE GYPSUM STORAGE BUILDING WAS DECONSTRUCTED, THE TIEDBACK WALL REMOVED AND THE FAILED SLOPE EXCAVATED TO REMOVE THE SOIL OVERBURDEN. THE GYPSUM STORAGE BUILDING WAS RECONSTRUCTED FURTHER NORTH WITHOUT ANY ISSUES AT A TOTAL MITIGATION COST OF OVER $10 MILLION.

THE PAPER FOCUSES ON THE UNIQUE CHALLENGES OF THE GEOLOGIC SETTING AT THE SITE, THE SHORTCOMINGS OF THE DESIGN PHASE GEOTECHNICAL EXPLORATION, THE IMPORTANCE OF IDENTIFYING LANDSLIDE GEOMETRY WHEN ASSIGNING DESIGN WALL LOADS, THE VALUE OF AN ADEQUATE SITE VISIT, THE IMPORTANCE OF VERIFYING THE ANCHORS ARE INDEED INTO THE TARGET MATERIAL AND NOT JUST DRILLED TO A SET LENGTH AND A DISCUSSION ON WHAT HAPPENED AND HOW IT WAS ADDRESSED.

INTRODUCTION

As part of the Coal Combustion Residuals (CCR) regulations, many coal fired power plants are installing equipment to handle and dispose of fly ash and gypsum both are by-products of the burning of coal to generate electricity. At some coal-fired power plants sulfur dioxide emissions are controlled by flue gas desulfurization (FGD) systems that use limestone or lime to react with the gaseous sulfur forming calcium sulfate or calcium sulfite commonly called synthetic gypsum. This project included a new pipe conveyor, gypsum storage building and a gypsum Reclaimer. To create room for the new facility the Owner elected to excavate into an adjacent hillside and build a retaining wall.

The Owner recommended a Geotechnical Engineering Firm (GEF) to the Design Build DBC (DBC) who retained the GEF to perform a comprehensive geotechnical report of the whole project including the gypsum storage building and the proposed retaining wall. The DBC retained a geotechnical specialty DBC to design and construct the wall.

GEOLOGY

The project site is located along the Ohio River in Northern Kentucky in the Outer Bluegrass Physiographic region. Bedrock is late Ordovician aged containing various percentages of interbedded non-durable shale and limestone. The project site straddles the Pleistocene and Holocene age Ohio River alluvium and the adjacent hillside formed by the eroding Kope Formation. The Kope Formation is comprised of about 80 percent non-durable shale with the remaining comprised of thin, fossiliferous limestone beds. The Kope Formation is a major contributor to the region's claim as being one of the most landslide prone areas of the country.

Dalrymple (1968)[1] devised a schematic of the geology of the project site which is reproduced below. The Dalrymple sketch (Figure 1) shows the transitional nature of the geology. The project site straddles the Colluvial Foot Slope and the Alluvial Toe Slope. The colluvium forms from the weathering of the shale and is predominantly low-plasticity clay with

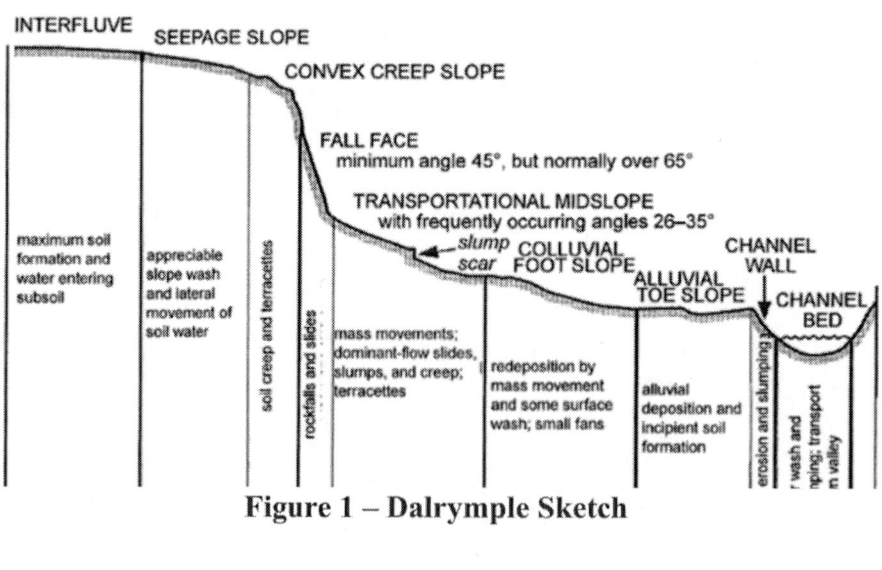

Figure 1 – Dalrymple Sketch

limestone cobbles and slabs throughout. A key feature is the convex shape of the slope forming a prominent foot slope zone with a scarp zone at the sharp transition. In this area, the Fall Face is usually formed by the Fairview and Grant Lake formations. These formations overlie the Kope Formation and display significantly less non-durable shale (less than 50 percent). Subsurface water is a significant contributor to slope instability in the Kope Formation. Discrete wet zones are often observed on the hillside as water bleeds from the hillside. The Kope weathers to a distinct brown clay with limestone slabs. It quickly transitions to gray shale with limestone below the weathered zone. Most of the seepage zones and slides occur within the colluvium, the brown weathered rock zone or at the interface between the weathered and unweathered bedrock.

The landslide mechanism is predominantly progressive translational slides that progress uphill producing the characteristic hummocky topography. Rotational slides are not as common but do occur when the soil thickness is thick. In summary, the geologic setting is well documented with obvious characteristics that can be readily identified by field and topographic observation.

The gypsum storage building is 130 feet wide and 450 feet long. Initial construction activities for the gypsum building included site grading including constructing the building pad for the gypsum storage building pad. The site grading required making a cut into the base of the north facing hillside and constructing a soldier pile and lagging tiedback wall. The wall height was predominantly between 12 feet and 18 feet, tapering down to five feet at the ends of the wall. Soon after constructing the tiedback wall, foundation and slab construction began. By the time of our involvement, the perimeter grade beam foundation was essentially complete, the slab placed, the rails for the Reclaimer installed, the Reclaimer installed on the rails and the steel erection was taking place. The construction progress is shown on Figure 2.

Figure 2 – Construction Progress

Figure 3

While finishing the west end of the south grade beam, the construction team noticed that a section of the grade beam that was placed earlier was bowed in Figure 3. Why was the grade beam bowed and not straight? Initial opinions placed blame on a survey error in laying out the grade beam. Others opined it may be poor construction methods. The DBC took a 4-foot level and checked the plumbness of the soldier piles, they were still essentially plumb so this temporarily ruled out wall failure. It was at this point that the DBC contacted S&ME to evaluate why the grade beam was bowed. From our site visit it was clear this was no survey error nor was it poor construction. A survey error would not produce a bow shape but a straight line that was skewed to the perpendicular as depicted in Figure 4.

Figure 4

Reconnaissance of the hillside above the wall revealed a tension crack in the hillside that lined up well with the bowed section of the grade beam. We also noted the hummocky terrain comprised of a series of narrow benches spaced about 75 feet apart all the way up the hillside. In our opinion the benches represented progressive head scarps from historic landslide activity. At the top of the hill we observed a fifteen foot high vertical wall of limestone with interbedded shale. This was the Fall Face (see Figure 1 - Dalrymple Sketch) and confirmed our suspicion.

We advised the DBC that we believed the hillside was sliding and caused the bow in the grade beam. At our request the DBC installed survey points along the top of the wall and began monitoring those points.

We needed to quickly find the failure surface before we could complete our evaluation and develop a mitigation plan. We installed fourteen slope inclinometers with nine on the hillside above the tiedback wall and five in the gypsum building area. Figure 5 depicts the location of the slope inclinometers. SI13 is not shown and it was located north of the building area. The initial tension crack was located between SI4 and SI5. The bow in the grade beam is near SI3.

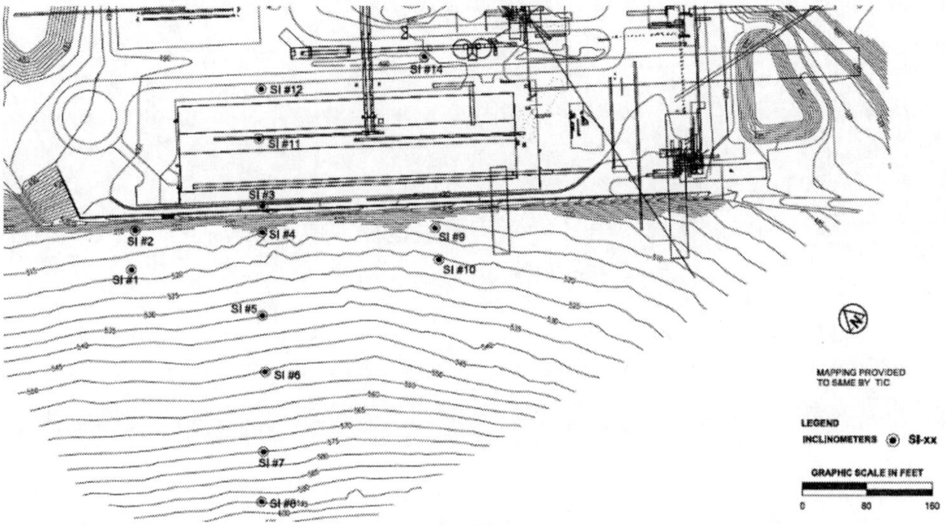

Figure 5

After four months of monitoring the survey points along the top of the wall, the wall had moved 25 inches toward the river with no slowing of the movement. Yet the wall remained remarkably plumb with no sign of distress. The bulge in the wall movement lined up with the bow in the grade beam. The wall movement is depicted graphically in Figure 6.

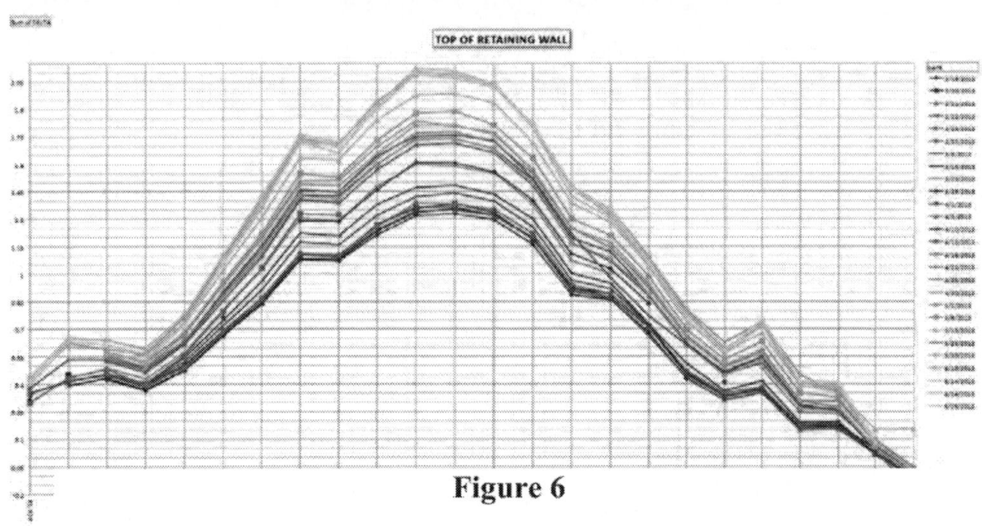

Figure 6

The slope inclinometer data showed that the entire hillside was creeping downhill with the failure surface transitioning from the colluvium in the upper part of the slope to the weathered rock zone at the lower part of the slope and extending under the wall. Figure 7 shows the failure surface relative to the soil profile. As if the whole hillside moving was not enough, the failure surface at the wall was over 30 feet deep! No wonder the wall was still intact, it was riding the slide downhill.

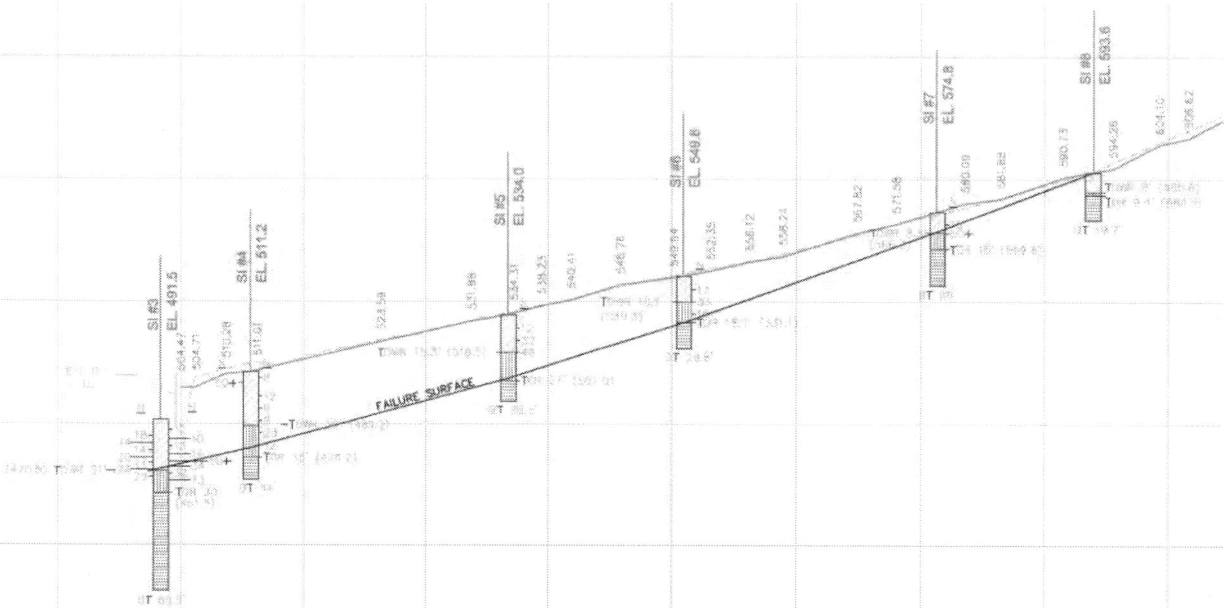

Figure 7

TECHNICAL REASONS THAT LED TO THE WALL FAILURE

Three technical errors contributed to the wall failure and one primary cause that had nothing to do with technical decisions. The three technical errors are:

- Failure to Identify Key Geologic Influences
- Incorrect Earth Pressure Diagram Used for Wall Design
- Soldier Piles and Anchors Were Too Short

Failure To Identify Key Geologic Influences

Failure to appreciate the influence of geology was a critical oversight in the design of the wall. Both site observations and understanding the geologic setting provide important clues about future performance to the Designer. For example, the hummocky topography that included a series of old head scarps that had weathered to form narrow flat benches up and down the hillside. We observed that the deer used these a paths to walk across the hillside. Further clues was the presence at the top of the slope of a fifteen foot high head scarp that represents the Fall Face predicted by Dalrymple. Both of these features are clearly seen in the LiDar topographic survey in Figure 8 but were not evidently recognized by the designers.

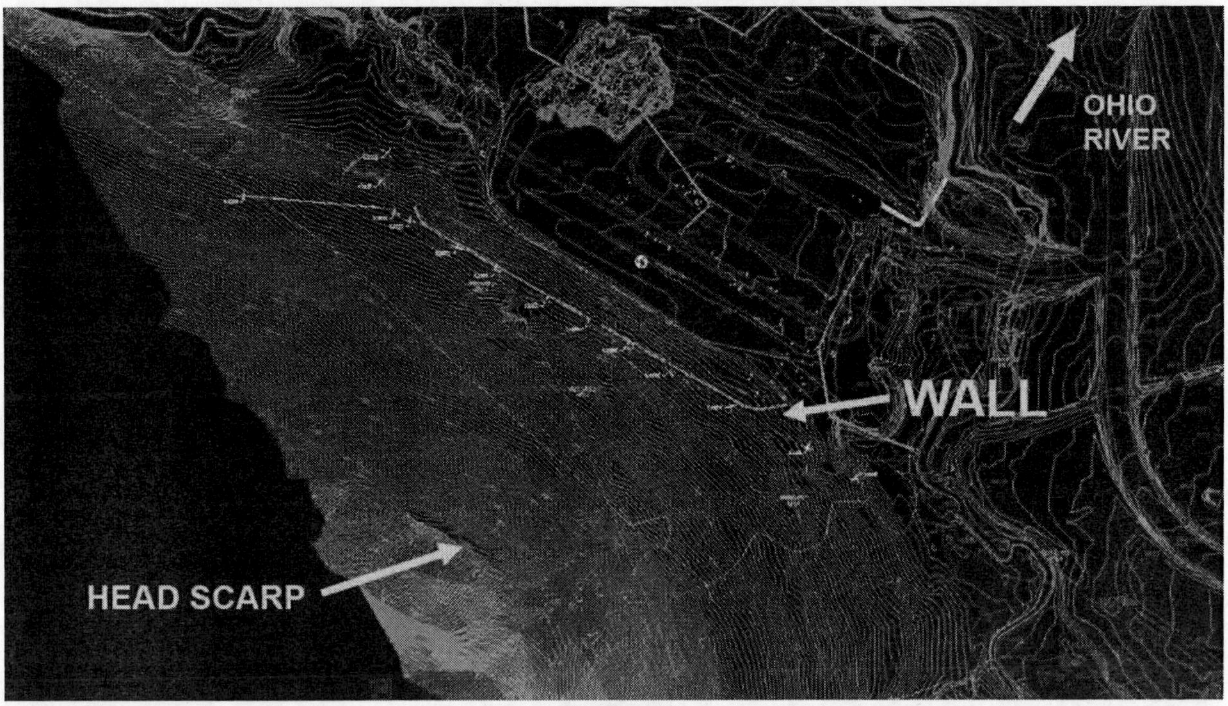

Figure 8

The Geology section of the original design report by the GEF discussed characteristics of the geologic formations including the likely presence of a weathered rock zone as well as the likely presence of colluvium at the foot slope. However, the wall design did not consider the presence of either of these characteristics at this site nor take into account the influence either of these had on the design. Correctly identifying the inherent geohazards of this hillside certainly would have warranted installing slope inclinometers on the hillside to check for pre-existing slope movement that would affect the wall design.

Incorrect Earth Pressure Diagram Used for Wall Design

The failure to identify the evidence of pre-existing landslides associated with the hillside resulted in the under-prediction of the design earth pressure. The GEF used a classic Rankine earth pressure distribution for the earth pressure but in reality they should have used a Landslide distribution. The difference is illustrated below. Note the original design on the left showing the earth pressure distributed only along the face of the wall in Figure 9. The actual earth pressure is shown on the right in Figure 10 as with the earth pressure distribution extending down to the failure surface as indicated by the red arrow.

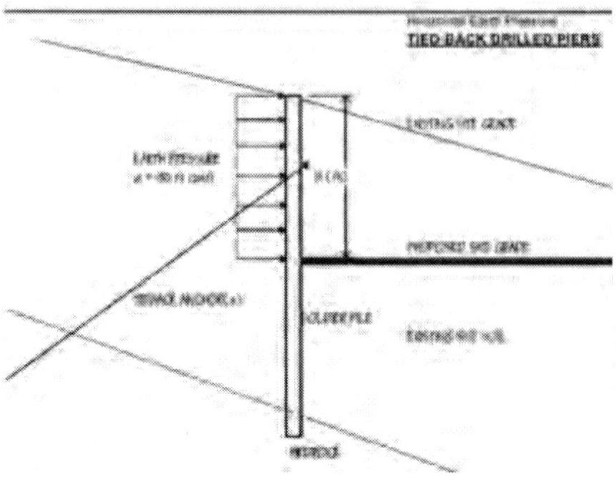

Figure 9

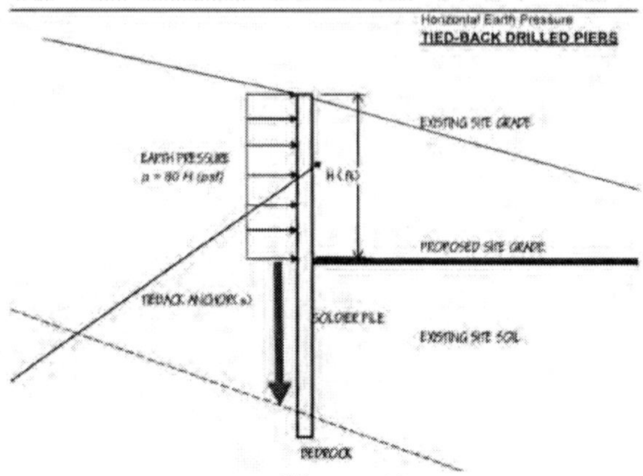

Figure 10

This difference is significant. The recommended Design Load from the GEF was 80H, where H is the height of the wall as shown on the graphic reproduced from the report. The actual load was computed to be 200H with H being from the top of the wall to the failure surface.

Soldier Piles and Anchors Were Too Short

Not identifying that the wall needed to be designed for landslide forces was in itself enough to doom the wall. However, in plotting the as-built geometry of the soldier piles and anchors with the boring data, we quickly recognized that construction issues were also contributory to the wall movement. The soldier piles did not reach bedrock but terminated in the weathered rock zone. The failure surface passed under the bottom of the soldier piles. The 45 foot long anchors terminated within the weathered rock zone without reaching the bedrock or terminated less than five feet into bedrock. The anchors that terminated within the weathered rock zone were well short of the failure surface. The anchors that made it to bedrock only extended a few feet beyond the failure surface. This explains why the wall could move over twofeet yet still look as if nothing happened. Figure 11 on the following page is one of our working drawing sections representing the actual constructed conditions at one of the cross-sections.

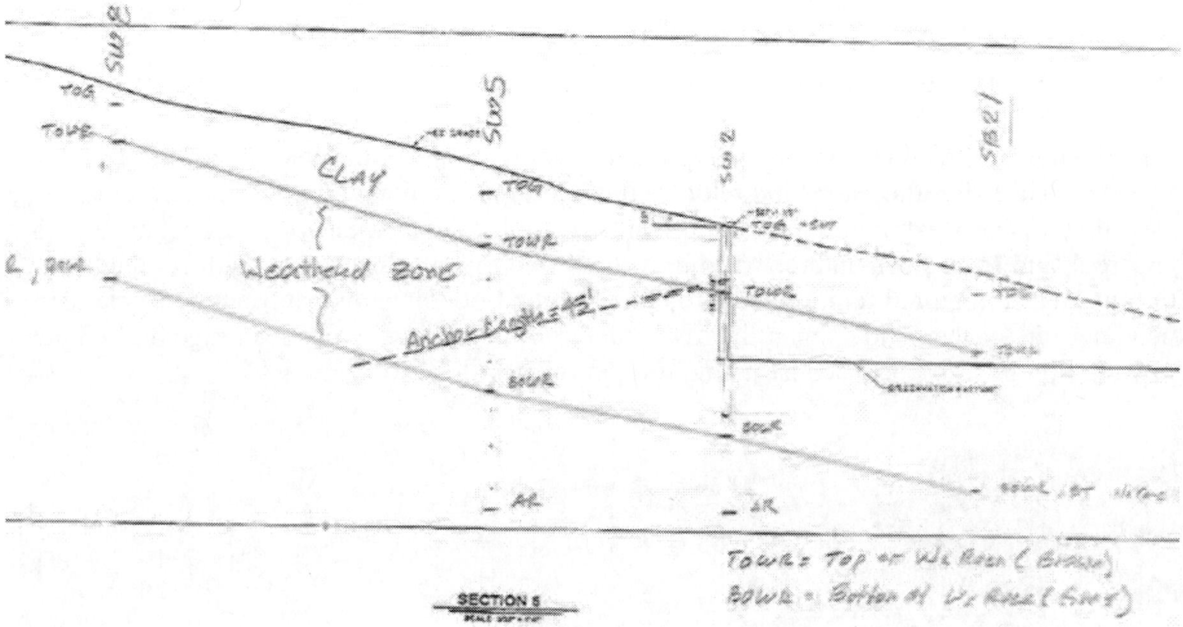

Figure 9

LESSONS LEARNED

- Research and understand the geologic setting of your project. Make sure your exploration program seeks to identify which of the characteristic geologic hazards are present at your site.
- Perform a thorough site reconnaissance with qualified personnel.
- Make sure your design takes into account the site observations, the inherent characteristics of the geologic setting and specific hazards identified in the exploration program.
- Someone has to own the wall design.

The primary reason for the wall failure was there was no ownership of the wall design. The GEF produced their report as part of a For Informational Purposes Only submittal to the Design Build team. The geospecialty wall contractor produced his own design. Because the wall was such a minor component of the large, complex, and fast paced construction project, no one owned the quality component of the construction. The design and construction took place in isolated pods disconnected from each other with little accountability for the design or construction.

SO HOW WAS THE SITUATION RESOLVED?

After looking at several options including reconstructing the wall the selected option was to remove the slide risk altogether and relocate the building. To that end the wall was removed, all the colluvium and weathered rock from the hillside in the active slide zone was excavated down to bedrock and hauled off, and the building was shifted toward the river away from the hillside. This option also required removing the foundation, the floor slab, dismantling the Reclaimer and Reclaimer rail system, and dismantling the structural steel for the gypsum storage building as shown in Figure 12 before reconstructing it all in the new location.

Figure 10

REFERENCES:

- Conference Proceedings
    1. Dalrymple, J. B. (et al.) A Hypothetical Nine Unit Landform Model. *Zeitschrift für Geomorphologic 12*, Fig. 1. P.62, 1968.

Poisson's Ratio Assessed from Ultrasonic versus Load Test

Sherif Elfass
Research Associate Professor
University of Nevada Reno
1664 N. Virginia St, MS258
Reno, NV 89557
(775) 784-6664
elfass@unr.edu

Evan Saint-Pierre
Geotechnical Engineer
Shannon and Wilson 400 N 34th Street, Suite 100
Seattle, WA 98103
(206) 632-8020]
esaintpierre@shanwil.com

Robert Watters
Professor
University of Nevada Reno
1664 N. Virginia St., MS172
Reno, NV 89557
(775) 784-6069
watters@unr.edu

Gary Norris
Emeritus Professor
University of Nevada Reno
1664 N. Virginia St, MS258
Reno, NV 89557
(775) 331-5582
norris@unr.edu

Prepared for the 70th Highway Geology Symposium, October, 2019

Acknowledgements

The effort presented herein was part of a project sponsored by the Nevada Department of Transportation (NDOT). The objective of the project was to improve Strain Wedge Model (SWM) program capabilities (AASHTO 2014) in analyzing the performance of laterally loaded drilled shafts embedded in cemented soils. A new material model for the partially cemented soils was developed using the results obtained from unconfined compressive strength (UCS) tests. The model has been implemented and verified against field tests.

Disclaimer

Statements and views presented in this paper are strictly those of the author(s), and do not necessarily reflect positions held by their affiliations, the Highway Geology Symposium (HGS), or others acknowledged above. The mention of trade names for commercial products does not imply the approval or endorsement by HGS.

Copyright Notice

Copyright © 2019 Highway Geology Symposium (HGS)

All Rights Reserved. Printed in the United States of America. No part of this publication may be reproduced or copied in any form or by any means – graphic, electronic, or mechanical, including photocopying, taping, or information storage and retrieval systems – without prior written permission of the HGS. This excludes the original author(s).

ABSTRACT

Fifty-three cored specimens from block samples of partially cemented soil (caliche), collected from the Las Vegas valley, were tested for their Unconfined Compressive Strength (UCS). Deformations of sixteen of these samples were recorded using a compressometer as well as Linear Variable Displacement Transducers (LVDTs). The compressometer, which is more commonly used on concrete cylinder samples, has the advantage in rock sample testing that lateral as well as axial deformation can be measured. Consequently, the equivalent linear secant Poisson's ratio can be evaluated ($\mu = -\varepsilon_L / \varepsilon_a$) with increasing stress level over the course of the UCS test. All samples were subjected to ultrasonic wave velocities determination prior to UCS tests. Compressional (V_p) and shear (V_s) wave velocities were measured using a commercial Ultrasonic Pulse Wave Transducer. The lab velocities were obtained in order to compare with field values from geophysical tests on site. However, from these lab wave velocities, the Poisson's ratio at zero strain/load were computed and compared with the variation in Poisson's ratio with increasing stress level based on deformation measurements from the UCS test. Such comparison is the subject of this paper.

INTRODUCTION

The Las Vegas region is known for its cemented soil stratigraphy. Cemented soils (commonly referred to as caliche) can vary widely with sudden changes in thickness, hardness and depth (Stone et al., 2001). The subsurface cemented soils are layered with various levels of cementation, with some horizons being more competent than engineered concrete, while others have fractures and clay seams making them less competent. Seismic methods have the potential to be a more economical and non-intrusive for determining the lateral extent and thickness of cemented soil over a large area (Sirles and Viksne, 1990). The same velocities used to establish subsurface topography might be useful when correlated with laboratory cored sample velocities in relating to the stress-strain behavior from UCS tests on such cored samples. In turn, the modeled material stress-strain behavior would then be used for foundation analysis.

Poisson's ratio is the negative of the ratio of transverse strain (ε_L) to the axial/vertical strain (ε_a) in an elastic material subjected to a uniaxial stress. It offers the fundamental metric by which to compare the performance of any material when strained elastically. Poisson's ratio plays an undeniably important role in the elastic deformation of rocks and rock masses subjected to static or dynamic stresses. Furthermore, its effects emerge in a wide variety of rock engineering applications, ranging from basic laboratory tests on intact rocks to field measurements for in situ stresses or deformability of rock masses. Therefore, information on various aspects of Poisson's ratio can be beneficial for rock engineering, e.g. it is a required computational input for the numerical stress analyses.

As part of a larger study (Saint-Pierre, 2018), a comparison was made of the relationship between Poisson's ratio assessed with changing stress level in the course of the UCS test and that assessed from laboratory assessed shear and compressional wave velocities prior to loading in the UCS test. This comparison is presented herein.

SAMPLE COLLECTION AND PREPARATION

As part of the larger study, samples representing different cementation levels had to be collected. Since in situ coring was out of the scope of the project, obtaining block samples from shallow excavations was identified as a viable alternative. Several construction projects in Las Vegas were identified and communication with the contractors was established. After evaluating several options, a site was selected and sample collection was coordinated with the contractor.

Details of the location and field collection of block samples, their transport to the Rock Preparation Lab of the Mining Engineering Department at the University of Nevada Reno, the coring of samples and further preparation of the 53 specimens to be tested are presented by Saint-Pierre (2018).

ULTRASONIC VELOCITY TESTING

Wave velocities were obtained using a Proceq Pundit PL-200 Ultrasonic Pulse Wave Transducer. The velocities were measured with 54, 150 and 250 kHz transducers for the compressional wave velocity, V_p, and 250 kHz for the shear wave velocity, V_s. It was decided that the lower frequency for V_p would be best since the higher frequencies are slightly distorted by voids and cavities (Ott, 2017). ASTM D2845-08 states that the aggregate in the sample should not exceed 10% of the specimen length. Given the nature of the material, this was not possible for many of the specimens.

Each test was performed using couplant gel to ensure good contact between the transducer components and the flat ends of the core specimen. The shear wave transducers required a special, much thicker couplant to adequately transmit the S-waves.

Careful consideration was taken to ensure the transducers were not in contact with material (aggregate pieces) that might have a higher velocity than the core specimen. Marks were made at each end of the core specimens to mark their mid-point so the transducers could be lined up as accurately as possible. Each specimen was also marked on its lateral surface with lines representing 120 degrees axial rotation. After the velocity was measured the first time, the specimen was rotated 120 degrees about its axis and measured again and this repeated for a third measurement. Once all three measurements were taken, the direction of propagation was reversed (end for end), and the same three measurements were taken. Averages were then recorded for each direction and the average of the two averages was recorded as the value for that specimen. Figure 1 shows the setup used in the test.

The gain was adjusted on the console to ensure the best delineation of first arrival. However, as the gain was increased, the pre-arrival noise also increased. To eliminate the ambient noise and ensure a strong first arrival, the gain was increased to where the ambient noise was noticeable, then decreased twice (Ott, 2017). Adjustment to the gain was made so that the first arrival wave peaked at 20% amplitude to ensure the observed wave forms were pronounced and unmistakable.

When measuring the P-waves, the equipment established the first arrival automatically. On the other hand, when measuring S-waves, this selection had to be done manually. Because shear waves are linear, one of the transducers would be rotated 90 degrees causing the shear waves to disappear, thus confirming that the wave interpreted to be an S-wave was indeed that. Figure 2 presents waves observed for specimen A-4-1 as an example.

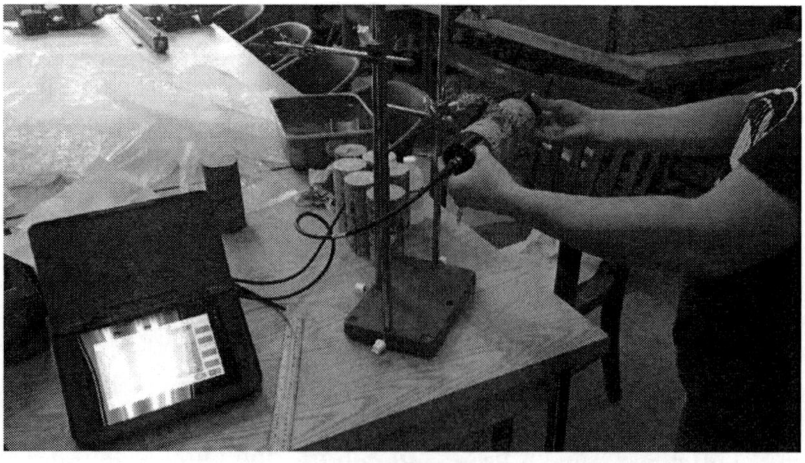

Figure 1 - Ultrasonic Testing Configuration

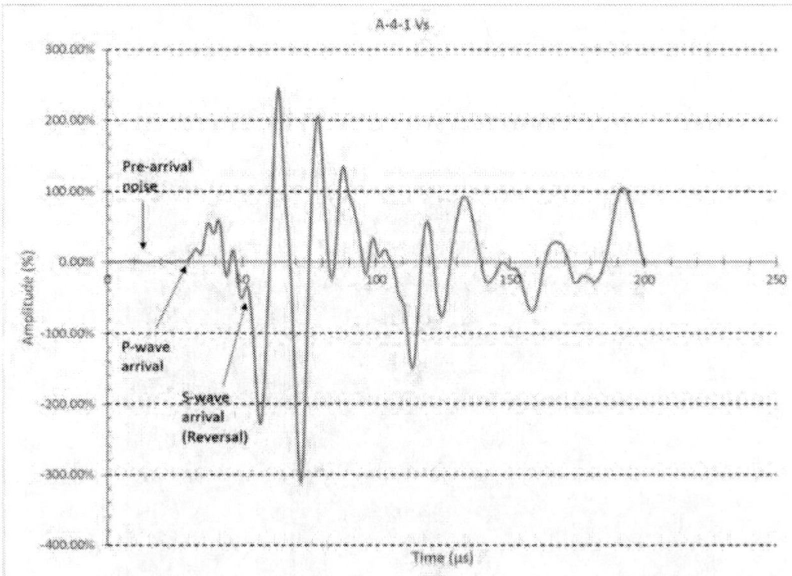

Figure 2 - Seismic Velocity Plot from Excel using Exported Data of Specimen A-4-1

UNCONFINED COMPRESSIVE STRENGTH TESTS

Unconfined Compressive Strength (UCS) test was performed using a Tinius Olsen displacement controlled, hydraulically powered load frame, The equipment uses manual controls with digital output. Compressive strength testing was conducted in accordance with ASTM D7012. The strain rate was maintained as constant as possible and failure of test specimens never occurred before two (2) minutes or after 15 minutes.

While LVDTs were employed throughout to record vertical sample deformation, it is with 16 of these tests that a compressometer was also used that is of interest here. The compressometer, which is used in tests of concrete samples, is not a piece of equipment that is typically used in the testing of rock samples. It records deformation in the central portion of the sample, away from the sample ends, thus reducing the end effects. Measurement is acquired in a single axial plane, from a hinged clamp equipped with a displacement sensor at one end. Since the axial displacement is measured from one end, a correction factor to the raw data should be applied per ASTM C469-14. This factor is based on the compressometer geometry and is equal to 0.505 for the compressometer used in this project. A significant advantage of the compressometer is that it also records lateral deformation of the sample. Figure 3 shows the mounted compressometer as well as the traditional LVDTs.

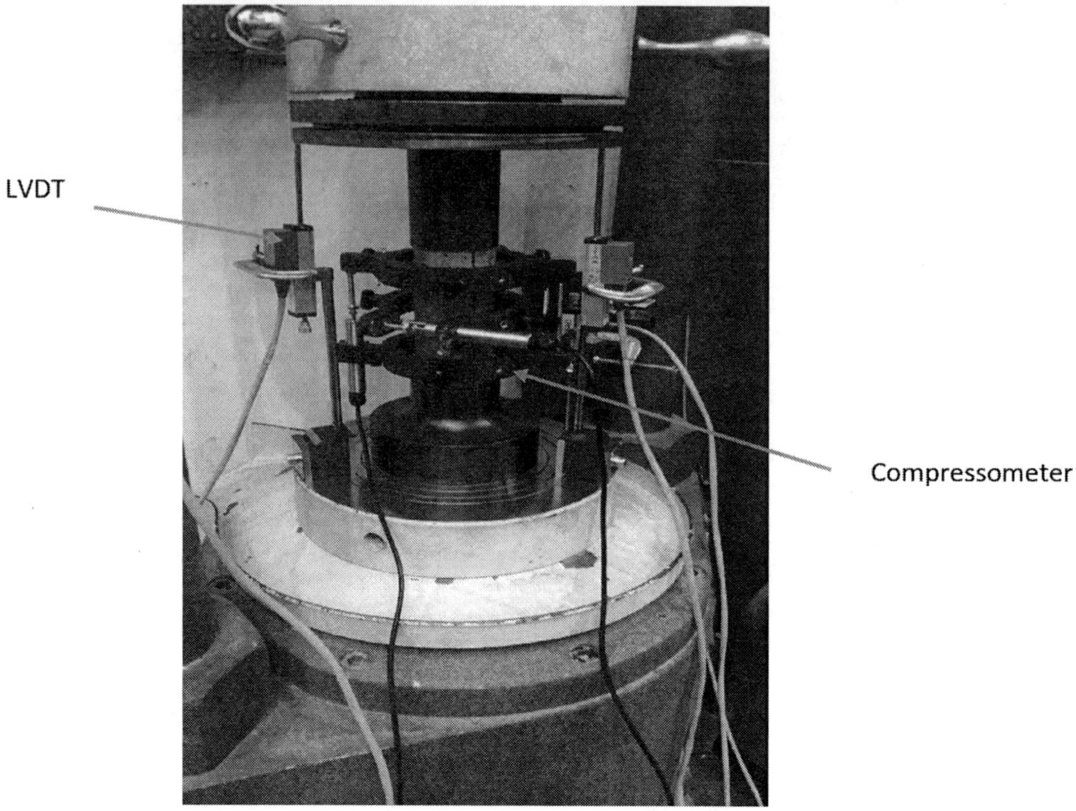

Figure 3 - Specimen with Mounted Compressometer Surrounded by Series of LVDTs

A steel blank was placed above and below the specimen to ensure the platens did not contact the axial and lateral displacement sensors of the compressometer. Once the hydraulics were engaged, the bottom plate advanced upwards. Specimens were loaded beyond failure to better establish and understand the failure mechanism.

POISSON'S RATIO DETERMINATION

The laboratory measured P- and S-wave velocities were used to calculate the sample's Poisson's ratio (μ) from Eq. 1 as provided in ASTM D2845.

$$\mu = \frac{(V_p^2 - 2V_s^2)}{2(V_p^2 - V_s^2)} \quad (1)$$

It is noted that this value represents Poisson's ratio at zero load. Furthermore, Eq. 1 represents a constrained (confined) condition as occurs in the field for geophysical tests under level ground conditions, the same as that in an oedometer test in the lab. This is not the unconfined condition observed in Figure 1 for the wave velocities obtained in this study. For unconfined conditions, for which Young's modulus, E, is ρV_p^2 and shear modulus, G, is ρV_s^2, where $E = 2(1+\mu)\,G$, it follows that Poisson's ratio becomes Eq. 2 (Das and Ramana, 2011),

$$\mu = \frac{(V_p^2 - 2V_s^2)}{2V_s^2} \quad (2)$$

By contrast, during the course of the UCS test, recorded values of vertical and lateral deformation obtained from the compressometer were used to assess increasing vertical/axial (ε_a) and lateral (ε_L) strains that develop under the applied axial stress (σ_d) and its associated stress level, SL (= σ_d /q_u where q_u is the peak stress, the unconfined compressive strength). From concurrent vertical and axial strains, the secant value of Poisson's ratio is given by Eq. 3 (Lambe and Whitman, 1969).

$$\mu = -\varepsilon_L / \varepsilon_a \quad (3)$$

However, this Poisson's ratio is not necessarily constant and typically changes as failure is approached (i.e. SL=1 at q_u) considering the dilation of the sample due to the opening of the fractures. Unfortunately, due to micro variations in lateral versus axial behavior of such nonhomogenous material (i.e. larger aggregate pieces and voids of varying size in the sample matrix), the instantaneous ratio of $\varepsilon_L/\varepsilon_a$ jumps around, yielding a wide band of its value, especially at low SL. Figures 4a through 4c show the variations in ε_a and ε_L at very low SL and the instantaneous Poisson's ratio with SL, for sample C-1-6. Another detail to notice is the near vertical variation in Poisson's ratio as SL approaches 1 as fractures open in dilatant response as the sample approaches failure. Of course, elastic theory would not have μ exceed 0.5 but elastic theory does not recognize extreme dilatant behavior in soil and rock.

While Figure 4a through 4c present the variations recorded for sample C-1-6, they represent a typical response recorded for all samples. Such response may also be due in part to fractures closing as load is reapplied after having been released due to sample retrieval and preparation. By contrast, assessing best fit equations for separate variations of ε_a and ε_L with SL, the ratio of the two equational values at the same SL was determined and Poisson's ratio assessed versus SL in this manner. Figure 6 presents the best fit equations for the variations of ε_a and ε_L with SL for sample C-1-6, as an example. It is noted that in many cases it was necessary to terminate the fitting short of SL = 1 because of the difficulty in finding best fit equations that would include data beyond where they were terminated (i.e. as SL approaches 1).

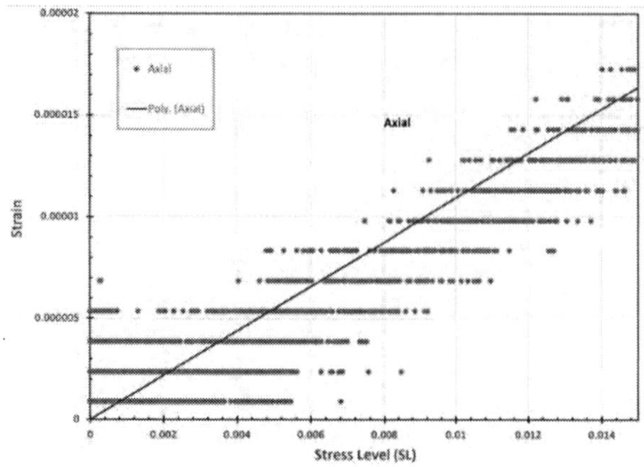

Figure 4a - Variations in Axial Strain (ε_a) with SL at low SL.

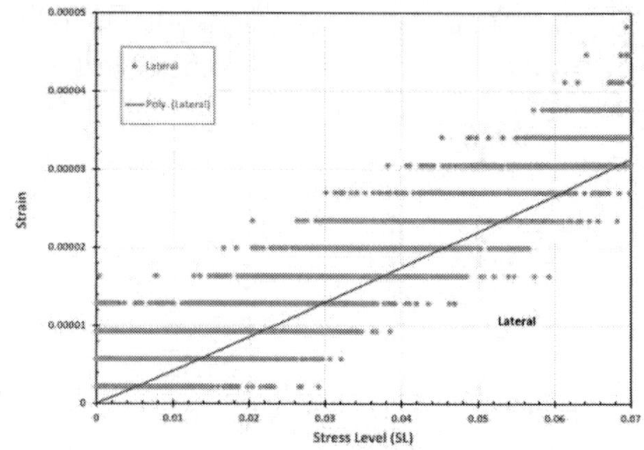

Figure 4b - Variations in Lateral Strain (ε_L) with SL at Low SL.

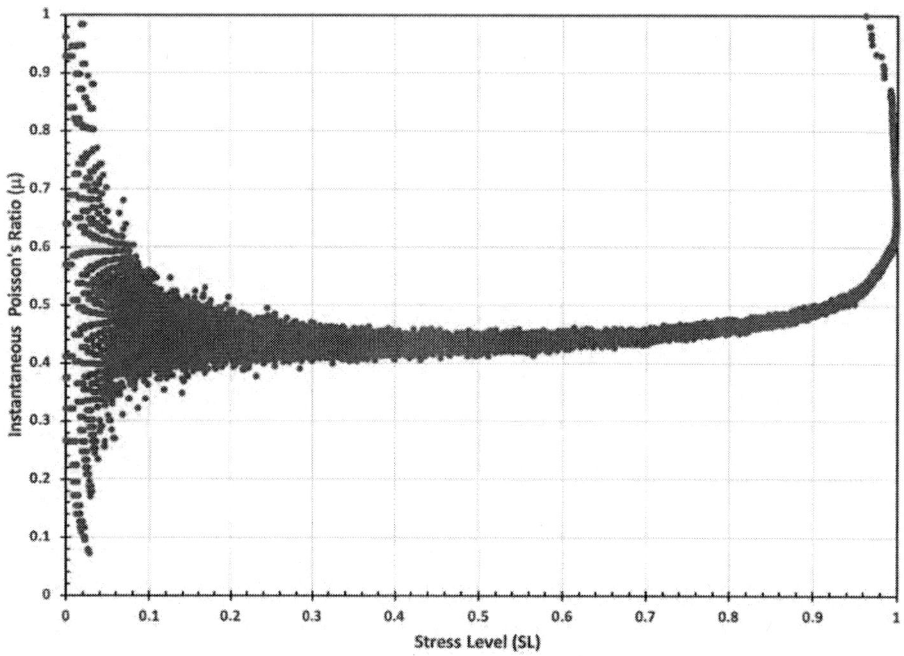

Figure 4c - Variation in Instantaneous Poisson's Ratio with SL

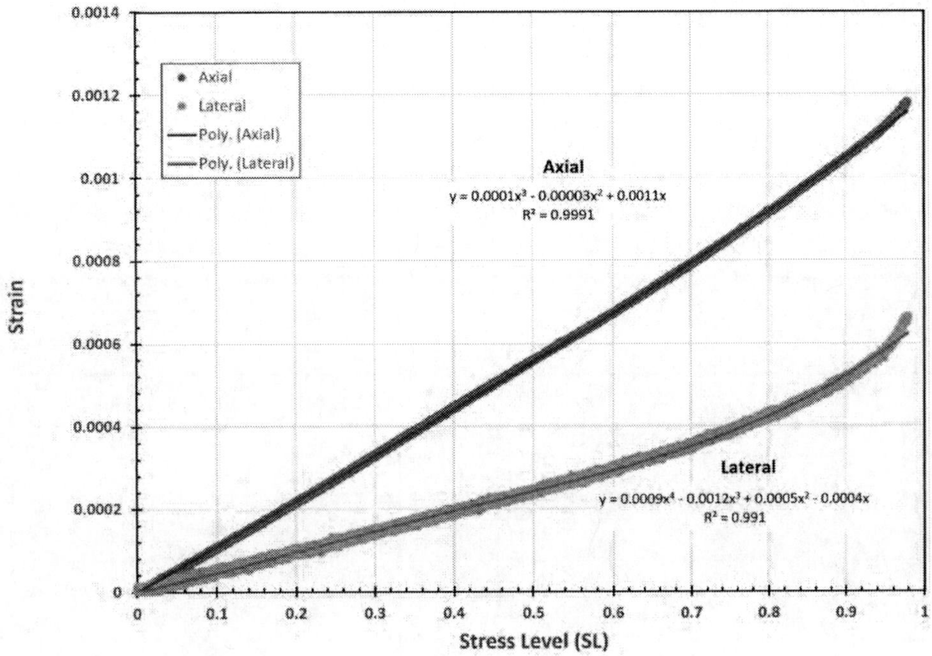

Figure 5 - Best Fit Equations for the Variations of ε_a and ε_L with SL for Sample C-1-6

Figure 6 presents the results of Poisson's ratio assessed using the aforementioned technique for 16 tests where deformations were recorded with the compressometer. As it can be judged, Poisson's ratio obtained from separate best fit functional variations of εa and ε_L show a continuous variation with SL, even approaching SL = 0. Values of Poisson's ratio at discrete values of SL (0, 0.5 and 1) taken from the various tests of Figure 6 are compared with those from wave velocity measurements (Eqs. 1 and 2) at SL = 0 in Table 1. Table 1 shows a summary of Poisson's ratio values for all the samples tested arranged in decreasing value of unconfined compressive strength (q_u).

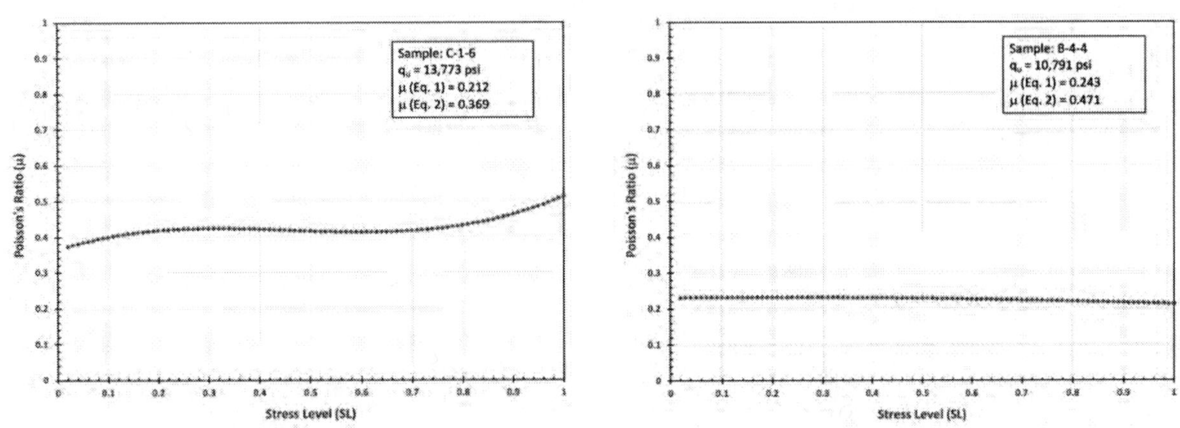

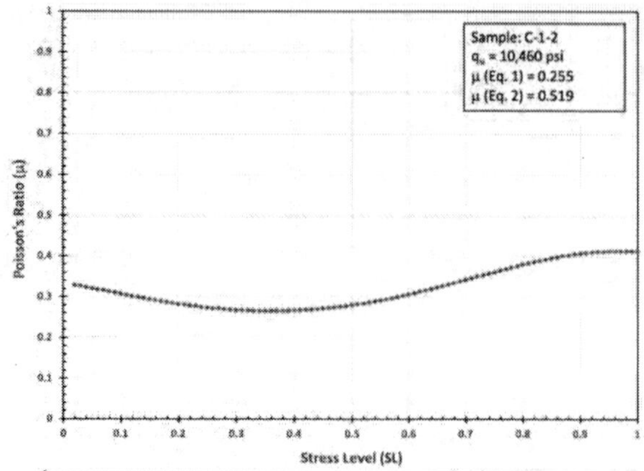
Sample: C-1-2
qu = 10,460 psi
µ (Eq. 1) = 0.255
µ (Eq. 2) = 0.519
Poisson's Ratio (µ)
Stress Level (SL)

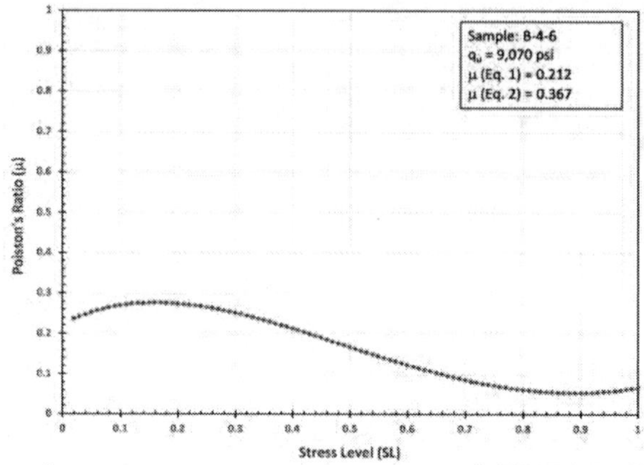
Sample: B-4-6
qu = 9,070 psi
µ (Eq. 1) = 0.212
µ (Eq. 2) = 0.367
Poisson's Ratio (µ)
Stress Level (SL)

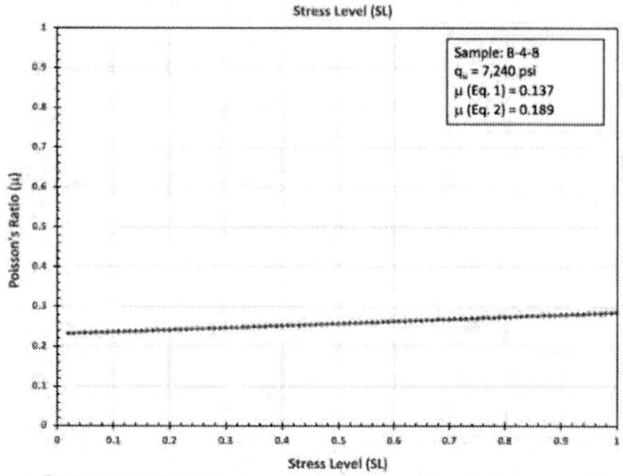
Sample: B-4-8
qu = 7,240 psi
µ (Eq. 1) = 0.137
µ (Eq. 2) = 0.189
Poisson's Ratio (µ)
Stress Level (SL)

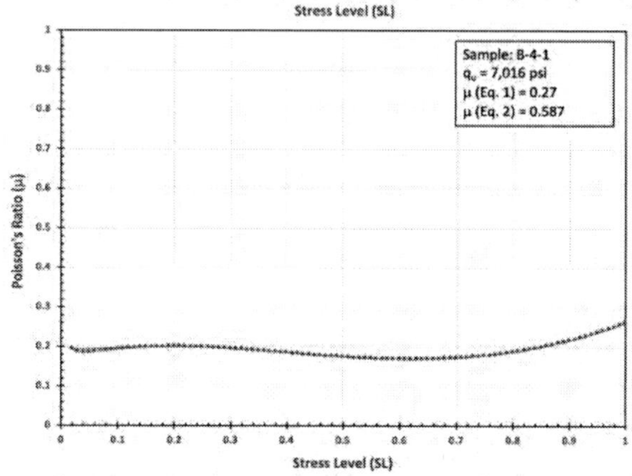
Sample: B-4-1
qu = 7,016 psi
µ (Eq. 1) = 0.27
µ (Eq. 2) = 0.587
Poisson's Ratio (µ)
Stress Level (SL)

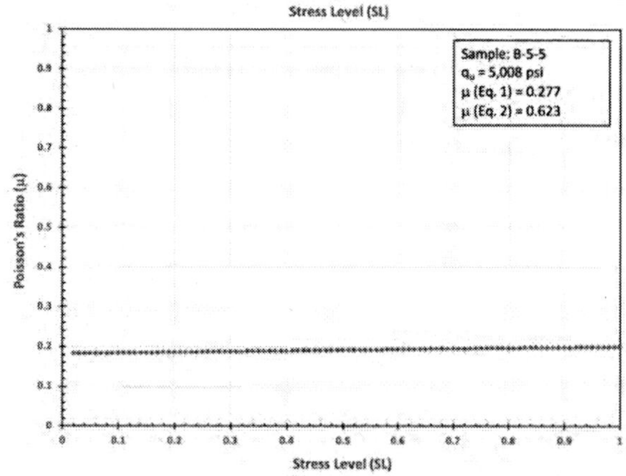
Sample: B-5-5
qu = 5,008 psi
µ (Eq. 1) = 0.277
µ (Eq. 2) = 0.623
Poisson's Ratio (µ)
Stress Level (SL)

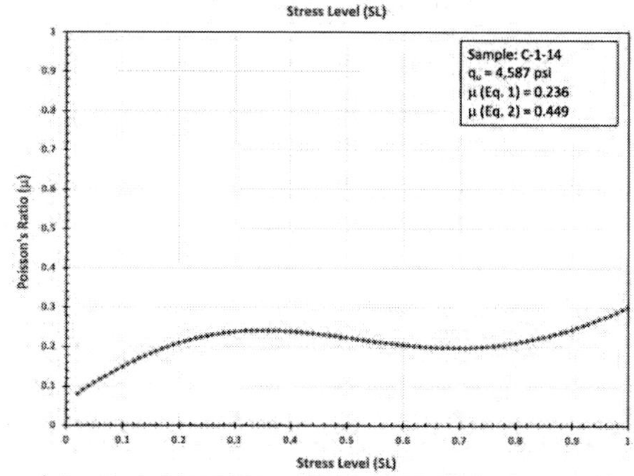
Sample: C-1-14
qu = 4,587 psi
µ (Eq. 1) = 0.236
µ (Eq. 2) = 0.449
Poisson's Ratio (µ)
Stress Level (SL)

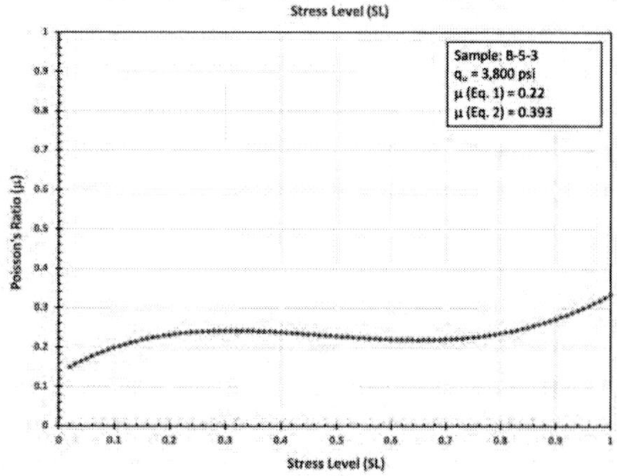
Sample: B-5-3
qu = 3,800 psi
µ (Eq. 1) = 0.22
µ (Eq. 2) = 0.393
Poisson's Ratio (µ)
Stress Level (SL)

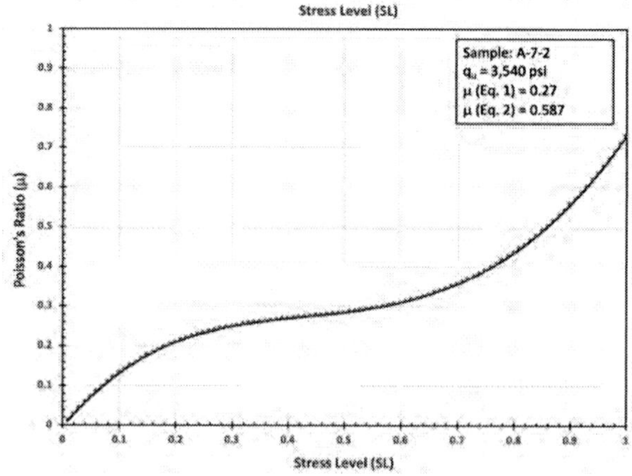
Sample: A-7-2
qu = 3,540 psi
µ (Eq. 1) = 0.27
µ (Eq. 2) = 0.587
Poisson's Ratio (µ)
Stress Level (SL)

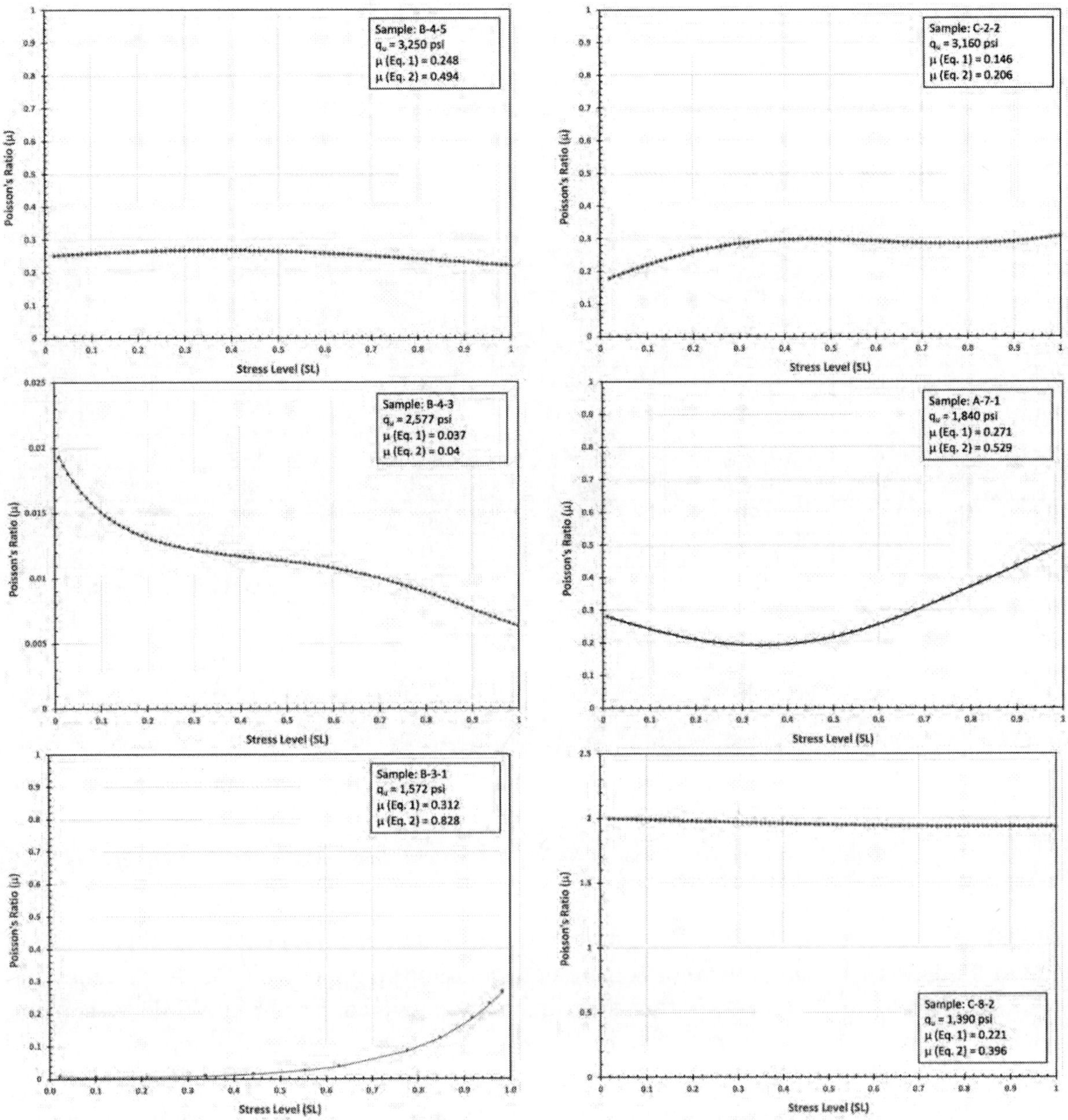

Figure 6 - Poisson's Ratio Assessed from Separate Best Fit Functional Variations of ε_a and ε_L

Table 1 - Summary of Poisson's Ratio Assessed by the Three Techniques

Sample ID	q_u (psi)	Poisson's Ratio				
		Confined (Eq. 1)	Unconfined (Eq. 2)	From UCS SL=0	From UCS SL=0.5	From UCS SL=1*
C-1-6	13,773	0.212	0.369	0.55	0.48	0.6
B-4-4	10,791	0.243	0.471	0.23	0.22	0.21
C-1-2	10,460	0.255	0.519	0.33	0.27	0.41
B-4-6	9,070	0.212	0.367	0.22	0.15	0.06
B-4-8	7,240	0.137	0.189	0.23	0.255	0.285
B-4-1	7,016	0.27	0.587	0.22	0.18	0.26
B-5-5	5,008	0.277	0.623	0.18	0.19	0.2
C-1-14	4,587	0.236	0.449	0.08	0.22	0.3
B-5-3	3,800	0.22	0.393	0.14	0.22	0.34
A-7-2	3,540	0.27	0.587	0	0.3	0.75
B-4-5	3,250	0.248	0.494	0.25	0.27	0.22
C-2-2	3,160	0.146	0.206	0.17	0.29	0.31
B-4-3	2,577	0.037	0.04	0.02	0.012	0.006
A-7-1	1,840	0.271	0.529	0.29	0.21	0.5
C-7-1	1,582	0.257	0.529	0.06	0.12	0.15
B-3-1	1,572	0.312	0.828	0.01	0.02	0.27

* Value assessed as SL approaches 1 but not exactly at 1

DISCUSSION/CONCLUSIONS

It should be noted that Eqs.1 and 2 for μ reflect V_p response in a homogenous isotropic elastic medium. Equation 1 is for confined or constrained conditions perpendicular to the direction of wave travel, while Eq. 2 is for V_p in the axial direction of an unconfined rod. While V_p is just distance divided by travel time, the value of V_p differs depending on the condition of perpendicular constraint; it is faster in the confined/constrained condition than the unconfined. V_p is equal to the square root of E/ρ for the unconfined rod, while it equals the square root of D/ρ in the oedometer test or the elastic half space of the field. D is the constrained modulus (= σ/ε in the direction of compressional wave propagation) which based on elastic theory is related to Young's modulus, E, as given by Eq. 4 (Lambe and Whitman, 1959).

$$D = \frac{E(1-\mu)}{(1+\mu)(1-2\mu)} \tag{4}$$

In the case of the cemented soil samples tested here, the material is not truly homogenous. Furthermore, it is not really elastic. With unbonded soil, it is known that the equivalent linear (or secant) values of G and E vary with the imposed stress and resulting strain. Such shear and Young's modulus reduction means that the resulting V_s and V_p experience a decrease with increased stress. It is the cementing present in the samples tested here that begs the question of how elastic-like is their response as compared to non-cemented soil. Poisson's ratio as assessed from Eq. 3 throughout the UCS test is the more realistic evaluation of an equivalent linear Poisson's ratio that is likely stress level dependent. The trouble with this approach is the wide band in strain that results at any particular SL, requiring the best fit equational approach to evaluation of μ with SL that was demonstrated here. In particular, μ exceeds 0.5 as SL approaches 1 in many cases. Such response does not comport with elastic theory and is likely due to dilatant behavior along micro and macro fissures/fractures in the sample. As judged by the best fit equations for horizontal and vertical strains versus stress level, the response is not linear, in which case we are assessing an equivalent linear (i.e. secant) Poisson's ratio variation.

It should be pointed out that the Poisson's ratio from best fit equations should only be considered up to the SL at which the best fit equations were assessed. As should be pointed out, the fit was terminated in some cases short of SL = 1 because of the difficulty in finding suitable equations beyond a certain point. Nevertheless, if the material is to be modelled under load, it is such variation in equivalent linear of secant Poisson's ratio that should be used, not that from wave velocity obtained at no load.

REFERENCES

AASHTO. (2014). AASHTO LRFD Bridge Design Specifications. Washington D.C.

Das, B.M. and Ramana, G.V. (2011), Principles of Soil Dynamics (2nd Edition), Cengage Learning, pp. 67-81.

Lambe, T.W. and Whitman, R.V. (1969). Soil Mechanics, John Wiley and Sons, p. 153.

Ott, T. (2017). Personal Communication with Evan-Saint Pierre.

Saint Pierre, E. (2018). The Development of a Material Model for Engineering Behavior Characteristics of Cemented Soils for the Las Vegas Valley. Thesis. University of Nevada, Reno.

Sirles, P. and Viksne, A. (1990). Site Specific Shear Wave Velocity Determinations for Geotechnical Engineering Applications. Geotechnical and Environmental Geophysics, Investigations in Geophysics, 5. 121-131.

Stone, R. C., Luke, B., Jacobson, E., & Werle, J. (2001). An overview of engineering with cemented soils in Las Vegas. In Luke B. Jacobson E. Werle J., eds., Proceedings: 36th Annual Symposium on Engineering Geology and Geotechnical Engineering (pp. 135-144).

Helicopter Sluicing for Rockfall Risk Mitigation in Response to the 2016 Kaikōura Earthquake

Rori Green, P.E.
Rori Green Consulting Limited
Christchurch, New Zealand
+64 21 289 9688
rori@rgconsulting.co.nz

Stuart Finlan, CPEng IntPE CEng MICE MASCE
New Zealand Transport Agency
24 Anzac Parade
Hamilton 3240, New Zealand
+64 21 985 135
stuart.finlan@nzta.govt.nz

Prepared for the 70th Highway Geology Symposium, October, 2019

Acknowledgements

The work described in this paper is based on the efforts of many professionals involved in the Kaikōura Earthquake recovery efforts. The team includes engineers, geologists, rope-access technicians, pilots, earthworks contractors, and project owners. The authors gratefully acknowledge their generosity in sharing their experience and insights. The authors also thank the following organizations for their support:

The University of Canterbury Quake Centre
New Zealand Transport Agency
KiwiRail
North Canterbury Transportation Infrastructure Recovery

Disclaimer

Statements and views presented in this paper are strictly those of the author(s), and do not necessarily reflect positions held by their affiliations, the Highway Geology Symposium (HGS), or others acknowledged above. The mention of trade names for commercial products does not imply the approval or endorsement by HGS.

Copyright Notice

Copyright © 2019 Highway Geology Symposium (HGS)

All Rights Reserved. Printed in the United States of America. No part of this publication may be reproduced or copied in any form or by any means – graphic, electronic, or mechanical, including photocopying, taping, or information storage and retrieval systems – without prior written permission of the HGS. This excludes the original author(s).

ABSTRACT

The November 2016 M7.8 Kaikōura earthquake resulted in excess of 50 landslides that directly impacted the key transportation corridor on New Zealand's South Island. In December 2016, the New Zealand government made the decision to restore road and rail service by the end of 2017. This meant that a significant effort was needed to remove unstable debris from earthquake-damaged slopes and enable relatively safer access for reinstatement works. Helicopter sluicing was selected as a key mitigation method for several landslides.

Several thousand hours of helicopter sluicing were performed during the emergency response and recovery works. The sluicing efforts were generally considered very successful and they were critical in achieving project deadlines. The overall sluicing programme is described, and site-specific examples are presented to show the range of outcomes.

INTRODUCTION

The November 2016 M7.8 Kaikōura earthquake caused significant damage to transportation infrastructure located in the northeast of New Zealand's South Island. Part of the damage was due to nearly 1 million cubic metres of rock falling onto the Main North Line (MNL) railway and State Highway 1 (SH1) from more than 80 landslides, cutting off a major transportation corridor and isolating the town of Kaikōura and surrounding rural communities. SH1 and the MNL railway form a critical transportation link serving New Zealand's second largest city (Christchurch) and the wider South Island.

By the end of 2016, the New Zealand Government made the decision to form an alliance to undertake work to restore the coastal transportation corridor. NCTIR, the North Canterbury Transport Infrastructure Recovery, is an alliance partnership between the New Zealand Transport Agency (NZTA), KiwiRail and four major construction contractors *(1)*. The alliance team consisted of up to 1700 people from more than 100 organisations. They were given the challenge of re-opening the corridor by the end of 2017.

THE EARTHQUAKE

The 14 November 2016 M7.8 Kaikōura earthquake was a complex event that involved rupture along multiple faults. Figure 1 shows the area most affected where significant ground shaking occurred. The event was felt throughout most of New Zealand. Fault rupture propagated northwest from the epicenter; surface ground rupture was observed along at least 20 faults spanning a distance of about 100 km.

Due to the significant ground shaking, more than 10,000 landslides were generated over an area of about 10,000 km^2 (*2*). The area affected by landslides is shaded red in Figure 1.

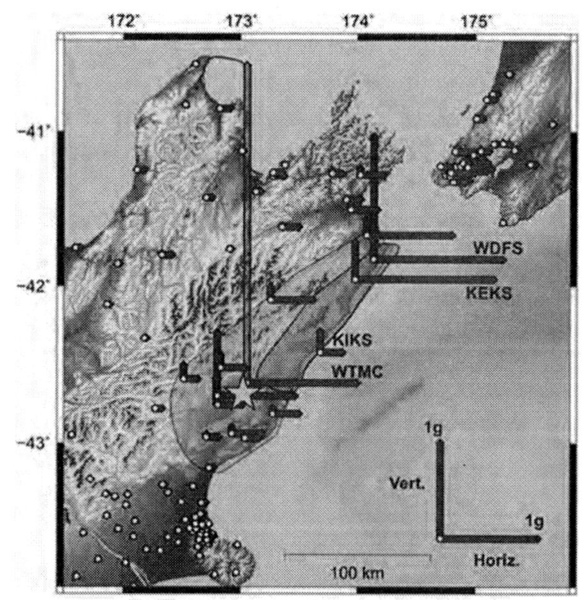

This graphic shows peak ground accelerations in the horizontal and vertical directions recorded by GeoNet strong motion instruments during the magnitude 7.8 Kaikoura earthquake of November 2016. The epicentre of the earthquake is shown as a yellow star. The vertical value of 3g recorded at Waiau in North Canterbury has unusual characteristics which scientists are still investigating. The approximate extent of mapped landslides is shown as pink shaded areas, with the majority occurring in the darker shaded region.

Figure 1 – Area affected by November 2016 M7.8 Kaikōura earthquake (*3*)

Effects on the Transportation Corridor

More than 80 landslides either directly affected or occurred upslope of the transportation corridor. Due to the landslides, as well as damage to road and rail infrastructure, access to Kaikōura (Figure 2) was completely cut off for 2-3 weeks until the Inland Route was re-opened as part of the Civil Defense emergency response works. Access via the main state highway (SH1) south of Kaikōura was re-established by mid-December 2016, but only during daylight hours due to increased rockfall risk from the tens of landslides along the 7 km southern coastal section. The 14 km-long northern coastal section of SH1 was more severely damaged and was not re-opened until mid-December 2017, 13 months following the earthquake.

Until SH1 was fully re-opened, traffic between Christchurch and Picton was detoured via an alternate route, which increased the travel time by 2 or more hours above the pre-earthquake 4.5-hour travel time via SH1. This route is a 2-lane road over a mountain pass; it had to accommodate higher traffic volumes (about 4000 vehicles per day), including an increase in heavy vehicles transporting freight that normally would have travelled by rail. The impacts of the diversion were a significantly reduced level of service as well as significantly increased transportation costs.

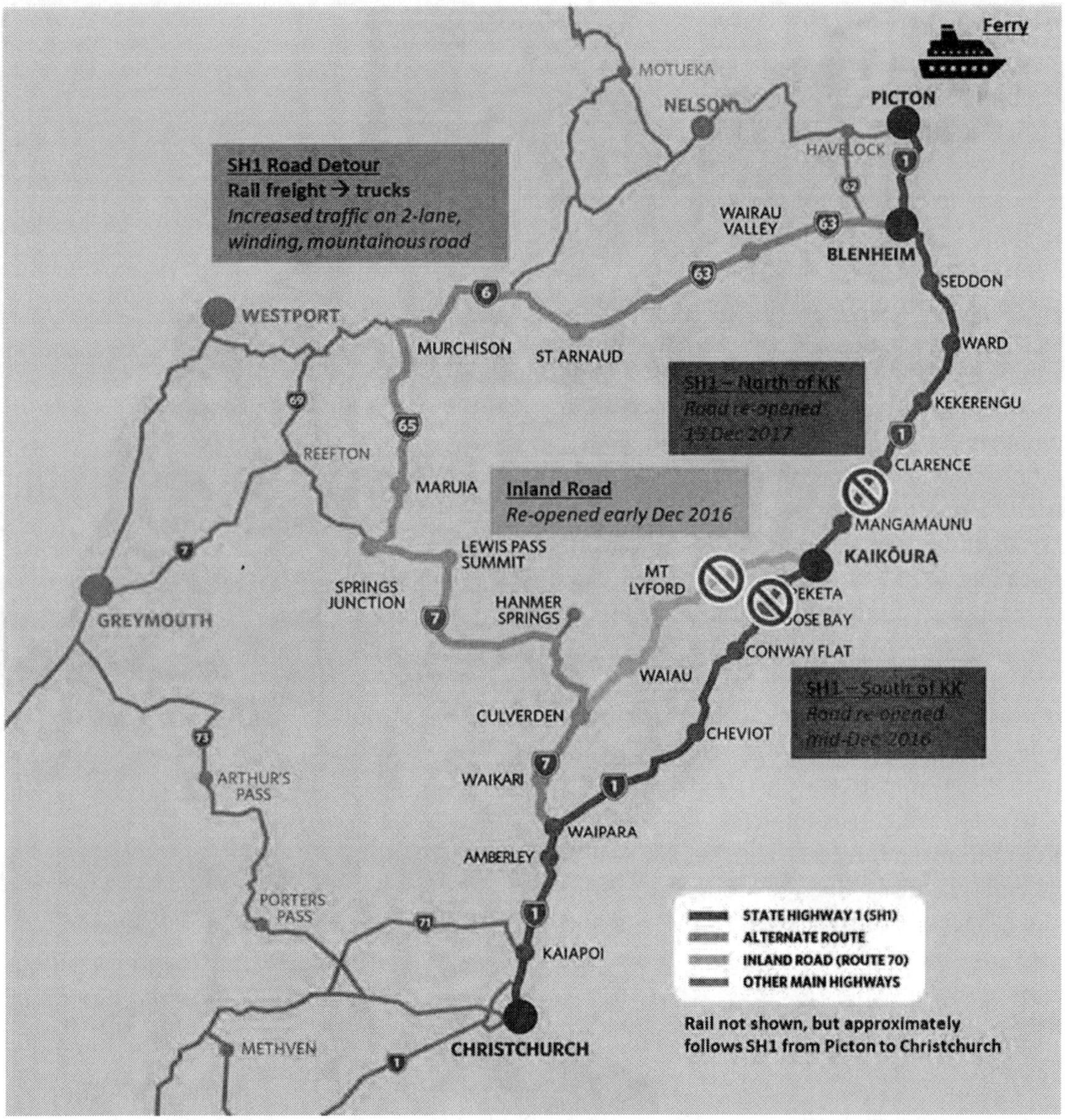

Figure 2 – Transportation network affected by Kaikōura earthquake (base map from NCTIR)

Time, more so than cost, was initially the key driver in selecting methods to mitigate the slope hazards. Landslide debris had to be removed so that work could begin to re-instate the road and rail within the narrow coastal corridor. Helicopter sluicing was selected as important tool to help achieve the tight project timelines.

Helicopter sluicing (Figure 3) has been used in New Zealand to clear unstable debris from slopes since at least the 1990's (*4*), although relatively little has been published about its use. The sluicing efforts undertaken in response to the Kaikōura earthquake are, by far, the most extensive use of this rockfall risk mitigation technique in New Zealand.

SLUICING FOLLOWING THE KAIKŌURA EARTHQUAKE

Sluicing was used in the early days of Civil Defence emergency response as part of the efforts to re-establish road access to Kaikōura via the Inland Road and SH1 south of Kaikōura (the south). It was used for periods of a few hours up to about 3 days to remove the loosest debris present on the slopes. It was typically used initially in headscarp areas to allow safer access for rope-access technicians undertaking scaling works; once rope-access teams were able to establish on the slopes, sluicing was undertaken in conjunction with scaling activities under the direction of site geologists and rope-access technicians. This methodology was used at more than 10 sites along the Inland Road and SH1 south. Once the roads were re-opened to traffic, sluicing (and scaling) activities had to be coordinated with road closures, which limited its further use in these areas.

Figure 3 – Helicopter sluicing, Kaikōura (photo from NCTIR)

Due to multiple major landslides north of Kaikōura (the north), the corridor was closed for a longer duration; loose debris had to be removed from the slopes to allow for safer access for earthworks teams to establish construction access roads. It was not until around mid- 2017 that construction road access was fully opened along the northern corridor. Because of the extent of the landslides and the duration of the road closure, more extensive sluicing works were undertaken; the majority of the sluicing was completed at seven sites.

Many thousands of hours were spent on sluicing efforts on the project – with significantly more (> 90 %) being undertaken in the north.

Sluicing in the North

Operations

Intensive sluicing was undertaken for about 5 months. At the height of the sluicing operations, up to 13 helicopters operated from a staging area established in a farm paddock as near as practical to the landslides. NCTIR set up an Air Operations Plan based on procedures in use by Fire and Emergency New Zealand for aerial firefighting, and engaged an Air Operations Manager to oversee helicopter movements.

A range of helicopters were used in the sluicing efforts, sourced mainly from the South Island. The smallest helicopter used was a Bell Jet Ranger (206B) with a 400 liter bucket (for small, short-duration jobs) and the largest was a Bell UH-1L (Vietnam-era "Huey") with a 1500 liter bucket. The bulk of the sluicing was done by B2 and B3 "Squirrel", and BK117 helicopters operating with 900 to 1100 litre buckets. Helicopters operated individually, or in circuits of up to 5 helicopters; it was found that more than 4 to 5 in a circuit reduced efficiency due to waiting time.

Site geotechnical personnel ("site geos") established sluicing plans with pilots daily due to the rapidly changing nature of the site conditions; rope-access technicians participated in the process where sluicing operations were undertaken in conjunction with scaling (Figure 4).

Figure 4 – Sluicing and scaling operations on an earthquake-damaged slope north of Kaikōura (photo from R. Musgrave)

Outcomes

An overview of some of the outcomes is illustrated via " before and after" photos in Figures 5 and 6. The photos show five major sluicing sites before and after mitigation works were undertaken; the mitigation works included a significant sluicing component together with varying levels of rope-access supported scaling and a limited amount of blasting.

Sluicing was considered, but rejected, at several sites where it was judged it would likely worsen the stability of the slope. This was particularly where landslide debris was situated above potentially erodible material, such as in-situ colluvium; it was also rejected as an option where there was a potential to cause further upslope regression of the landslide (Figure 7)

Figure 5 – Five of seven major sluicing sites north of Kaikōura before mitigation (photos from NCTIR)

Figure 6 – Five of seven major sluicing sites north of Kaikōura after mitigation; photos taken 7 to 11 months post-earthquake (photos from NCTIR)

Figure 7 – Example of site where sluicing was considered, but rejected due to potential to negatively impact slope stability (photo from NCTIR; LiDAR image from NCTIR GIS webviewer)

CASE EXAMPLES

To illustrate the range of outcomes, additional information about sluicing activities is presented via three brief case examples. These were selected as examples of:

- Successful outcome with a small effort (Conway Bluffs)
- Successful outcome with a substantial effort (Slip 7, Chutes B and C)
- Challenging issue that arose from sluicing efforts (Millie)

For simplicity, the information is presented in a series of tables and photographs. Information about the geologic setting is provided for overall context.

Geologic Setting

Most of the project site is situated within "greywacke" bedrock of the Early Cretaceous Pahau Terrane; this term is used locally to describe interbedded (strong) sandstone and (weak) mudstone. The bedrock is slightly to moderately weathered and is typically highly fractured and blocky, often with open jointing; it was significantly loosened by shaking associated with the Kaikōura earthquake, particularly along ridge crests. Bedrock is overlain in most areas by a layer of colluvium that is typically up to 1m thick on ridge tops, increasing in thickness downslope.

Numerous ancient landslides are present within the project area; these are believed to be slow-moving or inactive. Many of the recent earthquake-induced landslides have formed within larger, ancient landslides. These recent landslides tend to be shallow, translational debris slides involving vegetation and colluvium. At a few locations, landslides occurred within bedrock, particularly where unfavorably-oriented jointing forms large rock wedge failures. (*5, 6*)

Example 1 – Conway Bluffs

Figure 8 – Conway Bluffs before (photo from NCTIR)

Table 1 – Conway Bluffs	
Aspect	Description
Site Description	40 to 60 m- high slope situated immediately above road; blocky to highly fractured sandstone and mudstone bedrock; slope angle 45 to 60 degrees
Hazard Description	Shallow, translational landslide involving very thin surficial soil/vegetation layer and limited portions of underlying bedrock
Project Context	Slope situated above Inland Road (one of three roads to Kaikōura); all were blocked due to earthquake damage (landslides and road infrastructure damage); urgent time-driven works to re-establish access to Kaikōura as soon as possible
Reason for Sluicing	Remove loose rock and soil debris from slope to reduce immediate rockfall risk to rope-access scaling teams and road-users
Sluicing Effort	About 3 days; 1 helicopter; 1000 litre bucket
Outcome to Date	Loosest debris cleared from slope via sluicing and scaling; there were subsequent reports of rockfall, especially during heavier rainfall events (when road tended to be closed); additional scaling and sluicing undertaken about 18-months post-earthquake as part of programmed mitigation works; rockfall activity of slope has significantly decreased

Figure 9 – Sluicing at Conway Bluffs; rope access technicians circled (photo from NCTIR)

Figure 10 – Conway Bluffs post-mitigation, 3 weeks after earthquake (photo from NCTIR)

Example 2 – Slip 7, Chutes B and C

Figure 11 – Slip 7, Chute C headscarp area before mitigation (photo from NCTIR)

Table 2 – Slip 7, Chutes B and C

Aspect	Description
Site Description	250 m-high slope situated above rail and road; dilated, highly-fractured greywacke overlain by up to 3 m of colluvium; competent mid-slope bluffs form wedge-shaped debris chutes; slope angle 40 to 50 degrees, with steeper bluffs
Hazard Description	Shallow, translational landslide through colluvium and upper layer of dilated bedrock; ridgeline shattered due to earthquake shaking and has persistent tension cracks; failure volume about 60,000 m^3; fallen (sandstone) boulders up to 10m in dimension observed at slope toe
Project Context	Slope situated immediately above rail and road; geometric constraints for rail due to tunnel located adjacent to site; original slope toe needed to be uncovered to reinstate rail; construction access road not completed until 6 months post-earthquake, therefore no road closures required for sluicing/scaling works; time-driven works to re-establish northern rail access to Kaikōura by August 2017 (rail freight re-opening) and road access by December 2017 (public road opening)
Reason for Sluicing	Remove loose debris from slope to reduce rockfall risk for rope-access scaling teams and earthworks (debris removal) teams
Sluicing Effort	About 5 months; up to 5 helicopters (average 3) operating in circuit; 900 to 1500 litre buckets
Issues Encountered	Trees/tree roots occasionally caught up and binding rock debris; some trees able to be removed via grappling hook; debris occasionally choked in narrow, mid-slope chute; had to be cleared, usually with a few sluice buckets; care needed to avoid undercutting headscarp area, in part due to temporary relocation of nationally important, earthquake-damaged fibre-optic cable along ridge crest
Outcome to Date	Considerable volume of loose, disturbed material cleared from slope via sluicing and scaling (10's of thousands of m^3); subsequent rockfall activity relatively low compared with project-wide rockfall activity, especially during significant rainfall events; minimal additional on-slope rockfall risk mitigation works have been required to date

Figure 12 – Slip 7, Chute C sluicing from top-down to enable safer access for rope-access scaling teams (photo from R. Musgrave)

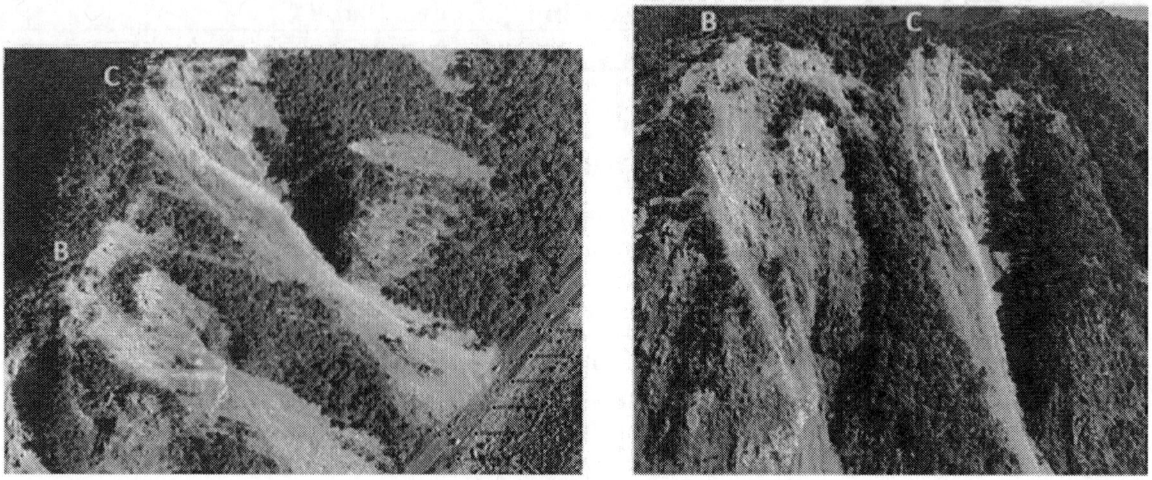

Before mitigation

After mitigation

Figure 13 – Slip 7 Chutes B and C before and after mitigation (aerial images from NCTIR GIS webviewer; oblique aerial photos from NCTIR)

Example 3 – Millie

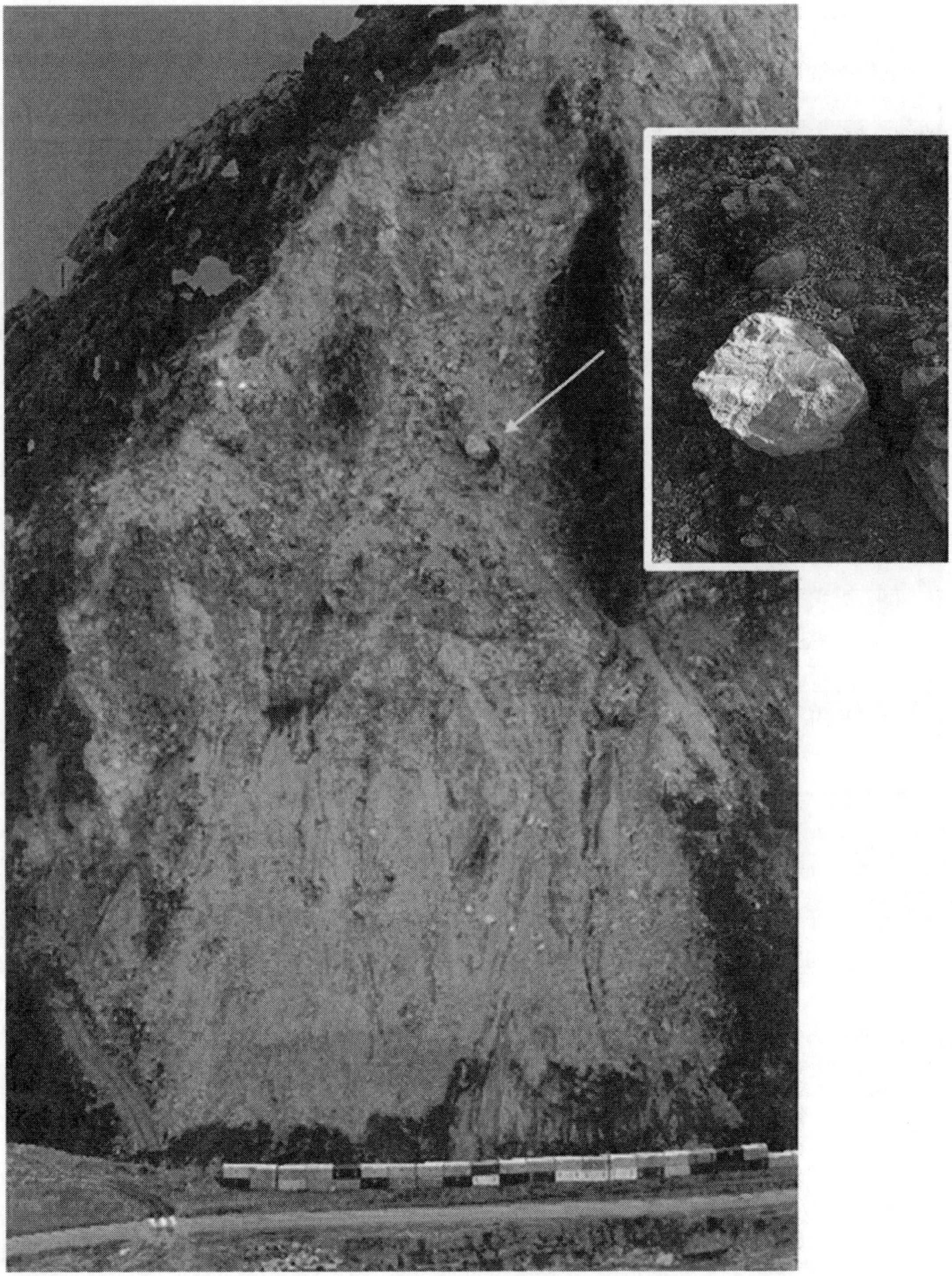

Figure 14 – Overview of slope post-sluicing, Millie indicated (inset photo and 3D photogrammetry image from NCTIR)

Table 3 – Millie	
Aspect	Description
Site Description	210 m-high slope situated above rail and road; dilated, highly-fractured greywacke overlain by up to 2 m of colluvium; occasional rock bluffs; slope angle 40 to 50 degrees, steeper in bluff areas
Hazard Description	Millie, a 50 m^3 boulder located mid-slope that has been partially undermined by sluicing efforts; maximum visible dimension 5-6 m
Project Context	Several similar-sized boulders were successfully removed from slope via sluicing; this was largely by removing surrounding (supporting) finer-grained debris, sometimes with assistance from the rope-access scaling teams where judged safely accessible. Water from sluicing efforts created flow channel adjacent to Millie, partially removing some of the surrounding debris; Millie was then targeted for removal by sluicing
Reason for Sluicing	Remove Millie
Sluicing Effort	Millie targeted for several days On the wider slope (Slip 7, Chute A), intermittent efforts over about 5 month period up to 5 helicopters (average 3) operating in circuit; 900 to 1500 litre buckets;
Issues Encountered	Millie unable to be successfully removed via sluicing; very large boulder left in a potentially precarious position; visually very scary; rope access teams not able to safely access for further assessment or removal
Outcome to Date	Millie successfully removed via blasting; considerable effort expended considering range of options for safe removal and executing the selected removal plan

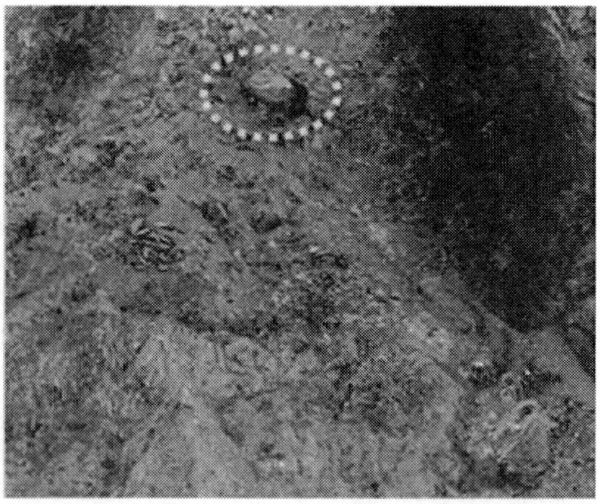

Figure 15 –Millie before (above) and after sluicing (photo and 3D photogrammetry image from NCTIR)

SUMMARY

Overall, the helicopter sluicing programme was judged to be very successful in quickly and safely mitigating slope instability hazards from areas that were not otherwise safely or easily accessible. The project deadlines could not have been achieved without sluicing.

There were a few challenges and less-than-successful outcomes, and arguably there are a few sites where decisions might have been made differently with the benefit of hindsight. Realistically, it is hard to know whether a different decision would have been made at the time given the available information; it is also difficult to know whether a different decision would have led to a better outcome in these few cases.

Helicopter sluicing is potentially a very useful tool in the rockfall risk mitigation toolkit. As with any solution, thoughtful consideration needs to be given to its appropriateness for the site conditions, the risks involved (financial, life-safety), and the project drivers (time, cost, safety). There is a real potential to end up with an unfavourable outcome and to drain the budget while reaching that outcome.

ON-GOING WORK

The information presented in this paper has been collated as part of a project aimed at documenting recent experience with two separate aspects of rockfall risk mitigation – sluicing and temporary rockfall protection – following the 2011 M6.3 Christchurch and 2016 M7.8 Kaikōura earthquakes. One of the project drivers is to document experience and lessons learned for use on future projects to aid decision-making, while both repeating successful practices and improving on those that were less successful. The project addresses in more detail the logistical considerations and methodology around sluicing, as well as presenting a number of case studies. Delivery is anticipated by the end of 2019.

The work is being funded by New Zealand Transport Agency and the University of Canterbury Quake Centre.

REFERENCES:

1. New Zealand Transport Agency. *Kaikoura Earthquake Response*. https://www.nzta.govt.nz/projects/kaikoura-earthquake-response/. Accessed May 2018.
2. C. Massey, D. Townsend, E. Rathje, K. E. Allstadt, B. Lukovic, Y. Kaneko, B. Bradley, J. Wartman, R. W. Jibson, D. N. Petley, N. Horspool, I. Hamling, J. Carey, S. Cox, J. Davidson, S. Dellow, J. W. Godt, C. Holden, K. Jones, A. Kaiser, M. Little, B. Lyndsell, S. McColl, R. Morgenstern, F. K. Rengers, D. Rhoades, B. Rosser, D. Strong, C. Singeisen, M. Villeneuve; Landslides Triggered by the 14 November 2016 Mw 7.8 Kaikōura Earthquake, New Zealand. *Bulletin of the Seismological Society of America*. doi: https://doi.org/10.1785/0120170305. 2018.
3. GNS Science. *Kaikoura quake produced strongest ground shaking in NZ, new research shows*. https://www.gns.cri.nz/Home/News-and-Events/Media-Releases/strongest-ground-shaking-in-NZ. Accessed April 2018.
4. Paterson, B. R. (1998). Effects and Mitigation of Rock-Fall Hazard SH73 Arthur's Pass Highway, South Island, New Zealand. *Roading Geotechnics 98: New Zealand Geotechnical Society 1998*, Auckland, July 1998, 235-245
5. R. Justice, G. Saul and D. Mason. Kaikoura Earthquake Slope Hazards – Risk Mitigation and Network Resilience. New Zealand Geomechanics, Bulletin of the New Zealand Geotechnical Society, Inc. December 2018, issue 96, pp. 26-37.
6. NCTIR. *Geology of Kaikoura Transport Corridor – Kaikoura North*. Internal report 100001-CD-GT-RP-0002. 24 April 2017.

Information about the three Case Examples has been collated from site notes, photographs, conversations with project personnel, and project reports; these have not been individually referenced.

Settlement Monitoring of a Trial Embankment in Philadelphia

Determining Site Specific Parameters for Large Embankment Construction

Sarah McInnes, P.E.
Pennsylvania Department of Transportation, Engineering District 6-0
7000 Geerdes Boulevard
King of Prussia, PA 19406
610-205-6544
smcinnes@pa.gov

Michael Yang, Ph.D., P.E.
Michael Baker International
300 American Metro Boulevard - Suite 154
Hamilton, NJ 08619
609-807-9534
myang@mbakercorp.com

Geoff Stryker, P.E.
STV Incorporated
1818 Market Street, Suite 1410
Philadelphia, PA 19103-3616
717-471-0600
geoffrey.stryker@stvinc.com

Seth Mascho, P.G.
Susquehanna Civil, Inc.
50 Grumbacher Road, Suite 10
York, PA 17406
717-846-7151
smascho@sqcivil.com

Prepared for the 70th Highway Geology Symposium, October 2019

Acknowledgements

The authors would like to thank The U.S. Federal Highway Administration; PennDOT Engineering District 6-0; Michael Baker International, STV Incorporated; Susquehanna Civil, Inc., James J. Anderson Construction Company, Inc.; and TRC Companies, Inc for their work on the various aspects of this project.

Disclaimer

Statements and views presented in this paper are strictly those of the author(s), and do not necessarily reflect positions held by their affiliations, the Highway Geology Symposium (HGS), or others acknowledged above. The mention of trade names for commercial products does not imply the approval or endorsement by HGS.

Copyright Notice

Copyright © 2019 Highway Geology Symposium (HGS)

All Rights Reserved. Printed in the United States of America. No part of this publication may be reproduced or copied in any form or by any means – graphic, electronic, or mechanical, including photocopying, taping, or information storage and retrieval systems – without prior written permission of the HGS. This excludes the original author(s).

Settlement Monitoring of a Trial Embankment in Philadelphia

Determining Site Specific Parameters for Large Embankment Construction

Author: Sarah McInnes, P.E., Pennsylvania Department of Transportation, Engineering District 6-0, 7000 Geerdes Boulevard, King of Prussia, PA 19406, phone: 610-205-6544, smcinnes@pa.gov

Coauthor: Michael Yang, P.E., Michael Baker International, Inc., 300 American Metro Boulevard, Suite 154, Hamilton NJ 08536, phone: 609-807-9534, myang@mbakerintl.com

Coauthor: Geoff Stryker, P.E., STV Incorporated, 1818 Market Street, Suite 1410, Philadelphia, PA 19103-3616, phone: 717-471-0600, geoffrey.stryker@stvinc.com

Coauthor: Seth Mascho, P.G., Susquehanna Civil, Inc. 50 Grumbacher Road, Suite 10, York, PA 17406, phone: 717-846-7151, smascho@sqcivil.com

PennDOT is presently reconstructing the I-95 Betsy Ross Bridge Interchange in the Bridesburg section of Northeast Philadelphia. This $880 million reconstruction includes replacement of the existing mainline, roadway and ramps. An innovative measure is being undertaken to eliminate structures at several ramps and replace them with roadway on fill. Staged construction, settlement considerations and existing deep foundations complicate the design. The benefits of replacing these structures with roadway on fill include considerable savings of construction costs and future maintenance efforts. Additionally, the use of on-site regulated fill from the project in the embankments will eliminate the cost of regulated fill disposal.

Evaluation of the elastic settlement from the proposed embankment and retaining walls over the in-situ sandy and non-plastic silty soil is a key factor to determine whether ground improvement measures will be needed. Significantly different elastic modulus values can be obtained using typical estimation methods based on Standard Penetration Test (SPT) N-values, Cone Penetration Testing (CPT) or relative density. To accurately assess the site-specific settlement, an instrumented, ±30-foot high test embankment is being constructed near a proposed I-95 ramp embankment. Instrumentation includes settlement plates, inclinometers, extensometers and piezometers. Vertical settlement at the ground surface of the test embankment at selected locations inside and outside the embankment is being monitored with settlement plates and settlement profiles. Inclinometers and extensometers were installed at different locations to identify the vertical and horizontal deflections at different depths. As the installation of test embankment continues, the field observed settlement data are compared with the theoretical settlement values, which are calculated using soil moduli estimated from both the SPT and CPT results. Based on the comparison, the estimated soil moduli that yield theoretical settlement results closest to the field data will be used at other ramp locations under similar subsurface conditions within the same project corridor.

INTRODUCTION & PROJECT DESCRIPTION

PennDOT's multi-billion-dollar I-95 corridor reconstruction effort in PennDOT's Engineering District 6 (comprised of the five-county region in Southeastern Pennsylvania) includes replacement of the existing mainline viaducts, roadway and ramps in Philadelphia from Vine Street to the Cottman Avenue Interchange. One of five sections of the corridor reconstruction is the Betsy Ross Bridge Interchange Reconstruction (referred to as section BRI) northeast of Center City in the Bridesburg and Port Richmond neighborhoods of Philadelphia. This design section, constructed in the 1960s and partially completed in 1973 & 1997, includes the Betsy Ross interchange, which connects I-95 to the Betsy Ross Bridge and New Jersey Route 90, and its associated ramps and connections to surface streets. This geometrically complex, three-tiered interchange services 160,000 vehicles per day and is the connection point for the Betsy Ross Bridge and major Philadelphia arterial routes including Aramingo Avenue and Richmond Street. The BRI project section comprises five design sections, BR0 and BR2 through BR5, and the overall goals of the project are to improve traffic movement between I-95 and the Betsy Ross Bridge, complete ramps and connections not previously constructed, increase capacity of the mainline and increase service life of the structures. The subject of this paper are Ramps B, D and H in construction section BR2 which connect NJ-90 and the Betsy Ross Bridge with I-95 southbound. The two locations discussed in this paper are referred to as Ramp BD and Ramp BH.

Figure 1: Section BR2 Shown in Red

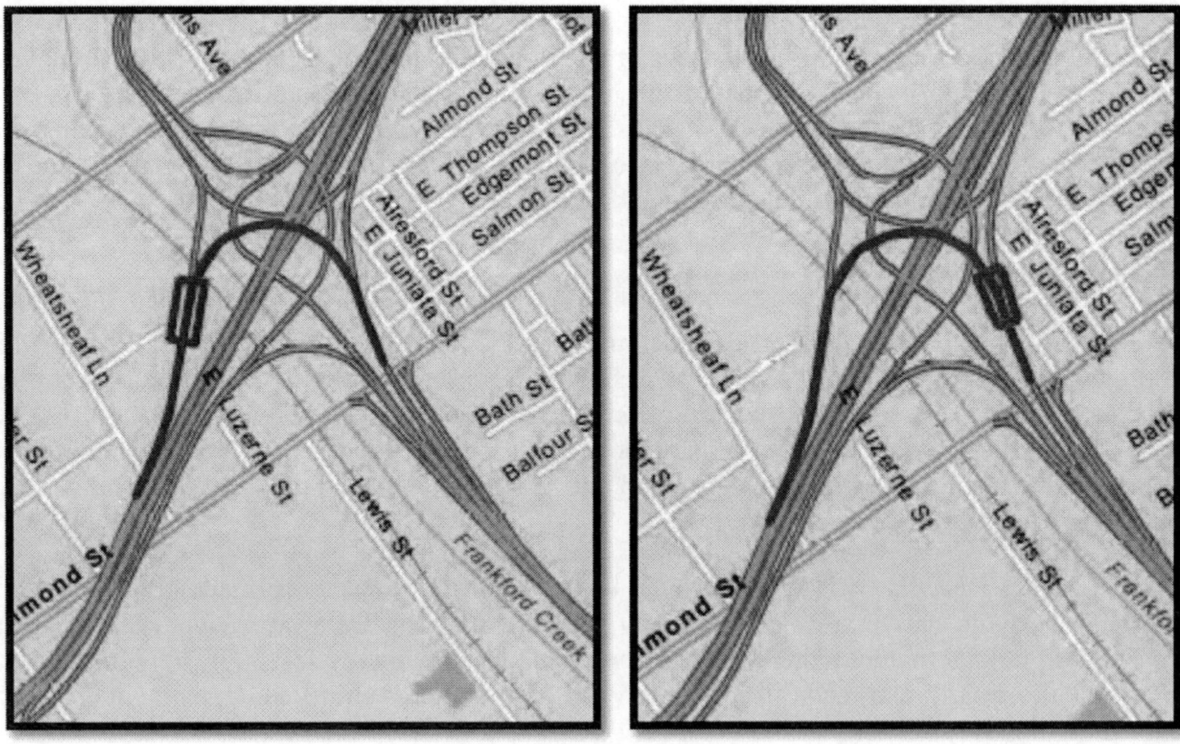

Figure 2: Location of Ramps BD (left) and BH (right)

Throughout the design process of the entire BRI project, the department and design team have considered replacing structures with roadway on fill where possible. Reasons for this include long term cost savings associated with eliminating bridge maintenance and replacement costs and using on-site regulated fill for embankment construction rather than paying the additional cost of off-site disposal. As a result, the design team began to investigate options for placing existing Ramps BD & BH on fill. Several factors were assessed including cost, staged construction and constructability, volume of regulated fill, and impact to existing foundations, however the most critical element required for success of this option was adequately predicting settlement of the existing loose soils. As a result, the department and design team partnered with a contractor from another I-95 construction project (section GIR) to construct an instrumented test embankment to measure site-specific settlement and back analyze the data to determine site-specific elastic moduli of the soil. A location adjacent to Ramp BH was determined to be ideal due to its location within the project limits and proximity to the proposed roadway on fill. Additionally, the department was able to recognize savings on the GIR project because regulated fill from GIR was used to construct the embankment and did not have to be disposed of.

SOILS AND GEOLOGY

Physiography & Topography

The project site is located in the Lowland and Intermediate Upland section of the Atlantic Coastal Plain Physiographic Province. It is located near the boundary of the Coastal Plain and the Piedmont

Physiographic Province, which is known as the Fall Line. The topography is characterized by flat upper terrace surfaces cut by shallow valleys of very low relief and the Delaware River floodplain. The area is underlain by unconsolidated to poorly consolidated sand and gravel deposits over complexly folded and faulted metamorphosed sedimentary and igneous rocks, primarily schist and gneiss. The drainage patterns are dendritic. Consequently, the project site is underlain by mostly unconsolidated to poorly consolidated sand and gravel.

Soil Survey

The Soil Survey Map of Bucks and Philadelphia Counties indicates the predominant soil within the project area is Urban Land (Ub). This designation represents highly variable and disturbed materials, generally including fill, resulting from previous construction and various land uses over time. Urban structures and works cover so much of this land type that identification of the soils is not practical. Most areas have been smoothed and the original soil material has been disturbed, filled over, or otherwise destroyed over time.

Adjacent to the Delaware River, the project area consists of loose man-made fills of various materials overlying native soils deposited by the action of the Delaware River. They consist primarily of granular material intercepted by lenses of clayey and silty soils. The uppermost strata are, for the most part, man-made fills. Sand and silt dominate the stratified deposits, interspersed with lenses of clayey soils.

Regional Geology

Figure 3 presents a portion of the Pennsylvania Geological Map (Philadelphia and Camden Quadrangles) with the project location indicated. As shown, the project site is mapped as being underlain by the Quaternary-aged Trenton Gravel (Qt). According to the Pennsylvania Geologic Survey and described in the *Engineering Characteristics of the Rocks of Pennsylvania*, the Trenton Gravel formation consists of gray to pale reddish-brown, very gravelly sand with interbedded, cross-bedded sand and clay-silt layers. These interbedded layers form a wedge that begins at the Fall Line and thickens toward the southeast. The Trenton Gravel is deeply weathered and composed of outwash and alluvium that consists of weathered gravel of granite, sandstone, gneiss, siltstone, and quartzite.

The Trenton Formation consists of gray to pale reddish brown, very gravelly sand with interbedded, cross-bedded sand and clay-silt layers. Thickness of the formation is about 30 feet. The formation is well bedded and cross-bedded. It has no apparent fracturing. It is deeply weathered and contains weathered gravel composed of granite, sandstone, gneiss, siltstone, and quartzite. These low-lying gravels occur at about 20 feet above sea level within the Delaware River Valley.

Below the Trento Formation lies the Wissahickon Formation, Oligoclase-Mica Schist Member composed of quartz, feldspar, muscovite, and chlorite. The oligoclase-mica schist variation is more coarsely crystalline than the associated albite-chlorite variation, excessively micaceous, and feldspar is more abundant. The estimated thickness of this formation is 8,000 to 10,000 feet. Bedding is fissile to thin and steeply dipping in most places. Fracturing is well developed and highly abundant. Joints are for the most part irregular, poorly formed, widely spaced, steeply dipping, and open. It is moderately resistant to weathering and often highly weathered to a moderate depth.

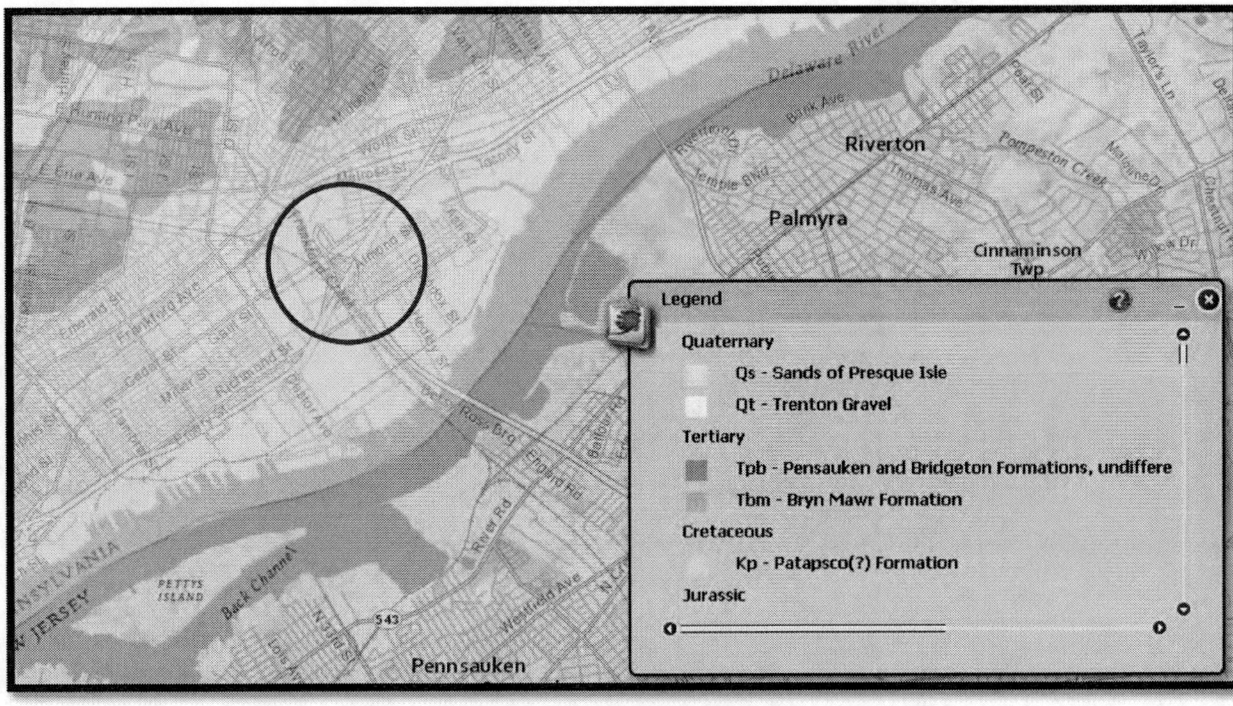

Figure 3: Geology Map

SUBSURFACE INVESTIGATION

The investigation for the BR2 design began with a review of available data from historic subsurface investigation programs conducted for this section of I-95 in the 1960s and 1980s. Seven borings were taken in the vicinity of Ramp BH and four in the vicinity of Ramp BD. Based on this review and experience with the potential for clay layers, organic silts and running sands encountered on other I-95 sections it was recognized that standard SPT test borings would not be sufficient. In addition to the traditional SPT borings, Shelby tube samples and cone penetrometer testing were specified. The subsurface investigation in the vicinity of Ramp BH consisted of eleven SPT test borings, five offset borings for Shelby Tubes and five cone penetrometer test (CPT) locations. At and near the Ramp BD location the investigation consisted of 21 SPT borings, nine CPTs and one Shelby Tube sample.

At Ramp BH, two CPT borings were drilled next to SPT test borings. Figure 4 presents the energy corrected SPT blow counts (N_{60}) from test borings and CPT borings. Commercial CPT interpretation software CPET-IT was used to obtain the equivalent blow counts. By comparing the CPT derived blow counts and measured blow counts from the drill rig with an automatic hammer it is concluded that the Energy Transfer Ratio (ETR) of the SPT drill rig was approximately 95%. "Calibration" using CPT results shows the advantage of providing continuous soil properties over the blow counts provided by test borings.

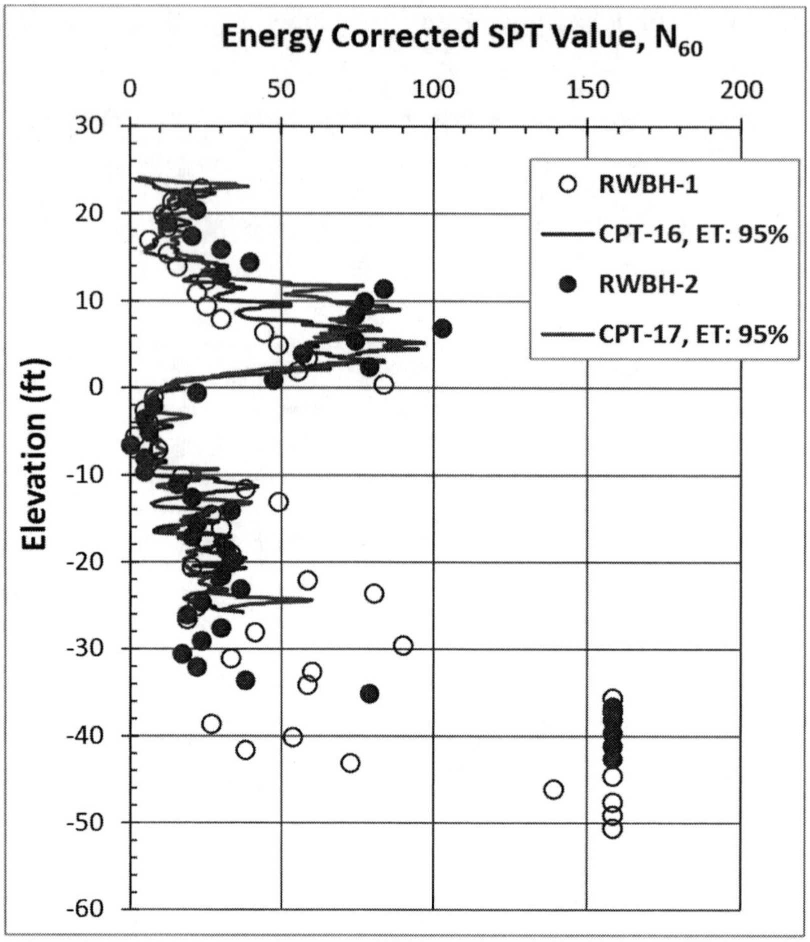

Figure 4: Comparison of SPT N_{60} Values with CPT Derived Blow Counts

The laboratory testing program for both locations included a series of tests to confirm classification, measure typical index properties, and evaluate compressibility, strength, and corrosion potential.

SUBSURFACE CONDITIONS

Overall, the subsurface conditions at both ramps were generally consistent. Figure 5 summarizes the subsurface conditions obtained from both SPT and CPT borings at Ramp BH general area.

Ramp BH area is relatively flat. The original ground surface elevation varies between 20 ft and 23 ft. The competent rock, which is defined as the bedrock that can be cored and has a variable recovery, is generally below elevation -50 ft. As shown in Figure 5, the soil above the bedrock can be classified in five (5) soil layers based on the field-tested blow counts, and material properties from the laboratory test results. The soil stratigraphy is presented in Table 1.

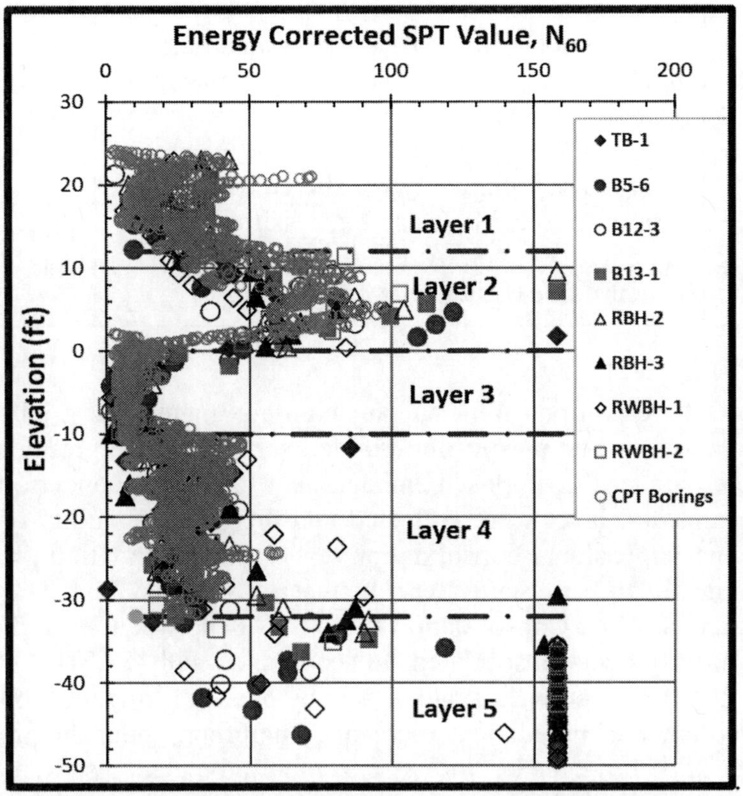

Figure 4 Subsurface Conditions at Ramp BH Area

Table 1: Summary of Subsurface Conditions

Soil Layer	Bottom Layer EL. (ft)	Soil Descriptions	Average N Value	Standard Deviation of N
1	~12.0	Loose to medium dense Silty Fine Sand fills. Average grain size D_{50}: 0.2 mm, fine content (passing #200 sieve): 30%-35%.	14	6.4
2	0.0	Very dense poorly graded alluvial Sand with Silt and Gravel. Average grain size D_{50}: 2 mm, fine content: 10%-15%. Out of 5 CPT borings, refusal was encountered in 3 borings.	43	19.3
3	-10.0	Stiff alluvial Lean Clay (plastic index less than 15) or non-plastic Silt. Sand content is usually less than 10%.	6	5.5

4	-32.0	Medium dense alluvial poorly graded Sand and Silt. Average grain size D_{50}: 0.5 mm, fine content: 15-20%.	19	1.25
5	-50.0 Bedrock	Very dense completely weathered bedrock (saprolite soil). Soils can be classified as Silty Sand with Gravel. Average grain size D_{50}: 1.0 mm, fine content: 10-15%.	Split spoon refusal	Not calculated

The average groundwater elevations at the site are at approximately Elev. 0.0 feet. As indicated in Table 1, Layer 3 is a low to non-plastic fine-grained soil below the groundwater. The potential time related settlement under the proposed embankment load are a concern. Undisturbed Shelby Tube samples from this soil layer were collected and the compressibility of this soil layer was evaluated through one-dimensional consolidation test with consolidation pressure applied to 32 tsf. The tested pre-consolidation pressures were between 5.8 tsf to 6.3 tsf. These pre-consolidation pressures correspond to an Over Consolidation Ratio (OCR) between 3.5 and 4.5, which indicates the Layer 3 soil is heavily over-consolidated. In accordance with AASHTO Section C10.6.2.4.1 (AASHTO 8th Ed., 2017), the settlement rate of heavily over-consolidated soils can be considered as an elastic-type settlement (immediate settlement). Therefore, under the proposed embankment load, the overburden soil above the bedrock is expected to be elastic settlement. It is recognized that the settlement within the embankment will be practically completed within one month after the final embankment height is reached.

Current PennDOT practice considers allowable total settlement for retaining wall and bridge structures as limited to one inch. Total anticipated settlement larger than one inch in the modular or mechanically stabilized earth retaining walls might be permitted if measures that could mitigate the differential settlement, such as, full height slip joints, are implemented. However, current PennDOT practice requires soil improvement or column supported embankments if the anticipated total settlement is greater than three inches.

In current design practice, the settlement of flexible footings/embankments is estimated by using Hough Method (AASHTO Equation 10.6.2.4.2-2, 2017) or using software FOSSA (Samtani & Nowatzki, 2006), which is the commercial version of the FHWA funded free software EMBANK. The advantage of the Hough method is that settlement values are directly related to the SPT blow counts and soil type. For the proposed embankment height of Ramp BH of 30 ft, the anticipated maximum settlement using Hough method is approximately 5 inches. Calculations of embankment/retaining wall settlement are also commonly performed using the FOSSA software. This program estimates elastic settlement based on the Boussinesq's equation. In addition to the soil stratigraphy information (soil layer thickness and unit weight), two parameters, elastic modulus and Poisson's Ratio, are required. The Poisson's Ratio of different soils does not usually vary significantly. However, the elastic modulus of different soils varies significantly, even for the same soil type and density. The current AASHTO (Table C10.4.6.3-1) has two different methods to quantify the elastic modulus values of different soils: from the empirical relationship of soil types and densities (consistent) and from the overburden and energy corrected SPT blow counts $N1_{60}$. Robertson (2009) also recommended a method of estimation of elastic modulus of soils using a CPT approach. Another elastic modulus relationship with energy corrected SPT blow counts (ETR 55%) proposed by Bowles (Table 5-6, 1997) is also widely used in practical engineering projects to estimate the settlement. In the Ramp BH area, soil properties for the settlement analysis for each soil layer are presented in Table 2.

Table 2 Soil Settlement Properties from Different Approach

Soil Layer	Soil Description	Assumed Unit Weight (pcf)	Poisson's Ratio	Various Modulus Values (ksi)			
				AASHTO $N1_{60}$	AASHTO Soil Type	Bowles N_{55}	CPT (Robertson)
1	Silty Fine Sand	120	0.25	3.1	1.5	2.0	8.0
2	Sand with Silt and Gravel	125	0.35	11.6	10.0	6.8	18.0
3	Lean Clay or Silt	105	0.32	0.5	1.0	0.7	5.0
4	Silty Sand	125	0.35	2.3	4.0	1.6	17.5
5	Silty Sand with Gravel	130	0.40	20	20	20	20

As shown in Table 2, the elastic modulus varies significantly using different approaches. The wide range of modulus values for the same soil types results in a significant variety of settlement prediction results. The magnitude of the settlement value will directly affect the decision to require ground improvement or even deep foundations. For Ramp BH, with a maximum embankment height of 30 ft, the predicted settlement using CPT derived modului resulted in a maximum settlement of slightly greater than one inch. No ground improvement is necessary if the actual settlement is close to the predicted value of about one inch. However, using AASHTO recommended modulus values, a maximum settlement of more than four inches is calculated and ground improvement is required.

TEST EMBANKMENT & INSTRUMENTATION

Calculated settlement using various methods resulted in inconsistent results at both ramp locations. Settlement at Ramp BD was calculated to be between 3.6 and 4.4 inches and settlement at Ramp BH between 2.6 and 6.1 inches, all settlement is predicted to be elastic. As a result of the variable results and elastic moduli and known inherent conservatism with correlated elastic moduli from tables, it was decided to construct a test embankment and directly measure settlements. The proposed embankment heights for Ramps BD & BH are 50 ft. and 30 ft. respectively. Based on available space, constructability and access requirements at the test location a 40 ft. wide and 30 ft. high embankment with wire mesh temporary MSE walls on three sides could be constructed adjacent to the proposed Ramp BH roadway on fill (Figure 6). Instrumentation included settlement plates, vertical and horizontal extensometers, inclinometers and vibrating wire piezometers (Figure 7).

Settlement plates (SP1 and SP2) were at the ground surface below the test embankment. Six vertical extensometers were installed at different locations below the ground to measure the effect of test embankment loads on the vertical deflections at various depths outside of the test embankment footprint. The horizontal and vertical deformation inside the embankment during construction was also monitored through horizontal and vertical extensometers. To capture the potential lateral movement of existing foundations adjacent to the embankment four inclinometers were installed at varying distances from the test embankment. Vibrating wire piezometers were installed in the Layer 3 low to non-plastic fine-grained soils to monitor the groundwater head changes under the embankment loads.

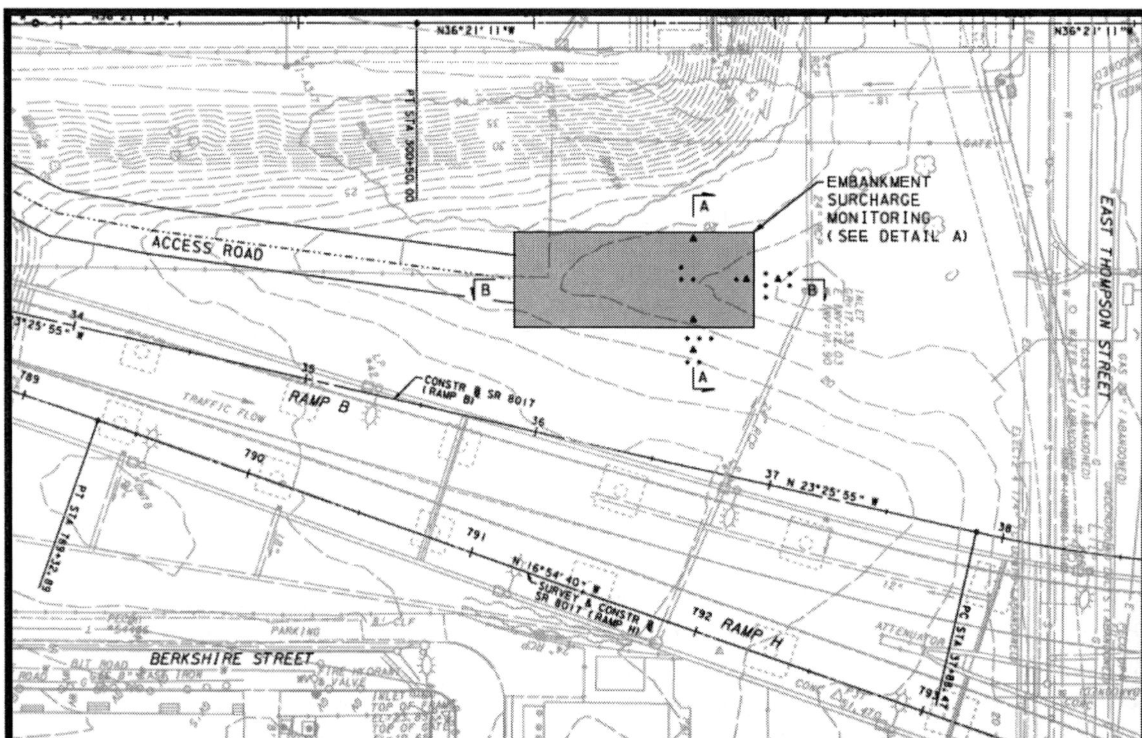

Figure 6: Test Embankment Location Adjacent to Proposed Ramp BH

Additionally, a test boring was taken within the embankment footprint, including laboratory testing of the soils, a proctor was run on the regulated fill material and nuclear gauge readings were taken. The horizontal and vertical deformation inside the embankment during the construction was also monitored through the horizontal and vertical extensometers. To capture the potential lateral movement on the existing footing adjacent to the embankment, four inclinometers were installed on the original ground with different distance to the test embankment. Vibrating wire piezometers were installed in Layer 3 low to non-plastic fine-grained soils to monitor the groundwater head changes under the embankment loads.

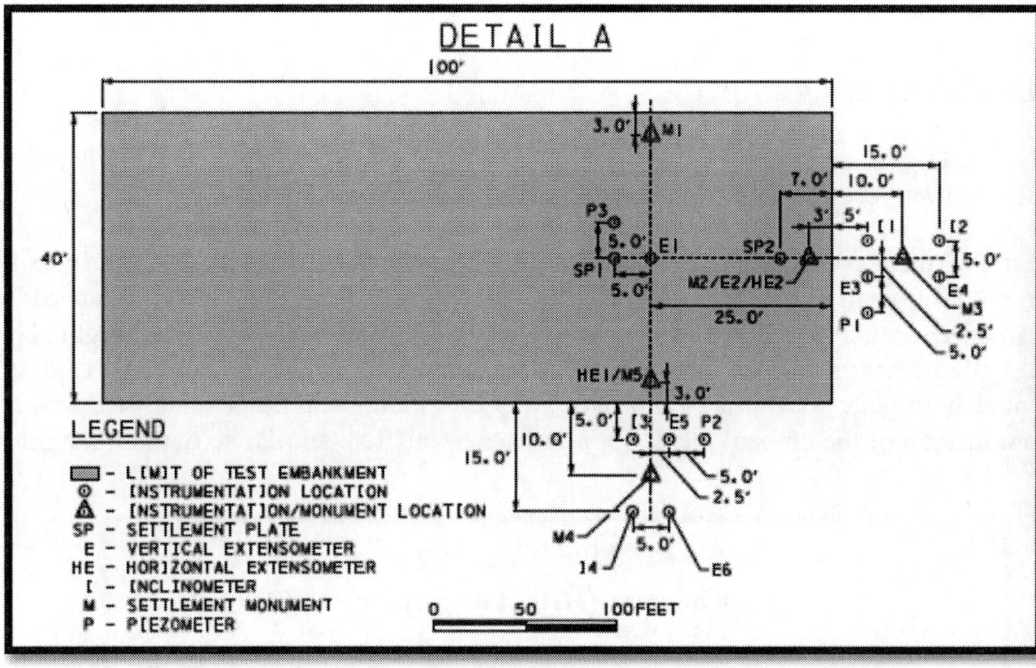

Figure 7: Test Embankment Instrumentation Plan

Figure 8: Test Embankment Construction Adjacent to Proposed Ramp BH

The estimated cost of embankment construction is $600,000 with a deduction of $210,000 because the regulated fill used in embankment construction did not have to be disposed of from the GIR project.

INSTRUMENTATION RESULTS

Currently, the test embankment is roughly 2/3 of its final planned height of 30 ft. The following are the observation results from the current instrumentation results.

Prior to embankment construction, two settlement plates were installed at the centerline of the test embankment 20 ft from both sides of the wall on the surface of original ground. Plate SP1 is 30 ft away from the end and Plate SP2 is 7 ft from the edge of the wire mesh MSE wall. Figure 8 presents observed settlement with embankment height. At the current embankment height of 18 ft, observed settlement at both plate locations is approximately one inch. Although Plate SP2 was only 7 ft away from the end of the embankment, both plates demonstrate similar settlement magnitudes.

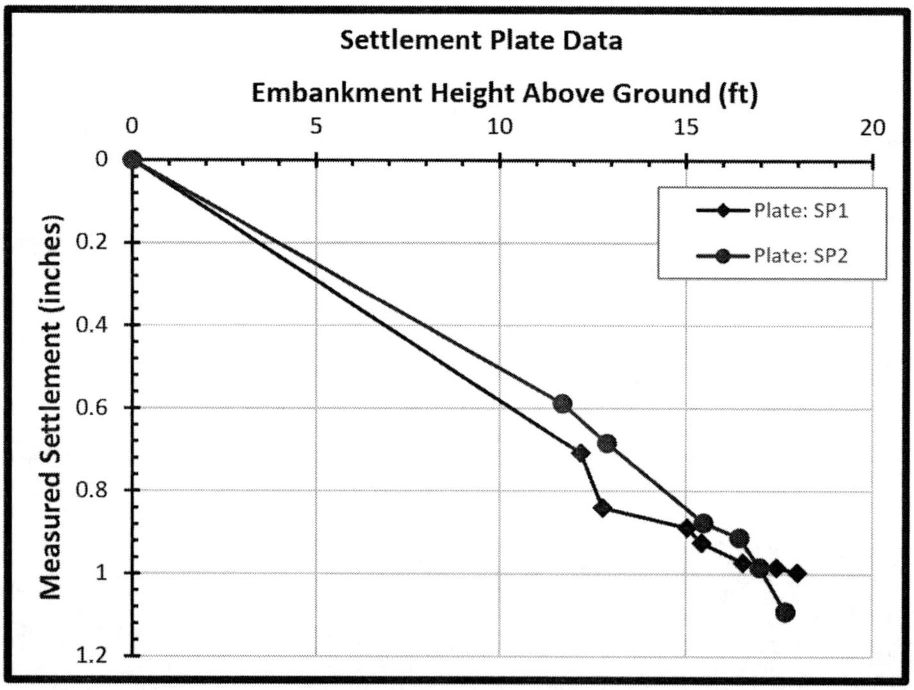

Figure 9: Observed Settlement with the Test Embankment Height

The vertical extensometers as well as inclinometers were installed at distances of 5 ft and 15 ft away from the end or side of the temporary wire mesh wall. The settlements registered at various depths below the ground are very small and it is generally less than 0.1 inches. No significant lateral soil movements were observed from the inclinometers.

All three piezometers at different soil layers have no significant groundwater head changes which suggests no pore water pressure build-up in any of the soil layers instrumented. All deformations are therefore, elastic deformation.

TEST SETTLEMENT ANALYSIS

Elastic settlement calculation in the FOSSA software program is based on the elastic solutions of stress in an infinite half-space. It computes the magnitude of the elastic and consolidation settlements resulting from roadway loading conditions (embankments, retaining structures, and all forms of permanent and temporary loads). As indicated previously, the elastic modulus values vary significantly from different resources. Different soil parameters listed in Tables 1 and 2 were used to estimate settlements across the embankment. The layer thickness, unit weight and Poisson's Ratio of each soil layer are identical in each analysis and the only variable parameter is the elastic modulus of the soil layer based on different approaches. The settlement analysis results are summarized in Figure 9.

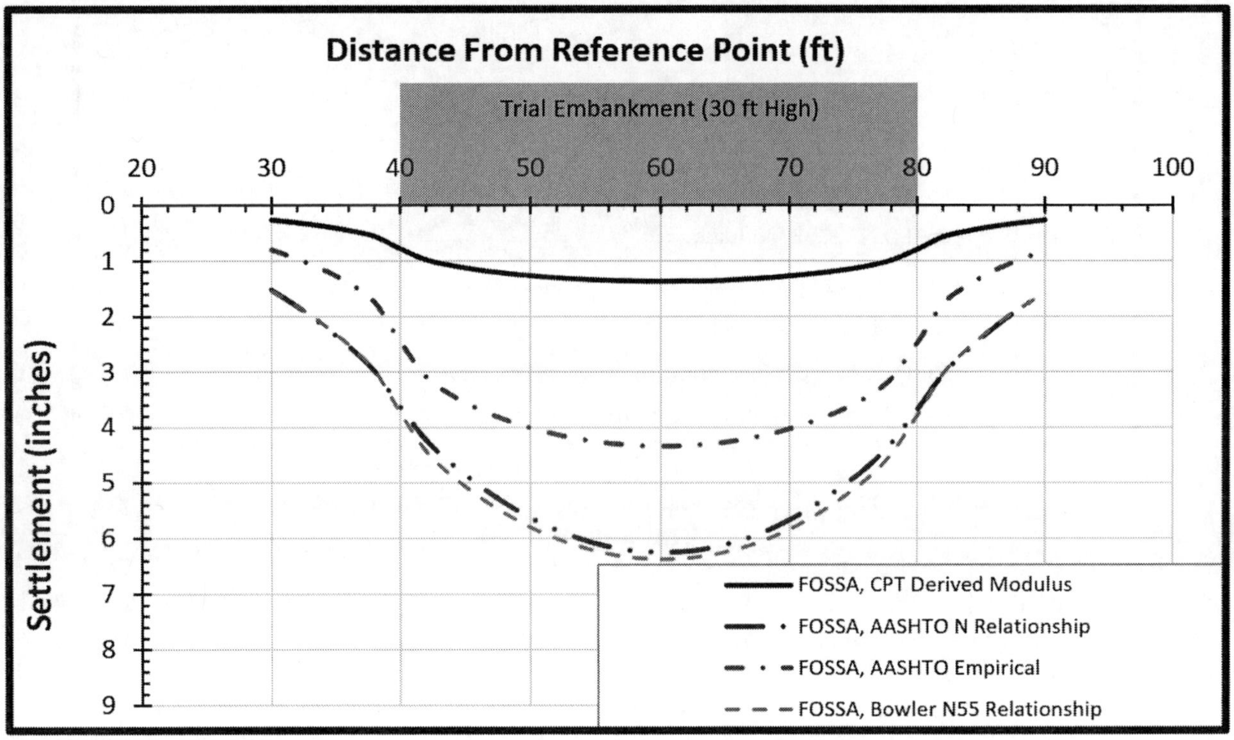

Figure 10: Predicted Settlement Values Across the Test Embankment

As indicated in Figure 9, the maximum settlement values vary significantly ranging from 1.4 inches using modulus derived from CPT results (Robertson, 2009) to 6.4 inches using AASHTO $N1_{60}$ relationship (AASHTO, 2017) and Bowles' N_{55} relationship (Table 5-6, 1997). The construction of this instrumented test embankment provides a valuable opportunity for the engineering professionals to select elastic moduli values that can provide more reliable settlement values. Figure 10 present the maximum settlement values using different approaches at the center of the test embankment with different embankment heights. The observed settlement at settlement plate SP1 is also plotted in this figure for the comparison purposes.

Figure 10 indicates that the modulus values derived from CPT results have a better overall prediction of ground settlement under the test embankment load. Compared with modulus values in the first four soil layers (except for the saprolite soil layer 5) listed in Table 2, the CPT derived modulus is 3 to five times higher than those recommended by current AASHTO (2017) and Bowles (Table 5-6, 1997).

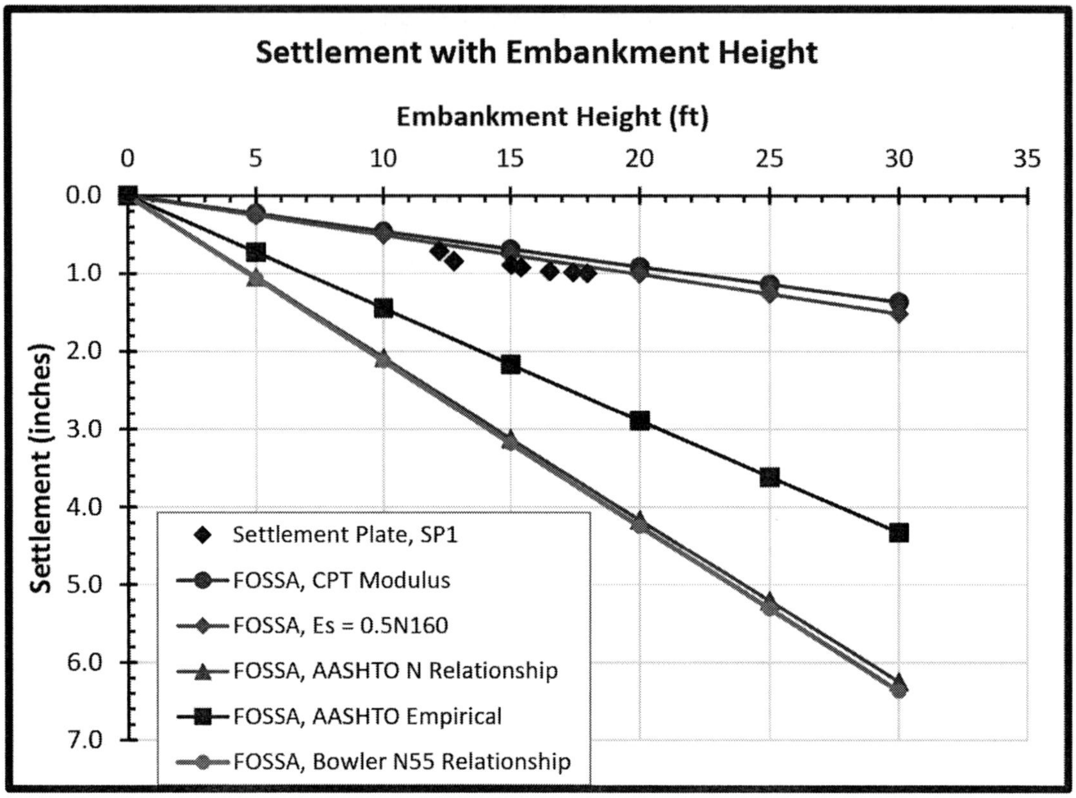

Figure 11: Comparison of Observed and Predicted Settlement at Center of Test Embankment

Although the site-specific CPT derived modulus is more reliable for each soil layer and can be used reasonably to estimate the settlement under the embankment loads, the SPT N values are still the predominant investigation tool used in the greater Philadelphia area because CPT data are not always available for all transportation projects. Therefore, a relationship of elastic modulus that correlates the SPT N-values is desirable. Following the AASHTO process, the new relationship between elastic modulus and the overburden and energy corrected $N1_{60}$ is developed by comparing the observed settlement values over the analyzed values. It is concluded that a simple linear relationship between elastic modulus (E_s) and $N1_{60}$ can be expressed as:

$E_s = 0.5\ N1_{60}$ (for the current study, a hammer energy transfer ratio of 95% is assumed)

The settlement analysis result using the new $N1_{60}$ relationship is also presented in Figure 11. The settlement at Ramp BD using the test embankment "calibrated" modulus values is calculated to be around two inches due to the thicker Soil Layer 2 and stiffer Soil Layer 3.

Based on the validated settlement analysis, it is possible to accurately predict the settlement magnitudes at Ramps BD and BH. After discussing this approach with all disciplines among the owner and design consultants, expensive ground improvement or column supported embankment is not required and regulated fill soil can be used in embankment construction. It is estimated that the potential cost saving is approximately $20 million and a schedule saving of at least 6 months can be achieved.

ROADWAY ON FILL DESIGN

The ability to obtain site-specific settlement data and predict the behavior of the soil profile impacts both design and construction. Originally both ramps were to be constructed using staged construction however presently Ramp BH can be closed and detoured eliminating staged construction at that location. Because of this requirement, the existing foundations at Ramp BD would require a retrofit as the existing cast-in-place concrete piles would not have the capacity to carry the additional vertical and lateral loads from embankment placement beneath the ramps while safely carrying traffic. Also, due to right-of-way constraints, retaining walls are required at both ramps. The specification of precast modular (PM) walls is precluded for settlements greater than one inch per PennDOT requirements. It was felt by the department, design team and design management that the predicted settlements were conservative, and it was more likely actual settlement would be one to two inches, hence the test embankment. After discussion with the structural engineers and reviewers it was agreed that if actual settlement were two inches or less, PM walls could be constructed with slip joints to allow movement during construction. It was also understood that all settlement would be elastic and post-construction settlement would be minimal and within required tolerances.

CONCLUSIONS

The current PENNDOT embankment and retaining wall could only tolerable limited total as well as differential settlements. Higher predicted settlement in the roadways and retaining walls will require soil improvement method(s) to mitigate the larger settlement. To accurately predict the settlement, even elastic settlement, is a challenge due to the significant variation of modulus values from different sources. The construction of this instrumented test embankment provides a rare opportunity to validate the elastic settlement and especially to validate the elastic modulus values. The following conclusions can be made:

- The current AASHTO recommended elastic modulus values are very conservative for the alluvial soils in the greater Philadelphia area. The best fitted modulus values are generally three to four times higher than the current recommended value. An over-conservative estimate may sometimes adversely affect an otherwise reasonable engineering solution for a potential geotechnical concern.
- CPT derived elastic modulus values using Robertson's approach results in a better settlement estimation and can be used in alluvial soils in the greater Philadelphia area. When the CPT results are not available, the approximate relationship between the elastic modulus and the overburden and energy corrected SPT $N1_{60}$ can be used for the settlement analysis.

- Due to the relative uniformity of the subsurface conditions, the site-specific elastic modulus parameters can be used with high confidence across the BRI project site going forward with roadway on fill designs, specifically for the critical mainline section BR3 & BR4.

- The construction of the test embankment results in a significant cost saving by verifying embankments can be constructed without ground improvement or column supported embankment. Further, regulated fill materials can be placed underneath ramps BD and BH, which result in a further cost savings.

REFERENCES

1. American Association of State Highway and Transportation Officials, "AASHTO LRFD Bridge Design Specifications," 8th Edition, 2017.

2. Berg and Dodge, "Atlas of Preliminary Geologic Quadrangle Maps of Pennsylvania", Pennsylvania Bureau of Topographic and Geologic Survey, 1981.

3. Bowles, "Foundation Analysis and Design," 5th Edition, The McGraw Hill Companies, Inc., 1997.

4. Geyer, McGlade and Wilshusen: "Engineering Characteristics of the Rocks of Pennsylvania", Pennsylvania Bureau of Topographic and Geologic Survey, 1982.

5. Pennsylvania DCNR, "The Geology of Pennsylvania", 1999.

6. Robertson, "Interpretation of Cone Penetration Tests – a Unified Approach," Canadian Geotechnical Journal, Vol. 46, N. 11, pp 1337-1355, 2009.

7. Samtani and Nowatzki, "Soils and Foundations, Reference Manual, Vol. I", FHWA-NHI-06-088, 2006.

TRIGGERING MECHANISMS OF THE LANDSLIDE AND ROCKFALL EVENTS OF THE HISTORIC FEBRUARY 2019 RAINFALL EVENT IN THE TENNESSEE VALLEY

David A. Freistaedter, P.E., P.G.
North Central Regional Engineer, GeoStabilization International®
Chattanooga, TN USA
855-579-0536
david.freistaedter@gsi.us

Michael W. Laney, P.E., G.E., P.Eng.
Deputy Chief Engineer, GeoStabilization International®
Olathe, KS USA
855-579-0536
michael.laney@gsi.us

Prepared for the 70th Highway Geology Symposium, October, 2019

Acknowledgements

The authors would like to thank the individuals/entities for their contributions in the work described:

Martin Woodard, PhD, PE, PG – Geostabilization International®
Jody Robinson, EIT – Geostabilization International®

Disclaimer

Statements and views presented in this paper are strictly those of the author(s), and do not necessarily reflect positions held by their affiliations, the Highway Geology Symposium (HGS), or others acknowledged above. The mention of trade names for commercial products does not imply the approval or endorsement by HGS.

Copyright Notice

Copyright © 2019 Highway Geology Symposium (HGS)

All Rights Reserved. Printed in the United States of America. No part of this publication may be reproduced or copied in any form or by any means – graphic, electronic, or mechanical, including photocopying, taping, or information storage and retrieval systems – without prior written permission of the HGS. This excludes the original author(s).

ABSTRACT

Usually, the Tennessee Valley experiences between 4 and 5 inches of rainfall during the month of February. Rainfall data obtained from the National Oceanic and Atmospheric Administration (NOAA) indicates that this region received rainfall amounts of 10 to 20 inches statewide in the month of February 2019 alone and this was on top of unusually large precipitation amounts the previous November through January. Monthly rainfall records were broken in Nashville, Crossville, and Knoxville. This rainfall was a result of a stationary front that stalled over the Tennessee Valley for nearly a week during the last half of February.

This rain combined with over-saturated conditions triggered an unexpected number of landslides and rockfall events impacting transportation routes within the Cumberland Plateau, Ridge and Valley, and Appalachian Mountain regions of Tennessee; as well as surrounding states including much of Kentucky to the north and parts of North Carolina to the east. These events, along with significant flooding, caused a state of emergency to be declared by the Tennessee Emergency Management Agency (TEMA).

The number of geohazard events impacting transportation routes in this region dramatically increased in comparison to what is typically expected; on the order of up to 10 times what is generally encountered.

This presentation will discuss the geomorphology and failure mechanisms of these landslide events that were triggered by the increased rainfall and discuss some of the solutions installed to mitigate the slope failure at one of these sites.

INTRODUCTION

The state of Tennessee, representative of most areas in the south-eastern United States with regards to climate, generally receives abundant rainfall throughout the year with the lack of a dry season as is experienced in other parts of the country. The climate is generally temperate, with warm to hot summers and cool winters with occasional cold air moving in from the north and west.

During the winter months, much of the state experiences daytime high temperatures ranging from 40 to the low 50 degrees Fahrenheit, with low temperatures at night near freezing, experiencing colder temperatures atop the Cumberland Plateau and other mountainous areas. Typically, 4 to 5 inches of rainfall is common, many times falling as heavy downpours. However, in the winter it is not uncommon for a sustained rainfall to set in, with light to moderate precipitation experienced consistently over days or even weeks with some breaks in between.

Below elevations of 3,000 to 4,000 feet above mean sea level (AMSL) trees generally lose their leaves around late November to early December, leaving the trees bare to early April in most locations. High elevation areas (above 4,000 ft AMSL) may see this period one month earlier to one month later each year. Areas of Tennessee, specifically mountainous areas which cover the eastern third of the state are classified as Appalachian Temperate Rainforest, especially on windward slopes. Other areas of Tennessee are not far removed from this, with over 50 inches of rainfall per year being normal, with areas in the mountains as high as 80 to 90 inches per year.

One factor that must be considered is that Tennessee is mainly covered with deciduous forests where the trees and brush lose their leaves in the fall. Most vegetation in the state goes dormant during the period between November and April, except for small plants that thrive in the understory of the forest, thriving on light which rarely makes it to ground level during times of thick vegetation. Figures 1 and 2 below illustrate how these forests can appear in the summer (Figure 1) as opposed to the wintertime (Figure 2).

Figure 1: Typical Woodlands in Summer

Figure 2: Typical Woodlands in Winter

Regarding climate and environment, the dormancy of vegetation and lack of sustained wind should be considered. During this time of the year where temperatures are lower, and precipitation remains high much of the thick vegetation goes dormant combined with a thick blanket of leaves and partly decaying organic matter on the forest floor. Due to the leaves and organic material on the forest floor the commonly experienced intense rainfall events do not run off quickly and saturate the soil.

Furthermore, this region lacks sustained wind such as is experienced in many states from Texas to the west. Without wind, water has a better chance of seeping into the soil over time and creating a wetting front that will move down due to gravity, increasing the weight of the overburden soils while decreasing the shear strength. Sustained wind is more effective for drying out soils than sunlight alone as can be observed with soils that are being scarified on a jobsite prior to compaction during grading. Dormancy of vegetation and lack of sustained wind are two contributing factors that aid in groundwater levels rising during this time of year, generally peaking in late winter to early spring.

"Climate Change" has been an important topic of discussion over the last decade or more. We have all observed prolonged weather conditions that deviate from what most would consider normal. Only a few years ago, parts of Tennessee experienced up to and over 100 days without rainfall and many areas began to turn yellow and brown by July or August, which is an uncommon experience. We have also witnessed wildfires in the mountains of Tennessee and near Gatlinburg which seem more appropriate in the western United States, where Wildfires are a common part of life especially in the summer and early fall.

As previously discussed, in recent years the western Appalachian Mountains have also been inundated with sustained and intense amounts of rainfall not seen since records began in the area (official record keeping began in the early 1800s in the eastern part of the United States). Whether or not climate change is occurring on the scale portrayed by main-stream media, deviations in what we consider "normal" need to be accounted for when designing and installing effective landslide and rockfall mitigation systems.

Figure 3 below represents the "Month to Date Observed Precipitation" for the Tennessee Valley on February 27th, 2019. Note that the Cumberland Plateau and Valley and Ridge Province lies at the epicenter of this sustained period of rainfall.

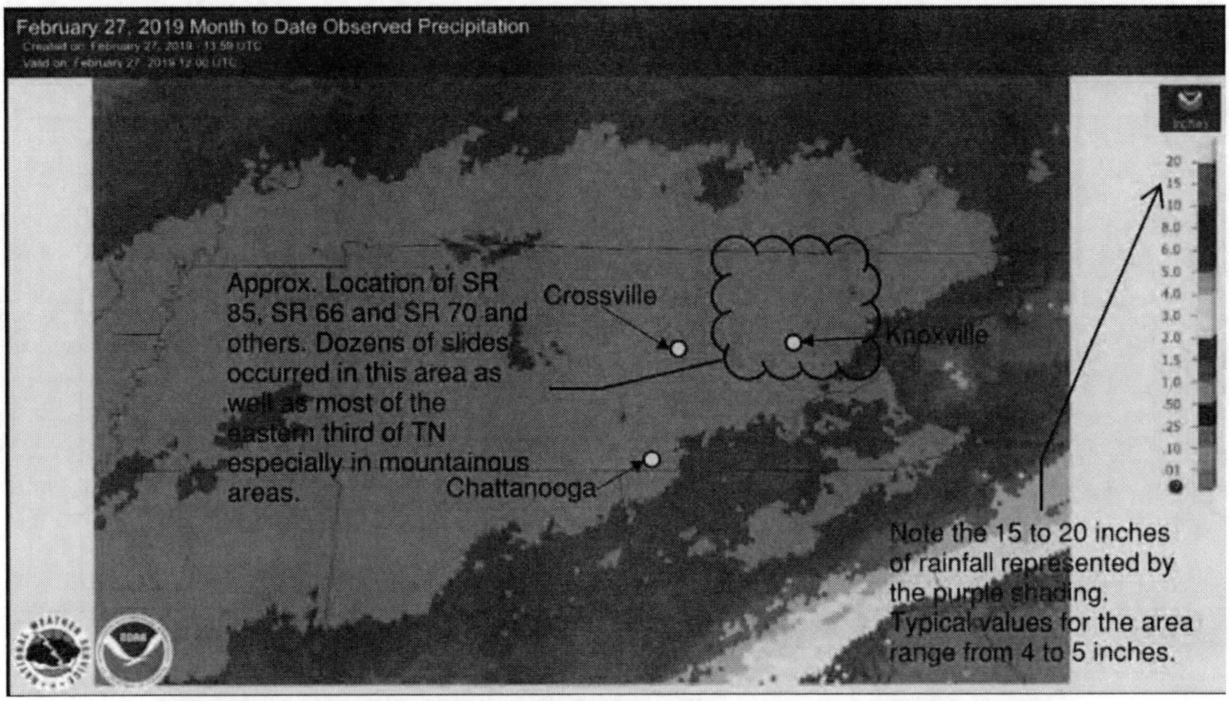

Figure 3: February 27, 2019 Month to Date Observed Precipitation

Weather patterns from the southwest brought significant amounts of Gulf of Mexico moisture northward which then interacted with the boundary of the cooler and drier air over the Appalachian Mountains for several days. This condition resulted in a prolonged period of heavy rain and flooding throughout middle Tennessee from February 19 through early February 24. In combination with the heavy rainfall that previously fell earlier in the month, and the unusually wet winter season so far, the result was widespread flash flooding and river flooding. Numerous homes and businesses flooded resulting in significant economic losses and several water rescues by emergency responders. New monthly rainfall records for February were recorded at many locations in Tennessee including Nashville and Crossville, with recordings of over a foot of rain. By the end of February most of the state of Tennessee had received between 10 and 20-inches of rain during this time period (Reference 1).

To understand the magnitude of rain that fell in this region Figure 4 shows February 2019 precipitation as a percent of normal precipitation. This shows that approximately 95 percent of the state of Tennessee received between 300 and 400 percent of normal precipitation for the month.

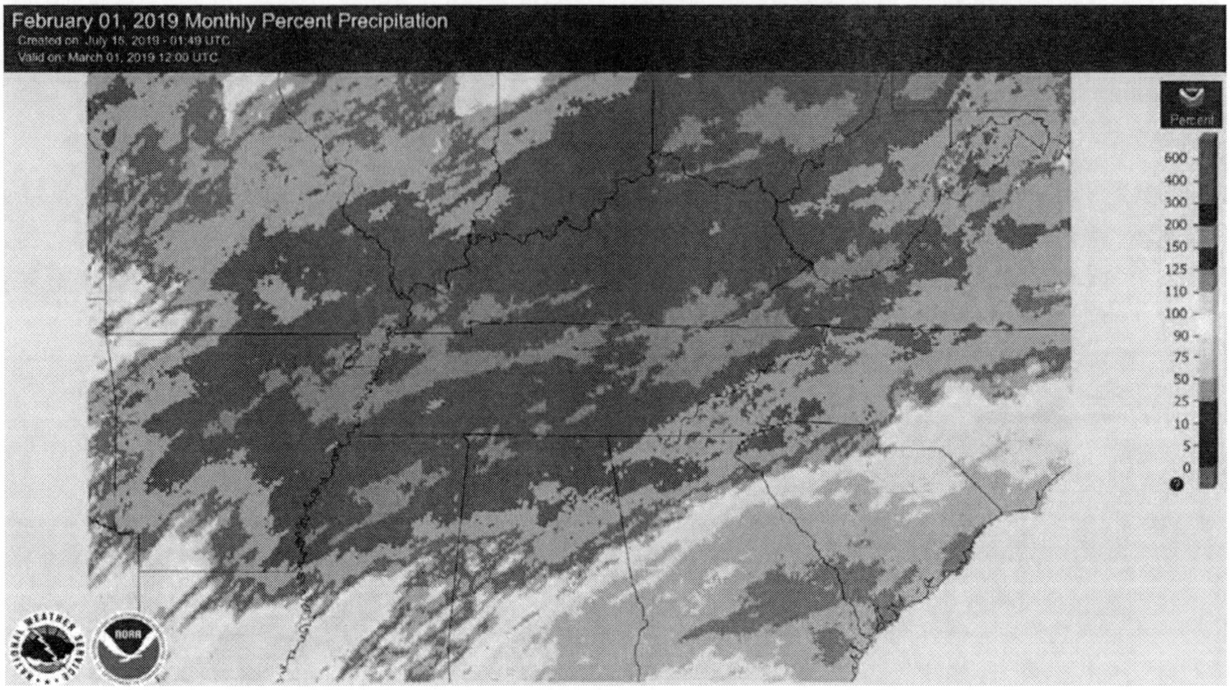

Figure 4: February 2019 Monthly Percent Precipitation

IMPACT OF RECORD-BREAKING RAINFALL

Due to the significant rainfall and saturated ground conditions of this event a significant increase in the number of geohazard events impacting transportation routes in the region occurred. Figures 5 and 6 represent landslides and rockfall events which occurred during late February. Dozens of significant landslides and rockfall events impacted the eastern two-thirds of the state causing government and private agencies to be stretched thin and resorting to unconventional methods in dealing with the sheer number of roadway disasters throughout the state. For example, State Routes 70 and 66 in Hawkins County, Tennessee, major state routes, were shut down completely. Figure 5 is a photo of one of the failures.

Typically, February is one of the slower months for landslide activity in eastern Tennessee, with weather dependent landslides picking up around March and April. From an operations standpoint, peak number of crews were forecasted to be running at the height of construction season which is typically around June. However, with the number of landslide events the number of crews at or greater than the typical June period, both public and private sector, had to be mobilized several months earlier. Crews had to be mobilized from outside of the region and hiring efforts needed to be expedited.

Figure 5: Aerial View of SR 70 Hawkins County Post-Slide, taken from a WVLT-TV drone survey. Two men in two separate vehicles drove over the edge of the landslide during the early morning hours of February 21, 2019 with one fatality.

Figure 6: Site Photos of SR 262 from February 25, 2019. At this point, the roadway became impassable and was completely shut down.

GENERAL GEOMORPHOLOGY AND FAILURE MECHANISMS OF LANDSLIDES AND ROCKFALL ENCOUNTERED WITHIN THE REGION

The Appalachian Plateau with its steep hillsides, thick soil cover, and precipitation of 35 to 50 inches per year, with the greatest amounts occurring in late winter and early spring has long been recognized as an area of major landslide severity (Reference 2). As illustrated in Figure 7 below, the Appalachian Mountains have a very high susceptibility for landslide activity (Reference 3).

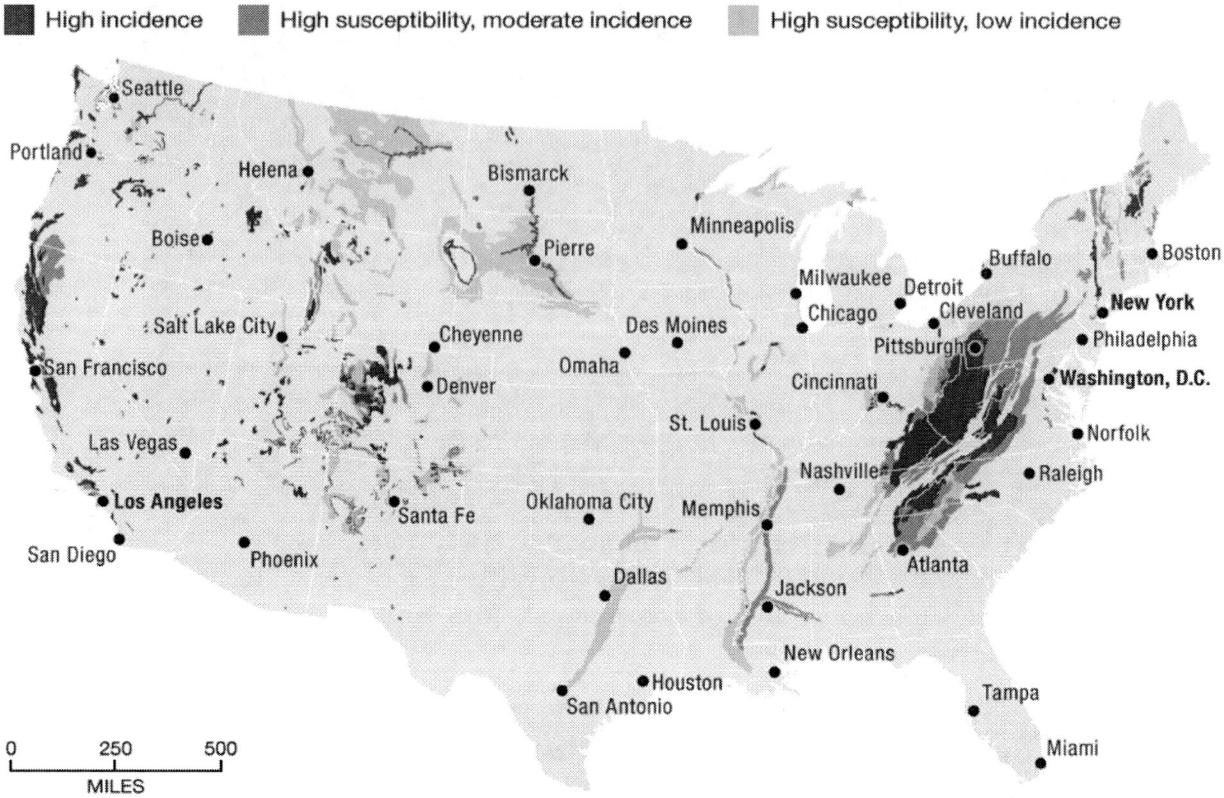

Figure 7: United States Landslide Susceptibility Map

Many areas of Tennessee average in excess of 50 inches of precipitation per year with elevated areas receiving over 60 inches per year and isolated areas of 80 to 90 inches per year not uncommon in the Great Smoky Mountains.

Gravity, weathering and erosional processes continuously work to bring slopes into equilibrium by essentially flattening them out over time. Rate of slope movement depends on slope geometry, strength of native soils, rock, and man-made fill, level of water saturation of these materials in the slope (which increases the weight of the materials) and changes in stress conditions which can be attributed to cutting, filling, traffic and other surcharges. In some areas seismic activity will create short-term instability of slopes as a consequence of resulting ground accelerations.

There are three common causes of landslide movement: 1- removal of lateral support, either by erosion or excavation, 2- surcharge loading by filling on slope, and 3- changes in groundwater conditions (Reference 2).

Natural slope geometry along the slopes of the Cumberland Plateau, Ridge and Valley Region, and Appalachian Mountains seem most susceptible to the first two causes. Slopes become unstable when driving forces outweigh resisting forces, and in the case of late February 2019, slopes became unstable largely due to the third most common cause where groundwater conditions changed dramatically due to near complete slope face saturation over an extended period.

The oversaturation of the slopes caused significant changes to slope conditions. Included in this are dozens of identified roadway sections that had been creeping almost imperceptibly for several years that suddenly mobilized into mudslides, landslides, and rockfalls. The most common cause of rockfalls was rock toppling. Toppling primarily occurred due in large part to saturation of joints and tension cracks in the rock masses and increased hydrostatic pressures on rock structures.

Below in Figure 8 are three images (Reference 4) illustrating the difference between rotational and translational landslides, as well as rock topples.

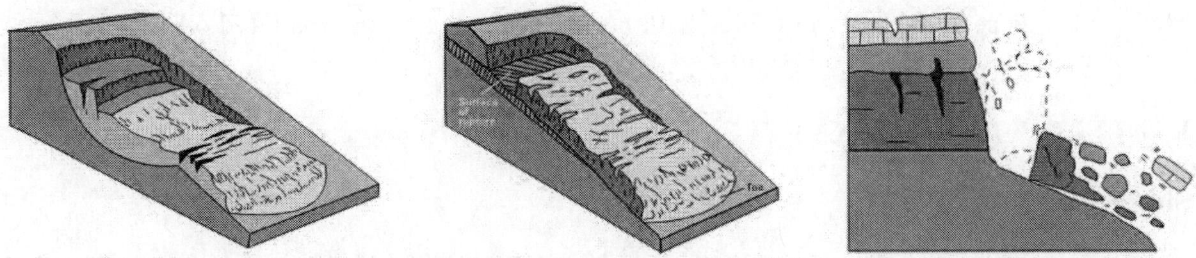

Figure 7. Schematic of a rotational landslide. (Schematic modified from Reference 9.) Figure 9. Schematic of a translational landslide. (Schematic modified from Reference 9.) Figure 5. Schematic of a topple. (Schematic from Reference 9.)

Figure 8: Rotational and translational landslide, and rock block topple, respectively.

Record rainfall resulted in an unusually high number of landslides on saturated soil slopes overlying rock in Tennessee and surrounding areas earlier this year. These events can be well compared to the 2011 Umyeonsan landslides in the vicinity of Seoul, Korea which have been well studied and documented (Reference 5). Dr. Jeong et al, had many interesting conclusions related to these types of slope failures that are discussed below.

Triggering mechanisms of natural slopes frequently comprise a complex interaction between hydrological and geotechnical processes, which in turn depends on irregular topography, hydro-geotechnical properties, boundary conditions such as permeability and weak soil/rock layers, and the initial state of the slope. The main reason for these types of slope failure is the loss of matric suction (the surface tension of water holding the grains of soil together) resulting in the loss of effective stress as water infiltrates the soil. (Reference 5)

Matric suction is the pressure dry soil exerts on the surrounding soils to equalize the moisture content in the overall block of soil. When there is an increase in water pressure, such as a sudden rise in the water table, the increase in pore pressure will result in these particles separating and a sudden reduction in soil shear strength ensues. The resulting shear movement of the soil will further drive up the pore pressure, until the system again reaches equilibrium and the pore pressure dissipates allowing grain-to-grain contact between the soil particles again.

Another triggering event due to increased hydrostatic pressures is at the interface between a pervious (coarse grained soil) and relatively impervious (clayey soil and/or bedrock) where the system is in equilibrium due to frictional resistance. However, a sudden increase in pore pressures at this interface due to a sudden addition or rise in groundwater above the impervious zone will result in a sudden decrease in the shear resistance due to friction, and the slide will translationally fail along this surface.

As the rainwater soaked into the mountainsides over the course of several days, it created a wetting front that increased the saturation of the underlying soils at depth. The increase in hydrostatic pressure above the impermeable layers increased the pore water pressure at this interface, decreasing the normal force and thus decreasing shearing resistance along the slide plane.

In addition, the fully saturated condition experienced by some soil particles within the landslide mass caused those materials to become weak and loosely consolidated, as can be seen by loss of matric suction/surface tension, resulting in rotational failures.

The result was several large landslides with negative impacts to the road and other infrastructure in the Cumberland Plateau and Ridge and Valley regions.

LANDSLIDE REPAIR EXAMPLE

Site Reconnaissance

The landslide at SR 262 MP 8.8 in Jackson County, TN impacted the outboard shoulder and both travel lanes following approximately 3 to 4 inches of rain over an approximate 12-hour period. A site reconnaissance was carried out to observe, map out, and collect field data from the failure zone. Based on field observations the heavy rainfall triggered approximately 3 and 5 feet of lateral and vertical movement in the roadway, respectively. (We note that the lateral movement was measured as the cumulative width of tension cracks observed within the roadway prism.) The landslide was likely a shallow rotational failure comprising a poorly consolidated colluvium with underlying bedrock. This type of mechanism essentially represents a localized failure condition where a soil that is generally in a drained state experiences localized transient pore-water pressures due to abrupt changes in hydraulic boundary conditions previously discussed.

Figure 9 below portrays a site sketch, from our February 25, 2019 site reconnaissance, showing excessive ground movement, slumping and tension cracking within the landslide complex. The site, located about 30 minutes northwest of Cookeville, TN is located well within the area of extreme rainfall as shown on Figure 3. Historically, the site showed signs of slope failure and roadway repairs were planned when the February 19 through 24 heavy rainfall temporarily shut down work. Upon arrival back to the jobsite on February 25, the roadway conditions, as shown in Figure 6 were noted by GSI and their affiliates. The initial intent of the roadway repairs was to maintain one lane of travel with temporary stoplights. However, after the heavy rainfall event, the entire roadway had to be closed due to the severity of the landslide.

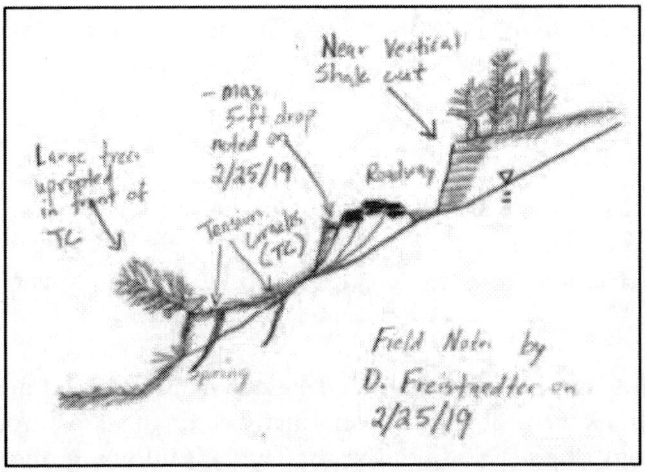

Figure 9: SR 262 MP 8.8 Field Notes Dated February 25, 2019

Permanent Slope Stabilization Elements, Strengths, and Sequencing

The slope stabilization plan consisted of soil nails, shotcrete facing, and a geosynthetic reinforced soil (GRS) structure. Soil nails and piles consisted of hollow steel bars. The soil nails were installed at -15 degrees from the horizontal. Nail embedment lengths were up to 30 feet. The soil nail facing was reinforced shotcrete.

Pile elements were integrated into the slope stabilization system to mitigate the landslide and the roadway shoulder. The pile elements consisted of two rows with batters ranging between 3 and 15 degrees off vertical. Embedment lengths for pile elements were up to 6 feet in rock.

The design grout-ground bond stress for the soil nails was approximated based on local experience and test results and ranged between approximately 4 and 12 psi. The ultimate yield strength for the soil nail and pile elements was approximately 97 kips.

The GRS structure geometry proposed at the site was variable, but generally, the wall was approximately 10 feet tall with widths varying between approximately 12 and 15 feet.

Slope Stability Analysis

We completed slope stability assessments of the failed slope and preliminary repair solution. The slope model geometry was developed using field and client developed cross sections. The goal was to obtain a minimum factor of safety of 1.3 and to recover on average, about 15 feet of roadway width. Given this criterion, a soil nail wall and GRS wall were chosen to stabilize the landslide and roadway.

Our stability assessments were completed using the two-dimensional slope stability software SLIDE 8.0 by RocScience. The SLIDE program performs two-dimensional limit equilibrium analysis to analyze slope stability and to determine a factor of safety (FS) against global failure. The FS against failure can be generalized as the ratio of forces resisting slope movement (e.g., soil strength, soil mass, etc.) and the forces driving slope movement (e.g., gravity, earth pressure, and earthquake shaking). A back analysis to obtain conditions just prior to failure was performed. In a back analysis the FS value is assumed to be equal to or less than 1 and indicates a condition where the shear stresses required to maintain equilibrium in the slope reach or exceed the available shear resistance.

We analysed stability of the slope stabilization under static (non-seismic) conditions. Due to the mapped peaked ground acceleration at the site being relatively low, seismic stability analysis was not performed. Doing so allowed us to select appropriate design shear strengths for the repair model.

Results from our stability analysis are included below in Figures 10 and 11. In Figure 10, the interpreted field conditions were obtained from our back analysis as indicated by the limit-state condition along the failed slope (FS of approximately 1). Following, the permanent stabilization system indicated in Figure 11 shows the repaired slope is stable under static conditions with a local FS of approximately 1.3 through the repair section and a global FS of 1.2.

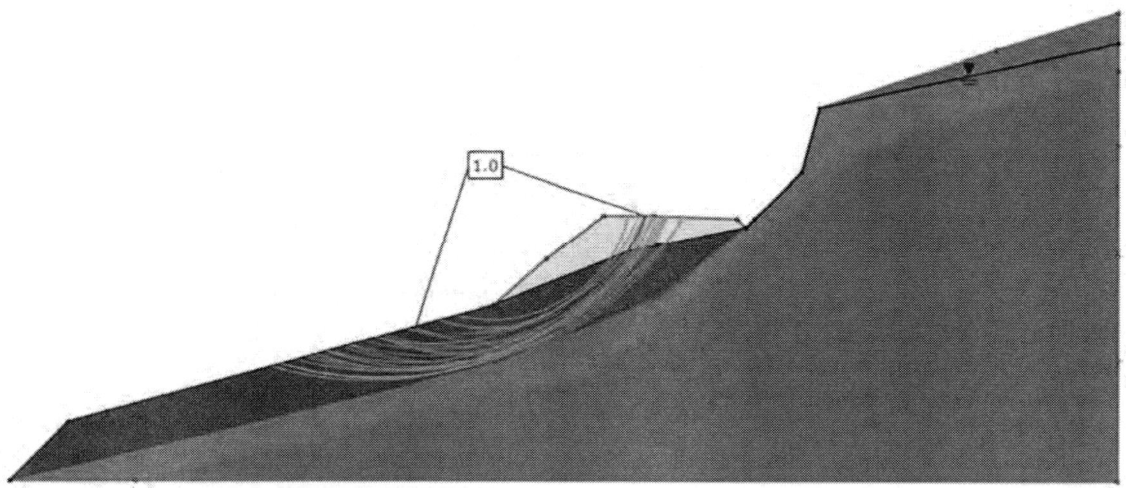

Figure 10: Back Analysis Factor of Safety

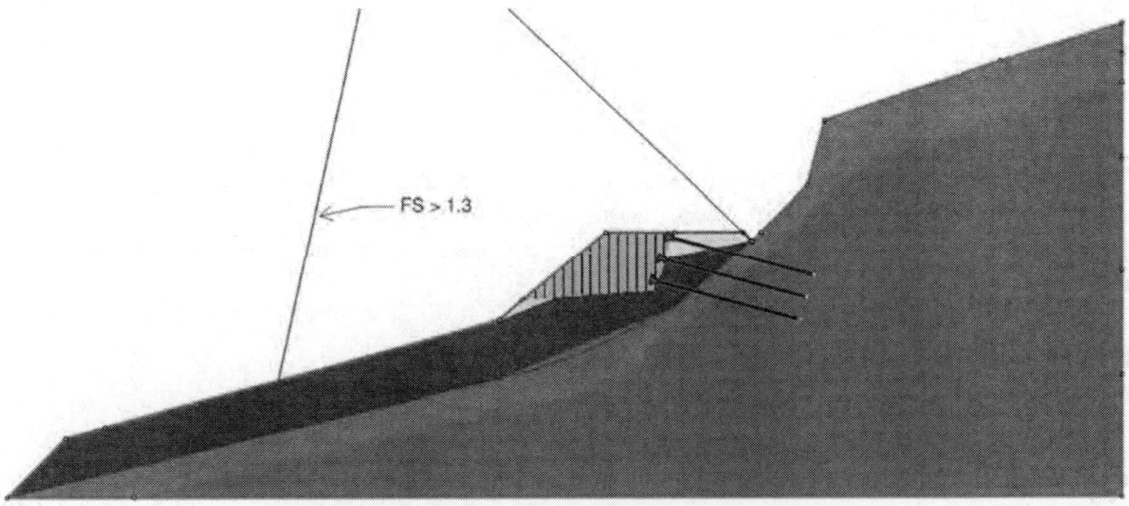

Figure 11: Repair Analysis Cross Section

Figure 12 below shows a portion of the repair at SR 262 MP 8.8 near end of construction. The full repair took approximately 3 months to complete, beginning in late February and ending in late May. However, crews worked long hours and 7 days straight initially to meet the Client's requested deadline of being able to open 1 lane of traffic after 7 days of work. Many people had to take long detours and school bus and trucking routes were heavily impacted due to the complete road closure. Being able to open the roadway to one lane of traffic within 7 days greatly minimized the impact to the public by turning a long detour into a short delay with timed traffic lights creating on average about 5 to 10 minutes of delay as opposed to up to 2 hours.

Figure 12: Photograph portraying repair near end of construction at SR 262 MP 8.8

SUMMARY

A driving factor in the documented slope failures impacting infrastructure in Eastern Tennessee and surrounding states was not primarily a result of weight or lubrication of the slide mass that caused the typical failures that are seen in our region, but an increase in localized transient pore-water pressures. The example presented in this paper was only one of several on-going repairs that have resulted from the rain events of February 2019.

REFERENCES

1. National Oceanic and Atmospheric Administration (NOAA), 2019, National Weather Service Website, *Late February 2019 Historic Rainfall and Widespread Flooding*, 2019, Accessed June 28, 2019. URL: https://www.weather.gov/ohx/lateFebruary2019flooding#

2. Gray, Richard E., 2008, The Appalachian Coalition for Geological Hazards in Transportation, "Landslide Problems on Appalachian Colluvial Slopes.", DiGioia, Gray & Associates, LLC., August 5, 2008,

URL: www.marshall.edu/cegas/geohazards/2008pdf/Session2/02_LANDSLIDE PROBLEMS ON APPALACHIAN COLLUVIAL SLOPES.pdf

3. NPR, 2014, NPR Website, *Map Warns of Patches Susceptible to Landslides*, March 26, 2014, Accessed July 29, 2019, URL: https://www.npr.org/sections/thetwo-way/2014/03/26/294829981/map-warns-of-patches-susceptible-to-landslides

4. Highland, Lynn M, and Peter Bobrowsky, 2008, *The Landslide Handbook - A Guide to Understanding Landslides*, United States Geological Survey. URL: https://pubs.usgs.gov/circ/1325/

5. Jeong, S.; Lee, K.; Kim, J.; Kim, Y. Analysis of Rainfall-Induced Landslide on Unsaturated Soil Slopes. *Sustainability* 2017, 9, 1280. URL: https://www.mdpi.com/2071-1050/9/7/1280/pdf

Precision Presplitting – Changes to Design Methodology Based on Young's Modulus

Anthony Konya M.S.
Precision Blasting Services
P.O. Box 189
Montville, OH 44064
(440)-823-2263
Anthony@idc-pbs.com

Dr. Calvin J. Konya
Precision Blasting Services
P.O. Box 189
Montville, OH 44064
(440)-474-6700
Konya@idc-pbs.com

Prepared for the 70th Highway Geology Symposium, October, 2019

Disclaimer

Statements and views presented in this paper are strictly those of the author(s), and do not necessarily reflect positions held by their affiliations, the Highway Geology Symposium (HGS), or others acknowledged above. The mention of trade names for commercial products does not imply the approval or endorsement by HGS.

Copyright Notice

Copyright © 2019 Highway Geology Symposium (HGS)

All Rights Reserved. Printed in the United States of America. No part of this publication may be reproduced or copied in any form or by any means – graphic, electronic, or mechanical, including photocopying, taping, or information storage and retrieval systems – without prior written permission of the HGS. This excludes the original author(s).

ABSTRACT

Precision Presplitting is a relatively new method of blasting a presplit blast which utilizes changes to the explosive load and spacing of blastholes based on the rock properties and structural geology. In recent years this technique has been utilized by numerous state Departments of Transportation, the U.S. Army Corp of Engineers, and dozens of private construction and mining companies as it easily facilitates the breakage of weak rocks to form stable highwalls with minimal to no slow-zones. This paper will discuss the variations to the explosive load and spacing based on the Young's Modulus of the rock. This paper will be showcasing real-world projects for which Precision Blasting Services (PBS) was the government or contractors blasting consultant which utilized this technique to effectively split from weak and weathered rock to strong, competent granites, along with encountering formations with multiple types of rock and loading methods for each. In addition to this, the methods for developing Hybrid Blasting Specifications for the Precision Presplit will be discussed.

INTRODUCTION

The use of explosives in rock blasting began on February 8, 1627 in the Oberbiberstollen of Schemnitz in Hungary which was designed and fired by Caspar Weindl (Guttmann, 1892) utilizing black powder. With the success of this first blast, the Hungarian Mine Tribunal quickly had this information disseminated through Hungary and the utilization of explosives in underground mining spread and by 1673 the technology had spread throughout the underground Hungarian mining industry (Brown, 1673). Blasting then spread throughout the world where it was introduced in Germany in 1700 and Sweden in 1724 for blasting in the mining industry. With the massive improvements in rock fragmentation, especially in hard rock which could not be mined except with fire setting, the construction industry soon began employing the use of black powder blasting. The first underground construction tunnel developed with the use of blasting is documented to have occurred 1679 to develop the Malpas Tunnel in Languedoc, France. By 1696 blasting had begun being used on surface blasting in construction for the development of roadways on the Abula Pass in Switzerland (Guttmann, 1892).

Since explosive first began being used in mining and construction, engineers and scientists have been developing theories to better understand how these explosives work and break rock in an attempt to improve the efficiency of blasting. These efficiency improvements have been in ways to improve explosives through chemical formulations and manufacturing processes, develop ways to better design blasts to increase fragmentation and heaving of the muckpile, and to reduce the environmental factors of blasting; such as ground vibration and overbreak of blasts. The overbreak of a blast, or breakage beyond the design line, has been of concern since these early construction projects as overbreak often leads to raveling of rock and incompetent walls. This overbreak can also cause increases in the speed of weathering and water penetration behind the slope which can lead to slope failures. In projects where concrete is used this overbreak requires the additional use of concrete which can increase project costs by millions. This is also of concern in underground workings where poor blasting along the perimeter of the blast will lead to large pieces of rock hanging on the back and ribs of the excavation which become immediately dangerous to workers in the area and add significant cost to remove or bolt.

In an effort to prevent overbreak a technique known as presplitting was introduced and worked by breaking a smooth line between holes in a rock. This technique began its use before explosives were ever introduced into construction; in places such as Egypt wooden wedges and soaked in water, causing expansion and pressure in the borehole. In northern climates rock was broken in a similar manner; were boreholes could be filled with water, the water was then frozen and crack between the blocks in the winter. This would create large, smooth blocks that could be used in construction and leave smooth back walls. Both of these ancient techniques involved drilling closely spaced boreholes and applying a pressure inside of the borehole to causes fractures between holes. In ancient times this pressure was often slow building, and in some cases would take months to fully split.

It was not until the 1950s where the first mention of explosive presplitting were utilized on the Lewiston Power Plant (Langefors & Kihlstrom, 1973) as part of the Niagara Power Project. In order to accomplish this, boreholes would be drilled from 12' to 52' depending on the bench height and loaded with detonating cord and partial cartridges of dynamite. These boreholes were 3 inches in diameter and fired before the main blast, with all presplit holes being fired at the same time utilizing instantaneous delays. This produced excellent results and essentially eliminated all overbreak on the project. This technique has then had widespread use in the mining and explosive industry to minimize overbreak from blasting.

THE MECHANICS OF PRESPLITTING

The mechanisms of production rock blasting are critical to understand because the same forces that apply to rock blasting also apply to presplit blasting. The way an explosive applies force and causes breakage to a rock do not change. Today's blasting industry understands that shock breakage in production blasting is impossible and that the gas pressure in the borehole causes breakage, however, the theories behind the mechanism of a presplit has not been well defined and relatively few studies have been completed on presplit formation. In today's blasting industry both shock breakage (Zhang, 2016) and gas pressure breakage (Konya & Konya, 2017) is presented in modern technical papers. Many have also argued that the mechanism behind the presplit is unimportant or academic, which may be true for the traditional case of presplitting which remains the same under almost all circumstances. However, with the advent of Precision Presplitting the mechanism behind a presplit is of importance as changes to dimensions such as the spacing of boreholes and explosive load in a hole are designed to meet the structural geology and rock properties. Without an understanding of the mechanisms behind a presplit formation a strategic design to eliminate overbreak while allowing for smooth breakage is impossible.

The first large scale explosive presplit was produced on the Niagara Power Project which was completed in 1962. This project was based in dolomite and limestone with a single layer of shale near the bottom of the excavation and had to have smooth walls in order to properly pour concrete. During the project, numerous methods of controlled blasting were attempted including Line Drilling, Line Drilling with explosive loads in every third hole, Modified Cushion Blasting, Decks of Dynamites throughout the borehole, and finally presplitting. It was reported that the only method that produced satisfactory results to minimize overbreak was the presplitting which was accomplished by taping 1 ¼" by 4" sticks of dynamite on Primacord every 12 inches. The boreholes were 2 ½" to 3" in diameter and spaced 24 inches apart and stemmed with crushed gravel. This resulted in increased rock excavation and a reduction in scaling by a factor of 10. Additionally, the project had significant savings on concrete costs and increased safety as the walls were cleaned smooth (Paine, Holmes, & Clark, 1961).

At the time, the project was designed based on the gas pressure generated by the explosive. The engineers assumed that if the gas pressure was kept below the compressive strength of the rock, they would avoid crushing the rock around the borehole. In order to create a break between boreholes the belief was that the borehole pressure had to be above the tensile strength. While this was a bit of a rudimentary theory at its time, the project was completed and the presplit functioned extremely well. Following the project, presplitting was widely accepted as the best and most cost-effective method of overbreak control.

Based on this theory, researchers of presplitting both in a laboratory and practical setting began looking into the decoupling of charges, or the reduction of the diameter of the explosive compared to the diameter of the borehole. This was done to decrease the dynamic gas flow on the borehole wall and to reduce the gas pressure in the borehole (Konya, Britton, & Lukovic, 1987) preventing large compressive strengths which would lead to overbreak (Day, 1982). However, this increase in decoupling ratio also led to minimal shock pressure transmission into the rock mass due to large impedance mismatches between explosives and air, then air and rock.

With the large increase in research of shock breakage in rock blasting, many authors began to investigate possible effects of shockwave collision between boreholes to develop tensile zones and causing presplit formation (DuPont, 1975; Crosby & Bauer, 1982). This theory was widespread due to the popularity of the DuPont Blasters Handbook and it is still circulated amongst many leading organizations today (International Society of Explosive Engineers, 2016) and researchers (Salmi & Hosseinzadch, 2014). This theory was heavily disputed and shown in numerous studies of the day and it was shown that the shockwave has almost no correlation between the dynamic shockwave and the presplit formation, with numerous studies showing that the quasi-static gas pressure in the borehole was responsible for presplit formation (Konya C. , 1973; Worsey P. , 1981; Worsey, Farmer, & Matheson, 1981; Daehnke, Rossmanith, & Kouzniak, 1996). Additional studies were conducted utilizing a propellant charge, Pyrodex, to fire a presplit blast. These propellant charges produced no shockwave as they deflagrate, not detonate (Akhavan, 2011), which completely isolated the gas pressure as the only working energy. Using the same principles as in traditional presplit design (Konya C. , 1980), the propellant charges produced the exact same results as a presplit blast that was fired with detonating explosives (Konya, Barret, & Smith, 1986). This proved that presplit mechanisms on a full-scale blast had no reliance on the shockwave generated by detonating explosives.

This led to the development of a Precision Presplit style of blasting, where extremely light loads of detonating cord are utilized to prevent all breakage except for the breakage between boreholes (Konya C. , 1982). This design utilized closely spaced borehole of 24 inches or less, to minimize the impacts of rock structure on the presplit (Worsey P. , 1984; Worsey & Qu, 1987; Tariq & Worsey, 1996). As this design methodology has begun widespread use, new empirical research into the explosive loading based on the rock properties has been developed (Konya & Konya, 2015; Konya & Konya, 2016; Konya & Konya, 2017).

This method of Precision Presplitting has effectively zero shock energy to form a fracture after accounting for impedance mismatches (Cooper, 1996), non-ideal detonation (Cook, 1974), and attenuation of the shockwave in the rock mass (Spathis & Wheatley, 2016). It has then been theorized that the mechanism behind the presplit formation is due to large hoop stresses which are generated between the boreholes causing a fracture, with no advancement of the fracture from gas penetration (Konya & Konya, 2017) and today this is the dominant theory of how presplitting functions with a strong correlation between borehole pressure and presplit formation, while no correlation exists between shock pressure and presplit formation (Konya & Konya, The Development of Pressure to Young's Modulus Models for Precision Presplit Blasting, 2019) and currently unpublished research by the authors showing a correlation between these hoop stress fields generated from gas pressure and split formation. The mechanisms of presplitting are:

1. The explosive detonates, causing a shockwave to propagate into the rock. This shock wave may cause initial micro-fractures on the borehole wall. This shockwave is of insufficient magnitude to cause major fracturing of rock, and with the almost infinite burden, will not cause any tension spalling.

2. The gasses within the borehole begin to expand, putting a pressure on the borehole walls. With the proper explosive load, this pressure will cause hoop stresses to form between two boreholes, causing a fracture to form.

3. The expanding gasses will extend into the fractures, causing an opening of the fractures and expand the fractures to the surface as the gas begins to blow-out.

4. If the proper amount of explosive and stemming is used, the explosive will blow the stemming out of the borehole and the gas pressure will be released through the top of the borehole (Choked Flow Gas Theory)

PRECISION PRESPLITTING: FIELD APPLICATION

Precision Presplitting is a new method of presplit blasting which utilizes very closely spaced boreholes, oftentimes 12" apart to 24" apart, with light charge loads, typically less than 700 grains per foot (0.1 lbs/ft) of detonating cord. Traditional presplitting is typically accomplished with 0.3 pounds per foot of 'presplit powder' and borehole spacings of 36" which is nearly identical to the Niagara Power Project and the advent of presplitting. While traditional presplitting may work in strong rocks with minimal structure, it often fails to produce proper results in weaker rocks or rocks which have significant structure. This often leads blaster's and contractors to state that "the geology does not allow for presplitting" when in actuality the presplitting is not proper for the geology. Specific instances are possible were any presplit will not properly function, such as when attempting to presplit a wall which is 15° from the dominant joint set or were loose bedding is steeply dipping into the cut.

Precision Presplitting is the only known, reliable method to achieve smooth walls where a majority of the half-casts (half borehole marks) remain after blasting. Because of this, the use of Precision Presplitting has replaced traditional presplitting in almost all areas on U.S. Army Corp of Engineer Projects. In addition to this, Precision Presplitting has been recommended and used on U.S. DOT blasting projects since the 1990s to ensure competent walls with minimal rock bolting or meshing (Konya C. , 1990). With the widespread adoption of Precision Presplitting, the authors have been developing an engineering based approach to design a Precision Presplit in any rock type based on the rocks properties and the structural geology. The highlights of this research is presented in this paper.

Figure 1 - Original Precision Presplit Testing with Variations to Spacing

KONYA PRESPLIT FACTOR

From the previous definition of the mechanics of presplitting, it is shown that the Young's Modulus has a relationship to the ability to presplit a rock. The authors then set out to answer, what is this relationship? Through both practical and theoretical testing, the authors have determined that rocks ability to presplit can be categorized using a presplit factor. This presplit factor is inversely proportional to the Young's Modulus of the rock and follow the formula in equation 1.

Equation 1 - Konya Presplit Constant

$$K = \left(\frac{40579}{E}\right)^{0.625}$$

Where: K = Konya Presplit Factor
E = Young's Modulus (GPa)

This presplit factor has been tested on numerous rock types and is assumed to hold true for almost all rocks and rock types, however, certain rocks (super brittle and extremely elastic-plastic) have not been tested but are expected to follow different mechanics due to the release/consumption of energy. This presplit factor may also change with excessive jointing, and methods to account for jointing are discussed below. Table 1 has values for the presplit factor for the average rock of different rock categories.

Table 1 - Konya Presplit Constants for Rock Types

Rock Type	Konya Presplit Factor
Granite	76
Limestone	86
Shale	116
Sandstone	127
Siltstone	201

In many cases, blasters and engineers will discuss the load for presplitting based on the 'split-factor'. This is similar to the powder factor for production blasting, but instead is the explosive load per area of presplit face. Just as powder factor is a poor design tool, the split factor is also a poor design tool. This is because this split-factor does not use any terms to derive and does not take into consideration the spacing, stemming, timing, or actual required explosive load. This split-factor also does not take into consideration the actual breakage or 'split' (fracture) produced. Generally, this split factor is taken as whatever is currently working at the mine or construction project and the powder per area is made equal throughout the blasting operation. As will be shown further in this paper, explosive load does not vary linearly with the spacing, using split factor to design assumes that this relationship is linear. It is the authors' recommendation that this split factor be used as nothing more than an economic or comparison tool, and not used for actual design purposes.

SPACING OF BOREHOLES

The next consideration in the design of the presplit is the spacing that will be used between boreholes, center to center. This is of extreme importance, as even properly varying the explosive load with the spacing will produce different fracturing. This is because:

1. The rock is non-homogenous and has joints and fractures that will cause fractures to deviate and in some cases stop fracture growth.

2. As the distance between boreholes increases, the proportion of spacing to distance to free face is decreased. This can result in additional breakage around the borehole.

Jointing and other discontinuities between two boreholes of a presplit will cause increased backbreak and a worse wall, and the joint frequency between boreholes will be one of the major limiting factors of the maximum spacing. In previous studies (Tariq & Worsey, 1995) studying the effects of discontinuities on presplit spacing, it was concluded that:

1. Increased discontinuities can help facilitate a presplit, however, a poor presplit facture is produced

2. A larger joint frequency between boreholes enhances the effects of cratering of the borehole, a worse presplit fracture is produced

3. For a single joint between boreholes, cratering occurs when the joint is farther than 8% of spacing and up to 20% of spacing between holes

From this research, one can see that the joint frequency between two boreholes is critical for the production of a smooth fracture. High jointing frequency decreases the explosive load needed (this is accounted for in the presplit factor, through the young's modulus) and it increases the potential for backbreak and cratering. In cases of high joint frequency, spacing should be reduced in order to reduce the effects of fracture widening and deviation based on the jointing. This can be done by limiting the actual number of joints between boreholes and varying the explosive load. In another paper it is stated "Of the most importance; the presence of discontinuities at less than 60 degrees to the proposed pre-split line tends to cause poor line definition. If the angle is less than 15 degrees, pre-split blasting has no visible effect on slope profile over bulk blasting." (Worsey P. , 1984). It also states that the pre-split fractures will intersect discontinuities at approximately right angles. This demonstrates not only the importance of spacing, but also the orientation of the presplit in relation to the natural jointing of the rock mass.

If a smooth presplit fracture must be obtained in extremely jointed rock, the authors have used a spacing of 12" (0.30m) to reduce the effects of the discontinuities. In rocks that are extremely massive with minimal jointing or other discontinuities, the spacing of the pre-split can be expanded much further to decrease the cost and maintain similar walls.

Increasing the spacing on a Precision Presplit can be done in rock with a low joint frequency, however considerations must be taken towards final wall conditions and distance to free burden. The presplit mechanism works because the distance between holes (spacing) is very small compared to the distance to the free burden. As the spacing increases and the explosive load is increased, the circumferential stresses towards the free face are increased, this can result in breakage towards the free face, or other production holes within the blast pattern.

In addition to this, by increasing the spacing the explosive load is increased. As the explosive load is increased, the total volume of gas in the borehole is increased. This can result in very large pressures in the borehole and fracturing into the solid rock mass. In addition to this, the large volume of gas attempting to escape the borehole and fractures can cause breakage at the top from the high velocity, dense gas shearing along the rock mass.

Spacing can be properly expanded and a similar fracture can be maintained, as long as the spacing is within reasonable distance, the rock is competent, and the explosive load and decoupling ratio are adjusted accordingly. Ultimately, the spacing will have to be determined after a thorough analysis of the rock and structure in the area; the desired results of the blasting; and test blasts. On any major project the authors recommend that a series of tests are done, possible in a single blast, were the spacing is varied to allow the project owner to determine 'satisfactory breakage'. This will not necessarily be an engineering procedure, but good management and a proper specification will allow the owner to ensure that the contractor will complete this procedure. The contractor must have a good degree of understanding on how Precision Presplitting works and how to estimate what the final spacing will be to properly estimate the cost of blasting before the project begins.

EXPLOSIVE LOAD FOR A PRECISION PRESPLIT

With the ability to calculate the presplit factor of different rocks one can now approach the design of a precision presplit with an engineering approach. The next factor to discuss will be the calculation of the explosive load, assuming a spacing has been selected from the information above.

The explosive of choice for a Precision Presplit is detonating cord. This is because detonating cord provides a consistent, controllable amount of energy throughout its length. Detonating cord is also easy to work with, with proper techniques it is simple to load, and explosive load variations can be made throughout the borehole to account for different rock types in a single blast (figure 2).

Figure 2 - Precision Presplit in Four Rock Types (Grundy, Virginia)

The explosive load with detonating cord, is calculated in grains per foot (grams per meter) where 7,000 grains is equivalent to one pound of explosive. Using the Konya presplit factor (K) and the spacing between boreholes, one can then calculate the explosive load to be used from equation 2.

Equation 2 - Precision Presplit Explosive Load Equation

$$D_{ec} = 7000 * \left(\frac{S}{K}\right)^2$$

Where: D_{ec} = Explosive Load (grain/ft.)
S = Spacing (inches)
K = Konya Presplit Factor

The use of the presplit factor takes into account the rock types, depositional environment, and the jointing/discontinuities throughout the rock, as long as proper sampling procedures have been used to quantify the Young's Modulus of the rock.

This equation has then been run for the average type of rock from multiple different classes and has been graphed in figure 3. This can then be used by taking the designed spacing for the rock and finding the appropriate line for the rock type (or similar rock type).

For example, if one is to use a 36-inch spacing in an average limestone, they would find 36 inches on the x-axis and go to the limestone line. Then match this point to the y-axis and get the optimal explosive load, for limestone with a spacing of 36 inches. The explosive load for this would be approximately 1200 grains per foot.

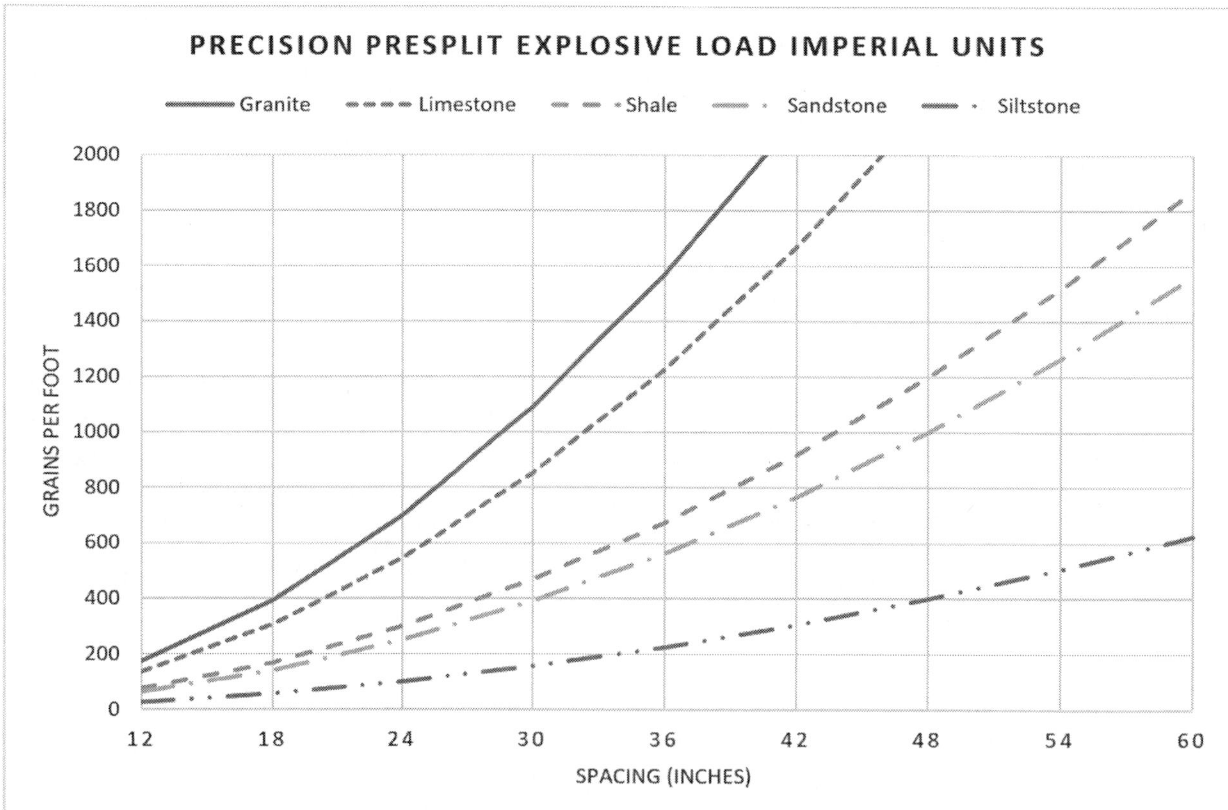

Figure 3 - Explosive Load Variations for Varying Spacing and Rock Type

ENVIRONMENTAL FACTORS OF PRECISION PRESPLITTING

The use of any type of blasting typically results in up to five 'environmental aspects' or considerations. These are:

1. Ground Vibration
2. Air Overpressure
3. Overbreak Control
4. Dust
5. Flyrock

In general, the ground vibration from a 'normal presplit' is up to five times that of a production blast. This occurs because of infinite confinement and no free face or relief. In previous research (Konya & Konya, 2016) it has been shown that Precision Presplitting has vibration comparable to that of production blasting. It can be concluded that Precision Presplitting will produce less ground vibration than that of a 'normal' presplit. To predict ground vibration from a Precision Presplit, the following equation can be used:

Equation 3 – Ground Vibration Prediction (Imperial Units)

$$PPV\left(\frac{in}{s}\right) = 26.79 * (Scaled\ Distance)^{-0.92}$$

In addition to prediction of the ground vibration, the air overpressure from a Precision Presplit can also be calculated (Konya & Konya, Airblast Prediction Equations for Construction Blasting, 2015):

Equation 4 - Air Overpressure Prediction (Imperial Units)

$$AP(dB) = -17.8 * LOG\left(\frac{D}{\sqrt[3]{W}}\right) + 153.7$$

Where: D = Distance from Blast (ft)
W = Weight of Charge per 8ms Delay (lbs)

When considering the impact of overbreak control, Precision Presplitting performs better than other methods creating a consistent plane without overbreak in weaker seams/areas. The dust and flyrock considerations for this technique will be similar to that of a normal presplit blast. When properly completed flyrock should be minimal however considerable dust will be present after the blast. In the past, the use of blasting mats to minimize flyrock or dust produced from the blast. However, the authors recommend this on a special case-by-case basis as it can lead to a large amount of gas pressure remaining in the borehole and causing cratering, and backbreak, along the presplit line. This should be determined through a joint evaluation from the project owner and the owner's blast consultant.

HYBRID SPECIFICATIONS

Drilling and blasting projects often present unique challenges, especially for specification writers, because of several unaccountable factors. In many areas, drilling and blasting contractors may not have the proper knowledge or experience to handle a project, yet as the government it is difficult to disqualify the contractor from bidding on a project. Drill and blast also have few applicable quantitative methods, unlike almost all other construction activities, and this makes control of a project extremely difficult for construction management personal.

For these and many other reasons, it is apparent that performance specifications do not work to keep minimal risk on the owner while allowing the owner to have control over a project to ensure the contractor uses proper blasting techniques. In addition to this, many agencies do not have the personal with proper experience to write prescriptive specifications nor would they want to take on the risk associated with a prescriptive specification. The authors have worked for years with the various state DOTs, Federal Highway Administration, and the U.S. Army Corp of engineers to develop specifications that provide control and proper performance without taking on risk of the contractors operations by the owner. From this experience, the authors have developed a new specification method which is known as the Hybrid Specification for use in drilling and blasting, which is used on almost all major rock excavation projects and is the top specification used by the Army Corp of Engineers for drilling and blast (U.S. Army Corp of Engineers, 2018).

The uniqueness of a Hybrid Specification comes in the development of performance criteria for the results of blasting and the limitations that the contractor must abide by in the development of their means and methods. This still has the contractor develop the means and methods of the project and take the liability of their means and methods, while ensuring proper performance goals are achieved. However, it allows the owner to have a say in what the contractor cannot do; for example, the specification may limit the contractor to using no larger than 1.5" diameter product on lifts between 10 feet and 20 feet. This is because it is well understood that using larger diameter product would lead to uncontrollable blasting on this size lift, by giving the contractor a maximum limit the owner can ensure that acceptable performance will be achieved and the contractor won't be asking for change orders because "the geology here is so unique that their normal means and methods won't break it to performance requirements." An example of this can be seen by looking at the State of Alaska Blasting Specifications which the authors helped to re-write. The previous specification was a performance specification (Alaska Department of Transportation and Public Facilities, 2017) and resulted in the inability to ensure that contractors properly completed the projects. The re-write of the specification was a hybrid specification (Alaska Department of Transportation and Public Facilities, 2018) which has given the state the ability to achieve better performance and control of projects.

The Precision Presplitting that is specified on a project should be developed utilizing this Hybrid Specification approach in order to ensure that contractors properly utilize and execute the

Precision Presplit. For example, the State of Alaska DOT utilized the following in their guide specification for Precision Presplitting:

"Precision Presplitting. Design precision presplit holes with a spacing from 12 inches to 24 inches on center. Adjust the spacing as necessary after evaluating the results of Test Blasts. Do not exceed an explosive charge of 0.1 pounds of explosive per foot of blasthole. Use fractional presplit cartridges as described in presplitting above, or use strands of detonating cord. Do not use cartridges greater than 1/3rd the diameter of the precision presplit hole. A larger charge is acceptable in the toe of the precision presplit hole. Do not use bulk explosives in precision presplit holes."

This specification tells the contractor what cannot be done on the project, without selecting the contractors exact means and methods. The limits that are put in place are developed to ensure that the project will achieve the performance requirements. This gives the owner the ability to mitigate change-orders based on poor drilling and blasting by the contractor and claims of 'geology' while keeping the risk of poor performance on the contractor's selected means and methods.

CONCLUSION

Precision Presplitting is regarded as the best method for overbreak control on construction projects by the US Army Corp of Engineers, Federal Highway Administration, and many State Department of Transportations. It utilized closely spaced holes with lightly loaded holes to achieve much better breakage in weak rock or rocks that have a lot of structure when compared to traditional presplitting. In this paper, the authors have discussed how gas pressure is the major factor in presplitting mechanics; the factors effecting the spacing of holes and how to mitigate geologic influence on a presplit; how to properly design the explosive load of a borehole based on the spacing and rock type; and how to properly develop a Hybrid Specification to ensure contractors properly utilize Precision Presplitting on projects.

REFERENCES

Akhavan, J. (2011). *The Chemistry of Explosives.* Cambridge: RSC Publishing.

Alaska Department of Transportation and Public Facilities. (2017). Standard Specifications for Highway Construction. 94-95.

Alaska Department of Transportation and Public Facilities. (2018). 203-3.02 Rock Excavation. *Section 203 Excavation and Embankment.*

Brown, E. M. (1673). Brief Account of some Travels in Hungaria as also some observations on Gold, Silver, Copper, Quick-filver Mines, Baths, and Mineral Waters. London: T.R.

Cook, M. A. (1974). *The Science of Industrial Explosives.* Salt Lake City: IRECO Chemicals.

Cooper, P. (1996). *Explosive Engineering.* New York, New York: VCH Publishers Inc.

Crosby, W., & Bauer, A. (1982). Wall Control Blasting in Open Pits. *Mining Engineering, 34*(2), 155-158.

Daehnke, A., Rossmanith, H., & Kouzniak, N. (1996, June). Dynamic Fracture Propagation due to Blast-Induced High Pressure Gas Loading. *2nd North American Rock Mechanics Symposium.*

Day, P. (1982). Controlled Blasting to Minimize Overbrak with Big Boreholes Underground. *Proceedings of the Eighth Conference on Explosives and Blasting Technique*, 264-274.
DuPont. (1975). DuPont Blasters Handbook. DuPont.
Guttmann, O. (1892). Blasting: A Handbook for the Use of Engineering and Others Engaged in Mining, Tunnelling, Quarrying, etc. London: Charles Griffin and Company.
International Society of Explosive Engineers. (2016). *Blasters Handbook 18th Edition.* International Society of Explosive Engineers.
Konya, A. J., & Konya, C. J. (2017). Precision Presplitting. Proceedings of the Society for Mining, Metallurgy, and Exploration.
Konya, A., & Konya, C. (2016). Precision Presplitting Optimization. Proceedings of the 42nd International Conference on Blasting and Explosive Technique.
Konya, A., & Konya, C. (2017). Precision Presplitting. Proceedings of the Society for Mining, Metallurgy, and Exploration.
Konya, A., & Konya, C. (2017). Precision Presplitting - Explosive Variations with Spacing. *Proceedings of the 43rd International Conference on Explosive and Blasting Technique.*
Konya, A., & Konya, C. (2019). The Development of Pressure to Young's Modulus Models for Precision Presplit Blasting. *European Federation of Explosive Engineers.* Helsinki: EFEE.
Konya, A., & Konya, C. J. (2015). Airblast Prediction Equations for Construction Blasting. *International Society of Explosive Engineers 41st Proceedings.*
Konya, A., & Konya, C. J. (2016). Precision Presplitting Optimization. *Proceedings of the Forty-Second Annual Conference on Explosives and Blasting Technique* (pp. 65-74). Las Vegas: International Society of Explosive Engineers.
Konya, C. (1973). High Speed Photographic Analysis of the Mechanics of Presplit Blasting. *Proceedings of Sprengtechnick International.*
Konya, C. (1980). Pre-split Blasting: Theory and Practice. *Preprint AIME*, 80-97.
Konya, C. (1982). Seminar on Blasting to the Ohio Laborers Union. Mont Vernon, Ohio.
Konya, C. (1990). *FHWA Rock Blasting Manual.* Federal Highway Administration.
Konya, C., & Konya, A. (2015, November). Precision Presplitting Optimization. *8th Drilling-Blasting Symposium*, (pp. 1-10). Istanbul, Turkey.
Konya, C., Barret, D., & Smith, J. E. (1986). Presplitting Granite Using Pyrodex, A Propellant. *Proceedings of the Twelfth Conference on Explosive and Blasting Technique*, 159-166.
Konya, C., Britton, R., & Lukovic, S. (1987). Charge Decoupling and Its Effect on Energy Release and Transmission for One Dynamite and Water Gel Explosive. *Proceedings of the Thirteenth Conference on Explosives and Blasting Technique.*
Langefors, U., & Kihlstrom, B. (1973). *The Modern Technique of Rock Blasting.* Uppsala: Almqvist & Wiksell Informationsindustri AB.
Paine, R., Holmes, D., & Clark, H. (1961, June). Presplit Blasting at the Niagara Power Project. *The Explosives Engineer*, 72-91.
Salmi, E., & Hosseinzadch, S. (2014). A Further Study on the Mechanisms of Pre-splitting in Mining Engineering. *App Mech Mater, 553*, 476-481.
Spathis, A., & Wheatley, M. (2016, July). Dynamic Pressure Measured in a Water-filled Hole Adjacent to a Short Explosive Charge Detonated in Rock. *Blasting and Fragmentation Journal, 10*(1), 33-42.
Tariq, S., & Worsey, P. (1995). An Investigation into the Effects of Joint Frequency and Spatial Positioning on Pre-Splitting. *International Society of Explosive Engineers*, (p. 287).

Tariq, S., & Worsey, P. (1996). An Investigation into the effects of some aspects of jointing and single decoupled blast holes on pre-splitting and boulder blasting. *Rock Fragmentation by Blasting - Fragblast 5*, 438.

U.S. Army Corp of Engineers. (2018). *Blasting for Rock Excavations.* US Army Corps of Engineers.

Worsey, P. (1981). Geotechnical Factors Affecting the Application of Pre-Split Blasting to Rock Slopes. University of Newcastle Upon Tyne.

Worsey, P. (1984). The Effect of Discontinuity Orientation on the Success of Pre-Split Blasting. *10th Annual Society of Explosive Engineers Conference on Explosives and Blasting Technique*, 197-217.

Worsey, P. (1984). The Effect of Discontinuity Orientation on the Success of Pre-Split Blasting. *International Society of Explosive Engineers*, (p. 197).

Worsey, P., & Qu, S. (1987). Effect of Joint Separation and Filling on Pre-split Blasting. *13th Society of Explosive Engineers Conference on Explosives and Blasting Technique.*

Worsey, P., Farmer, I., & Matheson, G. (1981). The Mechanics of Pre-splitting in Discontinuous Rocks. *22nd U.S. Symposium on Rock Mechanics*, 205-210.

Zhang, Z.-X. (2016). Rock Fracture and Blasting Theory and Applications. Elsevier Inc.

Optimization of Rockfall Hazard Assessment for the I-90 Snoqualmie Pass Corridor-Snow Bridge Project

David P. Findley, LG, LEG
Golder Associates Inc.
18300 NE Union Hill Road, Suite 200
Redmond, WA 98052
(425) 883-0777
dfindley@golder.com

Kyle E. Obermiller, PE, LEG
Golder Associates Inc.
18300 NE Union Hill Road, Suite 200
Redmond, WA 98052
(425) 883-0777
kobermiller@golder.com

Prepared for 70th Highway Geology Symposium, October, 2019

Disclaimer

Statements and views presented in this paper are strictly those of the author(s), and do not necessarily reflect positions held by their affiliations, the Highway Geology Symposium (HGS), or others acknowledged above. The mention of trade names for commercial products does not imply the approval or endorsement by HGS

Copyright Notice

Copyright © 2019 Highway Geology Symposium (HGS)

All Rights Reserved. Printed in the United States of America. No part of this publication may be reproduced or copied in any form or by any means – graphic, electronic, or mechanical, including photocopying, taping, or information storage and retrieval systems – without prior written permission of the HGS. This excludes the original author(s).

ABSTRACT

The Snoqualmie Pass rockfall hazard assessment area is a roughly ¼ mile section of Interstate 90 located in the Snoqualmie Pass area of the High Cascades of Washington State along the shoreline of Lake Keechulus. The subject area involves the replacement of a 500-foot long snowshed built in 1950 with two 1,200-foot-long avalanche bridges, one bridge for east-bound traffic and one bridge for west-bound traffic. Each bridge will accommodate three lanes of traffic. The bridges will span active avalanche chutes that allow future avalanche events to pass beneath the road deck. The project is part of a much larger 15-mile corridor improvement project from Snoqualmie Pass to the vicinity of Easton, WA aimed at improving driver safety and reliability while reducing traffic congestion, and avalanche closures in the winter. The overall purpose of the rockfall hazard analysis was to evaluate the potential for rockfall to impact the avalanche bridge structure. Parameters used in the rockfall models such as source areas, surface roughness, block size, and coefficients of restitution were based on field observations made during the field investigation. The initial rockfall models were developed using "as-designed" configurations and United States Geological Survey (USGS) topography. Following the construction of the rock slope cut, the cut slope geometries were updated using terrestrial LiDAR data and additional analyses were performed to evaluate potential changes in the rockfall trajectories and energies due to differences between "as-constructed" and "as-designed" slopes.

A rockfall event that occurred during construction provided the opportunity to evaluate the rock fall model input assumptions and calibrate the parameters used through back-analysis. The results of the rockfall event back analysis resulted in a reduction of the predicted in the kinetic energy, velocity, and percentage of rockfall with the potential to impact the westbound avalanche bridge structure.

This paper will illustrate the evolution of the rockfall model, the challenges with back-analyses of the rockfall event, and the differences between the as-constructed sections with both the assumed and calibrated restitution coefficients.

INTRODUCTION

Golder Associates Inc. was retained by the design-build team of Atkinson and Jacobs Engineering to provide rock slope geotechnical design services to support the design and construction of two avalanche bridge structures located within the I-90 corridor from Snoqualmie Pass in the High Cascades of Washington State to the vicinity of Easton, Washington as shown on Figure 1.

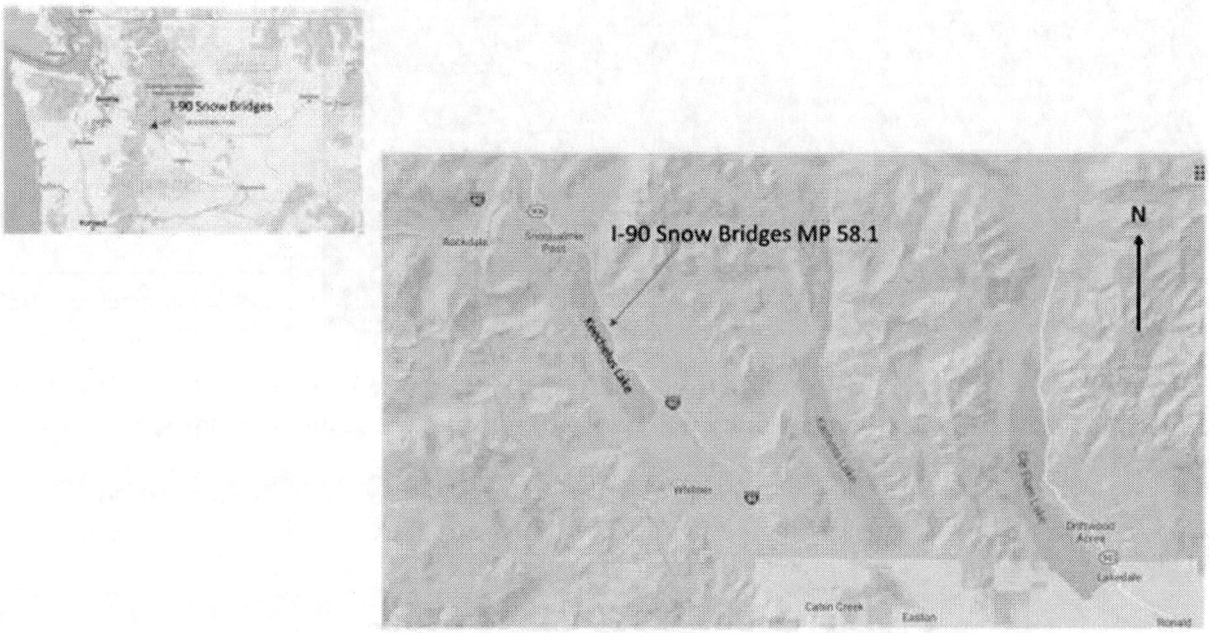

Figure 1 – I-90 Snow Bridges Location.

The snow bridge project was one design element within the entire corridor improvement project to increase capacity, safety, and reduction of avalanche related road closures during the winter months. The corridor is located along the east shore of Keechelus Lake, adjacent to a steep, west-dipping bedrock slope as shown on Figure 1. The required road width for the corridor improvement project was achieved through significant rock cuts along the inboard (west-bound) side of the right-of-way. Adverse dipping bedrock structure within the tuffaceous bedrock required significant slope reinforcement.

The snow bridge concept was the result of WSDOT's Cost Reduction Incentive Program (CRIP). The CRIP process is a method for contractors to propose alternative designs to the original design concepts if cost reductions are realized as well as achieving the original design objectives. The WDSOT design concept was to replace the original reinforced concrete snow shed built in 1950 with a much larger snowshed structure that would span all six lanes through the avalanche chute section. The 1950 snow bridge only spanned the west-bound two lanes. Frequent closures were common during the winter months in order to clear snow and avalanche debris that traversed over the west bound lanes only to accumulate and close the east-bound lanes as shown on Figure 2.

Photos courtesy of WSDOT

Figure 2 – View in the Westbound Direction with 1950 Snow Shed on the Right Covering the WB Lanes Leaving the EB Lanes Susceptible to Avalanche Closures.

The Atkinson/Jacobs CRIP proposal consisted two bridge structures; one for the west-bound lanes and the second for the east-bound lanes. The bridge structures would allow future snow avalanches to pass beneath the roadway rather than over the top via a snow shed.

Golder was subcontracted to the design-build team to provide geotechnical design and construction support for the rock excavations along the inboard side to provide the necessary clearance between the west-bound bridge structure and the bedrock slope. Additional services included rock fall analysis to evaluate the potential for rock fall impacting the bridge structure and columns-the topic of this paper.

PROJECT GEOLOGY

The project site is situated along a north-south oriented segment of I-90 between the base of a steep, west-facing slope and Lake Keechelus. The native slopes in the vicinity of the proposed cut range from 30 degrees (58 percent) to 40 degrees (84 percent). Areas of pronounced bedrock exposure are locally steeper. The slope extends from an approximate elevation of 2,560 feet (ft.) adjacent to the existing I-90 shoulder to the crest of the slope situated at an approximate elevation of 3,700 ft. The ground surface is vegetated with mixed density mature second growth timber. Vegetation density varies from open in the snow accumulation zones high on the slope and avalanche chutes to denser tree cover lower on the slope. The subject slope contains seven well developed avalanche chutes.

The bedrock lithology forming the ridge slope has been mapped by Tabor and others *(1)* as a tuff member of the Oligocene age Ohanapecosh Formation. In the project region, the Ohanapecosh

Formation is characterized as well bedded tuff, breccia, and minor flows of highly altered andesite, basalt, and dacite. The Ohanapecosh Formation is locally overlain by alpine glacial deposits of till, outwash, colluvium, and talus.

The slopes located within and above the project are formed on the southwest-facing limb of a northwest trending anticline. This structural geologic framework results in a dip-slope geometry where bedding planes within the Ohanapicosh Formation dip in the general direction of the slope dip direction, that is toward the I-90 corridor as shown on Figure 3.

Based on the site information collected, the rock on site can be described as: unweathered to slightly weathered, widely to very widely jointed with some closely jointed zones, dark grey porphyritic, fine to course crystalline, strong (ISRM R4) to very strong (ISRM R5) meta-welded lapilli dacite tuff.

Well developed, persistent joint sets control the kinematic stability of the rock mass. The most dominant joint set is a planar joint set dipping out of or parallel to the slope. See Figure 3. These joints are persistent over distances of tens of feet. Secondary intersecting near vertical joints form local wedges.

Figure 3 – Persistent Dip-Slope Joints Exposed in the Original 1950 Era I-90 Rock Cut.

The tuffaceous bedrock is overlain by a variable thickness of colluvium. Limited drill hole information and field observations suggest colluvium thickness ranging from 0 ft. at rocky outcrops and avalanche chute bottoms to about 10 to 12 ft. thick in the intervening ridges between avalanche chutes.

There is an existing cut slope that runs along the entire length of the proposed bridges. We assume that this cut was originally created during road and snow shed construction in 1950. The existing cut slopes range from about 40 degrees (84 percent) to 50 degrees (119 percent) at the north portions of the snow shed where talus/colluvium was encountered, to near vertical at the south end of the proposed bridge section where bedrock was encountered. This existing cut provides good exposure of overburden material, which is likely similar to material that will be encountered during the construction of proposed cuts for the bridges. The proposed cut will cross the paths of several avalanche chutes.

Evidence of older rock fall events was observed in the forested slope located above the proposed cut. Rock fall originating above the forested slope was stopped by the mature, second growth timber as shown on Figures 4 and 5.

Figure 4 – View in the West-Bound Direction. Both WB and EB traffic is temporarily on the new EB snowbridge structure.

Figure 5 – Bedrock Rockfall Debris Stopped by Second Growth Timber.

ROCK FALL ANALYSIS

The computer code RocFall (*2*) was used to model and evaluate rock fall potential originating from the slope adjacent to the proposed snow bridges. RocFall is a statistical analysis program designed to analyze risk of slopes with the potential to produce rock fall, providing such information as trajectory envelopes of modeled rock fall, energy, velocity, and bounce heights.

Factors affecting rockfall analysis include the following:

- Slope geometry
- Roughness of the slope surface
- Rolling resistance
- Restitution characteristics of the normal coefficient of restitution (Rn) and tangential coefficient of restitution (Rt) of the slope materials
- Geometry of the rock particles
- Density of the rock
- Vertex variation (x and y coordinates of vertices between slope segments)
- Initial rock block velocity (horizontal, vertical and rotational)

In modeling the rock fall potential for the snow bridge site, two source areas were evaluated:

1. Rockfall originating from near the crest of the proposed cut-referred to as the low energy source area, and
2. Rockfall originating higher on the slope near the crest of the ridge (around elevation 3,700 ft.) referred to as the high energy source.

In addition, two sizes of rock masses were modeled: a smaller rock assumed weight 1,400 pounds (lbs) and a larger rock mass with an assumed weight of 23,000 lbs.

These two source areas were evaluated along five cross sections through the project section. One of the sections located near Pier 5 (project Station 8358+25) is shown as representative of the process output on Figure 6 through Figure 9. Table 1 summarizes the rockfall simulation at Pier 5 (Sta. 8358+25).

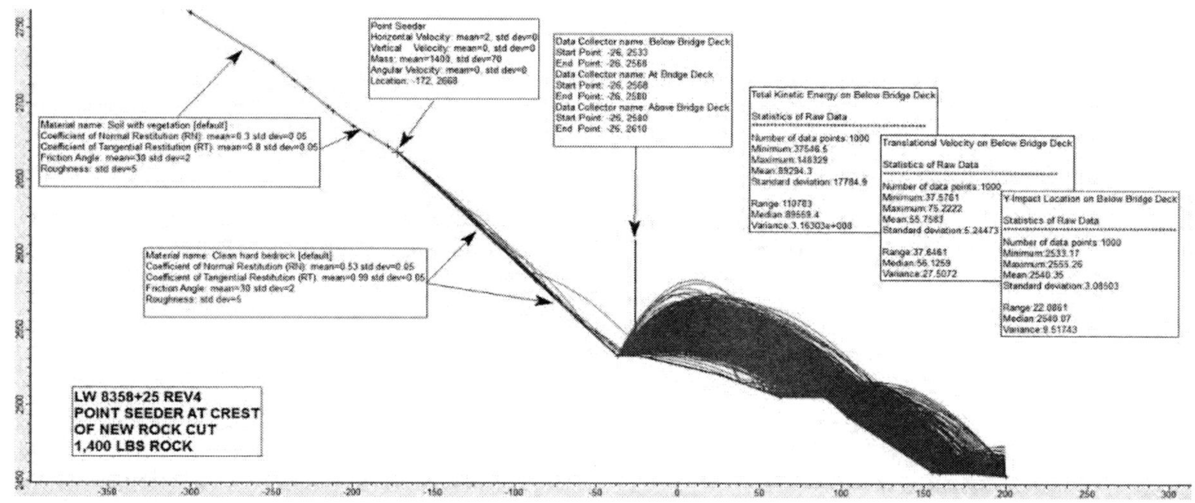

Data Collector	Maximum Bounce Height [feet]	Maximum Kinetic Energy [ft-lbs]	Maximum Translational Velocity [ft/sec]
Below Bridge Deck	22.26	1.5E+05	75.2
At Bridge Deck	n/a	n/a	n/a
Above Bridge Deck	n/a	n/a	n/a

Figure 6 – Top of Cut Low Energy Source, Small Rock.

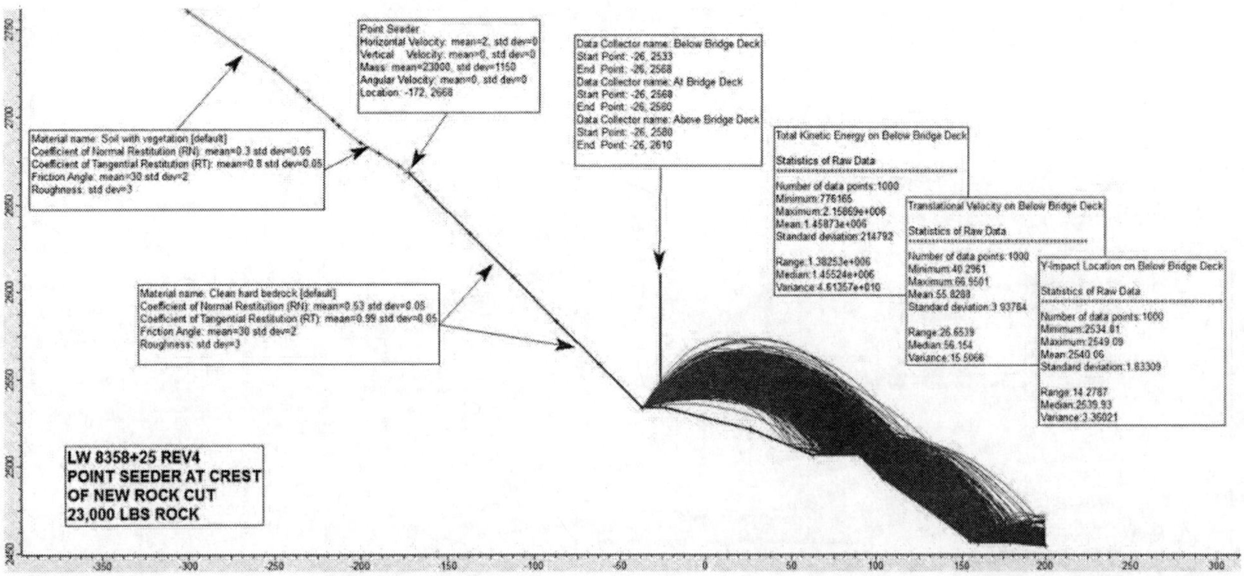

Data Collector	Maximum Bounce Height [feet]	Maximum Kinetic Energy [ft-lbs]	Maximum Translational Velocity [ft/sec]
Below Bridge Deck	16.09	2.2E+06	67.0
At Bridge Deck	n/a	n/a	n/a
Above Bridge Deck	n/a	n/a	n/a

Figure 7 - Top of Cut Low Energy Source, Large Rock.

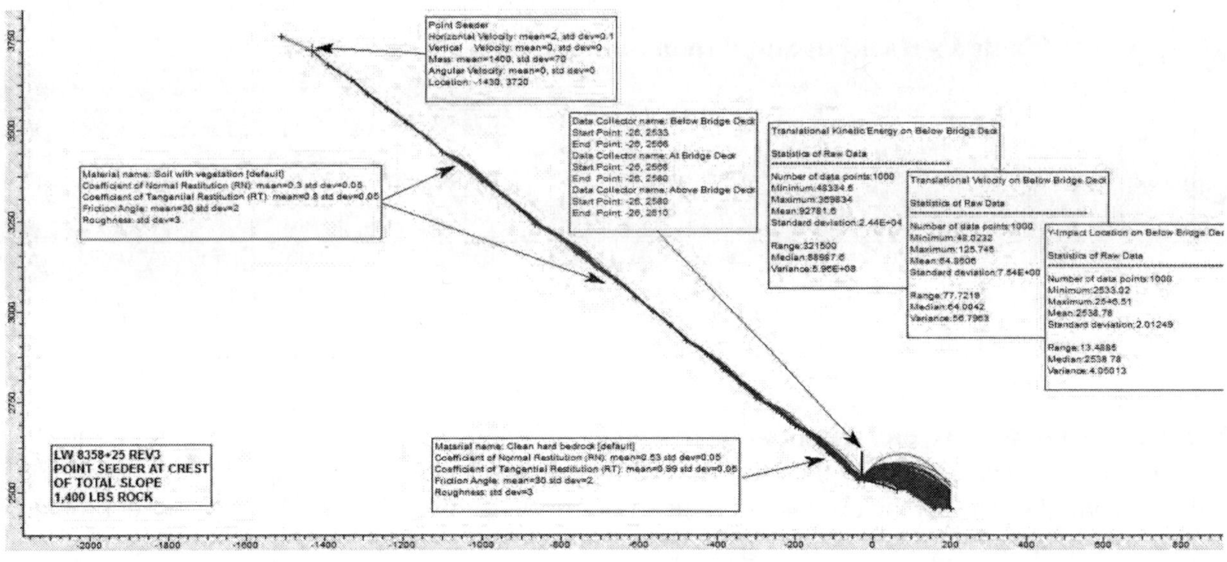

Data Collector	Maximum Bounce Height [feet]	Maximum Kinetic Energy [ft-lbs]	Maximum Translational Velocity [ft/sec]
Below Bridge Deck	26.86	4.4E+05	124.2
At Bridge Deck	n/a	n/a	n/a
Above Bridge Deck	n/a	n/a	n/a

Figure 8 – Top of Slope Crest High Energy Source, Small Rock.

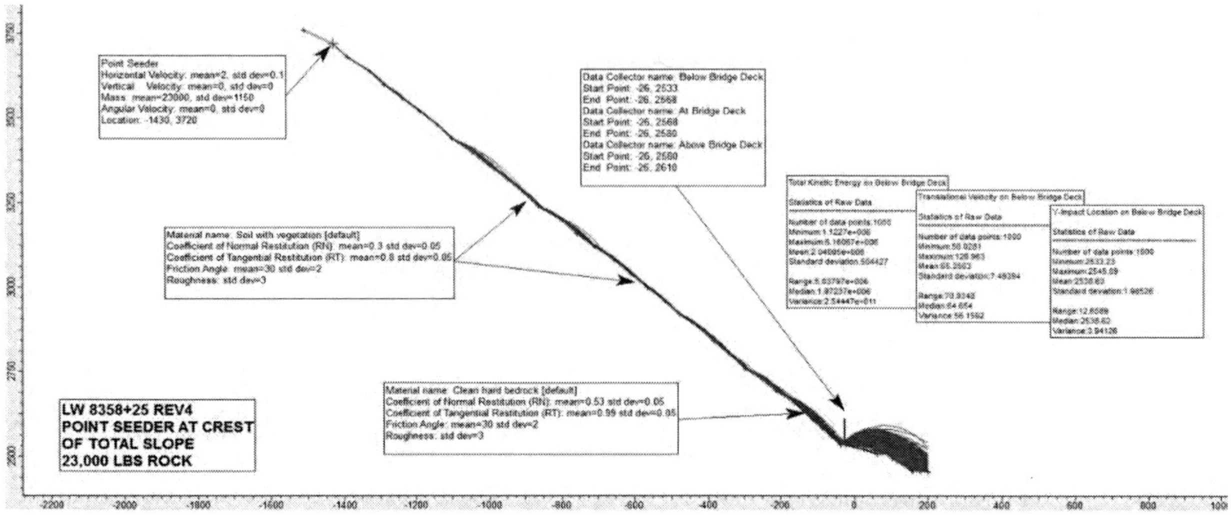

Data Collector	Maximum Bounce Height [feet]	Maximum Kinetic Energy [ft-lbs]	Maximum Translational Velocity [ft/sec]
Below Bridge Deck	16.04	5.3E+06	61.3
At Bridge Deck	n/a	n/a	n/a
Above Bridge Deck	n/a	n/a	n/a

Figure 9 – Top of Slope Crest High Energy Source, Large Rock.

Table 1 - Rockfall Simulation Summary Pier 5 (Sta. 8358+25)

Simulation Cross Section Location	Pier Number(s)	Simulation Seed Point	Rock Size (1)	Data Collector											
				Below Bridge Deck (i.e., Piers)				At Bridge Deck				Above Bridge Deck			
				Percentage of Rocks Passing (2)	Maximum Bounce Height [ft]	Maximum Total Kinetic Energy [ft-lbs]	Maximum Translational Velocity [ft/sec]	Percentage of Rocks Passing (2)	Maximum Bounce Height [ft]	Maximum Total Kinetic Energy [ft-lbs]	Maximum Translational Velocity [ft/sec]	Percentage of Rocks Passing (2)	Maximum Bounce Height [ft]	Maximum Total Kinetic Energy [ft-lbs]	Maximum Translational Velocity [ft/sec]
LW 8358+25	Pier 5	Crest of New Cut	Small	100	22	1.48E+05	75	n/a	n/a	n/a	n/a	n/a	n/a	n/a	n/a
			Large	100	16	2.16E+06	67	n/a	n/a	n/a	n/a	n/a	n/a	n/a	n/a
		Crest of Natural Slope	Small	99.9	27	4.39E+05	124	n/a	n/a	n/a	n/a	n/a	n/a	n/a	n/a
			Large	100	16	5.25E+06	61	n/a	n/a	n/a	n/a	n/a	n/a	n/a	n/a

Except for two of the five cross sections analyzed, the rockfall simulations indicate rockfall (both large and small) emanating from the crest of the proposed rock cuts, as well as the crest of the total slope, passes beneath the proposed bridge. For cross section LW 8359+50, the simulations of rockfall emanating from the crest of the total slope condition indicate 0.3% of the large rockfalls pass over the guard rail. For cross section LW 8361+50, the simulations of rockfall emanating from the crest of the total slope condition indicate 2.4% of the small rockfalls and 0.2% of the large rockfalls pass over the guard rail.

A summary of maximum rockfall bounce heights, translational kinetic energy and translational velocities from the five cross sections are summarized in Table 2.

Table 2 - Summary of Maximum Rockfall Bounce Heights, Translational Kinetic Energy and Translational Velocities

Data Collector	Maximum Bounce Height [feet]	Maximum Kinetic Energy [ft-lbs]	Maximum Translational Velocity [ft/sec]
Below Bridge Deck	54.08	7.8E+06	143.53
At Bridge Deck	76.95	6.7E+06	126.79
Above Bridge Deck	83.23	5.8E+06	123.48

Sensitivity Analysis

Golder conducted a parametric or sensitivity analysis of the coefficients of restitution (tangential and normal) for the two slope materials used in the rockfall analyses (vegetated slope and bare bedrock). One section located adjacent to the Pier 5 (LW 8358+25) was selected for the sensitivity analysis. To accommodate the influence of mini-benches, two types of slopes were analyzed: simple (no bedrock benches); and complex (bedrock benches). For each slope type, point seeders were placed at the crest of the proposed rock cut slope and at the crest of the total slope, about 1,250 ft. above the bridge location. For each analysis, the initial input parameters used in the preliminary rockfall analysis were held constant, and only one falling rock was simulated. One input parameter was adjusted up or down from the initial condition, and the effect of the rockfall bounce height, translational velocity and translational kinetic energy were then measured at the downslope data collectors. Standard deviations were set to zero for all the input parameters.

Simple Slope Condition

For the simple slope condition with the point seeder placed at the crest of the **new rock cut**, results indicate the modeled slope is most sensitive to bedrock Rt and Rn, where variations in these parameters are linearly related to rockfall bounce heights. Increases in Rn lead to increases in bounce height; increases in Rt lead to decreases in bounce height. Variations in velocity parameters lead to very small variations in bounce height. Variations in vertex variation also lead to very small variation in bounce height. Translational kinetic energy is roughly linearly related to rockfall mass, while bounce height and translational velocity are independent of rock mass.

For the simple slope condition with the point seeder placed at the crest of the **total slope**, the results indicate the modeled slope is most sensitive to vegetative slope Rt, above 0.9, where increases in this parameter lead to large increases in bounce heights. Increases and decreases in Rn lead to very little variation in bounce height. Variations in velocity parameters lead to very small variations in bounce height. Variations in vertex variation also lead to very small variation in bounce height. Translational kinetic energy is roughly linearly related to rockfall mass, while bounce height and translational velocity are independent of rock mass.

Complex Slope Condition

For the complex slope condition with the point seeder placed at the crest of the **new rock cut**, results indicate the modeled slope is most sensitive to bedrock Rn, where variation in this parameter is linearly related to rockfall bounce heights. "Noise" in signals is indicative of the effect of benches on bounce trajectories. Increases in Rn lead to increases in bounce height. Variations in velocity parameters lead to very small variations in bounce height. Variations in vertex variation below 3% lead to very small variation in bounce height. Translational kinetic energy is roughly linearly related to rockfall mass, while bounce height and translational velocity are independent of rock mass.

For the complex slope condition with the point seeder placed at the crest of the total slope, the results indicate the modeled slope is most sensitive to vegetative slope Rt, above 0.8, where increases in this parameter lead to large increases in bounce heights. Increases and decreases in Rn lead to very little variation in bounce height. Variations in velocity parameters lead to very small variations in bounce height. Variations in vertex variation from 1 to 10% lead to oscillating variations in bounce height, most likely related to the bedrock benches. Translational kinetic energy is roughly linearly related to rockfall mass, while bounce height and translational velocity are independent of rock mass.

The coefficients of restitution are difficult to characterize in lieu of actual measured rockfall data (e.g., trajectory, impact locations, bounce heights and rock mass) that can be used to calibrate the model. The sensitivity analyses indicate the rockfall model is most sensitive to the bedrock coefficients of normal and tangential restitution, and to the vegetative slope coefficient of tangential restitution. The model is less sensitive to vertex variation, and weakly sensitive to initial rockfall velocities (horizontal, vertical and rotational). Translational kinetic energy is directly proportional to rock mass, while variations in rock mass have no effect on bounce height or horizontal velocity.

BACK ANALYSIS OF OCTOBER 2013 ROCKFALL EVENT

A rock fall event occurred on October 3, 2013 during construction that originated well above the top of cut and produced rock debris that reached the westbound lane of I-90. The rock fall event provided a unique opportunity to calibrate the rock fall model with real world, site specific data.

A field reconnaissance was completed to assess the October 3, 2013 rockfall event. The source area of the rock fall was identified. The source area was a single boulder located approximately 245 ft. (slope distance) from the crest of the cut slope and based on field observations the single boulder broke up into smaller blocks on the decent down the slope. Field observations and measurements included impact locations, tree scar heights, boulder locations, slope length and angle, and path orientation. An approximate rock fall debris envelope was determined as shown on Figure 10.

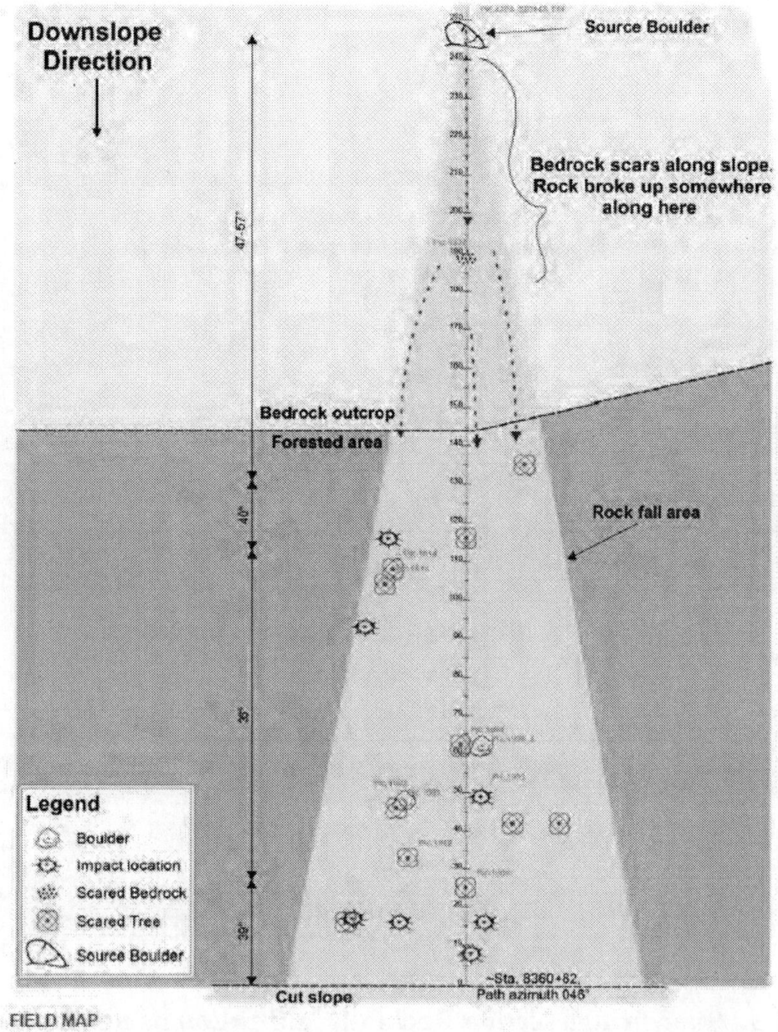

Figure 10 - October 3, 2013 Rockfall Event Field Reconnaissance Observations.

The previous rock fall analysis had used computer code default values for the various input parameters to model the rock fall using RocFall (*2*). The back analysis used the updated, commercially available software RocFall v5.014 by Rocscience (*3*). For the back analysis, Golder initially used both the Lump Mass and the Rigid Body Impact Mechanics (RBIM), models available within the software. The results based on both models were assessed and it was concluded that the RBIM model (that considers rock shape and size) appeared to produce more realistic rockfall trajectories based on limited field observations, compared to those based on the Lump Mass model (Figures 8 and 9). See Figure 11 for rockfall calibration using the RBIM model.

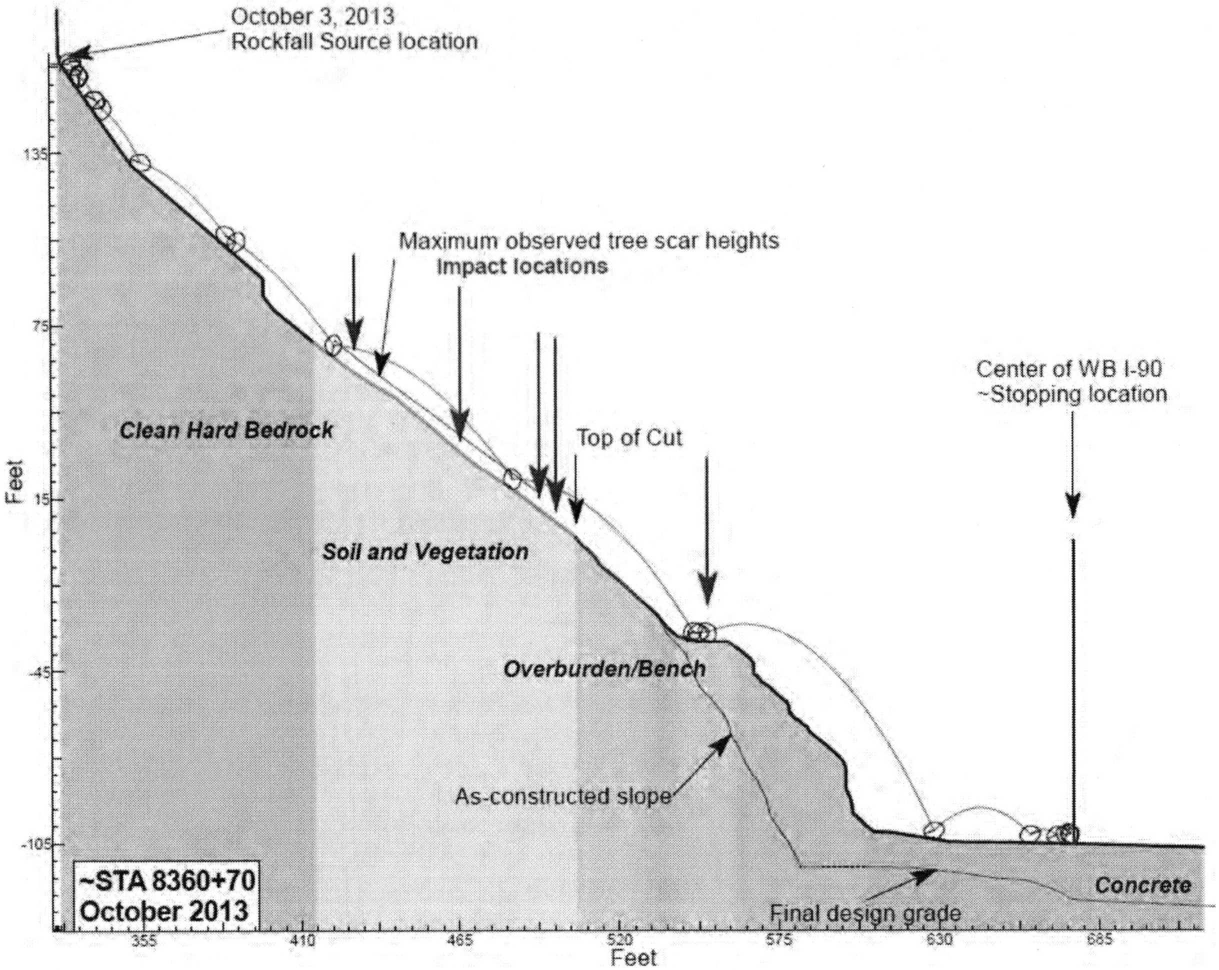

Figure 11 – I-90 Snowbridge section Rockfall Calibration using Rigid Body Impact Mechanics (RBIM).

The RBIM model was used to implement the back-analysis results in evaluating rockfall potential in two selected areas using the terrestrial LIDAR based as-built topography. The two as-built areas included one that was over-excavated and a second that under -excavated.

Input Parameters

Topography

The topography used during previous analyses was updated with terrestrial LIDAR data collected by WSDOT June 16, 2014 in the vicinity of the rock cut. Topography upslope of the rock cut was obtained from the topographic quadrangle map published by the United States Geological Survey (USGS) (*4*). The terrestrial LIDAR figure was provided by WSDOT on June 19, 2014. Variations in slope topography less than the topographic interval in the published USGS maps, i.e., 20 ft., were not incorporated into the rockfall models. Steeper local conditions than that modeled will likely produce higher energies and potentially higher bounce heights/trajectories. A topographic map derived from ground based or airborne Lidar will produce a more accurate estimate of the slope geometries above the current site topographic mapping limits (i.e., above approximate elevation 2,750 ft.). These data are not available. To account for uncertainty in the slope surface, a vertex variation of 5% was applied to the x and y coordinates between each slope segment in the model.

Roughness

The Slope Roughness in the RocFall Slope Material Library allows for variability in the local surface angle of segments of the slope. In the Rocscience RBIM, slope roughness is defined by spacing and amplitude, and statistical distributions can be assigned to each parameter. The spacing is the distance between amplitudes, while the amplitude is the distance from the base slope. As recommended by Rocscience, the amplitude was equal to zero, so that the roughness is more evenly distributed around the base slope. A standard deviation of 20% was applied to both material roughness properties to account for slope roughness variability. The material slope roughness is based on observations made near the vicinity of the slope cut and may not reflect conditions further away. The material roughness parameters are summarized in Table 3.

Table 3 - Material Roughness Parameters

Material roughness		Mean	Distribution	St. Dev	Rel. min	Rel. Max
Soil with Vegetation	Slope Roughness Spacing	3	Normal	0.2	0.6	0.6
	Slope Roughness Amplitude	0	Normal	0.2	0.6	0.6
Clean Hard Bedrock	Slope Roughness Spacing	1	Normal	0.2	0.6	0.6
	Slope Roughness Amplitude	0	Normal	0.2	0.6	0.6

Boulder Dimensions, Shape, and Rock Type

The boulder shape and mass used in the analysis were based on field observations made in 2012. The boulders observed along the slope were slightly tabular with length to width ratios slightly greater than 1; average ratio of 1.4. To be conservative, a square shaped boulder with rounded corners was modeled. The assumed boulder masses were the same as those used in the previous analyses. The large rock block was 23,000 lbs and the small rock block was 1,400 lbs. Table 4 summarizes the rock fall source rock properties assumed for the October 3, 2013 back analysis.

Table 4 - RocFall Source Rock Properties

<u>**Source Rock**</u> Shape: Super Ellipse^4 Mass (lbs): Large – 23,00 Small – 1,400 Density (lb/ft^3): 165	 Example of boulder caught in forest, 2012	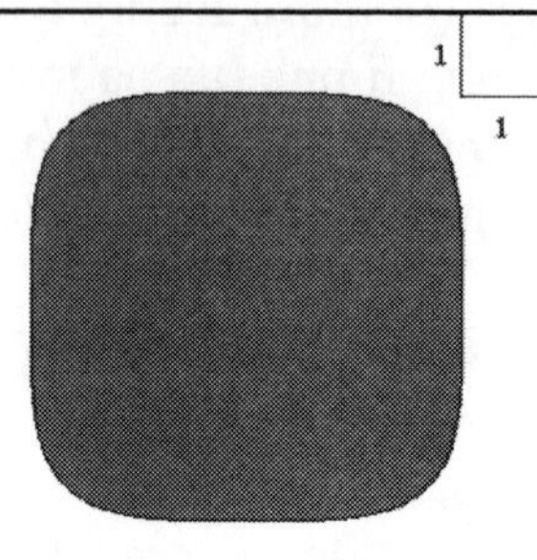 Shape selected in RocFall

Note: The boulders used in the October 2013 back-analysis have a different shape than those observed during the 2012 slope reconnaissance. Rockfall, from the October 2013 event, caught in the timbered forest where rhombus in shape.

Rockfall Source Location

Like previous analyses, a total of 1,000 rocks were "rolled" for each cross-section from the source area (a point seeder) located either at the crest of the proposed rock cut (described as low energy) or from the top of the natural slope (described as high energy). The seeders located along the crest of new cut were adjusted from the 2013 analysis to reflect as-constructed conditions; generally relocated to or just above the crest dowel. The 2015 sections were only evaluated for sources areas originating at the crest of new cut, except where sections coincided with those evaluated in 2013.

Slope Materials

Consistent with previous analyses, two surface conditions were modeled for the analysis: 1) rock exposed as a result of cut slope production, and 2) soil with vegetation to approximate the forest floor and colluvium layer above bedrock. Each of these surfaces has different coefficients of restitution. The coefficient of restitution is a measure of the energy transference between the ground surface and the moving boulder. The analysis includes the coefficient of normal restitution (Rn) and the coefficient of tangential restitution (Rt) for each material. Coefficients for the soil with vegetation was back analyzed from a rock fall event which occurred in October of 2013. Coefficient values for the clean hard bedrock and material friction values were reference values contained in RocFall (*5*, *6*). The material coefficients are summarized in Table 5.

Table 5 - Slope Material Coefficients of Restitution

Properties	Soil with Vegetation	Clean Hard Bedrock
Normal Restitution	0.25*	0.53
Tangential Restitution	0.55*	0.99

Note: The "*" symbol denotes the parameters that back analyzed.

Representative rockfall runs using the back analyzed data from the October 2013 rockfall event are shown on Figures 12 and 13.

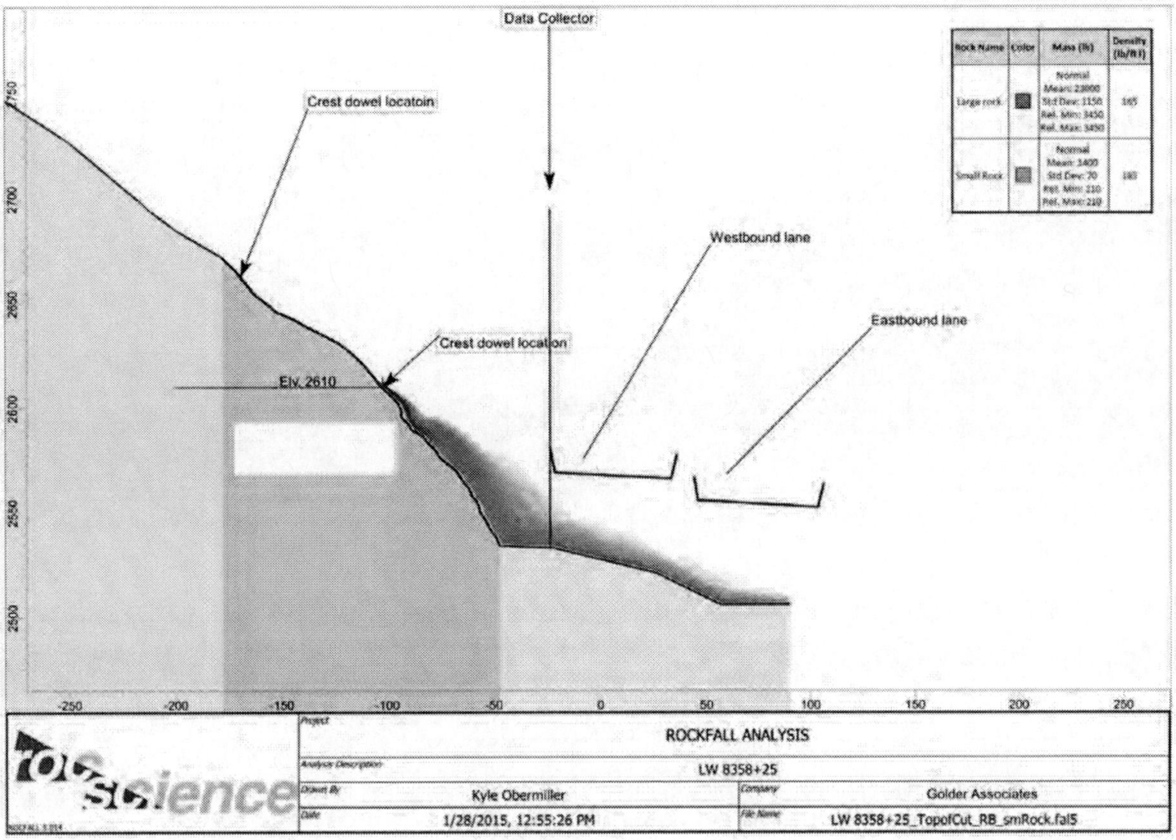

Figure 12 – Rockfall Analysis for Sta. 8358+25 for Small Rocks.

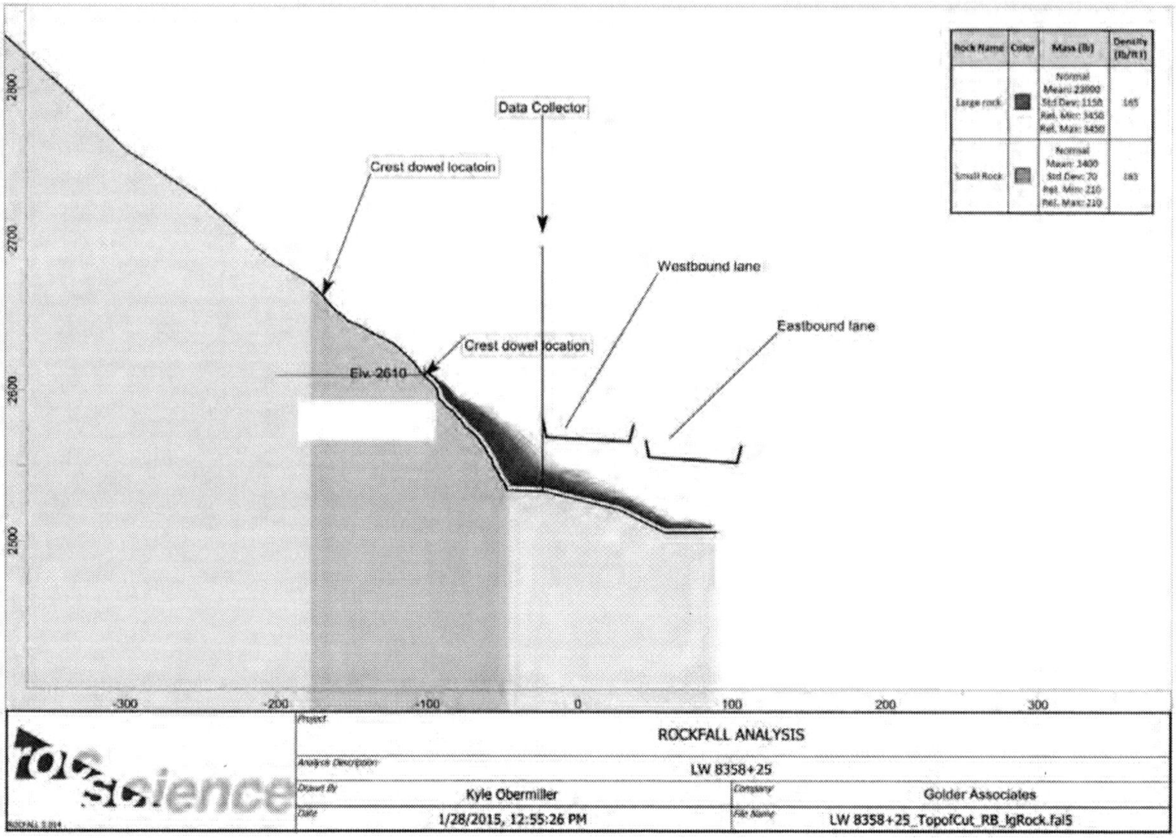

Figure 13 – Rockfall Analysis for Sta. 8358+25 for Large Rocks.

Rockfall Analysis Results

Upon completion of the back-analyses of the October 2013 rockfall event, we were able to apply the back-calculated coefficients of restitution, for the material upslope of the rock cut, and re-analyze all sections previously analyzed in 2013 and all under/over excavated slope analyzed in 2015 for a total of ten sections in the vicinity of the westbound snow bridge. In general, the results of the 2016 updated analysis show a similar percentage of rockfall impacting or passing under/over the bridge, however, the rockfall maximum translational velocity, and the rockfall maximum total kinetic energy all along the snow bridge sector have decreased (Figures 14 and 15).

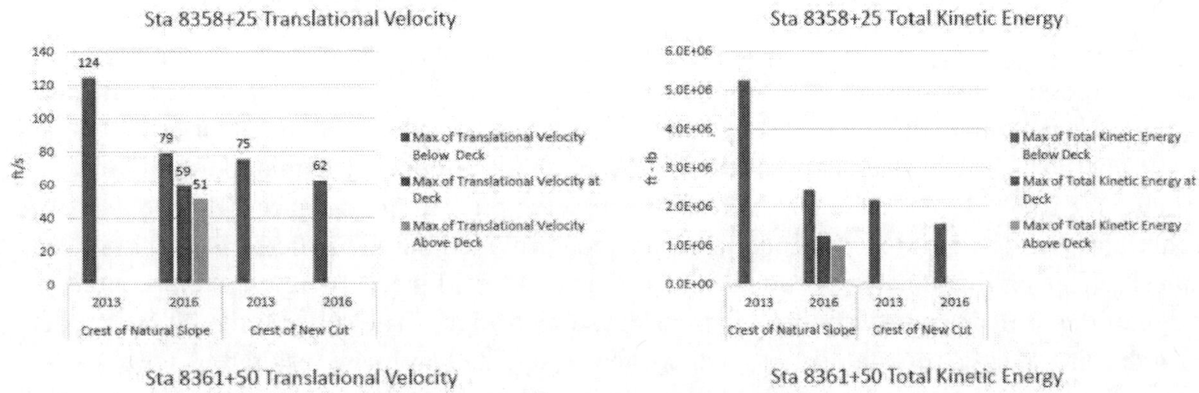

Figure 14 – Comparison of Translational Velocity and Total Kinetic Energy at Sta. 8358+50.

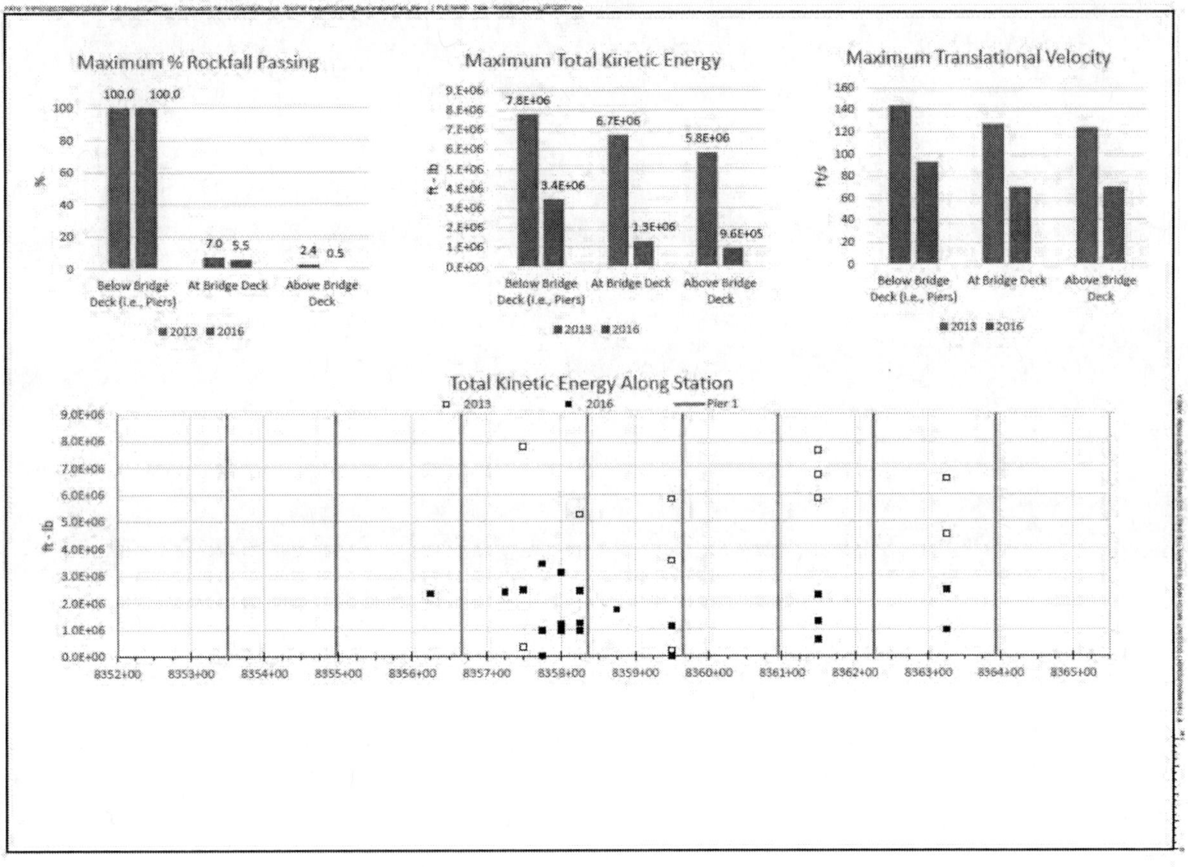

Figure 15 – Summary of 2016 Rockfall Data with Calibrated Model.

SUMMARY AND CONCLUSIONS

The results of this study demonstrate the importance of obtaining site-specific information whenever possible to calibrate rock fall modeling software. In the absence of site-specific data, default rock fall model input parameters tend to over-emphasize rock fall trajectories and total kinetic energies. The back analysis of the 2013 rock fall event provided an opportunity to calibrate the rock fall modeling that had been completed prior to the October 2013 rockfall event. Preliminary design input recommendations had been provided based on the earlier rockfall analysis. This original analysis provided the anticipated total kinetic energies to support the design of the drilled/socketed bridge piers for the snow bridge. The calibrated rock fall model using the site-specific rock fall data as well as newer, updated software resulted in a 50 % decrease in the total kinetic energy for the bridge structure.

REFERENCES

1. Tabor, R.W., V.A. Frizzel, Jr., D.B.Booth, and R.B. Waitt. *Geologic Map of the Snoqualmie Pass 30 x 60 Minute Quadrangle, Washington*, United States Geological Survey Geologic Investigations Series Map I-2538, 2000.
2. Rocscience Inc., May 14, 2012, RocFall – Risk Analysis of Falling Rocks on Steep Slopes, Version 4.057.
3. Rocscience Inc., July 31, 2015, RocFall – Risk Analysis of Falling Rocks on Steep Slopes, Version v5.014.
4. United State Geological Survey. Stampede Pass, Washington 7.5 minute Topographic Map, 2017.
5. Rocscience, Inc. *Dynamic Friction and Rolling Friction Coefficients Table*. Online: https://www.rocscience.com/help/rocfall/webhelp/baggage/Dynamic_Friction_Rolling_Friction_Table.htm. Accessed 10/2/2016.
6. Rocscience, Inc. *Rocscience Coefficient of Restitution Table*. Online: https://www.rocscience.com/help/rocfall/webhelp/baggage/rn_rt_table.htm. Accessed 10/2/2016.

Estimating Rockfall Kinetic Energy as a Function of Rock Mass

John D. Duffy P.G., C.E.G.

Yeh and Associates

391 Front Street, Grover Beach, CA 93433

(805) 440-9062

jduffy@yeh-eng.com

Prepared for the 70th Highway Geology Symposium,

Acknowledgements

Many colleagues and friends supported me during these many rockfall tests. Too many to list them all. But I would like to acknowledge a few who without them these tests may not have ever occurred.
From my California Department of Transportation, Marvin McCauley, Duane Smith and Ralph Fitzpatrick.
Robert Thommen and Bruno Haller from Geobrugg.
Howard Ingram from Hi-Tech Rockfall
Larry Pierson from Landslide Technologies (Retired Oregon Department of Transportation)

Disclaimer

Statements and views presented in this paper are strictly those of the author(s), and do not necessarily reflect positions held by their affiliations, the Highway Geology Symposium (HGS), or others acknowledged above. The mention of trade names for commercial products does not imply the approval or endorsement by HGS.

Copyright

Copyright © 2019 Highway Geology Symposium (HGS)

All Rights Reserved. Printed in the United States of America. No part of this publication may be reproduced or copied in any form or by any means – graphic, electronic, or mechanical, including photocopying, taping, or information storage and retrieval systems – without prior written permission of the HGS. This excludes the original author(s).

ABSTRACT

Data gathered from 685 individual rock rolls conducted on colluvial slopes and rock slopes ranging in vertical height from 20 to 132 feet, and slope angles from 34 to 63 degrees, have been analyzed to develop a method to predict maximum rockfall kinetic energy as a percentage of the potential energy of the rock at the initiation point and the rock mass. Rock rolling test methods, characteristics of the test sites, and data collected from 16 unique rock rolling test sites are discussed. Data collected from 28 rock-rolling test series were used to define four easily obtainable field variables: rock mass, slope height, slope angle, and nature of the slope (defined as rock or colluvium), which are used to calculate rockfall energies. Results of the analyses are presented as an energy estimation chart that defines the ratio of kinetic energy to potential energy as a function of rock mass. The energy estimation chart is presented in a "practitioner-friendly" format.

Introduction

The data analyzed in this report consists of over 28 years of compiled testing of flexible rockfall fences and catchment area designs throughout the world. Six hundred and eighty-five rock rolls were analyzed from 28 rockfall experiments at 16 different test sites to determine the ratio of kinetic energy at the base of the slope to the potential energy at the initiation point of the rockfall (Figure 1).

Rockfall analyses measure kinetic energy by combining rotational energy and translational energy. The principal of conservation of energy requires that the kinetic energy of the rock cannot exceed the potential energy the rock had at its initiation height before the fall, per Equation 1 (where m=mass, v=translational velocity, I=moment of inertia, ω=angular velocity, g=gravity, and h=vertical slope height).

$$\frac{1}{2}mv^2 + \frac{1}{2}I\omega^2 \leq mgh \qquad \textbf{Eq. 1}$$

Figure 1: Flexible Rockfall Fence Tests, California Department of Transportation, Grey Slip Test Site, 1989.

The data analyzed in this report correlate two key pieces of information: 1) mass of the rock and height of the slope at the initiation point, (i.e., potential energy 2) the translational and rotational velocities and maximum moment of inertia of the rock about the rotational axis, (i.e., kinetic energy) at the base of the slope. Test sites vary in slope and rock characteristics, but the essential elements of the test setups were the data were collected using high speed and video cameras to determine the time of travel and number of rotations the rock made between defined distances marked on the slopes, and the use of a load cell that measured the mass of the rocks before or after the rock roll.

Field Test Data

The test sites were located in California, Oregon and Switzerland. The slopes ranged in height from 20 to 132 feet (V), with slope angles that ranged from 34 to 63 degrees. Table 1 summarizes the locations and names of these sites.

Table 1: Summary of Rock Rolling Tests and Test Sites

Location		Date	# Rx Rolls	Test Site	Slope Angle (degrees)	Slope Height (feet)	Material Type
02-MOD-299-PM54.25	CA	1985	21		45	63-70	Fractured Rock
02-MOD-299-PM55.5	CA	1985	29		35	23-71	Collivium
02-PLU-70-PM62.0 Slope A	CA	1985	25		52	15-44	Fractured Rock
02-PLU-70-PM62.0 Slope B	CA	1985	20		50		Fractured Rock
02-SIS-5-PM21.5	CA	1985	18		50	58-79	Fractured Rock
02-SIS-5-PM22	CA	1985	11		52	31-46	Fractured Rock
03-NEV-20-PM40.2	CA	1985	30		48- 38	22-55	Colluvium
03-SIE-49-31.17	CA	1985	10		50	53-80	Fractured Rock
03-SIE-49-PM31.20	CA	1985	8		48	51-90	Colluvium
03-SIE-49-PM31.23	CA	1985	10		55	41-81	Fractured Rock
03-ALP-88-PM14.0	CA	1985	14		45	58-88	Glacial Till
TEST 1 CT	CA	1990	13	Gray Slip	34	132	Colluvium
TEST 2 CT	CA	1990	30	Gray Slip	34	132	Colluvium
TEST 3 CT	CA	1990	18	Gray Slip	34	132	Colluvium
TEST DTWM CH	CH	1992	4	Oberbucsiten	45	88.5 - 113	Rock
TEST Chain Link CH	CH	1992	7	Oberbucsiten	45	113	Rock
Test Brugg LE CT	CA	1993	19	Shale Pt	37	66	Colluvium
Test HiTech LE CT	CA	1996	10	Shale Pt	37	66	Colluvium
Test HiTech ME CT	CA	1996	31	Shale Pt	37	66	Colluvium
Cuesta Grade Chain Link Tests	CA	1998	11	Shale Pt	37	66	Colluvium
Cuesta Grade Gawk Screen Tests	CA	1998	15	Shale Pt	37	66	Colluvium
Cuesta Grade Jersey Barrier Test	CA	1998	26	Shale Pt	37	66	Colluvium
Cuesta Grade Tests	CA	2000	18	Cuesta Grade		66	Colluvium

Location		Date	# Rx Rolls	Test Site	Slope Angle (degrees)	Slope Height (feet)	Material Type
ODOT Rockfall Catchment Guide	OR	2001	51	Krueger Quarry	63	20	Rock
ODOT Rockfall Catchment Guide	OR	2001	96	Krueger Quarry	63	50	Rock
ODOT Rockfall Catchment Guide	OR	2001	78	Krueger Quarry	63	20	Rock
ODOT Rockfall Catchment Guide	OR	2001	48	Krueger Quarry	53	20	Rock
Smart Rock Tests 2013	CA	2013	4	Shale Pt	37	60	Colluvium

Statistical Analysis

Huang and Turner, 2013, statistically analyzed 567 rock rolls to determine the ratio of kinetic energy at the base of the slope to the potential energy at the initiation point. The objective of the statistical analysis was to identify an acceptable upper limit of the ratio of kinetic to potential energy that could then be used to calibrate a more sophisticated model and provide empirical evidence to support design work (Huang, Turner 2013).. Analysis of the data yielded valuable results when considering the potential uses of the empirical data provided in this report.

In order to present the data in a fashion that illustrates trends and allows for fitting of known distributions, Huang and Turner developed a normalized frequency diagram of the rockfall data, with the x-axis representing the bin sizes of the ratio of kinetic energy to potential energy. The sample mean and standard deviation of the data was determined and used to find a best-fit known statistical distribution (Huang, Turner 2013). Table 2 provides a summary of the data for those parameters. The log-normal distribution provided the best fit to the sample data, is shown in Figure 1.

Table 2: Summary of data parameters and calculated values used to define the log-normal distribution (Huang, Turner 2013).

Parameter	Kinetic Energy/Potential Energy (%)
Sample mean μ	29.42
Sample standard deviation σ	20.81
Log-normal λ	3.18 (ln(KE/PE))
Log-normal ξ	0.64 (ln(KE/PE))

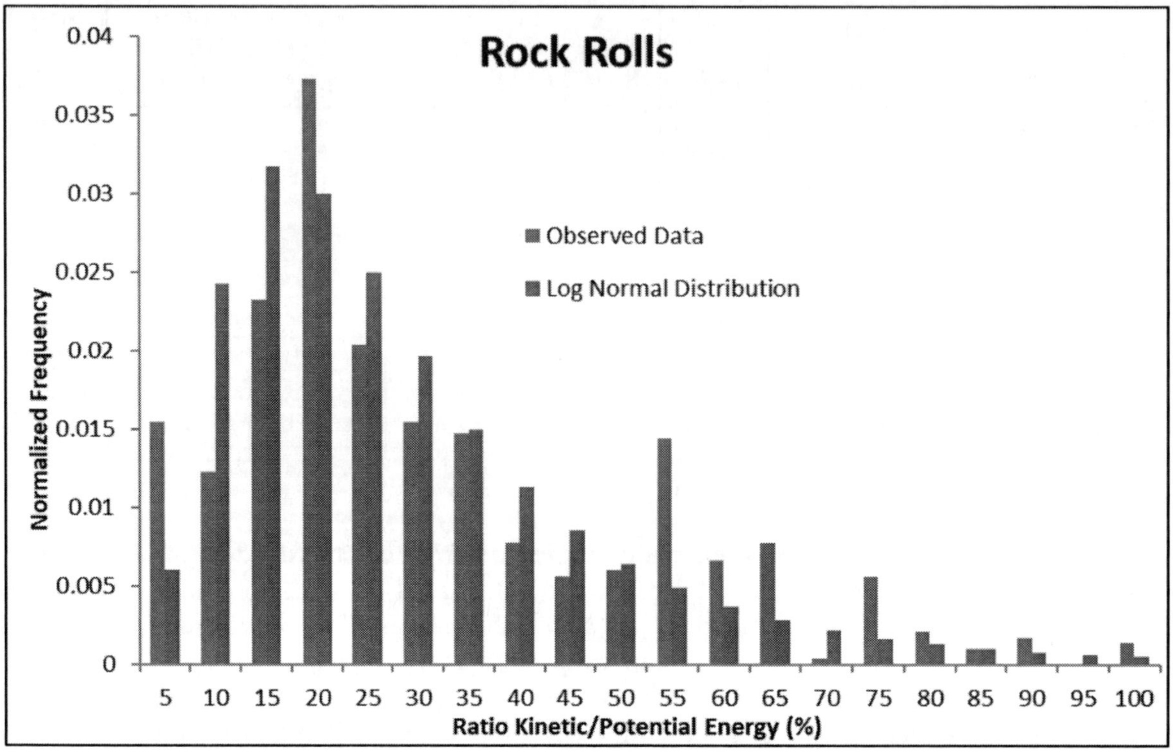

Figure 2: Normalized frequency diagram of the sample rockfall data (blue), plotted versus the log-normal distribution of the data defined by the sample mean and standard deviation of rockfall data (red) (Huang, Turner 2013).

Without any experience in performing modeling or statistical analysis, a designer can measure the slope height and rock mass of potential rockfalls in the field to get an estimate of the potential energy, and use that information with the chart to determine a design kinetic energy at grade by considering an acceptable percentage of rockfalls that could potentially exceed the energy ratio predicted by the distributions. The actual data was compared to the log-normal distribution considering 10% and 5% exceedance, where for the measured data, 90% of the observed rockfalls had energy ratios of 60.9% or less, so 10% of rockfalls could potentially exceed energy ratios of 60.9% (Huang, Turner 2013). The results for the 90% and 95% interval are summarized in Table 3.

Table 3: Summary of analyses (Huang, Turner 2013).

Distribution	Interval	Kinetic Energy/Potential Energy (%)
Measured Data	90%	60.9
Measured Data	95%	72.8
Log-normal	90%	54.5
Log-normal	95%	68.9

Energy ratio values predicted by the log-normal distribution are reasonably close to the results of the measured data. As with many best-fit distributions, the tail of the best-fit distribution varies slightly from the actual data, as can be observed in Figure 2. The tail is typically where the critical analyses needed for design take place; considering acceptable limit for exceedances are likely between 90-99%. Despite the variation in the tails of the distributions, the log-normal distribution could be used to make preliminary estimates of energy ratios when considering the variability in the random variables from Equation 1(Huang, Turner 2013).

Estimating Impact Energy

Smaller rocks were observed to be more likely to approach the full limit of their initial potential energy during a rock rolling event. As the rock weight increased for a given slope height, the initial potential energy also increased, but there was a reduction in the KE/PE ratio for larger rocks. The reduction observed for the larger rock was likely due to the impact mechanics of a rolling rock on a slope. As rocks roll and bound down a slope, energy is dissipated into the ground upon each impact. The heavier the rock the greater reduction in energy ratio will occur during ground impact. The data in this study indicate that lightweight rocks lose less energy during their trajectory than heavier rocks. This does not mean that heavier rocks have less impact energy at the base of slope than lighter rocks, only that the percentage of kinetic energy that developed during rock rolls relative to the initial potential energy was less for heavier rocks than for lighter rocks. This data also indicates the rockfall energy at the base of these slopes for all rock rolls never exceeded 500 kilojoules regardless of the rock weights that were tested. This finding could be significant because when considering the available rockfall barriers on the market today, 500-kilojoule barriers are very common "off the shelf" systems.

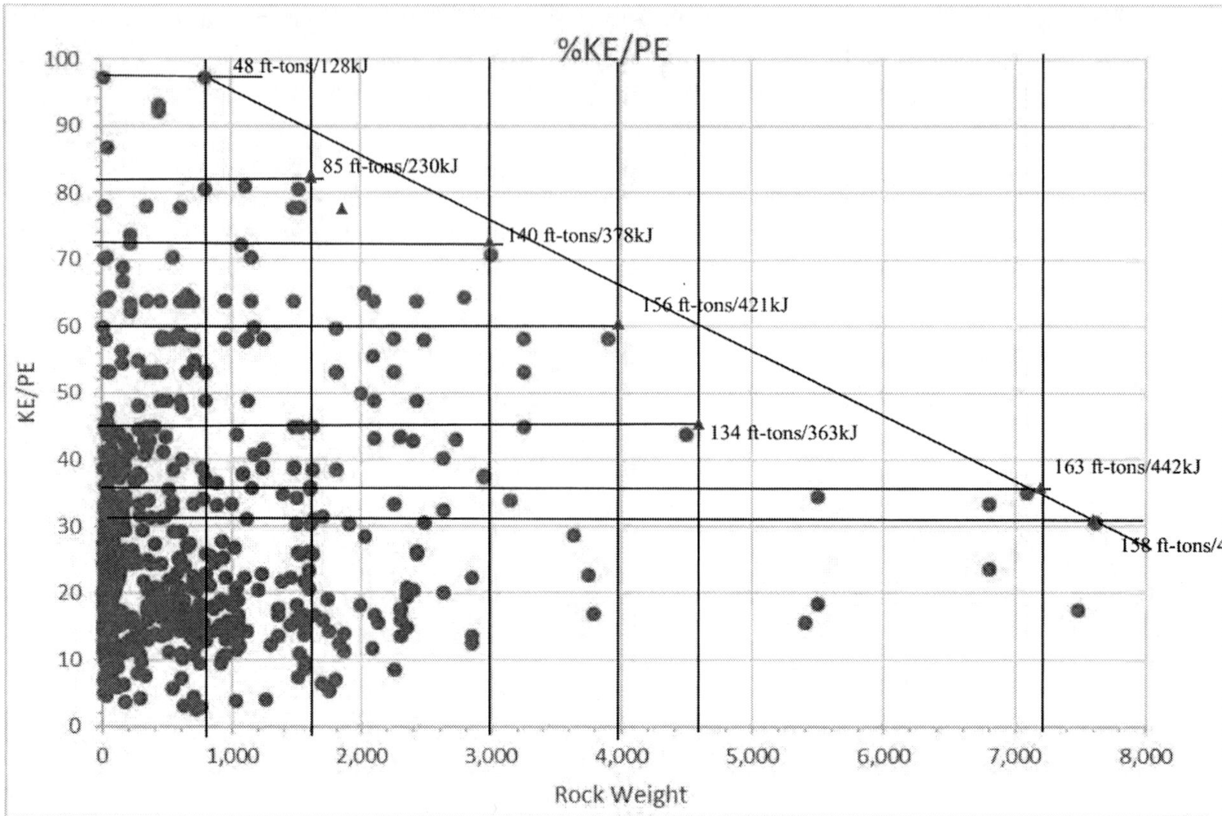

Figure 3. Percent KE/PE vs Rock Mass. Rocks ≤ 8,000 Pounds and Slope Heights ≤ 130 feet.

A rockfall barrier design requires an estimate of the total kinetic energy at the base of the slope. The designer typically utilizes a comprehensive field investigation supported by computer modeling. The findings presented herein can be used to support that work to check modeling results or support rock rolling site tests. During advanced planning or emergency work an energy estimate can quickly be established with a few simple field measurements (vertical slope height and rock weight).

Equations 2.0 and 3.0 provide a simplified method to estimate kinetic energy at the base of a slope for the test slopes in Table 1. Equation 2 expresses the energy ratio (in decimal form) as a function of rock weight. The ratio can then be used in Equation 3 to calculate the energy at the base of the slope (KE_T) at the limits indicated on Figure 3. These calculations can be made using only the rock weight and the vertical slope height to the rock.

$$KE/_{PE} = .001\ (Rock\ Weight) + 26 \leq 1 \qquad \textbf{Eq.2.0}$$

$$KE_T = \left(KE/_{PE}\right) \times PE \qquad \textbf{Eq 3.0}$$

Conclusion

Time and effort spent to test and record data for real-world rock rolls has provided valuable design information for selection and design of rockfall mitigation measures. Relatively inexpensive and simple test setups provide empirical support for the results of sophisticated computer models and can be used to calibrate the analysis results to accurately represent field conditions. Statistical analysis yielded a useful tool for designers as presented in Tables 2 and 3. Limited field and design work can be used to provide a reasonable estimate of rockfall impact energy as a percentage of potential energy. Commercially available flexible rockfall barriers are typically available in energy increments of 100 to 500 kilojoules (kJ); designing to the nearest 100 kJ using the data presented is possible for most projects and may even eliminate the need to run computer models. More in-depth statistical analyses of the data is possible, and consideration of rock mass, slope height, slope angle, slope hardness, and slope roughness could be used to develop design tables similar to Table 2 for commonly encountered field conditions.

REFERENCES

Duffy, J.D. 1992. *Field Tests and Evaluation of Flexible Rockfall Barriers*, Brugg Cable Products, Oberbuchsiten, Switzerland, 79 p.

Duffy, J.D. & W. Hoon. 1996a, *Field Tests and Evaluation of Hi-Tech Low Energy Chain Link Rockfall Fence,* Report No. CA/05-96-01. California Department of Transportation, San Luis Obispo, California, 39 p.

Duffy, J.D. & W. Hoon. 1996b. *Field Tests and Evaluation of Hi-Tech 50 and 70 foot-ton Rockfall Fence,* Report No. CA/05-96-02, California Department of Transportation, San Luis Obispo, California, 42 p.

Duffy, J.D. & W. Hoon. 1998. *Field Tests and Evaluation of Chain Link Fence, Gawk Screen, Fence and Jersey Barrier*, Report No. CA/05-98-01, California Dept. of Transportation, San Luis Obispo, California, 56 p.

Duffy, J.D., 2012, *Summary of Worldwide Rockfall Tests*, A.A. Balkema Publishers, Rotterdam, The Netherlands.

Duffy, J.D., & Chris Jones, 2000, *Cuesta Grade Temporary Rockfall Fence Tests,* California Department of Transportation, Memorandum, San Luis Obispo, Ca.

Huang, L., Turner, R., 2013. *Statistical Analysis of Rockfall Energy Dissipation*, California Polytechnical State University, San Luis Obispo, California.

Kane, W.F. & J.D. Duffy. 1993. *Brugg Low Energy Wire Rope Rockfall Net Field Tests,* Technical Research Report 93-01, The University of the Pacific, Dept. of Civil Engineering, 34 p.

McCauley, M., et. al., 1983, *Rockfall Mitigation*, California Department of Transportation, Sacramento, California.

Pierson, L.A., Gullixson, C.F. & R.G. Chassie. 2001. *Rockfall Catchment Area Design Guide,* Final Report SPR-3(032), Research Group, Oregon Department of Transportation, Salem, Oregon, 77 p.

Smith, D.D. & J.D. Duffy. 1990. *Field Tests and Evaluation of Rockfall Restraining Nets*, Final Report No. CA/TL-90/05, California Department of Transportation, Sacramento, California, 138 p.

Turner, Jack & J.D. Duffy, 2013, *Smart Rock Rolling Tests*, Senior Project, California Polytechnical State University & California Department of Transportation, San Luis Obispo, California.

The meandering Mundo Mud Pot:

Or how Salton Sea tectonics affect international trade

Dean G. Francuch, PG, CEG, R. Travis Deane, PE, GE, and Carol Zamora
Shannon & Wilson, Inc.
664, West Broadway
Glendale, California 91204
818-539-8400
dgf@shanwil.com
rtd@shanwil.com
clz@shanwil.com

Prepared for the 70th Highway Geology Symposium, October 2019

Acknowledgements

The authors would like to thank to thank the individuals for the contribution tin the work described:

Dr. David Lynch, PhD.
Thule Scientific
22914 Portage Circle Drive
Topanga, California 90290
dave@caltech.edu

Disclaimer

Statements and views presented in this paper are strictly those of the author(s), and do not necessarily reflect positions held by their affiliations, the Highway Geology Symposium (HGS), or others acknowledged above. The mention of trade names for commercial products does not imply the approval or endorsement by HGS.

Copyright Notice

Copyright © 2019 Highway Geology Symposium (HGS)

All Rights Reserved. Printed in the United States of America. No part of this publication may be reproduced or copied in any form or by any means – graphic, electronic, or mechanical, including photocopying, taping, or information storage and retrieval systems – without prior written permission of the HGS. This excludes the original author(s).

ABSTRACT

The Union Pacific Railroad's California to Texas mainline parallels the eastern shore of the Salton Sea, where approximately 70 trains per day carry international goods from Asia to the interior United States. Geothermal activity created by extensional tectonics between the south end of the San Andreas fault and north end of the Imperial fault creates mud pots and mud volcanos throughout the area by venting of carbon dioxide gas. The gas pushes groundwater laden with Pleistocene age lake sediments to the surface.

Aerial photos suggest that the Mundo Mud Pot may have originally formed prior to the 1950s, 1100 feet northeast of the railroad. Since 2004, a newly created mud pot developed a crater shaped caldera approximately 60 to 80 feet in diameter, 400 feet northeast of the railroad. In 2016, after a series of earthquakes in the Brawley Seismic Zone, the mud pot began moving in a southwesterly direction towards the railroad tracks.

In May 2018, the railroad was forced to cope with the mud pot and its water discharge as the mud pot had moved to within 80 feet of the mainline. Geophysics and geotechnical methods were used to focus on geologic structures allowing the movement of the mud pot. Methods used to mitigate the discharge and erosion include levee construction, sheet piles, dewatering wells, and riprap. Train traffic has continued to move by constructing two new detour tracks to either side of the mud pot, as it continues to gradually move under the original railroad alignment.

INTRODUCTION

The Yuma Subdivision of the Union Pacific Railroad (UPRR) is the main artery for the railroad's transcontinental operations from the ports of southern California to the midwest and eastern potions of the continental United States. Consisting of two mainline tracks on the eastern shore of the Salton Sea in southeastern California, this portion of the UPRR passes through a tectonically-active area at the southern end of the San Andreas Fault. In addition to abundant earthquakes, the area is pulling apart as part of an active spreading center extending northwest from the Gulf of California, generating Pleistocene to Recent volcanic activity. The volcanic activity and magma heats calcareous sediments at depth deposited from the ancient Colorado River, releasing carbon dioxide gas to the surface. This gas along with minor amounts of hydrogen sulfide emanates to the surface through clay and mud-rich Pleistocene lake deposits, mixing with near surface groundwater creating numerous mud pots and mud volcanos.

As the gas migrates towards the ground surface, it passes through deep groundwater aquifers, carbonating the water and pushing it upwards under pressure much like a carbonated beverage. Upon reaching the surface, clay-rich water and gas are discharged with a slurry-like consistency. If the mud slurry flows out onto the surface relatively evenly, it is typically termed a mud pot. If the mud deposits form an elevated ring or ridge around the discharge area, it is termed a mud volcano. Mud pots are common around the Niland, California area and several miles to the northwest along the UPRR tracks.

Figure 1-Active static mud pots located about 1000 feet north of the Mundo site.

A series of mud pots in the area have been studied by scientists for over a decade (Lynch and Hudnut, 2008). The mud pots may appear and disappear over time but typically remain stationary (Figure 1). The mud pots generally lie along the trend of the Wister Fault, which may be a southeast extension or an inactive segment of the San Andreas Fault. One mud pot on this trend, designated W9 by workers studying these features, was located 4.5 miles northwest of

Niland, California, 250 feet northeast of the UPRR tracks at Milepost 662.65, on the south side of Gillespie Road, about 600 feet due east of California Highway 111. Described as a large, shield-like mud spring similar to a mud pot, this feature went largely unnoticed by the UPRR and general public until late 2017.

A UNIQUE PHENOMENON

UPRR began experiencing difficulties with the W9 mud spring when muddy water surged out of it in October 2017. The mud-water slurry flooded the field and ditches on the east side of the UPRR tracks. The ditches carried the slurry to the southeast, where it entered Gravel Wash, a dry arroyo that flows southwest towards the Salton Sea under the UPRR tracks and parallel Highway 111 under separate bridges. The shallow gradient allowed sediment carried by the mud slurry to settle, accumulating in the ditches and Gravel Wash, blocking flow and diverting additional flows of the mud slurry discharge towards the tracks (Figure 2).

Figure 2-Southeast view of the Mundo mud spring in October 2017. The ponding of water is due to the plugging of the bridges below both the UPRR tracks and Highway 111 to the right. Image courtesy of the Union Pacific Railroad.

The UPRR attempted to divert flow and ponding from the mud spring by excavating the mud accumulating in the ditches and Gravel Wash. However, the drainages would rapidly become blocked again as new flows of mud slurry from the mud spring flooded the area, re-depositing sediment into the channels.

In May 2018, UPRR requested Shannon & Wilson to evaluate the site conditions and develop recommendations to mitigate the mud spring. Before arriving at the site, Shannon & Wilson studied Google Earth historic and current aerial images, which showed the mud spring location. The mud spring was first visible in 2005 images and remained in relatively the same location through the latest 2016 aerial images (Google Earth, 2016). When Shannon & Wilson arrived on site, the mud spring was about 125 feet closer to the UPRR tracks than shown on the Google Earth October 19, 2016 aerial image (Figure 3).

Figure 3-Mud spring in May 2018, approximately 80 feet west of the Union Pacific tracks. View to south.

The team contacted Dr. David Lynch, who co-authored many of the studies of the mud pots in the area, including studies at mud spring W9. Dr. Lynch reached out to his colleagues studying mud pots and mud volcanos worldwide, and after consulting with them, concluded that the movement of the mud spring was a rare phenomenon without precedence. Dr. Lynch joined the project team to assist in devising measures to mitigate the impacts of the mud spring on the UPRR tracks.

EMERGENCY RESPONSE

The mud spring has carved a roughly 75-foot wide path towards the southwest since 2016, which was filled with the discharge mud slurry, creating an oblong-shaped pond (or caldera) about 150 feet in length (Figure 4). In early May 2018, the mud spring was located in the southwest portion of the caldera pond. Agitation of the pond water caused by the mud spring discharge created small waves that were actively eroding the caldera edges towards the southwest as the mud spring moved. The erosion was generally 25 feet beyond the mud spring location. During our May 4, 2018 site visit the mud spring was approximately 80 feet from the eastern main line track and the eroding caldera edge was 60 feet away. In mid-May, the mud spring movement accelerated towards the southwest, moving over 20 feet to the southwest in a one-week period. The rapid movement towards the mainline tracks spurred rip rap placement along the west edge of the advancing caldera to slow the erosion caused by the mud spring surface agitation. UPRR was concerned that future rapid movement could compromise the main line tracks in less than a month, closing the tracks and forcing UPRR to detour train traffic for hundreds of miles.

In the initial assessment, Shannon & Wilson recommended implementing one or more of the following options:

- Breaching the caldera pond on the east side to allow the mud slurry discharge to drain into fallow crop fields;
- Stockpiling rip-rap between the mud spring and tracks in an attempt to slow the erosion of the caldera;
- Installing instrumentation to monitor for track settlement;
- Evaluating options to construct a temporary detour (shoofly) track to divert traffic around the mud spring;
- Driving sheet piles between the mud spring and tracks in an attempt to slow the erosion;
- Install ground improvements below the tracks that would densify the underlying subgrade, potentially diverting the mud spring flow around and to the west of the improved soil, and/or;
- Drilling relief wells in an attempt to depressurize subsurface gas and divert the mud spring.

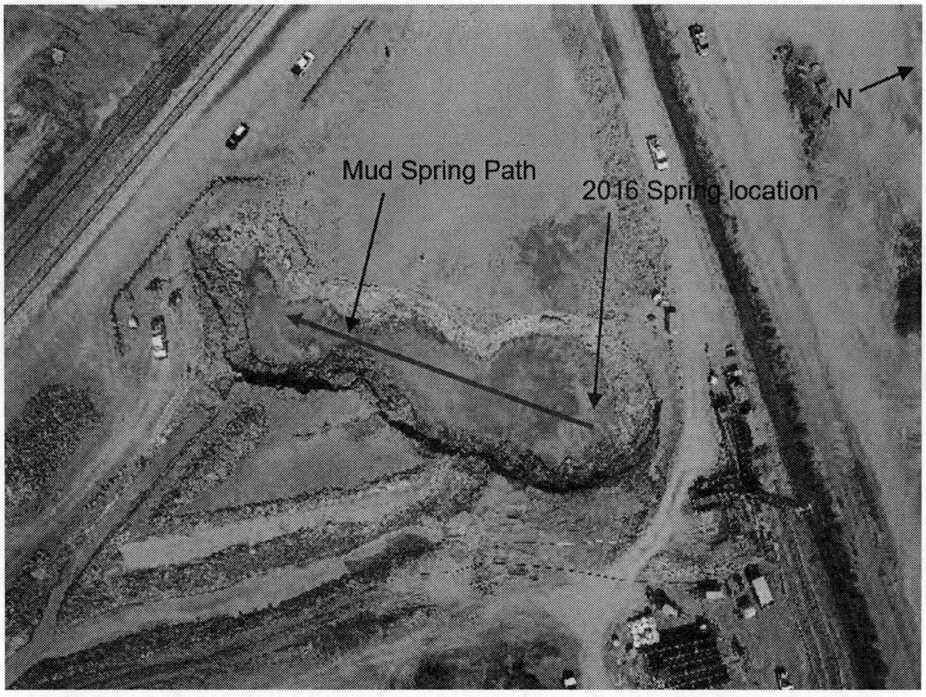

Figure 4-Path of mud spring from its location in October 2016. Image courtesy of D. Lynch.

The team proceeded with placing rip rap on the west side of the mud spring caldera and mobilizing a contractor to drive sheet piles. UPRR also began studying options for a shoofly track to divert train traffic around the mud spring. Shannon & Wilson also studied options for installing ground improvements, relief wells, and possibly building a bridge over the anticipated path of the mud spring.

GEOLOGICAL EVALUATION

Before many of the above options could be implemented, more information on the underground processes driving the mud spring was needed.

For most geologic hazards, we would evaluate the subsurface conditions by drilling borings to investigate the subsurface conditions or mechanism. However, the gas pressure observed in the mud spring and nearby mud pots deterred us from starting with drilling due to the risks of blow-outs. Therefore, we initially turned to geophysical methods that could image the subsurface from the ground surface. We consulted with our geophysical subconsultant, GeoVision of Corona, California, to determine which geophysical methods would best image the subsurface "plumbing" powering the mud spring. Based on the subsurface site conditions and near surface groundwater the project team performed the following:

- Electrical Resistivity Imaging (ERI)
- Spontaneous Potential (SP)
- P-wave Seismic Refraction
- Multi-channel Analysis of Surface Waves (MASW)
- P-wave Reflection
- High Resolution S-wave Reflection

We performed the geophysical imaging on up to four lines, designated Lines 1 through 4 (Figure 5).

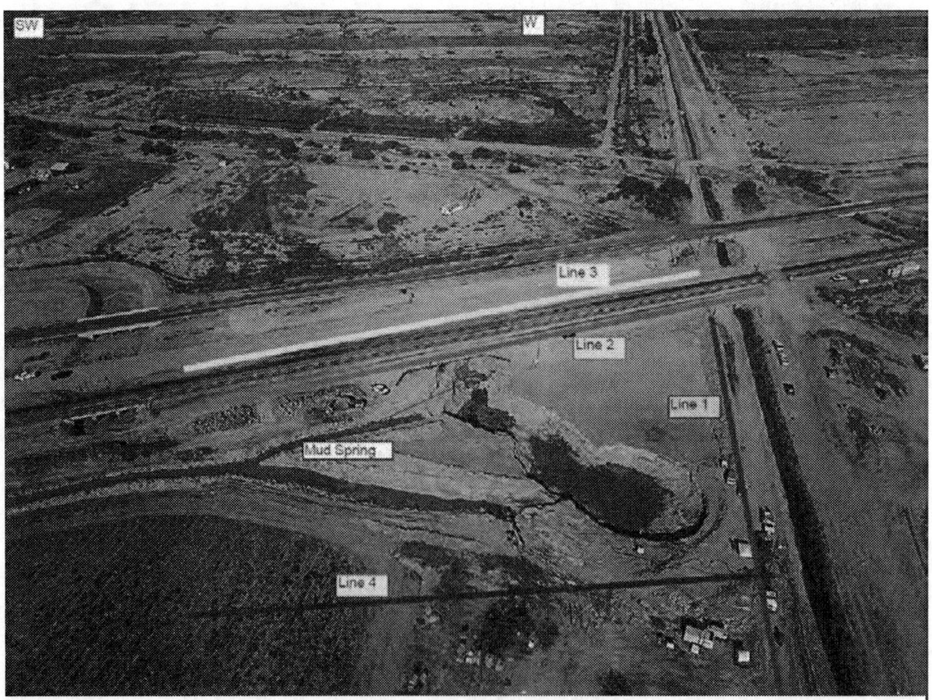

Figure 5-Locations of Geophysical section lines, relative to the caldera and the UPRR tracks. View to the west.

The geophysical methods had two purposes:

- Detect subsurface structures (e.g., faults, density variations) that could explain the mud spring trend, and
- Image the surrounding subsurface conditions to predict future movement of the mud spring.

Of the methods utilized, the High-Resolution S-wave Reflection provided the best quality results of the subsurface conditions. Images collected using this method indicated a relatively horizontal subsurface profile in the upper half of the imaged depth. The subsurface profile indicated two distinct, near-horizontal dense layers (Figure 6). These dense layers were relatively continuous except for linear, near-vertical disruptions that were interpreted as faults breaking through these layers. Below the deeper dense layer, the subsurface images indicated a more chaotic profile than the upper horizontal beds. The chaotic profile was interpreted as upwelling of groundwater that was focused at a parabolic-shaped structure that aligned with the

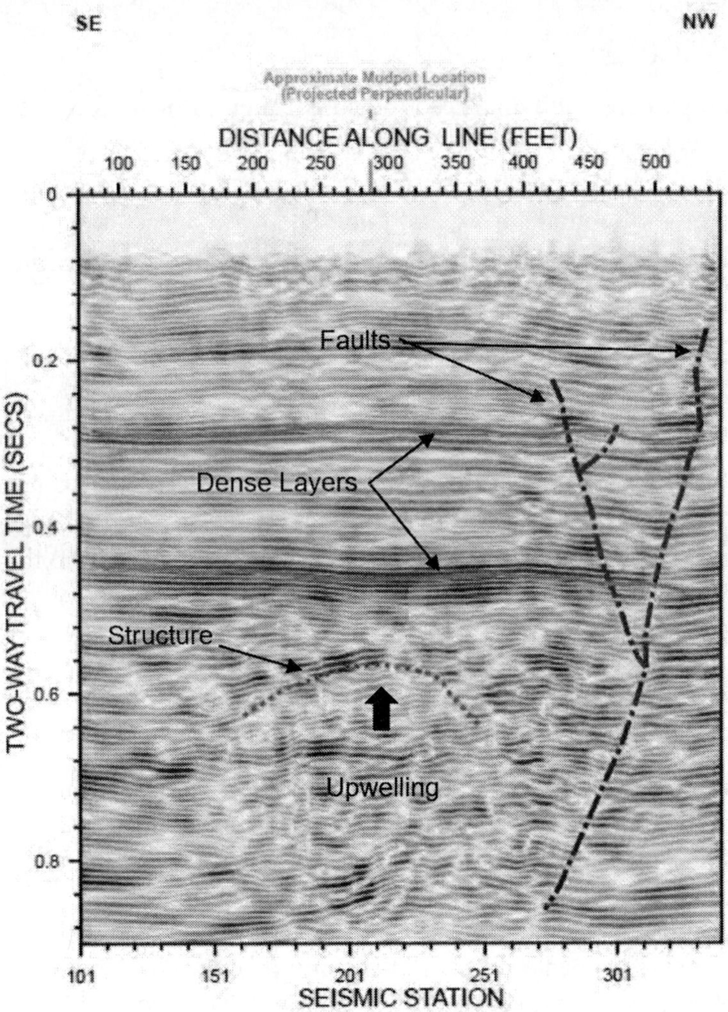

Figure 6-Geophysical S-Wave Seismic Reflection Line 2 along the east side of the Union Pacific track alignment with interpreted fault and upwelling structure.

path of the mud spring. The depth of this parabolic structure below the mainline tracks was about 200 feet below ground surface (GeoVision, 2018). The parabolic structure was more pronounced at Line 1 paralleling Gillespie Road to the north of the mud spring. The structure became less pronounced at Line 3 located between the UPRR tracks and SR 111.

Shannon & Wilson reviewed bridging over the mud spring path with either ground improvements or a steel structural bridge. In order to evaluate these measures, we attempted cone penetration tests (CPTs), which would provide near continuous vertical information on the subsurface profile. We initially chose the CPT exploration method because the CPT probe holes would be less susceptible to a blow-out than a larger diameter soil boring. The original target depth with the CPTs was 200 feet, but a dense layer at about 75 feet deep resulted in CPT probe refusal. We assumed this dense layer was the upper dense layer detected in the geophysics at a two-way travel time of about 0.3 seconds. Railroad personnel also reported that steel piles driven for recently constructed railroad bridges in the immediate area also encountered refusal at about 75 feet below the ground surface, indicating a through-going horizon that may be acting as a capping layer to the gas.

MITIGATION MEASURES

As Shannon & Wilson gathered and analyzed the subsurface information, the mud spring continued periodic movement towards the UPRR tracks. Rip rap placed to slow the erosion of the caldera was successful in slowing the surface erosion, but it rapidly disappeared into the mud spring presumably through the mud spring conduit. The team agreed that a better short-term solution was needed to slow or temporarily arrest the erosion in order to protect the railroad right-of-way.

Surface Protection

We developed a working theory on the mud spring to devise short-term measures to slow its progress. The origin of the carbon dioxide gas is likely thousands of feet below the ground surface to the southwest of the actual mud spring location based on the location of the nearest volcanic activity. The gas was likely escaping along a northeast-southwest trending structure (likely a fault) based on trends of local faults related to recent seismic activity (Persuad and others, 2016). While escaping towards the surface, the carbonated water discharge was actively eroding the soil in front of its path. Using this theory, we assumed that driving sheet piles in front of the mud spring's path would slow the surface erosion thereby protecting the railroad, allowing development time to implement other measures. While the soil would easily erode from wave action, steel sheet piles would not be subject to near surface erosion.

We began driving the sheet piles in early June 2018 and by mid-July, the sheet piles were installed for a width of 100 feet across the projected trend of the mud spring. We also drove sheet pile wing walls at either end of the main sheet pile line in attempt to "corral" the mud spring erosion at the surface. The depth of the sheet piles was about 75 feet, where the piles encountered refusal conditions at a depth similar to the CPTs and previous UPRR bridge piles. After completion of the piles, a near surface dewatering program using dewater pumps placed into the caldera was started to decrease water height, which also reduced erosion to either end of the sheet pile limits.

Subsurface Containment

The mud spring approached the sheet piles shortly after their installation. We were concerned that the mud spring would pass below the sheet piles unimpeded and strike the railroad tracks 40 feet away. Fortunately, the sheet piles kept the mud spring surface erosion contained, allowing us to proceed with measures to maintain train traffic and attempt to depressurize the gas powering the mud spring.

We attempted to depressurize the mud spring by tapping into the gas reservoir below the site. These relief wells would be drilled to depths of several hundred feet with a target of the upwelling structure identified in the geophysical sections and cased to allow the gas to vent, thereby reducing the gas flow feeding the mud spring. The first relief well (B-1) was positioned behind the path of the mud spring. This relief well was located along the mud spring trend near its 2016 location prior to the beginning of movement, but far from the infrastructure of the railroad, Highway 111, and parallel utilities given the experimental nature of the drilling. Below about 300 feet, the drilling encountered a pocket of gas that resulted in a blow out of water over 100 feet in height. While the gas pressure eased with time, it remained relatively high (approximately 170 psi) such that we decided to install permanent casing and screening to allow the gas to escape.

With a successful test, we attempted two additional relief wells (B-2 and B-3) closer to the existing mud spring location. Because of the proximity of the railroad tracks, we could not place a relief well directly along trend of the mud spring, and it was decided to install relief wells to the northwest and southeast sides of the active mud spring location. The second relief well (B-2), located less than 100 feet northwest of the existing mud spring, was drilled to a depth of 800 feet. During drilling, we did not encounter significant gas or groundwater despite the close proximity of the active mud spring. We eventually abandoned the hole as it did not have a discernible impact on the mud spring activity.

While we were drilling the relief wells, UPRR constructed a shoofly track at the west edge of their right-of-way. The purpose of the shoofly track was precautionary as we assumed the mud spring would eventually cross the sheet piles if the relief wells were unsuccessful. UPRR also continued to pump the water out of the caldera to limit further erosion and evaluate the bottom of the pond. We exposed a muddy bottom that was about 25 feet below the existing ground surface, but there was no evidence of a fissure or fracture in the area that the mud spring had previously passed through. The railroad attached a flowmeter to the pumps and we began recording daily measurements of the pumped slurry directly from the mud spring, which averaged about 40,000 gallons per day.

The mud spring, contained behind the sheet piles for almost four months, actively eroded out the soil supporting the sheet piles at depth. We observed the sheet piles tilting towards the mud spring, and we were increasingly concerned that the mud spring was undermining the sheet piles. Following the poor results at relief well B-2, we attempted a third relief well located to the southeast, with the hope that we would encounter the gas reservoir. This third well, drilled less than 100 feet southeast of the mud spring, showed more promise than the second well. We encountered significant gas pockets to a depth of 400 feet (Figure 7). From our recent experience, we had devised a method to reduce the blow outs observed in the first relief well

Figure 7-Mud and gas blow-out during drilling of well B-3 at a depth of approximately 400 feet.

such that we could progress deeper. However, as we started to deepen the relief well to tap into the gas, the mud spring breached the sheet pile containment.

SHEET PILES BREACH

As the mud spring eroded out much of the soil supporting the lower portion of the sheet piles, causing the piles to tilt towards the caldera. In mid-August, we observed a tension crack open up on the railroad track side of the sheet piles followed by visual deflection and sinking of the central sheet piles. We installed a tiltmeter on the sheet piles that allowed us to monitor the magnitude of the tilt. In September, the tilting accelerated towards the caldera (Figure 8).

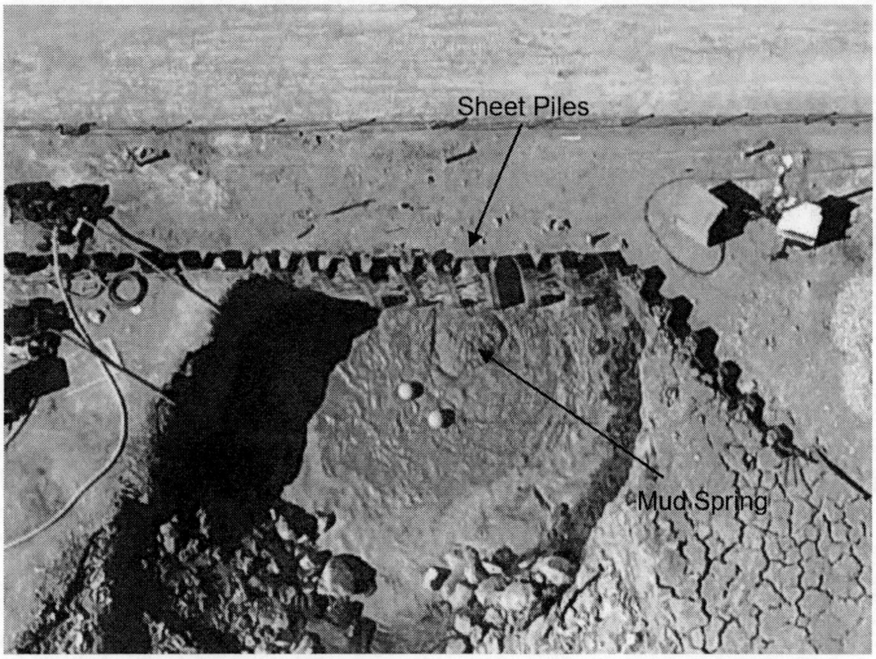

Figure 8-Aerial drone view of mud spring adjacent to the sheet piles prior to the breach. Image courtesy od D. Lynch.

We stabilized the sheet pile tilting by placing rip rap into the caldera adjacent to the sheet piles in mid-September. To reduce the potential for plugging of the mud spring conduit from the rip rap, a 5 foot diameter steel pipe was driven vertically at the mud spring location. The additional rip rap placement arrested the tilting for approximately two weeks.

On October 2, 2018, the tiltmeter recorded an anomalous movement of the sheet piles. We noted that certain sheet piles had been sinking up to 1 inch relative to its neighboring sheet pile. Most alarming was a spike in the mud discharge, which was three times the normal daily flow at about 120,000 gallons for that day (Figure 9). This was the highest amount of mud discharge recorded to date during our monitoring. A similar spike occurred on February 9, 2019, for unexplained reasons.

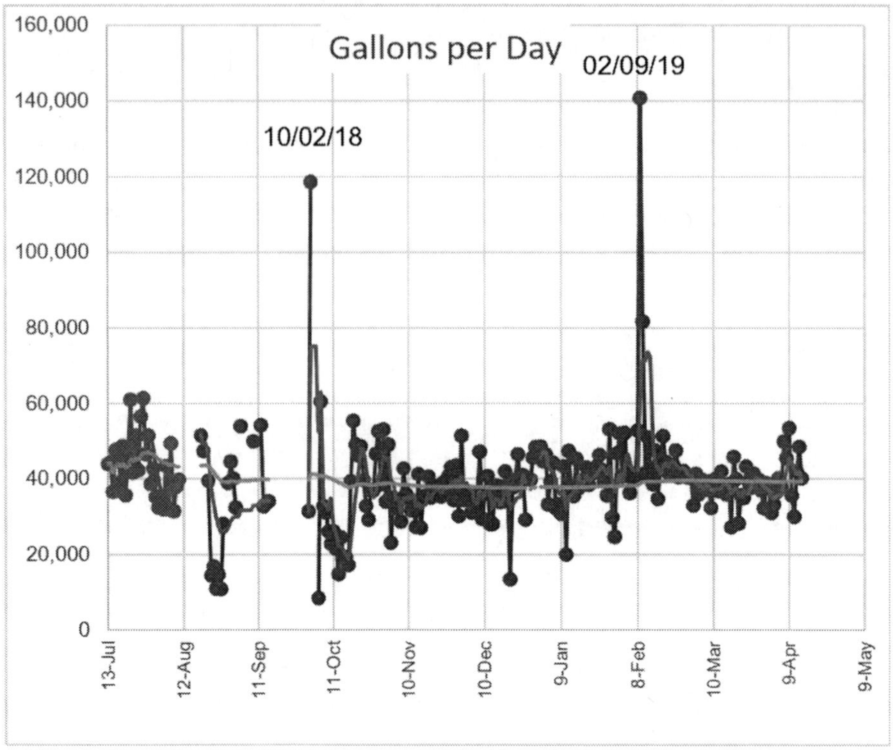

Figure 9-Daily volume if pumped slurry from mud pot caldera.

The following day, we observed water and gas emanating from the interlock joint between two sheet piles. Also, the sinking of the sheet piles accelerated from what was observed previously. We concluded that the mud spring was likely beneath the sheet piles.

We further concluded that breaching of the sheet piles by the mud spring was in progress and a work train was dispatched to deliver rip rap to the site. The rip rap would be placed at either end of the sheet piles in an effort to contain the erosion of the emerging mud spring between the sheet piles and tracks. In the early afternoon of October 4, a fountain of dust driven by gas was being expelled on the track side of the sheet piles in the new mud pot location. The ground between the sheet piles and the main line tracks failed by heaving upwards and then collapsing, forming a sinkhole between the sheet piles and tracks (Figure 10).

The work train, parked on the near track, was safely removed and all train traffic stopped in the rail corridor. The sinkhole was about 70 feet wide and about 25 feet deep. Through the

Figure 10- Aerial drone images of the mud spring and caldera both pre- and post-breach of the sheet pile containment structure. Images courtesy of D. Lynch.

remaining afternoon into the evening, the sinkhole began to fill with water. The mud spring disappeared from the caldera on the opposite side of the sheet piles after the sinkhole formed. Later in the evening, the mud spring re-emerged at the bottom of the sinkhole.

UPRR track crews and contractor began realigning the mainline tracks to the western shoofly track constructed in mid-August. Less than 12 hours after the gas explosion created the sinkhole, the UPRR trains began operating on the western shoofly track and western mainline. However, because of the high volume of traffic on the double track mainlines, UPRR wanted a second shoofly track constructed. As we had not observed the mud spring "back-tracking" to the northeast during movement, we recommended backfilling the caldera northeast of the sheet piles and placing the second (eastern) shoofly track at that location.

Contractors for the UPRR backfilled the entire caldera area to the east of the sheet piles with rip rap and soil. The eastern shoofly track became operational later in October. While functional, the eastern shoofly track required weekly tamping of the ballast due to settlement of the fill. The sheet piles, which had previously been tilting to the east prior to the sinkhole formation, began tilting west towards the sinkhole, adding to the settlement of the eastern shoofly track. Despite these issues, the railroad was operating trains on the eastern and western shoofly tracks with the mud spring located in the main line tracks alignment (Figure 11).

Figure 11-View of the mud spring within the original main line track alignment. Original tracks are cut on the opposite wall of the mud pot caldera. View to the southeast.

POST BREACH OPERATIONS

Since the breaching of the sheet piles UPRR has decided to utilize both shoofly tracks and wait for the mud spring to clear the mainline track alignment. Relief well drilling operations were discontinued as they appeared to have no effect on the mud spring discharge. From our past observations, we estimated the rate of the mud spring movement between 10 to 15 feet per month prior to the encountering the sheet piles. If this rate holds true for the future movement, we anticipate the mud spring will clear both mainline tracks by late 2019.

As the sheet piles are acting to support the eastern shoofly track, we have placed rip rap in the sinkhole to buttress them much like our previous effort to support of the sheet piles on the opposite side. Because of the close proximity of the mud spring to the sheet piles, the rip rap was periodically swallowed by the mud spring conduit. As rip rap was lost, the sheet pile tilting would accelerate towards the sinkhole, destabilizing the eastern shoofly track. Starting in early November through December 2018, the contractor mobilized six times to place additional rip rap to buttress the sheet piles. In early 2019, the contractor placed additional rip rap in the sinkhole an additional four times. The last rip rap placement took place on mid-February 2019. While the sheet piles have continued to gradually tilt towards the sinkhole, the rate of tilt has steadily slowed such that additional rip rap has not been placed through the end of April 2019.

FUTURE OPERATIONS

With the sheet piles stabilized and the mud spring moving away from the rip rap buttress, a final repair for the mainline tracks has been designed. The concept developed is a repair system that directs gas and water that may form new mud pots behind the mud spring's path away from the mainline tracks and into the lateral ditches on either side of the original mainline tracks.

One of the challenges posed by the repair is timing with railroad operations. Once moved off the mainline tracks, the mud spring will threaten the western shoofly track. The western shoofly track has served as the main route of the diverted train traffic given the instability of the eastern shoofly track. A narrow window of time will exist between reconstructing the mainline track before losing the western shoofly track. Therefore, the repair plans will need be in place and ready to implement based on the position of the mud spring, which has demonstrated erratic movement in the past.

Once the mud spring has cleared the main lines and tracks are restored, the western shoofly track will be taken out of service. Between the railroad and highway remain buried utilities including fiber optic lines. A petroleum pipeline that was located west of the railroad was rerouted to the east of the mud spring in early 2019. The fiber optic lines will be placed above grade on temporary power poles to span over the mobile mud spring.

The final critical infrastructure in the mud spring's path is Highway 111 owned by the California Department of Transportation (Caltrans). Caltrans engineers devised concepts for protecting the highway, but after reviewing the costs, we understand they will likely close the highway, or detour traffic around the area. Should the mud spring continue its trend towards the southwest, a gravel parking lot and county road remain in its path. Beyond that are fields and duck ponds before encountering the Salton Sea, a distance of about 2 miles (Figure 12).

Scientists from NASA, the Jet Propulsion Laboratory, and several universities have studied this unusual phenomenon since we brought the mud spring to the attention of various scientists, engineers, and geologists of the mud spring in June 2018. Much of the testing and data gathered by the UPRR team will aid these researchers.

Figure 12-Aerial drone image of the Mundo mud spring site showing the current shoofly railroad tracks around the mud spring and adjacent Highway 111. State owned park land and parking lot are on the west side of Highway 111. Image courtesy of D. Lynch.

For the UPRR team accustomed to working with natural hazards such as landslides, debris flows, and washouts, the Mundo mud spring has been a unique experience. The project team resisted the temptation to directly affect the mud spring (such as plugging the mud spring hole) as we were uncertain of the consequences of those actions. Instead, we accommodated the mud spring's movement through the UPRR right-of-way while maintaining operations on the critical rail lines.

REFERENCES

GeoVision, Geophysical Survey – Mudpot Investigation, Gillespie Road and Highway CA-111, Mundo, California Project No. 18203, June 24, 2018.

Google Earth, Image date 10/19/2016, Image Copyright 2019 Maxar Technologies.

Lynch, D. K. and Hudnut, K.W., The Wister Mud Pot Lineament: Southeastward Extension or Abandoned Strand of the San Andreas Fault?, *Bulletin of the Seismological Society of America* vol. 98 (4), pp. 1720-1729, August, 2008.

Persaud, P., Ma, Y., Joann M. Stock, J.M., John A. Hole, J.A., Gary S. Fuis, G.S., and Han, L., Fault zone characteristics and basin complexity in the southern Salton Trough, California, *GEOLOGY* vol. 44 (9); 2016, pp. 747–750.

An Innovative Solution for Debris Flow Barriers:

Better Performance with Less Maintenance

Chiara Morstabilini
Maccaferri Innovation Center
Via Ipazia 2
Bolzano (BZ, Italy), 39100

Luca Gobbin
Officine Maccaferri
Via Kennedy 10
Zola Predosa (BO, Italy), 40069
l.gobbin@maccaferri.com

Marco Luigi Deana
Officine Maccaferri
Via Kennedy 10
Zola Predosa (BO, Italy), 40069
m.deana@maccaferri.com

Daniele Lepore
Mares
Via Fiume Giallo 3
Roma (Italy), 00144

Prepared for the 70th Highway Geology Symposium, October, 2019

Acknowledgements

The authors would like to thank the individuals/entities for their contributions in the work described:

Aronne Armanini – University of Trento (Italy)
DICAM – University of Trento (Italy)

Disclaimer

Statements and views presented in this paper are strictly those of the author(s), and do not necessarily reflect positions held by their affiliations, the Highway Geology Symposium (HGS), or others acknowledged above. The mention of trade names for commercial products does not imply the approval or endorsement by HGS.

Copyright Notice

Copyright © 2019 Highway Geology Symposium (HGS)

All Rights Reserved. Printed in the United States of America. No part of this publication may be reproduced or copied in any form or by any means – graphic, electronic, or mechanical, including photocopying, taping, or information storage and retrieval systems – without prior written permission of the HGS. This excludes the original author(s).

ABSTRACT

The transformation of our natural environment due to human activity and ever growing population density has led to an increase in natural hazards (e.g. flooding, landslides and debris flow events) with consequent risks to infrastructure and human lives. Today, one of the major challenges for authorities and designers is to protect and manage these risks properly.

Debris flow events are becoming more dangerous, with higher magnitude and significant consequences. The solutions for protecting from debris flow events are several, but they are all strictly related to a precise and very detailed site characterization and characterization of the expected events.

The Maccaferri Innovation Center (M.I.C.) has researched and developed a new solution together with a new reliable design approach for debris flow protection based on accurate hydrological characterization of sites.

The new solution, called Mini Skirt Check Dam (MSCD), is an innovative weir aimed at diluting event flow rates by cutting the peak flows of the more destructive events. The conclusion of the research culminated in design installation of a MSCD in a basin located in Ottone (Italy): the MSCD was designed according to the brand-new design approach and based on a complete hydrological characterization of the boundary conditions. This paper presents a review of the research conducted and the new design approach to debris flow protection.

INTRODUCTION

Takahashi (1991) defines the debris flow as a phenomenon of transport of both liquid and solid particles that occurs in mountain areas characterized by a severe slope where the motion of the solid phase is driven by gravity. This phenomenon is not constant in time and space and, in addition, is extremely violent and impulsive. This last characteristic makes the forecasting and design extremely difficult because the rheology of the flow and its motion are still not completely characterized.

One of the most popular solutions for debris flow protection is flexible barriers. The main goal of these kinds of applications is to stop all the material flowing down the slope, or through a channel, at a given location. After each event, the barriers are partially or completely clogged by sediment, and requires maintenance for the barrier components after debris removal. Sometimes the barriers need to be completely replaced due to major damage because of the debris front impact. For these reasons, flexible barriers are usually utilized as a last protection means against debris flow events.

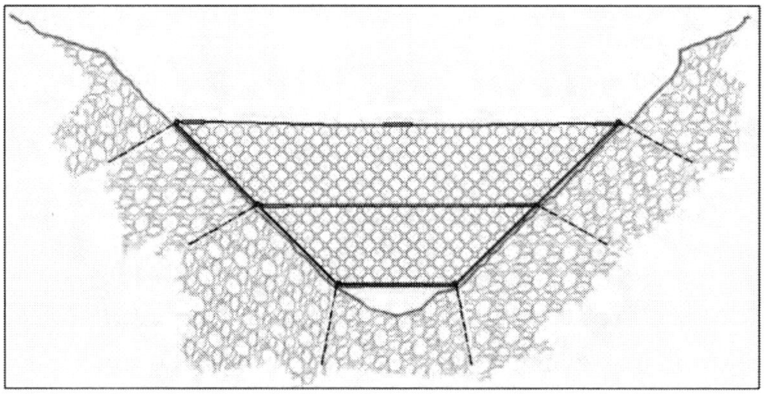

Figure 1: Typical drawing for debris flow barrier in channel. (Courtesy of Maccaferri)

Figure 2: Clogged open slope debris flow barrier requiring maintenance. (Courtesy of Maccaferri)

MINI SKIRT CHECK DAM

The Maccaferri Innovation Centre (M.I.C.) started a research program, in cooperation with the University of Trento, aimed at implementing a solution for debris flow protection without major maintenance operations. From the program a new solution has been identified with a product called Mini Skirt Check Dam (MSCD) [1]: a conceptual drawing for the MSCD is shown in Figure 3.

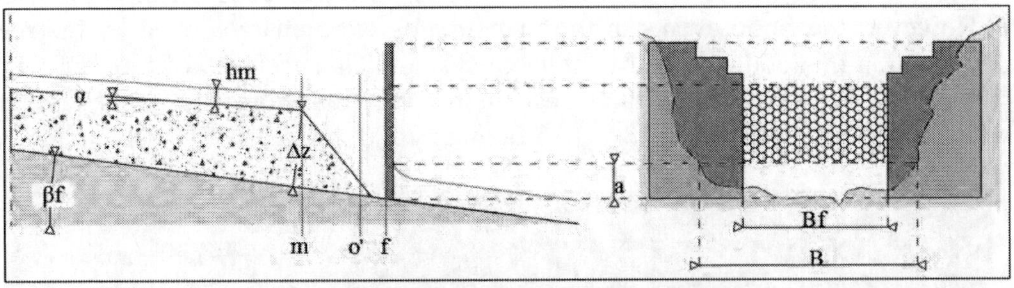

Figure 3: Example of a MSCD [2].

The research conducted in the M.I.C. demonstrated that the Mini Skirt Check Dam design is different from designs commonly used for traditional weirs because it considers the characteristics of the granular part of the debris materials, paying particular attention to big stones and floating logs. At the same time, it also demonstrated that a hydraulic approach could be very effective for designing this kind of application.

The MSCDs are designed using a formula that allows an accumulation of sediments in the netting; the sediments flow out of the structure after a wave peak of the biphasic material. Based on that peak, a solution is developed that ensures a complete self-cleaning system that discharges accumulated debris, considerably reducing maintenance. The dimensioning formula must be properly calibrated according to specific debris characteristics and the type of netting used by the system. Less concrete, better hydraulic properties and reduced maintenance give interesting characteristics to a Mini Skirt Check Dam from both performance and environmental points of view [4].

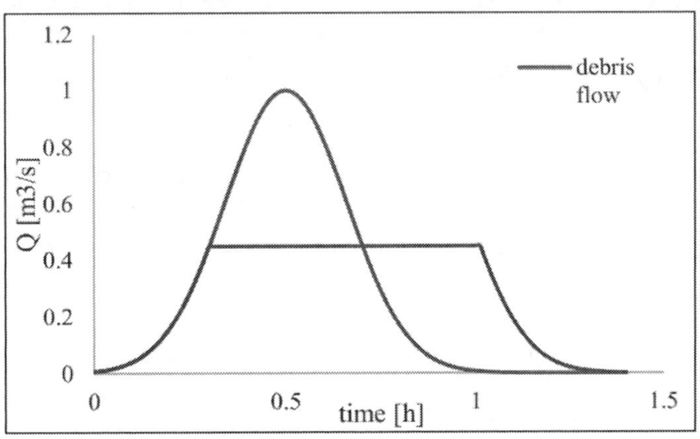

Figure 4: Flow rate per hour comparison for a site with and without a MSCD installation [2].

DESIGNING THE NEW SYSTEM

The first test installation project was chosen in the municipality of Ottone (Piacenza, Italy). The most challenging part of the project was the presence of a sewer made of masonry having a vaulted roof and two hydraulic sections of 1.30×1.40m and 2.00×1.50m in section. The drainage disposal capacity, assessed with a maximum filling level of the sections equal to 2/3 of the available height and considering clear water, was equal to $8m^3/s$ considering a multi-centennial flow rate. However, the flood events in the two streams were characterized by the presence of solids transportation that could plug the drainage system. This scenario of blockage by massive debris transport has occurred more than once in the recent history; the most relevant events happening on: 19th September 1953; 13th -16th October 2000; and 13th -14th September 2015.

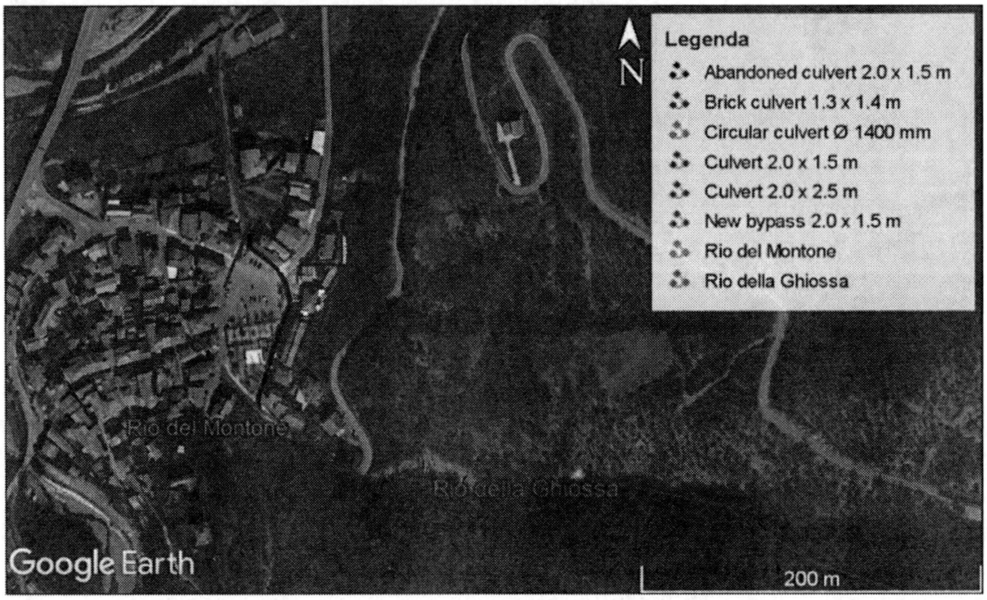

Figure 5: channel system below Piazza Vittoria after the first restoration [2].

The most critical section was identified using a Hec Ras model implemented in cooperation with the Politecnico of Milano. 88% of the 64 sections have a fixed geometry (sewer dimensions are known). Starting from this data, the peak of the flow rate was selected from previous studies (Menduni, 2016) and approved by the local administration.

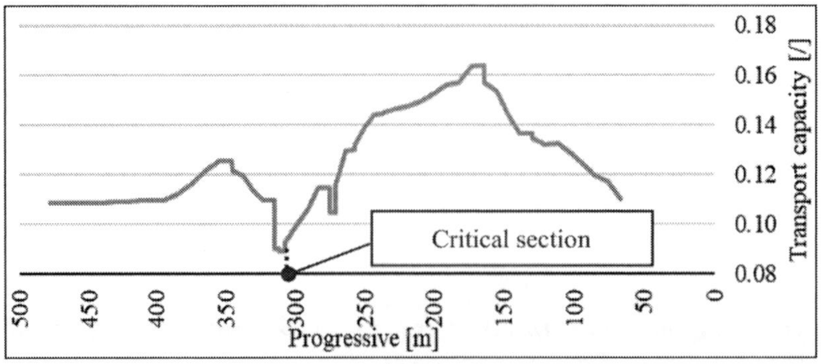

Figure 6: transport capacity based on Hec Ras model (Brunner et al., 2016).

The hydraulic modelling also included creation of a GIS model to obtain a complete curve of discharge over time, and it was made possible by using a tool called Peak Flow (Rigon et al., 2011). The calibrated function was adapted to the peak value of previous studies (Menduni, 2016) for the official test system design.

The total flow rate is represented by the sum of liquid and solid transport, and it was obtained by applying Takahashi criteria (Takahashi, 1991).

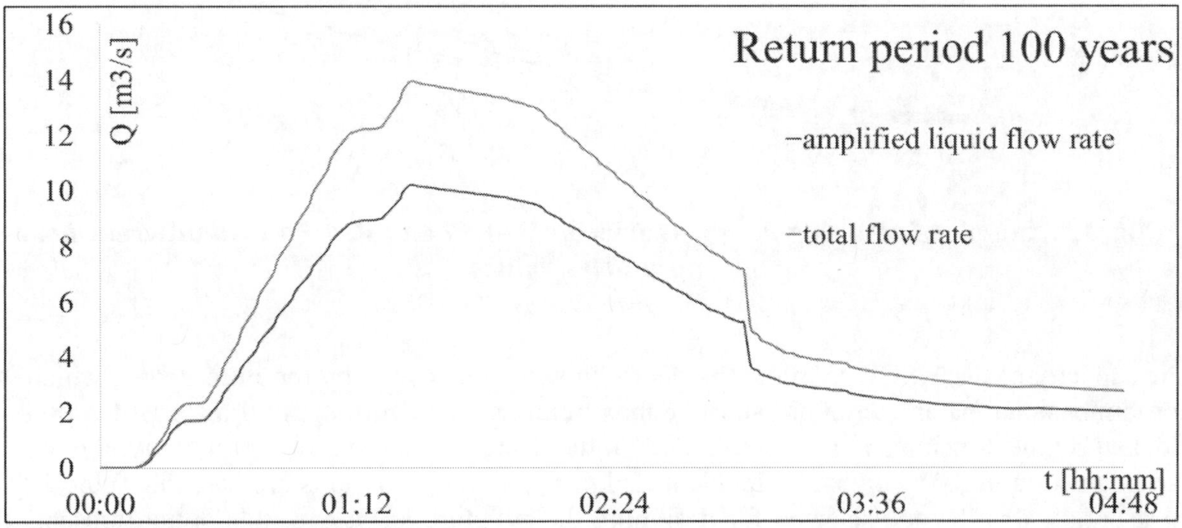

Figure 7: amplified liquid flow rate (blue) compared to the total flow rate (orange) [2].

These flow rate values were used to define the MSCD opening. With reference to Fig. 3, the design parameters to be determined were:

- *a*, the vertical dimension of the opening, i.e. how much the net is detached from the ground: this parameter could be found in the design abacus [3];
- *R*, the ratio between *B* (the average width of the undisturbed section) and *Bf* (the horizontal dimension of the opening, i.e. the distance between the two wings of the structure).

The design cross-section is 4.7 m wide, and two values of *Bf* were chosen (*Bf* = 3 m and *Bf* = 3.5 m). This approach led to two design options: one with ratio $R = 1.57$ and one with $R = 1.34$. The resulting design graphics for these two options are shown in Fig. 8.

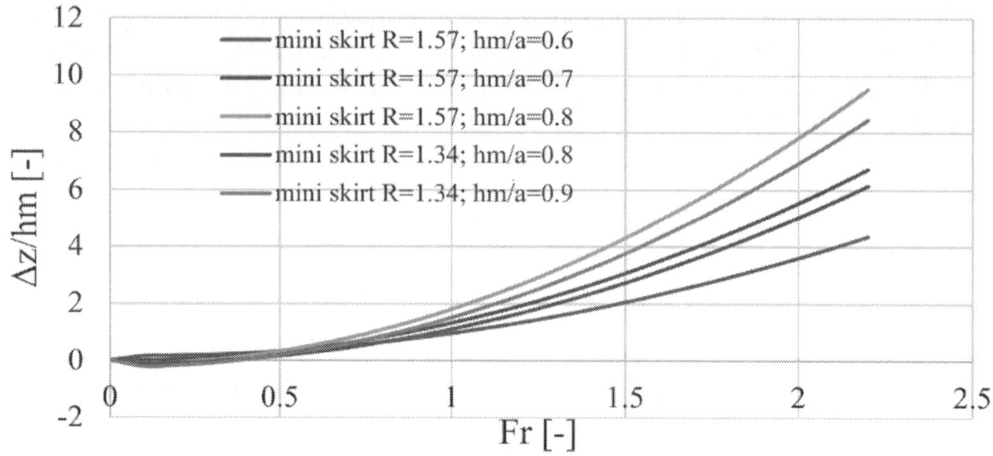

Fig. 1: Example of resulting design abacus for R=1.57 and R=1.34 with different *hm/a* parameter values

The capacity of the MSCD to dilute the debris flow is represented by the ratio $\Delta z/hm$, which is a non-dimensional parameter expressing the maximum deposit depth of solid material temporarily blocked by the structure. Based on that and on the design cross-section morphology, the MSCD can be maximum 2.0 m high and the identified possible design options are listed in Table 1. The third option was the one selected for designing the structure in Ottone: this option presents the biggest opening combined with the highest flow velocity. In this way, the opening obstruction related to the passage of small/medium size boulders is avoided, and the high velocity values support the self-cleaning effect, a peculiar characteristic of this kind of structures.

R [-]	a [m]	hm / a [-]	hm [m]	dz / hm [-]	dz [m]	dz + hm [m]	Fr [-]	U [m/s]	Q [m3/s]
1.57	0.35	0.70	0.25	6.72	1.65	1.89	2.20	3.41	3.42
1.57	0.30	0.80	0.24	7.04	1.69	1.93	1.90	2.92	2.50
1.57	**0.70**	**0.60**	**0.42**	**3.61**	**1.52**	**1.94**	**2.00**	**4.06**	**8.14**
1.34	0.35	0.80	0.28	6.14	1.72	2.00	2.20	3.65	3.66
1.34	0.30	0.90	0.27	6.22	1.68	1.95	1.90	3.09	2.66

Table 1: Final results of the design stage

INSTALLATION STAGE

The purpose of the project is to observe the performance of this innovative solution concept. For this reason, a monitoring system was also implemented and installed to collect data about the structure behaviour in case of debris-flow event.

The MSCD installation was carried out according to the design results: the interception structure, consisting of ring net panels combined with double twist mesh properly fixed on it, was extended to the ground on both sides in order to dimension and limit the opening.

For the job site in Ottone (Italy), a customized monitoring system consisting of three instruments was provided and installed on site including:

- A rain gauge for measuring the level of water over the time: this system was installed by the municipality before the start of the project. The data acquired were to be used to find a possible correlation between rain and pressure for setting a general design approach for future applications;
- Cameras: two high resolution cameras and one regular camera powered by three solar panels installed for recording the events; and
- Special shackles installed directly on the barrier; with the shackles connected by radio to a data logger placed near the solar panels.

The first step for the monitoring system installation was the positioning of the solar panels: a small area on the right side of the stream was levelled for installing the solar panels and connecting them to a control unit and to the concrete plinths. The pipe for connecting the control unit to the plinths was installed before and then covered by soil. Figures 8 and 9 show this phase and some details of the control unit.

Figure 8-9: installed solar panels and box containing the data logger.

The three monitoring cameras were installed on top of three steel posts. The posts were founded on concrete plinths having plastic pipes cast inside as conduits for connecting wires (see Figure 10 and Figure 11).

The cameras are located both upstream (one camera) and downstream (two cameras) the MSCD: the camera positions were set for having the complete overview of the MSCD and the expected debris flow event.

The internet connection for the entire system was checked and calibrated to ensure data transmission.

The calibration process will also involve other phases, (i.e. the energy supply and the efficiency of solar panels), that will be tested in a second stage of the project.

Figure 9-11: installation of the upstream camera and one downstream camera.

The last stage of the job consisted of the installation of four special shackles on the MSCD. The special shackles are directly connected to the data logger to record all the tensile variations acting on the monitored connections.

These shackles have the mechanical characteristics described in Figure 12.

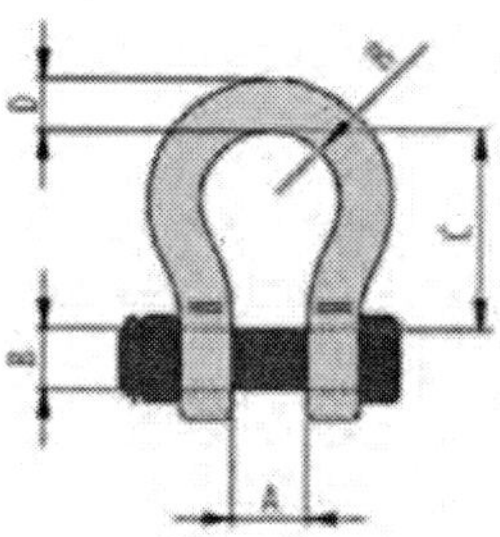

A	B	C	D	2R	weight	Max load
mm	mm	mm	mm	mm	kg	ton
47	32	108	28	75	3.87	9.5

Figure 12: special shackles characteristics.

After the installation, the monitoring system will require a calibration process over the next few months. This is necessary to have a functioning integrated monitoring system to facilitate post-analysis of acquired data.

CONCLUSION

The paper discusses the design of an innovative solution, called Mini Skirt Check Dam (MSCD), aimed at diluting event flow rates by cutting the peak flows of the more destructive events. The installation phase of a monitoring system is also presented.

In the near future the installation in Ottone will be constantly monitored for recording and analyzing the behavior of the MSCD over time, during and after the expected debris flow events.

Also, other MSCD structures are scheduled to be installed in other sites and then monitored for progressing the research and increasing the knowledge about debris flow events and this innovative solution.

REFERENCES:

- Conference Proceedings

    1. Morstabilini, C. and Deana, M. L. Laboratory test of an innovative check dam. In *7th International Conference on Debris-Flow Hazards Mitigation*, 2019.
    2. Morstabilini, C., Boschini, I., Zambrini, F., Menduni G., Deana, M. L. and Zorzi, N. Application of an innovative, low-maintenance weir to protect against debris flows and floods in Ottone, Italy. In *7th International Conference on Debris-Flow Hazards Mitigation*, 2019.
    3. Morstabilini, C. and Deana, M. L. Debris Flow: a new design approach. In *European Rock Mechanics Symposium: Geomechanics and Geodynamics of Rock Masses., v. 2*, 2018.
    4. Morstabilini, C., Ferraiolo, F. and Armanini, A. Mini Skirt Check Dam: an innovative apparatus against debris flow. In *IAHR*, doi: 10.3850/978-981-11-2731-1_251-cd, 2018.

- Federal Government Reports
    1. Brunner, G. W.. *HEC-RAS: River Analysis System: Hydraulic Reference Manual (5.0 ed.)*, US Department of Defense, Army Corps of Engineers, 2016.

- Journal
    1. Armanini, A., Fraccarollo L. and Larcher M.. Debris Flow. *Encyclopedia of Hydrological Sciences* v. 4(12), 2004, pp. 2173-2186.
    2. Takahashi, T.. Debris Flow. *International Association for Hydraulic Research*, 1991.
    3. Takahashi, T.. Mechanical characteristics of debris flow. *Journal of the Hydraulics* Division v. 104, no. HY8, 1978, pp. 1153–1169.
    4. Rigon, R., d'Odorico, P. and Bertoldi, G.. The geomorphology structure of the runoff peak. *Hydrogeology and Earth System Science*, doi:10.5194/hess-15-1853-2011, 2011.

- University Report
    1. Lepore, D.. Industrial products for water management. Department of Mathematics, Physics and natural science, La Sapienza University, Roma, 2019.
    2. Menduni, G.. Analysisof the hydrogeologic risk in Ottone related to the events of 13-14 september 2015 and residual risk, Department of Environmental Engineering of Politecnico of Milan, 2016.

Mitigation of Chronic Bluff Retreat with an Engineered Riprap Revetment BNSF Bellingham Subdivision, Whatcom County, Washington

Matthew Grizzell, EIT, PG
Senior Engineering Geologist, Shannon & Wilson, Inc.
1321 Bannock Street, Suite 200
Denver, Colorado 80204
mtg@shanwil.com

Tyler Stephens, PE
Associate Geotechnical Engineer, Shannon & Wilson, Inc.
400 north 34th Street, Suite 100
Seattle, Washington, 98103
tjs@shanwil.com

Prepared for the 70th Highway Geology Symposium, October, 2019

Acknowledgements

The authors would like to thank the individuals/entities for their contributions in the work described:

Walt Smith – BNSF
Glen Gaz – BNSF
Megan McIntyre - BNSF

Disclaimer

Statements and views presented in this paper are strictly those of the author(s), and do not necessarily reflect positions held by their affiliations, the Highway Geology Symposium (HGS), or others acknowledged above. The mention of trade names for commercial products does not imply the approval or endorsement by HGS.

Copyright

Copyright © 2019 Highway Geology Symposium (HGS)

All Rights Reserved. Printed in the United States of America. No part of this publication may be reproduced or copied in any form or by any means – graphic, electronic, or mechanical, including photocopying, taping, or information storage and retrieval systems – without prior written permission of the HGS. This excludes the original author(s).

ABSTRACT

The BNSF Bellingham Subdivision traverses coastal bluffs along Puget Sound in Whatcom County, Washington. In 1998, an approximately 50-foot-wide landslide involving an entire 75-foot-high bluff face developed about 3 miles north of Bellingham, initiating chronic bluff retreat to within 15 feet of edge-of-tie the following year. In 1999, Shannon & Wilson installed a proprietary vertical movement detector (VMD) instrumentation system, an owner-serviceable monitoring system connected to the BNSF signal to alert traffic to vertical shoulder movements or catastrophic shoulder failures. In early 2016, precipitation caused renewed bluff retreat to within 10 feet of edge-of-tie, in turn approaching, but not activating the VMD system. In response, Shannon & Wilson performed an emergency geotechnical site evaluation including geologic mapping, review of existing subsurface information, evaluation of slope stability, and development of conceptual mitigation options.

The bluff face exposes a bedded sequence of glacial deposits including (from top to bottom) glacial outwash deposits consisting of dense to very dense, cross-bedded sand and gravel with interbedded silt and clay layers overlying glaciomarine deposits consisting of stiff to very stiff massive clay with silt, sand, and gravel. The bluff includes at least two perched groundwater tables in coarse-grained outwash deposits above relatively impermeable clay beds in the glacial outwash and the glaciomarine clays. The perched groundwater horizons issued from the bluff face in two spring lines. A large industrial facility adjacent to the site likely increased seepage volumes due to effluent from a septic system drain field, and increased surface runoff from a large area of pavement that infiltrates upgradient of the site. Erosion of coarse-grained soils at the bluff face spring lines undermines the overlying soils, causing repeated debris flows. Coupled with wave erosion at the base of the slope, these debris flows steepened the face to an approximate average angle of 45 degrees after the storms, causing a cycle of bluff retreat and prompting renewed stability concerns.

Following analysis, Shannon & Wilson and BNSF determined that an engineered riprap revetment would provide an effective and economical means to arrest bluff retreat and buttress the slope. In addition to accommodating wave erosion forces, Shannon & Wilson's revetment design required considerations to accommodate environmentally sensitive tidal flats and constructability concerns due to the lack of a staging area between the railroad and the bluff face. Environmental concerns were accommodated by constructing a relatively steep (1.25H:1V) revetment slope. This slope angle was constructed by individually placing interlocking riprap boulders to form a fortified revetment face. The lack of staging area was accommodated by employing a large crane to transport equipment and revetment materials over the rail line.

INTRODUCTION

The Burlington Northern Santa Fe (BNSF) Bellingham Subdivision provides a critical freight and passenger rail link between Seattle, Washington and Vancouver, British Columbia. Chronic bluff retreat has impacted operations where the route traverses coastal bluffs along Bellingham Bay. A chronic location of bluff retreat capable of impacting railroad operations began developing in the 1990s near Milepost 99.2, or approximately 2.5 miles railroad north (northwest) of Bellingham, Washington (Figure 1). This case study documents geologic characterization, monitoring, and mitigation design efforts at this location beginning with a landslide and bluff retreat sequence in 1998 and culminating in the construction of a permanent riprap buttress and revetment in 2016.

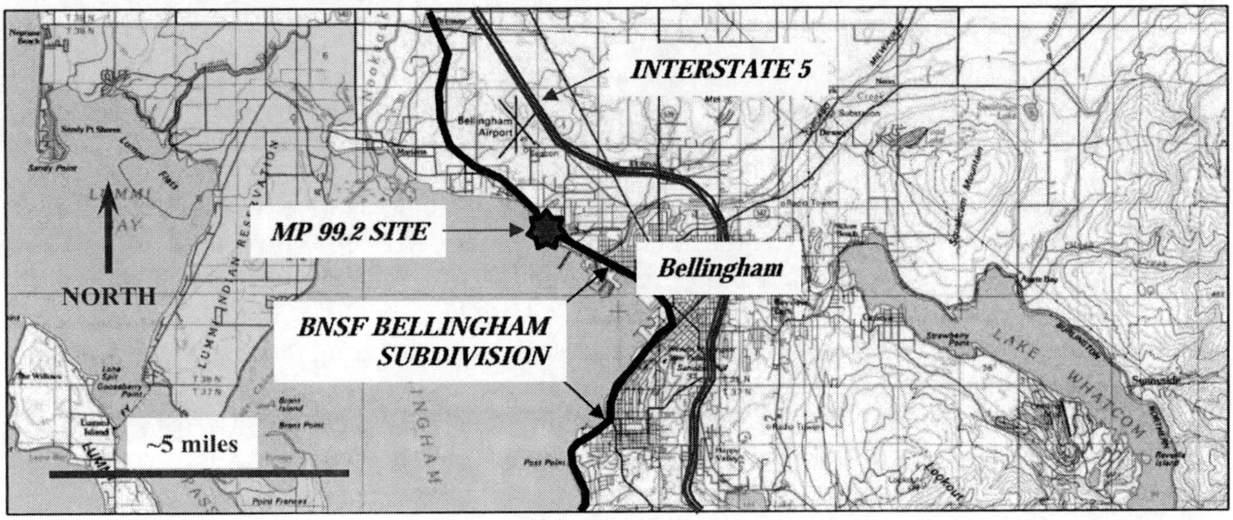

Figure 1 – Site Vicinity Map (Scale Approximate)

SITE DESCRIPTION

The BNSF Bellingham Subdivision line was constructed beginning in 1888 by the Fairhaven and Southern Railroad Company and subsequently sold in 1891 to the Great Northern Railway company, a predecessor railroad to the BNSF (Lewis, 1912) At MP 99.2, the BNSF Bellingham subdivision consists of a single main line with a roughly 1 percent grade to the south. The track is aligned on a short tangent section that passes along the top of approximately 70- to 80-foot-high coastal bluffs just east of the track. The coastal bluffs stand at overall angles on the order of 1H:1V (Horizontal to vertical), and in early 2016 regressed to within approximately 10 feet of the west rail along an about 75-foot-long reach of track. At the north end of the project area, the line transitions to curve that diverges eastward from the bluff crest. A Portland cement manufacturing plant covering about 20 acres is present to the east of the track (see Figure 2).

Based on a 1998 report by Shannon & Wilson, the Portland cement plant operates a septic system and drain field approximately 80 feet away from the site. This system likely contributes additional drainage into the subsurface. The plant also includes extensive pavements, some of which may drain toward the track and into the right-of-way ditch along the track. The right-of-way ditch drains to the southeast and discharges beneath a railroad bridge.

Figure 2 – Site Photograph view due northeast, June 2006. Note BNSF Bellingham Subdivision (yellow highlight), Cement Plant (top, center), MP 99.2 site (red shading), and Bellingham Bay (at bottom). ***Photo Courtesy of Washington Department of Ecology.***

Railroad stationing maps and historical county assessor's maps covering the area are the oldest readily available maps documenting the project site. These maps both indicate a 100-foot-wide ROW with the track situated 75 feet away from the bluff side ROW boundary (BNSF, ca. 1960 and Whatcom County Assessor, 1964). These maps do not clearly identify the bluff crest. However, it is presumed that the Fairhaven and Southern Railway company selected a ROW extending to the bluff crest in this area in 1888.

MP 99.1 Landslide (Early 1990s)

The first available documentation of mass wasting and bluff retreat affecting rail operations at the site are Shannon & Wilson project records indicating that a slope failure in 1991 or 1992 at approximate MP 99.1. In response to the MP 99.1 slope failure, Burlington Northern (a predecessor railroad to BNSF) reportedly placed side-dumped riprap to form a revetment and buttress. Aerial photos indicate two riprap covered slopes along an approximately 150-foot-wide section of the bluffs that likely correspond to this event.

MP 99.2 Landslide (April 1998)

In April 1998, Shannon & Wilson project records indicate that a second mass wasting event occurred at MP 99.2. The 1998 landslide occurred in the bluff slopes immediately northwest and adjoining the MP 99.1 riprap revetment. Reconnaissance observations suggested that the slide event caused the upper portion of the bluff to fail due to undercutting by wave erosion. The event reportedly caused the bluff crest to regress approximately 15 feet toward the rail, reducing the distance between the rail and bluff crest from 30 to about 15 feet in some locations. Based on review of historical aerial photos (City of Bellingham, 2019), this event appears to have denuded relatively mature vegetation from the bluff face.

Following the 1998 failure event, BNSF retained Shannon & Wilson to perform geotechnical studies at the site, including completing a geologic reconnaissance, drilling a 107-foot-deep geotechnical boring from track level, completing stability modeling with PCSTABL 5 and Mohr-Coulomb material parameters, and providing recommendations for mitigation and/or additional monitoring. Due to slope flattening that resulted from the 1998 failure event, stability modeling suggested that large, deep-seated failures extending beneath the track were unlikely to occur if surface drainage was maintained to limit stormwater infiltration.

Vertical Movement Detector (VMD) System Installation and Monitoring (1998 to 2016)

Between 1998 and 2000 Shannon & Wilson designed and installed a proprietary VMD instrumentation system at the site. The VMD system was funded in part by a Federal Railroad Administration grant awarded to Washington State Department of Transportation (WSDOT) to improve the performance and increase cost effectiveness of Amtrak traffic along the Bellingham Subdivision. The VMD was designed as an alternative to the previously deployed mercury switches, an existing landslide warning technology designed to activate track signal systems in the event of rotational ground movements. The VMD was designed to detect and activate track signals if downward vertical ground movement occurred at one or more VMD sensor locations regardless of sensor/ground rotation.

The VMD system at this location (Figure 3) consists of a line of partially liquid-filled chambers or "pots," each containing float switches mounted approximately 1 to 2 inches above the fluid surface. The chambers are connected in series via buried flexible tubing, and the system is connected to a fluid reservoir containing a substantially larger volume of the reference fluid. If any of the chambers moves vertically within the tolerance of the system (e.g. the distance between the float sensor and the fluid level in the chamber), the fluid level seeks a common hydrostatic level by flowing from the reservoir into the depressed chamber. The rising fluid level engages the float switch and activates electronics to activate the track signals and alert oncoming rail traffic. The VMD system also includes electronic sensors to detect fluid losses or catastrophic system failure, conditions that would also activate the track signals to alert approaching rail traffic.

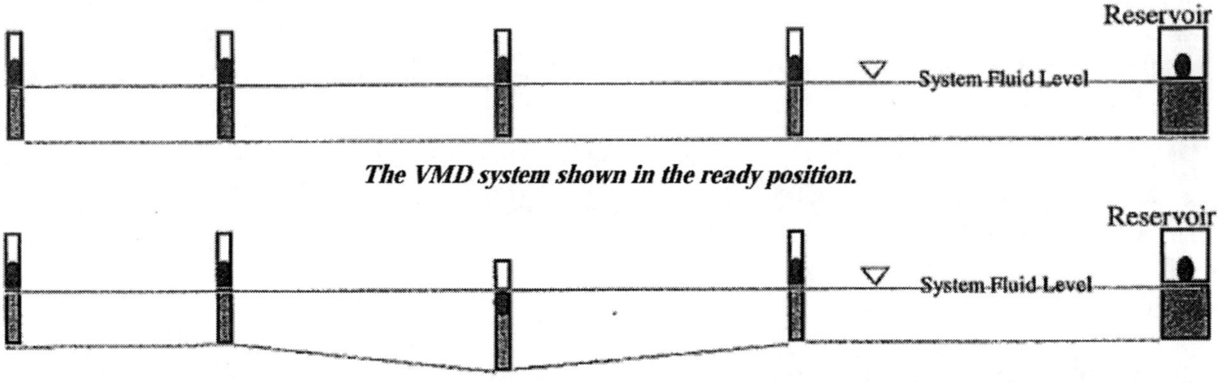

The VMD system shown in the ready position.

The VMD system is triggered with one (or more) of the sensors drops below the common fluid level.

Figure 3 – Schematic of Typical VMD System

The VMD was constructed parallel to a 150-foot-long section of rail with roughly equal spacing of 25 feet between individual pots (Figure 4). The VMD alignment is offset approximately 10 feet west of the western rail. Individual VMD pots were located between about 3 and 7 feet uphill of the bluff crest (hachured line at bottom of Figure 4).

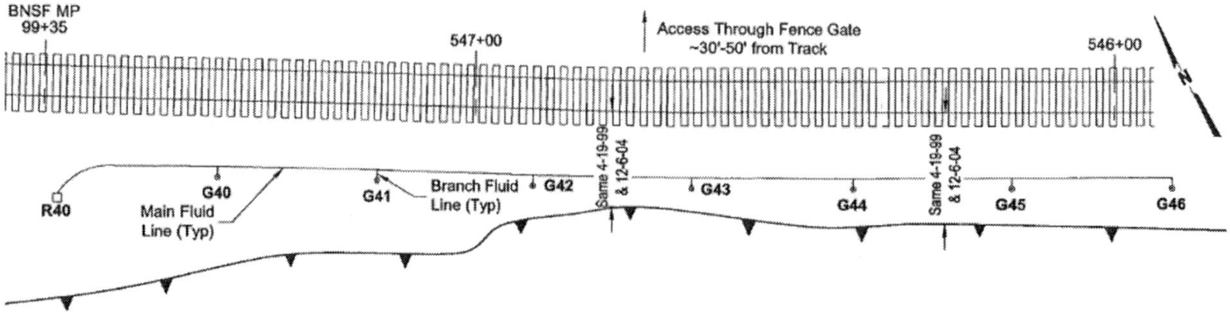

Figure 4 – VMD System Components and Bluff Crest, December 2004 (not to scale)

Based on a Shannon & Wilson report to BNSF dated December 2004, the VMD system at MP 99.2 operated as intended following installation with no false alarms. The same Shannon & Wilson report also suggests no measurable bluff retreat at select monitoring points between April 1999 and December 2004. No additional records are available describing operation of the VMD system or bluff retreat between December 2004 and February 2016.

Landslide, Bluff Retreat, and VMD System Damage (2016)

On February 16, 2016, BNSF requested that Shannon & Wilson visit the MP 99.2 site after a series of landslides and resultant debris avalanches caused renewed bluff retreat toward the tracks. The landslide events occurred after a series of winter rainstorms totaling about 3.5 inches of precipitation during the previous week, of which 1.6 inches fell on February 15 (Weather Underground, 2019) As shown on Figure 5, the 2016 landslides occurred along a 120-foot-long reach of the crest, causing retreat to within about 11 feet of the western track in some areas.

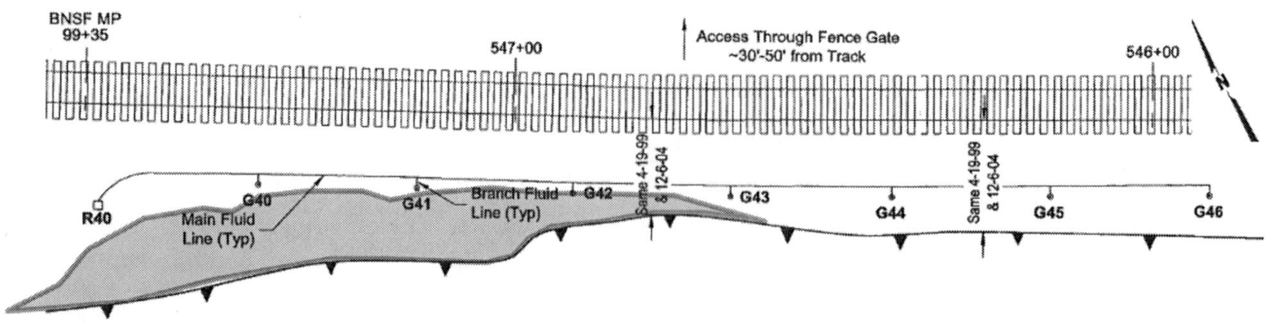

Figure 5 –VMD System and Bluff Retreat December 2004 to February 2016 (not to scale)

The bluff crest regressed to the edge of one of the pots (designated G42 – see Figure 5), causing downhill rotation, but not activating the switch and track signal. The bluff retreat was reportedly observed by a BNSF locomotive engineer familiar with the area. As shown on Figures 6 and 7, the debris avalanches covered much of the tidal flat along the toe of the bluff slope with up to 6 feet of saturated soil debris. Following the 2016 event, the upper slopes of the bluff stood at angles of up to about 70 degrees.

Figure 6 – MP 99.2 Bluff Crest and Debris Runout Following February 2016 Landslide Note rotated VMD pot (yellow arrow) and Debris Runout (Yellow Shaded).

Following the event, BNSF signal personnel inspected the VMD system to ensure it continued to operate. We understand that the VMD system required only minimal maintenance and has continued operating to the time of this writing.

SITE GEOLOGY

Regional Geologic Framework

The project site is located within the Cascadia Subduction Zone forearc at the northeastern edge of the Puget Lowland. The Puget Lowland is an elongated north-south depression positioned between the Olympic Mountains and the Cascade Range. The Puget Lowland is underlain a series of tectonically depressed sedimentary basins, including (from south to north) the Tacoma, Seattle, Everett, and Bellingham Basins (Kelsey and others, 2012). Beginning approximately 300 thousand years before present (ka), a continental ice sheet originating in the Coast Mountains of British Columbia (the "Cordilleran Ice Sheet [CIS]) made at least four cycles of advance and retreat into the Puget Lowland, including (oldest to youngest) the Defiance, Double Bluff, Possession, and Fraser Glaciation (Troost, 2014; Easterbrook, 2003). Three glacial maxima (or "stades") and one intervening retreat (or "interstade") are recognized within the Fraser Glaciation, including the Evans Creek Stade (a primarily alpine advance in Washington ca. 25 to 15 ka), the Vashon Stade (a regionally extensive advance ca. 18 to 13 ka) the Everson Interstade (ca. 12.2 to 11.7 ka), and the Sumas Stade (a less-extensive advance in the northern Puget Lowland ca 11.5 to 10 ka) (Weber and Kovanen, 2000, Porter and Swanson, 1998, Riedel and others, 2010, Easterbrook, 2003).

Figure 7 – MP 99.2 Bluff Slope Following February 2016 Landslide
Note rotated VMD pot (yellow arrow) and Debris Runout (yellow shaded).

The CIS reached its maximum extent ca. 16.9 ka, at which point the top-of-ice elevation on the "Puget Lobe" of the CIS at Bellingham was on the order of 5,000 feet (1,500 meters) above sea level (Porter and Swanson, 1998 – see Figure 8). Beginning at about 12.5 ka, the now-stagnant Vashon ice sheet had thinned sufficiently to allow seawater originating from the Straits of Juan de Fuca and the Pacific Ocean to float the ice sheet, forming an extensive ice shelf. As a result, a laterally extensive glaciomarine deposit consisting of massive clays with marine fossils and dropstones (gravel- to boulder-sized clasts that melted out of the ice shelf) was deposited in the northern Puget Lowland (Weber and Kovanen, 2000 and Easterbrook, 2003). Due to isostatic depression of the crust beneath the Vashon Lobe, sea level elevations at 12.5 ka were on the order of 400 to feet higher than current sea level. As the crust rebounded prior to about 11.7 ka, relative sea levels ultimately dropped to approximately modern levels. Of note, Weber and Kovanen (2000) and Easterbrook (2003) suggest that an additional, nearly 700-foot oscillation of relative sea level occurred between 12.5 and 11.5 ka during the Everson Interstade. The sea level oscillation inference is based on observation of a laterally persistent sequence of terrestrial sediments (Deming Sands) within the Everson glaciomarine drift sequence at elevations of up to 600 feet above present sea level. Easterbrook (2003) attributes the sea level oscillation to a combination of isostatic, eustatic, and tectonic variations of relative sea level.

Following re-emergence, a final advance of the CIS (the Sumas Stade) began about 11.5 ka. During the Sumas Stade, a piedmont glacier advanced from the Fraser River Valley to a position within about 5 miles (9 kilometers) northwest of Bellingham and the project site.

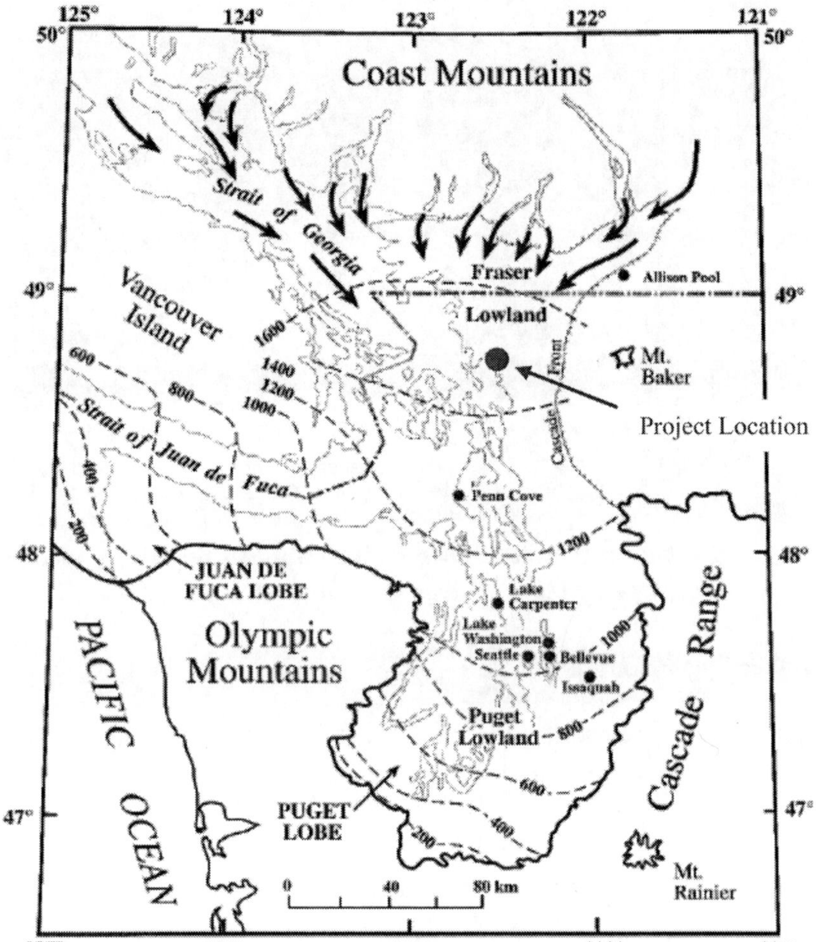

Figure 8 – Maximum Vashon Stade glacial extent, ca. 16.9 ka with ice elevation contours shown in meters above sea level (modified from Porter and Swanson, 1998).

As the Sumas Stade waned, meltwater issuing from the piedmont glacier flowed into Bellingham Bay through wide outwash channels that incised into the older glaciomarine deposits (Kovanen and Easterbrook, 2002). These outwash channels form topographically prominent valleys in the vicinity of Bellingham, including the underfit valley of Squalicum Creek adjoining the site to the Southwest (Figure 9).

Local Geologic Framework

Easterbrook (1976) mapped glacial and bedrock features in the Bellingham 1:62,500 quadrangle, including Sumas Outwash sands and gravels at the MP 99.2 project site and Everson Interstade age Bellingham Glaciomarine Drift north of the site. Examination of Light Ranging and Detection (LiDAR)-derived hillshade maps in the Bellingham area (Washington Department of Natural Resources, 2019) suggests that Easterbrook (1976) mapped the contact between Sumas Outwash and Bellingham Glaciomarine Drift along the northwest margins of the nearly 3-mile wide, topographically prominent mouth of the Squalicum Creek outwash channel (Figure 9).

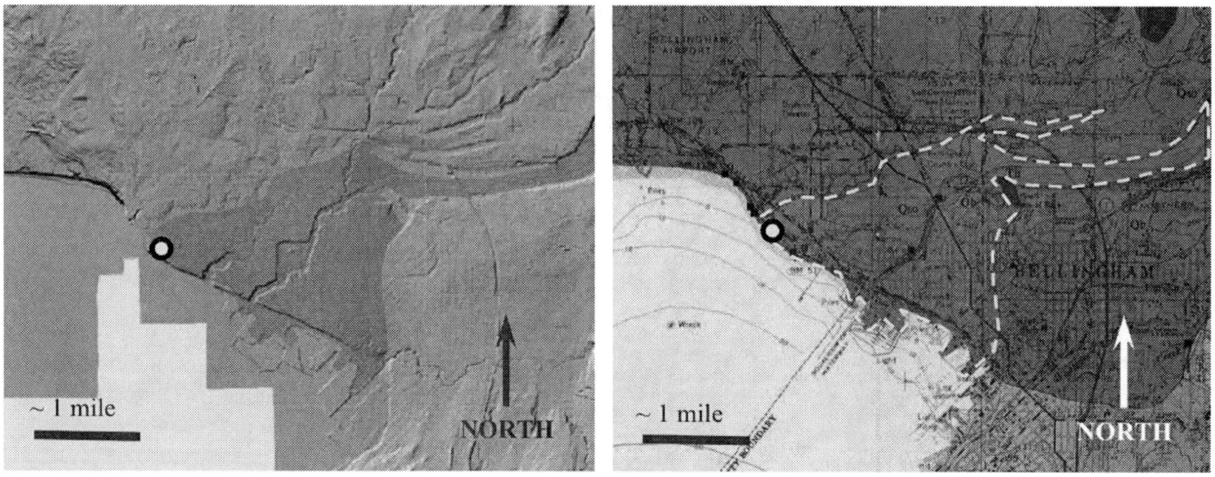

Figure 9 – Squalicum Creek Outwash Channel showing Sumas Stade outwash deposits within shaded regions (WDNR, 2019 and Easterbrook, 1976).

Site and Engineering Geology

During the February 2016 reconnaissance, Shannon & Wilson photographed and mapped the eroded bluff face. Figure 9 shows a field drawing of the bluff face with engineering geologic conditions. A 1998 boring log from track level indicated a very similar interbedded sequence consisting of (approximate thicknesses and cumulative depths below track level [btl] listed from top to bottom):

- 5 feet of stiff gray clay,
- 6 feet of loose to dense, brown sand with silt layers (to 11 feet btl),
- 9.5 feet of very stiff, brown-gray to light gray clay with scattered organics and slickensides near the bottom contact (to 20.5 feet btl),
- 37.5 feet of dense to very dense, black to gray, poorly graded gravel with silt and sand (to 48 feet btl),
- 22 feet of stiff to very stiff clay with numerous shell fragments and disturbance (to 70 feet btl or the approximate base of bluff),

- 31 feet of very soft to soft, gray, slightly gravelly to gravelly clay with sand (to 101 feet btl), and
- 5.5 feet of medium dense to very dense, gray silty sand (to 106.5 feet btl).

The field engineering geology observations were consistent with published mapping in the area. We interpret the uppermost 5-foot clay layer to be a fill deposit, with underlying Sumas outwash gravels and fluvial overbank deposits (about 5 to 50 feet btl) incised into the underlying Everson Interglacial-age Bellingham or Kulshan glaciomarine drift.

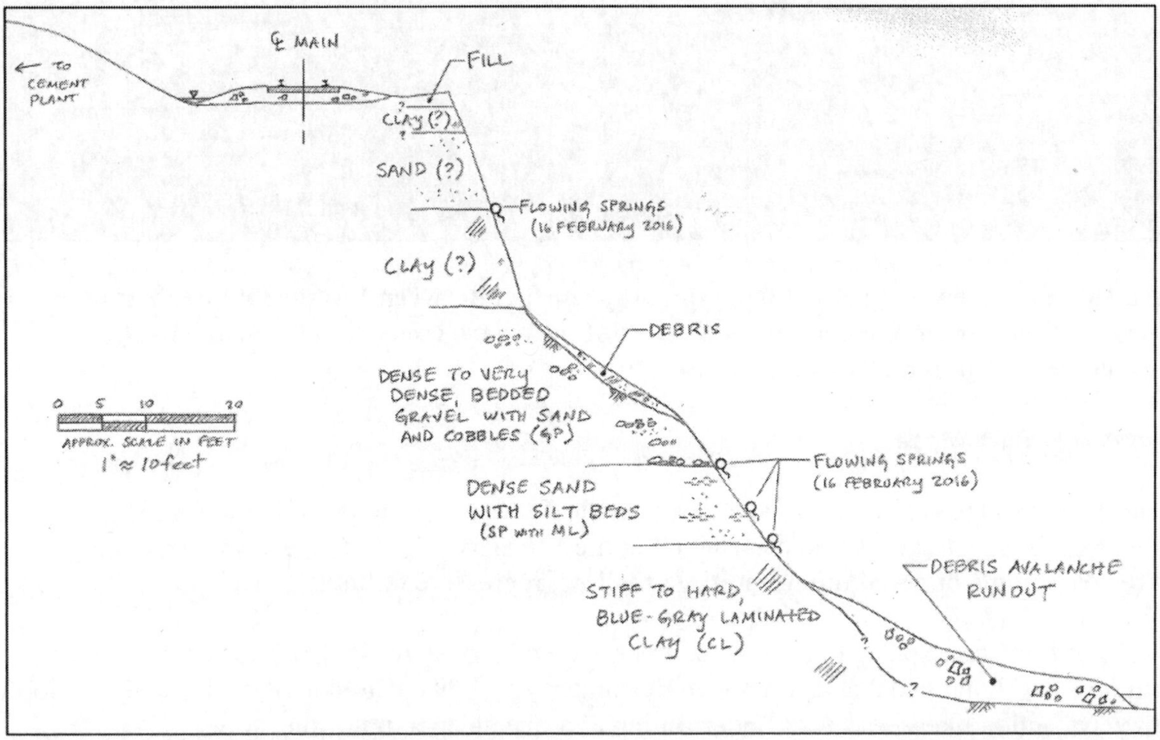

Figure 9 – 2016 Field Drawing of bluff face (modified from Shannon & Wilson project records). Ground surface profile from field measurements with a handheld clinometer and recreational laser rangefinder; geologic interpretations queried at inaccessible locations.

Hydrogeologic Observations

We noted flowing springs in the bluff face during the 2016 field reconnaissance, each of which occurred as a perched groundwater horizon where a sand or gravel beds overlie a relatively impermeable bed. The flowing springs occurred at three horizons within the Sumas Outwash deposits, including (1) the base of sand beds overlying a clay overbank deposit about 10 feet btl, (2) the base of an incised gravel bed overlying silt and sand at about 40 feet btl, and (3) the base of the silt and sand bed above the clayey glaciomarine drift at about 50 feet btl. We observed some springs flowing at estimated rates of several gallons per minute, with surface drainage rills extending to Bellingham Bay. We also noted that cohesionless deposits tended to erode back into the face at the location of the springs.

Figure 10 – February 2016 Site Photo showing contact between Sumas Outwash and underlying Bellingham Glaciomarine Drift (dotted where concealed by debris). Note incised channel geometry at base of outwash.

Interpreted Failure Mode

A combination of toe erosion and seepage in the bluff face result in periodic mass wasting events and resultant bluff retreat. It is unclear whether the February 15, 2016 mass wasting event occurred as a single large failure or multiple shallow, regressive failures.

Drainage of bluff face spring lines resulted in erosion of cohesionless deposits (which in turn undermines overlying beds) and softens/erodes underlying soils. Erosion at the base of the slope due to wave action likewise causes undermining and oversteepening of the slope.
Apparent cohesion of the materials allows steep slopes to temporarily form . With continued erosion and undermining, however, steep slopes eventually develop shear failures and fall to the base of the bluff as debris avalanches.

Extrapolated Bluff Retreat Rate

Shannon & Wilson measurements at this site suggest a maximum bluff retreat of about 13 feet between 1999 and 2016. While bluff retreat occurred abruptly as described above, these measurements indicate an annualized maximum bluff retreat rate of approximately 9 inches per year at the site.

If it is assumed that the west side of the original ROW was established at the bluff crest in 1888, and if it is assumed that the rail alignment did not appreciably change between 1888 and 2016, it follows that the bluff retreated a distance of 65 feet in 128 years. This suggests a maximum average rate of bluff retreat of about 6 inches per year. This figure considers the original bayside ROW width of 75 feet and 2016 measurements indicating the bluff crest about 10 feet west of the track.

MITIGATION DESIGN

In response to the February 15, 2016 mass wasting event, BNSF requested that Shannon & Wilson design a permanent, economical, and readily-constructible mitigation to arrest bluff retreat and stabilize the slope. Shannon & Wilson conducted a limited alternatives analysis to identify feasible mitigation concepts and identify the most favorable option for final design.

In consultation with BNSF, Shannon & Wilson evaluated the following mitigation options:

- **Realignment** – Realignment provided the most economical means of reducing risk to rail traffic but would not provide a permanent mitigation. Realignment would also require earthwork, reduce the ROW buffer between the track and the adjacent cement plant, and would require geometry considerations in adjacent areas to maintain curvature. Accordingly, we rejected realignment as a mitigation option.
- **Deep Drainage** – Deep drainage consisting of permeable drains drilled from track level or toe of bluff could provide an effective means to intercept and convey groundwater to the base of the bluff in a controlled manner. Shannon & Wilson implemented deep drainage for Burlington Norther in nearby areas of bluff retreat in the 1980s. While relatively economical, deep drains can become blocked or shear off if subsurface movements occur. Similarly, deep drainage provides no mitigation for wave erosion at the bluff toe and can thereby only provide supplemental mitigation. We rejected use of deep drainage for mitigation.
- **Engineered Retaining Structure or Wall** – We considered the feasibility of various types of retaining structures at the site, including (1) a partial or full-face soil nail wall, (2) a cantilever or tie-back soldier pile wall, (3) a tieback secant pile wall. These options would be relatively expensive due to the requirement to mobilize a specialty contractor and use of expensive construction materials. This option would also likely require construction of a revetment at the slope toe for wave erosion protection. Further, preliminary analysis indicated that long nails or tiebacks would be required for any feasible wall type. BNSF has historically not allowed structural elements to be placed beneath its tracks. For these reasons, we rejected a structural mitigation from further consideration.
- **Engineered Riprap Revetment and Buttress** – An engineered revetment and buttress would provide erosion resistance and substantially increase stability. A riprap buttress could be constructed by using readily available, relatively inexpensive construction materials (riprap and fill). Due to relatively simple construction equipment and techniques, a revetment and buttress could be constructed by a wide array of available contractors. If properly designed and constructed with hydrogeologic considerations, a revetment would also allow free drainage of groundwater springs in the bluff face. A revetment, however, presented considerable constructability challenges including (1) environmental considerations to place material at the edge of the tidal flat, (2) establishing construction access and materials staging in the cement plant, (3)

establishing construction access to the remote intertidal zone at the bluff toe, and (4) establishing a means to import materials to the bluff toe.

In consultation with BNSF, we determined that the riprap revetment provided the best combination of constructability, economy, and stability improvement. Accordingly, BNSF proceeded to issue a generalized request for proposal (RFP) to select a contractor capable of constructing a revetment, and retained A-CECO Equipment Company, Inc. of Redlands, California and their local subcontractor IMCO Construction, Inc. of Ferndale, Washington.

Revetment Analysis and Design

Shannon & Wilson completed a limited analysis to support revetment design utilizing existing subsurface information and reporting and topographic survey provided by BNSF subcontractor.

Project Survey

The Shannon & Wilson design utilized a photogrammetric topography survey. The photogrammetric imagery was acquired with an unmanned aerial vehicle by the BNSF contractor's survey subcontractor. The survey deliverables included a three-dimensional surface model of the project site. Due to the denuded condition of the ground surface in the area of greatest bluff retreat, the photogrammetric methods yielded an accurate survey with limited interference from vegetation (Figure 11).

Figure 11 – Photogrammetric Model Showing MP 99.2 Site (center bottom).

Analysis and Design

Shannon & Wilson completed limited analysis including review of existing stability analyses and evaluation of hydrology and hydraulics to support sizing of riprap and filter compatibility of fill. Based on this analysis, Shannon & Wilson specified a zoned revetment with a minimum 5-foot-thick zone of riprap on the outboard revetment face and a fill zone consisting of railroad ballast. The specification a readily available WSDOT specification for the riprap with a d_{50} (average diameter) of approximately 24 inches.

To accommodate environmental considerations, the revetment was designed to match the slope toe of the existing revetment to the south and natural slopes to the north. This toe geometry required placement of the riprap face at an overall angle of up to 1.25H:1V (approximately 39 degrees). The specification indicated that the riprap slope would require stacking individual riprap boulders in an interlocking fashion to form the relatively steep slope in a stable configuration. To limit toe erosion and provide additional resistance, the design included a minimum 5-foot-deep, riprap filled key trench along the toe alignment.

The top of the revetment was designed with a minimum 5-foot wide shoulder extending beyond the bluff crest. Because of the fixed toe alignment, the shoulder width required the relatively steep revetment face. However, the shoulder was necessary to allow sufficient width for the excavator to access the top of the revetment during construction. The shoulder provided the added the benefit of additional useful ROW width for BNSF following construction.

Shannon & Wilson prepared a simplified plan set utilizing four individual cross sections and a plan sheet derived from the photogrammetric survey (Figure 12). Concise specifications were incorporated directly into the plan set as dimensions and notes.

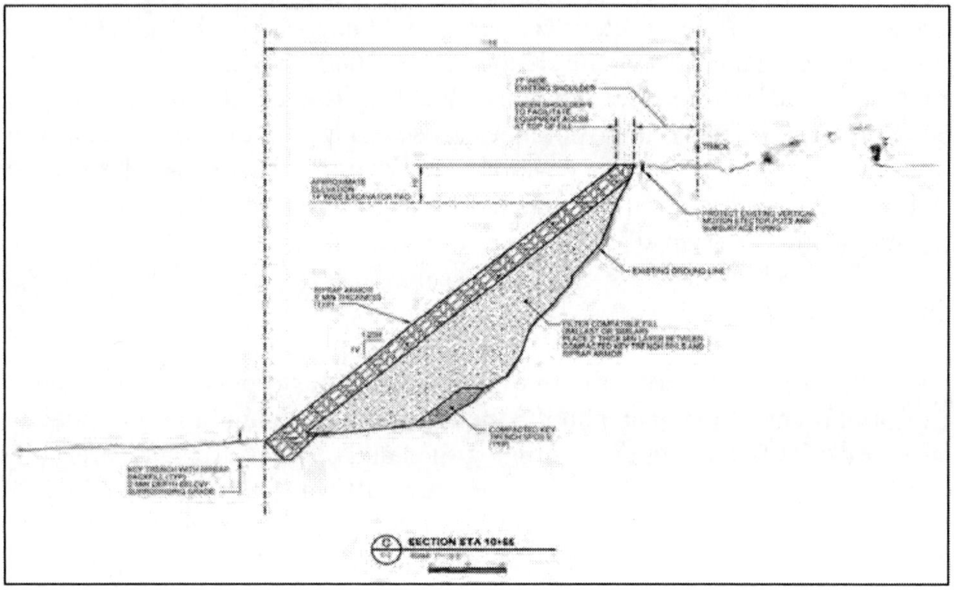

Figure 12 – Excerpt from Plan Set; note photogrammetrically derived topographic surface.

Site Access Considerations

Because there is no right-of-way access road along the line at MP 99.2, vehicles can only access the rail through the fence surrounding the cement plant property. Similarly, access to the base of the bluff is severely limited due to high tidal range and distant road access. The National Oceanic and Atmospheric Administration (2019) indicates that the daily tidal range in the area is on the order of 10 to 12 feet. While the subaerial tidal flat can extend nearly 1,000 feet from the bluffs at low tide, the tidal flat is inundated to the toe of the coastal bluffs at high tide. To access the base of the bluffs, vehicles or equipment would likely have to use the nearest road access at Squalicum Beach Park, approximately 3,700 feet to the southeast of the site. Access from the park would then require traversing environmentally sensitive tidal flats and passing beneath a wooden pedestrian pier and a process water discharge pipe from the cement plant, both of which extend approximately 2,000 feet into Bellingham Bay.

To accommodate staging and stockpiling access concerns, BNSF secured access to the cement plant property to establish a staging and stockpiling area to support construction. BNSF and Shannon & Wilson also determined that construction access across the tidal flat would not be feasible due to environmental consideration and limitations on equipment size capable of passing beneath the pier and process water pipeline. In their RFP, BNSF indicated that access to the base of the bluff would require a barge to deliver an excavator to the work site. BNSF also assumed that materials would be stockpiled in the cement plant and ferried to the revetment using a crane.

REVETMENT CONSTRUCTION

Mobilization

Revetment construction began on July 28, 2016 and was completed on August 29, 2016. The BNSF contractor determined that a relatively lightweight excavator (Deere 160C) could be placed on the beach using the same crane (Kobelco CK 2000 1F – see Figure 13). The crane would be maintained onsite to ferry construction materials to the bluff toe (Figure 14). This mobilization approach eliminated the risk and expense of barging the crane to the site. To avoid damage to the excavator by seawater, excavator mobilization occurred as tides receded. The contractor subsequently delivered riprap material onto the beach to allow the excavator to construct and move onto a "dry pad" above high tide.

Key Trench

The key trench was excavated during the first two days of construction. The trench exposed a thin veneer of littoral gravel less than about 6 inches thick overlying blue-gray glaciomarine deposits consisting of stiff clay with occasional gravel and cobbles. In some areas, the key trench was widened to 10 feet where soft and medium stiff clays were encountered. Key trench construction was timed to allow excavation during periods of low tide and to allow sufficient time for the contractor to backfill each segment of trench before demobilizing the excavator onto the "dry pad" at the end of shift.

The key trench was backfilled with a larger riprap gradation than required by the specification (12- to 60-inch-minus gradation) at the contractor's discretion. The contractor placed the largest riprap boulders at the base of the trench and smaller riprap individually placed to fill interstices. Consistent with the plan set, the contractor temporarily stockpiled key trench spoils at the base of the bluffs. Following substantial completion of the key trench, the contractor re-worked key trench spoils along the base of the bluff, creating a wedge that did not extend high enough along the bluff face to inhibit drainage along spring lines. At the suggestion of Shannon & Wilson, the contractor excavated three trench drains through the key trench spoils (backfilled with crushed aggregate gravel) to promote drainage.

Figure 13 – Mobilizing Excavator at Start of Construction

Revetment Construction

The contractor constructed the revetment in lifts, placing a pad of 8-inch minus fill compacted by the excavator followed by a facing of riprap. The contractor maintained grade and riprap thickness within design limits by mobilizing a Deere 160G excavator with GPS grade control. Shannon & Wilson observed progress to confirm materials and construction were completed in accordance with project plans and specifications. The contractor elected to use materials coarser than specification for the fill zone of the revetment. The fill materials consisted of 8-inch-minus quarry spall blended with recycled aggregate road base. Shannon & Wilson determined that these materials would meet filter compatibility requirements.

After approximately two weeks of construction, the revetment surface was at an elevation of about 55 feet above bluff toe, and the bench was becoming too narrow for the excavator to efficiently work. At this point, the contractor removed the excavator and used a heavier excavator with a longer reach to complete placement of the revetment from track level. By August 29, 2016, the revetment was substantially complete. Shannon & Wilson measured the final face at about 38 degrees, or slightly flatter than the 1.25H:1V design grade.

Figure 14 – Crane-Mounted Skid Bucket Ferrying Fill Materials to Bluff Toe.

CONCLUSIONS

The BNSF Bellingham Subdivision was initially constructed in 1888 along the coastal bluffs of Bellingham Bay. Due to wave erosion and mass wasting events, the bluff face retreated at an annualized average rate of about 6 to 9 inches per year. Following a mass wasting and bluff retreat event in 1998, the bluff crest retreated to within about 10 feet of the rail. In response, Shannon & Wilson completed geotechnical studies and installed an owner-serviceable, proprietary Vertical Movement Detector (VMD) system to integrate shoulder stability monitoring with BNSF signals. In February 2016, a series of storm events initiated a renewed sequence of mass wasting and bluff retreat that widened the erosion scar but did not trigger the still-functioning VMD monitoring system. In response, Shannon & Wilson designed a mitigation system consisting of an engineered riprap revetment. The revetment was selected following an alternatives analysis due to its favorable combination of economic feasibility, constructability, erosion resistance, and global stability improvement. Construction of the revetment was completed over an approximately one-month period. The revetment has performed favorably since completion of construction in August 2016.

Figure 15 – Partially Completed Revetment During Construction

Figure 16 – Completed Revetment Following Construction.

REFERENCES

Easterbrook, D.J., 2003, Cordilleran Ice Sheet glaciation of the Puget Lowland and Columbia Plateau and alpine glaciation of the North Cascade Range, Washington: in Easterbrook, D. J., ed., Quaternary Geology of the United States, INQUA 2003 Field Guide Volume, Desert Research Institute, Reno, NV, p. 265-286.

Easterbrook, D.J., 1976, Geologic map of western Whatcom County, Washington: U.S. Geological Survey, Miscellaneous Investigations Series Map I-854-B, scale 1:62,500.

Kelsey, H. M., B. L. Sherrod, R. J. Blakely, and R. A. Haugerud (2012), Holocene faulting in the Bellingham forearc basin: Upper-plate deformation at the northern end of the Cascadia subduction zone, J. Geophysical Res., 117, B03409, doi:10.1029/2011JB008816.

Kovanen, D.J. and Easterbrook, D.J., 2002, Timing and Extent of Allerød and Younger Dryas Age (ca. 12,500-10,000 ^{14}C yr. B.P.) Oscillations of the Cordilleran Ice Sheet in the Fraser Lowland, Western North America: Quaternary Research: 57, 208-224.

National Oceanic and Atmospheric Administration. *Tides & Currents, Annual Prediction Tide Tables for Bellingham Washington (9449211).* National Oceanic and Atmospheric Administration. https://tidesandcurrents.noaa.gov/noaatideannual.html?id=9449211. Accessed July 2, 2019.

Porter, S.C. and Swanson, T.W., 1998, Radiocarbon Age Constraints on Rates of Advance and Retreat of the Puget Lobe of the Cordilleran Ice Sheet during the Last Glaciation: Quaternary Research 50, 205-213.

Riedel, J., Clague, J., & Ward, B. (2010). Timing and extent of early marine oxygen isotope stage 2 alpine glaciation in Skagit Valley, Washington: Quaternary Research, 73(2), 313-323. doi:10.1016/j.yqres.2009.10.004

Troost, K.G., 2014, The Penultimate Glaciation and Mid- to Late-Pleistocene Stratigraphy in the Central Puget Lowland, Washington: Geological Society of America Abstracts with Programs, Vol. 46, No. 6, p.347.

Washington State Department of Natural Resources, Division of Geology and Earth Resources. *Washington Lidar Portal.* Washington DNR. http://lidarportal.dnr.wa.gov/#48.76585:-122.54001:14 Accessed on July 2, 2019.

Weather Underground. *Bellingham International Airport, Washington (Historical weather records for February 2016).* TWC Product and Technology. https://www.wunderground.com/history/monthly/us/wa/bellingham/KBLI/date/2016-2. Accessed July 2, 2019.

Weber, S., and Kovanen, D. J. (2000). Evidence for beach deposits in the Deming sand at Bellingham Bay, Washington and their implications for Late Pleistocene relative sea level. Geological Society of America Abstracts with Programs 32, 74.

ADVANCES IN ROCKFALL PROTECTION: A PRELIMINARY DESIGN TOOL FOR ATTENUATORS ESTIMATING ROCKFALL KINETIC ENERGY AS A FUNCTION OF ROCK MASS

Helene Hofmann, Corresponding Author
Geobrugg AG
Aachstrasse 11, CH-8590 Romanshorn, Switzerland
Email: helene.hofmann@geobrugg.com

Tim Shevlin, R.G.
Geobrugg North America LLC.
4676 Commercial Street SE, Salem, OR 97302
Tel:503-423-7258; Email: tim.shevlin@geobrugg.com

Prepared for the 70th Highway Geology Symposium, October, 2019

Acknowledgements

We would like to acknowledge the kind help of the WSL for lettings us borrow the high speed cameras and thank you to the RAMMS team at the SLF for lending us the rock motion sensors and the license for RAMMS::ROCKFALL and James Glover for all his help and knowledge during testing.

The authors confirm contribution to the paper as follows:

D. Wyllie, T. Shevlin, H. Hofmann, J.Glover – Study Conception and Design
T. Shevlin, D. Wyllie, H. Hofmann, J.Glover – Data Collection
T. Shevlin, D. Wyllie, H. Hofmann J.Glover – Analysis and Interpretation of Results
H. Hofmann, T. Shevlin – Draft Manuscript Preparation

All authors reviewed the results and approved the final version of the manuscript.

Disclaimer

Statements and views presented in this paper are strictly those of the author(s), and do not necessarily reflect positions held by their affiliations, the Highway Geology Symposium (HGS), or others acknowledged above. The mention of trade names for commercial products does not imply the approval or endorsement by HGS.

Copyright Notice

Copyright © 2019 Highway Geology Symposium (HGS)

All Rights Reserved. Printed in the United States of America. No part of this publication may be reproduced or copied in any form or by any means – graphic, electronic, or mechanical, including photocopying, taping, or information storage and retrieval systems – without prior written permission of the HGS. This excludes the original author(s).

ABSTRACT

Attenuators are a passive rockfall protection solution combining a flexible rockfall barrier Attenuators are a passive rockfall protection solution sometimes described as a flexible rockfall barrier with a prolonged draped tail. To date there is no formal solutions to their design. In comparison to classical flexible rockfall barriers where only translational kinetic energy is considered and the rockfall is stopped completely; attenuators present a design challenge wherein both the rotational and translational component of rockfall must be considered. In addition, Attenuator Systems are not stopping the rockfall, rather changing the trajectory and moderating the velocity. In order to develop a dimensioning concept that addresses these dynamics, it is important to fully understand the attenuation process. A joint research program between Wyllie & Norrish Rock Engineers, Ltd. in Canada, Geobrugg North America, LLC, and Geobrugg AG, Switzerland investigated this process. The loading of the system, the attenuation processes and the importance of the rotational component have been analyzed in a full-scale testing over a three-year period, on a site in British Columbia. This contribution provides insights into the analysis of the loading mechanisms acting on the attenuator system during rockfall impact. Rock motion dynamics are compared between those extracted from the accelerometer and gyroscope sensors embedded in the test blocks, high speed video analysis and rockfall simulations. These analyses provide the foundations with which a dimensioning tool has been created.

Keywords: rockfall, flexible protection system, attenuator, dimensioning, rotation

INTRODUCTION

Rockfall impact attenuators intercept rockfall trajectory, reduce potential bounce height, and dampen the rockfall velocity therefore attenuate the total kinetic energy of rockfall. A controlled guiding of the rock(s) to a designated collecting area is then possible avoiding costly clean-outs, as with standard flexible rockfall barriers. This type of low maintenance, passive rockfall mitigation system is increasing in popularity worldwide but no design guidelines exist. The loading mechanisms and the importance of the angular velocity and directional behaviour of the rocks upon impact have to date been largely unknown. In a joint research program between Wyllie & Norrish Rock Engineers, Ltd. in Canada, Geobrugg North America, LLC, and Geobrugg AG, Switzerland, considerable progress has been made in the understanding of the attenuation process and loading mechanisms of attenuator systems. Notably the importance of a rock's rotational component during impact is being analysed. Full-scale rockfall testing into attenuator systems was performed in 2015, 2016 and 2017 on a test site in British Columbia.

This contribution provides the results of the comparison of the acceleration and rotation components between the rock motion sensors, the video analysis and the RAMMS::ROCKFALL simulation and sets the basis for a dimensioning concept.

Rock Rolling

Full-scale one-to-one rock rolling testing has been central to the understanding of rockfall mechanics both for the development of rockfall protection systems as well as trajectory and impact models. The need to test and understand rockfall has a long history, some of the early efforts to control rockfall date back to the start of railway construction around 1834. Many rock rolling programs have since then been conducted. In the 1960s the US, Japan and Switzerland start with comprehensive rock rolling experiments (1; 2). Most recently in 2015, 2016, and 2017 rock rolling experiments were performed in Hope BC, Canada to test the capabilities of an attenuator system (3). For attenuators with their multi-dynamic interactions between rock and net along with the slope, the need for 1:1 rockfall tests is essential to understand the behaviour of rockfall trajectories and to calibrate simulation models. More importantly, the interaction between rockfall and a flexible protection structure has not been quantified yet and needs 1:1 rockfall tests for data collection in order to develop empirical relationships between the mesh and the rock.

Flexible Rockfall Protection and Attenuator Systems

Attenuators combine two long standing rockfall control methods, namely rockfall barriers and rockfall drapery. Flexible rockfall barrier systems are designed to intercept upslope rockfall and absorb the total energy of a rock impact until it has stopped. Whereas rockfall drapery is placed over an entire rock-mass to control rockfall that occurs within the drapery and direct them to a catchment area at the base of the slope (4; 5; 6; 7).

Attenuator systems therefore offer the interception function of rockfall barriers while, like rockfall drapery, further guide the rocks to a catchment ditch at the base of the slope instead of collecting them (Figure 1). Intercepting rockfall during freefall or after slope rebounds, the mesh system redirects the rockfall trajectory, reduces its bounce height, and can reduce velocity (8; 9). Both, the deformation of the netting at impact and the rock-ground contact during transport under the drape, dissipate a great quantity of energy (10). Attenuator systems are highly applicable to regions with a high rockfall frequency where it would be costly to often clean a standard rockfall barrier that retains rocks in its structure. Moreover, for situations where access for maintenance is difficult, attenuator systems offer a solution to rockfall control that delivers the maintenance needs to a more practical region at the base of a slope. Finally, attenuator systems offer the potential to enhance existing protection structures, such as a rockfall gallery for example, which does not meet the required height or energy level required to meet the actual rockfall hazard. The attenuator dissipates the kinetic energy of the rockfall to the design values of the other protection structure (8). Rockfall attenuators have mainly been applied since the 1990s in North America. Some testing was performed but no appropriate design guidelines exist for them (5; 11). To understand the ability of the attenuator to reduce bounce heights, kinetic energy of rockfall and its efficiency more one to one testing is needed.

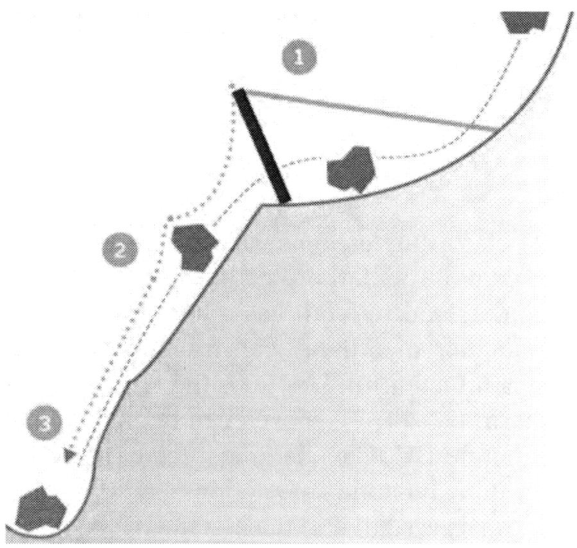

FIGURE 1 Simplified geometry of the attenuator tested in Canada (Geobrugg AG), showing the evolution of a block passing in an attenuator in three steps.

TEST SITE

The Nicolum Quarry in Hope, British Columbia, was chosen as a test site in February 2013, partly based on previous tests carried by the quarry owner, the British Columbia Ministry of Transportation and Infrastructure (MoTI) in the 1990's (12). The initial "proof of concept" full-scale attenuator testing series, performed in 2014 and 2015 by Wyllie and Geobrugg, confirmed the suitability of the Nicolum test site and the instrumentation systems utilized at that time (12). Two large test series were then conducted in January 2016 and subsequently in September 2017.

The slope is 60m high and near vertical with three inclined benches where a thin layer of soil covers the massive bedrock slope. Below the first gully, some rock debris has accumulated. The ground at the bottom of the slope is covered with a layer of soil as well. The rocks are released at the top of the slope with an excavator, approximately 16.5 ft (5 meters) above the ground. After testing in 2015, some trim blasting was undertaken to improve the hit rate on the attenuator system (12). After 2016 testing, the whole system was extended to a greater width to increase the hit rate on the mesh even more. Rockfall modelling contributed to this decision, which was confirmed successfully, during the testing in 2017. Here we present some selected results of the latest testing series.

Natural granitic blocks approximately up to 1.5 ft (0.45 m) in diameter and cubic reinforced concrete blocks 1.8, 2.5, and 3.28 ft (0.55, 0.75 and 1m) in diameter, with a housing for instrumentation, were used for testing. The concrete blocks' corners were painted black and their faces white, to enhance visibility in the videos.

Load cells were installed in all support ropes with two DAS systems (QuantumX MX840-B with eight channels and a HBM Spider system) on either side of the test site to accommodate 10 load cells (Figure 2).

The testing was recorded with two high speed cameras, kindly lent by the Swiss Federal Institute of Forest, Snow and Landscape (WSL) and several other cameras to cover most angles of view (front, side, top; see Figure 2).

FIGURE 2 Whole instrumentation setup on site. The view looks from the top down onto the slope and the attenuator location.

The main velocity analysis and observations of the rock-net interaction are performed using the data from the high-speed camera with a frame rate of 500 fps. The front view camera is used to document the impact location and the depth of field of the rock, allowing to calculate a correction factor for the side view video analysis.

Four rock motion sensors were used. One was from DTS, a micro slice accelerometer and gyroscope modular unit measuring tri-axial accelerations and rotations at 20 kHz. The sensor is placed into a custom housing and inserted into the test rocks centre of mass (3). The three other sensors were kindly provided by the SLF and were recording at 2 kHz.

ROCKFALL MODELLING

In order to gain insights into the rockfall behaviour at the test site, rockfall modelling was conducted using RAMMS::ROCKFALL (13). Importantly it permits a simulation of rotational behaviour and contact impact forces of rocks during runout (14). Calibration of the rockfall model for the test site was completed in a separate study (15). The rigid body rockfall code considers natural shape of rock blocks and has an extensive rock library to choose from (Figure 3). The simulations assisted in designing the test facility to optimise the placement of the attenuator system in the rock slope and provided valuable data of expected impact velocity distributions, along with angular speeds as governed by different rock shapes. Rockfall modeling was applied both to compare against the measured data and to investigate test site optimization for the most recent testing series in 2017.

Model Inputs

Input parameters were defined as the following:

- Rock shapes: An equant rock shape was chosen to represent best the cuboid form of the test blocks. A density of 145 pcf (2300 kg/m^3) was selected as this was representative of the onsite lithology along with the density of the reinforced prefabricated concrete blocks used for testing. For the concrete blocks a volume of 14.4 ft^3 (0.407 m^3) and dimensions x/y/z = 3.28 ft, 3.28 ft, and 2.75 ft (1.02 m, 0.98 m, 0.84 m) yielded a mass of 2,060 lbs (937 kg), which reflect a representative mass of the test bodies used (Figure 3).

- Topography was obtained with photogrammetric methods applying structure from motion (SFM) algorithms to obtain digital terrain model (DTM) of the test site. The soil types are defined in three categories depending on slope angle (approximately 0 to 15°; 15° to 40° and 40° to 90°) and are characterized by extra hard, hard and medium hard according to the user manual (Figure 4).

- Protection barrier: of interest for the analysis was to sample the dynamics of rockfalls at the location of the proposed barrier. In order to sample the rockfall dynamics at this location, an artificial wall was created in the DEM with GIS software which acted as a barrier upon which the data could be sampled along a profile line. With the sampling line the rockfall trajectories could be analyzed for the proportion of entering the region of the attenuator and those that potentially missed the structure, along with their dynamics (velocity, angular speed, impact force) at the point of contact.

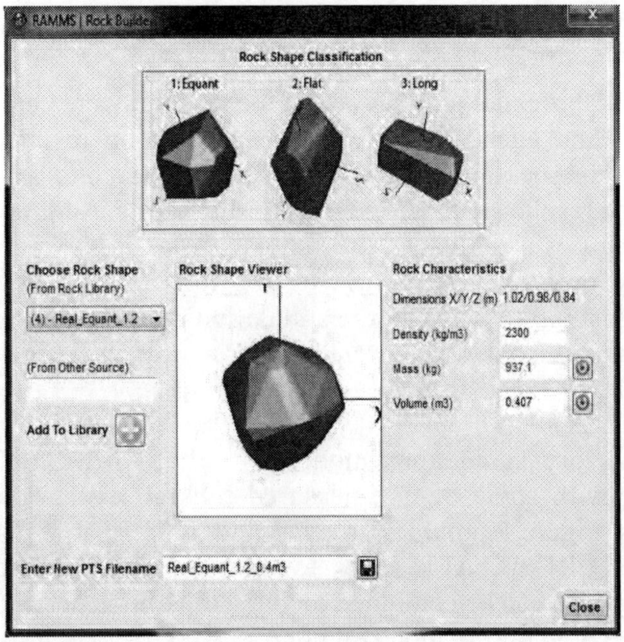

FIGURE 3 Rock shape library in RAMMS::ROCKFALL. Equant normal was chosen as most rocks and the concrete testing cubes resemble closest this shape. Mass and density are set in this example to fit the concrete test block of test T030, used throughout this contribution for comparison purposes.

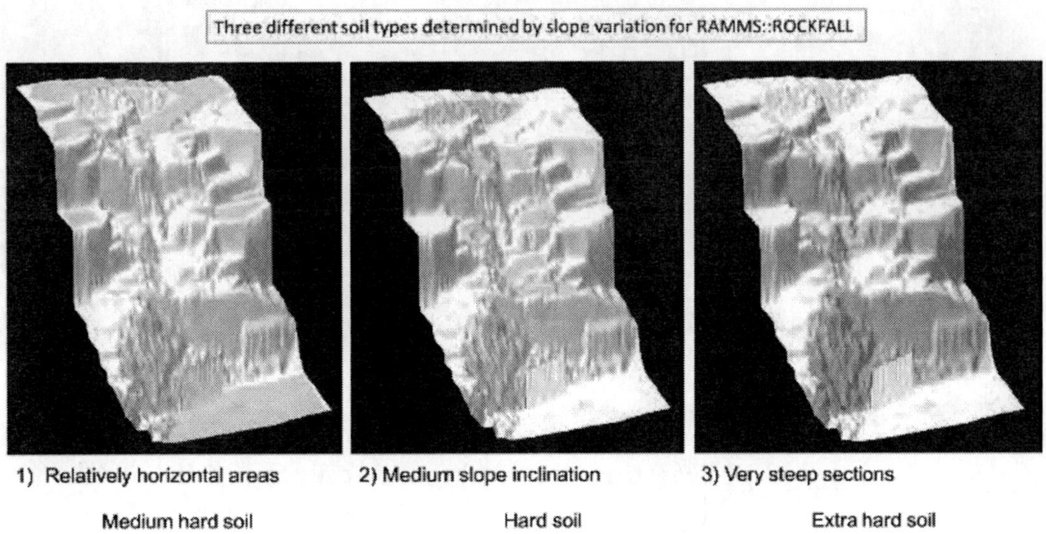

FIGURE 4 DEM and polygons chosen according to slope angle to define soil types. The steeper the angle the less soil cover does the granite slope have and the above-mentioned soil types in RAMMS::ROCKFALL correspond best to the field description.

Model Results

A total of 1000 rockfall simulations of the test site were conducted and examined for the rockfall hit rate into the attenuator barrier. The simulation results showed that 53.5% of the trajectories impacted the rockfall attenuator system. Compared to the experiments were a 57% hit rate was recorded, the results are close to the 2016 field tests. Additionally, the spatial distribution of the rockfall trajectories is similar in the simulations as in the recorded field tests. It is shown that 8.9% missed to the West and 19.2% to the East of the barrier. Notably the east misses demonstrate the closest parity between the simulations and field tests. On the other hand, 20.6% stayed on the slope or passed over the protection structure, which is almost double the percentage of the field test results.

Of the n=1000 trajectory simulations modelled with RAMMS::ROCKFALL, the trajectories showed congruence with some of the measured rockfall events during the field testing were selected for analysis. Figure 5 provides an overview of the spatial distribution simulated rockfall trajectories. RAMMS::ROCKFALL saves every calculated trajectory of one run and single trajectories of choice can then be combined on one DEM. In general, the trajectory distribution on the slope matches the field observation (Figure 6).

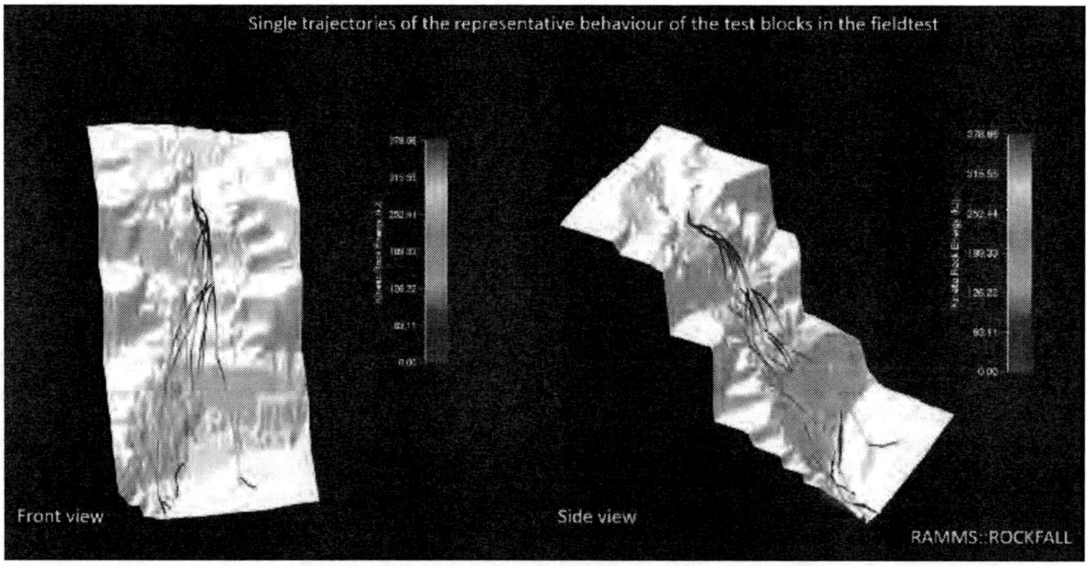

FIGURE 5 Single trajectories modelled, which resemble closely some eccentric behaviour observed while testing, confirming the accuracy of the model.

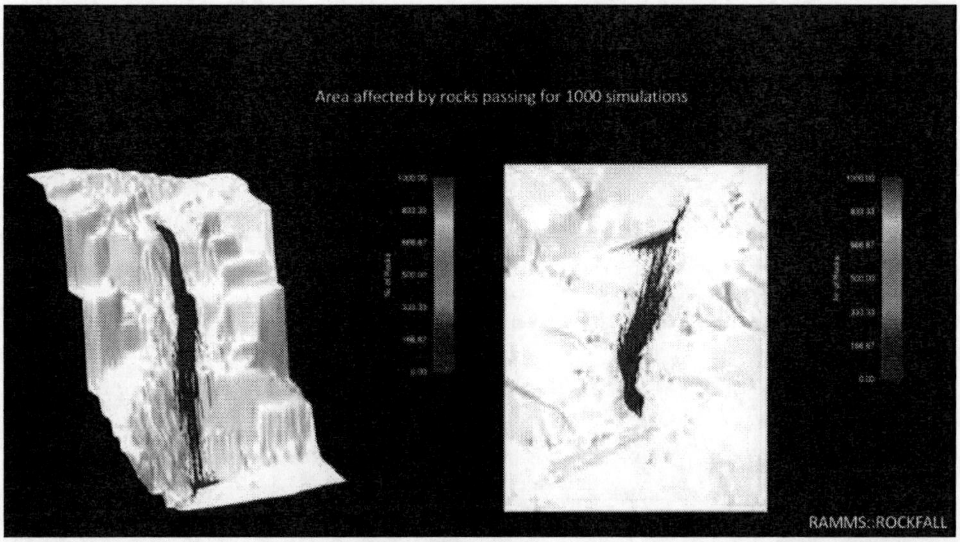

FIGURE 6 Area affected by passing rock, modelled with RAMMS::ROCKFALL for 1000 blocks.

In RAMMS::ROCKFALL velocity and total kinetic energy are always given for the whole trajectory when looking at 1000 trajectories at once. When interested in translational and rotational kinetic energy or the x, y, z, components of the angular velocity, up to hundred trajectories at one time can be studied. Many trajectories stop just short of the improvised barrier, therefore the summary statistics obtained are based on only a little number of impacts and account for a certain amount of error.

The translational velocities in RAMMS::ROCKFALL range between 0 and 98.5 ft/sec (0 and 30 m/s). This range is visible in Figure 7. The average maximum velocity is of 82 ft/sec (25 m/sec), getting close to the maximum values of the field test, but the frequency distribution ranges with most values placed around 59 to 62 ft/sec (18 to 19 m/sec), show slowerbulk velocities than from the video analysis.

The angular velocity in RAMMS::ROCKFALL for 1000 simulated trajectories range between 0 to 49.8 rad/sec (0 to 6 rev/sec); (Figure 8). The frequency distribution gives the main values ranging around 20 to 24 rad/sec (3 to 4 rev/sec).

It is notable is that the translational kinetic energies ranged between 0 and 485 kJ, the average maximum translational kinetic energies for 1000 simulated trajectories is 463 kJ with a standard deviation of 13.7. While the rotational kinetic energy ranges between 0 and 111 kJ with an average of 93 kJ and a standard deviation of 10.1. The rotational kinetic energy makes up to 20% of the total kinetic energy.

Total kinetic energy (Figure 9) being a function of translational kinetic energy (depending on velocity) and rotational kinetic energy (depending on angular velocity), its distribution is slightly underestimated with most values around 230 kJ whereas 300 kJ would be more realistic when compared to the field test.

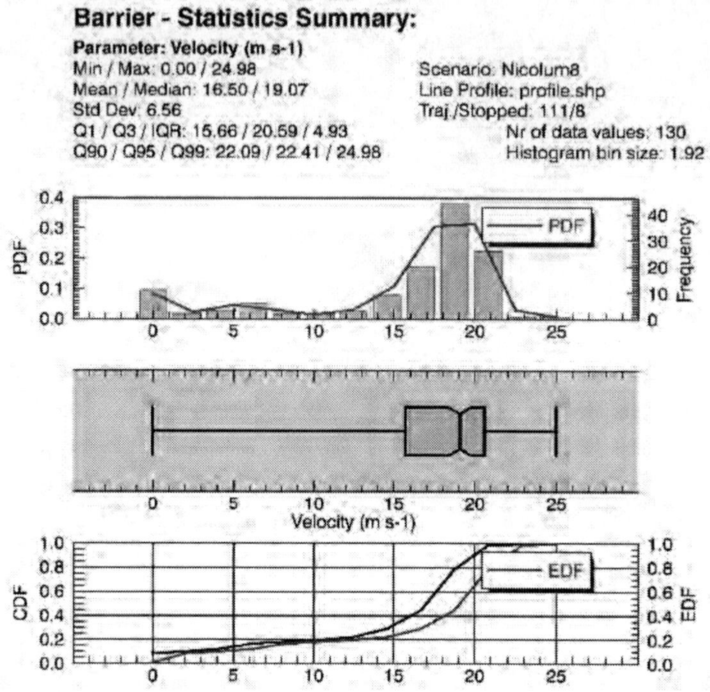

FIGURE 7 statistical distribution of the translational velocities modelled for 119 blocks with associated probability density function (PDF), cumulative distribution function (CDF) and empirical distribution function (EDF).

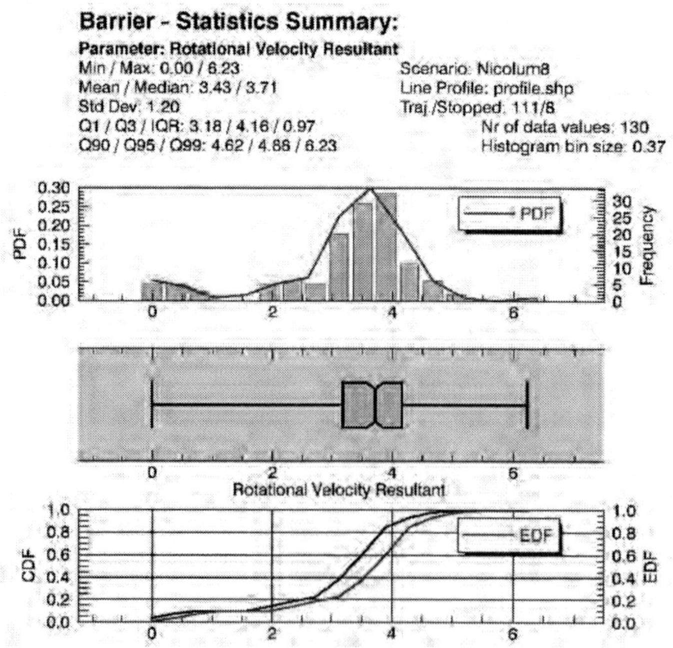

FIGURE 8 statistical distribution of the angular velocities computed for 119 blocks.

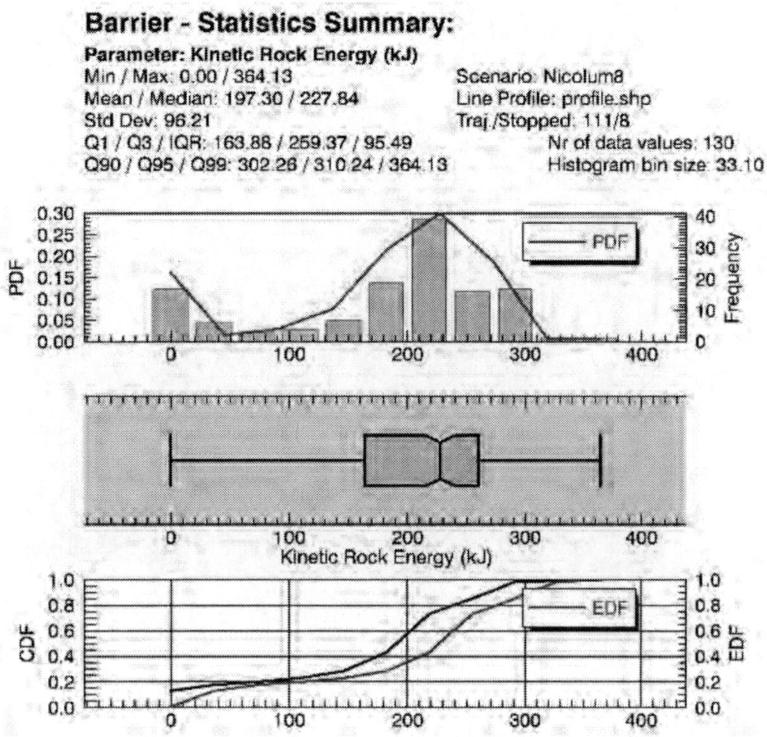

FIGURE 9 Calculated kinetic energies by RAMMS::ROCKFALL for 119 trajectories.

FULL-SCALE TESTING

Methodology

The impact velocity and angular velocity are measured from the videos, with the help of a video analysis software. In this case, Kinovea (16) was used as it is an open source software and relatively easy to handle as a beginner. Originally it is a sport motion analysis software but can be used for rock rolling experiments. Once a certain distance was calibrated (here the post of 26.25 ft (8m) length, the rock can be tracked automatically and manually, depending on lighting conditions, from first appearing in the frame all the way down to the ground through impact with the net (Figure 10). The velocity is then computed from the x and y points obtained from tracking and corrected for depth as described as in (17).

The rock motion sensor data is downloaded from the sensor using the proprietary software and processed to remove signal noise

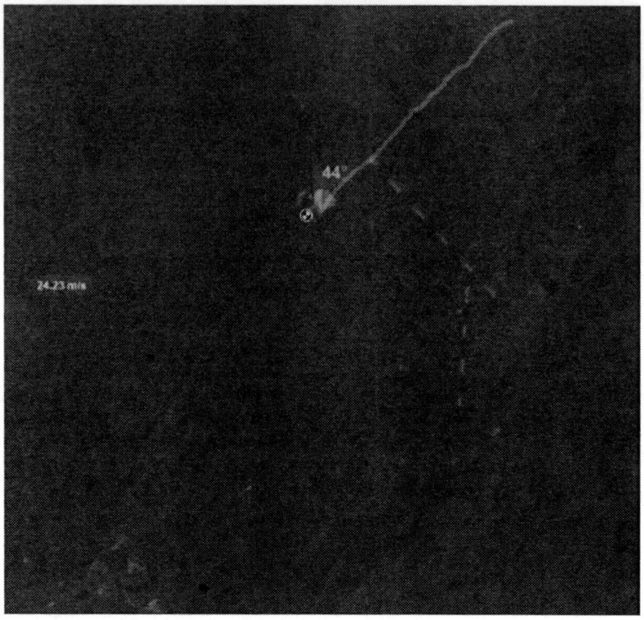

FIGURE 10 Tracking of a block throughout its fall with the software Kinovea.

Results

The velocity evolution throughout the fall is represented in Figure 11. Velocity at impact is of 88.5 ft/sec (27 m/sec) and decreases towards 19.7 ft/sec (6 m/sec) just above ground. This illustrates the attenuation process, it is observed how the block does not come to a full stop, but only attenuates its dynamics as the rock passes through the system.

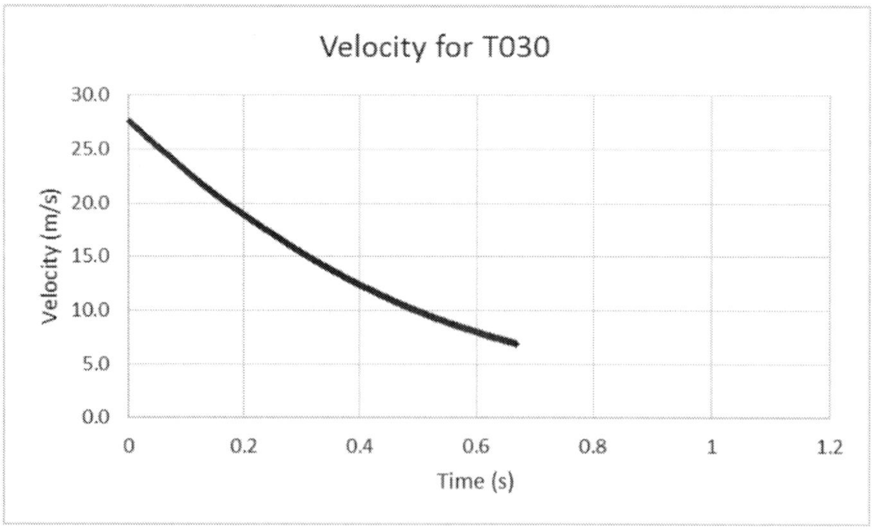

FIGURE 11 Velocity in m/sec for block T030 from impact with mesh until shortly above ground.

It is possible to compare the theoretical freefall of block T030 with its actual trajectory, illustrating the attenuation process in the perspective of distance travelled. Figure 12 shows how the trajectory of the block is intercepted, and its anticipated height is considerably dampened when impacting the attenuator instead of freefalling without any protection structure.

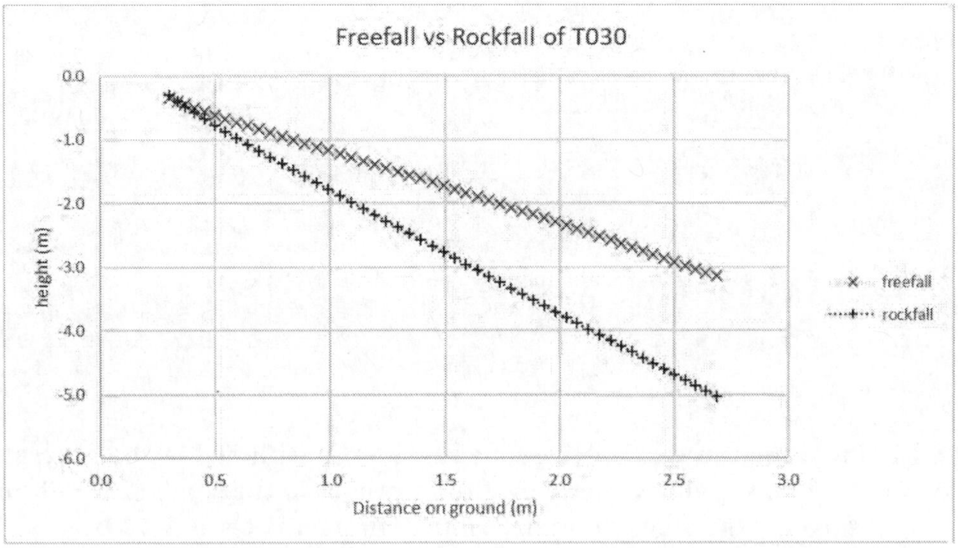

FIGURE 12 Comparison between theoretical freefall behaviour of block T030 versus the actual trajectory with impact of the protection structure from time of impact onwards.

The video analysis was also applied to measure the block's angular velocity. This was achieved by tracking given face of the block and marking every 90° rotation in the software. The time stamp of these frames then allows the computation of the angular speed in rad/sec. Figure 13 illustrates the evolution of T030 and T062 from impact with mesh onwards. The angular velocity extracted from the video analysis could then be compared with the measurements made with the gyroscope measurement of the rock motion sensor (Figure 14).

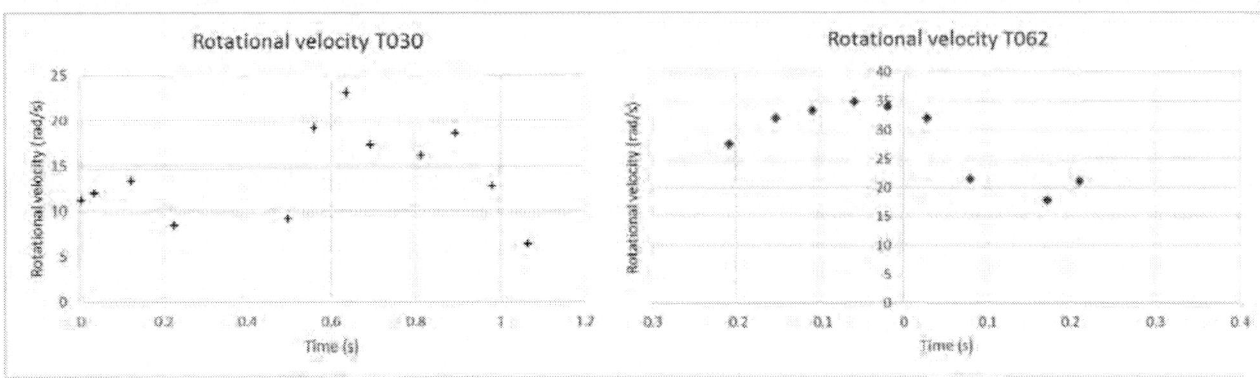

FIGURE 13 Angular speed (rad/sec) of block T030 (left) and T062 (right). Time of impact at T= 0s.

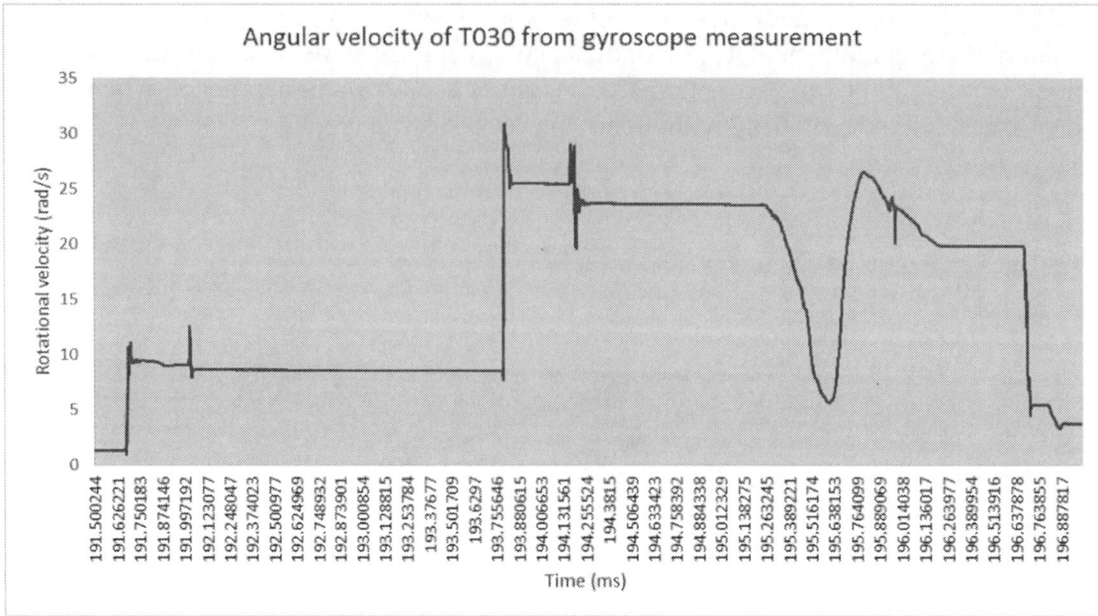

FIGURE 14 Plot of the three-axis gyroscope resultant positioned in the concrete blocks. Example of test T030.Angular velocity evolution ranges in the same order of magnitude then for the video analysis. Time of impact is t = 195.2 Ms.

The gyroscope measurement shows the same evolution of angular velocity to the video analysis results yield. Block T030 comes into the mesh with an initial rotation of approx. 10 rad/sec and 35 rad/sec, decreases and then increases again up towards 25 rad/sec before decreasing in steps (slightly visible as well on Figure 13) towards 0 eventually when reaching the ground. The T030 example shows well the problem with the sampling steps. Although the analysis of the video catches the trending rotation in freefall of 10 rad/sec, it misses the short acceleration before impact where the gyroscope indicates a velocity of 33 rad/sec (Figure 14). The left part of figure 13 and figure 15 show a similar evolution for test T062, therefore the angular velocities through video analysis seem to be coherent and the video analysis method can be applied when no rock motion sensor data is present.

Discussion

First analysis of the load cell data captures well the classic behavior of attenuator systems (Figure 16). The first peak corresponds to the initial impact of the block with the mesh and the second peak corresponds to the peak torque generated as the friction between the attenuator netting and the rock causes a reversal of the rocks rotational direction while rolling along the mesh (time steps match between the load cell graph and the video analysis where the reversal in rotational direction is observed).

Both the translational impulse and rotational impulse of the rock interacting with the attenuator system form the principal load cases attenuators should be designed for and are

the basis of the proposed attenuator design concept conserving momentum (18). These two load cases seem to correspond to the boundary conditions of the mesh.

1. Maximum translational impulse presents a puncturing risk through the mesh, and
2. A high rotational component of rockfall can shear the mesh open while being contained behind it.

Further, the load cell readings have confirmed these load cases which is forming the basis of the design principles for attenuators.

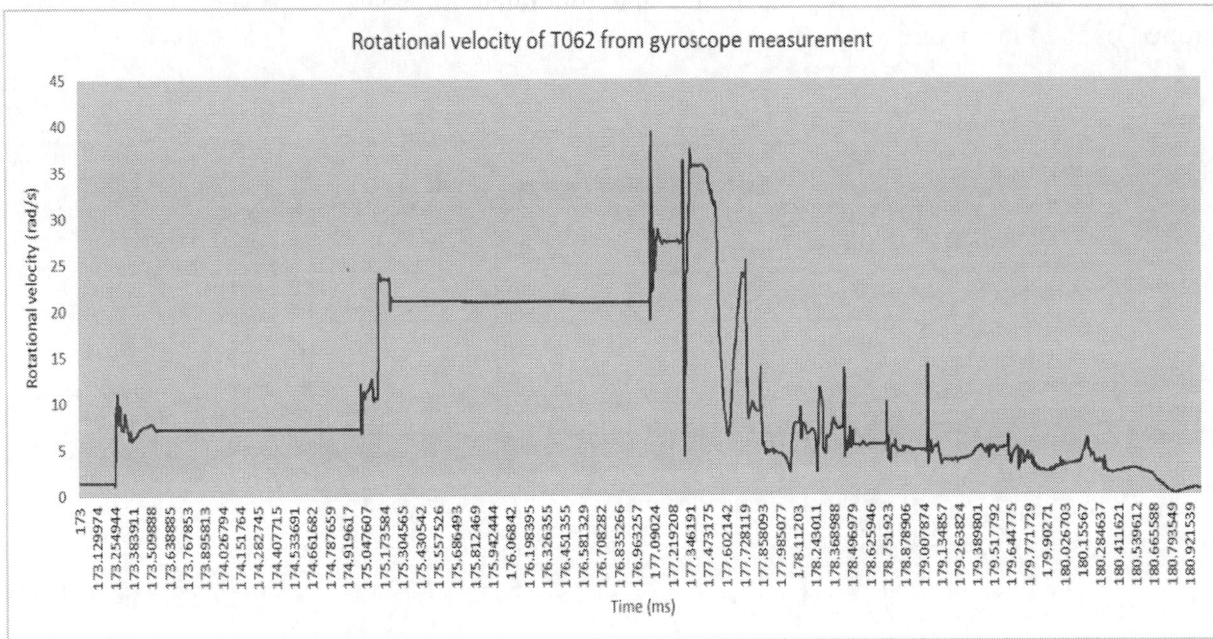

FIGURE 15 Plot of the three-axis gyroscope resultant, positioned in the concrete blocks. Example of test T062. Angular velocity evolution ranges in the same order of magnitude then for the video analysis. Time of impact at t = 177.4 ms

CONCLUSION

Attenuators are an interesting addition to flexible protection structures for rockfall hazard and a formal design procedure is important. To conclude, the rockfall simulations indicate similar results to the measured values and permit further insights into the full range of rockfall dynamics to be expected at the test site. Moreover, the simulation results assisted in developing test site design changes for the September 2017 tests in which the width of the attenuator system was increased. The widening of the attenuator test barrier successfully increased impact rate of the 2017 testing series. Rockfall dynamics are situated in a realistic range. The video analysis is associated with some error but in the case of angular velocity it is possible to compare the values with rock motion sensors. Although the resolution is not the same between the time steps of the video analysis and the 20kHz sampling rate of the rock motion sensor, it is possible to use the values obtained from video analysis for tests without a rock motion sensor recording, as the comparison between both is satisfactory.

Overall the combination of rockfall modelling and 1:1 real scale testing allows to understand rockfall dynamics better, as well as the general attenuation process. The proof of concept and advantages of attenuators have long been known, now the concept for a design approach is being assembled with these latest results, in order to build standardized systems. The necessity of high strength nets to cope with shearing load because of rotation is evident from the results obtained in this study. The dimensioning tool will be presented at the conference with a detailed explanation of the two-step verification of translation and rotational impulses for several types of rock properties found in nature.

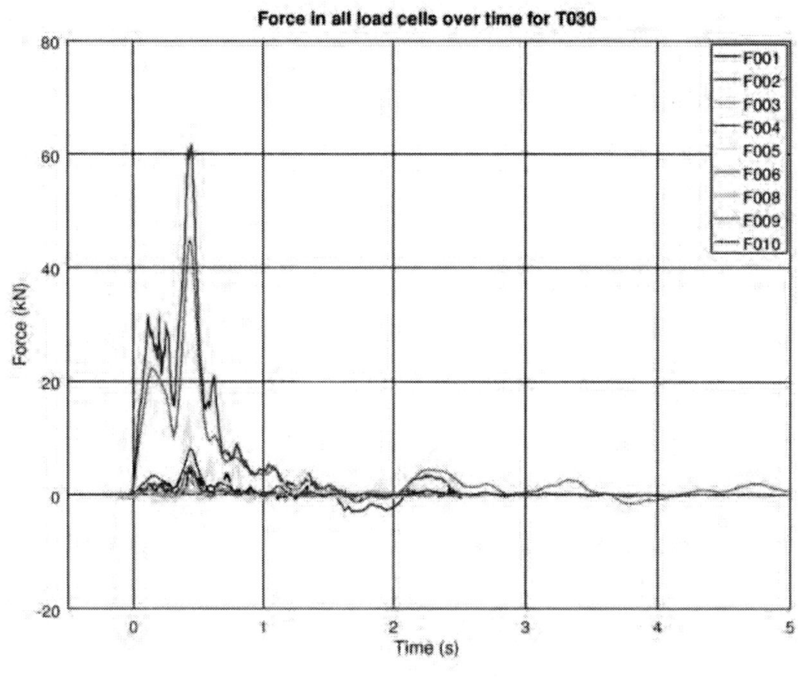

FIGURE 16 Two peaks recorded with load cells (F002 and F009 correspond to the top rope), t=0 corresponds to the time of impact.

REFERENCES

1. Duffy, J. and Glover, J. *A brief history of rockfall barrier testing.* 6th Interdisciplinary Workshop on Rockfall Protection, Barcelona, Spain, 2017.
2. Balkema, A. T., Westers, G., & Duffy, J. D. *Summary of World Wide Rockfall Tests.*
3. Glover, J., and Ammann, W. *Geobrugg internal testing report No. 1-a, Rockfall Attenuator Testing Nicolum Quarry, Hope BC.*, Global Risk Forum Davos, GRF, Switzerland, 2016.
4. Badger T.C. and Duffy J.D. *Drapery Systems In Rockfall: Characterization and Control*. Transportation Research Board, National Academy Press, Washington, D.C., 2012. pp 554-576.
5. Muhuthan, B., Shu S., Sasiharan N., Hattamleh O.A., Badger T.C., Lowell S.L., Duffy J.D. *Analysis and Design or Wire Mesh/Cable Net Slope Protection*, Washington State Transportation Center, Seattle, Washington, WA-RD 612.1, p. 267, 2005.
6. Wyllie, D.C. and Norrish, N.I. Stabilization of rock slopes. In Landslides: Investigation and mitigation Transportation Research Board, National Academy Press, Washington, D.C. pp 474-504.TRB Special Report 247, 1996.
7. Andrew, R.D., Barttingale R., and Hume H. Context Sensitive Rock Slope Design Solutions, Publication FHWA-CFL/TD-11-002. Federal Highway Administration, Central Federal Lands Highway Division, Lakewood, CO, 2011.
8. Glover, J., Denk, M., Bourrier, F., Volkwein, A., Gerber, W. Measuring the kinetic energy dissipation effects of rockfall attenuating systems with video analysis. *Conference Proceedings, Interpraevent*, 2012.
9. Glover, J.; Volkwein, A.; Dufour, F.; Denk, M.; Roth, A. Rockfall attenuator and hybrid drape systems - design and testing considerations. In: Third Euro-Mediterranean Symposium on Advances in Geomaterials and Structures. Third Edition, Djerba, 2010. pp. 379-384.
10. Badger T.C., Duffy J.D., Sassudelli F., Ingraham P.C., Perreault, Muhunthan B., Radhakrishnan H., Bursi O.S., Molinari M., and Castelli E. Hybrid Barrier Systems for Rockfall Protection, in A. Volkwein et al. (eds) Proceedings from the Interdisciplinary Workshop on Rockfall Protection, Morschach, Switzerland, 2008. pp. 10-12.
11. Arndt, B., Ortiz, T., Turner, K. Colorado's Full Scale Field Testing of Rockfall Attenuator Systems, Transportation Research Circular E-C141, Transportation Research Board, October 2009.
12. Wyllie et al. Attenuators for controlling rockfall: first results of a state-of-the-art full-scale testing program. Geovancouver 2016.
13. SLF, RAMMS ::ROCKFALL, retrieved July 2017 from: http://ramms.slf.ch/ramms/index.php?option=com_content&view=article&id=66&Itemid=93
14. Leine, R. I., Schweizer, A., Christen, M., Glover, J., Bartelt, P., & Gerber, W. Simulation of rockfall trajectories with consideration of rock shape. Multibody System Dynamics, 32(2), 241- 271, 2014.
15. Hofmann, H. Rockfall trajectories and dynamics analysis from 1:1 real rockfall

tests. MSc Thesis, University of Portsmouth, unpublished, 2017.
16. Kinovea, retrieved July 2017 from: http://www.kinovea.org
17. Glover, J. Rock-shape and its role in rockfall dynamics. , p.266, 2015.
18. Wyllie, D., Shevlin, T., Glover, J., Wendeler, C. Development of design method for rockfall attenuators. Proceedings of the 68th Highway Geology Symposium, 2017.

TDOT Counters Landslides Triggered by Floods of February, 2019

Robert Jowers, PE
Tennessee Department of Transportation
Materials & Tests Division – Geotechnical Engineering Section
6601 Centennial Blvd.
Nashville, TN 37243
(615)350-4133
robert.jowers@tn.gov

Prepared for the 70th Highway Geology Symposium, 2019

Acknowledgements

Bill Trolinger, Tennessee Department of Transportation (retired) – for hiring me, and inspiring my hunger for the geotechnical discipline of civil engineering.

Lori M. Fiorintino and Travis W. Smith, Tennessee Department of Transportation – For putting in the hours to tackle this awful flood.

Tennessee Valley Authority – for saving the State of Tennessee millions of dollars in flood damage year after year. And providing me power.

Disclaimer

Statements and views presented in this paper are strictly those of the author(s), and do not necessarily reflect positions held by their affiliations, the Highway Geology Symposium (HGS), or others acknowledged bove. The mention of trade names for commercial products doesnot imply the approval or endorsement by HGS.

Copyright Notice

Copyright © 2019 Highway Geology Symposium (HGS)

All Rights Reserved. Printed in the United States of America. No part of this publication may be reproduced or copied in any form or by any means – graphic, electronic, or mechanical, including photocopying, taping, or information storage and retrieval systems – without prior written permission of the HGS. This excludes the original author(s).

ABSTRACT

Over ninety-five landslides will be mitigated in middle and east Tennessee that were triggered by a 2019 February flooding event. The event created and continues to create tremendous strain on the Tennessee Department of Transportation's (TDOT's) customers, staff, and construction equipment resources; all this while programmed transportation improvements must continue to be rolled out with every scheduled letting. Seeking inspiration, the author reviewed proceedings developed for previous highway geology \ transportation geotechnical engineering conferences, including papers prepared by Mr. David L. Royster for guidance.

This paper and presentation will discuss a landslide categorization system, the type of landslide most common to Tennessee, an overview of landslide stabilization methods, and a discussion of landslide case studies that occurred in Tennessee due to this intense flooding. Lastly, proposed programmatic measures presently underway to better manage the geotechnical slope assets of TDOT will be discussed.

INTRODUCTION

The intense rains during the winter and spring months of 2019 were some of the heaviest historically recorded. This paper describes the landslides, mostly debris slides, but a few rock falls as well, that occurred at sites across the state in different geologic settings due to the flooded conditions. It also describes the corrective measures TDOT is taking to overcome these slope movements. This paper also discusses proposed programmatic measures that are being conceived to minimize future landslide risk. Our professional staff spent long hours preparing emergency relief contracts. TDOT Operations Maintenance forces mobilized continuously during this time, as well. Such disasters are never desired, but there are lessons being learned. When disasters strike, the best in people is brought out, but out of necessity.

GEOLOGIC SETTING

Tennessee is often referred to as having three "grand divisions", West, Middle, and East Tennessee (Hargett, 2018). Topographically however, Tennessee is divided into eight physiographic boundary regions (Miller R., 1974). Although precipitation records were set in West Tennessee in February 2019, the most severe slope movements occurred in Middle and East Tennessee, where the physiographic boundary regions Blue Ridge (Unaka), Valley and Ridge, Cumberland Plateau, Highland Rim, and Central Basin are located.

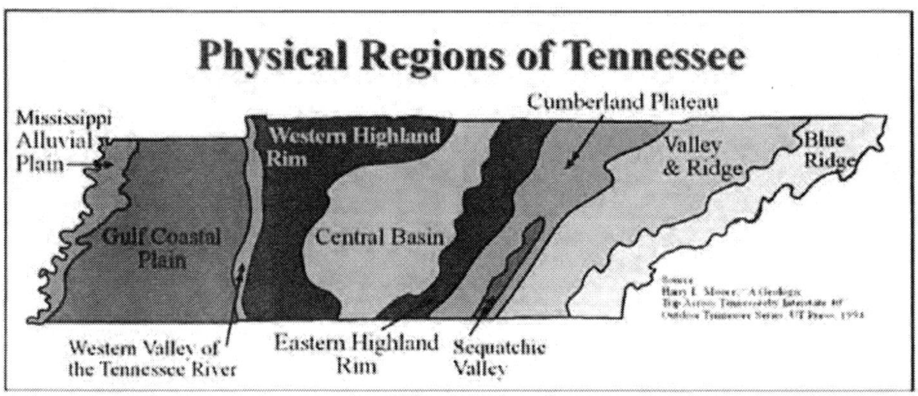

Figure 1: Geophysical Boundaries of Tennessee (Source, Harry L. Moore, "A Geologic Trip Across Tennessee Interstate 40", *University of Tennessee Press*, 1994.)

In size and scale, the most economically devastating landslides occurred at the physiographic boundary contacts between the Cumberland Plateau and the Valley and Ridge, and also the Cumberland Plateau and the Highland Rim. These physiographic boundary contacts often expose a long documented roadbuilding nemesis, the Pennington Shale or Pennington Formation (Royster, 1973, 1974, 1978).

In terms of sudden devastation, a recent debris slide on SR-70 in Hawkins Co. was particularly destructive. It is discussed later as one of the case studies. The escarpment of the debris slide was located in a roadway within the Valley and Ridge physiographic boundary, nearby to a switch back roadway curve. The saturated debris released violently, and was contained only by the topography of the same roadway section located downhill of the switchback.

FEBRUARY FLOODING

Tennessee has three separate watersheds: the Cumberland River, Tennessee River, and Mississippi River. Following the Great Depression, federal legislation funded works programs for a system of dams to be constructed by the U.S. Corps of Engineers (USCOE) and the Tennessee Valley Authority (TVA). This was indeed fortunate for flood relief in the cities of Chattanooga and Nashville. Prior to dam control, during the flooding of December 1926, the Cumberland River crested in Nashville at 56.2 feet, and the city flooding was catastrophic. Following dam regulation of water levels, the February 2019 flooding the Cumberland River crested in Nashville at only 40.9 feet, and the city experience no real flooding (National Weather Service, 2019).

Rainfall and weather patterns in the United States is monitored by the National Weather Service (NWS). Most counties in the state of Tennessee experienced the wettest February since precipitation records were kept. Since 1940, the weather station at the Nashville airport (BNA) in Davidson Co., for example, has a mean monthly February precipitation normal of 4.25 inches. The February 2019 total rainfall amount at the Nashville airport tripled the normal, arriving at the new record of 13.47 inches (National Weather Service, 2019). The NWS Nashville office summarized the weather event in the narrative captured below.

> A stationary frontal boundary stalled over or near the Tennessee Valley for nearly a week in mid to late February 2019. Persistent southwest flow aloft brought copious amounts of Gulf of Mexico moisture northward and interacted with this boundary for many days, causing a prolonged period of heavy rain and flooding throughout Middle Tennessee from Tuesday, February 19 through early Sunday, February 24. Due to the heavy rainfall that had already fallen earlier in the month along with the already unusually wet winter season so far, widespread flash flooding and river flooding resulted, with dozens of water rescues being conducted and numerous homes and businesses flooded. In addition, this heavy rainfall set new monthly rainfall records for the month February at many locations including Nashville and Crossville, both of which saw over a foot of rain. By the end of the month, nearly the entire state of Tennessee had received between 10" and 20" of rain in February 2019 (National Weather Service, 2019).

The Tennessee Valley Authority (TVA) manages dams along the Tennessee River. Figure 2 depicts the TVA facility where an engineering team huddles in a room to receive incoming hydrologic information, make tough data-driven decisions, and regulate the water levels of the Tennessee River. And there is also TVA coordination required with the USCOE engineers that regulate the Ohio River water stages. (National Weather Service, 2019 (4)).

Figure 2: TVA Headquarters where Water Levels are Managed (source Robert Jowers)

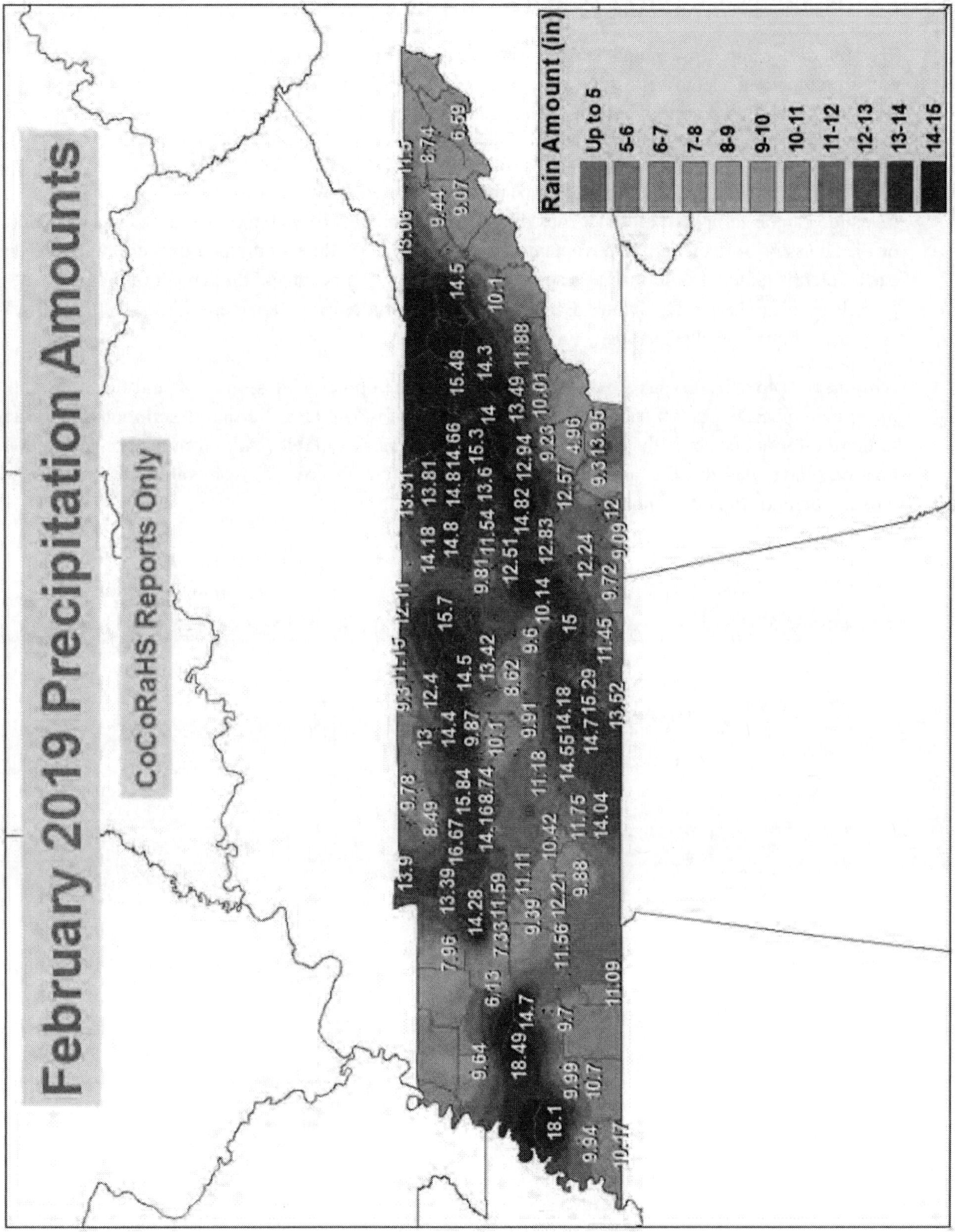

Figure 3: February 2019 Precipitation Amounts (Source NWS website)

LANDSLIDE TYPES AND CLASSIFICATIONS

When in discussion, Cruden and Varnes stress that the term, "landslide," should be categorized in a consistent method according to standard features and attributes. They stress consistently describing landslides with two descriptive nouns. The first noun describes the material type, while the second describes the type of movement (1996).

Material: In the Cruden and Varnes (1996) method of categorization, the material type name in a landslide can be composed of either rock or soil. The rock type of material is described as an intact hard or firm mass prior to the initial movement. The soil type of material is divided further into debris and earth. Debris material is defined as a soil where between 20 percent and 80 percent of the landslide mass particle sizes are larger than 2 mm. Earth describes material where 80 percent or more of the particle sizes are 2 mm or smaller (Cruden, Varnes, 1996).

Movement: Movement types are divided into either falls, topples, slides, spreads, and flows. And then, movement types are further subcategorized into modes of movement. Further descriptions can be added as more information about the landslide movement becomes available. And even further descriptions can be used such to describe the landslide's activity state (active, inactive, or reactivated), the rate of movement, and water content (Cruden, Varnes, 1996).

<table>
<tr><th colspan="2" rowspan="3">TYPE OF MOVEMENT</th><th colspan="3">TYPE OF MATERIAL</th></tr>
<tr><th rowspan="2">BEDROCK</th><th colspan="2">ENGINEERING SOILS</th></tr>
<tr><th>Predominantly coarse</th><th>Predominantly fine</th></tr>
<tr><td colspan="2">FALLS</td><td>Rock fall</td><td>Debris fall</td><td>Earth fall</td></tr>
<tr><td colspan="2">TOPPLES</td><td>Rock topple</td><td>Debris topple</td><td>Earth topple</td></tr>
<tr><td rowspan="2">SLIDES</td><td>ROTATIONAL</td><td rowspan="2">Rock slide</td><td rowspan="2">Debris slide</td><td rowspan="2">Earth slide</td></tr>
<tr><td>TRANSLATIONAL</td></tr>
<tr><td colspan="2">LATERAL SPREADS</td><td>Rock spread</td><td>Debris spread</td><td>Earth spread</td></tr>
<tr><td colspan="2" rowspan="2">FLOWS</td><td>Rock flow</td><td>Debris flow</td><td>Earth flow</td></tr>
<tr><td>(deep creep)</td><td colspan="2">(soil creep)</td></tr>
<tr><td colspan="2">COMPLEX</td><td colspan="3">Combination of two or more principal types of movement</td></tr>
</table>

Figure 4: Types of Landslides (Source: Highland, 2004) abbreviated version of Varnes' slope movement classifications (Varnes, 1996)

It is with certainty, the writer agrees with Royster (1978), in that the most common slope *movements* in Tennessee are best described as translational slides. And the material type is most often best described as debris in size. Shown below in Figure 5 is an image of a translational landslide.

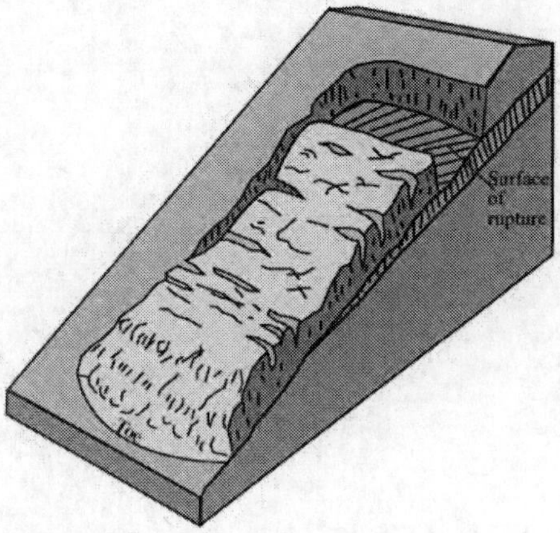

Figure 5: Translational Landslide (source Highland, 2004)

Example: Shown in Figure 6 is a roadside sandwich restaurant located in Chattanooga, TN along SR-8 (Signal Mtn. Road) at the foothill of a rather large ridge. The restaurant, destroyed by a landslide on February 23, 2019, would be best described by the Cruden and Varnes categorical method of landslide types, as an extremely rapid, translational, debris slide. The restaurant was closed, and there were no personal injuries reported, but the application for insurance coverage was denied (Pace, M., 2019).

Figure 6: Sandwich Restaurant before Landslide (source: google maps)

Figure 7: Landslide Destroys Sandwich Restaurant (source Chattanooga Times Free Press, Strickland, D.)

SLOPE INSTABILITY RISK FACTORS IN TENNESSEE

Perhaps the most significant factor for landslide risks in Tennessee is the nature of geologic deposition. Tectonic processes and subsequent weathering created Tennessee's landforms. Tectonic forces have caused beds to dip adversely toward the roadway, that often create "chutes" for debris to slide and flow, but also the opportunity for rocks to slide down into the roadway on an inclined surface plane.

The majority of slope instabilities occur on existing rural state routes. In the past, TDOT has had a policy whereby ownership of county rural roadways has been transferred to the State, and becomes a state route. And with this adopted ownership, the future maintenance of these rural routes is implied. Most of these rural former county roads did not meet the standard safety requirements of a state road. So, the process of increasing lane width, increasing shoulder width, and installing guardrail was required. This widening was commonly accomplished with no compactive effort, but by simply "side casting" from the top of slope whatever loose gravelly clay material was readily available. This condition led to a steeper slope; a condition the author and others refer to as an "over steepened" slope. Some locations in the Valley and Ridge, and Cumberland Plateau provinces have side-hill template roads with steep vertical grades, often built with "over steepened" slopes.

Lastly, colluvial material causes great landslide risks to Tennessee roadway slopes. When material on these slopes becomes infiltrated with water, the slopes often lose their stability. In particular, residuum of the Pennington formation, deposited by gravity as colluvium, is well known as being highly susceptible to landslide activity (Royster, 1978).

TRIGGERING MECHANISM

There are many slopes in Tennessee that possess the factors discussed previously, that lie waiting for an opportunity to move. The slopes need a landslide activation trigger. Though there could be many factors that contribute to a slope failure, Wieczorek argues that there is a single triggering mechanism that causes active slope movement (1996). Overwhelmingly, across the state of Tennessee the greatest trigger of landslides is intense rain. The intense rainfall events of late February fell upon a wet ground and infiltrated causing saturation and accompanying increases in pore-water pressures. This increase in pore-water pressures, accompanied with a decrease in cohesive strength, is generally considered the trigger of the many translational and rotational debris slides that are generated across the state (Wieczorek, 1996) (Cruden, Varnes, 1996).

LANDSLIDE STABILIZATION

The author's preferred semantic is the term landslide stabilization, but the term is synonymous with landslide mitigation, landslide correction, and landslide arrest, etc.

Royster (1978) wrote that avoiding a landslide occurrence or stabilizing an active landslide requires one or more of the following actions to be taken:

- *Relocation*
- *Removal*
- *Restraint*
- *Drainage*

Relocation: In most cases, the relocation of an existing state roadway route to a more favorable geologically stable site is impractical. It is more economical and straightforward to correct the landslide using one of the other actions listed, and retain the existing alignment. Historically, however, the author has witnessed the contract cancellation of a proposed SR-28, Bledsoe Co. four lane alignment improvement due to a number of significant landslides where the alignment climbed out of the Sequatchie Valley floor and encountered Pennington formation residuum. In addition, Royster (1964) describes a construction engineer's nightmare that must have been caused by the alignment location chosen for I-40, Cocke Co., where strata within the Unaka Mountains and Blue Ridge dip adversely toward the interstate roadway, and create the opportunity for rock slides. He describes another alignment that would have made a lot more sense, where the beds dipped away from the roadway. The present alignment of I-40 through Cocke Co. and Haywood Co., North Carolina, has encountered rock slide and translational debris slide landslide challenges since it was constructed in the early 1960's (Royster, 1964).

Removal: This action is the simple removal of the failed material in a manner that the continuance of traffic can be restored or a river can flow without restriction. The author witnessed recently a landslide (rock topple) remediation project on SR-25, Smith Co. whereby the contract duration was terribly excessive and the temporary traffic control terribly painful because of the great excavation volume required to obtain a proper rock fall catchment ditch. In the future, all other options will be exhausted before a wholesale excavation removal will be employed to correct a rock topple on one of the author's projects. Royster (1978) describes a full removal required (162,000 CY) of the 1972 Waterville debris translational slide on I-40, Cocke Co., where material slid downward along a "chute" wedge failure plane created by weathered dipping formation beds, into the Pigeon River. Royster (1978) goes on to write, that restraint measures for the Waterville slide were not considered practical at that time.

Restraint: Slope stability analyses require consideration of the forces that drive the landslide, against forces that restrain, or resist, the landslide. The author recognizes advancements over the past twenty years that have provided increased machinery power and efficiency, improved materials, and general innovation can provide systems with greater resistance to landslide movement than were thought possible in the past or, during initial construction of the interstate and state systems.

Advanced drilled systems that TDOT routinely employs include micropiles used laterally as soil nails, or vertically as articulated elements. Soil nails and micropiles mobilize passive resistance to restrain lateral landslide movement. Also, drilled and grouted steel anchors can be tensioned and used as active resistance to stabilize a slide. High strength wire mesh is another technology that can be installed quickly to arrest slope movement. Admittedly, many of these advanced systems of mobilizing landslide resistance can be costly, but when TDOT's "user costs" are considered, these advanced methods often prove economically viable.

The tried and true methods of landslide arrest that TDOT employed when initial construction of state and interstate roads were sliding, were used again during the recent 2019 flooding events. A graded rock buttress requires time-consuming excavation of a temporary slope before placing a final graded rock buttress to grade, but it is one of the go-to landslide corrections that TDOT employs. It is also observed, that the cost of repair due to the wear and tear of loaded trucks rolling on these rural roads must be a consideration.

Soldier pile and lagging walls, with or without anchors is another favorite landslide corrector in the TDOT tool box.

Drainage: Water introduced into a slope increases pore water pressure and subsequently decreases cohesive strength. Improvements to drainage must always be included in the landslide stability design.

Cross drains that are not functioning properly introduce water into a side hill template. When a side hill fill has noticeable a slope failure in the form of roadway cracking, etc., the cut ditches should be evaluated, and if drainage is not flowing properly, consideration should be given to lining the ditch with an impermeable geomembrane, especially if the ditch contains fractured rock or karst drainage. When roadway slope movements are occurring, it is the author's opinion that ditches should not be lined with concrete, because by it's nature, concrete is very susceptible to cracking, even without the introduction of slope movements.

Removal of subsurface water using horizontal drains has long been used as one of TDOT's strategies to stabilize landslides and the TDOT Geotechnical Engineering Section (GES) has recommended their use on most landslide stabilization projects. Subsurface water decreases cohesion, provides subsurface erosion, increases lateral pressure in fractures and joints, and increases pore-water pressure. All these have the effect of reducing the stability of cuts and embankments. TDOT initially used horizontal drains in 1972 on I-75 in Campbell Co. and with great success in stabilizing I-40, Cocke Co. too (Royster, 1973).

LANDSLIDES DURING CONSTRUCTION PHASE

Historically, many landslides were triggered during construction of state roads and federal interstates. It is the opinion of the author that the geotechnical, construction, and construction observation (quality assurance / quality control) professions have advanced, and it can be said, after reviewing the documents of Royster (1973, 1978, 1974, 1964) that fewer landslides have occurred during the past six years during the roadway construction, than occurred when I-75, I-24, and I-40 were being constructed. But it can also be argued that TDOT is simply not letting the volume of large scale grading projects that were being constructed during the time span reviewed. In review of Royster's works, it is believed the complex construction and engineering geology effort involved in the interchanges and bypasses of recent times, is small in comparison to the engineering geology and construction effort that must have occurred in the 1960's and 1970's when the interstates were being constructed.

2019 LANDSLIDES IN TENNESSEE ON EXISTING ROUTES

Due to intense flooding in 2019, there were over ninety-five separate roadway landslides that required stabilization. The cost of the flooding response will likely be over \$200M. The locations of these landslides are scattered very disproportionately across the ninety-five political counties of Tennessee. The landslides are scattered disproportionately because the great majority of slides occurred in Region 1 and Region 2, or Eastern grand division of Tennessee, the Valley and Ridge, Cumberland Plateau, and Blue Ridge physiographic regions. A great deal of the stabilization effort remains underway. As mitigations plans were developed for contract letting so rapidly, there are many contract administration challenges and the GES unit has been in the field providing support regularly.

Four high profile landslides are discussed in the narrative below. These four landslides were arbitrarily considered high profile to the author because of the economic cost to repair and inconvenience to TDOT's customers. Two of the four landslides are located within geologic strata Royster (1973) terms the "PC Complex", which is an expression he coined for slope movements due to the Pennington Shale residuum, colluvium, and water.

I-24 EB near MM 42 (LM 10.1), Davidson Co. (Pin 126626.00, CNT269, GES 1917618): A sudden translational debris slide occurred at the site in the early Sunday morning hours of February 24, 2019. The landslide risks included an extremely high cut slope on what the author considers an over-steepened slope. The toe of the high cut slope is behind an approximately twenty-foot high pre-split rock cut. The site is located in the Highland Rim geophysical region, just a few miles north of Nashville.

Sunday morning the author's team, roadway designers, TDOT leadership, surveyors, and a grading contractor met on site and made decisions on a plan necessary to get the interstate open again. The surveyors provided adequate ground information for the transportation engineers to design a temporary traffic control plan (TCP). The TCP consisted of grading work and pavement work necessary for a build out into the median to accommodate two southbound lanes of temporary travel. The interstate detour was unsafe and time consuming, so it was with great relief that the emergency traffic control plan opened to traffic on March 13, 2019.

The stabilization effort involved removing the slide material entirely, from the interstate and also the slope, and then restraining the slide with a graded rock buttress. The buttress was completed April 27, 2019.

Figure 8: I-24, Davidson Co., February 24, 2019 (source: WSMV-TV, Garbee, J.)

Figure 9: Looking southerly at I-24, Davidson Co. Debris-Slide, February 24, 2019 (source: Robert Jowers)

Figure 10: I-24, Davidson Co., Haul Road Being Built, late February, 2019 (Source: Hartman, J., TDOT)

Figure 11: I-24, Davidson Co., Temporary Traffic Control lanes in Place 3/13/2019 (Source: Hartman, J., TDOT)

Figure 12: I-24, Davidson Co., Buttress in Place 4/30/2019 (Source: Hartman, J., TDOT)

SR-85 approx. LM 20.5, Overton Co. (CNT154, Pin 128694.00, GES 6718498,6718718, 3/15/19 letting):

The site is located in a sparely populated area of the northwestern corner of the Cumberland Plateau physiographic province, but near the border of the Highland Rim. The landslide could be best described as a reactivated, slow to very slow, translational debris-slide. The landslide risk factors involved include primarily the residual, colluvial material of the Pennington Shale and Fentress formations, along with water. Royster (1973) coined the failure attributed to the "PC complex". The Pennington, as he described, was the "P", and the colluvium (Pennington\Fentress residuum) was the "C".

Landslides have affected this stretch of SR-85 through the decades, and it has been necessary to temporarily close lanes, and entire sections of SR-85 from time to time. Particularly the outside shoulder and lane of the side-hill roadway template as it displaces downhill. SR-85 has oversteepened slopes, tension cracks are common, and there are many switch-back horizontal curves.

The subject landslide was first reported in 1998, to the author's knowledge, but is assuredly ancient and has been sliding in terms of recent geologic time. Several mitigation proposals for slide stabilization were presented in 1998 including a graded solid rock buttress, but due to cost and relatively low traffic count, an option of re-grading, re-paving, and observation was selected. The movement was arrested, likely because the slide had reached a new driving forces/resisting forces equilibrium, and for a period of about twenty years the road was relatively stable, moving only following during intense precipitation.

Figure 13: SR-85, Overton Co.- Slide in late 90's, 6/15/1999 (Source: Hornal, G., TDOT)

The landslide was triggered by the February 2019 intense flooding and became active again. The road had to be closed and a time consuming detour of over fifty miles was posted. Plans were rapidly developed to mitigate the slide and reopen SR-85. The plans included reducing the pore pressures through installation of horizontal drains, and resisting the slope movement with construction of a rock buttress. Several smaller individual landslides along SR-85 in Overton Co. and Fentress Co. were also stabilized with concurrently in the contract using stabilization techniques that included reticulated micropiles, soil nails, and mechanically stabilized earth / geosynthetically reinforced soil.

Figure 13: SR-85, Overton Co.- Slow moving, Debris-Slide 3-4-2019 (Source: Williams, S., TDOT)

Bids were opened on March 15, 2019 and the award was to the lowest bidder for $12.9 M. As of this writing the project remains under construction and the road is closed. Substantial quantity overruns have occurred because of the nature of the debris slide translating, and also because of the absence of a thorough subsurface exploration program.

SR-70 LM 15.5, Hawkins Co.(CNT 131, 128644.00 GES 3717218): The first movement of the translational debris-slide was best described as moderate to rapid. The landslide risk factor present was an over steepened colluvium slope, but the trigger that caused movement was intense raiinfall.

The movement was reportedly first recognized the night of February 21, 2019 by maintenance personnel removing a fallen tree from the roadway. In mid-morning the author's team, roadway designers, TDOT leadership, and surveyors met on site and formulated a plan to take the site survey, work over the weekend developing construction plans, and let the project one early day in the upcoming week. The decision was immediately made that the roadway must be temporarily closed during construction and a temporary detour was to be implemented.

Figure 12: SR-70, Hawkins Co. Devastating Debris-Slide 2/21/2019 (Source: Robert Jowers, TDOT)

The stabilization application involved restraining a temporary slope face from the roadway surface downward. The temporary slope was temporarily soil nailed using a pinned geofabric\wire mesh face. With a stable temporary slope in place, a graded rock buttress could be installed. The project construction is still underway. During excavation, adversely dipping rock was encountered, and that appears to be the failure plane along which the colluvium slid.

I-40WB MM344-342 - Rockwood Mtn., Roane Co. (CN unavailable at this time), Pin 128729.00, GES 7324002): The site is within the Cumberland Plateau physiographical region, but near the contact with the Valley and Ridge physical region. The landslide could be best described as a reactivated, slow to very slow, translational debris-slide. The landslide risk factors involved include primarily residual and colluvial material of the Pennington Shale formation and water. The triggering mechanism was again the intense rainfall in the month of February, 2019.

A design solution is ongoing, but should involve removal, restraint, and drainage improvements. The driving forces from the colluvial material above the west bound lanes will be removed to the unweathered formations of the Pennington. Below the west bound lanes of I-40 an array of restraining ground anchors are being proposed drilled through the colluvium and into and tensioned against the unweateathered Pennington formations. An array of horizontal drains is proposed, as well.

The project engineer's bid estimate for just this small segment of the route is in the several tens of millions range. The amount of excavation will be substantial.

SR-2 (US41) between LM 25.20 and LM 26.40, Marion Co. (CNT229, Pin 128666.00, GES 5807417, $16.5M in 5-10-19 letting): The site is located near the contact of the Cumberland Plateau and Valley and Ridge physiographic region and another slope movement that Royster would have attributed to the "PC Complex". The slope movement could best be described as a reactivated, slow to very slow, translational debris-slide. The landslide risk factors involved include primarily residual and colluvial material of the Pennington Shale formation and water. The triggering mechanism was again the intense precipitation following, unusually wet winter months.

The solution will be to reduce pore water pressures by installing horizontal drains, and restraining affected sites using reticulated micropiles. The embankment slopes will be stabilized using soil nailing with a shotcrete surface.

The project was let to bid, and a contract for $16.5M was administered. The bid items included 40,000 CY in excavation, and approximately $10M for micropile landslide stabilization. The quantity of horizontal drains was 32,290 LF. The project activities were recently begun.

FUTURE STRATEGIES – CONCLUSION

RMP Adaptation to a General Landslide Program: TDOT's Rockfall Management Program (RMP) was reemphasized in 2016. The RMP, presently funded at $10M annually, identifies the state's sites that pose the most rockfall hazard risk, prioritizes a list of these high risk sites, and then programs these sites into rockfall mitigation projects. The author feels this funded RMP should be adapted to include landslides, as well. And moreover, it is felt the author spent too many precious resources devoted to the rockfall hazard risk in the state and failed to acknowledge the seriousness of the landslide risks of the state.

Following the floods of late February, 2019, the TDOT Asset Management Plan (TAMP) was executed in June, 2019. The TAMP ranked the risk due to *Rock Slides / Slope Failure* as the second highest risk, below *Flooding* (TDOT, 2019). Via the TAMP, TDOT leadership has tasked GES with exploring the potential of addressing the different forms of slope movement. TDOT leadership recognizes an organizational gap, because TDOT does not have a formal landslide program dedicated to mitigation. By not addressing all of TDOT's roadway slopes, landslides are exposing our customers to excessive risk.

The author is in agreement with the landslide classification criteria proposed by Varnes (1996), which states a rock fall should be categorized as a form of landslide. And the rock fall, can be broken down into a two noun description emphasizing the type of material, and the type of movement (*rock fall, rock topple, or rock slide*). Just like a landslide. If not for the general lack of understanding of Varnes (1996) naming convention of slope movements by the engineering geology community at large, the RMP could have, and perhaps should have, easily been termed the Landslide Management Program, or Slope Management Program.

The author acknowledges many state DOT representatives are present in the conference have implemented similar programs in the recent past, and plans to be actively engage these representatives during breaks in the HGS proceedings.

REFERNCES:

1. Hargett, T. (2018). *Tennessee Blue Book 2017-2018*, (2018). State of Tennessee, Nashville, Tennessee, pp. 539.
2. Miller, R. A. (1974). *The Geologic History of Tennessee*, State of Tennessee Department of Environment and Conservation, Division of Geology, Nashville Tennessee, pp. 3.
3. Royster, D. L. (1973). Highway Landslide Problems Along the Cumberland Plateau in Tennessee. *Prepared for Presentation at the Southeastern Section of the Geological Society of America Meeting – April 12, 1973, Knoxville Tennessee*, pp. 6.
4. National Weather Service (2019). *Late February 2019 Historic Rainfall and Widespread Flooding.* Retrieved from https://www.weather.gov/ohx/lateFebruary2019flooding
5. National Weather Service (2019). *February 2019 Rainfall Compared to December 1926 Rainfall.* Retrieved from https://www.weather.gov/ohx/February2019vsDecember1926
6. National Weather Service (2019). *NOWData – NOAA Online Weather Data.* Retrieved from https://w2.weather.gov/climate/xmacis.php?wfo=ohx
7. Royster, D. L. (1978). Landslide Remedial Measures. *Prepared for Presentation at the 37th Annual SASHTO Convention, October, 1978, Nashville, Tennessee*, pp. 18, 36.
8. Cruden, D.M., Varnes, D.J. (1996). Landslide Types and Processes. In *Special Report 247: Landslides-Investigation and Mitigation, (*Schuster, R.L., Turner, A.K. eds.) TRB National Research Council, Washington D.C., pp. 37-38, 52.
9. Pace, M. (2019, April 29). Subway owners won't rebuild after landslide as insurance claim rejected. *The Chattanooga Times Free Press*. Retrieved from https://www.timesfreepress.com/news/local/story/2019/apr/29/subway-owners-wont-rebuild-after-landslide-in/493607/
10. Royster, D.L. (1974). The Spring of '73 (An Overview of Recent Landslide Problems in Tennessee). *Prepared for Presentation at the 56th Annual Tennessee Highway Conference, Knoxville, Tennessee, April 25, 1974,* pp. 2-5, 7, 8.
11. Wieczorem, G.F. (1996). Landslide Triggering Mechanisms. In *Special Report 247: Landslides-Investigation and Mitigation, (*Schuster, R.L., Turner, A.K. eds.) TRB National Research Council, Washington D.C., pp. 76-81.
12. Royster, D.L. (1964). Highway Landslide and Stability Problems in Tennessee. *Prepared for Presentation at the 46th Annual Tennessee Highway Conference, April 9, 10, 1964.*, pp. 7.
13. American Association of State highway and Transportation Officials. (2008). *AASHTO LRFD Bridge Design Specifications*, Washington, D.C, Section 11 pp. 19.
14. Highland, L. (2004) Landslide Types and Processes. *USGS.* Retrieved from https://pubs.usgs.gov/fs/2004/3072/pdf/fs2004-3072.pdf
15. Royster, D.L. (1980). Horizontal Drains and Horizontal Drilling: An Overview. *Prepared for Presentation at the 59th Annual Meeting of the Transportation Research Board, January, 1980.*, pp. 1,2.
16. TDOT (Tennessee Department of Transportation). (2019). Transportation Asset Management Plan – 2019. Chapter 5, pp. 11. Retreived from https://www.tn.gov/content/dam/tn/tdot/maintenance/asset-management-office-/tamp/2019.1.0%20TAMP%20Report.Revised.8.26.2019.pdf

Application of Rockfall Simulation to Risk Analysis

Timothy J. Pfeiffer, P.E., G.E.

Foundation Engineering, Inc.
7857 SW Cirrus Drive, Bldg. 24
Beaverton, OR 97008
tjp@foundationengr.com

Prepared for the 70th Highway Geology Symposium, October 2019

Acknowledgements

The author would like to thanks Susan Conrad for her editorial assistance and friends and colleagues who have provided valuable insight into rockfall analysis and perceptions.

Disclaimer

Statements and views presented in this paper are strictly those of the author, and do not necessarily reflect positions held by their affiliations, the Highway Geology Symposium (HGS), or others acknowledged above. The mention of trade names for commercial products does not imply the approval or endorsement by HGS.

Copyright

Copyright © 2019 Highway Geology Symposium (HGS)

All Rights Reserved. Printed in the United States of America. No part of this publication may be reproduced or copied in any form or by any means – graphic, electronic, or mechanical, including photocopying, taping, or information storage and retrieval systems – without prior written permission of the HGS. This excludes the original author(s).

ABSTRACT

Rockfall simulation software was developed to provide a rational basis for the engineering of rockfall protection. Rockfall simulation is also now being used in the evaluation and quantification of rockfall risk. However, the limitations of the statistics and the inner workings of rockfall simulation are often inadequately addressed in risk analysis applications. In this paper, we discuss the limitations of using rockfall simulation in risk analysis and offer some suggestions for addressing the uncertainty involved in rockfall analysis.

Rockfall simulation uses Monte Carlo methods to simulate the variability observed in rockfall and provide a statistical model of rockfall. Statistics developed using Monte Carlo methods have the same limitations as any other statistical analysis. In addition, understanding the assumptions and bias built into the simulation algorithm is needed to interpret the simulation statistics and apply the results to risk analysis. Finally, no statistical model and analysis can overcome human fears and perception of risk, so the presentation of risk statistics must consider human perceptions.

This paper provides:

1. A look at how understanding the inner workings of rockfall simulation can inform the interpretation of rockfall simulation results applied to risk analysis
2. A discussion of statistical analysis applied to rockfall simulation and risk analysis
3. A brief discussion of risk perception and communication of geohazard risks.

The paper concludes with some ideas for improving the interpretation of rockfall simulation results.

INTRODUCTION

Rockfall simulation tools provide a means to evaluate potential rockfall using computer simulation rather than rolling a statistically significant number of rocks at every site. Since the 1980s, rockfall simulation has been used to provide an estimate of rockfall behavior for the design of rockfall protective structures. Rockfall simulation is also being used in rockfall risk evaluation. The purpose of this paper is to discuss the implications of applying simulation modeling to risk analysis.

The application of rockfall simulation to risk analysis differs from typical engineering design. Engineering design solutions nearly always include conservative assumptions and a factor of safety. In contrast, if conservative assumptions are applied to risk analysis, the result is higher calculated risks which may be misleading when compared to experimentally verified risks. In addition, risk analysis applications may use probability of occurrence values less than 1 percent, which is far lower than the accuracy the model justifies. Therefore, understanding the assumptions used to develop the simulation algorithm and the model accuracy is helpful when interpreting rockfall simulation data for use in risk analysis.

ROCKFALL MODELING AND VALIDATION TESTING

Probably the most notable observation of rockfall is the variability. It is obvious from watching even a few rocks on a relatively uniform slope that modeling rockfall will involve probability and statistics rather than a deterministic solution. Therefore, the results of experimental tests or simulation are probabilistic distributions rather than a deterministic solution. In rockfall modeling, this variation is simulated by randomly varying a parameter or parameters between limits. The simulation process using chance variables and random generated outcomes is often called Monte Carlo simulation after the famous gambling center. Monte Carlo simulation is applicable to a wide variety of problems involving random variables, variable distributions and complex systems (1). It is probably most known for its use in financial analysis. Monte Carlo simulation does not predict whether the market will go up or down, but it does provide a reasonable method to evaluate the potential distribution of outcomes based on past performance. Likewise, Monte Carlo methods used in rockfall analysis does not predict where or when a rockfall will occur, instead modeling provides a potential distribution of trajectories.

Rockfall simulation has come a long way in the last 30 years. However, regardless of the sophistication of the computer program, the simulation must calculate the complex interaction of the rock and the slope and provide a statistical model of the variation observed in the field. The interaction of the rock and the slope (the bounce) is complex and involves considering the material properties of the slope and rock, surface irregularities and variation, rotation of the rock, and vegetation. Considering the number of variables and complexity of the rock slope interaction, the resulting algorithm requires simplifying assumptions. These assumptions may result in significant variation between the simulated and the true condition. Therefore, the results have a built-in statistical bias.

The assumptions built into the algorithm and the selection of variables by the modeler reflect the end purpose of the analysis. For example: the Colorado Rockfall Simulation Program (CRSP) was developed to provide a rational basis for designing rockfall protective structures. Therefore, assumptions, calibration and validity testing were intended to provide realistic values for bounce height and velocity (2). The goal was to provide a reasonable match between the model and experiment for the 90 to 95 percent probability. Therefore, the model was not designed to model the behavior of all possible rockfall. A rockfall model intended for risk analysis would be designed and tested to closely match the behavior of all possible rockfall.

Validation

As discussed above, rockfall simulation involves modeling complex interactions and simplifying assumptions. Therefore, the validity of the model is based on comparing simulation results to field trials. Formal simulation model testing was developed for validating simulations designed to model complex electrical, computer and manufacturing systems. Simulation models are developed for specific purposes and the validity of the model should be based on how well the model satisfies the purpose. It is often costly and time consuming to validate a model for the entire set of conditions. Therefore, testing is limited to specific conditions of interest (3).

Quantitative simulation validity testing involves the numeric comparison of results between simulations and field testing. Field trials are relatively expensive and not applicable to all sites. However, validation testing also involves qualitative testing and comparison to historical information.

Quantitative Testing

In the case of rockfall, quantitative validation is typically limited to comparison of bounce height and velocity at specific points on the slope. However, more sophisticated collection of 3D trajectories is becoming available. An example of validation testing comes from a full scale rockfall testing conducted on an avalanche track in the French Alps (4). The experiment site was a 570 m long debris and talus cone following an avalanche track free of trees or large vegetation. One-hundred rocks averaging 0.8 cubic meters were started with a 5-meter initial drop. Table 1 provides a comparison between 1000 simulation runs using two simulation models and the 100 field trials.

Table 1. Comparison of results between field trials and two simulation methods for velocity and passing height (4).

Location	Data Source	Velocity m/sec			Passing Height		
		Mean	Std. Dev.	Max	Mean	Std. Dev.	Max
Elevation Line 1	Observed	12.5	5.2	28.1	1.4	1.1	5.0
	Method A	11.5	4.2	27.7	1.2	0.9	10.4
	Method B	12.7	4.3	30.3	1.4	1.0	11.0
Elevation Line 2	Observed	13.8	5.5	28.9	1.6	1.4	6.2
	Method A	10.9	4.7	29.6	1.2	1.0	15.5
	Method B	12.1	5.1	31.8	1.4	1.1	12.7

The field test results were compared to 3D modeling and found that the modeling provided a reasonable prediction of the rockfall trajectories. The distribution of velocities at two locations on the slope showed reasonable agreement for the mean, standard deviation and maximum velocities. While the average passing height of the field trials agreed reasonably well with the average simulation passing height, the maximum height values in the simulation models were more than twice the observed field values. This may be partially due to the larger samples size used in the simulation. However, bounce heights tend to follow a log-normal distribution, so the difference between the 90 percent probability and 98 percent is much greater than for a normal distribution and any bias has a more pronounced effect. This also illustrates the difficulty encountered when using maximum values. For field testing or simulation models, the maximum value may represent 1 in 100 field trials or 1 in 1000 simulation runs. However, for a limited sample size, the true probability of the maximum value is unknown and the maximum value could represent a 10% probability or a 0.1% probability.

The study from the French Alps also found that some of the test rocks came to rest soon after the initiation, which was not observed in the simulations. This situation was also observed during the validation testing of CRSP (5). Some rocks never got going or were not usable in a 2D model. These rocks were not included in the validation testing, because the intent of the simulation model was to provide a rational basis for rockfall mitigation, and these rocks did not represent rockfall that required mitigation.

Validation testing that finds the model is sufficiently accurate for the intended purpose and condition tested does not validate the model for all conditions and purposes. In the study from the French Alps discussed above, testing was limited to 100 rocks of similar size, started from the same height, and all on the same slope. Simplifying assumptions were incorporated into the testing and modeling. In rockfall hazard evaluations, slopes are often vegetated and all rockfall does not start from the same location with the same initial drop. Assuming the slope is rocky and free of vegetation is likely to be a conservative assumption for many sites and result in a statistical bias.

Conservative assumptions that may be built into rockfall simulation, validation testing, or parameter selection may include:

- Validity testing comparing mean values rather than extreme values
- Validity testing and simulation models for sparsely vegetated slopes
- Simulation modeling assuming all rocks stay intact
- Simulation modeling using maximum rock size or validation testing using one rock size
- 2D simulation models
- Selecting conservative parameters that increase values in the simulation results.

Any of these assumptions will result in biased output where the simulation model predicts higher bounces and velocities than would be observed for rockfall at the specific site. Conservative bias from conservative assumptions in engineering design is a widespread and a widely accepted practice. However, when applied to risk analysis, conservative bias will over predict risk and may be misleading when compared to risk data generated from historical population studies.

Qualitative Testing

Qualitative testing considers the question, does it look like rockfall? Therefore, validity testing of rockfall simulation should also include qualitative testing where an animation or rockfall trajectory is compared to personal observation. Some casual observation of rockfall reveals the following:

- Rocks don't bounce, but they do rebound on a slope.
- A spinning rock rolls further than a rock that is not spinning.
- Small rocks will stop amongst larger rocks, while larger rocks will travel further on a slope made of smaller rocks.
- Rockfall behavior varies widely even for similarly sized rocks following similar paths.

A valid rockfall model will need to satisfy these observations and generally appear to provide a reasonable result to someone who has observed rockfall. Figure 1 shows two hypothetical rockfall trajectories. Trajectory A may be appropriate for falling basketballs, but trajectory B looks more like rockfall observations. Note that at Points 1 and 2, both trajectories will have similar bounce heights and velocities. Qualitative comparison between the simulation and experience should be the first validation test that any simulation must meet.

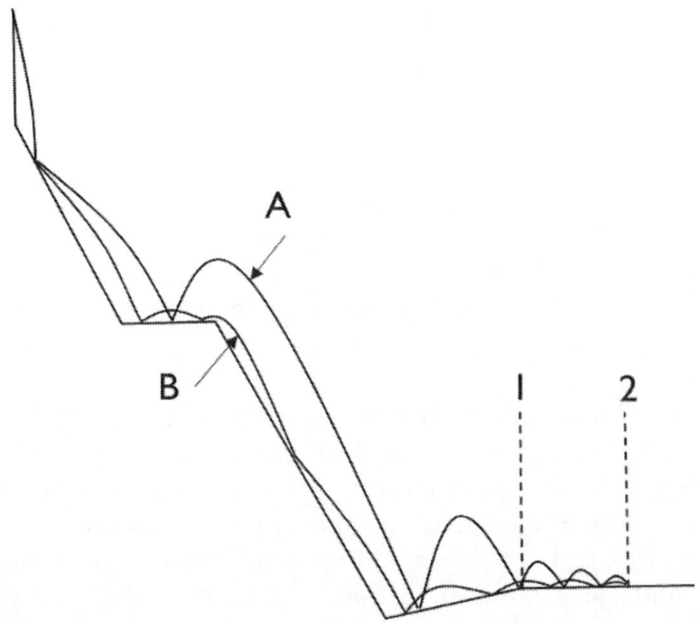

Figure 1. Two potential trajectories (A and B) with similar bounce height and velocity at points 1 and 2.

Historical Validation Testing

Historical validation testing compares the simulation results to the historic record or experience. Historical validation testing for rockfall is primarily a qualitative test because actual quantitative historical data is rare. Historical validation testing may used to assist the investigator's assessment of the model and input parameters validity for a specific site. Most commonly, the investigator compares the stopping points of rockfall observed at the site to the stopping points in the rockfall simulation. Sometimes there may be eyewitness observations, but these observations are often limited to a single event. While comparing the simulation results to field observation is essential, the sample size is generally very limited and incomplete. Reports of rocks on the road may miss rocks that rolled over the road, and the investigator must judge whether the rock observed represents an average rockfall or a low probability event.

For historical testing, the simulation results are considered reasonably valid if they include the set of observations available in the historic record. The model input parameters may be adjusted until the model is consistent with the available observations. The historical observation can provide a point of reference for evaluation of options. For example, mitigation options may be evaluated based on the relative effectiveness compared to the initial model that included the historical observation data. This method can be used to provide relative comparisons between mitigation options and the historical data.

STATISTICAL EVALUATION OF ROCKFALL SIMULATION

The limitations of rockfall modeling and validation testing described above, present a significant challenge to providing high quality data for a quantitative risk analysis. In addition to the bias from the difference between the true value and actual value, the large variation observed in rockfall and the limited field test sample size results in a relatively large confidence interval. This confidence interval represents the probability that the true value for the population is within the confidence interval of the sample assuming the sample does not include bias.

Bounce Height Statistics

As previously discussed, rockfall velocity tends towards a normal Gaussian distribution. However, in rockfall risk evaluations, height and roll out distance are often the parameters of interest. Rockfall height and roll out distance tend to more closely follow a log-normal distribution. The following provides a discussion of the implications of bounce height distribution on risk evaluations at low probabilities.

During the development of the Colorado Rockfall Simulation Program, the Colorado Highway Department rolled rocks down a 320-foot high by 400-foot long slope near Rifle, Colorado. Bounce height data was collected at two points on the sparsely vegetated slope. Figure 2 provides a histogram of the bounce height from the field testing and from the simulation model using the Colorado Rockfall Simulation Program. While there is substantial variation between the experimental data and the simulation data, the log trend lines appear similar. The simulation trend line shows higher bounce heights at low probability (5). This over estimation is an intentional conservative bias for software intended to be used in the design of mitigation features. However, this feature may be misleading if applied to risk analysis at low probabilities.

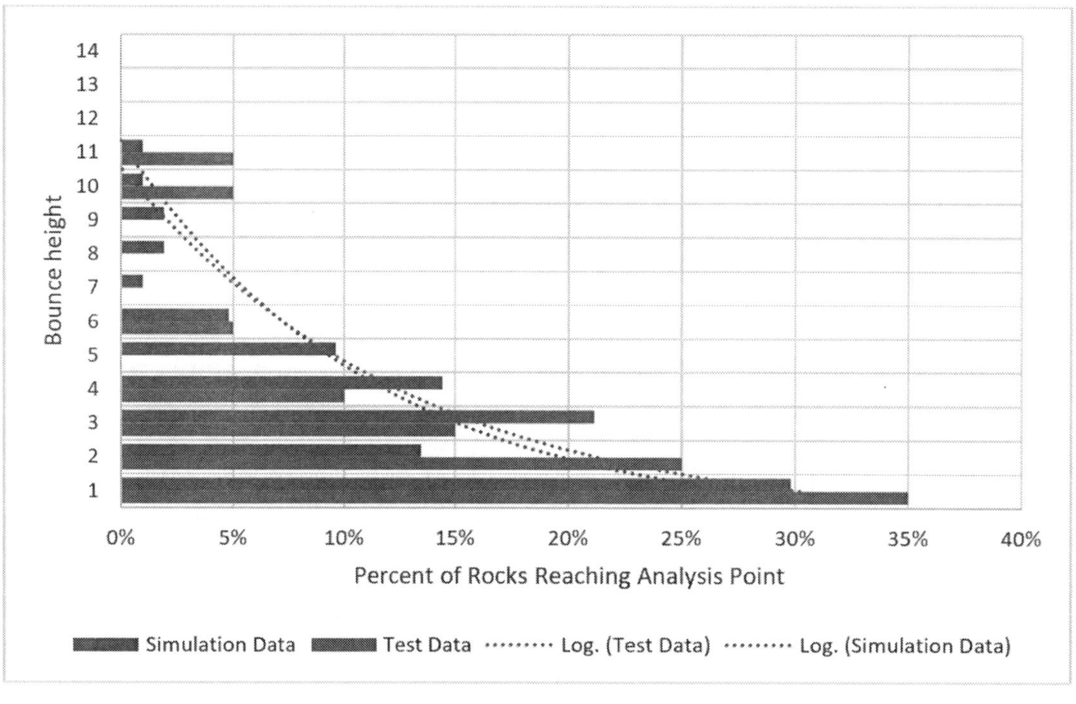

Figure 2. Comparison of bounce height from rockfall simulation data and field test results.

In the example above, the slope geometry limited the maximum bounce height. At a taller or steeper site, the impact of the log-normal bounce height distribution may be far more pronounced. Figure 3 illustrates how the conservative bias provides an 11-foot bounce height at 95% probability compared to a 10-foot experimental bounce height. However, the simulation bounce height at 98% probability is 20 feet compared to an experimental value that may be 15 feet. While the 20-foot bounce height may be possible, it likely represents far less than 2% of rockfall at the site and may provide misleading statistics if used in a risk analysis.

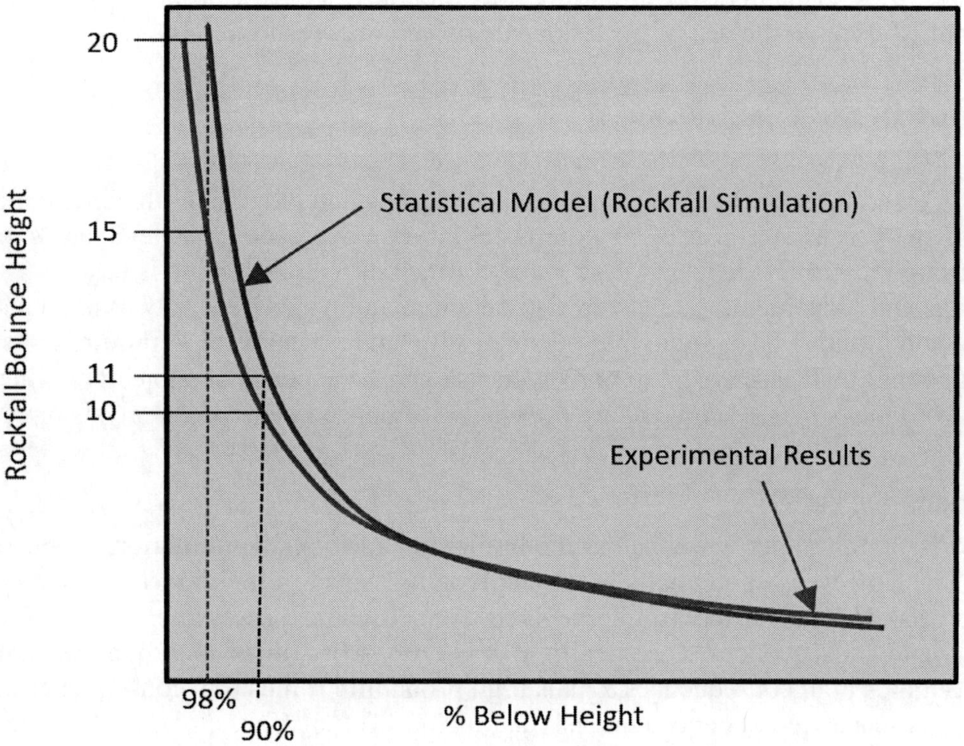

Figure 3. Comparison of bias for a log-normal distribution at low probability.

For a site with taller and steeper slopes, the bounce height difference between 95% and 98% can be dramatic. Table 2 compares the percent retained for a proposed 10-foot barrier verses the barrier height needed to retain 99% of the simulated rockfall. While the 10-foot high barrier is a viable option, 25 to 30-foot tall barriers would not be constructible at this site. While bounce heights of 25 to 30 feet may be possible at these locations, because of conservative bias in the model, the probability of a 25 to 30-foot bounce height may be far lower than 1%.

Table 2. Percent retained for a 10-foot rockfall barrier and barrier height needed for 99% retained based on rockfall simulation.

Section	Proposed Barrier Height	Percent Retained	Height for 99% Retained
3	10 feet	94.7%	30 feet
4	10 feet	92.5%	25 feet
8	10 feet	93.8%	13 feet

The assumptions that go into modeling the complex interaction of rocks on slopes may result in statistical bias where the distribution of simulation values is offset from experimental values. This bias may have a dramatic effect on the calculated values for log-normal distributions, by statistically overestimating the probabilities of low probability events. This bias gets multiplied through the rest of the risk analysis and may result in a bias in the resulting risk.

PERCEPTION OF ROCKFALL RISK

So far in this paper has looked at the assumptions and judgement that goes into rockfall simulation analysis used as part of an expert rockfall risk evaluation. The evaluation may be carried through to provide a quantitative value for the risk, but this value is based on a series of assumptions and judgements. However, risk on a human level is generally assessed at an emotional and intuitive basis following personal subjective criteria for assigning risks. Studies comparing expert risk assessment to perceived risk show the personal subjective criteria are far different from the expert system and are not easily influenced by expert analysis or opinions (6).

Risk Perception and Rockfall

So far in this paper, we consider the limitations of rockfall simulation applied to rockfall risk analysis. However, a potentially more challenging aspect of rockfall risk analysis could be how expert risk analysis differs from perceived risks. Technical assessment of risk focuses on probability and consequences. Technically, risk is probability times consequences, and low probability times high consequences equals high probability times low consequences. However, most people use a different concept of risk that include voluntariness, personal controllability, familiarity, and catastrophic potential, amongst other factors (7).

Voluntary, controllable, and familiar risks are generally perceived as lower risk than empirical data would indicate. Except for possibly rock climbers, exposure to rockfall is not voluntary, and it is not something that is user controlled. Rockfall injuries and fatalities are extremely uncommon and few people are aware of rockfall. Therefore, rockfall could generally be considered an unfamiliar risk. While rockfall can range from inconsequential to catastrophic, must people will see getting hit by falling rocks as catastrophic. Based on this model of risk perception, rockfall risks are likely to be perceived as a much higher risk than empirical data would suggest.

To mitigate the mismatch between perceived risks and calculated risks, calculated risk is often compared to documented historical risks. However, the comparison is between calculated values that may include substantial bias and empirical historic data that is presumably not biased. Therefore, the comparison may be misleading for calculated values that include conservative assumptions and bias.

Presentation of rockfall simulation results may inadvertently encourage misperception of rockfall risk. A figure showing the potential trajectory of rockfall impacting a site will tend to be perceived as a dangerous condition. However, an expert analysis may show the rock trajectory is a very rare event and the probability of impact and injury is very low. Figure 4 shows the trajectory of 1,000 simulated rockfall events. The caption states that the individual risk is calculated to be between $6x10^{-10}$ and $3x10^{-13}$ per year and the annual risk to all users is between $5x10^{-5}$ and $3x10^{-8}$. However, the figure clearly shows someone getting hit by a rock. Considering personal risk assessment is primarily an emotional reaction, a picture of someone getting hit by a rock likely has far greater impact on the perception of risk than the scientific notation in the caption.

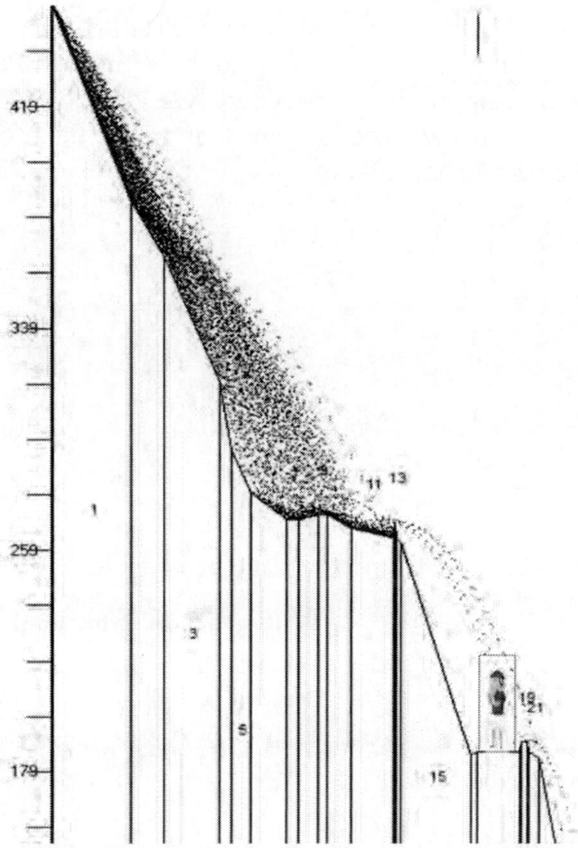

Figure 4. Simulated rockfall trajectories for a site where the annual risk to all users is calculated to be between $5x10^{-5}$ and $3x10^{-8}$ and the individual risk is between $6x10^{-10}$ and $3x10^{-13}$.

CONCLUSIONS

Rockfall modeling and simulation analysis includes judgement and assumptions. These assumptions and judgements are built into the simulation algorithms and site analysis. Understanding the potential bias and limitations of rockfall simulation modeling informs the interpretation of the results when used to evaluate risks. Some ideas to improve the interpretation of rockfall simulation results include:

- Understand the assumptions used in the rockfall simulation model and how these assumptions may generate statistical bias in risk analysis.
- Review the validity testing of the rockfall model and how the validity testing applies to the conditions being evaluated.
- Perform qualitative and historical validity testing for all sites.

- Observe scaling operations to develop the ability to qualitatively compare rockfall to rockfall simulation results.
- While simulation tools are improving, for the foreseeable future consider these tools as an aid to engineering judgement rather than a replacement.

Rockfall simulation analysis used in qualitative and quantitative risk evaluations can help inform decisions and compare options. However, comparing rockfall risk calculated from simulations to empirical data may be misleading and may not have the desired influence on the perceived risk. Alternatively, rockfall simulation and risk analysis can be used to compare the relative effectiveness of options at a site and select options that reduce the rockfall risk while considering cost, aesthetics, and constructability.

1. Mechanical and Aerospace Engineering, Probabilistic Engineering Design, Chapter 8 – Monte Carlo Simulation, Missouri University of Science and Technology, http://web.mst.edu/~dux/repository/me360/me360.html, Accessed June 7, 2019.

2. Pfeiffer, T.J. and Higgins, J.D., "Rockfall Hazard Analysis Using the Colorado Rockfall Simulation Program." Transportation Research Record, 1288, Transportation Research Board, National Research Council, Washington, DC (1990) pp. 117-126.

3. Sargent, R.G., "Verification and Validation of Simulation Models." Proceedings of the 2011 winter Simulation Conference, Phoenix, AZ, 2011, pp. 183-198.

4. F. Bourrier, L. Dorren, F. Nicot, F. Berger, F. Darve. "Toward Objective Rockfall Trajectory Simulation Using a Stochastic Impact Model", Geomorphology, Elsevier, 2009, 11 p.

5. Pfeiffer, T. J., "Rockfall Hazard Analysis Using Computer Simulation of Rockfalls." Master of Engineering Report, Colorado School of Mines, Golden, CO, 1989, 103pp.

6. Slovic, P. "Perception of Risk", Science, Vol 236, 1987, pp. 280–285.

7. Kasperson, R.E., Renn, O., Slovic, P., Halina S., Brown, J.E., Goble, R., Kasperson, J.X., Ratick, S., "The Social Amplification of Risk: A Conceptual Framework", Risk Analysis Vol. 8, No. 2, 1988.

LARGE-SCALE EARTHQUAKE-INDUCED LANDSLIDE REPAIR FOLLOWING NEW ZEALAND'S KAIKOURA EARTHQUAKES

Colby Barrett, P.E., J.D.
CEO, GeoStabilization International
4475 E. 74th Ave., Suite 100, Commerce City, CO USA
303-909-6083
colby@gsi.us

Derrick Hayes, M.S., P.E.
CEO, GeoStabilization International
4475 E. 74th Ave., Suite 100, Commerce City, CO USA
503-523-7306
derrick.h@gsi.us

Prepared for the 70th Highway Geology Symposium, October, 2019

Acknowledgements

Matt Hayes Photography - www.matthayes.co.nz, for the use of photos.
NCTIR for access to work sites throughout the rebuild zones.
South Pacific Helicopters for ongoing helicopter support.

Disclaimer

Statements and views presented in this paper are strictly those of the author(s), and do not necessarily reflect positions held by their affiliations, the Highway Geology Symposium (HGS), or others acknowledged above. The mention of trade names for commercial products does not imply the approval or endorsement by HGS.

Copyright Notice

Copyright © 2019 Highway Geology Symposium (HGS)

All Rights Reserved. Printed in the United States of America. No part of this publication may be reproduced or copied in any form or by any means – graphic, electronic, or mechanical, including photocopying, taping, or information storage and retrieval systems – without prior written permission of the HGS. This excludes the original author(s).

ABSTRACT

In November 2016, a 7.8-magnitude earthquake caused more than 100,000 cubic meters of rock and debris to tumble down and bury the transportation infrastructure below Ohau Point in New Zealand. All routes between the northern and southern portions were South Island severed. To undertake the rebuilding efforts, the New Zealand government established North Canterbury Transport Infrastructure Recovery (NCTIR) – an alliance to repair the road and rail networks between Picton and Christchurch. NCTIR's aim was to have the road open by Christmas 2017, but work also continued into 2018 to complete a number of safety improvements along the highway. The presentation will review the initial remediation attempts, the necessity to re-evaluate the original efforts, the need to incorporate techniques that were "outside-the-box" and different from traditional methods, and the logistics and planning that were required to complete the work safely.

INTRODUCTION

Shortly after midnight on Monday 14th November, a magnitude 7.8 earthquake struck the small South Island settlement of Waiau in North Canterbury, 100km north of Christchurch. The quake was the largest in New Zealand since the magnitude 7.8 Dusky Sound earthquake in 2009 (GNS 2016). While shaking was widespread with over 15,000 recorded 'felt reports,' the worst shaking occurred about 50 seconds after the quake rupturing started. The energy of the tremor progressed north over several minutes with surface rupture recorded on a total of 21 faults. The length of all the fault ruptures combined was close to 100km (GNS 2016).

The degree of ground shaking was high, recorded as Modified Mercalli Intensity Scale (MMI) Level 8 – Severe. This level of shaking causes considerable damage in ordinary buildings with partial collapse including fall of chimneys, factory stacks, columns, monuments, walls, etc. At a human level, people experience difficulty standing; while furniture and appliances shift. Geologically, the damage depends on the geological setting. In Kaikoura and the surrounding area, especially the coastal highway State Highway 1, north and south of the town, many of the slopes consist of over steepened weathered, fractured greywacke. During the quake, the shaking caused significant and widespread damage with a total of 26 major slips (and many smaller slips); closing both State Highway 1 and the Main North Rail Line between Picton and Christchurch.

The earthquake most severely impacted the coastal road and rail which hugs the coastline along a stretch of some 20 km north and south of the coastal tourist town of Kaikoura. These links serve as the South Island's primary rail and state highway routes. The transport corridor follows the coastal alignment and is situated on a very narrow bench at sea level. The coastal mountain range rises from sea level up to some 500 meters of elevation. The earthquake triggered a significant series of rock slides with some 40 major slides reportedly dislodging some 750,000 m^3 of rock and soil debris which buried many sections of the transport corridor. The town of Kaikoura was completely isolated for over a month until access south was re-established via an alternative inland route. During November and December, the initial recovery effort involved helicopter support to service Kaikoura and this effort was primarily supported by the coincidental presence of an international Navy operation which provided airlift and marine support from USA, Canada, Australia, and New Zealand.

The highway and rail north arteries were closed for 12 months and a reconstruction team, North Canterbury Transport Infrastructure Recovery (NCTIR), was quickly established as a government led alliance, tasked with the rebuild. A deadline of 12 months was established to have the full length of the alignment (southern and northern road and rail corridor) opened by Christmas 2017. The monumental task of the design-build recovery at an initial value of $1.5 billion and a workforce of up to 3000 personnel began in earnest in January 2017.

INITIAL REMEDIATION ATTEMPTS

Immediately following the earthquake, it was imperative to get an understanding of the situation; what damage had occurred to what infrastructure. Representatives from the two main transport networks, New Zealand Transport Agency (NZTA) (road) and KiwiRail (rail), and their respective engineers inspected the main routes by helicopter. At the same time, geologists and earthquake specialists from the Institute of Geological and Nuclear Sciences (GNS) and around the world were undertaking more specific scientific inspections. It soon became very apparent that the severity of this quake was far worse than initially thought, and the damage to infrastructure was significant.

Geotechnical engineers identified over 85 slope failures affecting the road alone, while GNS scientists reported 21 faults (later expanded to nearly 30 separate fault surface ruptures). The slope failures affecting the road were categorized into primary, secondary and tertiary failures depending on the level of effect to the road. Primary failures, over 40 in total, were those directly affecting the highway, i.e., presenting an immediate hazard such as debris on the highway, or debris subject to imminent failure. Secondary and tertiary failures would likely affect the highway at some point (following future trigger events) but weren't immediately of concern.

Engineers quickly determined that treatment of the failed primary slopes was going to be non-conventional. The slopes were surrounded by inaccessible terrain and too steep to allow treatment using excavators and bulldozers, as would normally be used in such circumstances. The slopes were also too active, and the area was at high risk of further ground shaking events through the expected typical aftershock sequences.

Heli-sluicing was quickly identified as the most appropriate primary means of treating the slopes while the ground shaking activity subsided; this would also allow time to consider more direct, permanent treatment options. Heli-sluicing involves the use of helicopters fitted with firefighting 'monsoon' buckets dropping around 1000 litres of water at a time directly onto the slope. In the case of Kaikoura, the buckets were filled from the ocean and water washed onto the slopes in a targeted manner. Geotechnical engineers monitored the sluicing operations directing the helicopter pilots to target areas.

Figure 1: Kaikoura heli-sluicing operations - www.matthayes.co.nz

At the peak of operations, thirteen helicopters were in use, and in some cases, up to seven machines focused on single areas. This repeated washing of the slopes released significant volumes of material. While not as effective as manual/mechanical scaling or blasting activities, in the area near Kaikoura heli-sluicing was deemed the only suitable (safe) option in the short term. Adding to reduced effectiveness was the fact that the slopes were dry with very little rainfall in the weeks preceding the earthquake (heli-sluicing is most effective on already saturated slopes where even small volumes of additional water can mobilize, or keep mobilized, saturated soils). The total volume of water used in the heli-sluicing operations exceeded 152 Million litres (the equivalent of over 60 Olympic-sized swimming pools) which had a profound effect on the slopes.
Sluice cannons were proposed initially but considered by NCTIR to be too ineffective and the decision was made to use helicopters. The benefit of using cannons is the delivery of high pressure and high volume. Heli-sluicing provides high volume, but only low-pressure water delivery. To improve the effectiveness of the helicopter operations, some operators, especially those from the Central Otago region of the South Island where heli-sluicing is quite common, used modified buckets with a funnel at the outlet to concentrate the flow of water and increase the delivery pressure.

After some weeks of heli-sluicing, several of the sites were able to be treated through more traditional manual scaling operations using geotechnical rope access techniques. The slopes were treated from the top down with scaling work focused on the head scarp area of the failures. This was to reduce the risk to the geotechnical rope access technicians by limiting their exposure to rockfall further down the slopes, while still managing to create safer work areas at the base of the slopes for protected machinery to operate.

Extensive manual scaling and blasting began in earnest initially in the southern zone (the area of State Highway 1 south of Kaikoura) where NCTIR saw an opportunity to reopen this section of the road before Christmas 2016. Rope access technicians were flown to the top of the slips and began clearing all large debris from the top down. Blasting was common on hard-to-move blocks, but high-capacity airbags were more commonly used as a fast and highly effective way to dislodge blocks weighing more than 20 tonnes.

Figure 2: Before and after extensive heli-sluicing (and some earthworks) on Slip 3 North

Shortly prior to Christmas 2016 the first heavy machinery was used to clear slip material from the road. The initial efforts used remote-controlled excavators, which were originally developed following the 2011 Christchurch earthquakes. This technique provided heavy machinery access into areas where ongoing rockfall presented too high a risk for human-operated machines, even with the most heavily armoured equipment. These efforts, clearing hundreds of thousands of cubic meters of slip material from the road, led to the southern zone reopening for the busy Christmas holiday traffic. Traffic resumption allowed a degree of normality to return to Kaikoura and boosted the local economy which is heavily dependent upon on tourism.

Heli-sluicing continued at various locations around Kaikoura through until May 2017 with an estimated cost of USD 1.2M per week. Manual scaling activity reduced during September and October 2017 with no further large-scale work undertaken by Christmas of that year. Earthworks were all but complete by Christmas 2017.

REEVALUATING THE ORIGINAL EFFORTS

Following Christmas of 2016, the focus moved onto designing and constructing more permanent stabilization solutions, as well as focusing on reopening the road in the northern zone (SH1 north of Kaikoura). The first durable design, for Slip 6 North Ohau Point, was released to invited tenderers in February 2017. This site was identified as the largest, most hazardous site on the network, and a potential choke point for rebuilding the road north of Kaikoura; due to its sheer size, the significant area requiring treatment, and the high degree of exposure from out-sized rockfall.

Figure 3: Ohau Point, December 2016 (photo courtesy Opus)

Hiway Geostabilization (HGS), a joint venture between Geostabilization International and the Hiway Group from New Zealand, were the successful tender winners with work starting mid-March of 2017. The solution for this slope was typical for almost all the remainder of slope treatments throughout the Kaikoura region and included pattern bolting and meshing of affected areas. Due to durability requirements, most of the slopes, including Ohau Point, were treated with high tensile steel wire TECCO mesh from Geobrugg which has a superior protective coating well-suited for coastal environments.

Completion of the installation at Ohau Point would allow safe access to the lower reaches of the slope that required additional remedial work, as well as significant debris clearance on the highway platform below. Before the meshes' and anchors' installation, significant manual scaling was needed to prepare the slopes and make them safe for access. Again, this was typical of all sites in the rebuild. The total area of treatment on the upper slope at Ohau Point was in the order of 5,600 m^2. The lower face, requiring manual scaling and localized rock bolting, had an area of around 12,500 m^2. The total area of slopes treated with this same or similar solution exceeded 50,000 m^2.

For Ohau Point, HGS proposed the use of low carbon, low phosphorous steel hollow bar product as an alternative to the proposed standard galvanized steel hollow bar. The reduced carbon content bar improves both the durability and ductility of the steel, which is critical in a coastal, high seismic environment.

Other sites throughout the rebuild involved very similar designs with local variations in dowel/nail depth and spacing. High angle, high risk, unstable slopes were predominantly treated with dowels/nails and high tensile steel wire mesh while low angle, lower risk sites used dowels/nails and mild steel wire double twist mesh. No sites were treated with dowels alone although several sites were treated with various draped mesh solutions.

In addition to the slope treatments, extensive rockfall and debris flow barrier installations were completed in the south with several long, moderate capacity (2000 to 3000 kJ) systems planned for construction in the north zone during 2018. The majority of the systems have been installed at or slightly above road level which reduced installation time and costs. In areas where space was available, rockfall and debris flow bunds were constructed. All these systems have already been proven effective with several requiring emptying and repair following recent (early 2018) rainfall-triggered slope failures. At these specific sites, no material reached the highway.

Figure 4: Rockfall barrier adjacent Raramai Tunnels

Figure 5: Rockfall/debris flow bund

THINKING OUTSIDE THE BOX

Several sites were remediated with unique solutions considered outside-the-box compared to more traditional slope remediation techniques. One of these is Site 14 South where treatment of the source area was deemed not to be cost-effective. A bespoke rockfall attenuator was designed for the site in combination with the early massive scaling and heli-sluicing operations. A sluice cannon was also used at this site, with great effect, to remove more unstable surficial material.

The installed rockfall attenuator is a modification of a Geobrugg rockfall barrier. The structure of the barrier is the same in regards to posts, base plates, and upslope anchors, but varies with a reduced number of lateral wire ropes (no bottom rope is used). The mesh and ring net of the main system is not attached at the bottom and is left to hang down the slope beyond the base of the normal rockfall barrier structure. This provides similar energy absorbing capacity of the original barrier, but allows the captured material to continue downslope to an area where it can be easily removed. These systems are ideal for sites where a high volume of material is expected to impact the system, or where access for future cleaning and maintenance is difficult.

This is the second installation of these high energy attenuators in New Zealand with two Geobrugg 3000 kJ and one 5000 kJ system installed on SH6 at Diana Falls (Haast Pass) following a significant landslide event in 2013. While these systems have required ongoing maintenance due to a very high energy environment, the road has not yet been affected by rockfall since the installation was completed.

Ohau Point was another site where non-traditional solutions were employed. During the early stages of the rebuilding process, the New Zealand Government pledged that up to NZ $2B would be spent to open this vital road and rail link. In addition to this funding pledge, the government pledged that the road would be open by December 15, 2017. The original programme for Ohau Point allowed for approximately three months of construction time including manual scaling, pattern bolting, and mesh installation. Understandably, the client required access to the road and beach platform below in the shortest timeframe possible to enable construction of a new seawall and highway.

Figure 6: Installing the drape on Ohau Point

Along with this increased pressure from the government and the people of New Zealand, significant weather delays, unfavourable winter conditions, and challenging & dangerous ground conditions were causing delays in completing the original design work. To accelerate access below Ohau Point, NCTIR requested HGS provide an alternative methodology/solution.

Some options to accelerate the programme for the original solution were proposed; however, after lengthy consideration, all were deemed to be unsuitable for various (mainly safety-related) reasons. HGS then started to think outside-the-box and suggested some alternative solutions including a polypropylene fiber-reinforced Gunite treatment of the entire slope.

After careful consideration including material availability and project schedule, the client accepted a ring net drape option to be installed as an addition to the existing permanent pinned mesh solution. Sequencing involved installing the drape system immediately and, in its entirety, with tensioned wire ropes across the bottom of the drape creating a closed system allowing access to the beach platform below, with the pinned mesh solution constructed later. The drape design included perimeter anchors connected with wire rope, a primary ring net drape over the full 5,600

m^2, followed by a secondary high tensile mesh drape over the top of the ring net. Lastly, wire ropes were installed across the base of the system and tensioned.

While in many situations the original high tensile mesh could be installed to act as a drape and no further work would be required, the large size of potential rock failures behind the mesh at Ohau would have risked compromising a high tensile system. The ring net was chosen to protect against more significant potential failures, with the high tensile secondary mesh installed to contain the smaller material. A 6-on-1 ring net was chosen as it maintains its width when lifted (i.e., doesn't choke in the middle) – which is critical when installation is conducted via helicopter. The ring net was installed first as it conforms to the slope better than the high tensile mesh. At the completion of the drape, the pattern bolts of the original design were installed directly through the drape with all components (ring net and high tensile mesh) which completed the permanent system.

LOGISTICS & PLANNING

Undertaking construction work of this nature is challenging at the best of times, even when resources, services, and transport options are abundant. Add to the mix the remoteness of the Kaikoura area, combined with severely restricted services following a major earthquake (including limited transport routes and essential emergency services), and it is easy to imagine the significant logistical exercise these projects became.

Planning for replenishment of materials and consumables required careful consideration of numerous factors, none the least the weather. With construction continuing through the winter, an average week saw the loss of 2-3 work days. The steepness of the terrain in the region means that

Figure 7: Installing the drape on Ohau Point

the majority of the slip sites have only helicopter access to and from the work site. This predestined air delivery of all staff, materials, tools, plant & equipment, and regular replenishing of consumables (such as water and fuel for grouting operations) - a very laborious effort.

While helicopters are great tools in the geotechnical construction industry, they have considerable downsides. They are highly dependent on weather, and the Kaikoura region is subject to regular low sea fog which inundates the coastal cliffs; as well as strong coastal winds. For safety reasons, work around Kaikoura was cancelled for the day if either of those conditions set in since the only emergency rescue option available is via helicopter and being left stranded on the top of the site overnight is not very appealing, nor safe, for geotechnical rope access workers.

With such irregular and uncontrollable delays inflicted on the construction programme, daily and weekly planning takes considerable time and effort. Coordination with local and national providers is essential. Even with the significant improvements to the slip sites, the road south of Kaikoura was regularly closed due to fresh activation of slips. This led to considerable delivery delays and local providers had only limited stock that needed to be continuously replenished.

For the redesign of Ohau Point, considerable challenges needed to be overcome. The design (by HGS) and review process were completed within one week, and perimeter anchor installation began immediately following. Sourcing of the 6-on-1 ring net resulted in a worldwide search, and eventually, the global supply was exhausted. Suppliers were found in North America and Western Europe, and due to the significant pressure from the government, the ring net was flown into New Zealand on specially chartered aircraft. The ring net installation began three weeks after the start of the perimeter anchor installation.

Further logistical issues arose with New Zealand's supply of shackles, required to connect the ring nets, being exhausted before even the first order was completed. The remainder of the product came from as far away as Europe as Australian suppliers scrambled to maintain stock. This replenishment in itself became a daily struggle, balancing daily quantities of ring net installation with urgent courier deliveries of bags of shackles.

While Ohau Point had significant challenges, these were combined with logistical problems from 40 other rockfall construction projects, over 60 km of road reconstruction, over 1 million cubic meters of earthworks, and rebuilding of 100 km of rail including collapsed bridges and damaged tunnels.

SUMMARY

The rebuild of the road and rail networks damaged after the Kaikoura earthquakes has been the largest such undertaking ever in the history of New Zealand. With four of the largest contracting firms in the country combining their resources to form the NCTIR alliance, and a plethora of sub-contractors ranging from bridge experts to rockfall specialists, at times up to 3000 workers have been involved in the rebuild.

Remedial measures have at times been somewhat rudimentary and at others entirely left field from more traditional solutions. During the initial response, many decisions were made on the fly in order to simply make the area safe for restricted access, and while longer term, permanent solutions were being designed. Add to this, political pressures requiring significant design changes only further complicated the already incredibly challenging task of working in this environment.

Thankfully, the target to get the road open on December 15th, 2018 was achieved to the relief of the nation.

Major project milestones:

- 19th November 2016 - Inland Road into Kaikoura open, 5 days after quake
- 12th December 2016 - SH1 south into Kaikoura open (restricted)
- 9th June 2017 - First rail into Kaikoura (from south)
- 8th September 2017 - SH1 north into Kaikoura, safe route under Ohau Point established
- 15th September 2017 - Rail open north and south of Kaikoura. The first train to pass through Kaikoura since the November earthquake.
- 15th December 2017 - SH1 Road Open. One year, one month, one day following the earthquake.

REFERENCES

- GNS Science, 2016. M7.8 Kaikoura quake the biggest since the Dusky Sound jolt in 2009 - 15/11/2016, *GNS media release*, 14/11/2016 7:49am
- FHWA, 2009. Use of Ring Nets for Slope Protection for Rockfall: End-of-Construction Report. Experimental Feature Study, Geotechnical Report C-7540, SR 28, Rock Island – Rock Slope Nettings.
- NCTIR, 2017. SH1- Ohau Point, the last slip in the red zone. September-November 2017 - milestones achieved. *NCTIR Newsletter, online release.*

Assessment of Unstable Rock Columns at the Tieton Royal Columns, SR-12, Oak Creek Wildlife Area, Naches, WA

William C. B. Gates, PhD, PE, LEG
McMillen Jacobs Associates
1101 Western Avenue, Suite 706
Seattle, WA 98104-2988
206-496-4829
gates@mcmjac.com

Dale Moore, PG, LG
McMillen Jacobs Associates
1101 Western Avenue, Suite 706
Seattle, WA 98104-2988
360-737-4762
dmoore@mcmjac.com

"Prepared for the 70th Highway Geology Symposium, October 2019"

Acknowledgements

The authors wish to acknowledge and thank Ross Huffman and Greg Mackey of Washington Department of Fish and Wildlife.

Disclaimer

Statements and views presented in this paper are strictly those of the author(s), and do not necessarily reflect positions held by their affiliations, the Highway Geology Symposium (HGS), or others acknowledged above. The mention of trade names for commercial products does not imply the approval or endorsement by HGS.

Copyright Notice

Copyright © 2019 Highway Geology Symposium (HGS)

All Rights Reserved. Printed in the United States of America. No part of this publication may be reproduced or copied in any form or by any means – graphic, electronic, or mechanical, including photocopying, taping, or information storage and retrieval systems – without prior written permission of the HGS. This excludes the original author(s).

ABSTRACT

Columnar volcanic flows of andesite and basalt crop out as rimrock along SR-12 from Rim Rock Dam to the junction of SR-410 in Eastern Washington. One of the rimrock areas have been appropriately named The Royal Columns which consist of sub-vertical columnar joints formed in the volcanic flows of the Tieton Andesite above the Tieton River. This is popular area for rock climbing enthusiasts.

Because of the popularity of the area, Washington Department of Fish and Wildlife (WDFW) became concerned about an apparent unstable column of andesite that appeared to be separating from the main cliff when climbers reported a new tension fracture gradually opening to at least 2 inches between the hanging wall and the footwall in 2016. In April 2017 the rock column collapsed and destabilized adjacent rock columns. Fortunately, no climbers were injured and WDFW immediately closed the climbing area until the rockfall threat could be investigated. WDFL asked McMillen Jacobs Associates (MJA) to investigate the recent rockfall. Most of the rock column had collapsed onto the bench and talus slope, however a residual rock wedge behind the original column of rock remained which posed a potential risk to climbers. The authors investigated the rockslope assisted by rope access techniques. The rock column appeared to initially have failed as a wedge and transitioned to direct toppling. The authors scaled approximately 3-CY of loose rock. After safety scaling was complete, the authors recommended the climbing area reopen for recreational use with the understanding, rockfall is an inherent risk.

INTRODUCTION

Location

Royal Columns is one of the less known rock climbing areas in Eastern Washington. Climbers started visiting the area in the late 1960s and really started frequenting the area in the 1980s. Present day, it is a very popular area for hiking and climbing. The Columns are directly across the Tieton River from the Oak Creek Wildlife Area headquarters, about 3 miles west of where State Route (SR)-12 splits off from SR-410 (Figure 1). By SR-410 it is about 2.5 hours from Seattle.

Figure 1- Location map for Royal Columns, Naches, WA. The project area is about 160 miles from Seattle, WA.

Geology

Royal Columns are part of the natural rimrock which borders the Tieton River at many locations along SR-12. The rock units are part of the Tieton andesite volcanic flows which cooled approximately one million years ago (1). Gusey and others (2) have postulated based on age dating that the Tieton andesites erupted from vents at Bear Creek Mountain near the core of the extinct Goat Rocks volcanic complex in the Cascade Mountains south of White Pass, WA; about 40 miles west and up river from the Royal Columns. In addition, the authors have suggested that this volcanic unit may be the longest know andesite flow in the world (2). The Tieton andesite is well known for its remarkable examples of columnar jointing. Royal Columns (Figure 2) is a great example where the rimrock forms clean classic sub-vertical columnar joints or colonnades that crop out above the river and US-12. Rockfall from the unstable columns is a common problem along US-12.

Figure 2 – The Tieton andesite rims the Tieton River canyon in many locations; because of the columnar structure, the local rimrock was named Royal Columns. This is a very popular rock climbing area in Eastern Washington. The Tieton River in the foreground.

BACKGROUND

Because of the popularity of the area, Washington Department of Fish and Wildlife (WDFW) became concerned about an apparent unstable column of andesite that appeared to be separating from the main cliff when climbers reported a new tension fracture gradually opening between the hanging wall and the footwall of the column in 2016 (3). Since US-12 was close by, WDFW contacted Marc Fish with Washington Department of Transportation (WSDOT) for advice on what to do. Marc referred them to Bill Gates with McMillen Jacobs Associates (MJA). Upon the referral, WDFW asked MJA to visit the site in 2016 to assess the unstable slope and provide recommendations for mitigation; but then postponed the trip because of budget issues. However, when the tension fracture appeared to be opening at a faster rate (from ¼-inch to over 2-inches) in the spring of 2017, they asked MJA to visit the site, ASAP. Before MJA scheduled visit, Gates received a call that at 0650 on Friday April 28th, 2017, the rock column collapsed and destabilized adjacent rock columns. Fortunately, no climbers or visitors were injured. WDFW immediately closed the climbing area until the rockfall threat could be investigated thoroughly. Most of the rock column had collapsed onto the bench and talus slope, however, a residual rock wedge behind the original column of rock remained which posed a potential risk to climbers (Figure 3).

FIELD ASSESSMENT

The authors mobilized to the Royal Columns climbing area in the Oak Creek Wildlife area Thursday May 11th of 2017 and met with WDFW to discuss the scope of the project. WDFW discussed the events leading up to the column failure, including interactions with a climber who frequented the area, who had been tracking the opening tension fracture behind the column before the block collapsed. Rock blocks from the column collapse were deposited between the toe of the slope and the meadow below (Figures 3 & 4). Blocks as large 7-CY were observed to have fresh broken and muddy surfaces suggesting they originated from the column failure (Figure 5).

Figure 3 – Photo on left identifies the present source area for the rockfall. The photo on the right shows one of the remaining andesite wedges after the column collapsed.

Figure 4 - Failure area and rockfall runout path. Note size of blocks and climbing anchors (rock bolt hangers).

Figure 5 – Bill Gates inspecting muddy surfaces on rock blocks linking detached blocks to failure area.

To evaluate the rock mass geology where the rockfall event occurred; the authors rigged a rappel line above the failed column and conducted a vertical scan line by rope access techniques to characterize the engineering geology (Figure 6).

Figure 6 – Dale Moore using rope access techniques to characterize the engineering geology and structure.

The overall approach to evaluate project area was comprised of five methods used to classify the quality and the stability of the rock mass. During the investigation the following methods were employed:

- Geomechanical rock mass classifications;
- Uniaxial compressive strength (UCS);
- Rock Quality Designation (RQD);
- Kinematic analysis; and,
- Tilt testing

Geomechanical Rock Mass Classification

As part of the field assessment, the authors developed the geomechanical rock mass classification for the project area. Two of the widely accepted classification systems are the Rock Mass Rating System (RMR) and the Geological Strength Index (GSI). RMR, also referred to as the geomechanics classification system by Bieniawski (4), is based on the algebraic sum of six rock mass property ratings, namely:

- Strength of intact rock material;
- Rock Quality Designation (Palmström's volumetric method);
- Spacing of discontinuities;
- Condition of discontinuities;
- Groundwater conditions;
- Orientation of discontinuities relative to the rock slope.

Field data was compared to tables published by Bieniawski (4) to estimate RMR. RMR can range from 0 to 100. From the ratings, rock class, corresponding descriptions, and engineering properties are assigned to the overall rock mass (Bieniawski's RMR classification). Bieniawski's RMR classification can be related to Hoek's Geological Strength Index (GSI). GSI = RMR_{89}' – 5 where RMR_{89}' has the groundwater rating set to 15 and adjustment for joint orientation set to zero (5). In addition, the GSI rating can be estimated directly from information collected from field mapping.

Uniaxial Compressive Strength

The authors used a geological hammer to indent or break rock specimens to estimate rock strength in the field. The results were compared to published tables by ISRM (6), Hoek and Bray (7) and Wyllie and Mah (8) on field estimates of rock strength. These values were converted to approximate uniaxial compressive strength of the rock.

Rock Quality Designation

To estimate the rock quality designation (RQD), the authors employed the joint volume relationship by Palmström (9), where RQD % = 115 – 3.3*J_v, where J_v is the total number of discontinuities per cubic meter. To evaluate J_v in the field, we measured a cleft of rock section in the x, y, and z directions and summed the discontinuities, then computed the RQD.

Peak Friction Angle

To estimate peak friction of the rock blocks, the authors completed ten separate tilt tests described by Kliche (10) (Figure 7). The highest and lowest peak friction values (60° and 32°) were removed from the data set and the average of the remaining data was assumed to be the peak friction angle, 40°.

Figure 7 – Dale conducting tilt tests of matching rock clasts to estimate peak friction.

SITE CONDITIONS

Rockfall occurs routinely from the Royal Columns as evidenced by talus deposits at the base of the columns and on the east side of the meadow below the climbing area. Column separation varies between 1 to 12 inches (Figure 8) along the length of the climbing area.

The Tieton Andesite mapped in the project area is dark gray on weathered faces, and light gray and black on fresh exposures. Geologic hammer tests indicate the intact rock strength varies from strong rock (R4) to very strong rock (R5), equating to 7,000 to 36,000 pounds per square inch (psi). The rock mass is columnar, with seven-foot diameter columns between 60 and 70 feet tall. The RQD calculated with Palmström's volumetric method is approximately 98.5%. Five discontinuity sets plus random joints are spaced between 1 and 4 feet. Discontinuities in the Tieton Andesite are generally rough to slightly rough with planar to undulating surfaces, and aperture widths between 1 and 12 inches. The rock mass is slightly weathered on exposed faces. In summary, the base RMR of the Tieton Andesite is estimated to be 89 suggesting Class I, very good rock. The GSI of the rock mass is about 84.

Figure 8 – Gates observing variable aperture widths between rock columns.

STABILITY ANALYSIS

The authors collected physical rock engineering information attendant to the rock slope, including dip and dip direction orientations of discontinuities. These data were plotted on stereonets using the computer program Dips® Version 7.0 by RocScience® (11). Where the stability of a rock slope is controlled by the structure of the rock mass, a Markland analysis is a well-documented and widely accepted design tool used to evaluate the rock slope stability (7). The information required to perform the analysis are the slope dip and dip direction, the orientation of the discontinuities within the rock mass, and the friction angle of the lithologies represented in the rock mass. The design team used a friction angle of 40 degrees for the Tieton Andesite based on tilt testing completed in the field. The design team considers the overall orientation of slope in the project area to be approximately 70 feet high, dipping 85° toward the west (285°).

The results of the Markland analysis show the columnar andesite at the Royal Columns is most likely to fail by direct toppling. In addition, observations in the field indicated an inclined basal plane of 55° (Figures 9 & 10) which further exacerbated the column instability by allowing the columns to slide on the basal plane in a wedge like fashion or special type of planar failure then topple. Of interest, the new Version 6 of Swedge® by RocScience® (12) allows the investigator to analyze the basal plane of a wedge. In this case, the arms of the wedge are sub-vertical (Joint sets 3 & 6) and the intersection lies on the basal plane (Joint set 2) which dips out of slope and exceeds the friction angle. Furthermore, for sliding and toppling to happen, the following conditions were met. The dip of the basal plane exceeded friction (55° > 40°) and the ratio of the base width of the column to the height (b/h) was less than the tangent of friction (7'/70' < tan 40°).

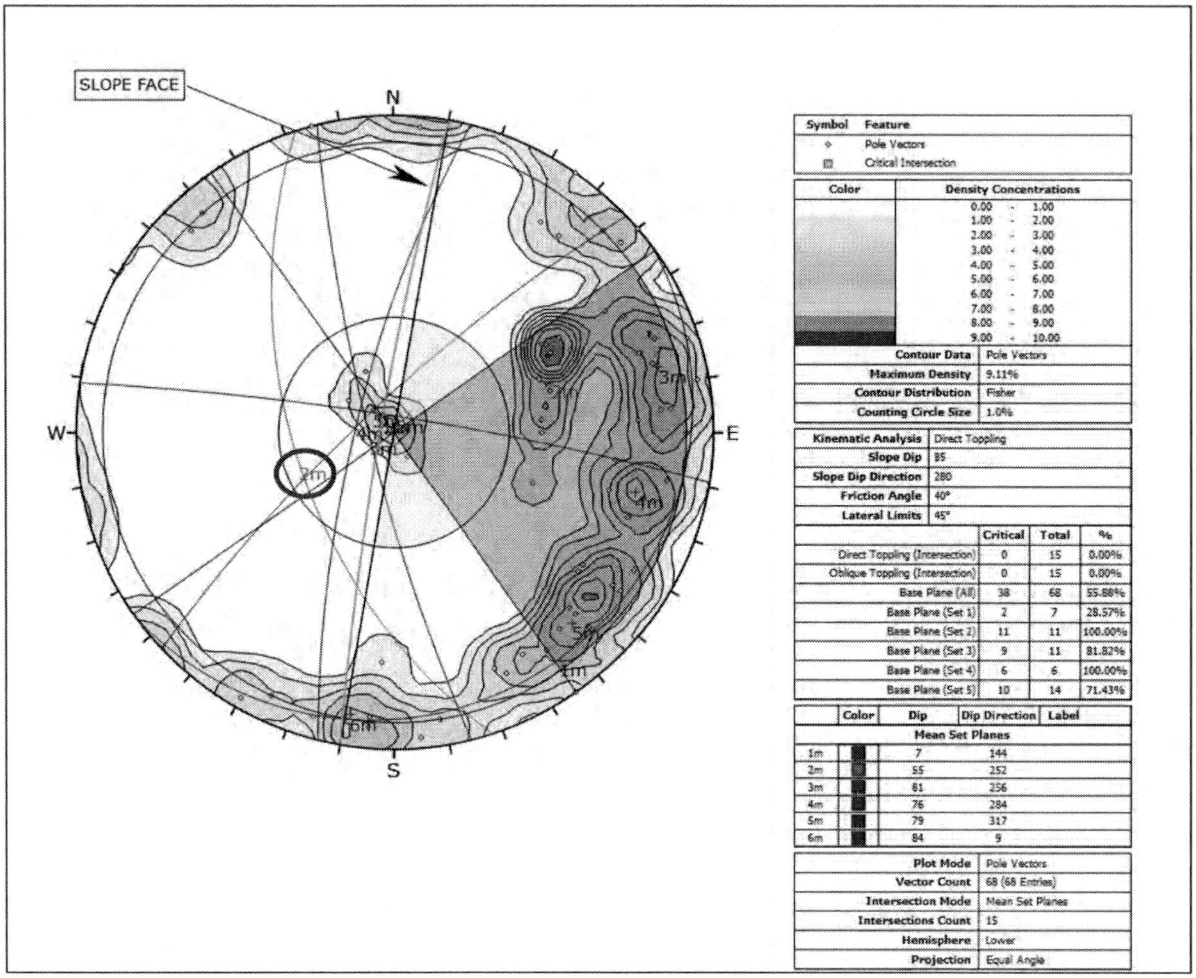

Figure 9 – Stereonet displaying kinematic analysis of direct toppling rock block failure, Joint set 2 is the basal plane on which the block failed. The joint faces of the neighboring columns intersect forming sub-vertical arms of unique wedges siting on the basal plane. This column failed by sliding on the basal plane then transitioning into direct toppling.

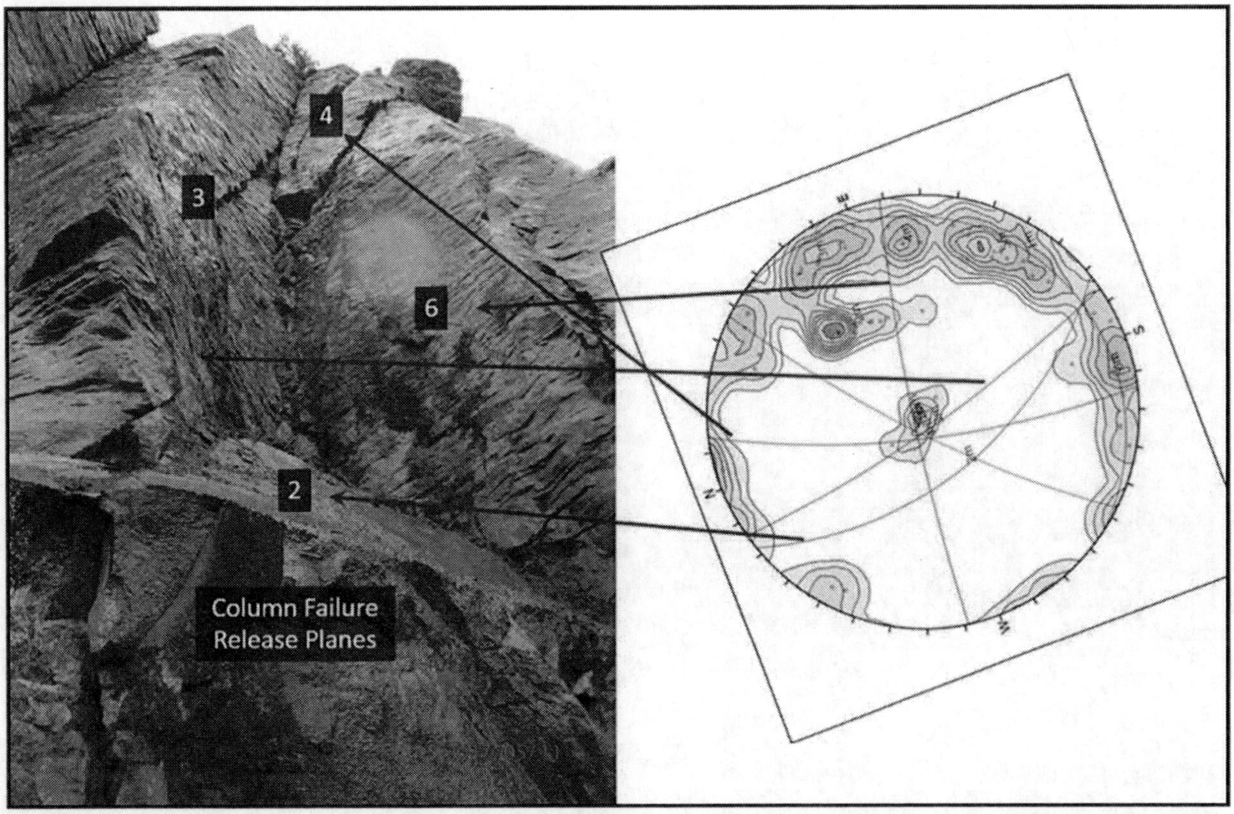

Figure 10 – Photo shows the release planes (3, 4, 6) and basal plane (2). Stereonet displays the mean set planes from Dips® analysis; note, the stereonet has been oriented to the slope. Joint number labels reference set numbers in Dips® analysis.

ROCK SCALING

The authors returned to the site on May 25th, 2017 to complete safety scaling in the project area. The goal of the safety scaling was to remove loose material from the residual rock wedge left behind after the column failure. The authors used steel hand scaling bars to pry loose rock from the residual rock wedge. Approximately 3-CY of material was removed near the base of the wedge (Figure 11). Safety scaling reduced but did not eliminate the rockfall hazard at the climbing area. After safety scaling, the authors recommended WDFW reopen the Royal Columns climbing area for recreational use (3).

Figure 11- Before and after safety scaling. The loose wedge of rock on the left was removed with a scaling bar.

CONCLUSIONS

The Royal Columns is a popular climbing area, in part, because of the columnar formations of the Tieton Andesite in that locality. Columnar volcanic flows are naturally susceptible to toppling failures due to their inherent geometry, tall columns on a narrow base. A history of rockfall at the Royal Columns climbing area is evidenced by the accumulation of rockfall debris in talus slopes on the east side of the meadow below the columns.

Column separation widths vary throughout the climbing area and were observed to be up to 12 inches wide. The failed column which prompted this study was observed to have progressive separation from adjacent columns prior to failure. Columns with inclined basal planes are at increased risk of failure from basal plane sliding. Column failure is expected to continue at Royal Column climbing area as the Tieton Andesite continues to age and weather, and column separation increases through time.

Safety scaling completed on May 25th, 2017 reduced but did not eliminate the hazard of rockfall in the climbing area. The authors expect rockfall to persist at the Royal Columns climbing area as the Tieton Formation continues to weather and additional joints form (3).

RECOMMENDATIONS

The authors recommended the Royal Columns climbing area be reopened for regular use. In addition, we recommend posted signage at the trailhead to include up to date information about recent rockfall and any additional hazards identified by WDFW (3).

REFERENCES

1. Hammond, Paul and Pat Pringle. *Selected geologic sites along US 12 between Yakima and the Cowlitz River Valley*: NAGT 2008 Fieldtrip http://www.centralia.edu/academics/earthscience/pubs/hammond_pringle_nagt2008.pdf

2. Gusey, D. L., P.E. Hammond and J.P. Lasher, *The Tieton andesite lava flows- of great length, but where did they come from?* Abstract, GSA Annual Meeting in Denver, Colorado, USA, 2016.

3. Gates, W.C.B. and Dale Moore, McMillen Jacobs Associates. *Royal Columns Rockfall Assessment, for Washington Department of Fish & Wildlife, Subject: Royal Columns Climbing Area Rockfall Reconnaissance and Stability Assessment*, May 31, 2017.

4. Bieniawski, Z.T. Engineering rock mass classifications. New York: Wiley, 1989, 251 p.

5. Hoek, E., and E.T. Brown, *Practical estimates of rock mass strength*. International Journal of Rock Mechanics and Mining Science, Vol 34, No. 8, 1997, pp. 1165-1186.

6. International Society of Rock Mechanics (ISRM). *Rock Characterization, Testing and Monitoring; ISRM Suggested Methods*. Pergamon Press, Oxford, UK, 1981.

7. Hoek, E., and J.W. Bray, *Rockslope Engineering*. Institution of Mining and Metallurgy, 1981, 357 p.

8. Wyllie, D.C., and C. W. Mah, *Rock Slope Engineering: Civil and Mining, 4th Edition*, Spon Press, 2004, 431 p.

9. Palmström, *A., Measurements of and Correlations between Block Size and Rock Quality Designation (RQD)*; published in Tunnels and Underground Space Technology 20, 1985.

10. Kliche, C.A., *Rock Slope Stability*: Tilt tests and Joint Roughness, 1999, pp. 38-41.

11. RocScience Inc. *DIPS, Plotting and Analysis Software for Geomechanical Structural Data*. Toronto, Version 7.0, 2016, Canada.

12. RocScience Inc. *Swedge, Plotting and Analysis Software for Geomechanical Wedge Analysis*. Toronto, Version 6.0, 2016. Canada.

OR213 MP 10 Spangler Hill Emergency Slide Mitigation & Monitoring

Max A. Gummer, P.E.
Oregon Department of Transportation
123 NW Flanders St.
Portland, OR 97209
(503) 731-8415
max.gummer@odot.state.or.us

Prepared for the 70th Highway Geology Symposium, October, 2019

Acknowledgements

The author would like to thank the individuals & entities for their contributions in the work described:

Stacy Stubblefield, P.E. – Oregon Department of Transportation
Brent Black, C.E.G – Landslide Technologies, Inc.
George Machan, P.E. – Landslide Technologies, Inc.
Phil Wurst, P.E., G.E. – Oregon Department of Transportation
Staff of ODOT Engineering Geology / Geotechnical Engineering
Staff of Landslide Technologies, Inc.
OSU RAPID Center

Disclaimer

Statements and views presented in this paper are strictly those of the author(s), and do not necessarily reflect positions held by their affiliations, the Highway Geology Symposium (HGS), or others acknowledged above. The mention of trade names for commercial products does not imply the approval or endorsement by HGS.

Copyright Notice

Copyright © 2019 Highway Geology Symposium (HGS)

All Rights Reserved. Printed in the United States of America. No part of this publication may be reproduced or copied in any form or by any means – graphic, electronic, or mechanical, including photocopying, taping, or information storage and retrieval systems – without prior written permission of the HGS. This excludes the original author(s).

INTRODUCTION

OR213 Cascade Highway South is a rural minor arterial highway located between Portland, Oregon and Salem, Oregon. The average annual daily traffic for 2017 was 15,300 which is an increase of 13% over the past six years. The highway is a critical north-south connector for communities outside of the Portland and Salem metropolitan areas.

In the winter of 2016-2017, record rainfall re-activated at least one ancient landslide within an evident landslide complex causing rapid and significant deflection of the roadway on the order of one foot in two weeks. An emergency subsurface investigation program was implemented and monitoring began. Following an official Emergency Declaration by the Governor of Oregon, both short and long term mitigations were implemented by the Oregon Department of Transportation (ODOT).

PROJECT TIMELINE

Staff members from ODOT Region 1 Geo/Hydro/Hazmat Unit (R1 Geo) were alerted to surface cracking in late February, 2017. Final horizontal drains and long term monitoring equipment were installed March, 2018. Figure 1 details the two schedule paths for this project, the Stage timeline predominantly on the upper half of the figure represents construction activities managed by ODOT Maintenance District 2B staff (D2B). The Phase timeline on the lower half represents design and analysis work conducted by Oregon Department of Transportation Geotechnical Engineers and Engineering Geologists (R1 Geo) and Landslide Technology, Inc. (LTI) staff.

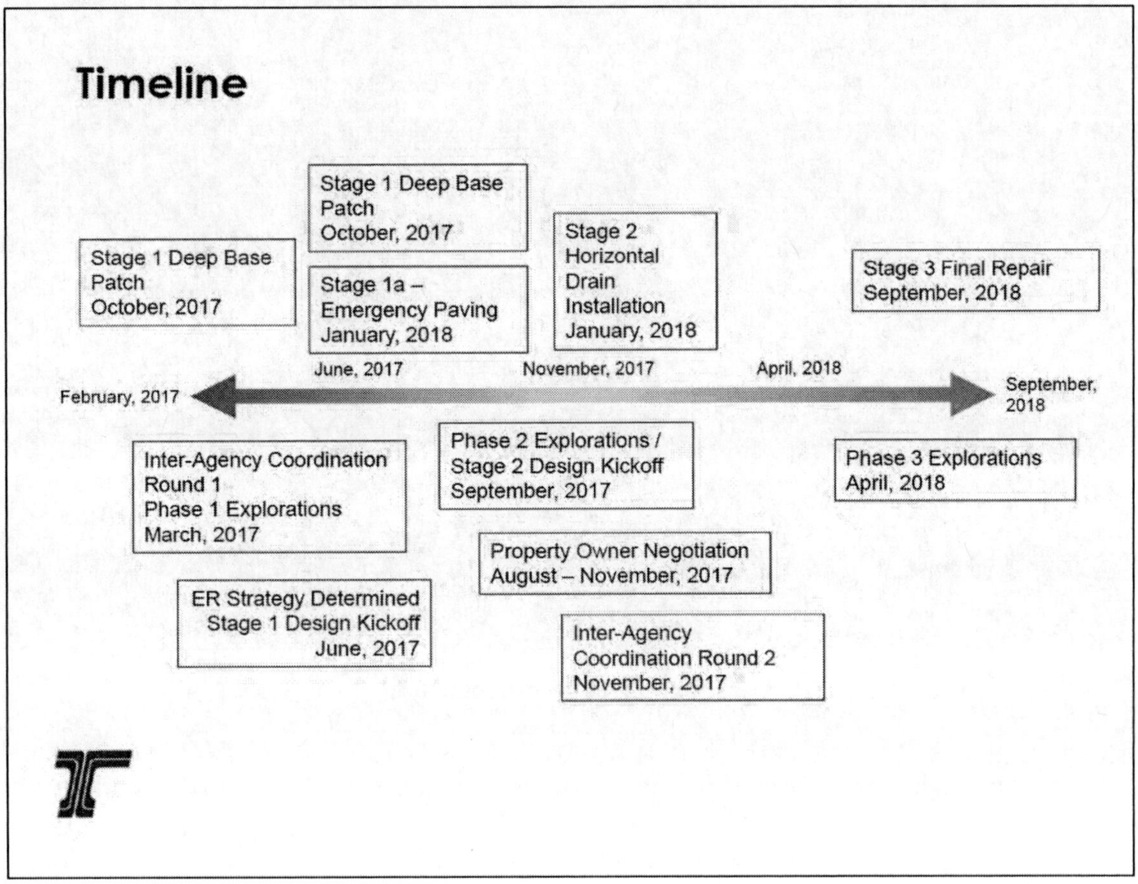

Figure 1: Project timeline

Two to three lurches of the landslide complex necessitated a larger scale repair as quickly as possible. Private property notification and logistics quickly became the critical path in terms of mitigation installation. Once final permit of entry negotiations were complete in November of 2017, long term mitigation was installed within four months.

PROJECT SETTING:

The project is located approximately 20 miles south south-east of Portland on the eastern edge of the Willamette Valley, as shown below in Figure 2.

Figure 2: Project Location

OR213 is a three lane highway that trends in a north south direction with the grade descending from north to south at the location of the landslide. The annual average daily traffic (AADT) at MP 8.9 to the north of the slide is 14,400. The highway has a lane configuration of one southbound lane and two northbound lanes, with an overall edge to edge pavement width of approximately 70 feet. At the location of the cracking, the embankment is estimated to have a maximum thickness of 25 feet to 40 feet.

At the time of initial crack development in February 2017, ODOT Geo staff consulted publically available LiDAR imagery published by the Oregon Department of Geology and Mineral Industries (DOGAMI) and observed evidence of ancient landslide activity, with residences built into the historic slide masses.

The highway is bordered to the east by a recently (2009) re-constructed gravel-lined ditch with a short soil slope inclining up to a vertical basalt rock outcrop. The highway is bordered to the west at the location of the pavement cracking by a descending embankment with a slope of approximately 1.5H to 1V for a vertical distance of approximately 30 feet. Beyond the toe of the embankment the inclination of the slope becomes highly variable with irregular, hummocky topography, typical of previous landslide activity. There is a 42-inch-wide culvert approximately 75 feet to the north of the pavement cracking that outfalls west of the highway and the water continues downslope within a natural drainage channel. The culvert was inspected via video as a result of the initial investigation and found to be in good condition. Following the full displacement of the landslide, the end run of the concrete culvert had separated 0.5 to 1 foot vertically. A seasonal spring and sag pond are present at the toe of the highway embankment. Flow from the spring has been relatively consistent seasonally, from dry to a rate of about one gallon per minute. The sag pond covered an area of approximately 160 square feet at the onset of movement in February, 2017.

GEOLOGIC SETTING

Published geologic maps of the area indicate the site is underlain by Pliocene-age basalt from the Boring Lavas and Miocene-Pliocene-age siltstone and sandstone of the Troutdale Formation (LTI, 2018). This portion of OR213 crosses onto and traverses a previously mapped, very large landslide complex, as shown below in Figure 3.

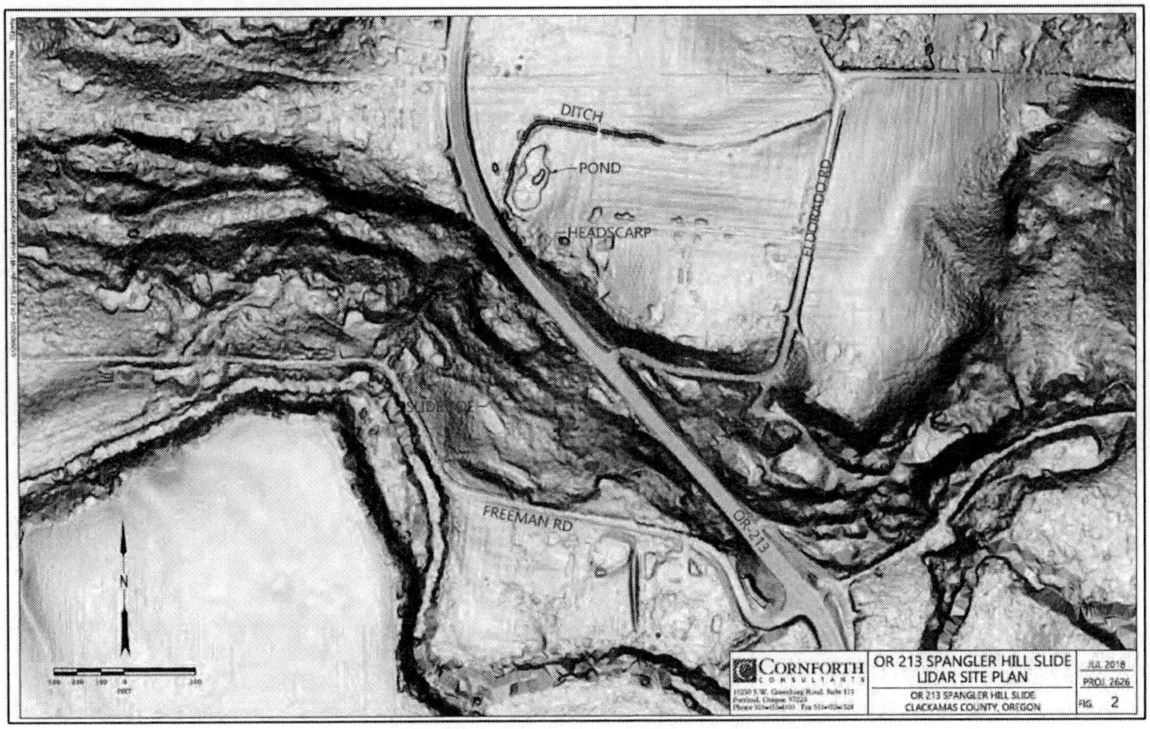

Figure 3: LiDAR imagery of slide complex

This landslide complex is one of several Quaternary landslide deposits mapped within the Milk Creek and Molalla River valleys (LTI, 2018). According to Ma, Madin, Duplantis, and Williams (2012), several of the valleys south of Oregon City have eroded into the Boring Lava plateaus, forming large landslide complexes.

The landslides are described as having developed at the contact between the overlying Boring Lava and the weaker underlying siltstone of the Troutdale Formation.

SUBSURFACE INVESTIGATION

R1 Geo staff conducted Phase 1 explorations to bracket the variability of the site. Four boreholes (TB-01-04) were installed in March of 2017. R1 Geo located the first borehole, TB-01, on the highway shoulder south of the crack on the displaced block. R1 Geo anticipated a total depth for TB-01 of 200-250 ft. in order to evaluate large scale mass failure of the area. After 10 working days, TB-01 was at a depth of 156 ft. due to installation difficulties associated with slope movement. R1 Geo advanced TB-02 at the toe of the regional slope within S. Freeman Rd, and TB-03 around mid-slope. R1 Geo installed 2.75 in. Slope inclinometer (SI) casing and vibrating wire piezometers (VWPs) in TB-01, -02, and -03.

Rapid slope movement quickly deflected TB-01, preventing readings after approximately June, 2017. Phase 2 explorations were conducted to replace the sheared SI in TB-01 as well as provide additional information regarding the slope movement geometry and ground water conditions. A total of three additional explorations were conducted with SI casing installed in one exploration and VWPs installed in all explorations.

LTI conducted Phase 3 explorations in April, 2018 following installation of the horizontal drains to complete the geologic modeling and install long term monitoring equipment. LTI installed six new borings at the site, TB-07 through TB-13.

Figure 4 below shows the locations of all the exploration phases.

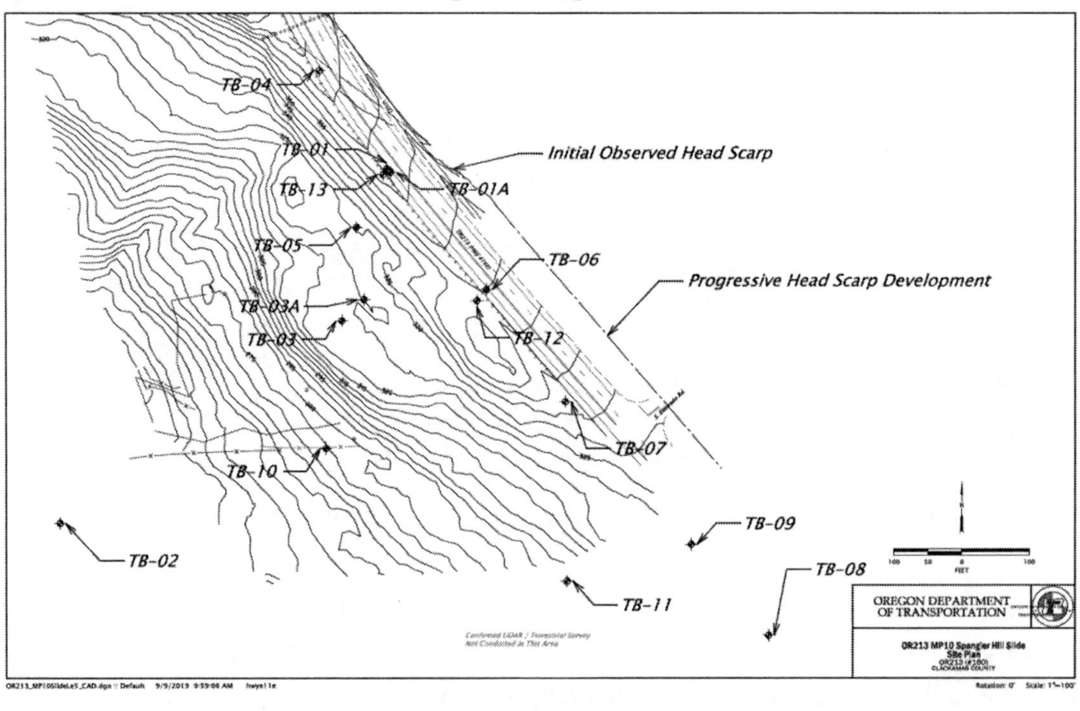

Figure 4: Site plan

Geologic units encountered at the project site, listed from youngest to oldest, include the following: Overburden soil (embankment fill, colluvium, and landslide debris), Boring Lava Rubblized, Boring Lava Basalt, and fine grained sedimentary rock of the Troutdale Formation. Locations of the borings are shown on Figure 4 and a brief discussion of each geologic unit is given below.

Overburden Soil. Overburden soils at the site consist of embankment fill, colluvium, and landslide debris. Embankment fill is interpreted to be present in borings TB-12 and TB-13 based on the fact that embankment fill was encountered in existing and adjacent borings TB-06 (9 feet of fill) and TB-01 (26 feet of fill). TB-12 and TB-13 were "quick" drilled to depths (75.5 and 42.0 feet respectively) well below embankment fill; consequently, no samples of fill were collected for analysis. The embankment fill encountered in existing TB-01 and TB-06 was characterized as silty to sandy gravel with cobbles and boulders.

Boring Lava Rubblized. The Boring Lava Rubblized consists of medium dense to very dense, sandy clayey boulders with some silt and coarse angular to subrounded gravel and cobbles. The unit is referred to as Boring Lava Rubble on ODOT's Summary Boring Logs. The boulders consist of medium hard to very hard (R3 to R5) basalt, up to six feet in diameter.

Boring Lava. The Boring Lava consists of fresh to slightly weathered, medium hard to hard (R3 to R4), gray, basalt. It is highly to slightly jointed (joint spacing between 1- to 6-feet) with three prominent joint sets: a dominant joint set is oriented from 30 to 50° from core axis; two secondary joint sets trend from 0 to 20° and 70 to 80°. All three joint sets display rough planar fracture surfaces with clay infilling. The basalt is finely phaneritic, finely vesicular, and contains occasional vugs up to 1.5 inches in diameter.

Troutdale Formation. The Troutdale Formation, alternatively referred to as the Molalla Formation on the ODOT R1 Geo Phase 1 summary boring logs, is a sedimentary unit of siltstone and sandstone interbeds that was encountered in every boring. This unit underlies all other geologic units at the project site. This unit consist of slightly to very highly weathered to decomposed, moderately to highly fractured, extremely soft to very soft interbeds of siltstone and sandstone. Some of the siltstone can be remolded to clay with hand pressure and displays fracture spacing between 2 inches and 4 feet. The Troutdale Formation contains numerous zones of diced and sheared texture, and numerous fractures displaying slickensides and polished surfaces. Occasional zones containing woody debris were encountered during drilling and occasional thin coal seams were encountered.

INSTRUMENTATION & DATA REDUCTION

Instrumentation at the project site consists of vibrating-wire piezometers (VWPs) and slope inclinometer (SI) casing. The VWPs were installed to measure piezometric pressures to determine groundwater elevations and the SI casings were installed to determine the depth and rate of landslide movement. All explorations except TB-04 have at least a VWP and SI casing installed in the borehole. All borings except for TB-04, -03A, and -05 have SI casing installed in the borehole. A selection of boreholes have single VWP's and the other borings have dual "nested" VWP's to measure ground water at discrete elevations. Boring designations, elevations,

SI casing depth, and VWP tip elevations are summarized in Table 1. Cumulative groundwater levels measured in the VWPs are shown below in Figure 5.

Table 1 – SI & VWP Installation Summary				
Boring	Ground Elevation (ft.)	SI Casing Depth (ft. bgs)	Piezo Depth (ft., bgs)	Piezo Elevation (ft.)
TB-01	362.8	154	98.0	264.8
TB-02	203.3	68	68.0	135.3
TB-03	323.8	98	73.0	250.8
TB-03A	323.1	N/A	42.0	281.1
TB-05	328.0	N/A	44.0	284.0
TB-06	280.0	86	82.0	198.0
TB-07	317.8	118	76.0 106.0	241.8 211.8
TB-08	306.0	132	69.0 115.0	237.0 191.0
TB-09	336.5	124	61.0 91.0	275.5 245.5
TB-10	274.4	82	23.5	250.9
TB-11	292.1	82	40.0	252.1
TB-12	348.0	122	90.0 108.6	258.0 239.4
TB-13	362.2	120	89.0 121.0	273.2 241.2

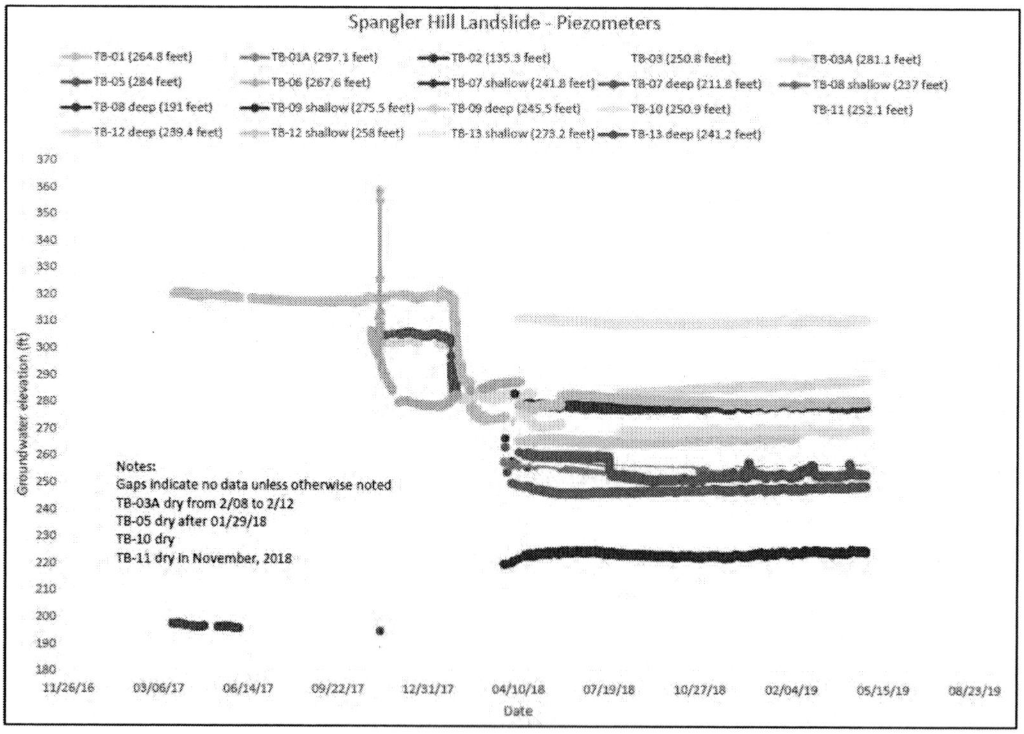

Figure 5: Cumulative groundwater monitoring data

As described previously, the horizontal drains were installed in January and February, 2018. The significant drops in hydraulic head shown in TB-01, -03A, -05, and -06 illustrate the effectiveness of the installed drains.

The figures below display representative movement profiles from Phase 1 and Phase 3 explorations.

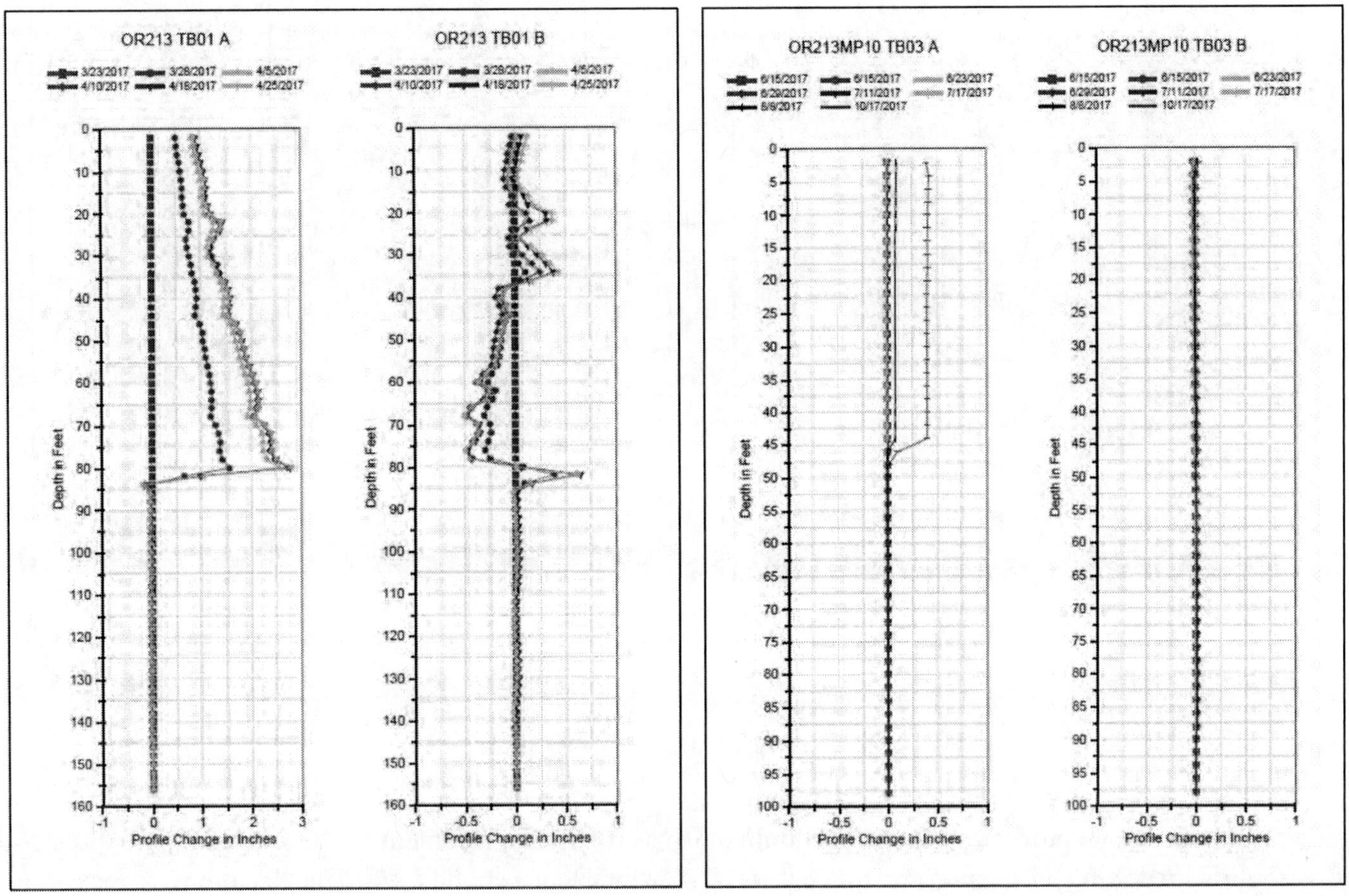

Figure 6: Phase 1 displacement profiles

Phase 1 displacement profiles, combined with the initial subsurface investigation, led R1 Geo and LTI to develop the design geologic model

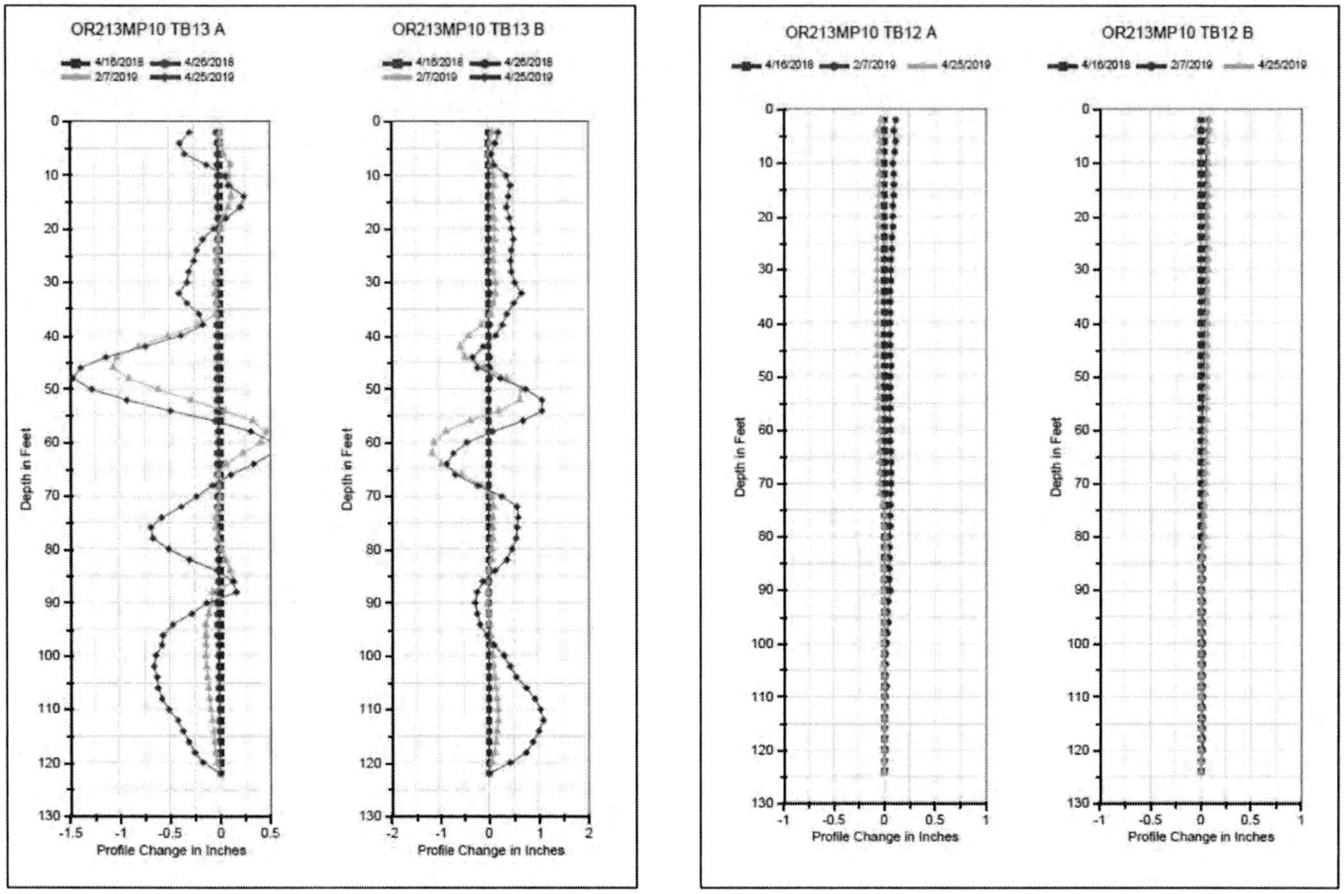

Figure 7: Phase 3 displacement profiles

Phase 3 displacement profiles indicate predominantly vertical displacement as a result of pore pressure redistribution, resulting in casing crushing at TB-13, which corresponds with the location of maximum displacement. Displacement profiles outside of the area of recent activity (TB-07 through TB-12) displayed little to no horizontal defection, confirming suspected failure mechanisms.

Full cumulative-displacement plots and select time-displacement plots for chosen inclinometer depths were analyzed in order to track movement rate which was correlated to factor of safety.

GEOLOGIC MODELING & DESIGN

Geologic cross sections were developed based on initial exploration results and observed SI displacement profiles.

Torsional ring shear tests conducted on visibly slickensided material gathered during Phase 1 explorations revealed a drained effective friction angle of 8.9°.

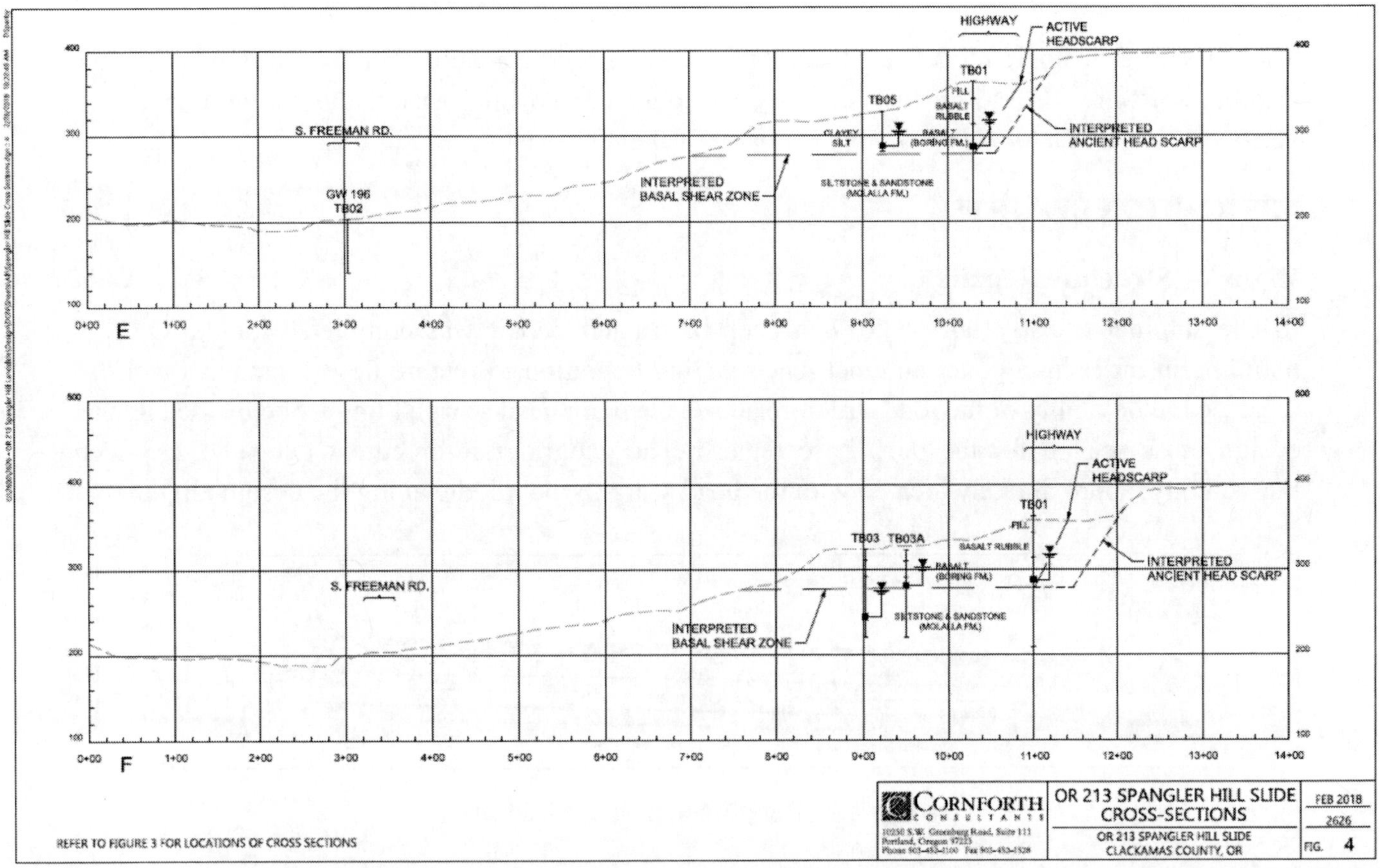

Figure 8: Interpreted Geologic Design Model

Traditional terrestrial observation indicated the main head scarp expressed across the road surface as shown in Figure 8.

The developed geologic model was then analyzed for a bi-linear failure surface in standard slope stability software. Based on the geometry and observed soil properties, the above model was verified to have a factor safety of approximately 1.0 under existing conditions.

MITIGATION EFFORTS

Highway Structure Repairs

In late summer to early fall of 2017, a deep base patch repair was completed for the OR213 mainline in order to conduct an emergency paving operation to restore design grade as well as increase the flexibility of the road structure across the main head scarp. Figure 9 below details the design cross section for the deep base repair. The geogrid reinforcement extended 235 feet horizontally, approximately located with the head scarp in the middle along the design alignment.

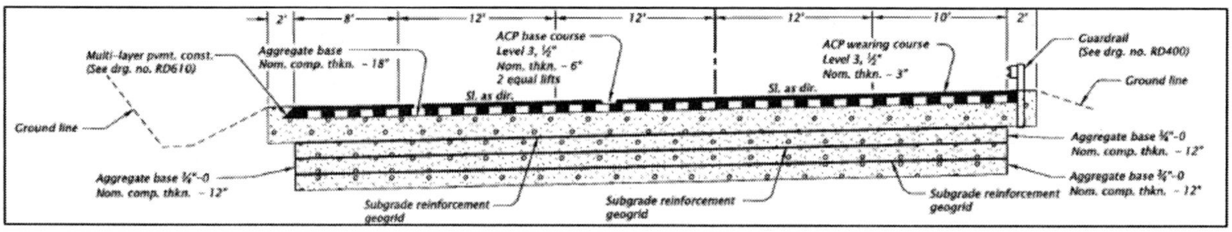

Figure 9: Deep base patch schematic

ODOT made the decision to keep at least two lanes of traffic open in each direction during construction. The staged construction presented difficulties in extending the geogrid reinforcement between excavation stages.

Following installation of the horizontal drains described below, the project team pivoted to restoring the grade of OR213 to as-built conditions. In the section south of the head scarp, the existing surface was approximately 3 ft. below original grade. Phase 3 instrumentation indicated that some displacement was still occurring. As a result, low density cellular concrete was specified to return the highway to as-built grades. Additional excavation was conducted in order to achieve a zero net stress condition with the addition of the increased backfill to meet grades. Forming and cure time for LDCC caused significant schedule impact, but the value of an additional driving force reduction appeared cost effective.

Slide Mitigation

Based on the interpreted geologic conditions and the observed instrumentation data, horizontal drains were confirmed as the most applicable mitigation method. This was based on a moderate reduction in ground water elevation of 5 to 10 feet, and a 15 to 20% increase in the factor of safety.

As can be seen in Figure 5, a ground water elevation decrease of more than 30 ft. was achieved in the most active area of the slide. Manual measurements of crack width and RTK survey observation indicated that the slide movement significantly decreased following drain installation.

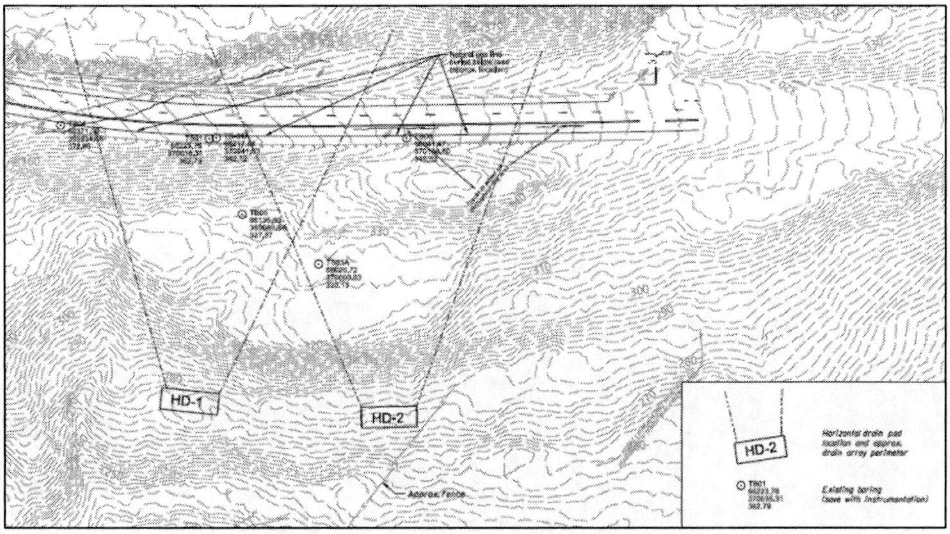

Figure 10: Horizontal drain layout

The property downslope of the highway is privately owned. In order to reach the optimal drain outlet elevation of 270 feet, the drill pads were forced to be located on the private property. Significant earth work and alteration of the property had to occur in order to get the drill rig into position.

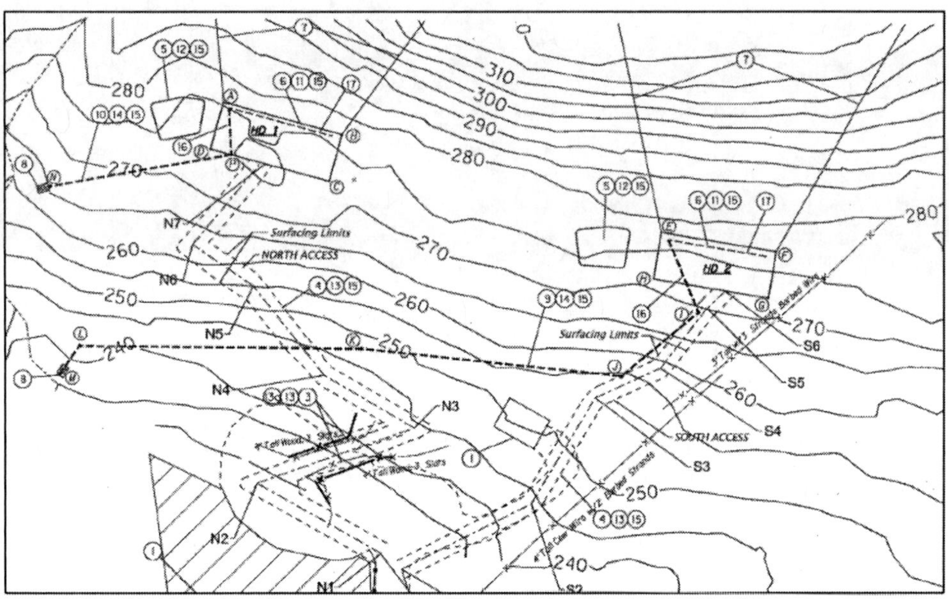

Figure 11: Detail shot of contract plans showing location of drill pads and access roads

Figure 10 above displays the negotiated general site layout with north being at the left, showing the staging area in the west, and the drill pads towards the east. Pond 1 was installed per plan (note

5 in northeast corner) while Pond 2 was shifted to the west and south of Pad 2 The access roads to the north and south as well as the drill pads required significant vertical excavation to adequately capture the anticipated drainage. The darker dashed lines in Figure 10 show the proposed path of the drainage pipe excavated and installed in difficult conditions. Initially, the access roads were to be constructed using native fine grained soil matrix from the overburden material once the boulders were removed. Since installation for the access roads started in January 2018, the saturated nature of the fine grained site soils prevented any functional construction. As a result, a contract change order was processed to allow the contractor to bring in pit-run cobbles.

Figure 12: Pad 1 excavation, looking south

Figure 13: Pad 2 excavation, looking south

Pond 1 was installed partially into the existing slope with the outboard side of the pond supported by a rapidly installed earthen berm constructed of native fine grained soil matrix. Due to the suspect stability of the outboard side of Pond 1, the specified volume of 20,000 gallons was not available. Once drilling for an individual drain reached design length and the steel drill string removed, the newly installed drain flowed at rates above 300 gallons per minute (gpm) of sediment-laden groundwater. After approximately 30 min. the flow rate subsided to approximately 10 gpm and the sediment load decreased. The limited amount of containment available at Pond 1 and the high flows of turbid flow created drainage outfall concern. The project team quickly identified a sediment control plan with settlement tanks and flocculants. With contractor assistance, the project team sized the equipment based on the rapidly fluctuating flow rates and timing. Figure 14 below displays the 20,000 gallon utilized for temporary storage and sediment removal. Ultimately three of the large tanks were needed to capture the flows from Pad 1.

Figure 14: Sediment control solution

UAS LIDAR SURVEY, CHANGE DETECTION ANALYSIS, FURTHER RESEARCH

Following the installation of the horizontal drains and Phase 3 explorations, the ODOT Research Division was able to contract with Oregon State University (OSU) to gather high quality LiDAR data using an unmanned aerial system (UAS) aka a quadcopter drone.

The additional analysis was conducted in part to verify assumptions regarding slide movement kinematics, as well as proof of concept for ODOT Geometronics unit for future projects. UAS gathered LiDAR was compared to traditional terrestrial survey techniques for accuracy verification. The collected data were then compared to existing LiDAR data previously acquired by DOGAMI. A change detection analysis was conducted between the two data sets that produced a displacement map of the landslide complex as shown below in Figure 15.

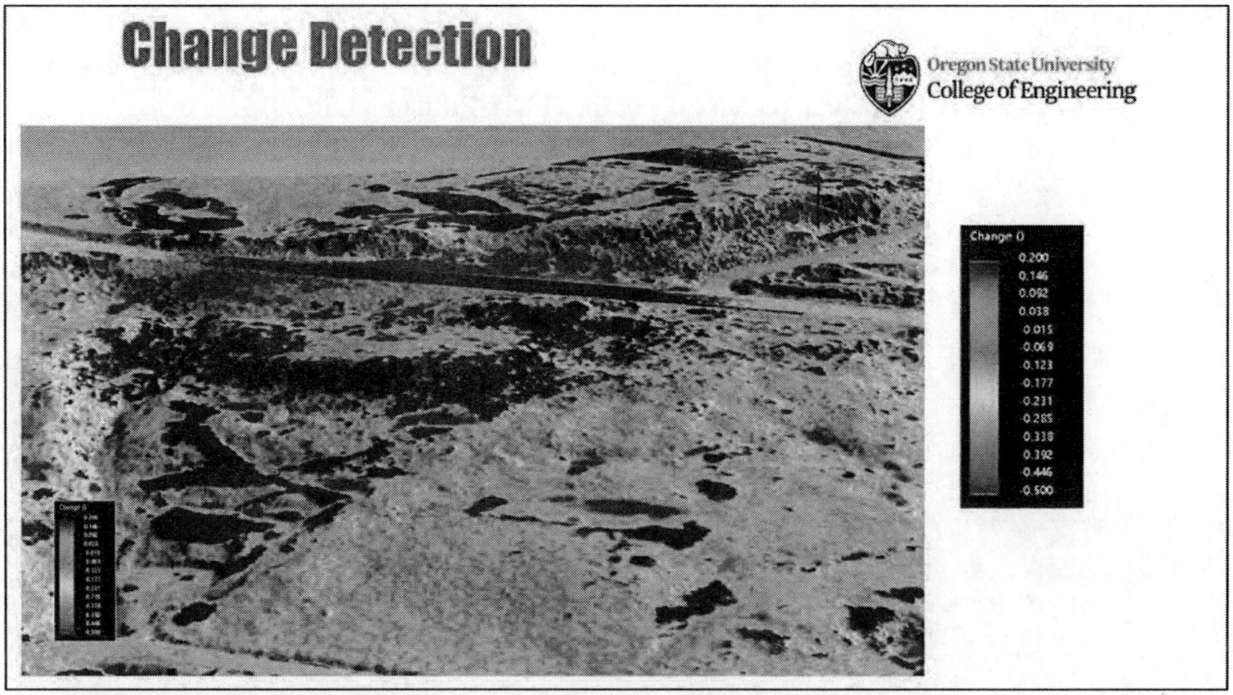

Figure 15: Displacement map based on change detection analysis with UAS LiDAR

The additional instrumentation and aerial surveys help to prove the efficacy of the horizontal drains. In addition, ODOT and OSU are currently developing an operations and maintenance (O&M) manual to track the long term performance of horizontal drains across the state, with the Spangler slide being one of the case studies. Output flow volume instrumentation is currently being installed at the site utilizing open source software and bespoke hardware.

CONCLUSION

Horizontal drains have been an effective mitigation strategy based on observed instrumentation readings and traditional terrestrial survey techniques. Long term performance of the drains will be monitored and incorporated into on-going research efforts.

UAS mounted LiDAR can be just as accurate as terrestrial LiDAR when sufficiently constrained. Aerial based surveys provide a broader coverage area that allow for more effective mitigation strategies.

REFERENCES:

1. Black, B., Machan, G., Ruby, Z. *Geotechnical Report, OR213: Spangler Hill Slide ER Project.* Landslide Technologies, Inc (LTI), Portland, OR 2019.

A Review of Scaling as Rock Slope Remediation Method

Gabriele Mellies
Stantec Consulting Ltd.
141 Kelsey Drive
St. John's, NL Canada A1B 0L2
1 709 576 1458
gabriele.mellies@stantec.com

John Nichols
Stantec Consulting Ltd.
1260 Boulevard Lebourgneuf
Québec City, QC, Canada G2K 2G2
1 418 626 1688
john.nichols@stantec.com

Maureen Matthew
Stantec Consulting Ltd.
40 Highfield Park Drive
Dartmouth, NS Canada B3A 0A3
1 902 468 7777
maureen.matthew@stantec.com

Prepared for the 70th Highway Geology Symposium, October, 2019

Acknowledgements

The authors would like to thank their clients for their support of this work:

Yves Lortie, Attila Fustos– Government of Bermuda, Ministry of Public Works
Zackary Modjeski – Defense Construction Canada
Steven Babstock – Parks Canada

Disclaimer

Statements and views presented in this paper are strictly those of the author(s), and do not necessarily reflect positions held by their affiliations, the Highway Geology Symposium (HGS), or others acknowledged above. The mention of trade names for commercial products does not imply the approval or endorsement by HGS.

Copyright Notice

Copyright © 2019 Highway Geology Symposium (HGS)

All Rights Reserved. Printed in the United States of America. No part of this publication may be reproduced or copied in any form or by any means – graphic, electronic, or mechanical, including photocopying, taping, or information storage and retrieval systems – without prior written permission of the HGS. This excludes the original author(s).

ABSTRACT

Rock scaling is defined as the activity of removing loose or unstable rocks from the face of a rock slope. It is commonly applied as a remediation measure to mitigate rockfall hazards, either as a singular measure or in combination with supplementary remedial actions. Scaling is often perceived as a simple task, not needing sophisticated skills, easily applied to all types of slopes, and requiring only simple tools or mechanical equipment. However, a recent review of a variety of scaling operations shows that the results are not always predictable or in accordance with the initial design. The scaling requirements are often not well understood, and risks of excessive scaling are frequently underestimated by the parties involved in scaling projects. Careful planning of scaling operations and discussions with all parties about the expected outcome, the uncertainties in reaching the agreed objectives and the remaining risks after scaling should be an essential part of the process. Contributing factors for a successful scaling project include a thorough understanding of the rock mass structure and expected failure modes, the selection of the right tools and equipment that suit the rock and the site conditions, and most importantly include the experience and knowledge of the scaling crew and the supervising personnel. In this work, several recent scaling projects carried out in a variety of locations and rock types were reviewed to evaluate the tools and techniques applied and to critically assess the success of the scaling operations.

INTRODUCTION

Rock scaling is a slope stabilization method, described as a type of excavation that consists of the removal of loose and unstable rock from slope faces which can be either individual rock blocks or rock masses or a layer of weathered rock on the face of a slope [Andrew and Pierson, 2012]. Scaling is commonly applied as a slope remediation or maintenance measure to various types of rock faces and rock slopes, such as highway cuts, natural slopes, open pit walls, tunnel portals and other areas of exposed rock, with the intention to remove rock that has the potential to fall. Andrew and Pierson, 2012 state an important distinction between rock that is loose and has the potential to fall and rock that is loose but does not have the potential to fall. A loose block may still be keyed in and even support the surrounding rock, and removal would rather have an adverse effect on the stability of the slope.

Scaling is utilized either as a single measure or in combination with other slope stabilization measures, such as rock bolting or installation of rockfall protection mesh, to address rockfall hazards and remediate potential or actual instabilities. As a single measure, scaling is often executed during regular maintenance work as proactive stabilization measure or as clean-up of a slope after a frost period or after a failure has occurred. It also is applied to mitigate rockfall hazards in areas where aesthetic or other limitations prevent the use of more aggressive or visible stabilization methods. In combination with additional measures, scaling is typically used to remove immediate hazards and prepare the slope for safe access for the work crew to facilitate the installation of further stabilization measures.

The outcome of a scaling operation is influenced by many factors. These factors must be considered and communicated during preparation for and execution of a slope remediation project to ensure that expectations are reasonable and achievable and that the limitations and risks of scaling operations are understood by all parties involved. Typical influencing factors are discussed in the following section.

ROCK SCALING – TYPICAL INFLUENCING FACTORS

Various factors impact the execution and success of scaling operations, including site specific, contractual, and human factors. Main factors include the following:

- *Rock Conditions:* The rock type and the rock conditions in the scaling area have a crucial impact on the scaling outcome. Different rock types require different scaling approaches and equipment and, as in most ground related projects, adverse rock conditions must be expected and prepared for.

- *Site Conditions:* Apart from the rock conditions, the general site conditions and constraints must be taken into account. Site conditions can include access limitations, aesthetic requirements, traffic restrictions, catchment conditions, adjacent infrastructure, etc.

- *Scaling Method and Equipment:* The method and equipment used for the scaling operation must be chosen based on the rock type, the accessibility of the area to be scaled and most importantly based on the intended outcome.

- *Experience of Scaling Crew and Owner's Engineer or Supervisor:* The experience and knowledge of the scaling crew is a critical factor for the implementation of a scaling operation, and the experience of the owner's engineer is crucial for the understanding of the contractor's work and for the proper communication between contractor and owner.

- *Safety of Scaling Personnel and the Public:* The most important factor in any scaling operation is safety. This includes the safety of the scaling personnel as well as the safety of the public.

- *Project Setup – Basis for Payment:* The basis for payment is often an area for discussion when tendering a scaling project. Typically, three payment options are discussed: payment on an hourly rate basis, payment as a lump sum for a specified scaling project, or payment on the basis of an area estimate.

- *Risk Management – Contractual Risks, Risks after Scaling and Acceptance of Risks:* Scaling operations require management of risks before, during and after construction. Risk management includes the risk of unforeseen rock conditions that become the basis for discussions, project changes and claims, risks that remain after implementation of scaling as well as readiness of the owner to assume risk.

PRACTICAL EXAMPLES OF SCALING PROJECTS

Black Watch Pass, Bermuda – Challenging Rock Type and Site Constraints

The Black Watch Pass is located on the north shore of Bermuda in Pembroke Parrish in close vicinity to the ocean. The two-lane roadway runs in north-south direction and was constructed as a box cut with an excavated rock face on both sides of the road (refer to Figure 1). The road cut was constructed in the 1930's and is a major tourist attraction in Bermuda's Pembroke Parrish and site of the major parade.

The rock cut has a length of approximately 300 meters, with sub-vertical (approximately 85 degree) walls that range in height from 3 to 24 meters. Approximately at the midpoint, the Black Watch Pass is crossed by a road that was built on an archway excavated into the rock. Except for a 20-meter long section at the north end of the eastern slope, the walls were excavated without benches. Pedestrian sidewalks are located at the toe of the rock cut with no catchment area or protection between the sidewalk and the rock faces.

Figure 1 – Black Watch Pass, Bermuda

The geology of Bermuda consists of Pleistocene epoch aeolianites or sandy limestone with layers of paleosols formed above a volcanic seamount consisting of basaltic rocks. The exposed rock at the project site is defined by many cavities and undermined areas that have formed due to differential weathering and erosion of the paleosols (refer to Figure 2).

Figure 2 – Cavities Formed by Erosion of Paleosols and Weak Sandstone

The 2018 slope stabilization project at the Black Watch Pass comprised vegetation removal from the rock faces and slope crests and manual scaling of loose rock and unstable existing masonry block walls, which had been part of stabilization measures carried out in previous stabilization efforts (refer to Figure 3). Further stabilization measures were limited to shotcreting and installation of rock anchors in selected areas.

The scaling was carried out from ropes, using scaling bars, hammers and blow pipes. Challenges included the varying strata of locally extremely weak rock that disintegrated into sand and the excessive vegetation growth along the crest. The weak rock had to be removed while not undermining more competent areas and the still intact existing reinforcement, such as the masonry walls.

Site constraints consisted of aesthetic requirements, which did not allow for more visible stabilization measures such as rockfall protection mesh, and lack of catchment at the toe of the rock faces, which necessitated thorough scaling to mitigate the remaining risk of subsequent rockfalls but without damaging the faces due to excessive scaling.

Figure 3 – Manual Scaling and Vegetation Removal

The roadway was closed for traffic during scaling and scaled material was dropped to the ground and subsequently removed off site. The scaling was monitored and directed by Stantec's engineer on site on behalf of the owner. Basis for payment was square meters of scaled area. This payment option proved to be reasonable for the work since the area was well defined and visible and allowed for an accurate estimate.

Cabot Trail, Cape Breton – Aesthetic Limitations and Work under Traffic

The Cabot Trail is a two-lane coastal highway that circumnavigates the Cape Breton Highlands National Park in Cape Breton, Nova Scotia, Canada. The construction of the highway required dozens of substantial rock cuts through the highland slopes.

Rock types are highly variable along the trail, but generally consist of high-grade metamorphic rocks (schist and gneiss) intermixed with plutonic rocks. The rocks of the Cape Breton Highlands can be heavily altered, highly fractured, and are exposed to extreme coastal weather patterns and resulting erosion. Typical slope configurations are shown in Figure 4.

Figure 4 – Cabot Trail, Cape Breton – Typical Slopes

Since 2015, a multi-year rock slope remediation program has been carried out along the Cabot Trail. Due to the high profile of the site and location within a national park, stabilization measures with low aesthetic impact were required to maintain the natural look of the rock exposure. The stabilization program included mainly manual scaling and installation of rock anchors; limited mechanical scaling and trim blasting were carried out. Concrete buttresses were installed locally to stabilize undermined slope areas. Some sections of the road with limited catchment area were protected against falling rock with low energy concrete barriers.

Scaling has been mostly carried out manually by rope access using scaling bars, air bags, and air pressure pipes. Mechanical scaling with long-reach excavators has been completed on some moderately high slopes where conditions enabled reach and allowed for safe and controlled rock removal. Typical scaling set-ups are shown in Figure 5.

Single lane traffic had to be maintained during construction and was managed with traffic light control systems. Where possible, scaling was carried out in off-peak seasons to reduce impact during peak tourist season. Protection of passing motorists and cyclists against falling rock was ensured with concrete barriers and mounted low impact fences, equipped with geotextiles that prevented fly rock to reach the travelled road lanes. In select circumstances, when potentially large areas of rock were released from the slope, all traffic would be stopped for brief periods of time.

Monitoring and direction of scaling activities was carried out by Stantec's engineers on behalf of the owner. Open communication between the owner's engineer and the contractor was essential to determine areas of concern, strategize the scaling activities and optimize the results. Challenges included zones of highly altered and fractured rock that sometimes led to over-scaling resulting in local zones of instability which were mitigated by either spot-bolting or, where possible, increased catchment areas. Site constraints such as aesthetic requirements and working under traffic required discussions about risk and future maintenance of the slopes and about acceptance of risk during construction and remaining risks after completion of scaling.

Figure 5 – Manual and Mechanical Scaling

Basis for payment for scaling were crew hours for both manual and mechanical scaling. Hours for measure were defined in the project specifications as time spent actively working on the slope and these hours were monitored by the owner's engineer and recorded in the daily work sheets. The method proved to be suitable for the project since an accurate estimate of the scaling area was difficult prior to the works.

Colwood Slope, British Columbia – Unexpected Rock Conditions

Located at Canada's Pacific Coast, the Colwood slope is located at the Canadian Forces Base on Vancouver Island, Canada. The rock slope forms the western boundary of the work yard and consists of a 130 m long slope with heights ranging from approximately 15 m to 25 m. The excavated rock face is irregular with slope angles varying between 45 degrees to near vertical. Figure 6 shows the mid-section of the slope at commencement of the project.

The geology at the site consists of fractured meta-igneous rock. The natural discontinuities in the rock mass were consistent in exposed areas. However, in areas with talus accumulation and dense vegetation growth an extensive network of fractures was observed that was inconsistent with the natural discontinuity sets and appeared to be the result of human impact on the slope.

The slope stabilization program for the site comprised scaling and installation of rock anchors and rockfall protection mesh. Scaling was carried out manually from ropes, and scaling tools were limited to handheld tools, such as scaling bars, pneumatic hammers, and blow pipes.

Figure 6 – Colwood Slope, British Columbia

The scope of the scaling operation was to remove loose rock from the slope in preparation for the installation of subsequent stabilization and protection measures. Challenges during scaling arose from the unforeseen poor rock conditions encountered in the middle section of the slope. The rock at both ends of the slope showed the expected natural discontinuities; however, the rock in the mid-section, although expected to be more fractured based on the pre-construction investigation, turned out to be of extremely poor quality and was heavily fractured, showing random fracture patterns, presumably the result of human impact such as blasting or other activities.

The mid-slope area caused significant scaling efforts in preparation for the subsequent anchor drilling and mesh installation; large amounts of loose rock were removed to prepare stable drilling locations and address safety concerns regarding access to the slope that were raised by the contractor. Further site constraints were the close proximity of the power lines, which had to be deenergized for the work, and other infrastructure and buildings that had to be protected against damages during scaling operations.

Monitoring and direction of scaling activities were carried out on behalf of the owner by Stantec's engineers on site. Issues and concerns were discussed on site between the owner's engineer and the contractor and communicated to all parties. When the unforeseen poor rock conditions were encountered, communication involving all stakeholders had to be intensified to properly address the situation. Recognizing the poor rock conditions and understanding the extent of these conditions and their impact on the project were crucial and establishing clear communication lines helped to get agreement on appropriate measures. The discussions allowed the team to determine areas of concern, strategize the scaling activities and optimize the results.

Figure 7 – Manual Scaling and Grubbing in Heavily Fractured Mid-Slope Section

Payment for scaling was made on a time and materials basis. Due to the encountered ground conditions, the costs and construction time for the scaling operation increased significantly. The method assigned the risk involved in the work, including unforeseen ground conditions, scaling method, experience of scaling crew and supervisor, etc., entirely to the owner.

Comparison of Scaling Project Examples

Table 1 shows a comparison of the three scaling project examples described above, including the challenges encountered and the outcome of the scaling work.

Table 1 – Summary of Project Examples				
		Black Watch Pass, Bermuda	**Cabot Trail, Cape Breton, NS**	**Colwood Slope, BC**
Site Location		Rock face along inner-city roadway.	Various rock slopes above scenic coastal roadway.	Rock face in work yard.
Adjacent Structures		Roadway and pedestrian sidewalks at slope toe.	Roadway at slope toe.	Work yard at slope toe. Buildings in close vicinity to slope. Power line along slope. Archaeological site at crest.
Slope Type		Excavated steep rock cut.	Natural rock slopes and slopes excavated for road construction.	Excavated steep rock face.
Slope Dimensions	Height	3 m to 24 m	10 m to 30 m	15 m to 25 m
	Length	300 m	50 - 250 m	130 m
Rock Type		Aeolianite, sandy limestone.	High grade metamorphic, plutonic.	Meta-igneous rock.
Rock Conditions		Weakly cemented limestone with paleosol of loose unconsolidated sediments.	Heavily fractured rock, generally foliated, moderate strength.	Fractured mass with signs of metamorphism and extensive unnatural fractures.
Access Conditions		Access from roadway and from slope crest. Crest had to be cleared for access.	Access from roadway and slope crest. Trails had to be cleared for access.	Access from work yard and from slope crest.
Catchment Area at Slope Toe		No catchment.	Catchment varies between wide ditches and narrow road shoulder; locally concrete barriers.	Limited catchment (paved edges of work yard).
Slope Remediation Measures		Scaling and selected rock bolting and locally shotcreting.	Scaling and selected rock bolting. Locally concrete buttress.	Scaling, rock anchors and rockfall protection mesh. Locally concrete buttress.
Scaling Method		Manual scaling from ropes.	Manual scaling from ropes and locally from lifting device. Limited machine scaling with long reach excavator.	Manual scaling from ropes.
Scaling Equipment		Scaling bars, blow pipe, hammers.	Scaling bars, air bags, blow pipe.	Scaling bars, pneumatic hammers, blow pipe.
Site Constraints, Requirements and Limitations		Aesthetic requirements (maintain natural appearance).	Aesthetic requirements (maintain natural appearance). Maintain single-lane traffic.	Protection of infrastructure and powerline at the base of the slope. Protection of archaeological site at slope crest.
Challenges Encountered		Extensive cleaning of loose rock. Avoiding over-scaling in loose sediment layers. Working in very hot environment.	Heavily fractured rock mass. Working adjacent to active traffic.	Very poor rock conditions in mid-section of slope. Highly fractured and unstable rock. Safety concerns during execution.
Payment Basis for Scaling		Square meter of scaled slope.	Crew hour rate. Payed for active scaling.	Time and Material.
Overall Project Outcome		Successful scaling operations within budget. Remaining risk of rockfall due to ongoing weathering. Regular maintenance and repetition of scaling required.	Successful scaling operations within budget. Remaining risk of rockfall due to ongoing weathering. Regular maintenance and repetition of scaling required.	Successful scaling operations; however, increased costs and extended work schedule due to unforeseen ground conditions. Remaining risk of rockfall mitigated with further stabilization and protection measures following scaling.

LESSONS LEARNED

Many lessons can be learned from successful projects, but it is usually the difficult or less successful projects that teach us the best lessons and help us understand what can go wrong, why it went wrong and which areas we can improve to deliver a successful project. Although most of these appear as common knowledge, the difficulties that arise during execution of a scaling project show that even the most experienced teams can face difficulties to successfully deliver scaling project. The following summary of thoughts and suggestions is based on the experience and lessons learned from recent projects; without claiming to be complete, the summary offers a comprehensive review of things to consider and areas for improvement when planning and executing scaling projects.

What are we dealing with?

Rock Conditions

Scaling in highly fractured or poor quality rock often leads to over-excavation, resulting in potentially damaged slopes and high costs. Removal of key blocks in more massive or blocky rock can destabilize areas above and trigger failures of initially stable rock. Undermining or creation of ledges, from which rock debris can bounce away from the slope during a later rockfall event, can occur due to layered rock with different rock strengths. In areas of very strong, competent rock, scaling of potentially loose material can be challenging and may require the use of more rigorous scaling methods such as supplementary tools (air bags, jack hammers) or heavy equipment.

When is enough?

Scaling in challenging rock conditions, such as highly fractured rock, can become a never-ending story. The hope for improving conditions or for finding the perfect clean rock face can sometimes result in over-scaling or even worsening of the slope conditions. It requires experience, discussions with all parties involved and sometimes courage to make the decision to either continue scaling or stop and accept that conditions will not further improve. This decision must be based on consideration of all information available and on clear communication and mutual agreement.

Where do we work?

Access Constraints

Site access constraints can limit the use of equipment for machine scaling or of lifting devices but may also hinder access for scaling crews on ropes. Access constraints can also refer to access to undermined or overhanging slope areas that are difficult to reach and pose a significant hazard when scaling manually from ropes or from lifting equipment. When rope access is used, available anchor points or possibilities for installing anchor points need to be investigated and specified. Areas that cannot be accessed during scaling, such as archeological sites or private properties, must be identified, since these may limit access to the scaling area.

Slope Toe Conditions

Conditions at the slope toe might require that scaled rock must be collected and removed from the site instead of being dropped to prevent damage to structures or blockage of waterways below the slope. This limits the scaling method and must be specified in the contractual documents.

Aesthetic Requirements

Aesthetic requirements might restrict the removal of vegetation or limit the scaling method and can necessitate a more thorough scaling when no further stabilization or protection measures can be installed. They may also limit the use of machinery to prevent significant scaling marks on the slope face. Aesthetic requirements have to be clarified and specified in the contract documents.

Traffic Conditions

When scaling adjacent to active road corridors, protection of the public is key. Wherever possible, roads and pathway below scaling areas should be closed for all traffic to protect the public. Scaling while maintaining open lanes for traffic limits the operations and catchment area at the slope toe and bears the risk that fly rock impacts the travelled lane and causes damage and/or severe injuries or death. Scaling under traffic requires adequate measures, such as exclusion zones, temporary barriers, fences, catchment areas, etc., to be in place to prevent overspill onto the travelled lanes during scaling and protect the public. Protection measures must be designed for the individual locations based on potential rockfall impact caused by scaling.

Traffic systems must be installed to regulate the traffic along a scaling area and to allow for complete closures during rock removal in critical areas where significant rockfall could occur with the potential to overflow the temporary barriers. Constant communication between the scaling crew and the ground crew is required to act in time and stop scaling when required. The risk related to scaling under traffic must be discussed with and accepted by the owner.

Adjacent Infrastructure

When scaling is carried out near infrastructure, such as roads, buildings, utilities, etc., protection measures have to be in place to prevent damage to these structures. These measures could include exclusion zones, barriers, protective covers, etc. The presence of sensitive infrastructure or property at the crest of the slope must be understood and may impact the degree to which scaling is carried out on a rock slope. In many cases, rock removal by scaling is the preferred option for remediation; however, in certain circumstances it can be preferable to use other stabilization measures in lieu of scaling to protect structures and preserve properties at the slope crest.

What do we want to do, and can we do it safely?

Scaling Method and Equipment

The method and equipment used for the scaling operation must be chosen based on the rock type, the accessibility of the area to be scaled and most importantly based on the intended outcome.

Machine scaling, using an excavator shovel or a long reach excavator, can be an adequate tool to scale without the need for personnel to access the slope and be exposed to the danger of falling rock. Machine scaling usually expedites scaling of larger areas and can create an evenly shaped clean face in a highly blocky rock mass by removing the outer layer of weathered rock or remove unstable individual blocks or rock masses from a slope. Machine scaling is, however, rather aggressive and often results in damages to the slope, over-excavation or unfavorably shaped slope crests. Limited reach of standard excavators can quickly result in undermining on high slopes that are better suited for long-reach equipment or manual scaling. Experience of the operator is crucial for machine scaling to prevent damages to the slope.

Manual scaling is slower but less aggressive and offers a variety of tools beyond the typical scaling bars, such as air bags for removal of large blocks or rock masses or hydraulic or pneumatic tools for rock splitting, to address various stabilization issues, and remove rock masses or individual blocks or boulders. Manual scaling can be performed from lifting devices or using rope access methods. However, scaling from lifting devices must be well considered based on the site conditions since it does not allow scaling from above the unstable rock and potentially exposes the scaler to falling rock.

Safety of Scaling Personnel

Whether manual or machine scaling, the scaling method and procedures must be such that safety of the scaling personnel is ensured. Scaling should always be performed from top to bottom. However, scaling manually from a lifting device cannot be performed downwards and includes the risk that scaled debris impacts the lifting device and the scaler. Manual scaling from ropes involves the risk that rock in already scaled areas can become detached while the scaler has continued further down the slope and now works in the path of the falling debris. This is particularly important in fractured, poor quality rock masses. Likewise, machine scaling might bring the operator and his machine into the failure path of the scaled debris.

What do we know?

Experience of Scaling Crew

Scaling crews need to have a general understanding of the rock conditions and the behaviour of the rock. This includes knowledge about typical failure mechanisms and the ability to identify if and how a rock block or rock mass could fail and what path the falling rock could take. The risk of damaging the slope due to aggressive machine or manual scaling or due to removal of key blocks should be understood.

Experience of Owner's Engineer or Scaling Supervisor

The scaling supervisor's experience is required to understand the contractor's work and concerns, direct the scaling operation and prevent over-scaling and damage to the slope. Lack of experience often prevents decision making on site and effective communication between the parties.

Definition of Required Experience

The required experience of the scaling crew and the scaling supervisor should be defined in the tender documents and documentation of this experience should be insisted on. If less experienced scalers work on a slope, supervision and control of their work should be performed by a more experienced team member to ensure all areas are properly addressed. Likewise, an experienced owner's engineer or supervisor should be on site who can ensure proper communication between the parties and is able to discuss technical aspects with the contractor and make educated decisions.

What are the Risks?

Basis for Payment

Typical payment options include: (I) payment on an hourly rate basis, (II) payment as a lump sum for a specified scaling job, or (III) payment on the basis of an area estimate. All three options have their advantages and disadvantages for both the owner and the contractor.

(I) Payment made on the basis of an hourly rate requires the owner, and typically their engineering consultants, to have a very good understanding of the rock and site conditions, the amount of scaling required and of the level of effort (number of hours) required for a scaling crew to perform scaling operations at a slope and achieve the intended objectives. This approach is not without its challenges as scaling crews have variable skill levels and/or methods that can impact the time required to carry out a project. Scaling time for payment must be clearly defined and accurately tracked to ensure mutual agreement between all parties. Uncertainties regarding rock conditions or access or inexperienced scaling crews might significantly increase the level of effort for reaching the intended goals and can either result in an extensive budget overrun or in an insufficiently scaled slope.

(II) Lump sum payment for a specified scaling project requires the contractor to well understand the rock conditions, site constraints and uncertainties which necessitates a good information basis on which a competitive bid can be made. Although site visits are valuable and recommended, they often cannot provide sufficient information and need to be supplemented by comprehensive geotechnical and geological data. Lack of information can bear either a great financial risk for the contractor when conditions are different from the assumed conditions or can result in increased costs for the owner due to conservative bids that cover potential risks.

(III) Square meter as basis for payment requires a reliable estimate of the area that has to be scaled. Surveying and aerial photographs of the slope provide means to define the scaling area. However, without direct access to the slope the scaling areas can be misinterpreted or varying conditions can be revealed after scaling has started. Underestimating the scaling area can result in insufficiently scaled slopes or significant change orders from a contractor and subsequent cost overruns to the owner. Overestimating the area might lead to re-negotiation of unit prices when the reduced square footage is significant.

Scaling on a time and material basis should be avoided. This option bears the risk of significant budget overruns due to unforeseen conditions, inexperienced scaling crews and misinterpreted scopes.

Sharing Contractual Risks

Scaling operations always bear the risk of unforeseen rock conditions that become the basis for discussions, project changes and claims. To reduce the risk of scope misinterpretations and budget overruns, the risk of changing rock conditions and the impact on the project must be discussed during project planning. Clear baselines need to be established for the contract documents to define which conditions must be expected and included in the bid by the contractor and which conditions would warrant a claim for changed conditions. Mandatory site visits with all stakeholders during tendering should be included in the contract documents to ensure a common understanding of the site conditions and constraints. Defining clear baselines for ground conditions that have to be expected and which conditions would justify renegotiations or claims can help to mitigate the financial risk for the owner as well as for the contractor.

Remaining Risks after Scaling

Remaining risks after scaling include rockfalls that occur shortly after a slope was scaled and increasing rockfall probability over time due to ongoing weathering and erosion of a slope.

Rockfall after Scaling: Scaling can be considered "controlled erosion" in which experienced professionals remove rock that would otherwise naturally fall under safe and controlled conditions. However, scaling is a form of ground disturbance and has the potential to accelerate instabilities where fresh rock surfaces are exposed to the elements. In some rock types that are susceptible to the effects of weathering and erosion, it is common to see an increased frequency of rockfalls in the first weeks or months after a slope has been scaled. The potential for these events to occur must be clearly communicated to the owner to prevent them from being unexpected.

Ongoing Deterioration over Time: Scaling is a temporary remedial measure, often paired with supplementary measures such as rock anchors, rockfall protection mesh, buttresses, etc. Over time, the effects of weathering and erosion (i.e., freeze thaw cycles, root jacking, water pressures, etc.) will result in renewed destabilization of the rock surface and additional loose blocks will form. Monitoring of slopes after remediation and regular repetition of a scaling operation are essential and clear communication with the owners regarding expectations for ongoing slope maintenance is key. The recommended scaling frequency varies significantly depending on the weathering rate of the rock and might typically range between 2 and 10 years [Andrew and Pierson, 2012]. If information about weathering rates is not available, best practice includes monitoring and implementation of more frequent scaling with the option to extend this frequency when more knowledge about the slope behavior is gained.

Understanding the Risk

Scaling will not eliminate a rockfall hazard and rockfall can occur even shortly after a slope was successfully scaled. Although the rockfall risk is typically reduced after scaling, the risk will increase over time and scaling must either be repeated, or other measures have to be put in place to address the remaining risk.

Acceptance of Remaining Risk

Risk management must be communicated throughout the project between all stakeholders. When discussing risk, clear definitions of the hazards and concise descriptions of probabilities and consequences should be established so that resulting terms such as "low risk", "medium risk", and "high risk" or "major and minor consequences" all share the same meaning for all parties. Examples of descriptors and descriptions for probability and consequence levels are provided in Pierson, 2012.

During project planning, risk tolerance must be discussed with the owner; the degree to which remedial measures will be carried out is a function of the owner's willingness and ability to accept a certain level of risk on their project site. Site or budget constraints or limitations such as aesthetic requirements might require acceptance of a higher risk. Requirements for post-scaling monitoring and/or maintenance as well as potential for rockfall after scaling must be made clear to the owner as part of the risk management.

What was expected?

Managing owners' expectations

The possibilities and limitations of scaling, the risk during and after implementation of scaling as remedial measure and the uncertainties that are inherent in scaling must be discussed with the owner during project planning stages. Engineers often misjudge the understanding of the presumably obvious and tend to take for granted that clients understand all technical terms. A precise understanding of the owner's needs and expectations is crucial to deliver a successful project.

Communication is key

Clear communication before, during and after the execution of a scaling project prevents misunderstandings and false expectations about the outcome and the remaining risk. Pre-construction meetings, daily updates and regular progress meetings, follow-up emails after field discussions to document findings and decisions and project end meetings should be standard communication items for every project.

Don't expect the impossible

Proper communication and education of all stakeholders are important to understand the possibilities and limitations of scaling regarding execution as well as durability. The understanding that not all loose rock can be scaled and that not all loose rock should be scaled is critical. It is also important to note that, despite scaling, rockfall hazards will likely remain. The communication that scaling is a temporary measure and the risk for rockfalls will increase over time is crucial to prevent misunderstandings about the outcome of a scaling project.

Taking scaling seriously

The perception that scaling is only a matter of removing some loose rock from a slope is unfortunately still present. However, scaling requires knowledge and a prudent approach to avoid damage to the slope due to improper scaling methods, aggressive scaling or removal of key blocks.

REFERENCES:

- Books
    1. Andrew, R. D. and Pierson, L. A. *Chapter 12: Stabilization of Rockfall*. In: Rockfall: Characterization and Control. Transportation Research Board of the National Academies, Turner, A. K. and Schuster, R. L. (Editors), Washington, D.C., 2012.
    2. Pierson, L. A. *Chapter 3: Rockfall Hazard Rating Systems*. In: Rockfall: Characterization and Control. Transportation Research Board of the National Academies, Turner, A. K. and Schuster, R. L. (Editors), Washington, D.C., 2012.

Geotechnical Asset Management, Collecting, Storing, and Disseminating Geotechnical Data to End-Users

Marc Fish
Washington State Department of Transportation Geotechnical Office
PO Box 47365, Olympia, WA 98504-7365
360-709-5498
fishm@wsdot.wa.gov

Tracy Trople
Washington State Department of Transportation Geotechnical Office
PO Box 47365, Olympia, WA 98504-7365
360-709-5594
troplet@wsdot.wa.gov

Jim Struthers
Washington State Department of Transportation Geotechnical Office
PO Box 47365, Olympia, WA 98504-7365
360-709-5409
struthj@wsdot.wa.gov

Prepared for the 70th Highway Geology Symposium, October, 2019

Disclaimer

Statements and views presented in this paper are strictly those of the author(s), and do not necessarily reflect positions held by their affiliations, the Highway Geology Symposium (HGS), or others acknowledged above. The mention of trade names for commercial products does not imply the approval or endorsement by HGS.

Copyright

Copyright © 2019 Highway Geology Symposium (HGS)

All Rights Reserved. Printed in the United States of America. No part of this publication may be reproduced or copied in any form or by any means – graphic, electronic, or mechanical, including photocopying, taping, or information storage and retrieval systems – without prior written permission of the HGS. This excludes the original author(s).

ABSTRACT

The collection of large quantities of geotechnical asset management (GAM) data presents unique challenges for the Washington State Department of Transportation (WSDOT), with respect to storing and then disseminating this data to its end-users. Within the GAM environment, the Geotechnical Office (GO) collects several large datasets that include subsurface data, laboratory test results, unstable slopes information, and the inspection records of constructed geotechnical assets. These datasets need to be safely stored, backed-up, and dispersed to the entire WSDOT workforce, other state agencies, consultants, and practitioners in ways that reduce, or minimize, the efforts of both our in-house geotechnical staff and WSDOT's Information Technology Office (ITO). Through the help of the ITO and a dedicated geographical information system (GIS) professional, some of these datasets are stored in enterprise databases (i.e. Sequel Server) by directly inputting the data using a customized web interface, or by "uploading" datasets from other databases or spreadsheets. Additional datasets that are collected in the field are stored in the "cloud" using IPad based applications such as ArcCollector and Survey 123. All these GAM datasets are accessed through ArcGIS Online customized web applications, and are viewed with previously available statewide datasets such as the highway inventory, 24k and 100k geology, topographic maps, orthophotos, CAD drawings, and airborne lidar. These applications were developed by our dedicated GIS professional, and are used to spread and display this data primarily through an on-line web browser. In some situations, this data is also viewed through more powerful customized desktop ArcGIS projects. This paper presents the types of GAM data the GO collects, how it processes and stores this data, and then how it disseminates this data to its end-users.

INTRODUCTION

The Geotechnical Office (GO) of the Washington State Department of Transportation (WSDOT) possesses an immense amount of Geotechnical Asset Management (GAM) information that includes subsurface and geological field data, laboratory soil and rock testing results, unstable slopes data, and constructed geotechnical asset assessments. This data needs to be safely stored, backed-up, and dispersed to the entire WSDOT workforce, other state agencies, consultants, and practitioners in ways that reduce, or minimize, the efforts of both our in-house geotechnical staff, and WSDOT's Information Technology Office (ITO).

WSDOT's GAM data is collected utilizing multiple methods/technologies including, GPS receivers, field notebooks, and IPads. Applications running on IPads, directly store data in the "cloud" that are accessed by other WSDOT programs or web applications. The large quantity of data, in conjunction with some very large aggregate datasets, and WSDOT internal IT protocol, create data storage and access issues that need to be overcome in order to ensure data integrity and usability by the end-user. Other geological data that is collected in the field (i.e. landslide morphology and subsurface data) is placed into WSDOT enterprise databases that are accessed through ArcGIS Online and served to others through web applications. WSDOT's geotechnical-based web applications and datasets are housed at locations where they are easily accessible (i.e. network server) and are backed up on a nightly basis. Some of the WSDOT web applications and GAM data are accessible only to WSDOT employees, while other applications and data need to be accessible to the public. This creates firewall issues that must be overcome. Also, with multiple end-users of these applications, software licensing issues can also become an issue.

The process starts with data collection. Through WIFI or cell modems (with or without VPN), data can be viewed and uploaded to the "cloud" or to WSDOT servers from the field. Most of WSDOT's GAM data are uploaded or stored into one of several enterprise databases (i.e. SQL server). Views are built, and the data is served to ArcGIS on a nightly basis as an automated process. This ensures that displayed data in a geographical information system (GIS) is never more than one-day old. Datasets from other entities (lidar, geology, landslide data, and roadways) are included in our web applications to provide additional detail and function to the applications. GIS within the GO is a full time job. Data routinely needs to be edited, and the web and desktop applications need to be maintained, so that data can be shared across the WSDOT platform and to the public. Additional work is needed in enhancing the SQL server databases and in automating data downloads.

THE DATASETS THAT WSDOT COLLECTS

Data is either collected digitally or in a field book. Data recorded in a field book is either entered into the computer each night or upon return to the office. This data is then "uploaded" into existing WSDOT geotechnical applications or programs, and, because it's eventually served out to the web, it can be accessed by others, both inside and outside of WSDOT. The following sections detail the datasets actively managed by the GO.

Constructed Geotechnical Asset Data

On an annual basis, the GO collects inspection data on constructed geotechnical assets (i.e. horizontal drains, type 1 & 2 slope protection, and rockfall fences) to determine if they are still functioning as designed, and if they are in need of maintenance. WSDOT has approximately 200 constructed geotechnical assets that need detailed inspections on 1, 3, or 5 year intervals. This inspection data is entered into the Survey 123 application that is running on an iPad, and the data is stored in the "cloud" (Figure 1). These survey forms are easily customizable to fit WSDOT's needs. No cell service is required to complete the survey, but to "upload" the data, or to access historical inspections or additional unstable slopes data, cell service and WIFI is required. Data is collected through drop down menus, "text" notes, and photographs.

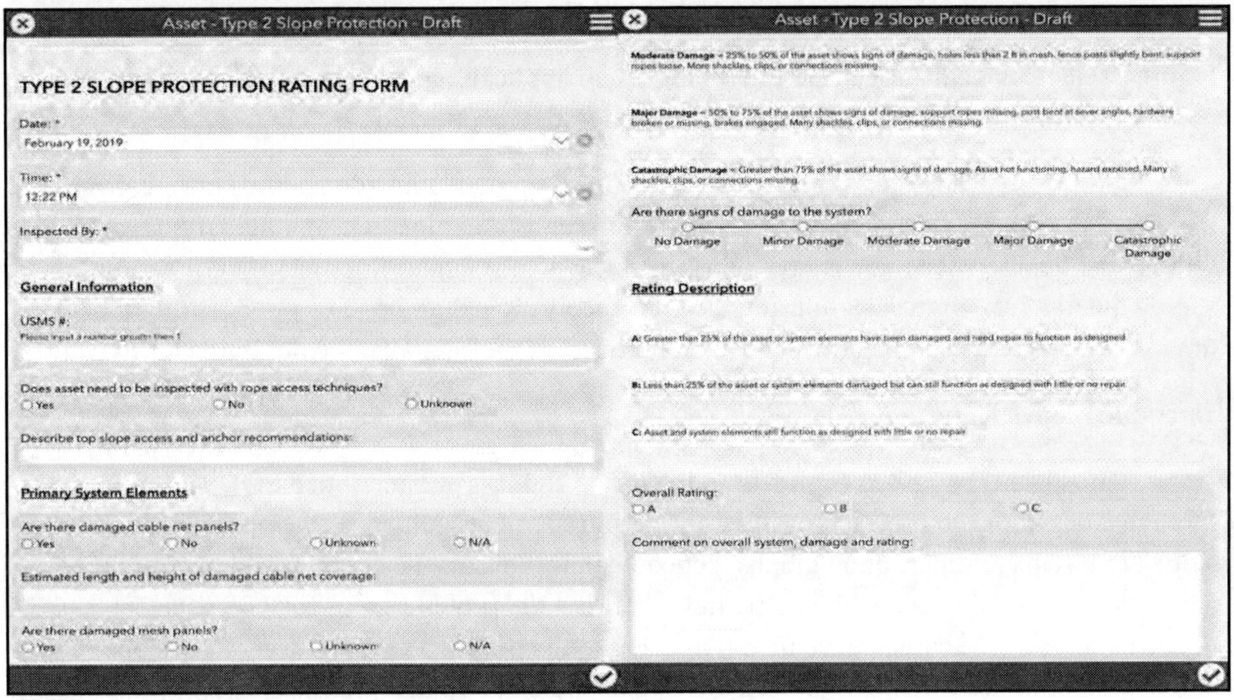

Figure 1: A portion of a type 2 slope protection inspection form in survey 123.

Landslide Mapping Data

Landslide data features (i.e. points, lines, or areas) are collected using either an IPad or a GPS receiver. When using an IPad, mapping is done on top of base layers (i.e. lidar, orthophotos, and topographic maps) or other previously collected GAM data (i.e. USMS data, landslide geomorphology, and constructed geotechnical asset information) (Figure 2). WSDOT internal datasets are only available with a WIFI connection. Photos and notes can be attached to any of the collected features as linked attachments.

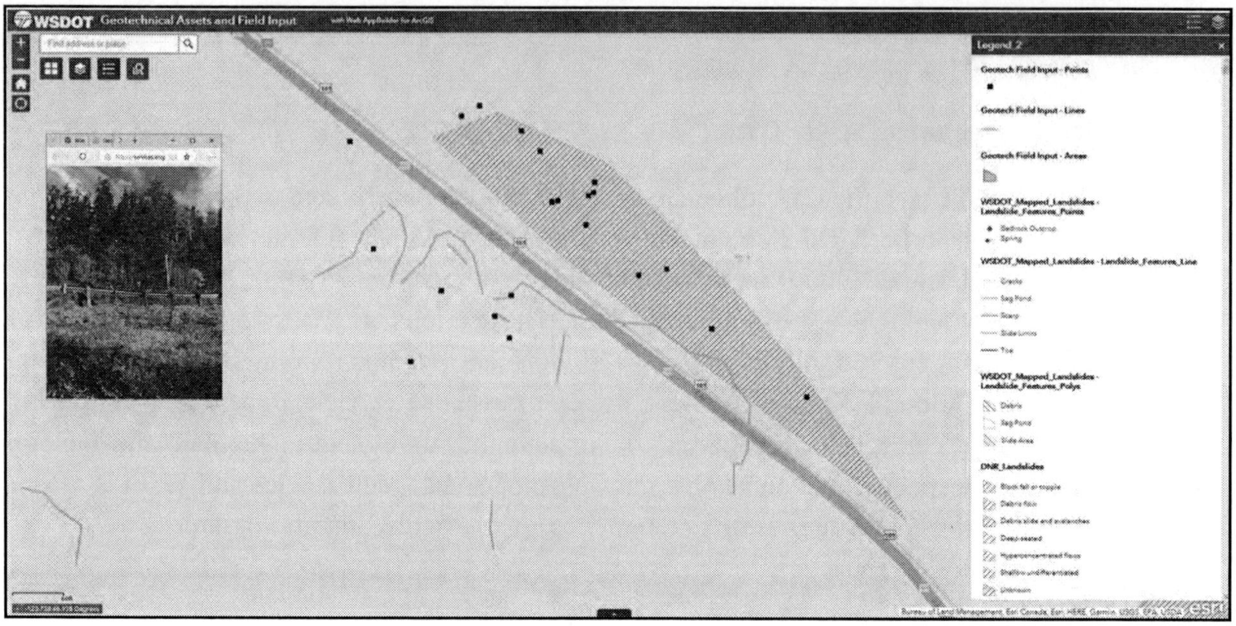

Figure 2: Geotechnical Assets and Field Input Web application showing landslide scarps and points with linked photographs that were mapped with ArcCollector.

Mapping-grade GPS receivers are often used to map landslide features or to collect locational data for geotechnical test borings. This data is downloaded from the GPS receiver into GIS layers. Position accuracy is less than 1 meter, which is superior to that of an IPad, but not survey quality. One drawback in using the GPS receivers is that they need to be brought back to the office so the data can be downloaded and processed, before it can be put into a GIS layer.

Unstable Slopes Data

This data is compiled from within the office and it is entered into a SQL server database through a desktop web browser, or it is collected in the field with an IPad (Figure 3). It includes slope type, status, ratings, photographs, conceptual designs, cost estimates, and benefit-cost analyses. To enter the data while in the field, WIFI with VPN is needed to access the database. A downside to using VPN is that connectivity back to the office is much slower. New slope ratings and photographs are collected for each unstable slope on 1, 3, or 5 year intervals.

Figure 3: Unstable slopes database accessed through a desktop computer or an IPad.

Test Boring Data

WSDOT's drill rig inspectors log holes using their field notebooks. Sample depths, blow counts, descriptions, and any drilling notes are recorded in the field book. Using a field notebook creates an extra step, as compared to directly entering the data into an IPad or a tablet, but it creates a hard copy of the data and allows for more detailed notes to be recorded. WSDOT established this protocol to provide flexibility, and to prevent loss of data due to the harsh working environment and potential damage to an IPad in the field.

HOW THE COLLECTED DATA IS PROCESSED

Following data collection, data is processed using a protocol that is dependent upon its type. The following paragraphs describe typical data processing performed prior to data storage.

Data collected using mapping-grade GPS receivers generally needs to be post processed to improve its accuracy. This is done through a process known as differential correction. In this process, man-made and natural errors are removed from the data that affect the GPS measurements.

At the conclusion of each test boring, or at the end of the project, the drilling inspectors enter their test boring data into a spreadsheet, and then email it to our GO. This spreadsheet is then imported into gINT (a software product that stores subsurface data and creates test boring logs) and a gINT project file and draft test boring log(s) are created for review by the project engineer/geologist.

Instrumentation data (i.e. groundwater levels and well design), and soil and rock testing results are entered into the project gINT file. Sample description are edited and new draft test boring logs are created. Before the geotechnical report is published, final edits and corrections are made in the project gINT database, and final test boring logs are created. The project gINT files are then uploaded to WSDOT's draft gINT enterprise database for use in GIS. Once a project has gone to advertisement, the gINT project is moved from the draft gINT enterprise database to the final gINT enterprise database.

HOW DATA IS STORED

The GO uses enterprise databases and geodatabases to store their GAM data. Whenever possible, data is entered into an enterprise database, such as SQL server. ArcGIS has the ability to directly connect to SQL server, as long as there is a spatial component in the data. Each enterprise database on WSDOT networks, is stored on a development, QA/QC, and a production server. These databases are backed up nightly. This leads to increased data storage needs that are approximately 3 times the original storage need for each enterprise database. Both the unstable slopes and gINT datasets are stored in an enterprise SQL server databases. Each night, database views are built that consist of the most recent data. These views are accessed by ArcGIS. GIS services are published to ArcGIS online, either on an internal (in-house) server, or an external (public) server. GAM data on the internal servers can only be accessed by WSDOT employees. For the public to access WSDOT's GAM data, the GIS service must be served to an external (public) server.

ArcCollector and Survey 123, used for initial data collection, store their data in the "cloud". This is convenient to the user, but trying to access this data for use in other applications is not easy, due to WSDOT firewall issues. GO personnel can manually access this data, but the process of an automatic download is still being worked out.

HOW WSDOT SHARES THE DATA

Most of the data that the GO has collected and has stored, is served out to others through web applications, web mapping applications, or sometimes through desktop ArcGIS. Through these applications, users can locate, view, query, and download specific geotechnical information. Permissions can be set, so only authorized users can gain access to specific data.

USMS Intranet Web Page

All unstable slopes data is entered, viewed, and edited using a web interface on the WSDOT intranet (in-house) (Figure 4). This data is accessible to all WSDOT employees. Permissions are set that grant certain privileges to each user. Viewing or searching for unstable slopes data along a state route can be done through either a "text" query or through a web map. Other data including SRview (WSDOT highway video logs), photographs, correspondence, and geotechnical reports are also available through the web application. Documents explaining WSDOT's unstable slopes program, user manuals, and published papers explaining how to enter and interpret the data are also available on the USMS website.

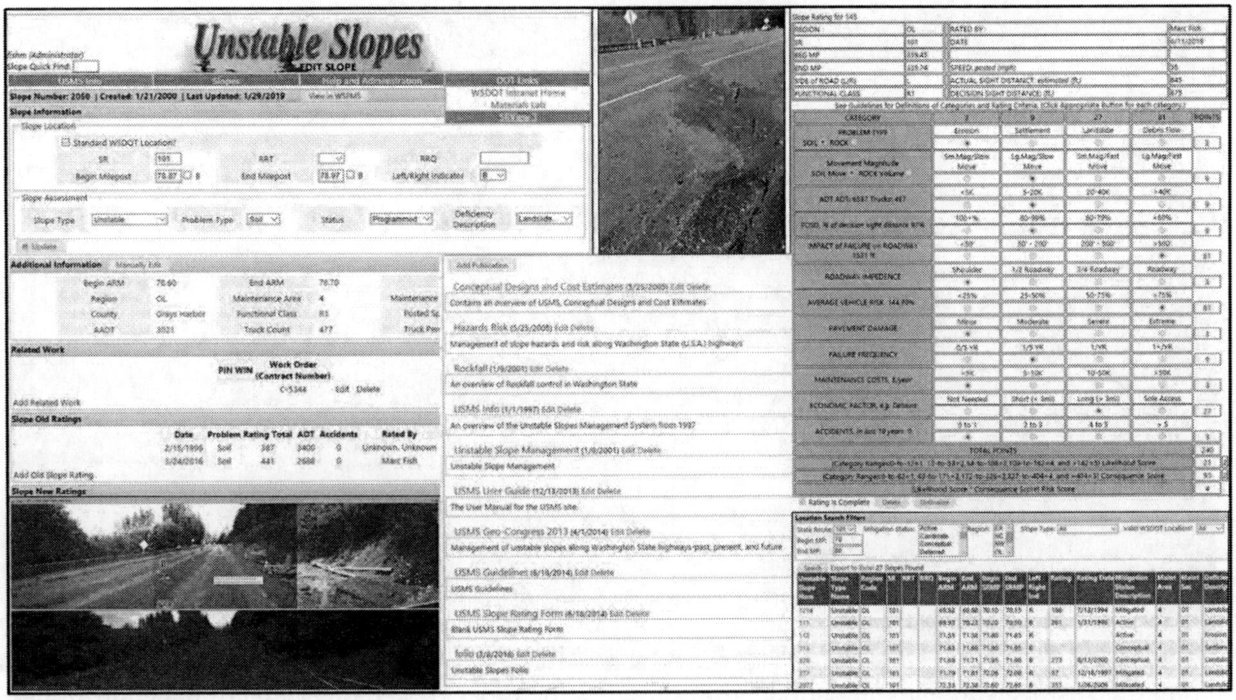

Figure 4: USMS Web Application showing "text" search functions, rating page, linked photos, downloadable documents, and the state highway video logs.

Web Mapping Applications

Unstable slopes data, mapped landslide morphology, geotechnical asset inspection reports, and detailed subsurface information are all available through several web mapping applications. Special viewing, editing, and administrator rights are assigned to all users. These web maps have been built for large mega projects and for statewide use. Built-in queries, layers, and base maps are available for each application. At this time, licensing issues limit the number of users, and all users need to be part of the geotechnical user group. Data that is housed inside the WSDOT firewall is not available to the public. Datasets that are available to the public are placed on a server outside the WSDOT firewall.

USMS Web Mapping Application

All of Washington's approximately 3400 known unstable slopes are displayed on a statewide map. They can be displayed by type, deficiency, status, or rating (Figure 5). As a user "zooms" in, the unstable slope "points" turn into "lines", and by "clicking" on a "point" or a "line", data about the unstable slope appears in a pop-up window. At the bottom of the window, a link is provided that takes the user directly to the USMS webpage for that slope. Different base maps can be displayed, and there are book marks that "zoom" the map to each individual county in the state.

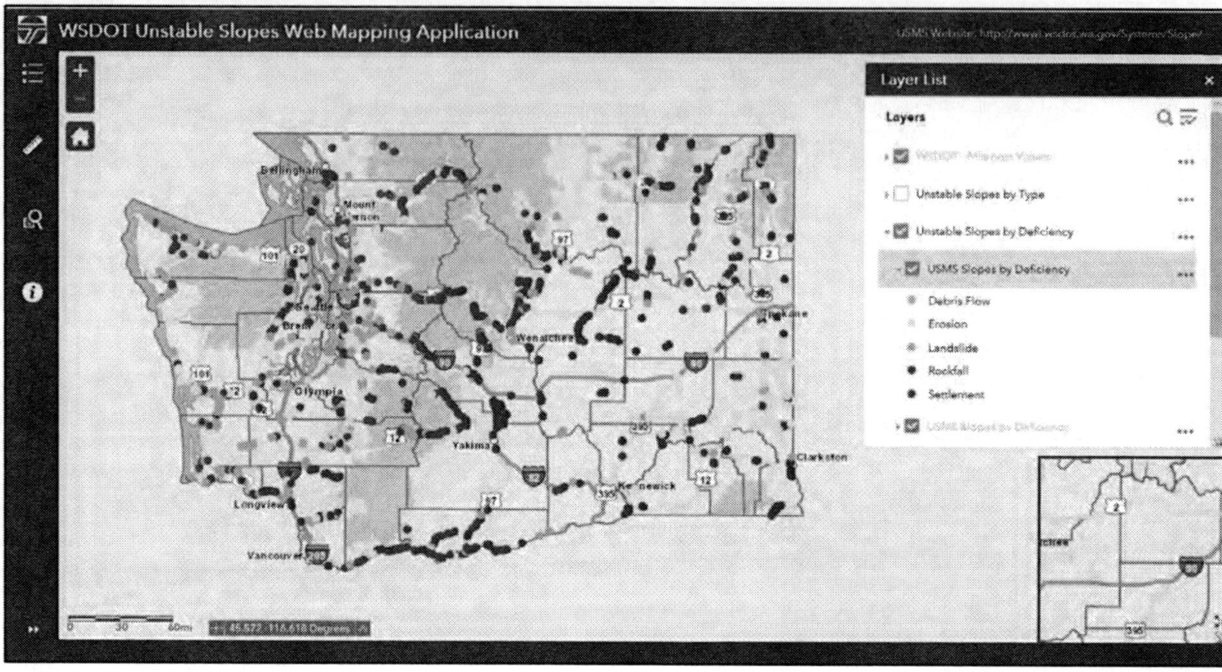

Figure 5: USMS web mapping application showing the types (based on color) of unstable slopes throughout the state.

Geotechnical Web Mapping Application

This web mapping application displays subsurface information and test boring logs (Figure 6). WSDOT has one application that is for public use (external), and another for in-house users (internal). The internal application shows all subsurface data, including draft data, which are not available in the public application. The historical data are composed of pdf version test boring logs that were acquired through the ECM portal (Oracle database), WSDOT's document archive database. The historical test boring locations were mapped by georeferencing historical plan sheets or by locating the stationing and offset along historical project alignments. The legacy boring data are older gINT databases, in a format different from that of the draft and final data, so only the point and project information is viewable in the GIS, with a link to the pdf test boring log.

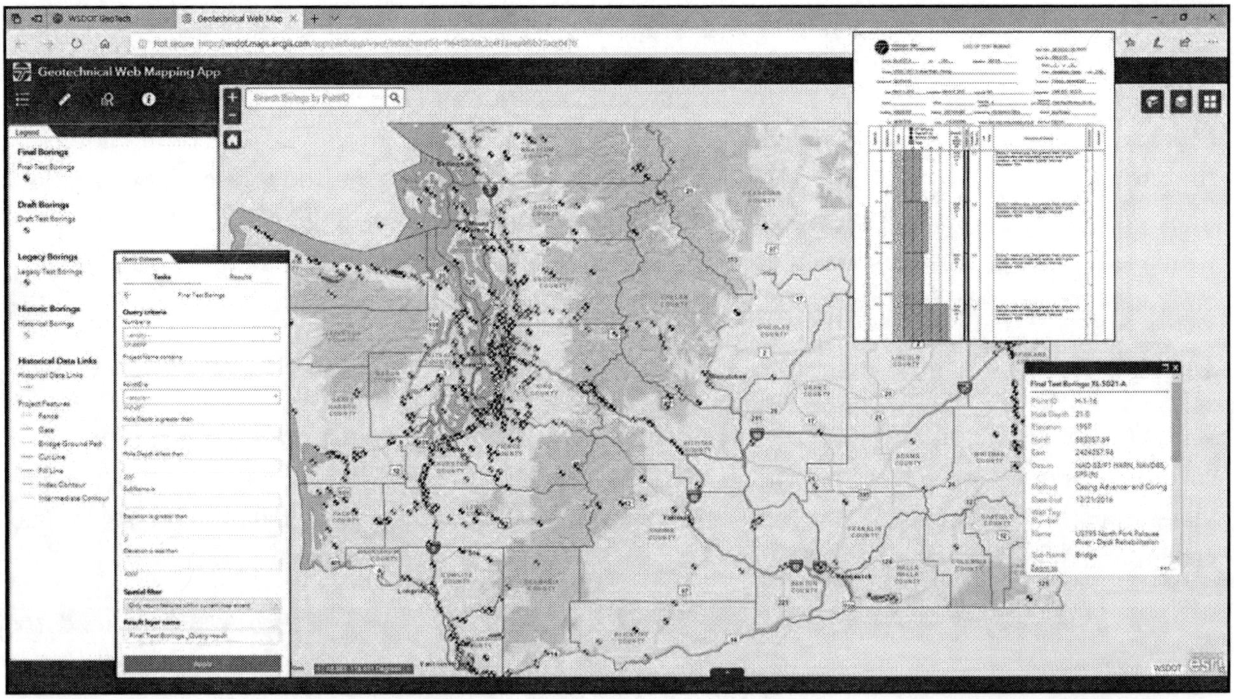

Figure 6: Geotechnical web mapping application showing legend, search function, pop-up window, and linked test boring log.

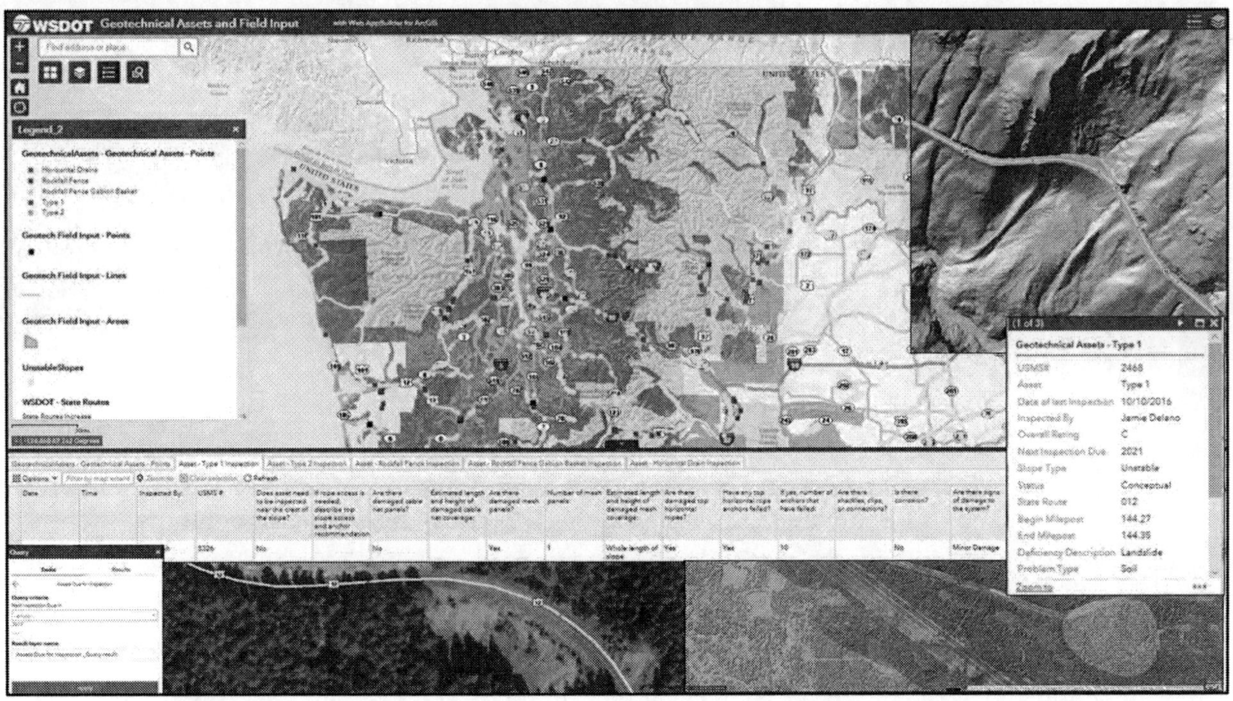

Figure 7: Geotechnical Assets Web Mapping application displaying types and location of the constructed geotechnical assets with the geology, lidar, and orthophoto layers, the legend, a pop-up window, and a search function.

Geotechnical Assets Web Mapping Application

This web mapping application displays the location of WSDOT's constructed geotechnical assets (i.e. horizontal drains, rockfall fences, and type 1 and 2 slope protection) (Figure 7). It is used by the GO to locate geotechnical assets that need a new inspection, and to track previous inspection results. It also has a couple of predetermined queries to help locate which assets needs to be inspected each year. Several other layers (USMS, mapped landslides, etc.) can be turned on to provide additional information.

Mega Project Web Mapping Application

These applications contain test borings (old and new), lidar, contour intervals, parcels, and specific design layers (Figure 8). This is an internal application, available to all WSDOT staff that are working on a specific project. This web mapping application is updated regularly and is used to make project specific decisions.

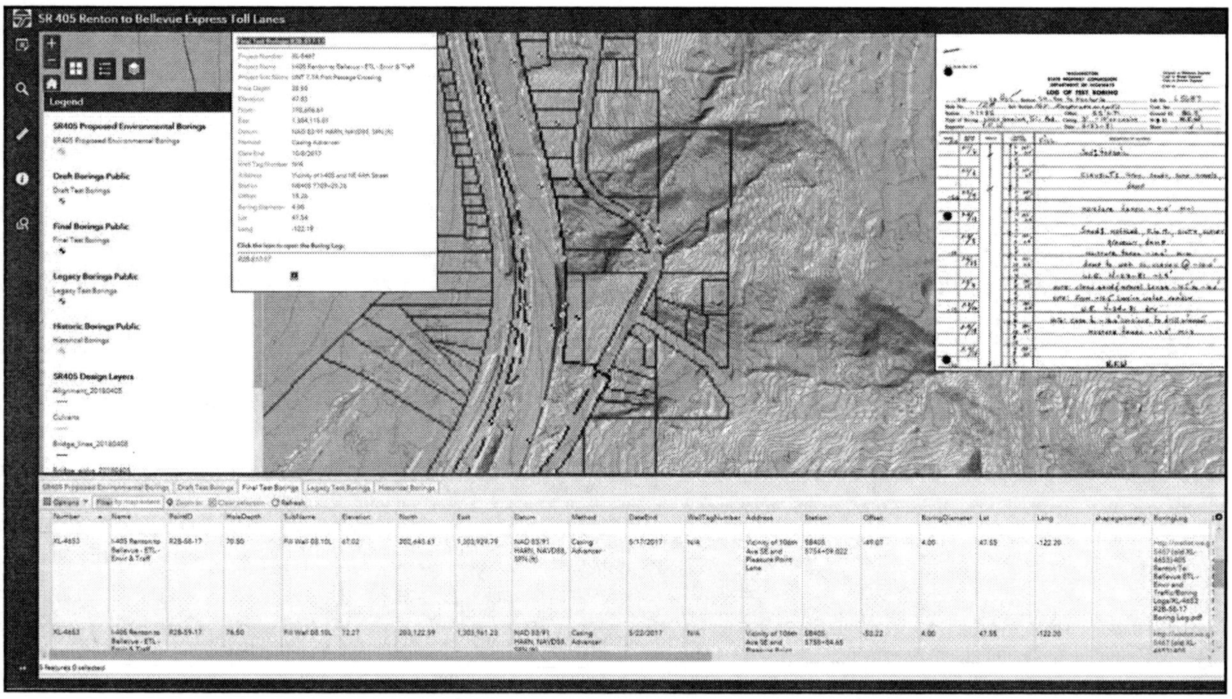

Figure 8: I405 Renton to Bellevue Mega Project. Data layers include parcels, contours, and the historic test boring logs, with the legend, a pop-up window, and the data window.

Desktop GIS

Some WSDOT employees have access to the full desktop version of ArcGIS. The GO has created several unique projects containing the same geotechnical/geological data that is used in the web mapping applications (Figure 9). These projects can be accessed from anywhere in the state, as long as the user is behind the WSDOT firewall. More functionality and data is available through desktop ArcGIS. A few drawbacks include the need for a software license, the knowledge of how to use the software, and the need for enhanced computer speed, depending upon the size of the datasets and data processing complexity. Additional licenses, provide special profile (i.e. cross section) and 3D capabilities to the software. Cross sections are used in change analysis, limit equilibrium analysis, and rockfall modeling (Figure 10) and 3D views are used to visualize topography (Figures 11).

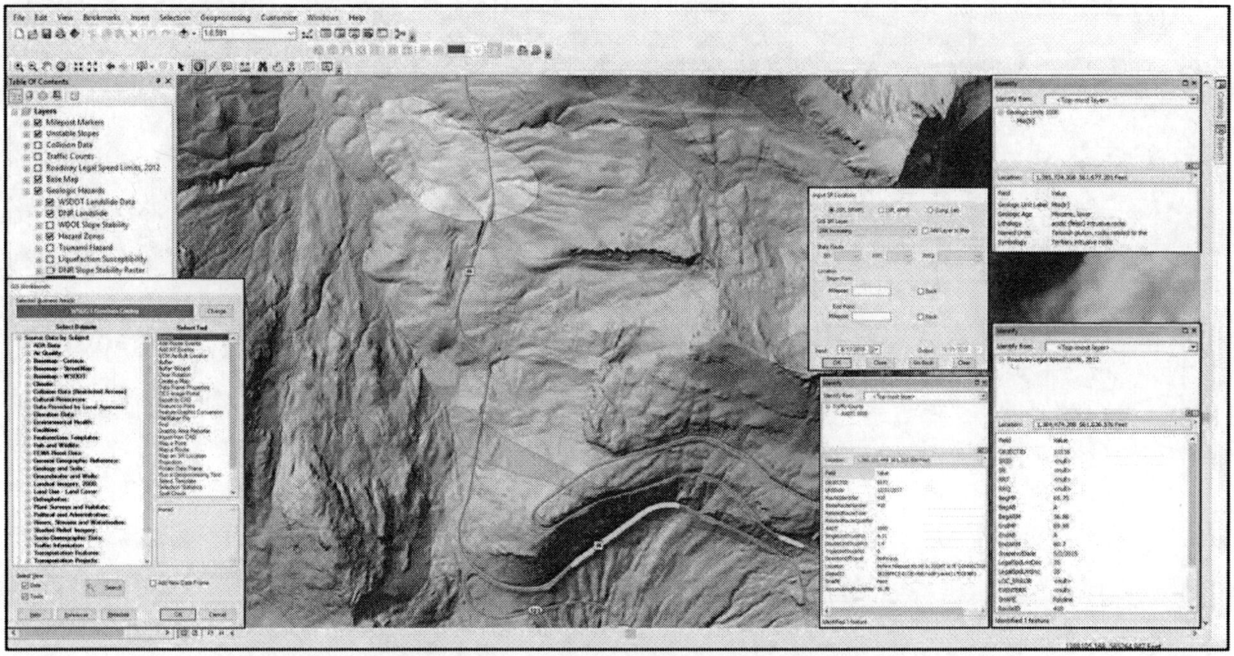

Figure 9: View of desktop ArcGIS showing the GIS workbench and map SR location functions, and the traffic count, speed limit, 100K Geology, and mile post marker layers.

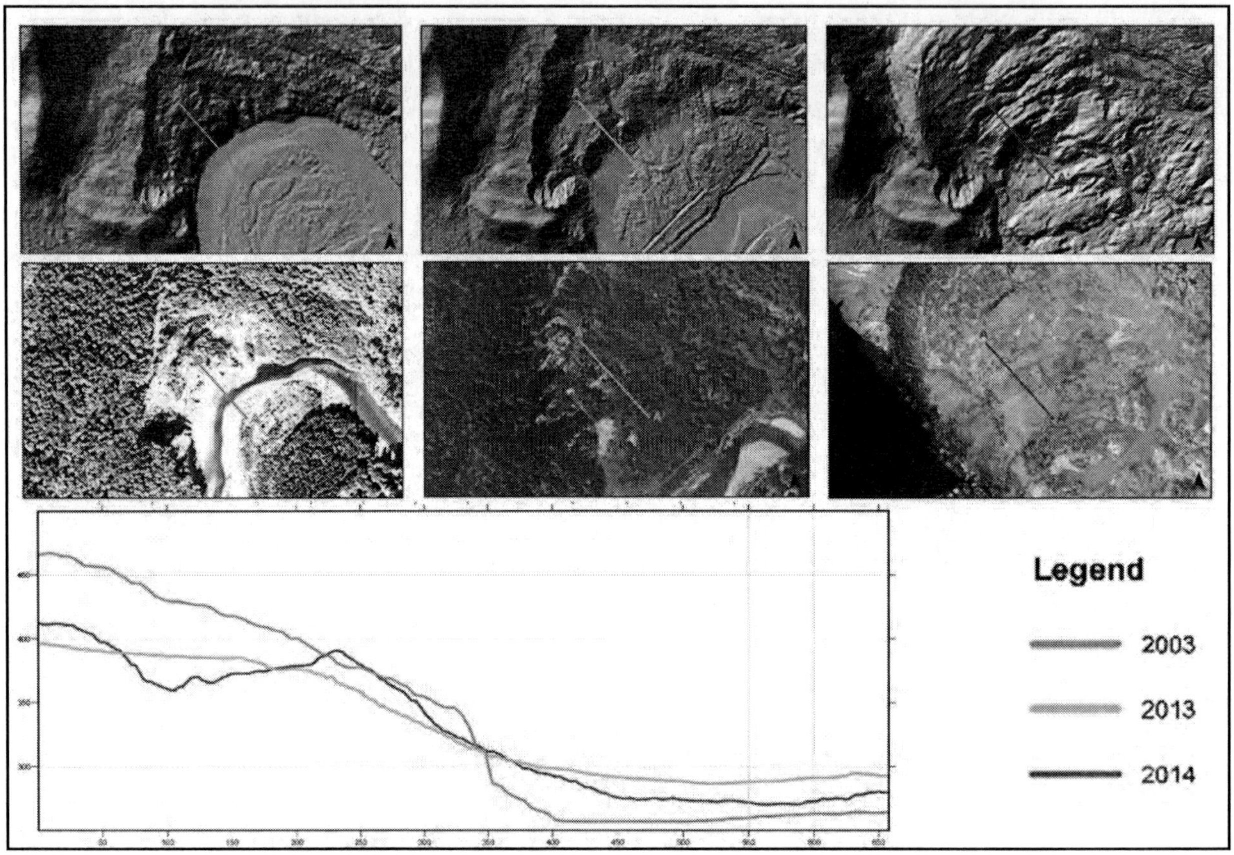

Figure 10: Cross sectional view of a landslide over a 3-year period.

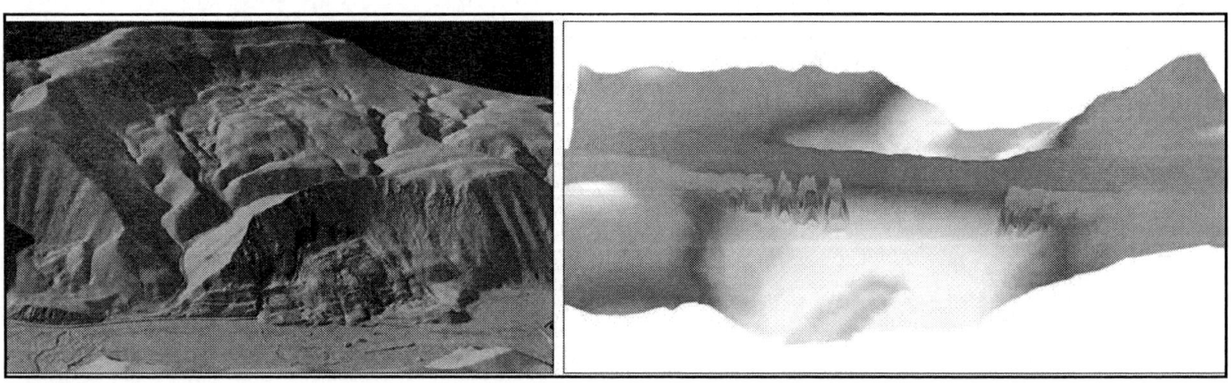

Figure 11: 3D views help to visualize topography at the site of an unstable slope or a project.

CONCLUSION

The GO has developed a methodology to disseminate GAM data to end-users through a visually pleasing technique. This method uses commercially available software with some minor programming by the ITO. For most needs and applications, the user only needs a web browser to access the data. Through their web browser they can locate, view, and download large quantities of GAM data including, unstable slopes data, test boring data, and geotechnical asset inspection records. The data is safely stored, backed-up, and dispersed in ways that reduce, or minimize, the efforts of both our in-house geotechnical staff, and the ITO. These applications were developed by our dedicated GIS professional over several years. During this time period, advances in GIS online mapping has helped to improve our office's ability to get data out to the end-user. This process is continually changing, and is being updated as new technology becomes available to the department. It has worked well, but more users need to take advantage of these web mapping applications that disseminates WSDOT's GAM data.

REFERENCES

WSDOT Geotechnical Assets & Field Input Web Mapping Application for WSDOT employees, https://wsdot.maps.arcgis.com/apps/webappviewer/index.html?id=e49a7fd663b245829fd937e6c90a86d2, accessed 6/19/19 2:32 PM PST

WSDOT Unstable Slopes Web application for WSDOT employees, http://webprod9.wsdot.loc/systems/slope/, accessed 6/19/19, 2:08 PM PST.

WSDOT Unstable Slopes Web Mapping Application for WSDOT employees, https://wsdot.maps.arcgis.com/apps/webappviewer/index.html?id=c3925b2bddad4337965c368991ad780d accessed 6/19/19, 2:10 PM PST.

WSDOT Geotechnical Web Mapping Application for WSDOT employees, https://wsdot.maps.arcgis.com/apps/webappviewer/index.html?id=f9645206fc2c4f33aea9f9b27acc0470, accessed 6/19/19, 2:20 PM PST.

WSDOT Public Geotechnical Web Mapping Application, accessed 6/19/19, https://wsdot.maps.arcgis.com/apps/webappviewer/index.html?id=03b9e16243ef480988e470c325e4f020, accessed 2:27 PM PST.

WSDOT Geotechnical Assets Web Mapping Application for WSDOT employees, https://wsdot.maps.arcgis.com/apps/webappviewer/index.html?id=c3925b2bddad4337965c368991ad780d, accessed 6/19/19, 2:10 PM PST.

WSDOT Renton to Bellevue Web Mapping Application for WSDOT employees, https://wsdot.maps.arcgis.com/apps/webappviewer/index.html?id=3662584d4fd44660941928c7314a96ea, accessed 6/19/19, 2:28 PM PST.

Investigation, Design and Construction of Rock Slopes and Foundations for the New Genesee Arch Bridge, Letchworth State Park, New York

Jay R. Smerekanicz, PG
Golder Associates Inc.
670 North Commercial Street, Suite 103
Manchester, NH 03101
(603) 668-0880
jay_smerekanicz@golder.com

Mark F. McNeilly, PE, D.GE
Golder Associates Inc.
570 Broad Street, 6th Floor, Suite 601
Newark, NJ 07102
(603) 668-0880
pingraham@golder.com

Jeffrey D. Lloyd, PE
Golder Associates Inc.
670 North Commercial Street, Suite 103
Manchester, NH 03101
(603) 668-0880
jeffrey_lloyd@golder.com

Prepared for the 70th Highway Geology Symposium, October 21-24, 2019

Acknowledgements

The author(s) would like to thank the individuals / entities for their contributions in the work described:

Norfolk Southern Railway
Kevin W. Johns, PE – Modjeski and Masters Inc.
Russell J. Marchese – Golder Associates Inc.

Disclaimer

Statements and views presented in this paper are strictly those of the author(s) and do not necessarily reflect positions held by their affiliations, the Highway Geology Symposium (HGS), or others acknowledged above. The mention of trade names for commercial products does not imply the approval or endorsement by HGS.

Copyright Notice

Copyright © 2019 Highway Geology Symposium (HGS)

All Rights Reserved. Printed in the United States of America. No part of this publication may be reproduced or copied in any form or by any means – graphic, electronic, or mechanical, including photocopying, taping, or information storage and retrieval systems – without prior written permission of the HGS. This excludes the original author(s).

ABSTRACT

Replacement of the Portageville High Bridge (High Bridge), spanning the scenic Genesee River Gorge in Letchworth State Park, New York, entailed the first true steel arch bridge structure constructed for the railway industry since the 1940s. Understanding the geologic and geotechnical conditions for rock excavation, stabilization, and foundation design were essential to the project's success.

The bridge lies above the Upper Falls within the southwest part of Letchworth State Park, often referred to as the "*Grand Canyon of the East*", about 40-miles southwest of Rochester, New York. The previous steel trestle bridge, built in 1875, was considered a Park icon and the new bridge had to accommodate the same historical and aesthetic context. By the 1990s, the High Bridge was showing its aged condition, functionally obsolete, and subject to both load and speed restrictions. Hence, its replacement was deemed necessary to improve operational efficiency of this important railway corridor.

The site geology consists of the Late Devonian Nunda Formation interbedded and closely jointed fine-grained sandstones and shales, overlain by glacial till, outwash, and colluvium. Significant site constraints complicated the new bridge's design and construction, including limiting activity that could adversely impact the river or Park, relocation of an existing park access road, and a right-of-way width limited to 75 feet. To address these constrains, the new bridge's design utilized cantilevered construction methods to construct the steel arch supported on skewback foundations, which required near vertical rock cuts up to 120-feet deep into the Genesee River Gorge sidewalls.

Design and construction of the new bridge's foundations required extensive geologic and geotechnical assessments, including rope rappel mapping of gorge walls and a detailed subsurface exploration program to provide a thorough understanding of geologic conditions at the project site. Geological engineering design evaluations addressed potentially unsafe soil and rock conditions which had the potential to affect the integrity of the new bridge, including: numerical modeling for the skewback foundations; rock slope design; rock excavation utilizing close-in, controlled rock blasting; rockfall hazard mitigation measures; rock slope reinforcement and stabilization; consideration of shale sulfide effects on the proposed skewback foundations; micropile design for approach pier foundations; and soil approach embankments.

Construction issues included controlling excavated rock from entering the river, minimizing blasting-induced vibrations on the previous bridge, installing spot rock dowels and shotcrete to stabilize overbreak from blasting, maintaining active rail traffic across the previous bridge, limited linear space for staging of construction materials and equipment, and use of a temporary tieback system used to assemble the cantilevered, arched bridge elements. Construction commenced in November 2015, and the new bridge opened to railway traffic in December 2017. Upon its completion, the new bridge was renamed the Genesee Arch Bridge.

INTRODUCTION

Replacement of the Portageville High Bridge, spanning the Genesee River Gorge, with a new steel arch structure (i.e., Genesee Arch Bridge) was a complex undertaking, and represents the first true steel arch bridge structure constructed for the railroad industry since the 1940s. Understanding the geologic and geotechnical conditions and challenges associated with the requisite rock excavation, slope stabilization, and foundation design were essential to the project success.

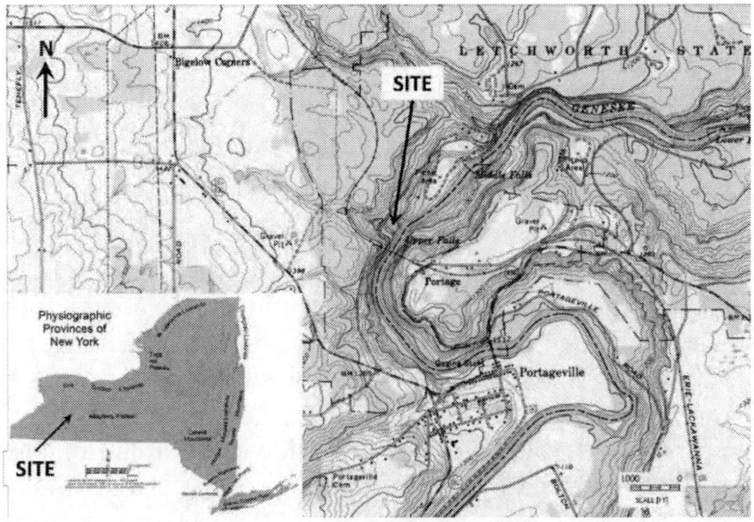

Figure 1 - Site location and physiographic map (USGS, 1976; NYSM, 2016)

The new Genesee Arch Bridge is located about 0.75-miles northwest of the Village of Portageville in Wyoming and Livingston Counties, New York, lying within the southwest part of Letchworth State Park (Park), about 40-miles southwest of Rochester, New York (Figure 1). This section of the Genesee River Gorge is commonly known as the "*Grand Canyon of the East*", and the High Bridge site is situated just upstream (i.e., above) of the Park's Upper Falls (Figure 2). The river has deeply incised the bedrock, creating a 240-foot-deep, steep, narrow gorge.

Figure 2 - View to north of the Upper Falls of Genesee River Gorge from the High Bridge

The former High Bridge was considered to be a Park icon by the public. Given this status, it was generally expected and understood that any new bridge would have to possess its own unique, iconic character and similar aesthetic qualities, while also restoring the Genesee River to its natural, unobstructed (i.e., pre-1851) condition. The High Bridge was considered both a vital and weakest link along Norfolk Southern's Southern Tier Line, which runs from Suffern, New York northwest to Buffalo, New York. By the 1990s, the High Bridge was showing its aged condition, functionally obsolete, and subject to both load and speed restrictions (i.e., 10-miles-per-hour), which created an undesirable bottleneck along an important railway corridor.

After it acquired the bridge in 1998, Norfolk Southern initiated plans to replace the High Bridge, and engaged Modjeski and Masters, Inc. to explore ways to improve the operational capacity of the High Bridge, including rehabilitation, strengthening, and replacement alternatives. Following a comprehensive study, Norfolk Southern decided to replace the High Bridge with a new single-track, 483-foot-long, 2-hinged spandrel-braced steel arch structure with three 80-foot-long approach spans flanking each side of the arch (total length of 963-feet), which was later renamed the Genesee Arch Bridge.

The alignment of the new Genesee Arch Bridge was set parallel to and offset 75-feet south of the former High Bridge, and it is supported atop two (2) reinforced-concrete skewback foundations (located about 80-feet above river-level). At its skewbacks, the distance from the top cord of the new railway trackage to the arch skewback bearing pins is 148-feet. Due to its span, alignment, and geometry, construction of the new bridge's arch skewback foundations required about 16,000 cubic yards of rock excavation, cut upwards of 120-feet into the existing, near-vertical river gorge sidewalls.

Design and construction of the new Genesee Arch Bridge foundations required extensive geologic and geotechnical evaluations, including geologic mapping on rope-rappel of river gorge sidewalls and a detailed subsurface exploration program to provide a thorough understanding of geologic conditions at the bridge site. In addition, geological and geotechnical engineering design analyses were conducted to evaluate ground conditions that could adversely impact the new bridge's integrity, consisting of:

- Arch skewback foundation design, including bearing capacity and settlement
- Numerical modeling of arch skewback settlements
- Evaluation of shale sulfides effects on arch skewback concrete
- Micropile designs for bridge piers and abutments
- Development of "*close-in*", controlled rock blasting rock excavation methods
- Evaluation of construction-induced vibrations on the old High Bridge structure
- Slope stability and rock slope stabilization / reinforcement designs
- Development of rockfall hazard mitigation measures

BRIDGE HISTORY

Between July 1851 and August 1852, the Buffalo and New York City Railroad, a subsidiary of the former Erie Railroad, constructed an 800-foot-long, 234-foot-high timber trestle bridge structure across the Genesee River Gorge, built atop sandstone pier foundations located just upstream of the Gorge's Upper Falls. At that time, the bridge became to be known as the Portage Viaduct (later renamed the High Bridge), and it provided breathtaking views of the Genesee River Gorge and its waterfalls downstream (north) of the bridge. On May 6, 1875, the Portage Viaduct was destroyed by fire. The bridge was subsequently replaced with an 819-foot-long, 245-foot-high wrought iron structure, built atop the prior bridge's repaired foundation piers, reopening to train traffic on July 31, 1875 (about three months following the fire).

In 1903, the 1875 bridge superstructure was upgraded to accommodate increased rail traffic loads, requiring replacement of its deck trusses and plate girders, which were supported atop existing wrought iron tower bents. The bridge's tower bents did not require replacement because they were originally designed to support two (2) tracks of rail traffic, and going forward the bridge would remain a single-track structure (Irwin and Johns, 2019). Figure 3 provides views of the original 1852 timber trestle and 1903 wrought iron bridge structures.

Figure 3 - Views to the south of 1852 (left) and 1903 (right) bridge structures

In 1958, the exposed rock slope faces beneath the bridge, which form the gorge sidewalls, were armored with shotcrete to reduce differential weathering and erosion (Erie Railroad Magazine, 1958). In addition, dental concrete was placed within open rock joints within the river channel to reduce the potential for scour around the High Bridge's foundation piers. In 1960, the Erie Railroad merged with the Delaware, Lackawanna, and Western Railroad, which became part of Conrail in 1976. In 1998, Conrail was jointly acquired by Norfolk Southern and CSX Transportation, and ownership of the High Bridge and its Southern Tier Line were then conveyed to Norfolk Southern.

REGIONAL AND SITE GEOLOGY

The project site is located on the northern edge of the Allegheny Plateau in western New York State. The Allegheny Plateau is situated on the northwest part of the Appalachian Plateau, which is on the west flank of the Appalachian Geologic Province. The Appalachians were uplifted during the Ordovician through the Permian periods during two separate tectonic collisions between the North American, European, and African continents, which occurred over a period from about 500 to 225 million years ago.

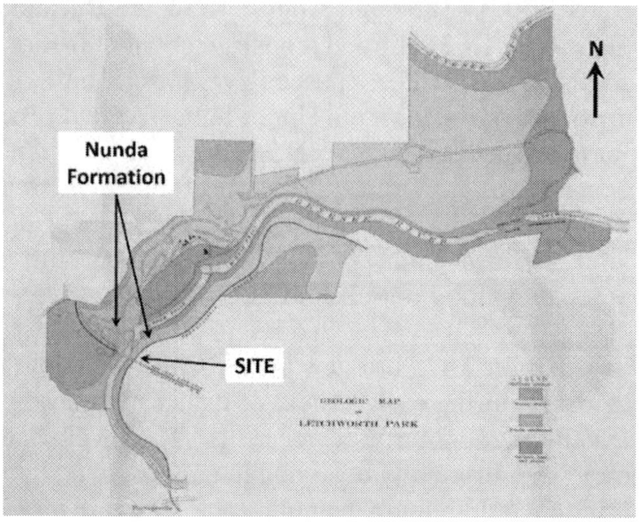

Figure 4 - Site bedrock geologic map (Clarke and Luther, 1908)

This region was later modified by four major glacial advances / retreats from about 2 million years ago to 6,000 years ago (van Diver, 2003). The Wisconsin glaciation (peaking about 20,000 years ago) was the final glacial surge, leaving the Valley Heads Moraine and a series of recessional moraines at the receding glacial front. Temporary meltwater lakes formed between the retreating ice sheet front and these moraines, establishing a base level for erosion. Rivers flowing toward these lakes down-cut into the underlying bedrock until they emptied into the lakes, where they deposited their sediment loads at the erosional base level. Lake levels dropped as the impounding moraines were overtopped and new spillways were established at lower elevations, creating new lower base levels which reactivated erosion and down-cutting (van Diver, 2003) to produce the series of waterfalls (Upper, Middle, and Lower Falls) downstream of the project site.

Bedrock within the region belongs to the Late Devonian Nunda Formation of the West Falls Group, which is about 400- to 950-feet-thick and characterized by horizontally bedded, fine-grained sandstones with interbedded shale layers (Figure 4; Rickard and Fisher, 1970; Clarke and Luther, 1908). The Nunda Formation is most likely a submarine fan deposit, as its sandstones are generally thick, massive to wavy / flaggy-bedded, have few primary sedimentary structures, terminate abruptly, and appear to have lobate forms (Jacobi et al., 1994). Overburden soils generally consist of glacial drift (till and outwash) and colluvium.

PROJECT CHALLENGES

The design and construction of the new Genesee Arch Bridge was a complex undertaking, which had to address and overcome multiple project challenges, including:

- Railway "right-of-way" limitations
- Proximity of the new bridge to the old High Bridge (i.e., 75-foot offset)
- Need to maintain rail traffic across the old High Bridge throughout construction

- Severely restricted site access, due to proximity of the new bridge to river gorge sidewalls
- Limited linear space for staging of construction materials and equipment
- Need to prevent adverse impacts on the Genesee River and Park
- No temporary or permanent structures allowed in the river
- Mitigate potentially adverse impacts (e.g., vibrations) on the old High Bridge

To overcome many of the above-noted project challenges, the design and construction of the new bridge considered and required implementation of "*top-down*" construction techniques, which involved accessing all foundation work areas solely from atop the river gorge sidewalls. This was achieved by using large cranes placed atop the river gorge sidewalls, which were used to transport (i.e., lower / raise / remove) all construction equipment, materials, excavation spoils, and personal required to complete the required foundation work.

On the west side of the river, the proximity of an existing Park access road and parking lot presented another noteworthy project challenge. This road provides access from the south, which is the busiest entrance to the Park, and it included a hazardous (i.e., "*blind*") curve within the vicinity of the old High Bridge. Figure 5 provides plan views of the west approach area, before and after new bridge construction.

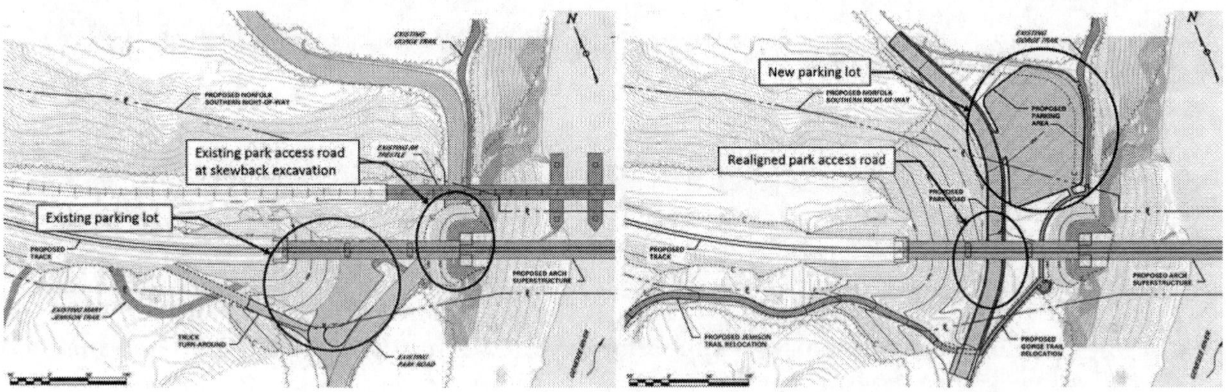

Figure 5 - Plans of west approach before (left) and after (right) new bridge construction

As illustrated in Figure 5, the new west railway approach embankment would directly impact an existing Park parking lot and west skewback excavation work would encroach and bisect the existing Park access road. To resolve these impacts, the solution was to relocate and realign the Park's access road westward, which would eliminate the "blind" curve condition and create a new, larger parking area.

SUBSURFACE CONDITIONS

Due to the proximity of the new bridge to the existing gorge sidewalls, site access issues restricted all subsurface exploration activities to only those accessible areas atop the river gorge sidewalls, which limited exploration options. As part of the new bridge's foundation design, a two-phase investigation approach was implemented, which involved:

- A geologic field reconnaissance (conducted in 2008), utilizing fixed-line, rappelling techniques, to collect geologic and geotechnical mapping data (e.g., lithographic descriptions and rock mass discontinuity measurements) for exposed river gorge sidewalls within vicinity of the new bridge (Figure 6).
- A subsurface exploration program (conducted in 2011; Figure 6), which included:
  - Four (4) inclined rock coreholes to collect rock core samples
  - Thirteen (13) soil borings along approach embankment alignments
  - Optical / acoustic borehole televiewer (OTV/ATV) surveys to collect rock mass discontinuity measurements (e.g., dip angle and direction)
  - Geotechnical laboratory testing on collected soil and rock samples

Figure 6 – Geologic mapping using rappelling techniques in spring 2008 (left); and inclined rock core drilling during winter 2011 on west side of gorge (right)

Due to safety and site access concerns, the Park also required that the access road be temporarily closed during all subsurface drilling activities, which required working in the middle of the winter. This exposed field crews to severe, extreme weather conditions, including frigid temperatures, high winds, and snow / ice, typical of western New York.

The surficial geology at the site consists of glacial till and outwash, overlain by colluvium. These sediments overly subhorizontally-bedded, gray, fine-grained sandstone / siltstones with interbedded shale layers, typically 2- to 8-inches thick, which account for about 1% to 3% the "as-drilled" rock mass. Between El. +1145 and +1155, an approximate 6- to 9-foot-thick layer of shale and/or shaly-sandstone lies at / near the bases of the east and west arch skewback foundations (see Figures 7).

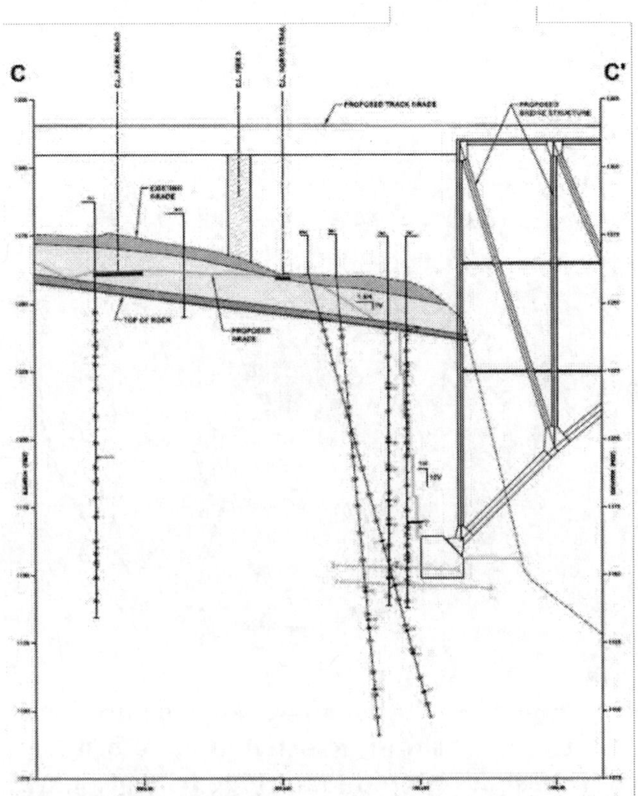

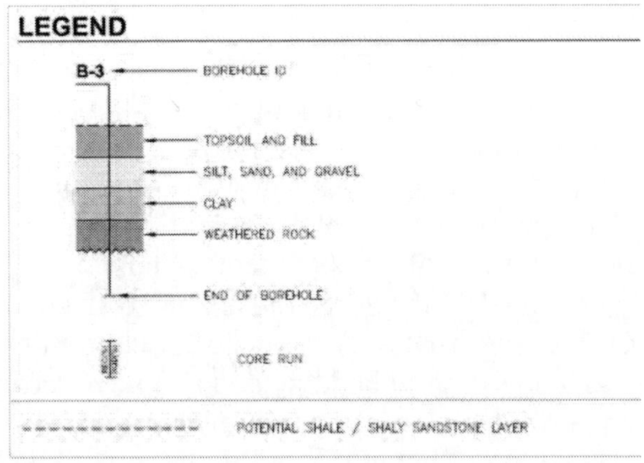

Figure 7 – Subsurface Profile of West Arch Skewback

This above-noted shale / shaly-sandstone coincides with a "*hard blue shale*" identified in Clarke and Luther (1908) and was also identified in the gorge sidewalls during the field geologic mapping program. As the shale was suspected of containing iron sulfides, petrographic laboratory testing was conducted to evaluate if the skewback concrete foundations could be adversely affected by pyrite-containing shale. The layer was found to contain upwards of 3% pyrite, which required over-excavation to prevent the proposed arch skewback foundations from being constructed directly atop pyrite-containing rock (Golder, 2009).

KINEMATIC ANALYSES

Due to the near proximity (i.e., on order of 10 to 20 feet) of the new bridge's skewbacks to the "*as-excavated*" rock slopes, a series of kinematic analyses, using the *Dips* (RocScience, 2012a) software analysis package were performed to evaluate the proposed rock cut slope stability, and establish maximum allowable rock cut slope angles. In addition, two-dimensional, limit-equilibrium stability analyses, using the *Slide* software analysis program (RocScience, 2012b), were conducted for evaluating overall stability of the proposed rock cut slopes (Golder, 2013).

In general, kinematic analyses utilize geometric methods to assess different modes (e.g., planar, wedge, and toppling) of rock slope instability and/or failure. In particular, these analyses entail using stereographic projection techniques to plot three-dimensional orientation data, such as slope geometry, orientations of rock mass discontinuities, and discontinuity strength values (e.g., internal friction), as two-dimensional representations, which can be used to identify the number and nature of potentially adverse rock mass discontinuities.

Discontinuity Data

The predominant discontinuity types measured in the field were bedding planes (503 out of 901 measurements or 56%), and bedrock joints (398 out of 901 measurements or 44%; see Figure 8). The discontinuity breakdown by collection method was 270 measurements obtained by rope rappel of the gorge (34%), and 631 via OTV/ATV (70%). Three major discontinuity sets, consisting of bedding and two near vertical and orthogonal joint sets were identified. Evaluation of both OTV/ATV and rappelling data indicates six additional minor discontinuity joint sets exist. Bedding discontinuities have a very low (< 3 ft) to high (30 to 60 ft) persistence, are spaced 0.1 to 10 ft, and have very tight (<0.1 mm) to cavernous (>1 m) openings. The shape of the bedding discontinuities is planar to stepped to undulating, and roughness ranges between rough to smooth. Infilling is common, and consists of silty sand, shale, clay, broken rock, and rare secondary mineralization (e.g., chlorite, talc, or gypsum). Joint discontinuities have a very low (< 3 ft) to very high (>60 ft) persistence, are spaced 0.1 to 25 ft, and have very tight (<0.1 mm) to extremely wide (10 – 100 cm) openings. The shapes of the joint discontinuities are planar, irregular, curved and stepped, and roughness ranges between smooth to very rough (and rarely polished and slickensided). Infilling is common, and consists of silty sand, clay, broken rock, and rare secondary mineralization (e.g., chlorite, talc, gypsum, and iron oxide).

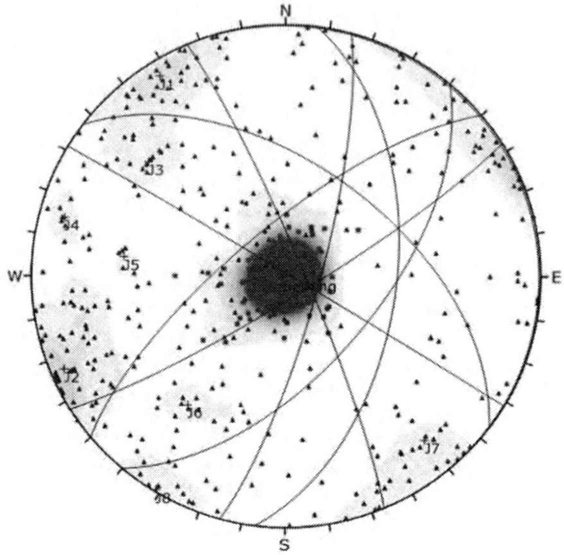

Figure 8 – Equal angle lower hemisphere stereonet showing distribution of bedding (red squares) and joint (blue triangles), with set poles and great circles (901 points, contour interval 0.63%)

In general, the nature of the OTV/ATV discontinuity data method does not allow for measurement of 2-dimensional characteristics such as surface shape and persistence, but the method does measure discontinuity asperity, and can allow for estimating joint spacing once joint sets are identified from preliminary stereonet analysis. The OTV/ATV data indicate the bedding spacing ranges from 0.01 ft to 16.5 ft, and the aperture ranges from very tight (<0.1 mm) to very wide (1 to 10 cm). The OTV/ATV data also indicate the joint spacing ranges from 0.02 ft to 25 ft, and the aperture ranges from very tight (<0.1 mm) to very wide (1 to 10 cm).

Figure 8 presents a representative polar distribution plot for the combined rappelling and OTV/ATV data, including set pole concentrations, and great circles of the eight discontinuity joint sets and bedding plane used in the kinematic stability analyses. The sets include poles representing the three major discontinuities that plot within the 1 percent contour interval, and the poles representing the six minor joint sets that plot within the 0.1 percent contour interval. Table 1 below provides a summary of the nine joint and bedding sets.

Table 1 – Discontinuity Set Summary

Set Identification	Discontinuity Type	Characteristic	Dip/Dip Direction
Bedding	Bedding	Bedding	01/058
J1 m	Joint Set 1	Major joint set	83/148
J2 m	Joint Set 2	Major joint set	84/067
J3 m	Joint Set 3	Minor joint set	60/130
J4 m	Joint Set 4	Minor joint set	79/105
J5 m	Joint Set 5	Minor joint set	54/097
J6 m	Joint Set 6	Minor joint set	54/037
J7 m	Joint Set 7	Minor joint set	74/320
J8 m	Joint Set 8	Minor joint set	89/031

Rock Strength Data

Laboratory values of rock peak and final friction angles along discontinuities range from 27.4° to 35.7° and 22.2° to 36.4°, respectively. The average peak and final friction angles are 29.4° and 29°. These averages are within the range of published values for sandstone and shale (25° to 35°, and 27°, respectively). The roughness of the discontinuity planes and the infilling material, if present, can increase or decrease the friction angle. While the field descriptions of the joints indicate rough to moderately rough surfaces and undulating to planar shapes, many of the joint and bedding plane discontinuities contain clay and shale, which can reduce the friction angle. As the shale occurs only in beds parallel or closely parallel to the bedding discontinuity set, and the bedding is nearly horizontal, an overall base friction angle of 32.5° was chosen for analysis, as this value better represents the base friction value of discontinuities within the sandstone.

Rock Slope Cut Angle Analysis

To assess the stability of potential rock cut slope angles as part of the foundation excavation for the new bridge, a series of planes were stereographically depicted to represent final cut slope angles of 79°, 82° and 84°, representing cut slopes of 5 vertical to 1 horizontal (5V:1H), 7.5V:1H and 10V:1H, respectively.

Potential Kinematic Sliding Modes

For each cut slope, planar sliding daylight envelopes were plotted. The stereonets also include a base friction cone, oriented about the center, and spaced 32.5° from the center. Any pole pertaining to a discontinuity plotting within the daylight envelope but outside of the base friction angle cone is free to slide (the daylight envelope allows for the planar sliding test, i.e., there must be free space for a planar slab to slide into). Poles plotting within the green shaded region represent planes which could be susceptible to planar sliding. See Figure 9 for example stereographic kinematic analyses for the west and east slope excavations.

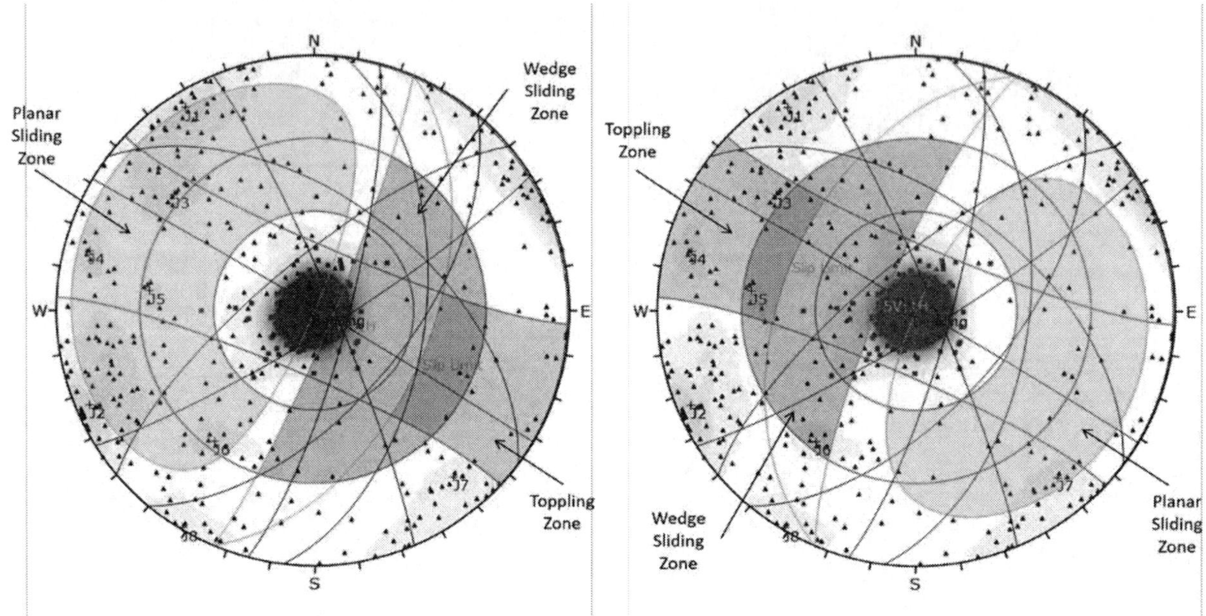

Figure 9 – Kinematics of west slope at 10V:1H (left), and east slope at 5V:1H (right)

Each stereonet also contains the slip limit (green great circle) defined as the slope angle minus the friction angle. For toppling failure to occur, the dip of the planes must be steep enough for slip to occur. Each stereonet also has a horizontal cone representing a kinematic bound envelope with limits of 20 degrees on either side of the dip direction of the proposed cut. Poles plotting outside this limit represent discontinuities that are likely aligned too oblique to be able to slide. Any pole plotting outside of the slip limit but within the 20-degree kinematic bounds represents a plane, which could topple kinematically. This zone is represented by the blue shaded region.

In wedge kinematic analysis, planes instead of poles are analyzed for the potential for failure. Therefore, the friction angle for the planes is measured from the horizontal, i.e., equator of the stereonet. An additional plane friction cone exists in the stereonets, oriented about the center, and spaced at $90° – 32.5° = 57.5°$. Intersections of planes occurring within this cone but outside of the cut slope angle represent the intersection lines of planes, which could cause wedge failures if kinematically possible. This zone is represented by the pink shaded region.

Table 2 below summarizes the controlling structural geology and potential kinematic sliding modes of the analyzed rock cut slope angles.

Table 2 – Structural Geology and Potential Kinematic Sliding Mode Summary

Foundation Excavation Slope	Cut Slope Orientation		Controlling Structural Geology	Potential Sliding Mode
	Dip	Dip Direction	Joint Set ID (1)	
West (5V:1H)	79	113	1, 2, 3, 4, & 5	Planar Wedge
West (7.5V:1H)	82	113	1, 2, 3, 4, & 5	Planar Wedge
West (10V:1H)	84	113	1, 2, 3, 4, 5, & 8	Planar Wedge

Foundation Excavation Slope	Cut Slope Orientation		Controlling Structural Geology	Potential Sliding Mode
	Dip	Dip Direction	Joint Set ID (1)	
East (5V:1H)	79	293	3, 4 & 7	Planar Toppling
East (7.5V:1H)	82	293	3, 4 & 7	Planar Toppling
East (10V:1H)	84	293	2, 3, 4, 7 & 8	Planar Wedge Toppling

(1) Major joint sets are in "Red".

The following sections provide further discussions of the controlling geologic structures and potential kinematic sliding modes for each excavation, and the design slope angles.

West Slope Excavation

As shown in Figure 8, the west slope excavation for all three rock cut slope angles is susceptible to all three modes of kinematic sliding (i.e., planar, wedge, and topping). The controlling geologic structures for planar sliding involve the major joint sets 1 and 2, as well as minor sets 3, 4, and 5. The controlling geologic structures for wedge sliding are the intersections of major joint set 1 and minor joint set 3, and major joint set 2 and minor joint set 4. No major joint set plots within the toppling region, but poles from random joints do plot within the toppling envelope; therefore, this mode of sliding was considered minor, but possible. Given the sliding modes exist regardless of the evaluated cut slope angles and rockfall mitigation measures (e.g., pattern rock dowels, rock drains, dental shotcrete, and drape netting) were incorporated into the project's rock slope stabilization design, the west rock slope excavation angle was designed to be 84° (10V:1H). To address potential toppling issues, a line of regularly spaced upper-brow rock dowels was also included, as part of the project's rock slope stabilization design.

East Slope Excavation

Figure 8 also shows that east slope excavation for all three rock cut angles is less susceptible to kinematic sliding. For the 5V:1H and 7.5V:1H rock cut slope conditions, the controlling geologic structures for planar and toppling sliding involve minor joint sets 3, 4 and 7. These are considered minor sets and the potential for these failures is low at these cut angles. For the 10V:1H slope angle, the controlling geologic structures for toppling and planar sliding involve the minor joint sets 3, 4 and 7; and for wedge sliding is the intersection of major joint set 2 and minor joint set 8. To reduce rock slope stabilization costs, the east slope was cut at an overall angle of 79° (5V:1H). As required for the west slope above, a line of regularly spaced upper-brow rock dowels was included to address potential toppling issues.

NUMERICAL ANALYSIS

Numerical modeling was performed using the two-dimensional, plane-strain finite-element program Phase^2 (RocScience, 2012c). Phase^2 simulates the behavior of soil and rock materials by representing geologic materials as an arrangement of elements and nodes that form a grid (or mesh) that can be adjusted by the user to fit the shape of the structure being analyzed. Phase^2

was used to make predictions (i.e., estimates) of the proposed arch skewback foundation deformations / displacements and contact stresses induced by the applied intermediate and final foundation loading conditions. The proposed arch skewback foundations were estimated to have incremental horizontal movements between 0.01- and 0.10-inch, and total vertical movements between 0.09- and 0.27-inch. These estimated movements were within the design tolerance.

ROCK CUT SLOPE STABILIZATION

Based on the findings and results of the slope stability, numerical and kinematic analyses, the following excavation, stabilization, and reinforcement methods were incorporated into the design:

- Rock Cut Slope Angles: East rock cut slope of 5V:1H; west rock cut slope 10V:1H to accommodate Park access road relocation.
- Rock Dowels: 40-foot-long, 1.375-inch-diameter (#11), Grade 150, galvanized steel thread-bars drilled and grouted within exposed, "as-excavated" rock slope faces on a 10-foot (typ.) staggered spacing pattern, at a 15-degree declination (i.e., below horizontal) angle. Shorter rock dowels were also included along the perimeter and tops of the excavations and as spot rock dowels, where needed.
- Rock Drains: 25- to 50-foot-long, "open", 3.5-inch-diameter drain holes, located about 5-feet above the proposed arch skewback foundations, installed at 10-degree inclination (i.e., above horizontal) angle.
- Rock Scaling: All exposed rock cut slope surfaces were scaled, as necessary, to physically remove loose, unstable rock blocks / fragments, which could potentially move down-slope and represent future rockfall hazards.
- Shotcrete: Where the shale / shaly-sandstone materials were observed within the "as-excavated" rock cut slopes, dental shotcrete was utilized, as needed, to reduce the degree and extent of differential weathering. In addition, structural shotcrete buttresses were used, as needed, in areas where the rock excavation created unsupported, over-hanging (i.e., potentially unstable) rock blocks, which could not be effectively removed by rock scaling methods or structurally stabilized by addition of spot rock dowels.
- Rockfall Drape: A flexible, interconnected drape netting system was installed over all "as-excavated" rock slope surfaces that present future rockfall hazard risks (i.e., slope areas where rockfall debris has potential to strike, impact the new bridge structure). The specified drape netting system included a combination of galvanized ring nets (primary drape) and double-twist wire mesh (secondary drape), which where anchored along the slope crest.

CONSTRUCTION

By December 2014, design and environmental permitting of the new bridge were completed. Construction began in February 2015 with clearing activities to provide site access and staging areas to allow construction of the new bridge. In fall 2015, Norfolk Southern awarded a $58-

million contract to American Bridge Company (AB) of Coraopolis, Pennsylvania to construct the new bridge (Irwin and Johns, 2019).

Figure 10 – Narrow, tight conditions for the west slope excavation

Figure 11 - Line drilling used for perimeter control for each rock excavation lift

Rock excavation activities, using "*close-in*" controlled rock blasting, were conducted from February to December 2016, removing about 15,750 cubic yards of rock. The blast design included unloaded, closely spaced vertical line drilling for perimeter control of each lift (see Figures 10 and 11). Typically, each blasting lift was limited to 15 ft in height. The Contractor was also required to mechanically excavate all rock within 10 ft of the proposed arch skewback foundation bearing surfaces.

Due to concern about how much construction-induced stress the old High Bridge could tolerate, construction of the new bridge required monitoring and adherence to strict, conservative construction vibration limits (i.e., 0.5-inch-per-second peak particle velocity above ambient conditions); and coordinating all rock blasting events with Norfolk Southern to maintain the flow of train traffic across the old High Bridge. Additionally, no rock was allowed to impact the river from blasting and excavation. Due to these tight constraints, the blaster used an electronic detonation system for

Figure 12 – Access to excavation areas with large cranes

the controlled, close-in blasting, along with anchored rubber-tire blasting mats to contain potential flyrock. In general, two to three blasts were needed for each lift. Excavation equipment consisting of remote-controlled hydraulic blasthole rigs and micro-excavators, along with the workers, explosives and other construction material, and the excavated rock, were lowered into or raised out of each excavation by two 200-ton cranes staged on both sides of the gorge (see Figure 12).

Figure 13 - Drilling rock dowels in the west slope excavation

As each lift was completed, the contractor installed pattern rock dowels (see Figure 13), per the rock slope stabilization design. Dental shotcrete and shotcrete buttresses were installed when each excavation reached full depth. Due to the presence of unsuitable pyritic shale materials at the base of the excavation (verified by rock coring as excavation activities approached foundation levels), the excavations for the arch skewback foundations were deepened to over-excavate the unsuitable shale materials and replace it with concrete, so that the arch skewback foundations would not bear directly atop such unsuitable pyritic shale materials and thus reduce risks associated with concrete degradation (see Figure 14).

Rock blasting and excavation activities were successfully completed without adversely impacting the old High Bridge, and train traffic was not interrupted throughout construction. Following completion of rock excavation activities, a series of rock drains were installed at the toes of each excavation, along with a ring-net drape, attached at the slope crests with wire rope anchors (see Figure 15). Overall, the east and west arch skewback concrete foundations were placed and ready for setting of the bridge's arch bearings, which were set by December 2016 and March 2017, respectively.

Figure 14 – Line drilling at the bottom of the east slope excavation to over-excavate pyritic shale and replace with concrete

Figure 15 – Completed drape netting and arch skewback bearings under construction in the east slope excavation

On December 11, 2017, the new bridge was opened to railway traffic, and AB subsequently started work to demolish the old High Bridge, which was subsequently removed by May 2018. On May 24, 2018, a formal bridge dedication ceremony was held, during which it was officially renamed the Genesee Arch Bridge (Irwin and Johns, 2019; see Figure 16).

CONCLUSIONS

Figure 16 – Completed Genesee Arch Bridge

While replacement of the High Bridge over the scenic Genesee River Gorge with a new steel arch structure is an engineering marvel and is a testament to the skill and expertise of the numerous engineers, scientists, and contractors that brought the new bridge from its initial concept through construction completion, its success would not have been possible without a thorough investigation of the geologic materials such a massive structure must bear upon. Modern approaches, including a 3-dimensional assessment of subsurface conditions using high-angle, rope rappel mapping techniques, inclined core borings, and downhole geophysics, along with assessment and mitigation of potential adverse geologic conditions, were vital for the project's success.

While the old High Bridge could not be saved, the project did incorporate one of the old steel foundation columns, mounted in the rebuilt overlook area on the west side (Figure 17), a fitting memento to the long history of the bridges built across the Genesee River Gorge.

Figure 17 – Monument to the old High Bridge

REFERENCES

Clarke, J.M. and Luther, D.D. (1908). Geologic Map and Description of the Portage and Nunda Quadrangles, Including a Map of Letchworth State Park, New York State Education Department, New York State Museum, Bulletin 118, January 1908.

Erie Railroad Magazine (1958). *Erie Spends $126,000 on Cliff to Keep Portage Span Safe*, October 1958, p. 8-9, and 27, Erie Railroad, Cleveland, Ohio.

Irwin, D.B and Johns, K.W (2019). *Replacing an Icon*, Civil Engineering, American Society of Civil Engineers, p.42-51.

Golder Associates Inc. (2009). Geotechnical Engineering Report, prepared for Modjeski and Masters, Inc., Norfolk Southern Railway Bridge No. 361.66, Portageville, New York, dated June 16, 2009.

Golder Associates Inc. (2013). Report on Preliminary Geotechnical Investigation, prepared for Modjeski and Masters, Inc., Norfolk Southern Railway Bridge No. 361.66, Portageville, New York, dated February 8, 2013.

Jacobi, R., Gutmann, Piechocki, A.I., Singer, J., O'Connell, S. and Mitchell, C. (1994). *Upper Devonian Turbidites in Western New York – Preliminary Observations and Implications*, in Landing, E., ed., Studies in Stratigraphy and Paleontology in Honor of Donald W. Fisher: New York State Museum Bulletin No. 481, p. 101-115.

New York State Museum (NYSM; 2016). NYS Physiographic Provinces of New York.

Rickard, L.V. and Fisher, D.W. (1970). Geologic Map of New York, Niagara Sheet, New York State Museum and Science Services, Map and Chart Series No. 15, Scale 1:250,000, dated March 1970.

RocScience (2012a), "Dips (ver. 5.109) Plotting, Analysis, and Presentation of Structural Data Using Spherical Projection Techniques", RocScience, Inc., Toronto, Ontario, Canada.

RocScience (2012b) Phase^2 (ver. 8.0) 2D Elasto-Plastic Finite Element Program, RocScience, Inc., Toronto, Ontario

RocScience (2012c) Slide (ver. 6.018) 2D Limit Equilibrium Slope Stability Analysis, RocScience, Inc., Toronto, Ontario, Canada.

U.S. Geological Survey (USGS; 1976). Portageville, New York, 7.5 Minute Series (Topographic Map), scale 1:24,000.

Van Diver, B.B. (2003). Roadside Geology of New York, MT, Mountain Press Publishing Company, Missoula, MT.

Route T, Ray County, Missouri Landslide Remediation Study, Design and Construction

John F. Szturo R.G. P.G
HNTB Corporation
715 Kirk Drive
Kansas City, MO 64105
816 472-1201
jszturo@hntb.com

Wayne A. Duryee P.E.
HNTB Corporation
715 Kirk Drive
Kansas City, MO 64105
816 472-1201
wduryee@hntb.com

Prepared for the 70th Highway Geology Symposium, October 2019

Acknowledgements

The authors would like to thank the individuals for their contribution in the work described

Missouri Department of Transportation

Disclaimer

Statements and views presented in this paper are strictly those of the authors and do not necessarily reflect positions held by their affiliations, The Highway Geology Symposium (HGS), or others acknowledged above. The mention of trade names for commercial products does not imply the approval or endorsement by HGS.

Copyright Notice

Copyright © 2019 Highway Geology Symposium (HGS)

All Rights Reserved. Printed in the United States of America. No part of this publication may be reproduced or copied in any form by any means – graphic, electronic, or mechanical, including photocopying, taping, or information storage and retrieval systems – without prior written permission of the HGS. This excludes the original authors.

Abstract

The study involved an evaluation of the existing failed slope and study of various stabilization measures. The primary purpose of the study is to assess the advantages and disadvantages of cost-effective remediation alternatives. In general, the existing slope is considered unstable and a roadway realignment, retaining wall system, or shallower slope is warranted to reopen Route T. Early on in the process, one of the primary factors identified was to have little to no impact to the BNSF operations and right of way. The major issue with the existing slope is its failure and movement of soil into a drainage ditch adjacent to the BNSF mainline tracks. The soil could block drainage and possibly foul the tracks. In the end, cooperation between MoDOT and the BNSF was required to facilitate repair of the slope and reopen Route T.

The existing slope along 1000 feet of the south side of Route T has failed. The head scarp of the slide was located very near the south edge of pavement. Many other stress cracks were evident in multiple places in the pavement indicating many areas in the roadway embankment are very near equilibrium and could fail with further triggers.

Maximum existing slope height is on the order of 30 feet. The toe of the present slope is positioned at the edge of the drainage ditch parallel to the BNSF double tracks. The slope has failed in multiple locations along the 1000 feet and the entire length should be considered for repair.

A total of seven different options were evaluated with pros and cons listed for each option along with an estimated cost.

1.0 Introduction

The Missouri Department of Transportation (MoDOT) closed Route T due to a slope failure that occurred in association with a high precipitation (4 to 5 inches) event covering several days in early April 2015. Route T is a rural two-lane asphalt roadway in Ray County between the towns of Fleming, and Camden, Missouri.

HNTB was contracted by MoDOT to perform a study of alternatives to reopen Route T. The primary purpose of this study wasto assess cost-effective alternatives to do so.

The following sections summarize the approach to the study, site conditions, assessment of each alternative, and recommendations.

Vicinity Map

2.0 Study Approach

The study involved an evaluation of slope stability as well as various retaining walls and roadway realignments to remediate the landslide. Consideration was given to constructability, cost, impact to BNSF RR and right of way, soil type, rock type, degree of weathering, slope height, MoDOT right-of-way, long-term maintenance, and other applicable factors that could influence the recommendations of the preferred system. In short, no work could be performed on or from the BNSF right of way and no additional right of way would be obtained by MoDOT opposite the slide.

The approach for the study was to initially determine feasible alternatives based on the applicable site constraints such as slope location, proximity to right-of-way limits, and impact to BNSF RR. HNTB personnel reviewed the proposed locations, elevations, cross sections, subsurface logs, and laboratory test information from exploratory borings provided by MoDOT in a preliminary report dated July 6, 2015 (see Appendix A).

3.0 Engineering Geology

Bedrock is not exposed, although the borings and geomorphology would indicate bedrock occurs in the near surface. Geologic maps and borings indicate the bedrock consists of Lower Pennsylvanian Age, thick layers of shale with layers and stringers of limestone. As is typical along the Missouri River bluffs, the overburden consists of loess, a wind deposit of angular interlocking silt sized grains. The borings describe the bedrock as shale, tan to gray, calcareous, soft to medium hard. The top few feet are typically weathered very soft. Limestone layers were encountered deeper in the shale.

The calcareous and clayey shale is impervious and acts as an aquatard. Water moves more readily through the loess and is concentrated at the contact of the shale. This concentration of water decreases the strength of the loess and facilitated the slope failure.

The north bank of the Missouri River was located adjacent to the site as indicated on the USGS map of 1894, 56 years later it is now nearly a mile to the south, probably the result of the flood of 1917 or flood control channelization. Route T was most likely constructed when the river was in its former position next to the roadway prior to 1917. Since the shale most likely outcropped very near the former steep surface adjacent to the Missouri River, the present roadway was likely cut into the side of the hill just enough to create a suitable roadway width. Material from the cut was probably pushed over and down the slope to an angle of repose at 2H: 1V or slightly steeper. The fill material was also most likely not compacted. Coal mining also occurred in the bluff during the 1800's to power steamboats.

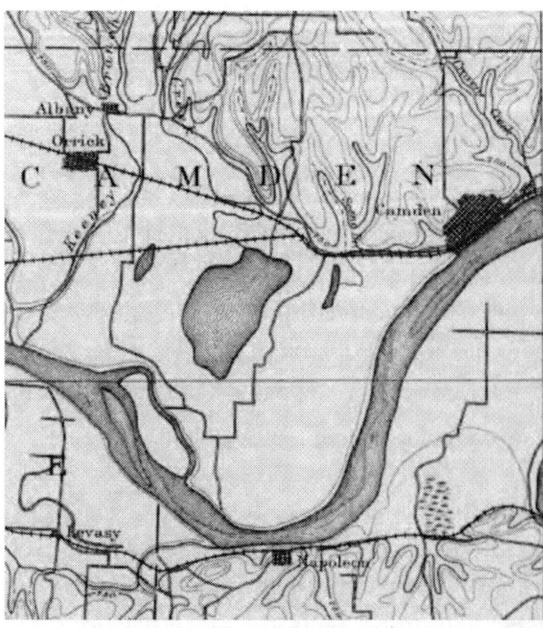

Missouri River 1894

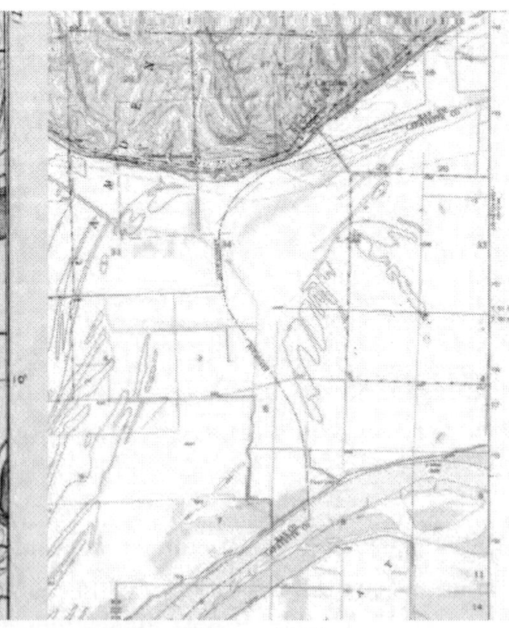

Missouri River 1950

4.0 Site Conditions and Constraints

Route T is located at the boundary of the flat Missouri River flood plain and the steep bluff. The river plain consists of alluvial sand, 100 feet +/- thick. In the near past, as per historic maps, the north bank of the Missouri River was located at the toe of the present slope of Route T. Route T is located in a sidehill cut in the bluff while the BNSF RR tracks are located in the flat alluvial flood plain of the Missouri River. The right of way at the study area is approximately 60 feet wide or 30 feet right and left of the roadway centerline. The right of way line is located approximately half way between the edge of pavement and the railroad drainage ditch.

The north (uphill) side of Route T is in a 15 to 20 feet deep vertical cut in loess. Also, at the east end of the study area and along the inside radius of the existing roadway, a significant amount of soil of the bluff has failed and is blocking the roadway. The downhill failure side of Route T has a steep, approximately 1H: 1V slope down to an existing ditch. The BNSF Railroad double tracks are located adjacent to the ditch. The slope failed into the ditch of the railroad during the event and the toe of the slide mass was removed to reopen the railroad. Lateral stress cracks are observed at various locations along the roadway from the centerline to the south edge of the existing pavement indicating instability under nearly all the existing roadway.

Approximately 400 feet of guardrail is constructed along the roadway. Due to the landslide, the existing soil has severely eroded and several guardrail posts are partially to fully exposed and are hanging in the air. Approximately ten wooden power poles for the overhead utility (ATT) are installed in the slope, a few of which are leaning due to the slope failure and will need to be removed. A waterline runs parallel to the north side of Route T through the project and will also need to be relocated.

The BNSF RR insists that no access or construction take place on their right of way. This includes no overhang from drill leads or crane swings. This restriction complicates nearly all practical solutions to the slope failure.

Slope as observed from BNSF

Slope as observed from BNSF

5.0 Remediation Alternatives – Approaches to Slope Stability

There are four basic ways to approach slope stability in this case.

A. Avoid
   a. Relocate the roadway
   b. Complete removal of the unstable materials
   c. Install a bridge
B. Reduce Driving Forces
   a. Change roadway line or grade
   b. Reduce the driving weight

C. Increase Resisting Forces
   a. Use a buttress or a toe berm
   b. Structural systems, retaining walls
   c. Anchors

D. Increase Internal Strength
   a. Drain water from the subsurface
   b. Use reinforced backfill
   c. Install in-situ reinforcement

The following options were carried forward from the basic approaches:

5.1 Option 1 – Do Nothing

Do nothing and leave Route T closed. Create a permanent detour or reconfigure state route system. Some maintenance will be required as the slope will continue to ravel and foul the BNSF right-of-way.

Pros:

- Minimal costs

Cons:

- Continued impact to RR
- Inconvenience to travelling public
- Increased emergency response times

5.2 Option 2 – MoDOT Proposal – Remove and Replace

The initial MoDOT recommendation was for a "Chimney Drain with Geogrid Reinforced Slope Repair".

The plan calls for excavating the slide material from near the centerline of the roadway to the toe of the existing slope, reconstructing the embankment with rock fill in the bottom 10 feet, and a Geosynthetic Reinforced Soil Slope (GRSS) above the rock fill.

Pros:

- Less construction complexity
- Proven MoDOT methodology
- Multiple contractors qualified to bid and construct

Cons:

- Cannot be constructed from existing roadway
- Impact to RR
- Requires substantial excavation and replacement with select material

5.3 Option 3 - Move Roadway to the North and Lower Grade

This option includes improving an existing substandard roadway geometry by locating it out of the area of instability. This option would involve possible addition of right of way, water line relocation, excavation of soil, establishing drainage, paving and guardrail. The backslope would be cut vertically in the loess and the fill slope toe would be located at the existing south right of way.

Pros:

- Improved roadway geometry
- Increased long term stability
- Minimal impact to BNSF RR
- Less construction complexity
- Multiple contractors qualified to bid and construct

Cons:

- Requires additional right of way
- May involve slight grading of material on BNSF right of way
- Requires offsite haul of large quantity of material

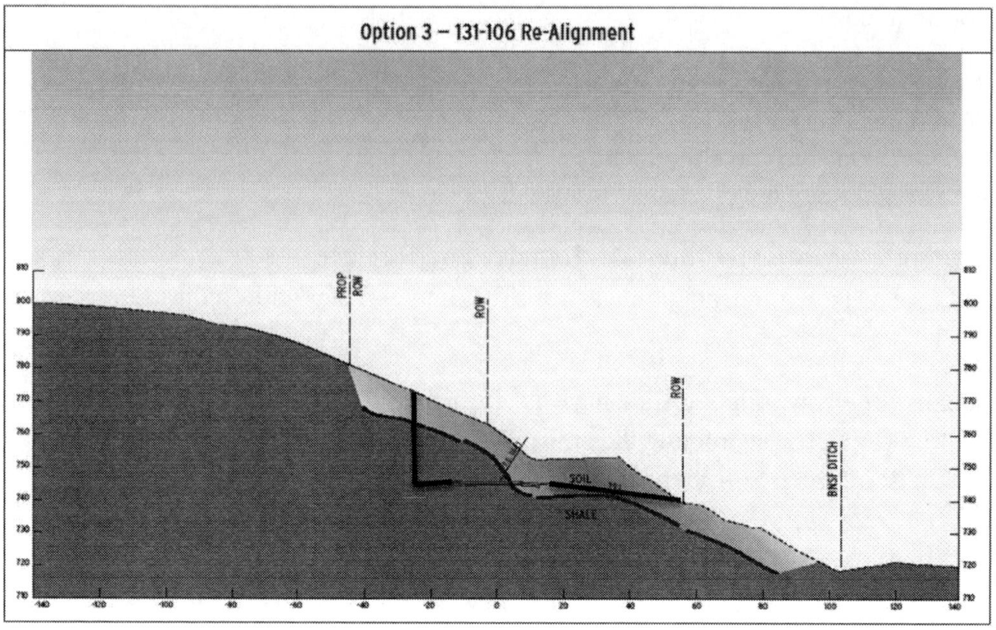

5.4 Retaining Walls

All the retaining wall options can be combined with various schemes of lowering roadway grade.

Option 4A – Retaining Wall – Concrete Crib (T Wall Proprietary) or MSE

Excavate failed soil material to shale bedrock and sufficiently embed into shale and construct a crib type concrete retaining wall system.

Pros:

- Readily available standard precast concrete segments
- Can be constructed from existing roadway

Cons:

- Construction (grading) may take place on BNSF right of way
- Requires a large deep excavation – haul material offsite
- Requires select granular backfill
- Not readily adaptable to variable bedrock topography.

Option 4B – 131 + 06

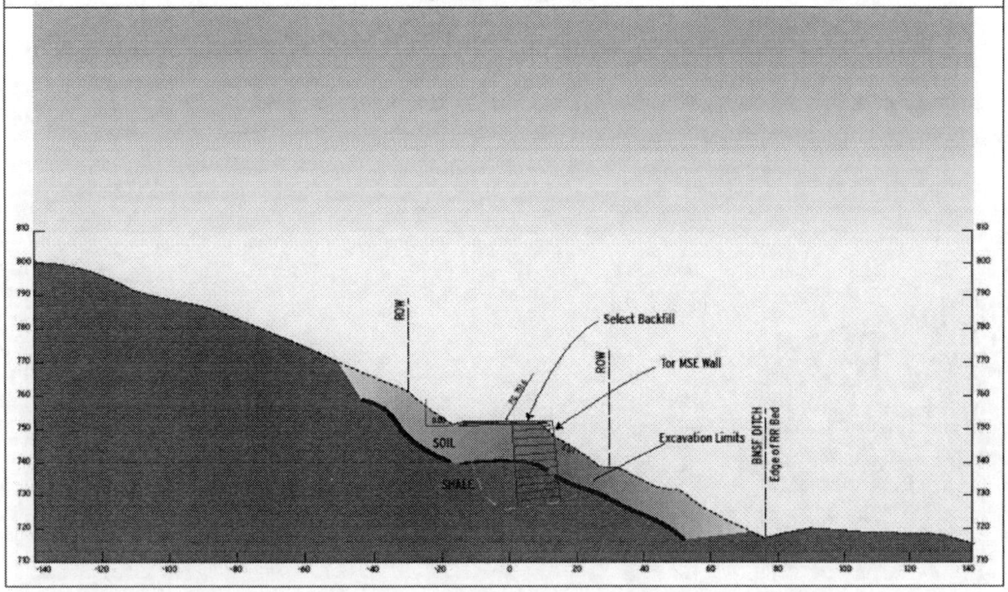

Option 4B – Retaining Wall - Soil Nail Wall

Construct a reinforced slope using soil nail construction. Construct a soil wall from the south edge of pavement to a stable elevation below top of shale bedrock.

Pros:

- Can be constructed top down on MoDOT right of way
- No large excavation and material haul off

Cons:

- Construction (grading) may take place on BNSF right of way
- Requires specialty contractor
- May require some adaptation where wall location is not entirely in cut section.

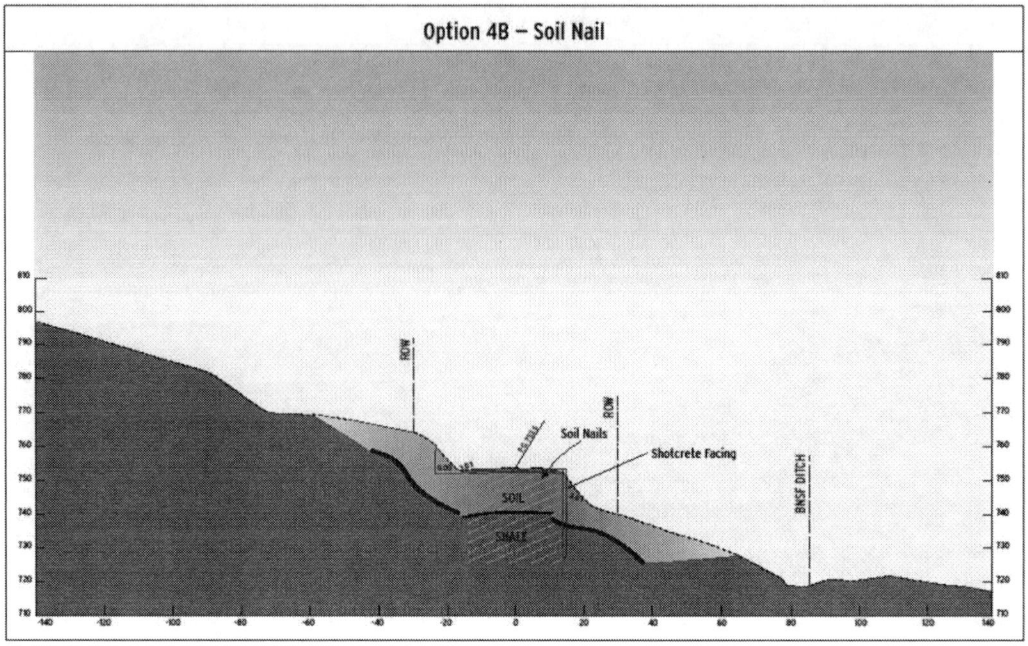

Option 4C – Retaining Wall Soldier Pile and Lagging

Construct a permanent structure retaining the existing Route T using a soldier pile and lagging system. Drill holes from the existing roadway, place piles, and backfill with concrete. Soldier piles will most likely require anchoring with tie backs into the shale bedrock. Precast concrete panels are then placed between the soldier piles and the soil graded between the wall and the BNSF RR right of way.

Pros:

- Can be constructed from existing roadway
- Multiple contractors qualified to bid and construct

Cons:

- Construction (grading) may take place on BNSF right of way
- Complex system
- Not readily adaptable to variable bedrock topography
- Greater risk of redesign and construction claims

Option 4C – Soldier Pile and Lagging

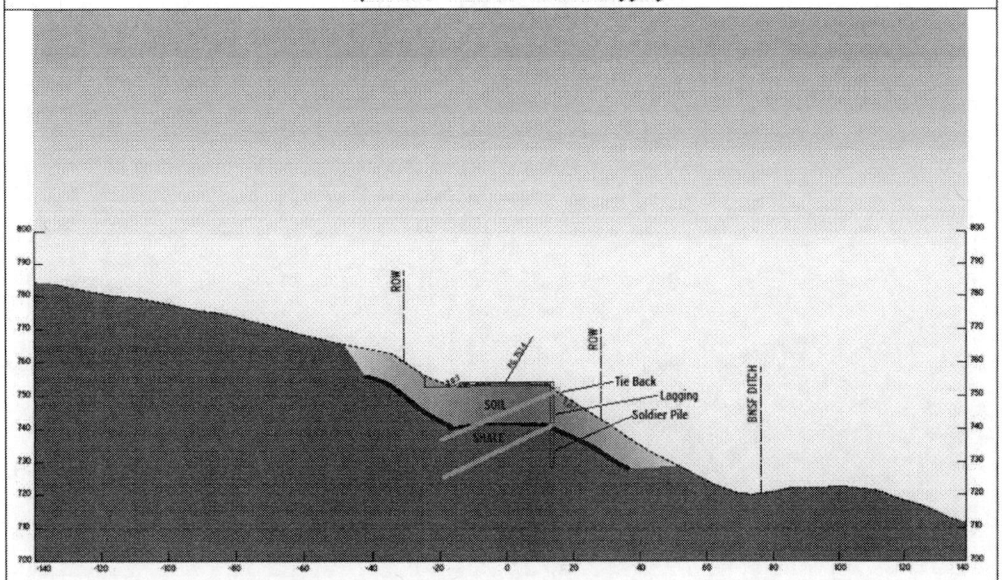

Option 4D – Sheet Pile Retaining Wall

Construct a continuous sheet pile retaining wall along the entire length. Sheet pile wall will require multiple levels of whalers and tiebacks.

Pros:

- Can be constructed from existing roadway
- Multiple contractors qualified to bid and construct
- Minimal excavation and haul off

Cons:

- Construction (grading) may take place on BNSF right of way
- Complex system requires specialty contractor for tiebacks
- Not readily adaptable to variable bedrock topography
- Greater risk of redesign and construction claims

Option 4D – Sheet Pile Wall

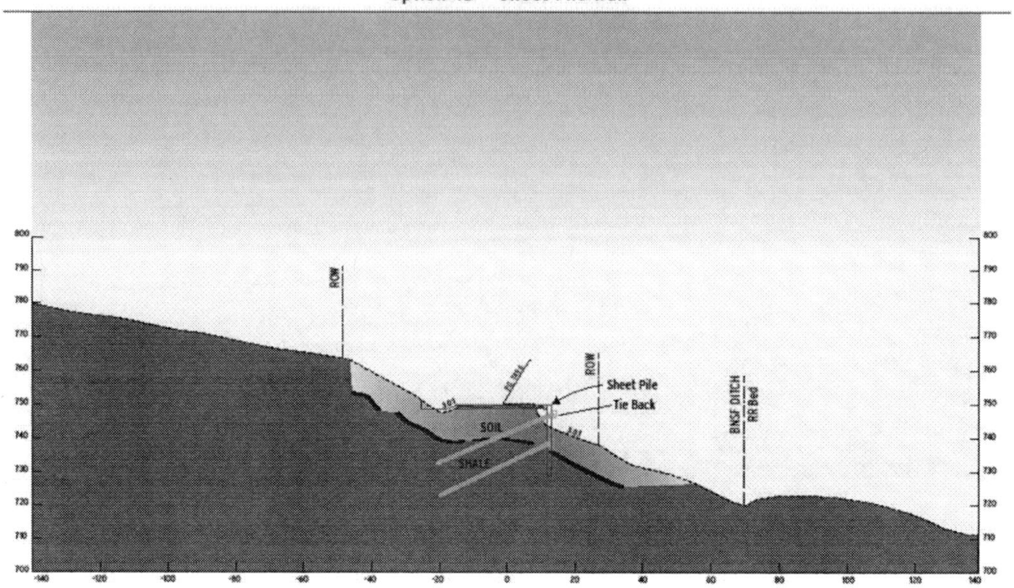

6.0 Stability Analysis

Four major slide locations along the alignment were evaluated for their stability.

An evaluation of the geometry of the slides, including the maximum height of the hill north of the roadway, the lateral extent of the slide, and the thickness of material described in the borings as highly weathered, extremely soft shale, determined the critical location for stability analyses. A resistance factor of 0.77 (Factor of Safety >1.3) was used to determine if a remediation alternative was effective. Options 3 and 4, as described in Section 6.0, were selected as viable options.

For Option 3, which involves re-aligning the roadway to the north and performing a cut into the loess slope, the stability of the slope above the roadway was generally evaluated and should be stable if the slope does contain loess. Global stability for this option has a factor of safety of 1.301 (resistance factor of 0.77).

For Option 4, involving various retaining wall types, only a general analysis with a simplified wall was performed to verify global stability. Internal stability and differences in wall configurations were not evaluated. Global stability for this option has a factor of safety of 1.710 (resistance factor of 0.58).

7.0 Comparison Matrix

Opt.	Description	Can be constructed from existing roadway	Add. ROW Required	Impact to Utility	Service Life (Years)	Impact to BNSF	Const. Difficulty	Cost ($Millions)	
								Low	High
1	**No Action**		No	Status Quo	N/A	Moderate	N/A		
2	**Remove and Replace** – Excavate and replace with rock	No	Yes on BNSF	Mod	100	High	Low	$2.00	$2.00
3	**Realignment** – Move roadway to the north and/or lower grade	Yes	Yes to North	High	100	Low	Low	$0.50*	$1.25*
4A	**Retaining Wall – T Wall/MSE** – Concrete crib or MSE type wall	Yes	No	Low	75	Moderate	Moderate	$1.58	$2.08
4B	**Retaining Wall – Soil Nail Wall** – Top down construction	Yes	No	Low	75	Moderate	High	$1.33	$2.08
4C	**Retaining Wall Soldier Pile and Lagging** – Install soldier piles with tiebacks and concrete or timber lagging	Yes	No	Low	75	Moderate	Moderate	$2.08	$2.83
4D	**Retaining Wall – Sheet Pile Wall** – Install sheet piles with whalers and tiebacks	Yes	No	Low	75	Moderate	High	$2.33	$3.46

8.0 Conclusions and Recommendations

All the remediation options require additional subsurface investigations to define bedrock surfaces, design parameters and final dimensions.

Option 4B Soil Nail Wall was selected by MoDOT and was recommended for final design. This option was preferred due to its ability to adapt readily to a variable top of shale bedrock and reduced risk of contractor claims and complicated redesign. The soil nail wall may be extended or shortened with minimal redesign or change order.

The re-alignment option requires the purchase of additional right of way and the haul off of all excavated material off site. The MSE/Concrete Crib wall option also requires extensive excavation and haul off of material. These two options may also require excavation up to 30 feet deep from the existing pavement, requiring temporary shoring or other staging.

The soldier pile and sheet pile wall also did not adapt well to a variable top of shale bedrock. The piles or sheets will have to be re-evaluated if lengthened. Both of these options require tie back anchors. If the wall is lengthened, additional tie backs may need to be added.

9.0 Wall Construction

MoDOT and HNTB prepared plans and specifications for a Contractor-designed soil nail wall. The specifications required a performance-based soil nail wall support system with certain criteria, such as minimum nail lengths and spacings set to meet external stability requirements for global stability. The Route T project was bid in June 2016 combined with another local slide project. As MoDOT wanted to get the roadway open by winter, the contract had a 90-day calendar completion time with a roadway opening date of December 2, 2016. The successful bidder's price for the Route T project was $ 2.49M. The successful Contractor immediately submitted a no-cost alternate proposal to change the soil nail wall to a soldier pile wall with an 8-inch finished shotcrete face. The Contractor felt the soldier pile wall could be constructed faster and was more adaptable to variations in subsurface conditions than a soil nail wall.

MoDOT held a series of meetings with the Contractor and his designer to discuss the soldier pile wall alternate and agreed on the requirements for design and construction of the soldier pile wall. After a number of back-and-forth design and plan submittals between the Owner and the Contractor, an approved set of soldier pile wall plans and specifications were finalized in early November 2016. The Contractor had started with site prep and ordering soldier pile and other materials ahead of final approval in order to make up for initial loss of time getting the alternate approved. The final soldier pile wall had a total length of 1165 feet with an exposed height varying from 4 to 21 feet. Soldier piles were spaced at 10-foot centers and consisted of HP 14x102 sections for cantilever heights up to 11 feet and HP 10x42 sections for tieback wall heights from 12 to 21 feet. Single tiebacks consisted of two and three 7-wire strand double corrosion protected tiebacks. Tieback lengths varied from 30 to 40 feet installed at a 20-degree rake. Tiebacks were required to stay on MoDOT right-of-way that was 46+ feet behind the wall face. All excavation and wall construction operations were required to stay within the narrow bench between the face of wall and property line with the BNSF Railway. A conventional soldier pile wall utilizes timber lagging between soldier piles with a cast-in-place concrete permanent face. The Contractor chose to excavate at the face in 5-foot lifts, install drainage strips, wire mesh and reinforcing bars, and place final shotcrete wall facing with trowel and screed finish before proceeding to the next lower level. Vertical drainage strips were connected to a lateral drainage collector pipe at the base of wall and outleted beyond the wall.

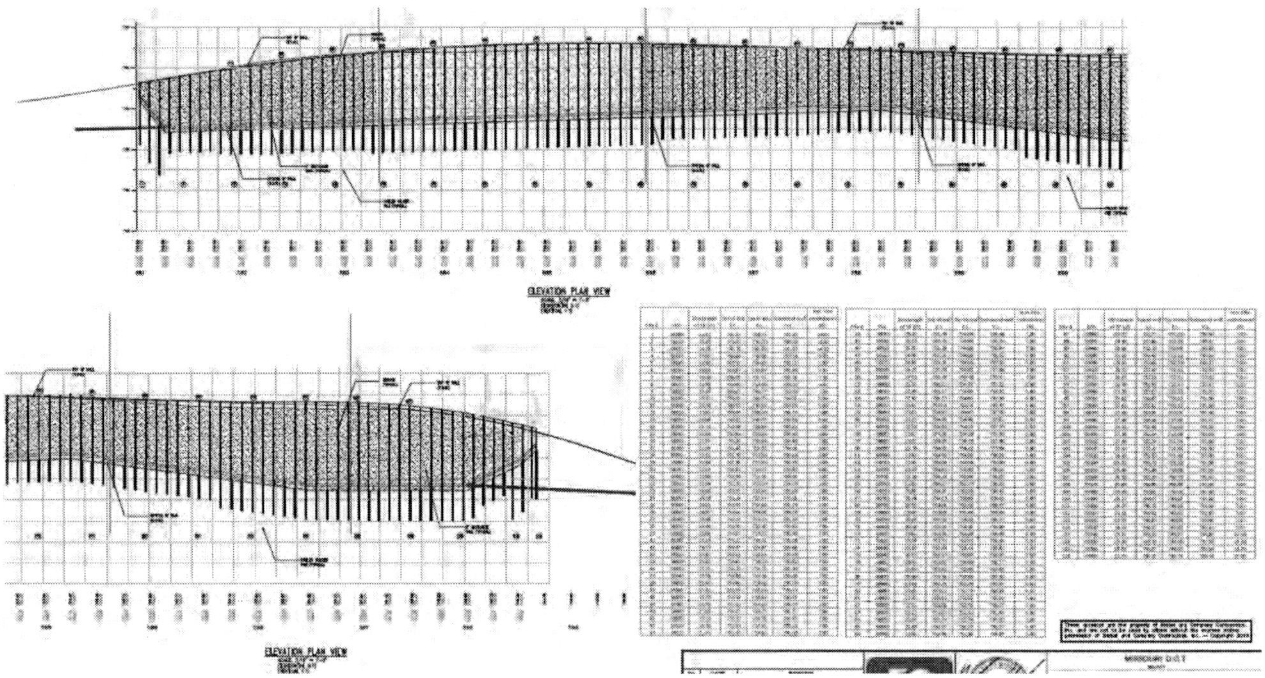
ELEVATION PLAN VIEW
ELEVATION PLAN VIEW
MISSOURI D.O.T

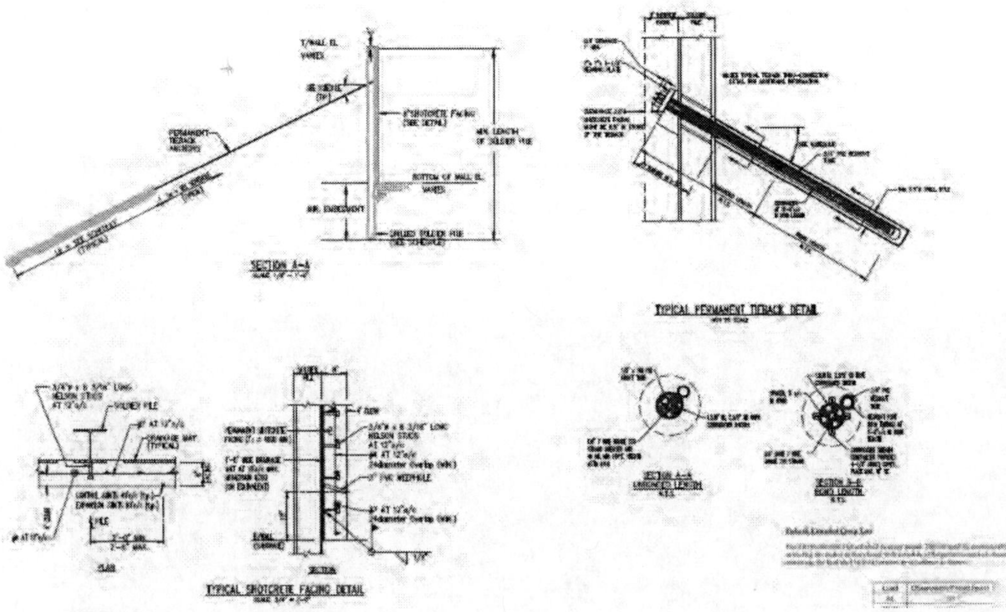

The soldier piles were installed by drilling an open hole from up top, inserting the pile, then backfilling around the pile with lean mix concrete (500 psi minimum) to allow for easier removal at the excavation face as the pile face was exposed. As excavation proceeded to tieback level, tiebacks were installed, stressed, tested, and locked off. Nelson studs were welded to the exposed pile face to support the shotcrete facing. Tiebacks were installed by drilling with continuous auger with air flush, then immediately inserting the tieback and filling the hole and tieback with grout from the bottom up. All tiebacks were either performance or proof tested in accordance with PTI recommendations for prestressed rock and soil anchors. The wall required a total of 118 soldier piles and 111 tiebacks. Tiebacks were stressed, tested, lift-off tested, then locked off at 90 percent of design load. One tieback was not able to be installed full length due to caving soil conditions. The tieback was tested and accepted as is.

Along about a 75-foot length of the wall, sand zones started to slough within a short period of time after exposure as lifts were excavated. Length of lift excavated was adjusted in this zone to shorten the time between exposure and shotcreting. The sloughing extended into the excavation and was filled with additional shotcrete as the shotcrete lift was placed. When the next lower lift was excavated, sand material from behind the upper shotcrete wall section sloughed out and left voids behind the shotcrete wall. These voids were filled by drilling a series of vertically-staggered holes through the shotcrete face and filling with grout from the bottom hole upward until grout flowed out of the next higher hole, then continuing to raise the grout level and fill voids to the top of void. Not all voids were filled by this procedure as surface collapse was observed behind the wall at several locations along the sand limits. A series of probing and excavation from the top behind the wall was performed to check for voids. Voids were filled with flowable fill from the top.

The wall construction was started in early November 2016 and completed in early January 2017. Wet mix shotcrete was cold-weather protected with blankets and heaters. MoDOT provided the resident engineering and daily construction inspection while HNTB Geotechnical section assisted with shop drawing, wall design, and tieback testing submittal reviews and observation of wall construction. Following final grading and paving, the roadway was opened to traffic on February 14, 2017, some 73 days beyond the contract date.

Finished Wall

Completed Slide Repair and Reconstructed Roadway

MANAGING RISK FOR WORKERS ON SLOPES FOLLOWING THE 2016 KAIKŌURA EARTHQUAKE, NEW ZEALAND

Rachel Musgrave
Heads Up Access Ltd
Christchurch, New Zealand
PO Box 41100, Ferrymead 8247
+64 21 296 4407
rachel@headsupaccess.co.nz

Leon Gerrard
Geosolve Ltd
Queenstown, New Zealand
PO Box 1780, Queenstown 9300
+64 204 076 4510
lgerrad@geoslove.co.nz

Prepared for the 70th Highway Geology Symposium, October, 2019

Disclaimer

Statements and views presented in this paper are strictly those of the author(s), and do not necessarily reflect positions held by their affiliations, the Highway Geology Symposium (HGS), or others acknowledged above. The mention of trade names for commercial products does not imply the approval or endorsement by HGS.

Copyright Notice

Copyright © 2019 Highway Geology Symposium (HGS)

All Rights Reserved. Printed in the United States of America. No part of this publication may be reproduced or copied in any form or by any means – graphic, electronic, or mechanical, including photocopying, taping, or information storage and retrieval systems – without prior written permission of the HGS. This excludes the original author(s).

ABSTRACT

Disaster recovery takes place in an abnormal environment. A defining tension exists between the need to rebuild quickly, but with careful deliberation. This tension poses risks for the health and safety of workers involved, at a time when risk levels are higher than normally encountered in the workplace. A key question is how to implement "good practice" health and safety procedures to protect workers in condensed timeframes that are distinctive post-disaster. This case study on the Kaikōura Earthquake will specifically address the demands placed on rope access workers involved in the reconstruction of the distributed transport network, the hazards encountered and how risk was managed.

Key findings are that the transition from disaster response to recovery is a crucial phase of reconstruction, during which clarification of expectations and information sharing benefit workers. Quantification of risk, including a consideration of societal risk, should be a process that is both transparent and inclusive of workers, according to the law and to "good practice". Preparation activities, such as pre-disaster training, planning and testing of emergency procedures can reduce risk.

Future research is recommended into reconstruction following the Kaikōura Earthquake to evaluate emergent safety culture and develop a model to improve risk communication through a multi-level organization to workers at field level. Improvements in the management of safety for reconstruction workers will allow for more effective and efficient recovery in future natural hazard events affecting critical lifelines and infrastructure, improving the resilience of transportation networks and communities in New Zealand.

INTRODUCTION / CONTEXT

The Kaikōura Earthquake

The M7.8 Kaikōura Earthquake on 14th November 2016 caused severe social, economic and environmental impacts in New Zealand. Ground shaking, surface rupture and thousands of co-seismic landslides damaged infrastructure on a regional scale, across the north east of the South Island and in Wellington. The township of Kaikōura was severely affected, suffering acute isolation during the peak tourist season. State Highway One (SH1) and the Main North Rail Line (MNL) are critically important strategic assets for New Zealand. Both were severely impacted by the earthquake (Fig.1 & 2), requiring closure for urgent repairs, lasting for 13 months.

Mitigation of the landslide hazard above the road and rail is ongoing; operational restrictions, such as single lane access and reduced speed limits still apply in places. The economic recovery of the Kaikōura and Hurunui Districts, the Canterbury and Marlborough Regions and the nationally important tourism and freight industries is directly reliant on a fully functioning and resilient transportation network *(1)(2)(3)(4)*.

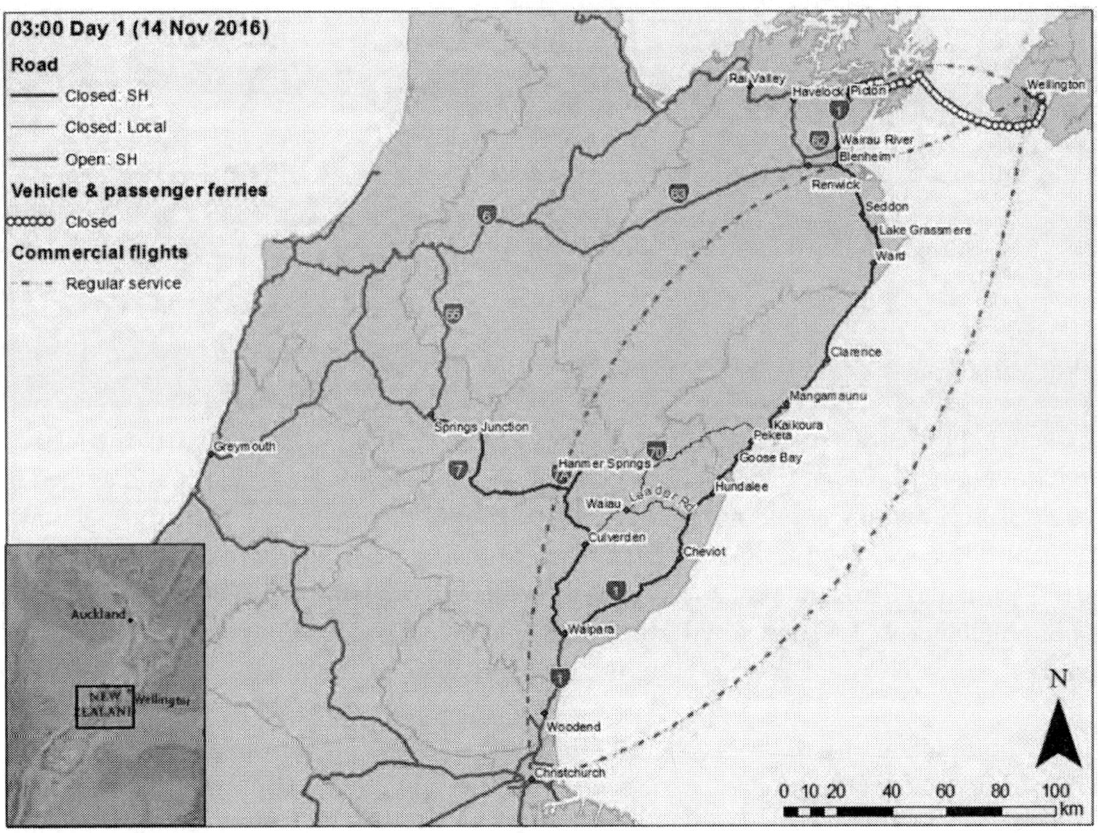

Figure 1 – State Highway Level-of-service on day 1 (14th November 2016) ***(1).***

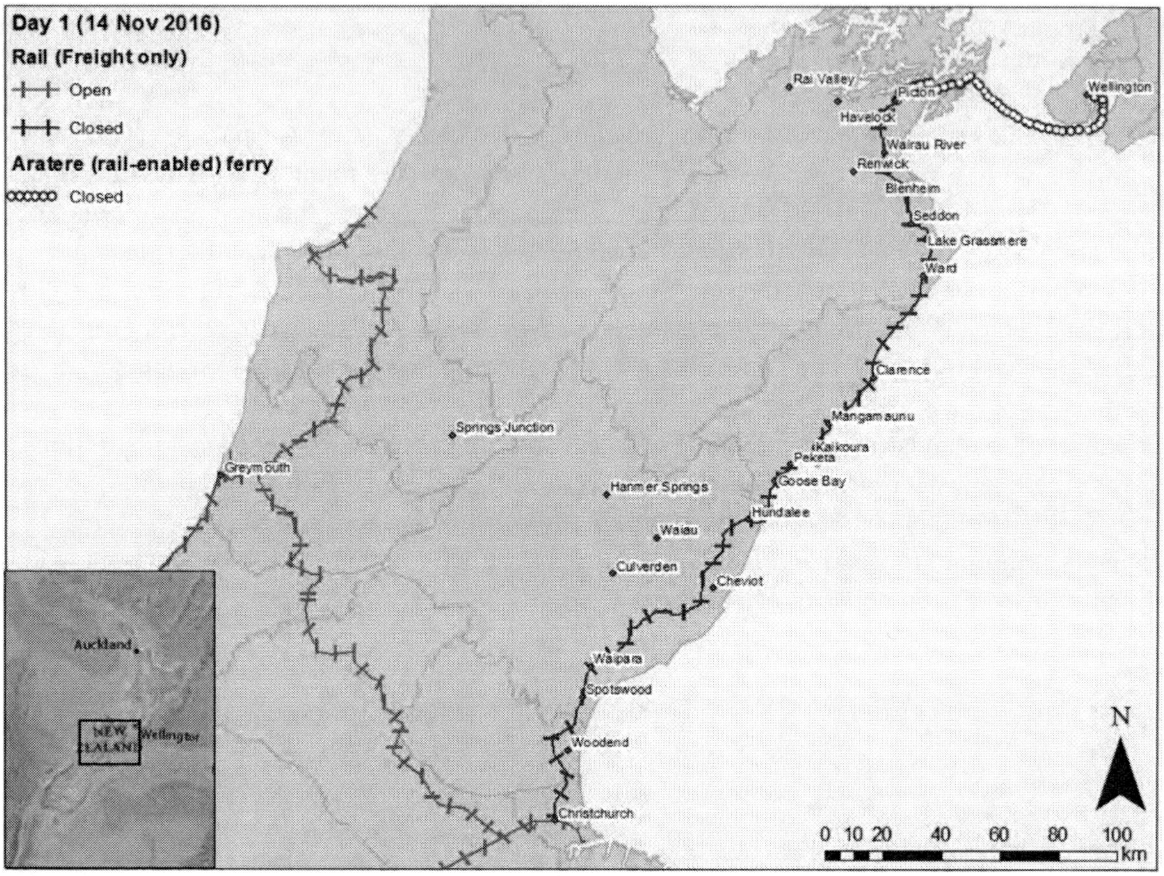

Figure 2 – Rail Level-of-service on day 1 (14th November 2016) *(1).*

NCTIR Alliance

Large-scale natural disasters require multi-agency responses. In December 2016, the New Zealand government passed the Kaikōura/Hurunui Earthquakes Recovery Act and agreed to fund the repair of SH1 and the MNL, north and south of Kaikōura. The North Canterbury Transport Infrastructure Recovery (NCTIR) was established to restore the transport network infrastructure between Picton and Christchurch. NCTIR is an alliance partnership between the New Zealand Transport Agency (NZTA) and KiwiRail (the asset owners) and four large construction companies (Fulton Hogan, Downer, Higgins and HEB) *(4)*.

Due to the complex nature of the reconstruction work, a large number of personnel with specialized rope access training and experience were needed on the slopes in Kaikōura to complete the scope of works in a timely manner. All the major rope access contractor companies operating in New Zealand and one Canadian company became involved in the reconstruction, as subcontractors to the NCTIR Alliance.

Risk Levels for Workers During Disaster Reconstruction

Reconstruction following a natural disaster is often large in scale, long in duration and complex in terms of the range of hazards, to which workers are exposed. An over-arching characteristic of disaster recovery is compression of infrastructure repairs in time and in a limited space. Both time and space compression have critical implications for protecting the health and safety of workers during the immediate and sustained phases of the response and recovery (Jackson et al., 2002; Johnson & Olshansky, 2016; Olshansky et al., 2012). During disaster reconstruction, workers are required to carry out critically important, urgent and dangerous work, at some personal risk. Even trained and highly skilled individuals increasingly have to cope with events of a scale larger than they would normally encounter. Inadequate training of some workers, due to the numbers required, results in situations and responsibilities being encountered, which fall outside their accustomed roles. There is often a more prolonged exposure to high-risk situations than they are equipped to deal with *(5)(6)(7)(8)*.

Intense pressure to open the transport network in the shortest possible time, contributed to the Kaikōura reconstruction being a high hazard industry sector with an elevated risk profile for workers on slopes. This paper identifies some approaches used by rope access workers during the reconstruction to implement "good practice" in health and safety standards and outlines risk management strategies used to manage the elevated risk levels, at a time where urgency to rebuild quickly was an overriding factor. Our recommendations for improvements form part of a long-term strategy to reduce risk during the emergency response and recovery phases for workers on slopes, where conventional means of access are not available following future large earthquakes where slope instability impacts critical infrastructure.

THE REGULATORY ENVIRONMENT

The Kaikōura Earthquake occurred in the year following a significant reform of New Zealand's Health and Safety at Work legislation, brought about by the Pike River Mine tragedy in 2010 and the subsequent findings of a Royal Commission of Enquiry. Post-disaster reconstruction in New Zealand is governed by two parallel pieces of legislation (and by risk management and "working at height" guidelines and qualifications).

The Health and Safety at Work (HSW) Act, (2015) governs health and safety at work, but recognizes that other New Zealand legislation may affect workers. The Act addresses such overlaps by providing that other legislation can be considered when deciding whether health and safety duties are being met. Where two pieces of legislation apply, the duty holder must follow both (Worksafe, HSWA Special Guide 2017). There is no distinction in the HSW Act (2015) between post-disaster times and "normal" times, yet an important distinguishing characteristic of disaster recovery is that it takes place in an abnormal environment *(9)*. Under the new Health and Safety at Work Act (2015) a framework for continual improvement includes appropriate scrutiny and review of actions taken by persons performing actions or exercising powers *(10)*.

The Civil Defence and Emergency Management (CDEM) Act (2002) creates a framework within which New Zealand can prepare for, cope with and recover from local, regional and national emergencies. The CDEM Act requires communities to achieve acceptable levels of risk by correctly identifying risks, adopting risk reduction management practices and provide for planning and preparation for emergencies, and for response and recovery (11). The CDEM Act (2002) does not specify particular health and safety environments for workers during or after emergencies, but does recognize that the safety, health and well-being of a "community" is an integral part of the generic recovery structure after a natural disaster. The "community" is not specified by the Act, but must surely include the workers who have been involved in reconstruction?

The AS/NZS ISO 31000:2009 Risk Management Principles and Guidelines outline good practice processes for managing risk in New Zealand. According to these guidelines every aspect of the risk management process needs to be systematic, transparent and inclusive, and facilitate continual improvement of an organization and a dynamic response to change. In the case of risk management for workers the priority should be protecting life and safety from harm. Agencies responsible for the safety of workers have an overarching responsibility to make good decisions about exposure of workers to known risks *(12)(13)(14)*.

"Working at height" is the term used to denote a preventative safety measure where work positioning is achieved using Personal Protective Equipment (PPE) to prevent a person from falling. "Working at height" methods allow the worker to access the place of work and perform tasks while suspended in areas where conventional means of access are not possible *(15)*. Workers must have either International Rope Access Trade Association (IRATA) or Industrial Rope Access Association of NZ (IRAANZ) qualifications (or both) in order to utilize "working at height" methods in the workplace in New Zealand.

Figure 3 - Rope access technicians (circled) scaling head scarp, Slip 7 north of Ohau Point, Kaikōura, Feb. 2017. Photo: R.Musgrave

Some occupations are unavoidably exposed to hazards that are in the nature of their jobs. The requirements of the role played by rope access workers involved in the Kaikōura reconstruction are such that it could not be performed without exposure to some risk (Fig.3). A lack of New Zealand-based rope access technicians with the relevant experience meant that many in the workforce were contractors from overseas, unfamiliar with New Zealand conditions, or were newly qualified technicians with no prior geotechnical experience.

HAZARDS

Working at heights is intrinsically hazardous; workplace accidents can have severe consequences. Worldwide, falls from height remain the most common cause of serious and fatal injuries in the workplace *(15)(16)*. Additional hazards to personnel are encountered in the geotechnical field. Environments can include falling rocks, toxic dust and unstable surfaces, as well as the frequent use of heavy machinery, drills and compressed air, which require intensive management processes. High levels of experience and supervision are needed to ensure that safe working methods are maintained *(17)*.

Aftershocks

A particular issue for workers in the Kaikōura reconstruction is the dynamic nature of the risk, which remains elevated due to the increased likelihood of aftershocks following a significant earthquake. Although the risk is expected to decline with time (as the aftershock sequence decays) the recovery effort may well be over by the time the probability of seismic events returns to background levels *(18)*. In the year following the earthquake, during the most intense phase of reconstruction activities, the probability of one or more M6.0-6.9 aftershocks in the Kaikōura area was initially estimated at 98% (extremely likely). This forecast was updated every 3 months; by February 2018 the probability estimates had fallen to 53% (Geonet, aftershock forecasts, 19thDec. 2016 & 5th Feb. 2018)). A large aftershock, if centered close to an occupied worksite on or below an unstable slope could have had severe consequences.

Tsunami

A locally generated tsunami is characterized by a short time interval between initiation and run up. Multiple tsunamis generated by either a fault rupture offshore, or by underwater landslides into the Kaikōura Canyon (or both) are possible following an aftershock near the Kaikōura coast. These types of tsunami have arrival times of between 10 minutes and 1.5 hours following an earthquake *(19)(20)*. Many occupied worksites on the coastal transport route were (and still are) situated close to sea level.

Post Seismic Rainfall-induced Landslides

A large earthquake not only triggers severe co-seismic landsliding but can also reduce the stability of slopes for a long period of time post-earthquake. The probability of recurrence of large-scale landslides is very high, as slopes have been weakened and fractured by recent seismic shaking *(21)(22)(23)*. In addition, critical rainfall thresholds for triggering landslides and debris flows decrease significantly (compared to the pre-earthquake thresholds), subsequently increasing the frequency of rainfall-induced landslides in regions affected by strong ground shaking *(24)(25)*. The seismically damaged slopes north and south of Kaikōura are now more susceptible to rapid failure in high-intensity rainfall events. Secondary effects such as rock falls, landslides and debris flows after heavy rain, have potential to cause significant problems for people working in the immediate areas where slopes have been seismically weakened.

THE ROLE OF ROPE ACCESS TECHNICIANS IN EMERGENCY RESPONSE AND RECOVERY

Rope access technicians were positioned on sites above the Inland Kaikōura Road (Inland Route 70) within days of the Kaikōura Earthquake. The value of rope access techniques to facilitate safer access for the opening of this critical lifeline was evident early on in the emergency response, as the slopes above the road were unable to be accessed by traditional means. Rope access workers were engaged to remove the critical hazards at the source by "scaling" (removal of loose rocks with crow bars) and were also used as "spotters" positioned on the landslides to observe initiation of movement and provide early warning. These actions were implemented to reduce the risk for other workers at road level who were clearing debris for emergency access, and with minimum disruption to New Zealand Defence Force convoys travelling the route daily to take essential supplies into Kaikōura.

On November 30th 2016 the Inland Kaikōura Road was provisionally opened to civilian convoys. Work then began on SH1 and the MNL, first south and then north of Kaikōura. Construction workers began to remove debris from the toe of the landslides, in order to facilitate access and begin repairs on the road and rail. Initially, the rope access technicians were providing a support role, reducing risk for other workers on the project (as in the response phase), and thus enabling important and urgent work below the earthquake damaged slopes to proceed. Later, the construction of temporary and permanent engineered rock-fall and landslide risk mitigation structures began on the slopes. This specialist activity requires a high degree of skill and experience (Figs. 4 & 5).

RISK MITIGATION FOR EMERGENCY RESPONSE

Aftershocks were occurring frequently at this time, creating a culture of extreme caution amongst rope access workers, who needed to descend into the zones of highest rock-fall hazard on slopes to perform tasks. Safety concerns had to be balanced with a commitment to assist in the emergency response and play what was considered to be a critically important role. Key safety considerations included:

- Limiting time spent and number of people in high risk zones, minimizing exposure to Individuals.

- The rope access teams employed a "top down philosophy" which dictates removal of rock fall hazard before descending below, and avoidance of areas with high hazard lower on slopes (where possible).
- Rescue systems were rigged prior to descent, with standby rescuers remaining at the top of slopes, to facilitate very rapid extraction of operators from the rockfall hazard zone if required.
- Only the most experienced and highly qualified rope access team was engaged in the response phase.
- The rope access team included two rope-access qualified engineering geologists who were able to report site observations to the ground-based geotechnical team at the time.

Figure 4 (left) - Construction of shallow landslide barrier, Slip 18 south of Kaikōura, Nov. 2017. Figure 5 (right) - Installing mesh by helicopter sling load, a high-risk activity, Slip 18. Heli-operations were often conducted over an open highway. Photo R. Musgrave

RISK MITIGATION FOR RECOVERY / RECONSTRUCTION

Avoid or Substitute

Where possible, operators avoided accessing the lower slopes of the landslides, by substituting alternative methods for removal of hazards (Figs. 6 & 7).

Figure 6 (left) - Air bags used to remove an unstable column of rock while operators retreat to a safer location, Slip 10, south of Kaikōura. Photo R. Musgrave. Figure 7 (right) - Sluicing to remove loose debris below abseiler, Slip 7, north of Ohau Point, Kaikōura. Photo R. Musgrave

Temporary (non-engineered) Risk Mitigation Structures

In order to begin the construction of engineered risk-mitigation design structures, it was necessary in some cases to first install temporary structures for the protection of workers required to spend long periods of time below significant rock fall hazard on the lower slopes of landslides (Figs. 8 & 9).

Additional Training

During the Kaikōura reconstruction, and due to the difficulties around access and safety, geotechnical professionals relied on observing slopes from a distance giving a broad overview and using information relayed by rope access workers about detailed ground conditions. Over the course of the reconstruction, some members of the geotechnical team became qualified for rope access to IRATA Level 1, allowing them to reach on foot difficult to access sites and observe slope conditions more closely (under the supervision of more experienced rope access operators).

Some rope access contractors also received additional training in basic structural geology, risk awareness, hazard identification and factors affecting stability on the earthquake damaged slopes. This enabled them to understand better the main factors that lead to slope failure, identify and report unsafe conditions and take appropriate action. Observing, monitoring and reporting slope conditions by all on site proved an effective way of managing risk in the work environment.

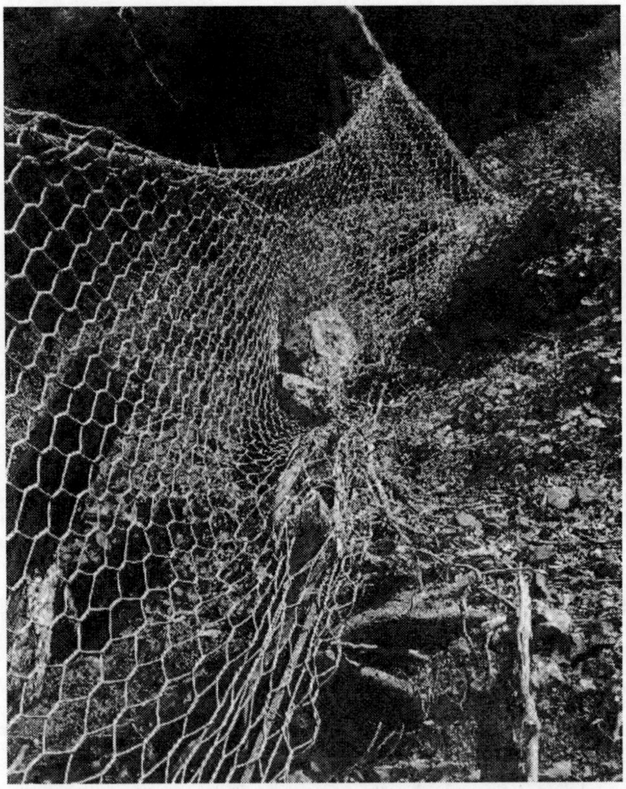

Figure 8 (right): Example of temporary rockfall mesh installed above an occupied worksite, Slips 18 &19, south of Kaikōura. This structure contained a ~100 m^3 failure which occurred during rainfall, June 2017. Figure 9 (left): Temporary rockfall catch fence above worksite, Slip 18. This structure was impacted 4 times by rocks ~1m^3 during the course of construction of a shallow landslide barrier directly below. Photos: R. Musgrave.

The NCTIR Rainfall Trigger Action Response Plan (TARP)

To manage the elevated risk to workers during rainfall, telemetered slope monitoring instruments and rain gauges were installed at numerous locations along the coastal transport route on sites most affected by slope instability. The NCTIR Rainfall Trigger Action Response Plan (TARP) was implemented as a predictive risk management tool in March 2017, to reduce risk for the travelling public and recovery workers. Decisions are made (using real-time monitoring of rainfall) to close worksites (and the road and rail), based on forecasted rainfall in relation to antecedent rainfall conditions, using a model developed *(26)* for determining rainfall-triggering thresholds for landslides in Wellington, New Zealand. Rope Access workers and the geo-technical team then used the model thresholds as a guide (along with ground observations and slope monitoring within their teams) for dictating avoidance of the slopes during or immediately after significant rainfall events.

Emergency Procedures

All qualified rope access teams have the training and capability to perform a rescue of an injured worker on ropes *(15)*. However, in January 2017, senior rope access contractors became concerned that the particular hazards encountered in the work environment on the slopes in Kaikōura, combined with the possibility of a large aftershock occurring, required specialist rescue training over and above "normal" rope access requirements. The possibility of multiple rock falls and slope failures in an aftershock could mean that many severely injured casualties would require rescuing simultaneously from different sites. It was felt that the capability to perform rescues in this scenario did not exist. The rope access teams requested that senior team members with specialist medic training also received "long line" rescue training where rescues are performed from underneath a helicopter, giving the option to perform rapid extraction of many injured persons from slopes if necessary.

Evacuation Planning

All rope access teams had their own evacuation plans in case of an emergency and identified "safe" places to muster, relevant to each worksite. Often, the safest means of egress from a worksite on a slope was identified as up to a muster point on a ridge, rather than down to road level. In January 2017 it was pointed out by rope access workers that the tsunami "safe" places and muster points identified in the current tsunami evacuation plan for worksites along the coastal route were at sea level, in the tsunami evacuation red and orange (must evacuate) zones *(27)*. Worker input prompted a broad scale review of evacuation plans.

DISCUSSSION

Assessing Risk

IRATA requirements specify that site-specific risk assessments be carried out, with input from all rope access team members, before work commences. These assessments are qualitative and use a risk matrix to assess the potential likelihood and consequences of hazardous events. Before commencing work, all tasks should be organized, planned and managed so that there is an adequate margin of safety to reduce risk *(15)*. Most experienced rope-access trained contractors are proficient in their work but lack formal training in geology or engineering geology. It is important that persons with training and experience in rockfall and landsliding are involved in assessing risk on site, because under or over estimating risk can affect the outcomes of the risk analysis *(28)*.

Quantifying Risk

Quantifying risk is useful because it allows a comparison of hazards and enables authorities and workers to prioritize risks in order to inform decision-making *(28)(29)(30)(31)*. In New Zealand, managing the risk from landslide hazards follows principles and guidance set in Australia by the Australian Geomechanics Society (AGS). The AGS recommends that some degree of quantification of risk is attempted in all cases, even if crude or preliminary, especially where loss of life is a possibility. This allows comparison with the acceptance criteria for loss of life, which is also quantified *(28)*.

Quantitative risk assessment is increasingly being used to inform government and private sector policy decisions in New Zealand. The Christchurch Earthquake Sequence (CES) set a precedent for its use. Individual annual fatality risk was the criterion for establishing upper limits of risk tolerability from rock fall and cliff collapse on the Port Hills. The Christchurch City Council then used this criterion to guide decision-making regarding the safe occupation of buildings below or on the edge of cliffs *(29)(31)*.

In 2012, two small eruptions from Mt. Tongariro produced multiple volcanic hazards in the Tongariro National Park, prompting closure of the popular Alpine Crossing track for 6 months. The key reasons for this extended closure were safety concerns for track users, however decisions had to made prior to the track opening, to determine whether the risk was tolerable, to allow Department of Conservation workers and GNS scientists access to the closed areas *(13)*. The period following the November 2012 eruption was a time of considerable uncertainty requiring a transparent decision-making process concerning access close to the active volcanic vent. Discussions with the New Zealand government agency responsible for health and safety in employment emphasized that there should be no compromise to staff safety standards by the Department of Conservation or GNS Science. Life safety risk mitigation was the paramount consideration, which had to be balanced against losses for the local, regional and national economy *(13)*. GNS scientists performed basic quantitative risk assessments within days of the eruption to analyze life safety risk from ballistic hazards, using an expert elicitation panel. This process balanced the urgent need to collect scientific data and repair infrastructure, with health and safety in employment regulations, in order to facilitate informed decision-making *(13)*.

In the Kaikōura situation, fatality risk for individuals is of primary consideration and should be quantified in order to manage overall risk in the workplace on slopes *(28)*. The entire risk management process must be transparent and inclusive of stakeholders at all levels according to New Zealand law, to risk management standards and good practice *(10)(13)(32)*.

Risk Evaluation, Establishing Criteria and Uncertainties

Risk evaluation assists with decision-making after risk has been assessed, to decide whether to accept or treat the risks and to set priorities for action. To make decisions, the level of risk is compared against criteria, to determine what is acceptable, tolerable or otherwise. The UK Health and Safety Executive (HSE) judges the tolerability of risk first in terms of the absolute levels of risk to individuals - only if individual risk is tolerable is it then reasonable to proceed. Individual risk is used as the primary measure of risk, but societal risk must also be considered *(33)*.

In general, higher risks are likely to be tolerated for workers in industries with hazardous slopes, than for society as a whole. Upper limits of tolerability of 10^{-4} per year individual fatality risk for members of the public and 10^{-3} per year for employees are suggested in the UK *(28)(31)*. In New Zealand, such criteria have not yet been firmly established at Government level and in the private policy sector with regard to the workforce.

A significant aftershock on a nearby fault, which ruptures at shallow depths has the potential to cause multiple rock falls, landslides and generate a tsunami on the coast near Kaikōura. In this scenario, significant hazards exist for people on worksites on (and below) steep slopes at sea level. Loss of lives is a real possibility. If the possibility of loss of lives exists, the probability that the incident might actually occur should be sufficiently low that relevant risk criteria are met (eg. probability x number of deaths $<10^{-3}$ for workers). This accounts for society's particular intolerance to events that cause many simultaneous casualties and is embodied in societal tolerable risk criteria *(28)*.

There will be an element of risk in all decisions that are made: zero risk is not achievable. Furthermore, it is not advisable to use quantitative risk estimates as the sole determinant for making decisions in light of the uncertainties in many estimates of risk. The assessed risk may span the acceptance criteria, requiring a high degree of confidence about what is tolerable when making decisions *(28)(31)(33)(34)*.

Exposure

The reconstruction following the Kaikōura Earthquake was a large-scale civil construction project for New Zealand. Since January 2017, over 7500 different workers have worked over 4,300,000 hours on 180 different worksites *(35)*. On Dec 15th 2017, State Highway 1 to Picton re-opened and the consequent traffic flow increased to approximately 5000 vehicles daily. This was a significant milestone for the Kaikōura community, the freight and tourism industries, the New Zealand government and the NCTIR alliance *(36)*.

The benefits of opening the highway were clear to all working on the project. However, conducting the repair works above an open highway with traffic passing below work-sites caused concern for rope access contractors. Many consequential stoppages lengthened the duration of the project and affected productivity, adding to the frustration for workers, stakeholders and the public. More importantly, this decision changed the level of risk workers were exposed to, as they were required to spend a longer period of time in the hazard zone that would be normally be acceptable to them.

The result of re-opening the highway while rope access work was still ongoing was that the temporal probability (of an individual being at a given location, given the spatial impact of the hazard) increased significantly for workers, because the scope of the project was increased by an unspecified period of time. An increase in temporal probability (exposure) increases the fatality risk for people working on or below slopes (and for members of the public using the road).

Testing of Emergency Plans

Under the HSW Act (2015), emergency plans must be prepared for each workplace prior to work commencing. There must be provision for the testing of evacuation plans and training and instruction given to workers. This is also a key component of IRATA risk management procedures *(15)*. The CDEM Act (2002) specifies that scheduling of training and exercises to validate plans falls under "Readiness" activities *(10)*. Pre-disaster plans can improve the speed and quality of post-disaster decisions. Organizations involved in recovery should plan and act simultaneously *(9)*.

The fast pace of the Kaikōura reconstruction and competition for limited resources made it difficult to prioritize the formulation of emergency plans when they were most needed early in the response and transition phases. The testing and refinement of emergency plans did not occur until the recovery phase was well underway.

Production Pressure

Worksafe recognizes that both physical and psychological factors are at play in the work place: deadlines create stress and fatigue amongst workers and can compromise efforts to maintain a work environment with acceptable levels of health and safety *(14)*. Since the Kaikōura Earthquake, a number of milestones have been heralded as major successes during the rebuild of the transport corridor. The scale of works completed or near completion would normally have taken many years during a time of standard operations or "business as usual".

Prolonged closure of SH1 and the MNL have incurred a high economic and social cost for New Zealand *(1)(2)(3)(4)*. High profitability and high health and safety standards can be complementary factors however it is important to acknowledge that tension can arise between different goals (profitability and safety) in specific decision-making processes. The heightened risk for workers on coastal slopes in Kaikoura had to be balanced against increased risk levels for users of the alternate route (State Highway 7) in the time that SH1 and the MNL were closed.

Cost-Benefit Analysis vs. the Cautionary Principle

Commonly, operating companies and regulators conceptualize risk as a product of frequencies and consequences, often by quantification. A cost-benefit analysis is a method for quantifying the advantages and disadvantages of different solutions and providing a basis for their prioritization and is used as the justification for implementing (or not) risk-reduction measures. A cost-benefit analysis is a useful tool, giving insights into risk and the considerations involved. However, there are limitations on its use as the results are conditional, based on a variety of assumptions. In the context of high-risk industry sectors, such as reconstruction after a natural disaster, it must be recognized that some benefits and costs (loss of life or injury) are exceedingly difficult to quantify in monetary terms. Any attempt to provide a comparison of costs and benefits in the context of major accidents with low probabilities of occurrence is nearly impossible (37).

The cautionary principle is a fundamental principle in safety management giving full weight to risk and uncertainty, thus representing an extreme safety perspective. According to this principle, in a context with uncertainty and risk, caution should be the ruling principle, by the implementation of risk-reducing measures, or by not starting an activity *(17)*.

RECOMMENDATIONS

Workers are an integral part of the disaster recovery community. The requirement to protect workers should be a primary consideration in a recovery done well in order to improve safety, capability and efficiency in processes and outcomes after future disasters. Future strategies for worker protection include:

- Establishment of a National register of trained and experienced rope access operators.
- Appropriate ratios of inexperienced to experienced operators maintained.
- Increasing the capacity for planning the transition between response and recovery and for the formulation and testing of emergency plans
- Workers and regulators require support mechanisms for decision -making during times of uncertainty.
- Assessing, quantifying and communicating risk through formal channels is critically important for the protection of workers.

Staff Training and Experience

Natural hazard events like the Kaikōura Earthquake will occur again in New Zealand, requiring input from contractors, consultants and stakeholders in a collaborative approach. Having a skilled New Zealand based workforce to call on in the event of a future emergency will increase the capability of regions to respond in a safe and effective manner. A register of all contractors and consultants, who have had experience with this type of geotechnical work in Kaikōura and following the Canterbury Earthquake Sequence (and have undergone additional training) would allow the experience gained during this reconstruction to benefit New Zealand in the future.

IRAANZ and IRATA qualifications are adequate for training people to work safely at heights. The training and certification are conducted in a controlled environment, which does not prepare technicians for the additional hazards encountered on unstable slopes in a seismically active area. Newly qualified rope access technicians require a high level of supervision by experienced operators. Appropriate ratios of inexperienced to experienced operators should be carefully managed in hazardous work environments.

IRAANZ and IRATA may be best placed to take the lead and develop a framework to formalize a register of rope access technicians trained and experienced in geotechnical work, and more specifically disaster response and recovery work.

Planning the Transition from Response to Recovery

Disaster recovery starts while the response is still active. The transition is a process, which should be planned, documented and communicated *(11)*. Increasing the capacity for planning the transition, by adding personnel and technical assistance is a solution to the tension between speed and deliberation that exists in disaster reconstruction *(9)*.

Workers require information and support during the transition, as the priorities for action differ during these phases. What is considered to be a tolerable level of risk for workers may change after a disaster. In the early phases of a response it may be appropriate to take risks if there is a reasonable chance of saving lives. At some point an incident must transition to recovery. At this point it is no longer considered appropriate to allow workers to expose themselves to significant risks to perform duties *(6)*. Where significant risks exist, adequate mitigation and protection must be in place according to law *(10)*.

Decision Making Support

Following disasters, decisions are made which may differ from those made during non-crisis (business as usual) situations. Crisis decisions are made in high-risk/low operating time environment with large uncertainties. Decision support mechanisms for regulators and operators have been shown to contribute to the prevention of major accidents *(37)*.

A suggested approach to clarify decision-making at practitioner level is based on the Fire and Emergency NZ (FEMNZ) "safe person" concept where an emergency responder "will, may, or will not" take risks, depending on the context (Fig. 10). The regulatory decision-maker should be able to take a dynamic approach, i.e. be able to give weight to an extreme economic perspective, or an extreme safety perspective, or a perspective on a continuum in between (according to the adopted risk management approach and the context) *(37)* (Fig. 11).

Communication Channels between Scientists, Engineers and Contractors

Knowledge and information sharing are critically important for successful worker protection after a disaster *(5)(6)(8)(38)*. Assessing risk should be an inclusive process involving workers representatives, experts, stakeholders and technical partners. Experts involved in risk assessments will need to collaborate beyond professional boundaries during recovery *(30)*.

An organizational model where the fast pace of disaster reconstruction is matched by equally fast-paced development of safety culture in organizations involved, would ensure that health and safety "good practice" can be implemented at all times in future disasters in New Zealand. This may be achieved through a communication network that transfers risk information quickly through the different levels of a complex organization, providing communication channels between recovery actors to facilitate information transfer, inclusion and transparency about known risks.

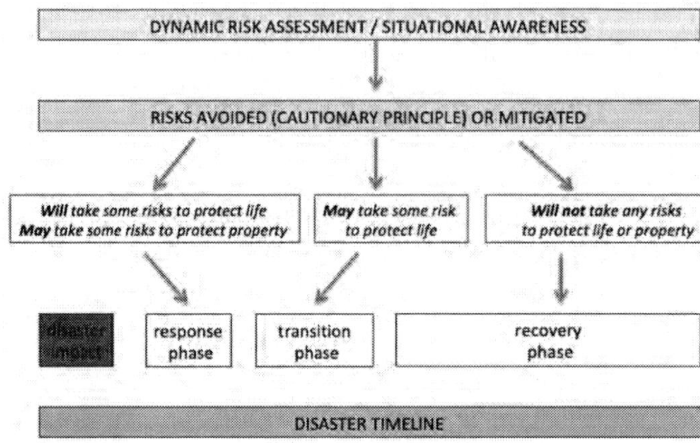

Figure 10 - Suggested decision-making approach for operators during different phases of disaster Practitioners require communication and support from regulators during transition between the phases. Where clear boundaries do not exist, risk levels and accident rates should be trending down. Image adapted from *(39).*

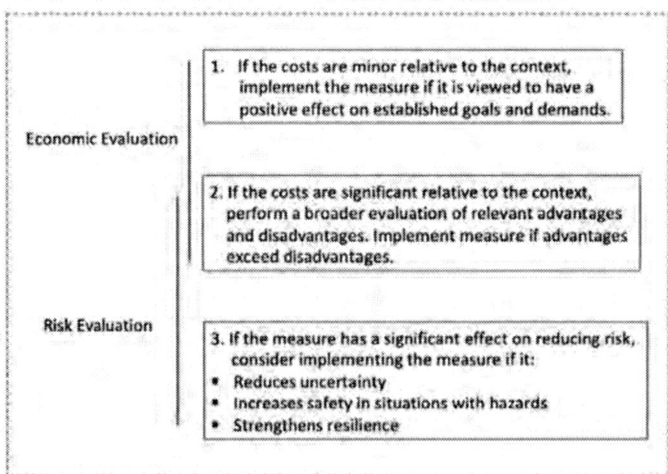

Figure 11 - Suggested decision-making approach for regulators and stakeholders during times of uncertainty. A dynamic approach is recommended, giving weight to either an economic or safety perspective, depending on the context, Image adapted from *(37).*

CONCLUSIONS

Following the Kaikōura Earthquake, the New Zealand Government, stakeholders, consultants and contractors faced considerable uncertainty and had to balance the tensions between urgency to rebuild and the need for due care with regard to health and safety concerns. Rope Access workers played an important role in the emergency response and recovery, reducing the long-term risk post-earthquake.

The Kaikōura reconstruction has highlighted the difficulty in maintaining a balance between reducing the risk to workers to tolerable levels and allowing nationally important strategic recovery works to proceed rapidly. Mitigating the risk to life safety is of upmost importance, although this will be balanced with contextual factors such as economic losses, political priorities, the well-being of affected communities and the increased risks associated with the use of alternate transport routes, especially where the distributed transport system lacks redundancy.

Improvements in the management of safety for reconstruction workers will allow for more effective and efficient recovery in future natural hazard events which affect critical lifelines and infrastructure, thus improving the resilience of transportation networks and communities in New Zealand. This report distills lessons from the Kaikōura reconstruction, where significant disaster recovery challenges led to different management approaches being used to reduce risk. The key findings of this study were:

- There was no formal process to effectively communicate information about hazards and risk on slopes to workers at field level.
- The planning and development of emergency procedures was often in a constant state of "catch up" to keep pace with the rapid reconstruction work.
- Additional training reduces risk for workers.
- Workers require support, communication and clarification of expectations from stakeholders and government during the transition from response to recovery.
- In Kaikōura the additional risks posed by ongoing seismicity and secondary natural hazards required additional input from experts with knowledge and experience in assessing, quantifying and managing these types of risks.

REFERENCES

1. Davies, A.J., Sadashiva, V., Aghababaei, M., Barnhill, D., Seosamh, B.C., Fanslow, B., Headifen, D., Hughes, M., Kotze, R., Mackie, J., Ranjitkar, P., Thompson, J., Triotino, D.R., Wilson, T., Woods, S., Wotherspoon, L. 2017. Transport Infrastructure Performance and Management in the South Island of New Zealand, During the First 100 Days Following the 2016 Mw 7.8 "Kaikōura" Earthquake. Bulletin of the New Zealand Society for Earthquake Engineering, 50(2) 2017.
2. Ministry of Transport. 2017. Economic Impact of the 2016 Kaikōura Earthquake, Ministry of Transport report no. 170217 Feb 2017, Market Economics Ltd. Auckland: New Zealand.
3. Mason, B. & Brabhaharan, P. 2017. Resilience assessment of state highways in New Zealand, Annual conference of the New Zealand Society of Earthquake Engineering (NZSEE) 2017.
4. New Zealand Government. 2016. Alliance established to rebuild Kaikōura coastal route. Retrieved from https://www.beehive.govt.nz/release/alliance-established- rebuild-Kaikōura-coastal-route (accessed 02/05/2017).
5. American Public Health Association Response to Disasters (APHA). 2008. Protection of Rescue and Recovery Workers, Volunteers and Residents Responding to Disasters. Policy Number 20069 Nov 2008. Retrieved from https://www.apha.org/policies-and-advocacy/public-health-policy-statements/policy-database/2014/07/18/14/10/response-to-disasters-protection-of-rescue-and-recovery-workers
6. Jackson, B.A. 2002. Protecting Emergency Responders: Lessons Learned from Terrorist Attacks. In RAND Corporation (ed). Retrieved from https://ebookcentral.proquest.com/lib/canterbury/detail.action?docID=227921.
7. Olshansky, R.B., Hopkins, L.D., Johnson, L.A. 2012. Disaster and Recovery: Processes Compressed in Time. Natural Hazards Review 2012 (13) 172-178.
8. Sim, M. 2011. Disaster response Workers: are we doing enough to protect them? Occupational and Environmental Medicine 68(5) 309-310.
9. Johnson, L.A. & Olshansky, R.B. 2016. After Great Disasters. How Six Countries Managed Community Recovery. Policy focus report PF041. Lincoln Institute of Land Policy.
10. Ministry of Business Innovation and Employment. 2015. Health and Safety at Work Act 2015. Ministry of Business Innovation and Employment, Wellington, NZ. Retrieved from http://www.legislation.govt.nz.
11. Ministry of Civil Defence and Emergency Management. 2002. Civil Defence and Emergency Management Act 2002. Ministry of Civil Defence and Emergency Management, Wellington. Retrieved from http://www.legislation.govt.nz.
12. ACC. Worksafe NZ. 2013. Safer. Healthier. Together. An Action Plan 2016-2019. Reducing Harm In NZ Workplaces. Retrieved from https://www.acc.co.nz/assets/business/action-plan-reduce-harm-NZ- workplaces.pdf
13. Jolly, G.E, Keys, H.J.R., Proctor, J.N., Deligne, N.I. 2014. Overview of the co-ordinated risk-based approach to science and management response and recovery for the 2012

eruptions of Tongariro volcano, New Zealand. Journal of Volcanology and Geothermal Research 286 (2014)184-207.
14. Worksafe NZ. Towards 2020. 2017. Progress towards the Government's Working Safer Fatality and Serious Injury Reduction Target, May 2017. https://worksafe.govt.nz/dmsdocument/1295-towards-2020.
15. IRATA International code of practice for industrial rope access (ICOP) 2014. IRATA International: England. Retrieved from https://irata.org/page/international-code-of-practice
16. Fleming, M. 2001. Safety Culture Maturity Model. The Health and Safety Executive (HSE) Offshore technology report 2000/049.
17. Industrial Rope Access in New Zealand (IRAANZ): Best Practice Guidelines. 2012. Industrial Rope Access Association of New Zealand in association with the Department of Labour. May 2012. Retrieved from http://iraanz.co.nz/technical-info/
18. GNS . Earthquake Forecast and Hazard Modeling. Retrieved from https://www.gns.cri.nz/Home/Our-Science/Natural-Hazards/Earthquakes/Earthquake-Forecast-and-Hazard-Modelling/M7.8-Kaikōura-Earthquake-2016.
19. Walters, R.A., Barnes, P., Goff, J.R. 2006. Locally generated tsunami along the Kaikōura coastal margin: [image] Part 1. Fault ruptures, New Zealand Journal of Marine and Freshwater Research, 40(1) 1-16.
20. Walters, R.A., Barnes, P., Lewis, K., Goff, J.R., Fleming, J. 2006. Locally generated tsunami along the Kaikōura coastal margin: Part 2. Submarine landslides. New Zealand Journal of Marine and Freshwater Research, 40(1) 17-28.
21. Huang, R. & Li W. 2014. Post-earthquake landsliding and the long term effects in the Wenchuan earthquake area, China, Engineering Geology 182 (2014) pp 111-120. Hunter, A. & Hendrickx, M. 2009. Rope Access Methods in Slope Risk Assessment and Remediation. Australian Geomechanics 44(2).
22. Qiu, H.Z., Kong, J.M., Wang, R.C., Cui, Y., Huang, S.W. 2017. Response mechanism of post-earthquake slopes under heavy rainfall, Journal of Seismology 2017: DOI 10.1007/s10950-017-9641-9.
23. Tang, C., Zhu, J., Xin, Q., Ding, J. 2011. Landslides induced by the Wenchuan earthquake and the subsequent strong rainfall event: A case study in the Beichuan area of China, Engineering Geology 122 (2011): 22–33
24. Lin, C.W., Liu, S.H., Lee, S.Y., Liu, CC. 2006. Impacts of the Chi-Chi earthquake on subsequent rainfall-induced landslides in central Taiwan, Engineering Geology 86: 87–101.
25. Zhang, S., Zhang, LM., Glade, T. 2014. Characteristics of earthquake- and rain-induced landslides near the epicenter of Wenchuan earthquake, Engineering Geology 175: 58–73.
26. Glade, T., Crozier, M., Smith, P. 2000. Probability Determination to Refine Landslide-triggering Thresholds Using an Empirical "Antecedent Daily Rainfall Model". Pure and Applied Geophysics 157 (2000) 1059-1079.
27. Environment Canterbury ECAN nd. [MAP] Tsunami evacuation zones. Retrieved from http://ecan.maps.arcgis.com/apps/Minimalist/index.html?appid=591062afb6b542abb247cc8d15a64855) (accessed Feb. 4, 2018).

28. Australian Geomechanics Society (AGS) 2000. Sub-Committee on Landslide Risk Management. Landslide Risk management Concepts and Guidelines, Australian Geomechanics 2000.
29. Massey, C.I., Yetton, M.D., Lucovic, B., McSaveney, M. J., Heron, D., Bruce, Z.R.V. 2012. Canterbury Earthquakes 2010/11 Port Hills Slope Stability: Pilot study for assessing life-safety risk from cliff collapse. GNS Science Consultancy report 2012/57.
30. Rovins, J.E., Wilson, T.M., Hayes, J., Jensen, S.J., Dohaney, J., Mitchell, J., Johnston, D.M., Davies, A. 2015. Risk Assessment Handbook. GNS Science Miscellaneous Series, v.84. Wellington, New Zealand: Massey University.
31. Taig, T., Massey, C., Webb, T. 2012. Canterbury Earthquakes Port Hills Slope Stability: Principles and Criteria for the Assessment of Risk from Slope Instability in the Port Hills, Christchurch, GNS Science Consultancy Report 2011/319
32. AS/NZS ISO 31000. 2009. Risk Management-Principles and Guidelines, Joint Australian New Zealand International Standard 2009.
33. Great Britain Health and Safety Executive (HSE) 2008) Societal Risk and land use planning. HSE Board Miscellaneous Paper Misc/08/14 Health and Safety Executive: UK.
34. Great Britain Health and Safety Executive (HSE) 2001. Reducing Risks, Protecting People: HSE's Decision Making Process. HSE Books: Sudbury.
35. Bell, S. 2018. Well Being and Mental Health in Kaikōura Rebuild.
36. New Zealand Transport Authority (NZTA). 2018. Kaikōura earthquake update. Retrieved from http://www.nzta.govt.nz/assets/projects/Kaikōura-earthquake-response/Kaikōura-earthquake-update-20180406.pdf (Accessed 23/4/2018).
37. Sorskar, L.I.K., & Abrahamsen, E.B. 2017. On how to manage uncertainty when considering regulatory HSE interventions. EURO J Decis Process 5: 97-116
38. United States Government Accountability Office (GAO). 2007. Disaster Preparedness. Better Planning Would Improve OSHA's Efforts to Protect Workers' Safety and Health in Disasters. Report to Congressional Committees 2007. Retrieved from https://www.gao.gov/products/GAO-07-193
39. FEMA 2017. [IMAGE]. Natural Disaster Recovery Framework. Retrieved from https://www.fema.gov/national-disaster-recovery-framework. (Accessed 14th May 2018).

A Geotechnical Perspective on Emergency Repairs: Case Studies from Federally Owned Roads

James Arthurs
Federal Highway Administration
Central Federal Lands Highway Division
12300 W. Dakota Avenue
Lakewood, CO 80228
720-963-3633
james.arthurs@dot.gov

Prepared for the 70th Highway Geology Symposium, October, 2019

Disclaimer

Statements and views presented in this paper are strictly those of the author(s), and do not necessarily reflect positions held by their affiliations, the Highway Geology Symposium (HGS), or others acknowledged above. The mention of trade names for commercial products does not imply the approval or endorsement by HGS.

Copyright Notice

Copyright © 2019 Highway Geology Symposium

All Rights Reserved. Printed in the United States of America. No part of this publication may be reproduced or copied in any form or by any means – graphic, electronic, or mechanical, including photocopying, taping, or information storage and retrieval systems, without prior written permission of the HGS. This excludes the original author(s).

ABSTRACT

Emergency repair work and restoration of access is a critical function of transportation agencies. The Emergency Repair for Federally Owned Roads (ERFO) program provides funding to Federal Land Management Agencies (FLMA) to repair or reconstruct facilities open to public travel damaged by a natural disaster or catastrophic failure. ERFO provides funding for emergency repairs (immediate access) and permanent repairs (restore facilities to their pre-disaster conditions). The Federal Lands Highway (FLH) division offices of the Federal Highway Administration (FHWA) frequently partner with FLMAs to design and deliver permanent repairs.

Initial site reconnaissance is typically performed by FLMAs and collected using a mobile app (Mobile Solution for Assessment Reporting (MSAR)) to create a Damage Survey Report (DSR). DSRs are used by FLMA and FLH personnel to evaluate and program repairs based on transportation priorities and available funding.

Permanent repairs under ERFO vary widely depending on the nature of the disaster, type of facility, and availability of funding. ERFO provides funding to restore the site to pre-disaster conditions; repairs balance technical feasibility with limitations on "betterment".

This paper presents a geotechnical perspective on recent repairs related to heavy precipitation that occurred in the Sierra Nevada in the winter of 2016 – 2017. Reinforced soil slopes, scaling, drainage improvements, and use of on-site materials are common ERFO methods and typical details used to deliver projects on an expedited schedule. The typical details approach promotes use of templates and trained cross-functional design teams, expedited contracting, and project specific solutions. Examples of repairs made and post-construction results are presented.

INTRODUCTION

Transportation infrastructure is vulnerable to damage from severe weather events, floods, earthquakes, and other natural hazards. After a major disaster, restoration of access for emergency crews is critical to prevent loss of life. Subsequent permanent repairs are essential to restore ordinary access to those affected by the disaster. Funding for this type of repair and restoration work is frequently provided by the Federal government and administered by local government agencies. In the case of Federal government owned facilities, the emergency funding is administered by the Federal Highway Administration (FHWA) through the Office of Federal Lands Highway (FLH).

When a qualifying emergency occurs, Federal Land Management Agencies (FLMA) apply for Emergency Repair for Federally Owned Roads (ERFO) funding to restore access and repair damaged facilities (FHWA, 2019). Each State is allocated a limited portion of ERFO funds. A qualifying emergency means that damage is caused by a widespread event, frequently with a disaster declaration. Small, localized failures do not normally qualify. Restoration of access typically consists of short-term repairs to provide emergency or administrative access. These repairs are typically completed by the FLMA using either internal staff or local contractors. Permanent repairs with the goal of returning the facility to pre-emergency conditions typically take place later, after some engineering analysis and design has been completed. FLH frequently takes the lead for design and delivery of permanent repairs.

Nearly all emergency repairs are subject to two major hurdles: time and budget. The nature of emergency work is that it is unplanned, and therefore does not fit into an agencies typical program planning. Loss of access puts pressure on agencies to deliver repairs as quickly as possible. Although many transportation agencies, including FHWA, set aside some funding for emergency work, it is typically not enough to meet all needs at all times. Therefore, agencies typically try to deliver emergency repair projects with minimal budgets, abbreviated design schedules, and accelerated procurements or contracting.

This paper presents typical geotechnical repair techniques along with the framework that Central Division of FLH (CFLHD) uses to evaluate and select between different options. Several case studies are presented to illustrate the repair techniques and decision framework.

ERFO Project Delivery

ERFO funding is administered by FLH, and the FLMA have several options with regards to design delivery and contract administration. The FLMA can deliver the project internally, contract an engineering firm, or partner with FLH. This paper focuses on projects that were designed and managed by FLH. These projects are managed from cradle to grave, meaning that FLH is responsible for programming, project design, and construction contract administration. This approach allows FLH to expedite delivery and manage design risks through continued involvement during construction. FLH frequently makes use of negotiated contracts rather than traditional sealed bid contracting as an additional tool to expedite construction and manage risk.

Initial site reconnaissance is typically performed by FLMA personnel once the site can be safely accessed. Snowbound roads, continuing volcanic eruptions, and other hazards frequently prevent immediate reconnaissance. FLH provides the FLMA's with a mobile app (the Mobile Solution for Assessment and Reporting (MSAR)) to record details of the site and to create a Damage Survey Report (DSR) (FHWA, 2019b). The DSR is the initial record of damage and is used by FLMA and FLH personnel to evaluate and program repairs based on transportation priorities, available funding, and scope of the repair work. It includes information such as size of the repair area and a rough estimate of repair costs for use in prioritizing and programming repairs. Later, the DSR is used to direct project scoping and data collection effort by FLH staff. An example repair sketch and cost estimate are presented in Figures 1 and 2.

Typical ERFO Projects

Immediate actions following damage vary depending on the nature and degree of damage, and the condition of the overall transportation network. Where damage is severe or there is a significant continuing hazard and detours are available, the first course of action is typically to close the road until permanent repair work is complete. If damage is less severe or the route provides critical access with no detours available, emergency access restoration work is performed. This work can include activities such as embankment reconstruction, debris removal, and rock slope scaling. This type of repair is typically performed directly by the FLMA using either government personnel or local contractors. Funding for this type of work is referred to as "quick release funds".

As previously stated, the collected DSR's are evaluated and prioritized, typically creating lists of projects that can be completed by FLMA personnel and projects that will require additional staff. For FLH delivered permanent repairs, site reconnaissance and project scoping is typically performed by the project team. DSR's are not necessarily prepared by engineers; therefore, this site visit provides technical experts with an opportunity to collect data and evaluate the site using their knowledge and experience. In addition, when DSR's are prepared, the number of potential project sites is very high, limiting the amount of time that can be spent at a single site.

When the FLH project team evaluates the site, they are able to spend additional time collecting measurements and characterizing the sites. Based on programming decisions made by the FLMA, site may be prioritized or completely eliminated. Data collected typically includes existing facilities and structures (signs, culverts, bridges, retaining walls, etc.), site geometry (failure area size, slope angles and heights, drainage patterns, etc.), and other more detailed observations about the type of damage that has occurred. This additional information is used to make judgments about the underlying mechanisms that caused the damage and what repair options will be most suitable. From a geotechnical perspective, this site visit allows an opportunity to document the on-site soil and rock materials, collect samples, and characterize the type of failure that occurred (i.e. landslide or rock slide, embankment failure, fault rupture, sinkhole formation, etc.). Due to the abbreviated schedule and limited budgets, subsurface investigations are not always performed. Where risk is low and involvement during construction can further reduce the risk, a subsurface investigation is not typically performed. Instead, conservative design parameters will be assumed with a plan to confirm assumptions with further observations during construction. Where risks are higher, such as with large landslides and structures, a subsurface investigation is more likely to be performed.

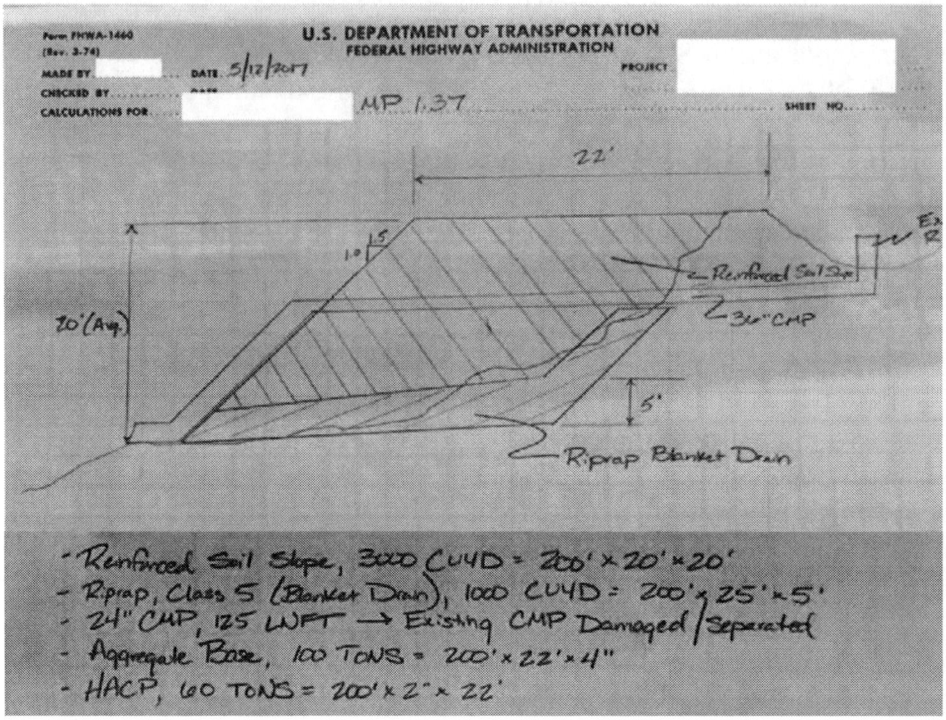

Figure 1 – Example Repair Sketch for a Recent ERFO Project

COST ESTIMATE for EMERGENCY REPAIRS				
Quantity	Unit	Item Description	Unit Price	Cost
1.00	lump sum	Hazard tree felling	$400,000.00	$400,000.00
1.00	lump sum	Scale upslope to remove hazard mater	$25,000.00	$25,000.00
Method	☐ Force Account ☑ Contract	Delivery Agency: CFL	Total Estimated Emergency Repairs	$425,000.00
COST ESTIMATE for PERMANENT REPAIRS				
Quantity	Unit	Item Description	Unit Price	Cost
6000.00	CY	Remove debris	$90.00	$540,000.00
5.00	per slide	Install MSE wall, prep foundation, a	$200,000.00	$1,000,000.00
1.00	lump sum	Reconstruct road prism	$200,000.00	$200,000.00
1.00	lump sum	Reconstruct chip seal road surface	$100,000.00	$100,000.00
26400.00	LF	Re-establish ditchlines	$10.00	$264,000.00
1.00	lump sum	Mobilization	$75,000.00	$75,000.00
1.00	lump sum	Traffic control	$40,000.00	$40,000.00
1.00	lump sum	Geotechnical investigation	$100,000.00	$100,000.00
1.00	lump sum	re-seed	$60,000.00	$60,000.00
Method	☐ Force Account ☑ Contract		Total Estimated Permanent Repairs - Construction	$2,379,000.00

Figure 2 – Example DSR from a Recent ERFO Project

After this initial site visit, members of the project team review their notes and then meet to discuss the proposed repair work. Once the team agrees on the conceptual plan for each site, engineering analysis is performed as needed to evaluate the specific design requirements. The team then prepares the contract plans, specifications, and estimate, which will be used for the repair work. Site survey and detailed topography are typically not available for ERFO sites. Most ERFO plans use typical sections and details with site sketches based on the limited measurements collected in the field. Rough quantities are calculated to be used as the basis for the contract. The final repair is field fit and based on a performance based methodology rather than strict adherence to the plans.

By statute, ERFO funding must be spent within two years of the triggering disaster (FHWA, 2019a). Due to this limitation, the contract development and award must follow an accelerated schedule compared to traditional project development. The contract is also frequently administered using a non-traditional method such as employing 8A contractors (small and disadvantaged businesses) and letter contracts. These contracting methods have the advantage of reducing the time required for advertisement and bid evaluation, compared to traditional low-bid contracting, and provide opportunities for small businesses to develop experience working with FLH.

ERFO TYPICAL DETAILS

Based on project experience, numerous "standard" repair options have been developed that are applicable to a variety of scenarios. These repair options are summarized on Tables 1Table 1 and 2 and discussed in the following sections. Table 1 presents repair methods that require at most limited design analysis, such use of design charts or simple empirical relationships. Table 2 presents repair methods that require more in-depth design analysis and/or the use of specialty contractors. The typical details create a shorthand for discussion between the geotechnical engineer and other project staff when various repair options are under consideration. A selection of these items are discussed in the following sections.

Conventional Earthwork

Conventional earthworks such as clearing debris from roads and rebuilding embankment fills is one of the most common repair methods employed for ERFO work, both for temporary and permanent applications. Given the goals and limitations of ERFO funded projects, reestablishing access by conventional earthworks is often the default repair option considered. Conventional earthworks may be the only repair method employed when the cause of the failure is not geotechnical in nature, such as from a culvert failure, and when construction space is not limited due to steep slopes or environmental restrictions. Conventional earthworks are also frequently combined with other methods that improve the resilience of the repair.

Conventional earthworks require little design work to include in plans and can be implemented by general contractors with convention equipment. However, conventional earthworks typically have the largest footprint of the various methods presented and are more likely to impact environment resources and right-of-way.

Table 1 – Common Emergency Repair Methods Requiring Limited Design

Repair Name	Applicable Scenario(s)	Materials Required	Equipment Required	Relative Cost	Typical Maximum Slope/Face Ratio (V:H)
Remove Debris	Cut slope failure	None	Loaders, excavators, haul trucks	Low	n/a
Rock Scaling	Cut slope failure	None	Pry bars, air bags, long reach excavator	Low to moderate	n/a
Embankment Repair	Embankment slope failure	Conventional fill	Loaders, dozers, haul trucks, compaction equipment	Low	1 : 2
Surface Drainage Repairs	Culvert/embankment blowout, scour, slope erosion, embankment slope failure	Conventional fill, pipe culverts and accessories	Excavator, haul trucks, compaction equipment	Low to moderate	n/a
Embankment Armoring	Embankment slope failure (due to scour/erosion)	Riprap, gabions, revetments, or other armor, separation geotextile	Excavator, haul trucks, compaction equipment	Low to moderate	1 : 1.5
Shoulder Stabilization	Embankment slope failure	Angular boulders, geotextile	Excavator, haul trucks, compaction equipment	Low to moderate	1 : 1
Deep Patch	Embankment slope failure	Granular fill, geosynthetic reinforcement	Loader, excavator, compaction equipment	Low to moderate	1 : 1.5

Table 2 – Common Emergency Repair Methods Requiring Detailed Design

Repair Name	Applicable Scenario(s)	Materials Required	Equipment Required	Relative Cost	Typical Maximum Slope/Face Ratio (V:H)
Heavy Scour Protection	Large washes and low water crossings	Concrete barriers, revetments	Excavator, haul trucks, compaction equipment	Moderate to high	n/a
Special Rock Embankment / Rock Buttress	Cut slope failure, creeping slopes, embankment toe protection, small landslides	Angular boulders, geotextile	Excavator, haul trucks, compaction equipment	Low to moderate	1 : 1.5
Buttress	Medium to large landslides	Conventional fill	Loaders, dozers, haul trucks, compaction equipment	Low to moderate	n/a
Reinforced Soil Slope (RSS)	Embankment slope failure	Granular fill, geosynthetic reinforcement	Loader, excavator, compaction equipment	Low to moderate	1 : 1.5 (vegetated) 1 : 1 (hardscaped)
Gabion Wall	Embankment slope failure, cut slope failure, stream erosion	Gabion baskets, stone fill	Excavator, haul trucks, compaction equipment	Moderate	4 : 1
Rockery	Embankment slope failure, cut slope failure, stream erosion	Angular boulders, drainage fill	Excavator equipped with hydraulic clamp or tooth	Moderate	4 : 1
Mechanically Stabilized Earth (MSE) Walls	Embankment slope failure	Granular fill, geosynthetic reinforcement, facing elements	Loader, excavator, compaction equipment	Moderate to high	18 : 1
Road Realignment	Embankment slope failure, cut slope failure, stream erosion, landslides	Typically, conventional fill, can also include retaining structures	Loaders, dozers, haul trucks, compaction equipment	Moderate to high	1 : 2 Or retaining structure as above

Shoulder Stabilization

The typical shoulder stabilization used in ERFO projects consists of rock with an average diameter of 29 inches. The rock is separated from the road embankment with a geotextile. The rocks are typically mechanically placed with a 1V:1H face and a base width approximately one-half of the height. The maximum height is typically limited to 5 to 8 feet. The shoulder stabilization is useful in cases where the outboard shoulder of a roadway has been lost and the steepness of the slopes below the roadway would require the construction of oversteepened or very tall embankments. They are most economical where an appropriate source of large, angular rocks are available near the project site. However, they can increase loading on the embankment slope that may be marginally stable. A typical shoulder stabilization detail is shown in Figure 3.

Drainage Repairs and Improvements

Water is frequently a driving force in geotechnical failures. Control of surface runoff is therefore both a critical aspect of repair work. Typically, it is also one of the relatively less expensive items and primarily consists of conventional earthworks to establish drainage ditches and installation of pipe culverts. Detailed analysis should be performed by an engineer for larger drainages, but for typical roadside drainage, typical roadside ditch and cross culvert details can be implemented without detailed design work. This item can therefore be included in plans with little cost to design and can be implemented by general contractor. One weakness of this item is that it may fail to address subsurface water concerns if groundwater is relatively deep, or the zone of infiltration is far away from the roadway. Surface drainage improvements outside of the roadway footprint are often difficult to implement due to steep topography or environmental impact restrictions.

Deep Patch

The deep patch repair consists of alternating layers of fill and biaxial reinforcement geosynthetic, either woven geotextile or geogrid, and fill. This design is intermediate between a typical subexcavation detail and a reinforced soil slope (RSS). Deep patch repairs are typically 5 to 15 feet deep and extend the full width of the roadway. The reinforcement is typically placed at 12-inch vertical spacing. The outboard slope of the deep patch is typically set between 1V:1H and 1V:1.5H. Deep patch is typically employed where embankment distress is present, but the embankment material has not been lost. The deep patch repair has several advantages including use of design charts, relatively shallow depth of excavation below the pavement section, and simple construction that can be performed by a general contractor. A typical deep patch detail is shown in Figure 4.

Special Rock Embankment/Rock Buttress

Where surficial creep of a cut or fill slope is of concern, a special rock embankment may be constructed. This detail consists of large rock that in inlaid into the slope. This provides additional resisting mass and protection from surface water erosion. The special rock embankment is most useful when sliding surfaces are relatively shallow, within about 5 feet of the surface and the rate of movement is slow. They are also limited to smaller features, typically with a maximum height of about 25 feet. Similar to other repair options, availability of durable boulders of appropriate size often governs the economy of this option.

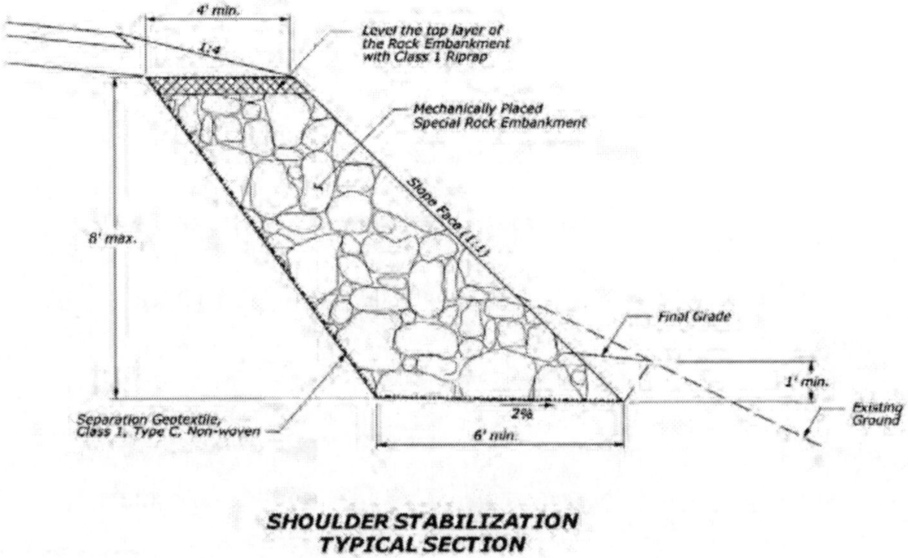

Figure 3 – Typical Shoulder Stabilization Detail by FHWA

Buttresses

Buttresses are a common landslide repair option that consists of constructing a fill at or near the toe of the landslide to increase forces resisting sliding. Buttresses need to be designed in order to be effective, but can be constructed by a general contractor using traditional earthwork methods. The size of a buttress is proportional to the driving force of the landslide that it is mitigating. A buttress for a large landslide can therefore have a very large footprint with commensurate costs and environmental impacts.

Reinforced Soil Slopes

Where the roadway embankment has been completely lost, a Reinforced Soil Slope (RSS) can be constructed to reestablish the roadway while limiting environmental impacts and import fill. RSS consist of alternating layers of select fill and geosynthetic reinforcement (Berg et al., 2009). Face angles can vary from about 26 degrees to 70 degrees, while vertical reinforcement spacing can vary from 8 inches to 2 feet. The reinforcement spacing depends on the backfill material and the slope angle. Where the slope is constructed steeper than 34 degrees (1V:1.5H), wrapping of the geosynthetic at the face is commonly employed. Relatively shallow slope faces are commonly seeded with native grasses; steeper slopes can employ a hard facing. RSS are taller than the typical deep patch repair and need more detailed analysis, but can still be constructed by a general contractor. A special rock embankment can be included to protect the toe of an RSS. A typical RSS detail is presented in Figure 5.

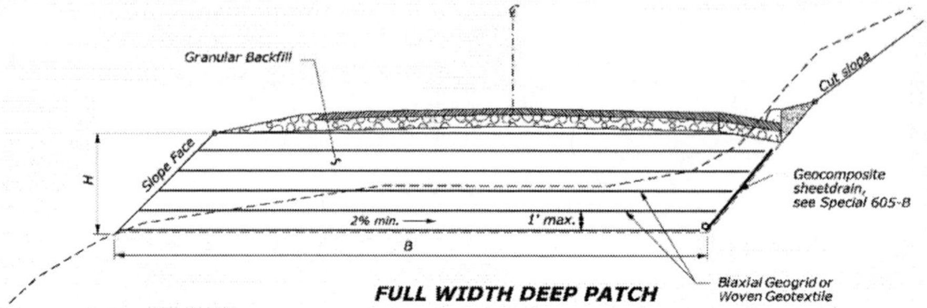

Figure 4 – Typical Deep Patch Detail by FHWA

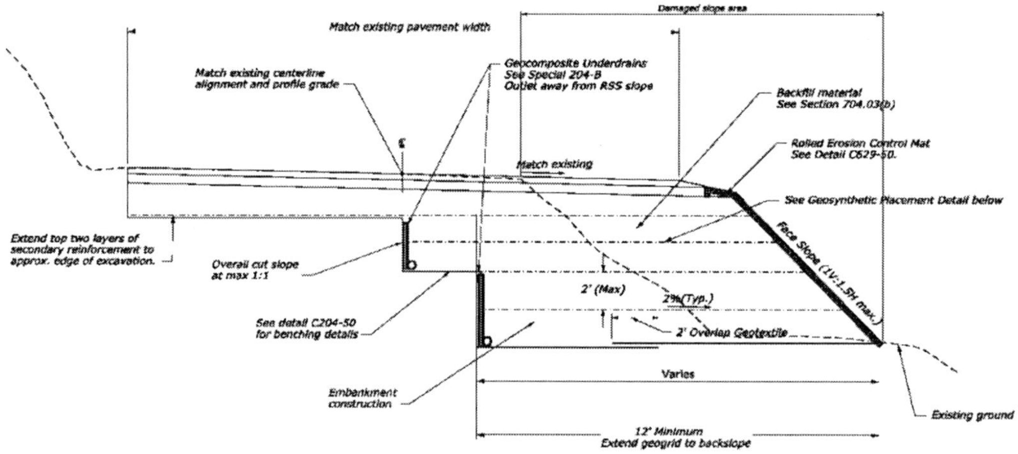

Figure 5 – Typical RSS Detail by FHWA

Gabion Wall

Gabions are typically 3-foot by 3-foot by 3-foot bins formed with twisted steel mesh and filled with a stone material. Other sizes are available from some manufacturers. The most common application for gabions is as a gravity retaining wall, although they can also be used for certain types of stream armoring and rockfall barriers. Gabions can be a cost effective option where suitable stone material is available near the project site. The stone fill for gabions is typically smaller than material used for rockeries or riprap, and are therefore more economical in certain regions.

Rockery Walls

Rockery walls are a special type of dry-stacked rock wall with construction standards developed by the FHWA (Mack et al., 2006). Rockery walls are designed as gravity walls and can be used as retaining structures or to buttress slopes. Rockery walls are more frequently used for cut slope applications, rather than supporting embankment fills and are typically limited to heights of approximately 12 to 15 feet, due to limitations of handling larger boulders. Rockeries are most cost effective when there is a rock source present on or near the project site. Design includes detailed analysis of the rockery as a retaining wall including global stability of the whole slope, including the wall. Rockeries do not require specialized contractors, but do require a skilled equipment operator to place the boulders. A typical rockery wall section is shown in Figure 6.

Mechanically Stabilized Earth

Mechanically stabilized earth (MSE) retaining walls consist of high quality backfill with geosynthetic or metallic reinforcement, and some type of rigid facing (Berg et al., 2009). Most projects on Federal lands, especially emergency work, make use of geogrid reinforcement and welded-wire basket facing. MSE walls are one of the higher cost options for repair of embankment failures, and are typically only deployed where other methods such as shoulder stabilization or RSS are not suitable. MSE walls require both detailed design analysis and specialty contractors to construct. On-site fill materials can be considered for MSE walls to reduce the need to import high quality backfill. Additional testing during both design and construction is critical to use of on-site materials to ensure that appropriate design parameters are used and that the material used meets the design.

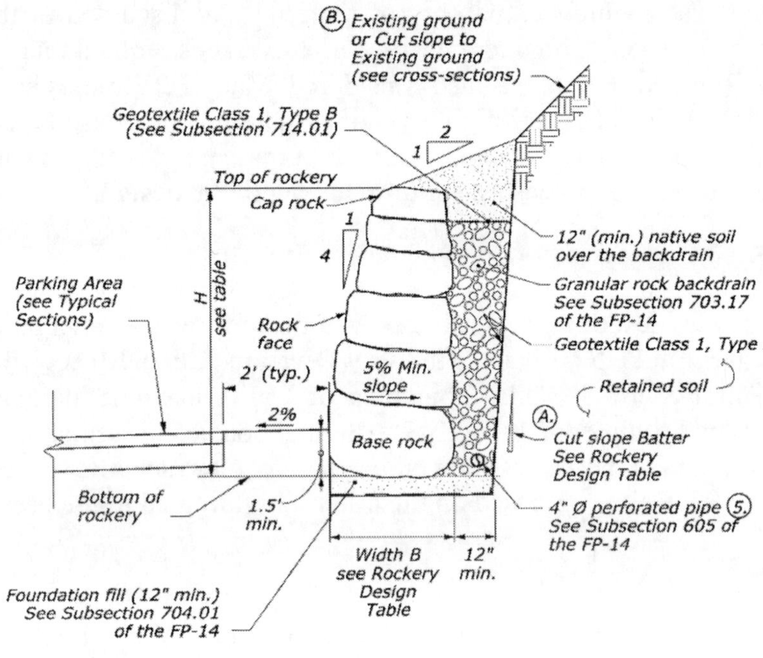

Figure 6 – Typical Rockery Detail by FHWA

Rock Scaling

In cases where rock fall or landslides leave a destabilized slope above the roadway, rock scaling can be performed to remove portions of the unstable material and improve short term safety of the roadway. For ERFO projects, rock scaling is typically performed by crews on rope using hand tools such as pry bars with occasional use of air bags. Scaling with machinery such as long-reach excavators is not typical due to the anticipated higher cost and longer lead time of mobilizing the specialized equipment. Although, rock scaling can improve short term safety, it is not a long-term solution. This must be kept in mind by the geotechnical professional and communicated to the facility owner.

APPLICATION OF STANDARD DETAILS

The standard details present many options for repair of various types of damage. The geotechnical engineer responsible for a project must select from these options the most appropriate repair technique. Criteria evaluated include technical feasibility, construction cost, environmental impact, design analysis required, construction complexity, and total (design and construction) schedule. These criteria together form a decision framework. Typically, technical feasibility is the first check that the geotechnical engineer makes, listing all of the options that will effectively repair the damage. Once a list of technically feasible options is compiled, the geotechnical engineer will discuss the remaining criteria with other members of the project team including other technical experts, environmental protection specialists, and the project manager. Typically, cost and schedule are weighted heavily, but other factors are also considered, especially as they affect cost or schedule. For example, a repair option with more environmental impact may cost less to construct, but would require a longer schedule in order to complete required environmental clearances.

Cost considerations also include availability of material. As discussed in the previous section, availability of materials such as boulders for rockeries and rock embankments, stone for gabions, and granular backfill for MSE walls, is also considered. Many ERFO sites are remote and costs to haul materials to the site can be very high. Marginal materials may also be considered for repair work, provided that conservative design assumptions are employed, and additional testing is performed during construction to ensure conformance with the design.

CASE STUDIES

The following section describes several case studies of recent emergency repair activities undertaken by the Central Federal Lands Highway Division (CFL). Most of these case studies are located in the central Sierra Nevada mountains of California with damage related to heavy precipitation events in the winter of 2016/2017. Although located in similar terrain and with similar failure mechanisms, each of these sites were unique and required specific repair methods. Selection of repair methods at each site was based on the criteria highlighted in the previous section.

Eleven Pines Road

Eleven Pines Road (Forest Road 14N08) is located in the Eldorado National Forest, approximately 18 miles east of Georgetown, California. The road descends from the intersection with Wentworth Springs Road, crosses the Rubicon River and then ascends the opposite side of the valley towards McCulloh Ridge Road. The forest around the site was previously burned by forest fires that destroyed most of the vegetation in the area. The link between forest fires and landslides is well documented (Steblein and Miller, 2019) and was likely a contributing factor in these failures.

In February, 2017, a series of heavy precipitation events led to damage to multiple locations along the first six miles of the route, including embankment failures, culvert blowouts, and loss of road base material (Figure 7). After an initial site visit by U.S. Forest Service (USFS) personnel to complete the DSR, the CFLHD project team performed a site visit to collect additional data after the snow had fully cleared from the site. Following the site visit, the project team spent several hours together comparing their notes and identifying the applicable repairs for each site. The embankment failures typically extended the full width of the roadway, were 100 to 300 feet wide and 20 to 50 feet deep. The slopes below the failures were relatively steep, with slope ratios between 1V:2H and 1V:1.5H. Based on these factors, RSS and MSE walls were the technically feasible alternatives. Conventional embankments would have required a significantly larger footprint to construct, and no other repair options were appropriate to the situation. RSS were selected due to their lower cost, ease of construction, and use of lower quality backfill compared to MSE walls. Other repair work included standard embankment construction, culvert replacement, drainage improvements such as headwall installation and erosion protection, and pavement reconstruction. The geotechnical engineer performed analyses to correctly size the RSS and assisted other personnel to compile the project plans and specifications.

Subsequent to the construction completion, heavy seasonal storms affected the area in the winter of 2019. One of the repaired areas moved as a result of the heavy precipitation, likely as a more deep-seated landslide rather than a relatively shallow embankment failure.

Figure 7 – Embankment Failure at Eleven Pines Road

Rattlesnake Road

Rattlesnake Road (Forest Road 05N14) is located in the Stanislaus National Forest, approximately three miles east of the unincorporated settlement of Camp Connell, California. The road climbs along a variably steep to gentle cross slope, following the course of the Little Rattlesnake Creek. Similar to Eleven Pines Road, heavy precipitation caused embankment failures at several locations along this road. In this case, stable rock was exposed in the slide scarp and the existing slopes were less steep (Figure 8). Therefore, conventional embankment, RSS, and MSE walls are the most technically appropriate repair options, and conventional embankment was selected due to the lower anticipated cost and faster construction. To improve final stability, the road alignment was shifted so the embankment would be supported by the bedrock.

Figure 8 – Embankment Failure on Rattlesnake Road with Bedrock Exposed

Bowman Lake Road

Canyon Creek crosses Bowman Lake Road (Forest Road 0018) in the Tahoe National Forest, approximately five miles east of the unincorporated settlement of Graniteville, California. At this site, heavy precipitation led to higher than normal outflow from Bowman Lake. This flow migrated from the main channel that Bowman Lake Road crosses via a bridge, to a side channel that the road crosses via a vented concrete low water crossing. Due to the higher than anticipated flows in the side channel, the low-water crossing was scoured and pushed down the channel several feet (Figure 9).

This type of damage does not directly incorporate the typical repair details presented earlier in this article, although the design decision process remains the same. Following the site visit, the project team met to discuss all technically feasible solutions, including replacing the low water crossing in-kind or replacing it with a bridge. Replacing in-kind was undesirable because the repair would lack resilience and be vulnerable in the future. Replacing with a bridge would be expensive and reduce the total number of repair projects that could be conducted. The project team then considered a repair option that would address the underlying cause of the failure: the shift of stream flow from the main channel to the side channel. An embankment fill was constructed at the low point in the channel bank to contain flows in the main channel so that they could safely pass under

Bowman Lake Road via the existing bridge. This repair option provided the best balance of cost, impacts, and technical effectiveness.

To ensure geotechnical stability of the fill, it was analyzed as a rapid drawdown condition based on the hydraulic model of the inciting event. This analysis showed that the channel banks were not likely to fail due to slope instability. Scour potential was evaluated by the CFLHD Hydraulics Engineer to design appropriate erosion control measures to protect the fill slopes. The damaged section of Bowman Road was reconstructed with an embankment fill crossing the side channel with a single corrugated metal pipe culvert to convey local flows in the channel.

Figure 9 – Damage to Low-Water Crossing on Bowman Lake Road

Blacksmith Flat Road

Blacksmith Flat Road (Forest Road 14N25) is located in both the Tahoe and Eldorado National Forests. The project site is located in the Eldorado National Forest, approximately six miles east of Foresthill, California. In this area, shallow bedrock has led to the creation of extremely steep valley walls. The bedrock slopes are typically marginally stable, leading to significant retaining devices including ground anchors and rockfall mesh being installed to protect the nearby Ralston Power House. At the project location, however, the bedrock is not intact. Instead, the hillslope is composed of what appears to be fault gouge material including large blocks of bedrock material in a matrix of silty to clayey sand. Seeps were observed in this portion of the slope after failure, in both the wet and dry seasons.

The March, 2017, failure at this location consisted of a relatively shallow debris slide that initiated in the slopes above the roadway (Figure 10). The debris blocked the roadway and redirected stormwater out of the roadside ditch, and concentrated the flow onto a point on the outboard shoulder, leading to scour of the fill embankment. Records from the water authority that operates the nearby powerhouse, show that the same area had previously failed in 1999. Due to limits of available funding and requirements of the ERFO program, the repair at this location was limited to clearing the debris from the road to reestablish access. Options such as ground anchors and horizontal drains were estimated to be too costly. A buttress would cost less, but space was not available due to the steep topography and river adjacent to the site. The large size of the slide and difficulty of slope access was also part of this decision. Limited scaling was performed using conventional track-mounted excavation equipment to remove unstable material from the slope. Following clearing of the roadway, no pavement was placed, under the assumption that additional debris material may slide onto the roadway in the near future. Construction was complete in November, 2018.

Raveling of the slope was noted after heavy rains in December, 2018, followed by a large failure of similar size to the original slide in February, 2019 (Figure 11). Repair of this new slide is underway, but is not covered by the ERFO program and not being designed or administered by FHWA. The reactivation of this landslide is indicative of the types of risk involved in emergency repairs.

Figure 10 – Slide on Blacksmith Flat Road in 2017.

Figure 11 – Slide on Blacksmith Flat Road in 2019.

SUMMARY AND CONCLUSIONS

Emergency repairs are critical to restore immediate access for emergency vehicles and later to restore access for various purposes including residential and commercial use. The ERFO program administered by the FHWA provides funding for such repairs and FLH provides engineering and contract administration support for FLMAs. Schedules for these projects are typically accelerated to meet both funding requirements and to restore permanent access as soon as possible. For this reason, a collection of common solutions is extremely useful to meet project schedules and deliver solutions that provide appropriate safety to the public.

Numerous repair options were discussed in this paper, providing appropriate methods for various types of damage, budgets, and impact concerns. Several case studies were also discussed, showing real world applications of the standard details. A summary of design and construction time, as well as project cost is presented in Table 3. This shows that Blacksmith Flat Road had the longest design time. For speed of contracting, Blacksmith Flat Road and Eleven Pines Road were packaged together, with the contractor starting work at Eleven Pines Road first. This allowed the CFLHD project team additional time to consider repair alternatives. This schedule also allowed additional time for natural processes to remove additional unstable materials from the slide area.

Table 3 – Summary of Case Study Projects

Project Name	Design Time	Construction Time	Approximate Construction Cost
Eleven Pines Road	150 days	120 days	$5,750,000
Rattlesnake Road	50 days	74 days	$1,000,000
Bowman Lake Road	260 days	100 days (planned)	$850,000 (bid award)
Blacksmith Flat Road	400 days	30 days	$1,500,000

The case studies present both successful applications, as well as potential shortfalls to some repair techniques. In at least a few of the case studies, limitations to the scope of repairs was a primary contributor to less successful results.

Lastly, management of institutional knowledge of these various repair options is critical. Over time, the standard details should evolve to meet a particular agency's needs and culture, based on the record of successes and shortfalls of the various repair types. In this way, the emergency repair standard details can be customized based on geologic terrain, risk management, material availability, and funding requirements or limitations. New staff can be quickly trained to use the standard details and a decision framework to maintain consistency between repair projects. The collection of standard details expedites plan development and reduces design costs so that more money is available for construction. Overall, the use of standard details and a decision framework improves an agency's ability to respond to emergency repairs.

REFERENCES

Berg, R.R., Christopher, B.R., and Samtani, N.C. *Design and Construction of Mechanically Stabilized Earth Walls and Reinforced Soil Slopes*. Publication Nos. FHWA-NHI-10-024 and FHWA-NHI-10-25. Federal Highway Administration, U.S. Department of Transportation, 2009

Federal Highway Administration. *Emergency Relief for Federally Owned Roads (ERFO).* https://flh.fhwa.dot.gov/programs/erfo/. Accessed on June 3, 2019.

Federal Highway Administration. *Mobile Solution for Assessment and Reporting (MSAR).* https://collaboration.fhwa.dot.gov/dot/fhwa/msar/default.aspx. Accessed on August 15, 2019.

Federal Highway Administration. *Plans, Specifications, and Estimate for Project CA ERFO TAHOE NF 18(1).* 2018.

Federal Highway Administration. *Plans, Specifications, and Estimate for Project CA ERFO 05N14(1).* 2018.

Federal Highway Administration. *Plans, Specifications, and Estimate for Project CA ERFO 14N08(1).* 2017.

Mack, D.A., Sanders, S.H., Millhone, W.L., Fippin, R.L., and Kennedy, D.G. *Rockery Design and Construction Guidelines*. Report No. FHWA-CFL/TD-06-006. Federal Highway Administration, U.S. Department of Transportation, 2006.

Steblein, P.F., and Miller, M.P. *Characterizing 12 Years of Wildland Fire Science at the U,S, Geological Survey: Wildland Fire Science Publications, 2006-17.* Open-File Report 2019-1002. U.S. Geological Survey, U.S. Department of the Interior, 2019.

Pfeiffer Canyon Bridge Failure Within the Context of Risk

Gresham D. Eckrich, P.E., P.G., C.E.G.
Yeh and Associates, Inc.
391 Front Street, Suite D
Grover Beach, California 93449
(805) 481-9590
geckrich@yeh-eng.com

Prepared for the 70th Highway Geology Symposium, October, 2019

Acknowledgements

Golden State Bridge: Will Reames
California Department of Transportation: Susana Cruz; David Galarza, P.E.; Doug Cook, P.G., C.E.G.; Ryan Turner, P.E., G.E.
California Polytechnic State University: Robb Moss, Ph.D, P.E.
Yeh and Associates: Judd King, P.E., G.E.

Disclaimer

Statements and views presented in this paper are strictly those of the author(s), and do not necessarily reflect positions held by their affiliations, the Highway Geology Symposium (HGS), or others acknowledged above. The mention of trade names for commercial products does not imply the approval or endorsement by HGS.

Copyright © 2019 Highway Geology Symposium (HGS)

All Rights Reserved. Printed in the United States of America. No part of this publication may be reproduced or copied in any form or by any means – graphic, electronic, or mechanical, including photocopying, taping, or information storage and retrieval systems – without prior written permission of the HGS. This excludes the original author(s).

ABSTRACT

Pfeiffer Canyon Bridge Failure Within the Context of Risk

Gresham D. Eckrich, P.E., P.G., C.E.G.
Yeh and Associates, Inc.
391 Front Street, Suite D
Grover Beach, California 93449
(805) 481-9590
geckrich@yeh-eng.com

The original Pfeiffer Canyon Bridge was a 3-span reinforced concrete box girder bridge constructed in 1968 that carried two lanes of Highway 1 across Pfeiffer Canyon near Big Sur in Monterey County, California. The famed scenic highway runs north-south along the Pacific coastline and generates approximately 1.7 million dollars per day in travel spending within Monterey County and neighboring San Luis Obispo County. A landslide mobilized by January and February 2017 storms displaced an interior pile column supporting the bridge. The California Department of Transportation (Caltrans) inspected the pile column and closed the bridge in late February 2017. Subsequent structural damage to the pile column and bridge deck required permanent closure, and demolition and replacement of the bridge using emergency funding sources. Caltrans contracted with Golden State Bridge to complete the 22 million-dollar ($22M) replacement project in approximately 8 months.

Bedrock at the site is mapped as mélange of the Franciscan Complex, which predominantly consists of a chaotic mixture of fragmented rock blocks embedded in a pervasively sheared matrix of clay shale. The sheared matrix typically weathers to a predominantly clay soil that is prone to landsliding. There are no geomorphic features indicative of landsliding in the topographic contours shown on as-builts for the original bridge, and there are no as-built documents indicating the designers were aware of the potential for landslides to impact the interior bridge supports. Therefore, the cause of the bridge failure (landsliding) was likely considered a low probability event by the original designers.

This study considers an idealized scenario where the potential for landsliding to damage the bridge was identified after construction of the bridge and before the bridge failure. Statistical analysis is performed to estimate the probability of landsliding and structural failure of the bridge pile as a result of landsliding. The probability of that outcome is multiplied by the costs of the outcome to estimate the risks to Caltrans and California businesses that benefit from Highway 1. The results of this risk analysis for Pfeiffer Canyon Bridge could be considered a case study for future evaluations of bridges (or other transportation assets) in the context of risks posed by natural hazards such as landsliding.

INTRODUCTION

The Pacific Coast Highway (Highway 1) runs north-south along the Pacific coastline of California. Caltrans constructed Pfeiffer Canyon Bridge in 1968 to carry two lanes of the highway across Pfeiffer Canyon near Big Sur in Monterey County. The highway was originally constructed in the late 1930's. A wagon road from the Post Ranch in Big Sur to Carmel was completed in 1889, and reportedly was only passable in the dry season. The location of the wagon road is not shown on historical maps, but it likely passed through the project vicinity and defined roadway grades that were eventually accommodated by the highway. The highway at Pfeiffer Canyon is located inland and approximately 1-mile northeast of the coast, on a northeast-facing slope that borders the Big Sur River watershed.

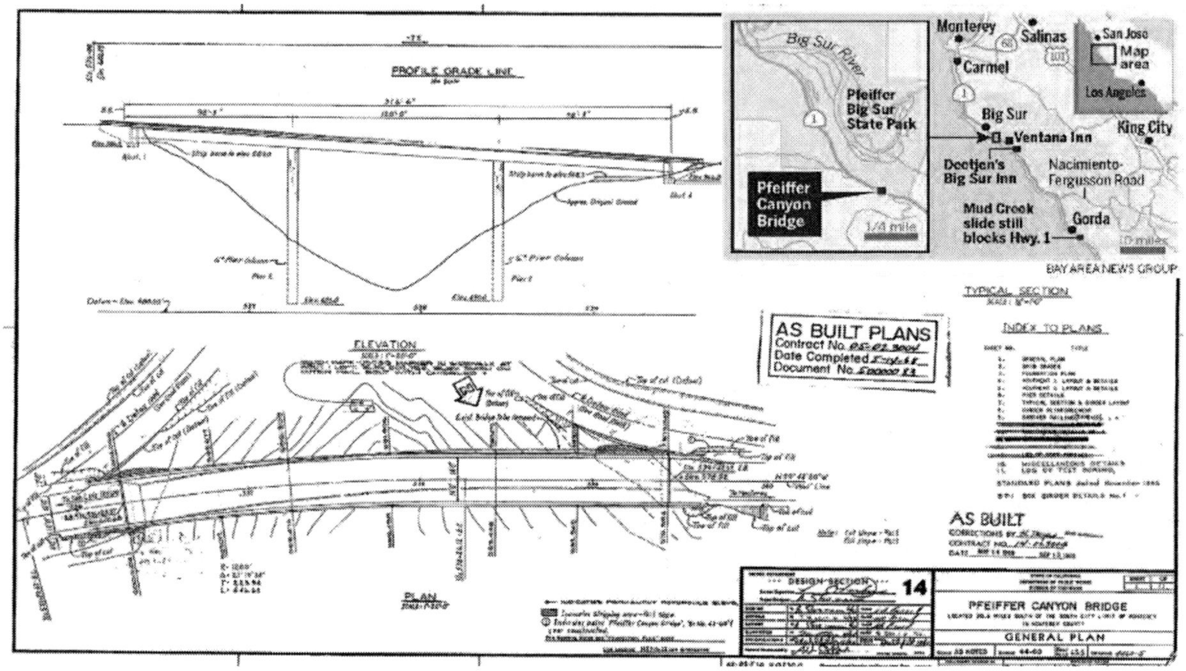

Figure 1: 1968 As-builts (Inset: Bridge Location [Bay Area News Group])

Figure 1 shows the 1968 as-builts and the bridge's location on Highway 1. The 1968 bridge was a 316.5-foot long, 3-span reinforced concrete box girder bridge oriented approximately southeast-northwest over the seasonal drainage that flows northeast through Pfeiffer Canyon to the Big Sur River. The two interior pier columns of the bridge were located on opposite drainage slopes and founded on single, 6-foot diameter reinforced concrete piles. Elevations at the site range from approximately elevation (el.) 498 feet at the canyon bottom to approximately el. 603 feet at the southeast end of the bridge. Slope inclinations below the bridge location range from approximately 1.5h:1v (horizontal:vertical) to 1h:1v. A landslide mobilized by January and February 2017 storms displaced the interior pier column (Pier 3) on the northeast-facing drainage slope. Figure 2 shows the site topography, the location of Pier 3, and the interpreted 2017 landslide head scarps. Caltrans inspected the pier column and closed the bridge on February 15, 2017.

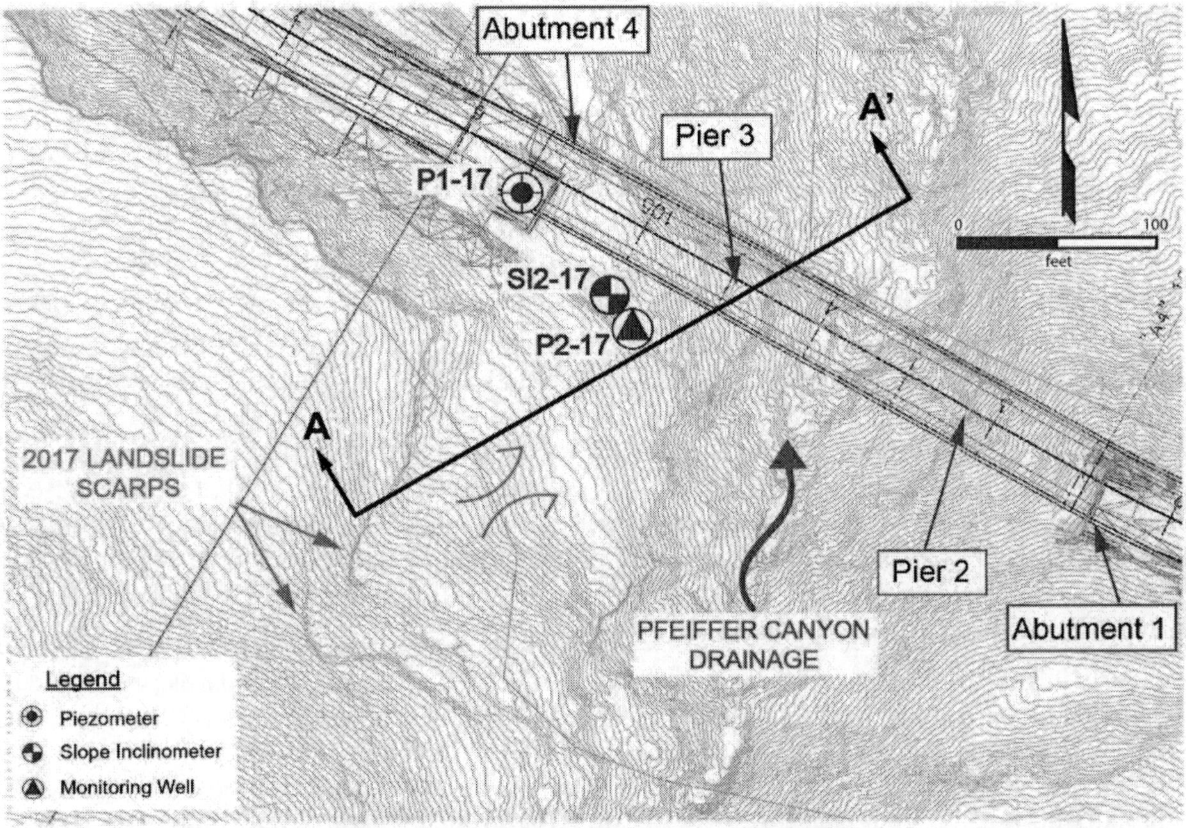

Figure 2: Pfeiffer Canyon Topography

Figure 3 shows photos of the displaced Pier 3 and the lateral scarp of the landslide immediately downslope of Abutment 4. Caltrans engineers considered structurally retrofitting Pier 3 before additional slope displacement and subsequent structural damage to the pier column and bridge deck required permanent closure, and demolition and replacement of the bridge using emergency funding sources. Caltrans contracted with Golden State Bridge (GSB) to demolish and construct

Figure 3: Facing east, displaced Pier 3 in foreground, Pier 2 in background (left); Facing west, lateral scarp of landslide downslope of Abutment 4 (right)

the new bridge, which was opened to traffic on October 13, 2017. GSB completed the $22M bridge replacement project in approximately 8 months.

Engineering Risk

The engineering lexicon defines "risk" as the product of probability of failure multiplied by the consequences of failure. The probability of failure is typically expressed as an order of magnitude within some time frame. The consequences of failure are often expressed as a monetary value. The risk value, in turn, can be expressed as a monetary value within some time frame. Risks are typically evaluated relative to risk thresholds defined by project stakeholders.

This study considers an idealized scenario where the potential for landsliding to damage Pier 3 was identified after construction of the bridge and before the bridge failure. There are no geomorphic features indicative of landsliding in the topographic contours shown on the 1968 bridge as-builts, and there are no as-built documents indicating the designers were aware of the potential for landslides to impact the interior bridge supports. Therefore, bridge failure caused by landsliding was likely considered a low probability event by the original designers. The new Pfeiffer Canyon bridge was designed as a single span structure so that interior supports would not be founded on the potentially unstable drainage slopes, thus reducing the probability and risk of landsliding to impact the new bridge.

Similarly, Caltrans retrofitted the bridge in 1996 to meet seismic design standards per 1983 American Association of State Highway and Transportation Officials (AASHTO) Bridge Design Specifications with Interims and Revisions by Caltrans. Caltrans seismic design criteria for the original design were not reviewed for this study, but it can be reasonably inferred the bridge was originally designed for a less conservative seismic hazard than the hazard recommended by current Caltrans guidelines, and that the retrofit reduced the risk of bridge failure due to strong ground motion.

A risk analysis was performed for the idealized scenario to estimate the risk of landsliding that could result in failure of the bridge and require mitigation consisting of bridge retrofit or replacement. This risk is then compared with the risk of structural failure and other potential natural hazards, including liquefaction, strong ground motion, debris flows, scour, and fault rupture. The results of the risk analysis are presented in a decision tree that could have been used in the idealized scenario as a basis for assessing the need for hazard mitigation and prioritizing mitigation alternatives. The branches of a decision tree map the sequence of events that could result in bridge failure and mitigation, and quantifies the risk for each hazard.

GEOLOGY

The bridge site is located within the Santa Lucia Range of the Coast Ranges geomorphic province. Figure 3 shows the surficial geology at the site as mapped by Dibblee (2007). The site is mapped as late Mesozoic Franciscan Complex (fs), which is described as a mélange of pervasively sheared, deep marine sedimentary and volcanic rocks, metamorphosed under conditions of high pressure and low temperature. The composition of the Franciscan Complex represents the subduction of oceanic plates and associated pelagic sediments deposited in a subduction trench during the Jurassic period. This tectonic setting resulted in the accretion of oceanic lithosphere (meta-basalt, fg) and near-source pelagic sediments dragged downwards with the subducting oceanic plate and subjected to high pressure-low temperature metamorphism producing metagraywacke, metasiltstone, chert and shale. The mélange is typically characterized as a bimrock (block-in-matrix rock). The sheared matrix of the mélange typically weathers to a predominantly clay soil that is prone to landsliding. Dibblee (2007) mapped a relatively large landslide (Qls) immediately east of the bridge site.

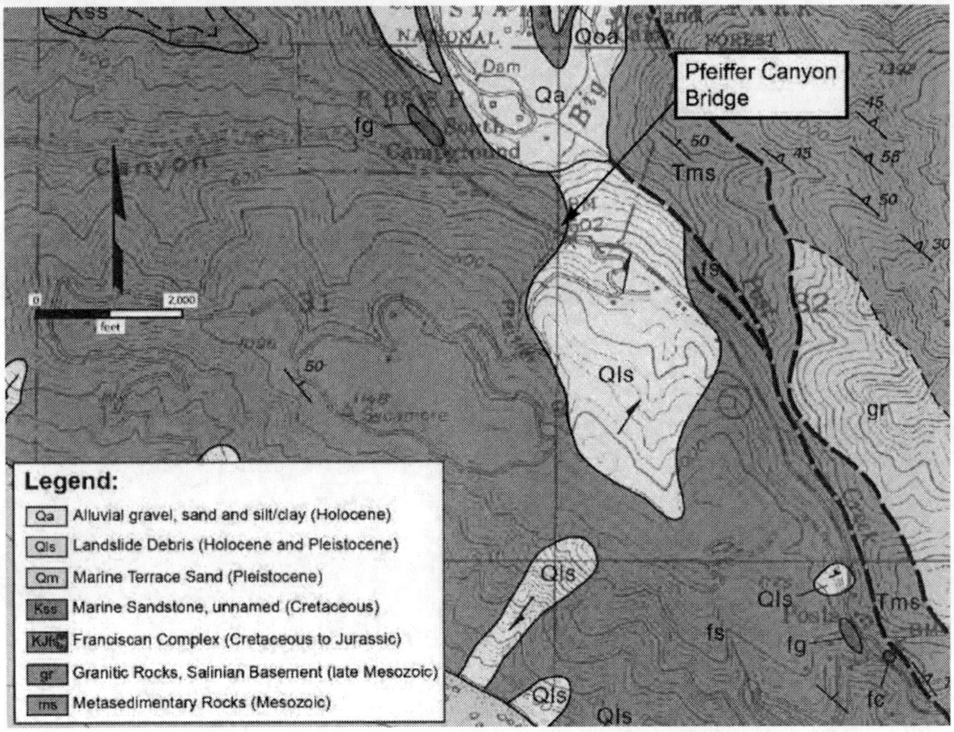

Figure 4: Geologic Map after Dibblee (2007)

SUBSURFACE EXPLORATION AND INSTRUMENTATION

Subsurface exploration for the 2017 emergency replacement project consisted of drilling six borings using mud rotary and HQ coring methods to depths ranging from approximately 47 to 122 feet. The borings were drilled approximately 1 to 3 weeks after the February 2017 landslide. The Franciscan Complex mélange encountered predominantly consisted of intensely weathered to decomposed, pervasively sheared shale with subangular to angular blocks of graywacke and shale. The blocks were chaotically distributed and ranged in size from sand to boulder.

Instrumentation was installed in five of the borings and consisted of slope inclinometers, piezometers, or monitoring wells. The locations of instrumentation on the northeast-facing drainage slope are shown in Figure 2. Slope inclinometer SI2-17 was installed in a boring drilled between Abutment 4 and Pier 3, as shown on Figure 2, within the interpreted limits of the 2017 landslide. Piezometer P1-17 was installed in a boring drilled near the northwest bridge abutment (Abutment 4), and monitoring well P2-17 was installed in a boring drilled between Abutment 4 and Pier 3, within the interpreted limits of the 2017 landslide.

INSTRUMENTATION AND RAINFALL DATA

Landsliding typically occurs in the Franciscan Complex during relatively wet winter months and is predominantly associated with elevated groundwater surfaces. Figure 5 shows Big Sur rainfall data[1] and groundwater data recorded by the piezometer P1-17 and monitoring well P2-17. Approximately 72.5 inches of cumulative rainfall was recorded in the Big Sur area between October 1, 2016, and April 30, 2017; the average cumulative rainfall for this period is approximately 44 inches. Approximately 47 inches of cumulative rainfall was recorded before Caltrans personnel observed the landslide damage to the bridge in mid-February (see Figure 5). Rainfall in January and February 2017 totaled 25.5 and 18.7 inches, respectively, approximately 16 and 10 inches greater than the averages for January (9.2 inches) and February (8.7 inches) in Big Sur.

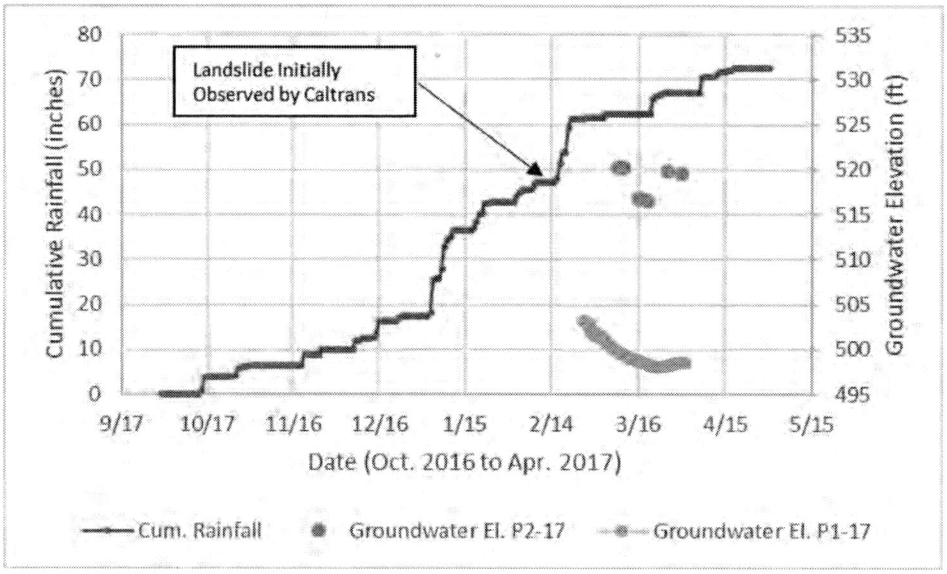

Figure 5: Big Sur Rainfall and Groundwater Data

Groundwater and seepage conditions within the Franciscan Complex bedrock are typically variable and associated with groundwater flowing along discontinuities, fractures, and relatively impermeable shear planes. The variability of groundwater conditions is illustrated in Figure 5, which shows the groundwater elevations in the monitoring well and piezometer, located approximately 80 feet apart, differed by approximately 17 to 22 feet during March 2017. Slope movement was evident in the slope inclinometer data at depths of approximately 35 and 55 feet, indicating there were multiple slide planes within the landslide mass. Thus, groundwater in P2-17 may have been perched above an interpreted landslide plane at a depth of approximately 55 feet (el. 517.5 feet). The slope inclinometer and groundwater data were used in slope stability analyses, including back-analysis of the 2017 landslide to estimate the shear capacity of the reinforced concrete pile supporting Pier 3.

[1]https://www.usclimatedata.com/climate/big-sur/california/united-states/usca2017/2017/4

The data in Figure 5 also show a difference in groundwater elevation response to rainfall between the two locations. Groundwater elevations in P1-17 and P2-17 rose by approximately 0.5 and 3 feet, respectively, following 4.7 inches of rainfall between March 21 and 25, 2017. Groundwater response was not considered in this study's risk analysis, but a more rigorous analysis would consider the probability of above-average rainfall, associated groundwater elevation response, and impacts to slope stability.

SLOPE STABILITY ANALYSES

Slope stability analyses were performed to evaluate the probability of failure along cross-section A-A' (see Figure 2), which is oriented along the same azimuth of movement observed in the slope inclinometer data (azimuth = 060) and includes the location of the damaged bridge pier column (Pier 3). The slope stability analyses evaluated potential failure surfaces (i.e., landslide planes) that intersected the 6-foot diameter reinforced concrete pile supporting the pier column, and potential failure surfaces located below the pile. The groundwater surface modeled in the slope stability analysis was based on the highest groundwater elevation measured in piezometer P2-17, approximately 3 weeks after the 2017 landslide.

Strength parameters for the Franciscan Complex mélange were developed on the basis of laboratory test results and statistical analysis of the test results shown in Figure 6. A total of 14 tests for direct shear strength or triaxial compressive strength using consolidated undrained loading were performed on mélange samples recovered from borings drilled in the bridge vicinity. Figure 5 shows an estimated mean friction angle of 33 degrees and a mean cohesion of 700 pounds per square foot (psf). Standard deviations of friction angle (4 degrees) and cohesion (220 psf) were estimated assuming a normal distribution for these parameters and using the "three-sigma rule" (Dai and Wang, 1992), which is predicated on the fact that 99.73-percent of all values of a normally distributed parameter fall within three standard deviations above and

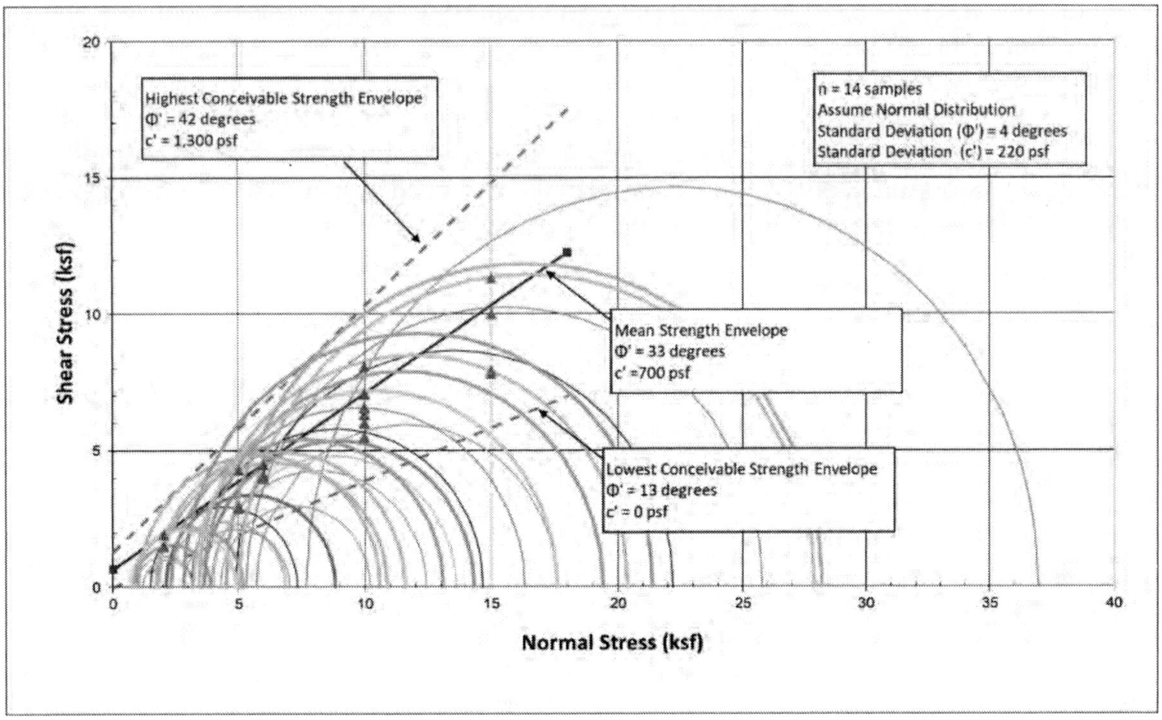

Figure 6: Results of DS and CU Laboratory Tests

below the mean value. The mean shear capacity of the pile at Pier 3 was estimated to be approximately 430 kips, with an estimated standard deviation of 55 kips, based on back-analysis of the 2017 landslide and a typical coefficient of variation (equal to the standard deviation divided by the mean) for the structural resistance of reinforced concrete (Ellingwood et al., 1980).

The results of the slope stability analyses indicate a probability of slope failure of 3.2-percent ($p = 0.032$), corresponding to a deterministic factor of 1.28 for the slope modeled. The estimated probability was used in the decision tree for this study's risk analysis to evaluate the risk of bridge failure posed by landsliding. The probability value was multiplied by four to account for the four bridge supports ($p = 0.032 \times 4 = 0.128$). This probability estimate assumes bridge failure would occur if any one of the supports was damaged. It should be noted the probability of slope failure would likely vary between the 4 supports and a more rigorous risk analysis would evaluate the probability of failure at each support.

It should also be noted that this probability does not represent the probability of the above-average rainfall during the 2016-2017 winter and associated groundwater response that was likely a predominant cause of the 2017 landslide. A more rigorous risk analysis would multiply the rainfall probability by the probability of failure and this would likely result in a lower (conditional) probability estimate. However, it may be reasonable to assume the groundwater elevations during the 2017 landslide event were higher than those modeled in the slope stability analyses, and the higher groundwater elevations would have resulted in a higher probability of failure that may offset the lower conditional probability of landsliding due to above-average rainfall.

RISK ANALYSIS

The risk analysis was performed to estimate the risk of multiple natural hazards and structural damage that could result in failure of the bridge and require retrofit or replacement. As noted above, the probability of failure and risk are typically expressed within some time frame. The risk analysis considered a time frame of 75 years, corresponding to the typical design life of bridges designed in accordance with AASHTO Load and Resistance Factor (LRFD) Bridge Design Specifications (2017). The risks are presented as the probability of failure multiplied by the consequences of failure, and would be used (in the idealized scenario) as a basis for assessing the need for hazard mitigation and prioritizing mitigation alternatives.

Probability of Failure

The results of the slope stability analysis performed for this study were used to estimate a probability of slope failure due to landsliding. The probabilities of structural failure and other natural hazards considered in this study were estimated from code-based values or based on subjective judgement. Caltrans design guidelines recommend a 5-percent probability of exceedance in 50 years for seismic design (e.g., strong ground motion), which corresponds to a return period of approximately 975 years. AASHTO's 2017 design guidelines recommend consideration of a flood event with a return period of 500 years for scour design, corresponding to an annual probability of exceedance of 0.2-percent probability of exceedance in any given year. The recommended probabilities were converted to probabilities within a 75-year time period assuming the occurrence of events is a Poisson process. It should be noted the 1968 bridge and 1996 retrofit may have been designed for different scour and seismic probabilities, however, this study assumed the probabilities recommended by current design guidelines for analysis brevity.

The probability of debris flows was estimated to be one-quarter the probability of landsliding (p = 0.032). Although there is no historical record of debris flows at the site, there is dense forest upslope of the site and the Big Sur area is generally susceptible to wildfires that could contribute to the potential for debris flows. The probability of liquefaction was estimated to be zero because the Franciscan Complex bedrock is generally not considered susceptible to liquefaction. The probability of fault rupture was estimated to be zero because the site is not within a designated Alquist-Priolo Earthquake Fault Zone and there are no mapped faults crossing the site.

AASHTO (2017) calibrated load and resistance factors for the Strength I Limit State based on a 0.02-percent probability of exceedance in 75 years, corresponding to a reliability index of 3.5. The risk analysis assumed the same probability for structural failure that would require emergency retrofit or replacement of the bridge. The strength limit state considers the stability or yielding of each structural element. AASHTO (2017) notes extensive distress and structural damage may occur under the strength limit state but overall structural integrity is expected to be maintained; therefore, the probability of structural failure may be lower than the value assumed for the risk analysis. The risk analysis also considered structural deterioration (or aging) defined as the gradual deterioration of the bridge's structural integrity over the design life of the bridge. The risk of a bridge's structural deterioration (or aging) is typically addressed by retrofitting or replacing the bridge at the end of its design life.

The probabilities considered in a decision tree column should sum to one. Therefore, the probability of structural deterioration during the design life of the bridge was estimated as the complement to the sum of probabilities of bridge failure due to structural damage or natural hazards (i.e., 1 – p[failure due to structural damage or natural hazards]). The probabilities of bridge replacement and retrofit were estimated for each hazard based on subjective judgment of the likelihood of a bridge retrofit effectively mitigating each hazard.

Consequences of Failure

The consequences of bridge failure considered for the risk analysis included the cost of bridge replacement or retrofit projects under emergency and planned circumstances, and the cost of lost travel spending for regional businesses in both Monterey and San Luis Obispo Counties, where Highway 1 is estimated to generate approximately $1.7M per day in travel spending for reginal businesses (Visit California, 2017). The risk analysis assumed a cost of $22M for the bridge replacement, reflecting the actual 2017 emergency replacement cost. The risk analysis considered a retrofit cost equal to one-half of the replacement cost ($11M).

The regional consequences of emergency projects, such as the 2017 bridge replacement, are generally greater than the consequences of a planned retrofit or replacement. Visit California (2017) performed a study to evaluate the economic impact of the Pfeiffer Canyon Bridge closure, and estimated approximately $554M in lost travel spending could be attributed to the emergency bridge replacement project, assuming the project would result in a road closure for approximately 320 days. As noted above, the emergency bridge replacement was completed and opened to traffic in approximately 8 months (i.e., 240 days). Therefore, the impacts of the emergency bridge replacement project were estimated to be approximately $416M by reducing Visit California's (2017) estimate by a factor of 0.75 (240/320 = 0.75) to account for the shorter duration. It was assumed that the road closure duration of the retrofit would be approximately half the duration of the replacement (i.e., 4 months or 120 days). Therefore, the impacts of the emergency bridge retrofit project were estimated to be approximately \$208M by reducing Visit California's (2017) estimate by a factor of 0.375 (120/320 = 0.375).

The costs and durations of emergency replacement and retrofit projects assumed for this study were simplified for analysis brevity. It should be noted the project costs and durations would depend on the hazard mitigated by the project and construction access. For example, the cost and duration of a Pier 3 retrofit for the 2017 landslide would have likely been greater than the cost and duration for the 1996 seismic retrofit.

The risk analysis assumed the consequences of a planned retrofit or replacement project to address structural deterioration would be less than the consequences of an emergency project to address bridge failure. The costs attributed to lost travel spending could be reduced, given sufficient time prior to construction, by providing a temporary detour during a planned retrofit or replacement project. Conversely, the time constraints of emergency bridge projects in steep mountainous terrain, like the 2017 Pfeiffer Bridge replacement project, often don't allow enough to time to provide temporary vehicular detours.

Caltrans' planned retrofit or replacement projects often involve construction stages to allow traffic through the existing route during construction, thereby minimizing traffic disruptions and the lost travel spending associated with those traffic disruptions. The risk analysis assumed a planned bridge replacement project would result in the equivalent of one month (i.e., 30 days) of lost travel spending and the impact was estimated to be approximately \$52M by reducing Visit California's (2017) estimate by a factor of 0.09 (30/320 = 0.09). The impact of a planned retrofit project was estimated to be approximately \$26M (i.e., half the estimated impact for a planned replacement project).

Decision Tree

Figure 7 presents the decision tree developed to evaluate the risk of bridge failure resulting from the hazards discussed above. The results of the analysis show that the risk of emergency bridge retrofit or replacement due to failure from landsliding over a 75-year design life is approximately \$53.2M (\$2.8M+\$50.4M), which is approximately \$18.4M greater than the risk of structural deterioration over a 75-year design life (\$11.6M + \$23.2M = \$34.8M).

The risks of bridge failure due to scour (\$9.1M+\$42.6M = \$51.7M) and strong ground motion (\$1.6M+\$28.9M = \$30.5M) may have been considered reasonable risk thresholds in this idealized scenario because the probabilities of these design events are defined by Caltrans guidelines. Therefore, Caltrans may have deemed the risk of bridge failure due to landsliding as unacceptable, relative to one or both of these risk thresholds. A replacement bridge would be designed to reduce the risk of landsliding and the risk of additional hazards, such as debris flows, at the same time. The replacement project would be planned to minimize the consequences of replacement and the associated risk (i.e., lost travel spending associated with traffic disruptions during construction).

The risk of bridge retrofit or replacement due to failure from landsliding is approximately \$22.7M greater than the risk of bridge failure due to strong ground motion. A similar risk analysis performed prior to the 1996 seismic retrofit of the bridge would have likely resulted in prioritizing the mitigation of landslide risk. However, as noted above, the original bridge was likely designed for a less conservative seismic hazard than the hazard recommended by current

Caltrans guidelines; therefore, the probability of strong ground motion resulting in bridge failure may have been greater in 1996 and the associated risk may have been greater than the $30.5M risk estimated for this study.

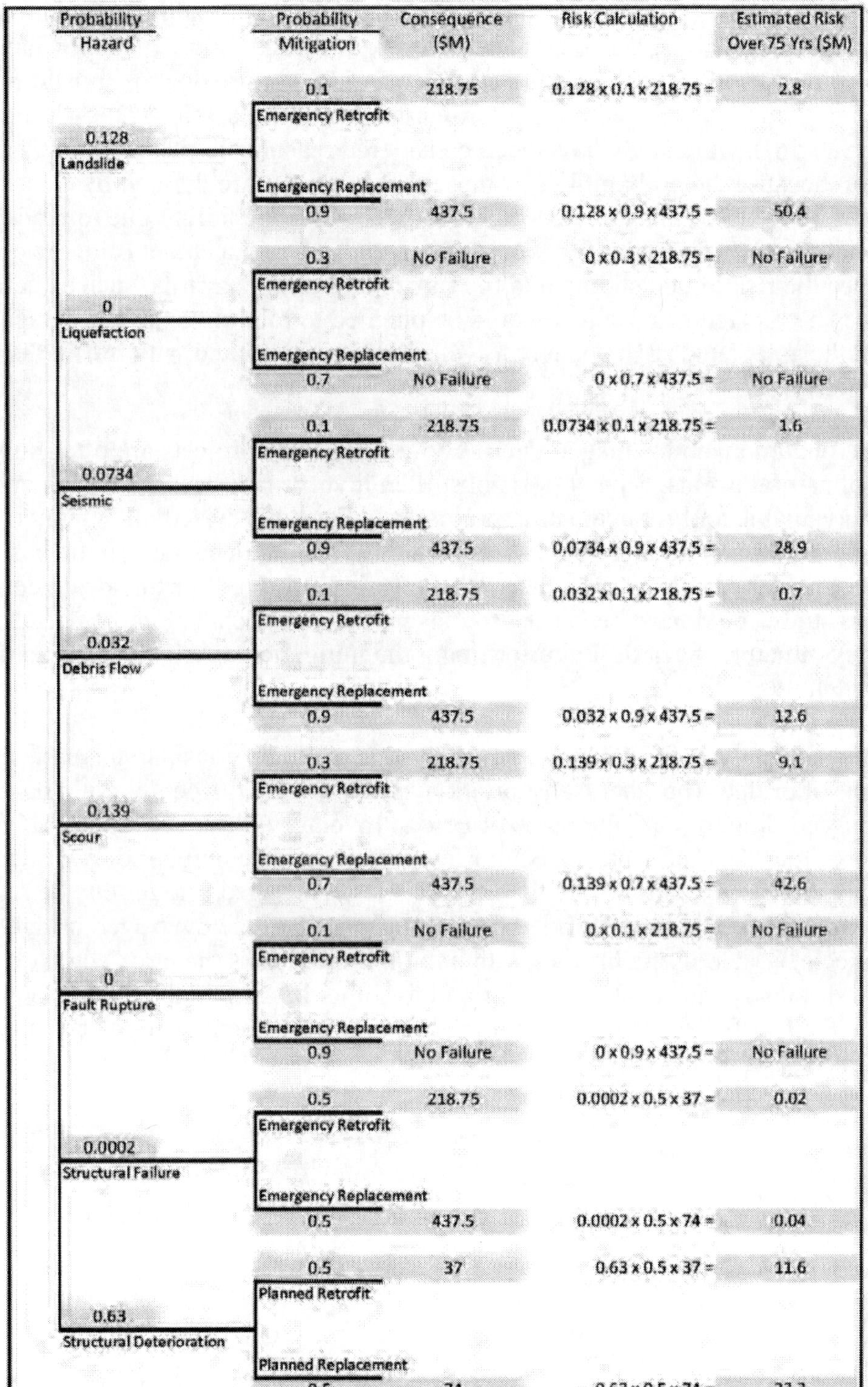

Figure 7: Risk Analysis Decision Tree

SUMMARY AND CONCLUSIONS

Statistical analyses were performed to approximate the distribution of shear strength for the Franciscan Complex bedrock and bridge pile shear strength to estimate the probability of landsliding and failure of the bridge as a result of landsliding. The probability was input as a decision tree branch in a risk analysis to evaluate the risk of bridge failure posed by landsliding. The results of the analysis indicate the estimated risk of bridge failure due to landsliding ($53.2M) is greater than the estimated risks of bridge failure due to scour ($51.7M) and strong ground motion ($30.5M), which may have been considered reasonable risk thresholds in this idealized scenario because the probabilities of these design events are defined by Caltrans guidelines. Therefore, Caltrans may have deemed the risk of bridge failure due to landsliding as unacceptable, relative to one or both of these risk thresholds. A replacement bridge would be designed to reduce the risk of landsliding and the risk of additional hazards, such as debris flows, at the same time. The replacement project would be planned to minimize the consequences of replacement and the associated risk (i.e., lost travel spending associated with traffic disruptions during construction).

Planning, design, operation, and management of civil engineering projects are often subject to the uncertainty of natural events. Project decisions often involve risk because of this uncertainty. Risk analysis is a valuable tool for evaluating project decisions in a systematic manner, with the goal of reducing uncertainty and thereby risk, and making rational decisions on the basis of objective criteria, such as risk thresholds. The risk analysis performed for the idealized scenario considered in this study could have been used to make a rational decision on the need for mitigating the risk of natural hazards and prioritizing the mitigation alternatives based on objective risk thresholds defined by Caltrans.

The results of the risk analysis for Pfeiffer Canyon Bridge could be considered a case study for future evaluations of bridges (or other transportation assets) within the context of risks posed by natural hazards. According to the Caltrans 2014 Bridge Inventory, there are 2,847 bridges in California that are older than the typical design life of 75 years. The number of over-aged bridges will likely increase at a faster rate than the existing over-aged bridges can be replaced or retrofitted. Risk analyses could be performed by Caltrans to prioritize over-aged bridge replacement projects by identifying bridges with natural hazard risks that are deemed unacceptable. Additionally, risk analyses performed for other bridges in the inventory may provide rationale (and sufficient time) for Caltrans to plan replacement or retrofit of a bridge that is younger than its design life.

REFERENCES

- American Association of State Highway Transportation Officials (AASHTO 2017), LRFD Bridge Design Specifications, 8th edition, American Association of State Highway and Transportation Officials, Washington, DC.
- Dai, S.H., and Wang, M.O. (1992), *Reliability Analysis in Engineering Applications*. Van Nostrand Reinhold, New York.
- Dibblee, Thomas W. (2007), Geologic Map of the Point Sur, Big Sur, and Pfeiffer Point Quadrangles, Monterey County, California, Dibblee Geology Center Map #DF-348, Scale: 1:24000.
- Ellingwood, B., Galambos, T.V., MacGregor, J.G., and Cornell, C.A. (1980). "Development of a Probabilistically-Based Load Criterion for the American National Standard A58." Special Pub. 577, National Bureau of Standards, Washington D.C. June.
- Moss, R. E. S. (2013), Applied Civil Engineering Risk Analysis. Shedwick Press.
- Singh, V.P., Sharad, K.J., and Aditya, K.T. (2007), Risk and Reliability Analysis: A Handbook for Civil and Environmental Engineers. ASCE Press.
- Visit California (2017), *California Highway 1 Closure Economic Impact Study Preliminary Presentation of Findings,* research prepared by Dean Runyan Associates and Destination Analytics for Visit California.

US 60 Pinto Creek Bridge Foundation Optimization with Emphasis on Cost and Constructability in a Challenging Geologic Environment

Daniel N. Fréchette, PhD, PE
Principal Geotechnical Engineer
Wood Environment & Infrastructure Solutions, Inc.
4600 East Washington Street, Suite 600
Phoenix, Arizona 85034
(602)-733-6029
daniel.frechette@woodplc.com

And

Patrice P. Brun, PE
Geotechnical Design Engineer
Arizona Department of Transportation, Bridge Group – Geotechnical Services
Infrastructure Delivery and Operations Division
205 South 17th Avenue, Room 217E, Mail Drop 613E
Phoenix, Arizona 85007
(602)-712-8099
pbrun@azdot.gov

Prepared for the 70th Highway Geology Symposium, October, 2019

Acknowledgements

The author would like to thank the individuals for their contributions in the work described:

Tony Freiman, PE – Wood Environment & Infrastructure Solutions, Inc.
Rafael (Rafe) Davis, PE – ADOT Bridge Group

Disclaimer

Statements and views presented in this paper are strictly those of the author(s), and do not necessarily reflect positions held by their affiliations, the Highway Geology Symposium (HGS), or others acknowledged above. The mention of trade names for commercial products does not imply the approval or endorsement by HGS.

Copyright Notice

Copyright © 2019 Highway Geology Symposium (HGS)

All Rights Reserved. Printed in the United States of America. No part of this publication may be reproduced or copied in any form or by any means – graphic, electronic, or mechanical, including photocopying, taping, or information storage and retrieval systems – without prior written permission of the HGS. This excludes the original author(s).

ABSTRACT

The existing US 60 Pinto Creek Bridge, located in a scenic part of rural eastern Arizona within the Tonto National Forest, is structurally deficient, functionally obsolete and needs to be replaced. The existing bridge crosses Pinto Creek, which is an incised canyon with steep rock slopes and narrow approaches. This segment of the US 60 is a vital transportation corridor for residents within the area and the regional mining industry, which require the existing roadway and bridge to remain operational during construction of the new bridge. The narrow approaches resulted in the proposed bridge being located within 30 feet of the existing bridge and almost immediately adjacent at one of the abutments. In addition, the existing ground varies in elevation up to 35 feet across a single foundation element due to the steep topography within the incised canyon, thus complicating construction due to limited access and the need to mobilize large equipment. Considerations evaluated during design included foundation type, construction equipment access, temporary slopes, temporary shoring, scour potential, historic mine workings, impacts to the environment and historical considerations, and construction costs. The design team evaluated a variety of foundation types including spread footings, small-diameter drilled shafts, and large-diameter drilled shafts (greater than 8 feet) to optimize the design for the new bridge. In the end, the robust design effort resulted in a cost effective and constructible design in the challenging rock environment that was able to satisfy all project stake holders.

INTRODUCTION

The existing US 60 Pinto Creek Bridge, located in a scenic part of rural eastern Arizona within the Tonto National Forest, is structurally deficient, functionally obsolete and needs to be replaced. The existing bridge is a nine-span, steel arch bridge constructed in 1949. The main arch span is 350 feet long with the total length of the bridge being 750 feet. The existing bridge spans Pinto Creek, which is an incised canyon with steep rock slopes and narrow approaches. This segment of the US 60 is a vital transportation corridor for residents within the area and the regional mining industry, which require the existing roadway and bridge to remain operational during construction of the new bridge. The narrow approaches resulted in the proposed bridge being located within 30 feet of the existing bridge and almost immediately adjacent at one of the abutments. In addition, the existing ground varies in elevation up to 35 feet across a single foundation element due to the steep topography within the incised canyon, thus complicating construction due to limited access and the need to mobilize large equipment.

GEOLOGIC SETTING AND GEOTECHNICAL PROFILE

Regional Geologic Setting

The project is located near the junction of the Mexican Highland and Sonoran Desert sections of the Basin and Range physiographic province, just to the southwest of the boundary between the Basin and Range and Transition Zone physiographic provinces (1). The project is located along the northern flank of the Pinal Mountains an area dominated by granitic intrusions. The project location is shown on Figure 1.

The rock unit exposed in the project area is the Schultze Granite which was intruded as a granitic stock during the late Cretaceous to early Tertiary Laramide orogeny. Many of these stocks in the region contain porphyry copper deposits that are actively mined and copper mineralization is present along many fracture surfaces. Most of the Schultze granite stock consists of porphyritic quartz monzonite with a fine- to coarse-grained groundmass and 1/2- to 2-inch-diameter feldspar phenocrysts. The granite is locally cut by aplite dikes and dikes of foliated granite.

Erosion of the bedrock units has resulted in the deposition of colluvium and debris slides (Qds) on hillsides and alluvium along drainages within the project corridors. Along some portions of the roadway, the bedrock units were cut and benched to provide a flat surface for roadway construction. The rock that was excavated from the cuts was typically wasted on the downslope side of the road commonly resulting in 10- to 30-foot-thick sections of embankment fill on the side of the road opposite of the cuts. The geologic units, in relation to Pinto Creek Bridge are shown on Figures 2 and 3.

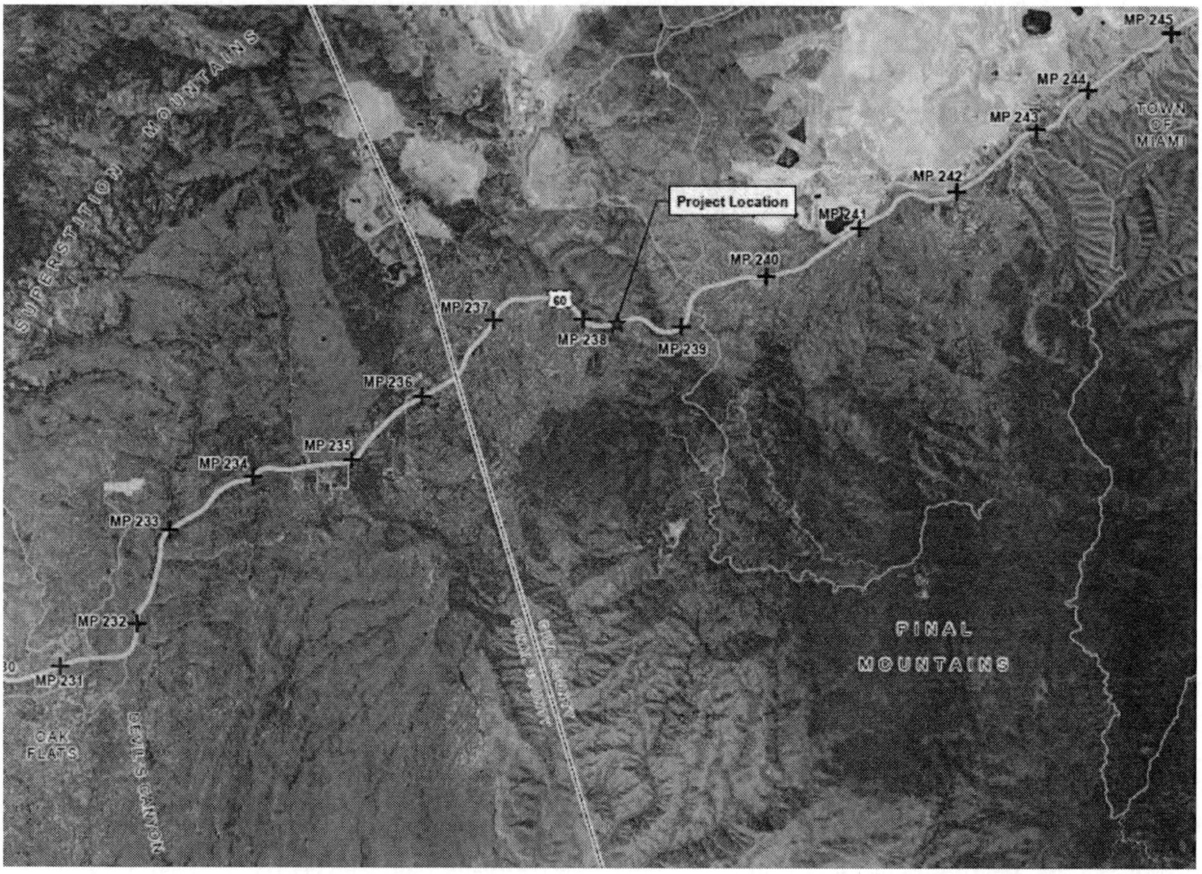

Figure 1 – Project Location Map of Pinto Creek Bridge Area

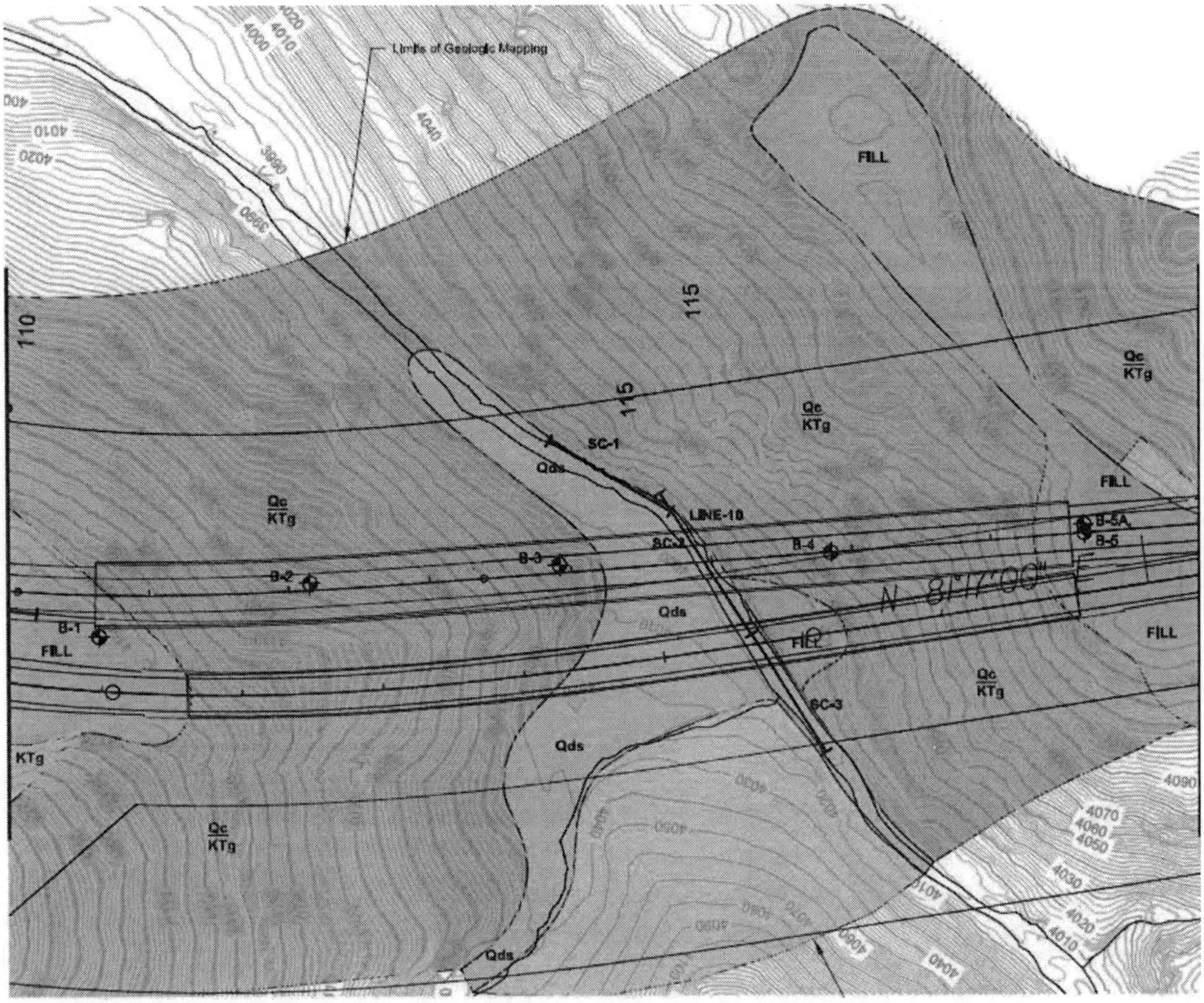

GEOLOGIC UNITS

FILL	Silty to clayey sand, gravel & cobbles with occasional boulders at the ground surface, nonplastic with medium plasticity zones & brown to gray.
Qal	Alluvial Stream Channel Deposits of sand, gravel & cobbles with sand zones, occasional boulders & clay zones, uncemented, nonplastic to low plasticity, light brown.
Qds	Debris slides of silty sand, gravel & cobbles with considerable boulders up to several feet in diameter, nonplastic & light brown. note: older debris slides appear weakly cemented.
Qc	Colluvium of silty to clayey sand & gravel with varying amounts of cobbles & boulders up to 3' in diameter, nonplastic to highly plastic & various shades of brown & gray.
KTg	Schultze Granite of fine grained to porphyritic texture with quartz-rich bands & alternating very hard & very soft zones. Shear zones, slickensides along fracture surfaces & hydrothermal alteration zones are common throughout the rockmass. The granite typically is various shades of speckled gray to yellowish-brown.

Figure 2 – Geologic Map of Pinto Creek Bridge Area

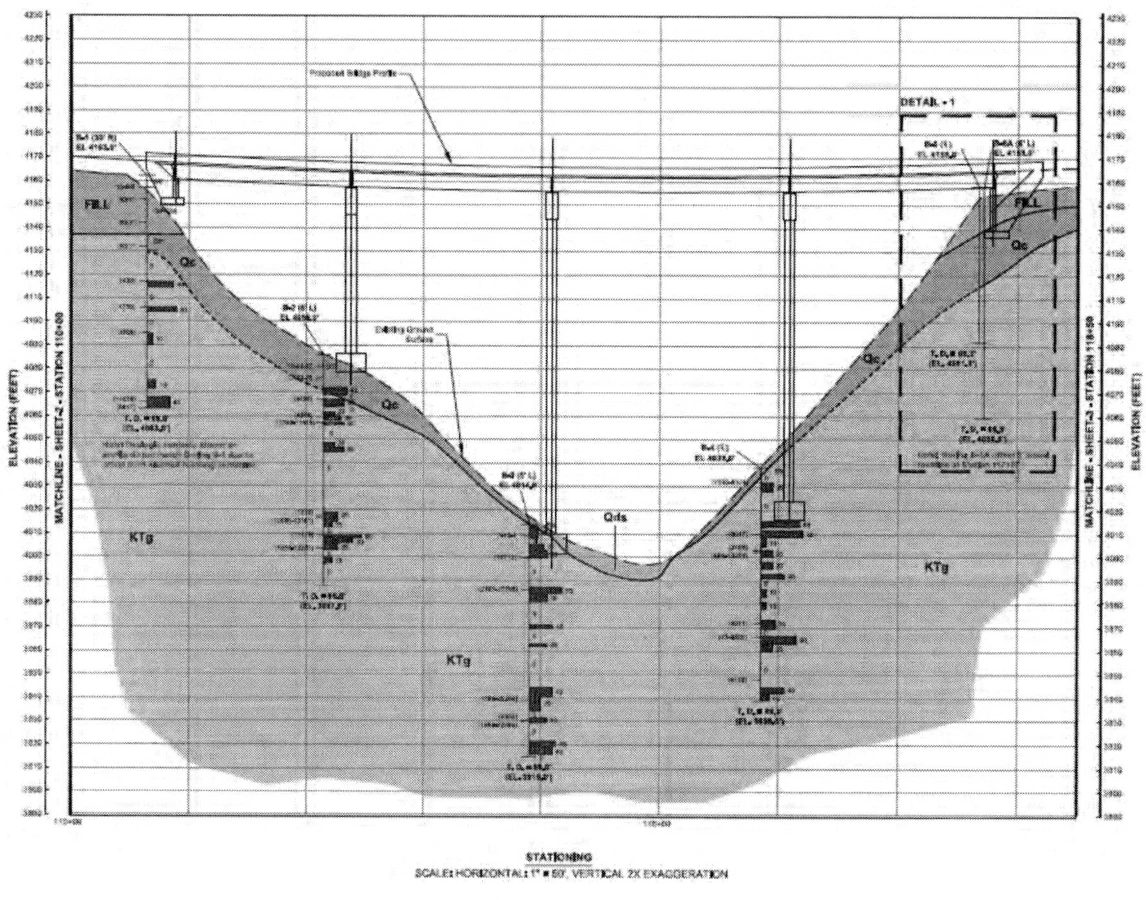

Figure 3 – Roadway Profile Showing Geologic Cross Section at Pinto Creek Bridge

Local Geologic Setting

Details of the units encountered at the Pinto Creek Bridge site are presented in the following sections.

Schultze Granite (TKg)

Schultze granite is exposed throughout the project corridor and is exposed in each of the six existing road cuts. Surface exposures of the granite are primarily slightly to moderately weathered, hard to very hard and gray to brown. The rockmass is characterized by two dominant joint sets. Joints typically are closely (2 inches to 1 foot) to moderately-closely (1 to 3 feet) spaced. Granite was encountered in all twelve of the borings advanced throughout the corridor. The boring locations are shown on the plans, profiles and cross sections on Sheets 1 through 3 along with rock quality designations (RQDs) of the core retrieved from the borings and results of point load index and UCS laboratory tests performed on selected core samples. The depth to bedrock along the corridor varies from one foot bgs to 45 feet bgs.

Colluvium (Qc)

Colluvial deposits occur on the hillsides adjacent to Pinto Creek and atop the planned cut section of the project, located east of the bridge. The colluvial deposits generally are derived from the in-place weathering of the underlying bedrock units and are of variable thickness across the corridor. A thin, 2 feet thick veneer of colluvium mantles the planned cut section area whereas upwards of 25 feet of colluvium was encountered along the planned bridge alignment. The colluvium generally consists of silty to clayey sand and gravel with varying amounts of cobbles and boulders up to 3 feet in diameter. Typically the colluvium is uncemented. Colluvial deposits along the new bridge alignment vary from nonplastic to highly plastic and generally are nonplastic along the cut section. The colluvium is typically described as various shades of brown and gray. Refraction seismic interpretations typically indicate p-wave velocities of 1,100 to 1,200 feet per second (fps) and s-wave velocities of 500 to 600 fps.

Alluvium (Qal and Qds)

Alluvial deposits occur along and within a couple of drainages within the project corridor. These deposits consist of channel deposits (Qal) and Qds. One small drainage located at the west end of the corridor contains Qal and the larger channel (Pinto Creek) contains Qds, which are probably underlain by Qal within the creek area. Qal are confined to the floors of the drainages and generally consist of poorly graded sand, gravel and cobbles with varying amounts of silt. Occasional boulders and isolated clay zones occur within these deposits. These deposits are uncemented, nonplastic to low in plasticity and light brown. Refraction seismic interpretations for the alluvium indicate p-wave velocities of 1,300 to 2,600 fps and s-wave velocities of 440 fps.

Qds deposits, which occur very locally at the base of steep hills, are a result of sliding or rolling rocks and other unconsolidated materials. These deposits generally consist of silty sand, gravel and cobbles with considerable boulders up to several feet in diameter. The older slides appear weakly cemented and generally are nonplastic and light brown.

Man-made Fill (Fill)

Most of the fill deposits within the project corridor occur as embankment fills along the roadway. In most cases, the bedrock units were cut and benched to provide a flat surface for roadway construction. The rock that was excavated from the cuts was typically wasted on the downslope side of the road commonly resulting in 10- to 30-foot-thick sections of embankment fill on the side of the road opposite the cuts.

The fill materials generally consist of sand, gravel and cobbles with varying amounts of clay, silt and boulders. The fill materials typically are uncemented and nonplastic, however, an isolated pocket of medium plasticity clayey material was encountered on the east abutment (Abutment 2). Fills encountered within borings are various shades of brown, gray and red.

Refraction seismic interpretations for the fill deposits were variable with p-wave velocities generally ranging from about 1,400 to 3,000 fps and s-wave velocities ranging from about 600 to 2,300 fps, with a few s-wave velocities ranging up to 3,000 fps.

Geotechnical Profile

The replacement bridge is a four-span, combination steel I-girder bridge with haunched sections at the pier. The new bridge alignment extends from Stations 110+63 to 117+58 (Sheet 2) and is located north (left of centerline) of the existing bridge. The new alignment is approximately 70 feet left of existing centerline at the west abutment (Abutment 1) and 45 feet left of centerline at the east abutment (Abutment 2). Six borings (B-1 through B-5 and B-5A) were completed within the bridge area. All six borings were advanced through unconsolidated deposits of fill and colluvium into bedrock. The depth to bedrock varied from 1 to 45 feet bgs. Profiles showing the contact between the fill and colluvial deposits and bedrock are presented on Sheet 2.

Fill materials along the planned bridge alignment were encountered at the ground surface at the new abutment locations. These materials generally are comprised of uncemented sand, gravel and cobbles with occasional boulders up to 18 inches in diameter and varying amounts of silt and clay. The fill deposits are likely derived from adjacent road cuts and the cobble- to boulder-sized materials are rock fragments excavated from the cuts, probably by blasting. Laboratory test results indicate these fill materials generally are nonplastic at the west abutment and nonplastic to medium plasticity at the east abutment. Standard penetration tests (SPTs) obtained within the fill materials yielded a range of N-values (summation of the second and third six-inch interval blow counts) ranging from six to refusal blow counts within the initial six-inch interval. Typically the softer fill zones were encountered at relatively shallow depths and the fill sections appeared firmer with increasing depths.

Colluvium was encountered at all of the planned pier and abutment locations for the new bridge alignment. Colluvial materials were encountered at the ground surface at the planned pier locations, at 25 feet bgs at the Abutment 1 and 20 feet bgs at the Abutment 2. The colluvium is described as uncemented, silty to clayey sand and gravel with varying amounts of cobbles and boulders up to 3 feet in diameter. Some thin clay zones were encountered within the colluvium at Abutment 2 locale (Boring B-5). Laboratory test results indicate the colluvial deposits along the new bridge alignment vary from nonplastic to medium plasticity. SPTs obtained within colluvial materials yielded N-values ranging from 42 to refusal within the initial six-inch increment.

Granite was encountered within all the borings completed along the new bridge alignment. Granite recovered from these borings is generally of poor rock quality and is characterized by a fine-grained to porphyritic texture. Quartz-rich bands are common within the rockmass and typically are various shades of speckled gray. The rockmass contains alternating zones of slightly to moderately weathered, hard to very hard rock intervals alternating with soft to very soft, decomposed zones throughout the profile. Some of the decomposed zones are comprised of clay and granite fragments. RQD values are commonly zero and generally less than 40. UCS test and point load index test results indicate highly variable rock strength values ranging from about 160 to 25,100 psi. Dry densities of core samples generally ranged from 152 to 162 pounds per cubic foot (pcf). Shear zones are common throughout the rock profile and typically associated with smooth to slightly rough, fracture surfaces with slickensides. Fracture surfaces typically are filled with either clay or lined with quartz.

The seismic refraction interpretations generally matched the results of the exploration borings with regard to the depth to bedrock. The estimated depth to bedrock along Line 11 and within Boring B-1, performed near Abutment 1 (west side of Pinto Creek) ranged from about 12 to 32 feet bgs. The bedrock contact slopes toward the bridge resulting in thicker sections of fill near the existing bridge abutment. The estimated depth to bedrock along Line 10 performed within the drainage bottom of Pinto Creek ranged from about 10 to 20 feet bgs. Interpreted p-wave velocities of the granite were highly variable generally ranging from about 5,100 to 11,700 fps and s-wave velocities generally ranging from about 2,600 to 4,800 fps.

SUBSURFACE INVESTIGATION

The subsurface investigation performed to characterize the geologic material at the project site consisted of geologic mapping, refraction seismic surveys, exploration drilling, laboratory testing, optical televiewer, piezometer installation, and scan lines. A brief overview of the each method is presented below. Prior to the subsurface investigation a review of existing data, consisting of reviewing published geological maps, aerial photographs, topographic maps, and as-built plans of the existing roadway and bridges was completed.

Geologic Mapping

Surficial geologic mapping of the project corridor included the delineation and characterization of soil and rock units and documentation of the general rock mass conditions exposed within the project corridors. The surficial distribution of soil and rock units within the limits of the Pinto Creek Bridge are shown on Figure 2. Documentation of rock mass conditions included rock type, hardness, weathering, and discontinuity characteristics pertinent to potential foundation conditions and slope stability. In addition, evaluation of existing cut slopes was performed consisting of descriptions of the soil and rock types exposed; measurements of the slope angles, orientations and heights of the cut slopes; and evaluations of the slope performance at existing slope angles.

Refraction Seismic Surveys

Eleven refraction seismic surveys were completed along the project corridor to assess geologic and geotechnical conditions within planned cuts and existing fill sections. Refraction seismic survey results were used to: aid in characterizing the subsurface geotechnical profile, provide general strength parameters for various subsurface materials, assist with developing earthwork factors, and provide information on the general rock mass conditions of the rock units. The refraction seismic surveys were 120-foot in length and utilized both compression wave refraction and shear wave refraction microtremor (ReMi) methods. One-dimensional (vertical) shear wave velocity profiles of the subsurface site soils were determined. The interpreted investigation for the seismic lines ranged from about 15 to 35 feet and 20 to 55 feet for the p-wave and s-wave, respectively.

Exploration Drilling

Twelve borings were drilled along the project corridor to characterize subsurface conditions of which six were completed along the new bridge alignment. The borings were drilled with portable drill rigs transported by helicopter. The borings were advanced using a HWT casing advancer, HQ- and NQ-sized, wireline, diamond-bit, rock coring system in a vertical orientation. Two of the borings along the planned cut section of the project were downhole scanned with an optical televiewer to identify discontinuities and other features, which were used for slope stability analyses. In addition, a piezometer was installed in one of the bridge pier borings to monitor potential water levels adjacent to the bridge alignment.

Laboratory Testing

Laboratory testing of materials collected from the borings and scan lines was performed to support engineering analyses. Testing included grain-size analysis in accordance with ASTM D2487, Atterberg limits (liquid limit and plasticity index) in accordance with ASTM D4318/ASTM D2487, subgrade R-value in accordance with ASTM D2844, moisture-density relationship (standard Proctor) in accordance with Arizona Test Method (AZ) 225, moisture content in accordance with ASTM D2216, pH, and minimum resistivity determinations in accordance with AZ 236, chloride content with AZ 736, and sulfate content AZ 733. Testing of rock core samples consisted of unconfined compression strength (UCS) in accordance with ASTM D2166 and point load index tests in accordance with ASTM 5731-08. The density of the core samples was determined using the volumetric method.

Scan Lines

Three scan lines were completed along the floor of Pinto Creek within the active stream channel to characterize the gradation of large particles on that could not otherwise be effectively sampled. Each scan line was 100 feet in length, as measured by a cloth tape laid out on the streambed. The size of the particle lying immediately below each foot "marker" on the tape was measured in the intermediate or b-axis (width), where the a-axis is a particle's longest dimension and the c-axis is the particle's thickness and shortest dimension. Each measurement was recorded for later analysis using a method based on Kellerhals and Bray (2). The purpose of the scan lines was to provide a particle size distribution to assist in characterization of the potential for scour at the bridge foundations.

INITIAL BRIDGE FOUNDATION DESIGN

Site Considerations and Constraints

Many variables were considered when deciding on the foundation type for the Pinto Creek Bridge: soil/rock conditions, scour concerns, presence of an abandoned mine, environmental impacts and historical considerations, aesthetics, presence of the existing bridge, and constructability. As part of the evaluation, consideration was given to the possibility to mix and match the foundation types across the bridge structure.

Soil/Rock Conditions

The geotechnical profile at the Pinto Creek Bridge consists of varying amounts of fill or colluvial soils overlaying the granite. The thickness of these soils at the abutment locations were between 30 and 45 feet. Pier 1 has approximately 16 feet of fill and/or colluvium, Pier 2 has 1 foot and Pier 3 has approximately 5 feet. The granite is predominantly moderately to slightly weathered with highly weathered to decomposed zones. The granite generally has a rock quality designation (RQD) value of zero or near zero due to a high degree of fracturing and clay in-fill zones. The exception is the slightly weathered zones with high RQD values. Due to the variability in fill depths and rock strengths, the elevation of the top of suitable rock varied for each bridge foundation element. The elevation of the top of suitable rock varied from elevations of 4,013 to 4,130 feet.

The general poor quality of the granite resulted in the drilled-shaft foundations being designed using the intermediate geomaterial (IGM) procedure within the highly fractured and variably weathered granite instead of rock sockets.

Scour Concerns

An initial analysis of the potential for scour was conducted for the project. The initial analysis concluded that there is no interaction between the 500-year flow and the proposed structure. Furthermore, channel contraction is not expected and long-term channel degradation and lateral migration were considered to be negligible. These aspects eliminate the significant contributors to scour.

Historic Mine Workings

The area surrounding the Pinto Creek Bridge has a long history of mining activity due to the presence of porphyry copper deposits within the Schultze Granite. A review of the record drawings for the existing bridge shows the presence of an underground mine adit beneath and near Pier No. 7 of the existing bridge, which is in the vicinity of Pier 3 of the new Pinto Creek Bridge, See Figure 4. Portions of the mine adit were grouted during construction of existing bridge, but only those portions in the immediate vicinity of the foundation for Pier No. 7.

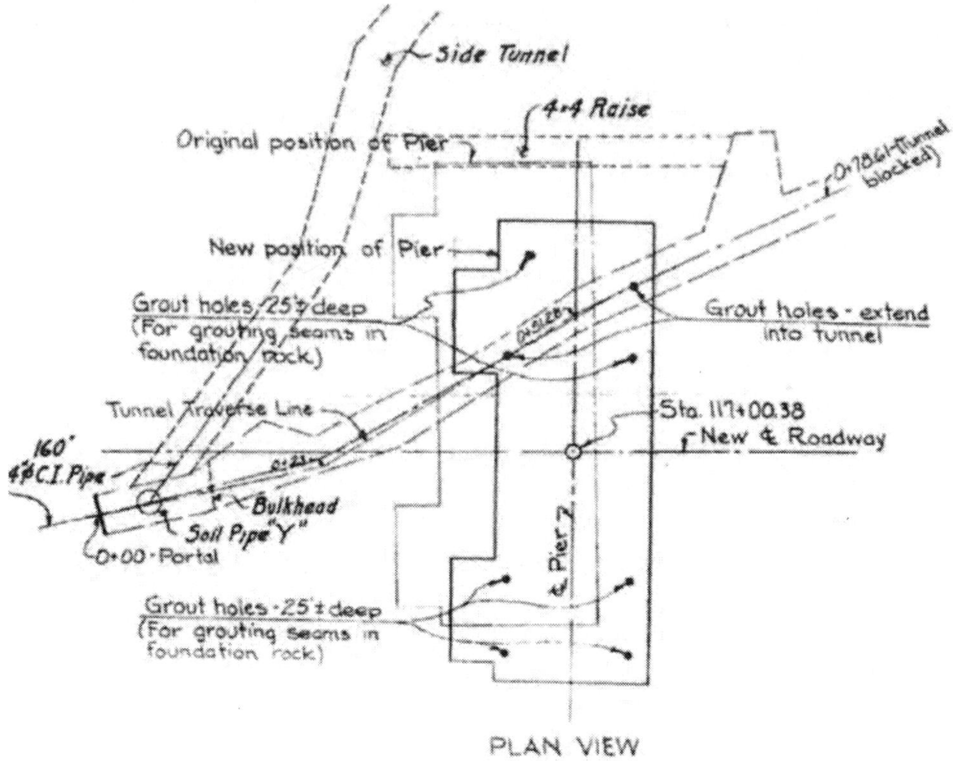

Figure 4 – Mine Adit beneath and near Pier No. 7 of the Existing Pinto Creek Bridge

Environmental Impacts and Historical Considerations

The Pinto Creek Bridge area is located in an environmentally sensitive area that most importantly includes an endangered species of cactus known as the Arizona Hedgehog. In addition, Pinto Creek is a waters of the United States requiring efforts to minimize the disturbance within the creek limits. Pinto Creek is listed on Arizona's list of impaired waters due to exceedances of naturally occurring dissolved copper (3).

The largest impact on the environment was anticipated to be the excavation limits for the foundation elements and temporary construction access roads to the bridge pier locations given the steep incised slopes of the canyon. During removal of the existing bridge, precautions to prevent the release of lead and asbestos during demolition will be implemented. Also, restrictions during construction may be required between September 1 and February 28 if bird nest sites are located in local trees.

Other site considerations include that the existing bridge is designated historic as per the Arizona State Historic Preservation Office while located along the designated Gila-Pinal Scenic Road as part of the network of America's Scenic Byways.

Aesthetics

The Pinto Creek Bridge is located within the Tonto National Forest. The Forest Service's project goal was to maintain the natural beauty of the area and minimize the impact of the construction and project on the surrounding area.

Presence of Existing Bridge

The US 60 is an important transportation corridor that needed to remain open during construction. Therefore, the existing bridge had to remain in place until construction of the new bridge was complete. The approaches to the existing bridge are narrow and didn't leave much room for an alternate alignment of the new bridge. The narrow approaches resulted in the proposed bridge being located within 30 feet of the existing bridge and almost immediately adjacent at one of the abutments. The close proximity of the existing bridge meant that the construction of the new bridge couldn't impact the foundations of the existing bridge.

Constructability

Ensuring the design could be constructed in light of consideration of all of the other controlling factors was a key consideration. These considerations including temporary construction access, temporary construction slopes, availability of contractors to perform the work among others. These were exacerbated by the steep terrain across each footing/pile cap, which resulted in as much as a 35-foot elevation change across the length/width of the footing/pile cap. This created a challenge for the construction of a spread footing or pile cap for a group of small-diameter drilled shafts.

Summary of Foundation Selection

Foundation types considered for the initial design of the bridge included spread footings, small-diameter drilled shafts (diameters less than 8 feet) and large-diameter drilled shafts (diameters equal to or greater than 8 feet). Small-diameter drilled shaft foundations were selected for Abutment Nos. 1 and 2 due to the presence of over 30 feet of fill soils at the bridge locations, which were not deemed suitable for supporting the bridge loads on spread footings at a shallow enough depth.

Spread footings were considered for the piers, but were quickly eliminated due to several factors including deep fill thickness at some locations, presence of the mine adit, proximity of the existing bridge to the proposed bridge, large excavations would disturb more area and increase the environmental impacts and decrease the aesthetics of the area however. Therefore, drilled shafts were selected as the preferred alternative. Large diameter drilled shafts were selected due to the high lateral and axial demands of the bridge piers. Since drilled shafts were the preferred alternative for some of the bridge piers, they were selected for all piers to simplify the construction and achieve savings associated with having a larger quantity to construct.

REEVALUATION OF BRIDGE FOUNDATION DESIGN

The project was advertised and ADOT received multiple bids. However, the low bid was significantly higher than the engineer's estimate and the department's programmed amount. ADOT completed a detailed evaluation of the bids and identified areas where they thought some cost savings could be recognized. Based on this, ADOT decided to reject all bids and regather the design team to reevaluate the design to identify cost savings with emphasis on the foundation design, specifically the large diameter drilled shafts at the piers. The unit prices for these elements of the project were higher than anticipated. Therefore, all of the initial assumptions used to identify large-diameter drilled shafts as the preferred foundation type were reevaluated.

Site Considerations and Constraints

Soil/Rock Conditions

The soil and rock conditions at the project site didn't change. However, more detailed analyses were completed to assess the feasibility of using spread footings at the pier locations including bearing capacity on or adjacent to slopes and slope stability analyses. These analyses confirmed that spread footings were not feasible at Pier No.1, but could be considered at Pier Nos. 2 and 3.

Scour Concerns

The initial scour analysis concluded that scour was negligible for the project site. The analysis was reevaluated considering the potential for using spread footing foundations for Pier No. 2, which is founded near the invert of Pinto Creek. The reanalysis confirmed the initial findings that scour was negligible. Scour concerns did not play a role in the final foundation selection.

Historic Mine Workings

A more detailed evaluation of the location of the existing mine adit located near the proposed Pier No. 3 was completed. This was done by overlaying the record drawings on to the project plans, See Figure 5.

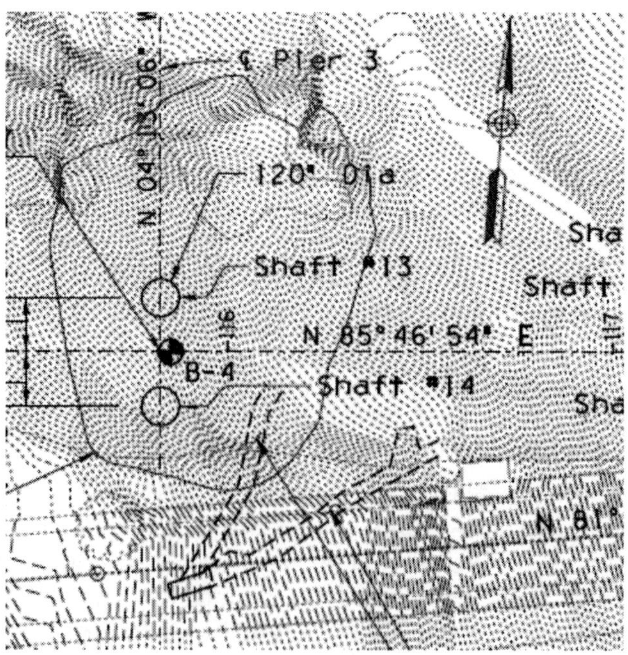

Figure 5 – Mine Adit near Proposed Pier No. 3

Based on the review there is a clear possibility the mine adit is beneath proposed Pier No. 3. Furthermore, the mine adit may have been exposed within the required excavation to construct a spread footing foundation at the proposed Pier No. 3. This would create an unsafe condition during construction that was unacceptable.

Environmental Impacts

Environmental impacts of using spread footing foundations, a group of small-diameter drilled shafts, or large-diameter drilled shafts was evaluated primarily through identifying the amount of disturbance that would be caused during construction. An evaluation was completed at each bridge pier based on the required footing/pile cap dimensions and bearing elevation. The impacts of these are shown in a Figures 6 through 8 considering temporary slopes of 0.5H:1V in the rock and 1.5H:1V in the colluvium and fill soils.

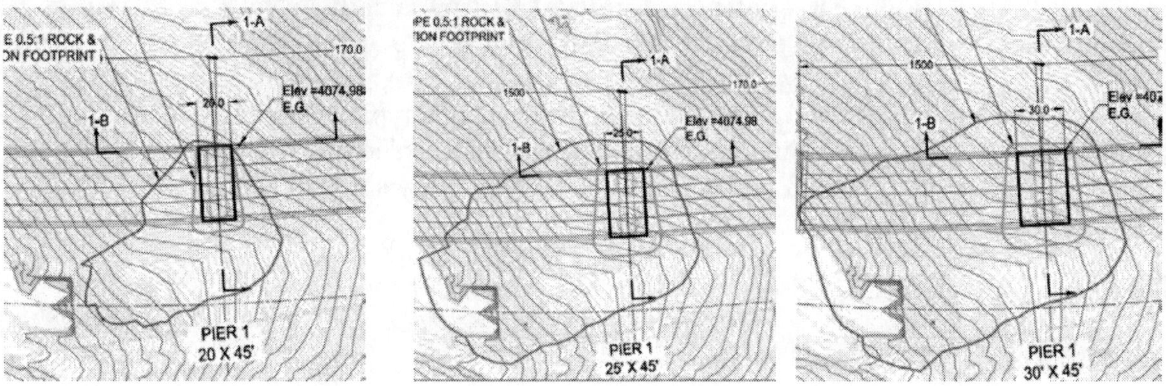

(a) (b) (c)

Figure 6 – Potential Limits of Disturbance Pier No. 1
(a) Large-Diameter Drilled Shaft (b) Small-Diameter Drilled Shaft (c) Spread Footing

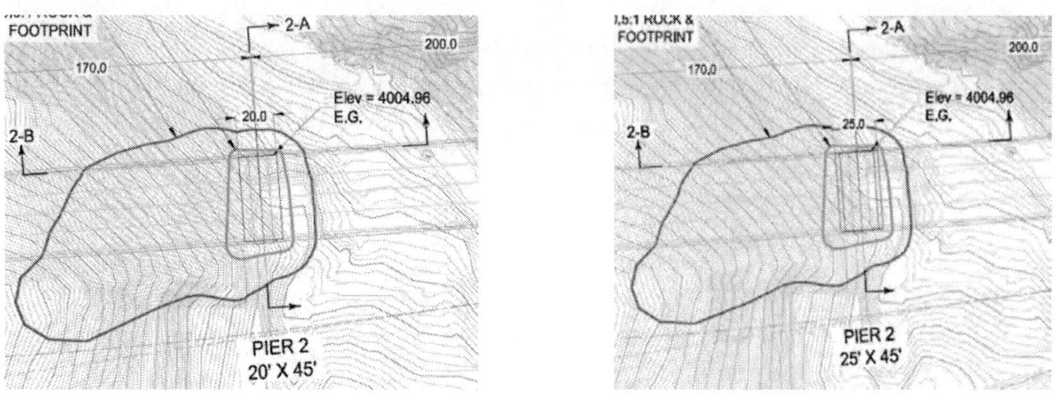

(a) (b)

Figure 7 – Potential Limits of Disturbance Pier No. 2
(a) Large-Diameter Drilled Shaft (b) Spread Footing

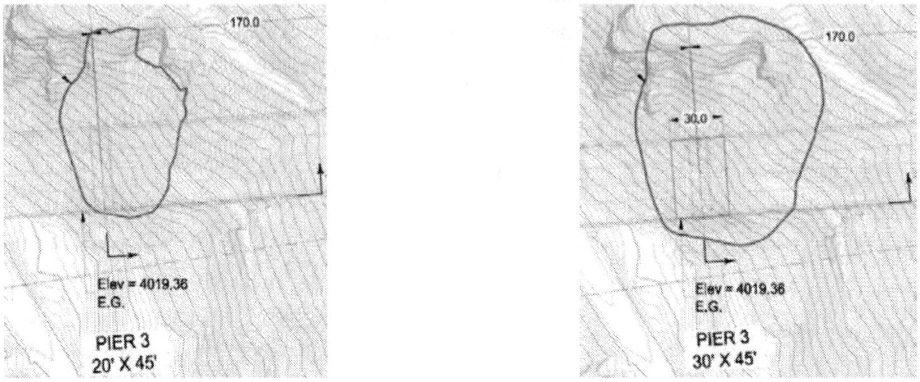

(a) (b)

Figure 8 – Potential Limits of Disturbance Pier No. 3
(a) Large-Diameter Drilled Shaft (b) Spread Footing

The spread footings and small-diameter drilled shafts typically had larger areas of disturbance due to a combination of large footing dimensions and/or lower bearing elevations as is shown in Figures 6 through 8. Alternatives using temporary soil nail shoring with steeper slopes to reduce the impacts were also evaluated.

Aesthetics

The impacts of using spread footing foundations, a group of small-diameter drilled shafts, or a pair of large-diameter drilled shafts was evaluated primarily through identifying the amount of disturbance that would be caused during construction similar to the environmental impacts. See Figures 6 through 8 for the areas of disturbance.

Presence of Existing Bridge

The impacts of using spread footing foundations, a group of small-diameter drilled shafts, or a pair of large-diameter drilled shafts was evaluated primarily through identifying the amount of disturbance that would be caused during construction similar to the environmental impacts. See Figures 6 through 8 for the areas of disturbance.

Constructability

The constructability considerations were unchanged during the reevaluation. However, these considerations were better identified and the costs associated with them were included in the overall cost analysis for the various alternatives.

Summary of Final Foundation Selection

Abutment Nos. 1 and 2

A group of small-diameter drilled shafts were selected for Abutment Nos. 1 and 2 due to the presence of significant thickness of fill soils unsuitable for the use of spread footings. The small-diameter drilled shafts were more cost effective than the large-diameter drilled shafts.

Pier No. 1

Spread footings – Highly fractured granite classified as IGM combined with steep slopes substantially reduce the bearing resistance to unfeasible levels for spread footings when taking into consideration the proximity of the slope to the footing. In addition, approximately 15 feet of colluvium on top of the highly fractured granite requires substantial excavation which envelopes the existing bridge foundations at the abutment and piers. Approximately 7,700 cubic yards of excavation were estimated to construct the spread footing without shoring.

Small-Diameter Drilled Shaft Group - A group of small-diameter drilled shafts would require approximately 6,700 cubic yards of excavation which would envelope the existing bridge foundations at the abutments and piers. A soil nail wall would be needed to keep excavation from undermining the existing bridge with an estimated cost of $360K. Furthermore, six 72-inch diameter drilled shafts with a 25-foot by 45-foot cap is estimated to cost more than the single row of 120-inch diameter shafts after accounting for excavation, shoring, and restoring slopes.

The shaft cap needs be embedded in firm undisturbed material to keep center-to-center spacing reduction factor at 1.0. If the shaft cap is embedded in loose colluvium, the center-to-center spacing reduction factors would make small-diameter drilled shafts cost prohibitive. This requires the shaft cap to have virtually the same excavation as the spread footing as it needs to be embedded into the fractured granite.

Large-Diameter Drilled Shafts – Requires minimal excavation to build 24-foot by 45-foot pad for drill rigs. Estimated excavation is 2,100 cubic yards without shoring and 600 cubic yards with shoring. Shoring may be needed to keep excavation from encroaching on the existing bridge foundations. Pier No. 3 requires a single row of 120-inch diameter drilled shafts so there is limited cost savings by using smaller shafts at this location as the larger drill rig needs to be mobilized to the project site.

Pier No. 2

Spread Footings – Spread footings may be considered since Pier No. 2 is in a flatter section of the canyon making slope stability a non-factor. Excavation is estimated at 4,000 cubic yards. Excavation and slope repair for spread footings are much greater than a single row of shafts. Scour is a concern with spread footings. However, historical photographs show little to no stream migration. Stringent scour countermeasures would be recommended as shallow spread footings are typically not used in waterways. The rock quality designation (RQD) of the highly fractured granite is far too low to classify as non-erodible.

Large-Diameter Drilled Shafts – Requires minimal excavation to build the 24-foot by 45-foot pad for drill rigs close to existing surface. Estimated excavation is 2,100 cubic yards without shoring and 600 cubic yards with shoring. Shoring maybe needed to keep excavation from encroaching on the existing bridge foundations. Pier No. 3 requires a single row of 120-inch diameter drilled shafts so there is limited cost savings by using smaller shafts at this location as the larger drill rig needs to be mobilized to the project site.

Pier No. 3

Spread Footing – The existing mine adit is in close proximity to Pier No. 3 and prevents spread footings from being an option at this location. The exact location of the mine adit is unknown due to the portal being covered up, but record drawings appear to put it within 10 feet laterally of a potential footing. A spread footing would require 7,700 cubic yards of excavation which chases up the canyon slopes increasing the area requiring restoration.

Small-Diameter Drilled Shaft Group – The existing mine adit is in close proximity to Pier No. 3 and prevents a small-diameter drilled shaft group from being an option. The exact location of the mine adit is unknown due to the portal being covered up, but record drawings appear to put it within 10 feet laterally of a potential drilled shaft cap.

Large-Diameter Drilled Shafts – Requires minimal excavation to build a 24-foot by 45-foot pad for the drill rig close to the existing surface. Estimated excavation is 1,900 cubic yards without shoring. This option reduces risk with reduced proximity to the existing mine adit.

COST EVALUATION

As noted previously, the project was rebid with the design team reevaluating the design to identify cost savings with emphasis on the foundation design. The process involved evaluating alternate foundation options at each pier location including: 120-inch diameter drilled shafts; 72-inch diameter drilled shafts, and spread footings. The evaluation process included site specific considerations such as excavation area (footprint) and overall volume, slope stability in excavation areas, shoring, soil nail wall (Pier No. 3 only), scour (Pier No. 2 only), and cost comparison of construction materials for the alternate foundation systems. The results of the analysis are presented in Table 1.

Structure	Cost Estimate Using Two 120-in. Dia. Drilled Shafts	Cost Estimate Using Six 72-in. Dia. Drilled Shafts	Cost Estimate Using Spread Footings	Cost Savings Using 120-in. vs. 72-in. Dia. Drilled Shafts	Cost Savings Using Spread Footings vs. 120-in. Dia. Drilled Shafts	Cost Savings Using Spread Footings vs. 72-in. Dia. Drilled Shafts	Considerations	Prefered Foundation System
Pier 1	$1,278,508	$1,409,373	$514,800	$130,865	$763,708	$894,573	SPREAD FOOTING NOT FEASIBLE STEEP SLOPES PROVIDE VERY POOR BEARING CAPACITY. REQUIRES MUCH DEEPER EXCAVATION, BACKFILL, AND RESTORATION.	120-in. Dia. Drilled Shaft
Pier 2	$1,539,508	$1,755,333	$456,133	$215,825	$1,083,375	$1,299,200	SPREAD FOOTING QUESTIONABLE SPREAD FOOTINGS ARE TYPICALLY NOT USED IN WATERWAYS. STRINGENT SCOUR COUNTERMEASURES REQUIRED IF SF CHOSEN.	120-in. Dia. Drilled Shaft
Pier 3	$1,423,508	$1,935,333	$489,867	$511,825	$933,641	$1,445,466	SPREAD FOOTING NOT FEASIBLE MINE ADIT IN CLOSE PROXIMITY. EXACT LOCATION AND LIMITS ARE UNKNOWN.	120-in. Dia. Drilled Shaft

Table 1 – Summary Foundation Cost Evaluation and Considerations

CONCLUSION

Construction costs for the planned Pinto Creek Bride along US 60 were too high above the programmed amount during the initial bidding process. Upon careful consideration, ADOT rejected the bids and reconvened the design team to reevaluate the project to identify potential cost savings with focus on the bridge foundation designs. The reevaluation included a variety of foundation types including spread footings, small-diameter drilled shafts, and large-diameter drilled shafts (greater than 8 feet) to optimize the design for the new bridge. Considerations evaluated during design included foundation type, construction equipment access, temporary slopes, temporary shoring, scour potential, historic mine workings, impacts to the environment and historical considerations, impacts to the existing bridge, and construction costs. The reevaluation resulted in the selection of the same foundation type for all the bridge elements, small-diameter drilled shaft groups at the abutments and large-diameter drilled shafts for the piers. In addition, an alternate option to use a spread footing at Pier No. 2 was included in the revised plan set with the thought that the contractor would determine which of the two options was the most cost effective. The project was re-advertised and resulted in savings to ADOT of almost $3 million with approximately $0.5 million coming from a reduction in the cost of construction of the large-diameter drilled shafts. Therefore, the reevaluation and the rebidding of the project was a success as it resulted in a more cost effective and constructible design in the challenging rock environment that was able to satisfy all project stake holders.

REFERENCES:

1 Menges, C.M and Pearthree, P.A., 1989. Late Cenozoic Tectonism in Arizona and Its Impact on Regional Landscape Evolution in Jenney, J.P and Reynolds, S.J. Geologic Evolution of Arizona, Tucson. Arizona Geological Society Digest 17, p 649-680.

2 Kellerhals, R. and Bray, D.I., 1971. Sampling Procedures for Coarse Fluvial Sediments. Journal of the Hydraulics Division, ASCE, Vol. 97, No. HY8, August. Pp. 1165-1180.

3 ADEQ, Arizona Department of Environmental Quality, April 2017, Pinto Creek Dissolved Copper TMDL – Salt River Watershed, Open File Report 16-05.

Rockfall, Rock Slope Stability and Risk: A Perspective from Managing Risks for Dams and Levees

Vanessa C. Bateman, PG, PE

US. Army Corps of Engineers

Prepared for the 70th Highway Geology Symposium, October, 2019

Disclaimer

Statements and views presented in this paper are strictly those of the author(s), and do not necessarily reflect positions held by their affiliations, the Highway Geology Symposium (HGS), or others acknowledged above. The mention of trade names for commercial products does not imply the approval or endorsement by HGS.

Copyright Notice

Copyright © 2019 Highway Geology Symposium (HGS)

All Rights Reserved. Printed in the United States of America. No part of this publication may be reproduced or copied in any form or by any means – graphic, electronic, or mechanical, including photocopying, taping, or information storage and retrieval systems – without prior written permission of the HGS. This excludes the original author

Abstract

Rockfall hazard classifications have continued to develop since the first systems were published to bring rigor to the assessment of rockfall hazard sites in order to rate and compare sites with one another. However, because these systems are used for management programs there is a need to take further steps to move these hazard rating systems from a hazard ranking to a true risk-informed basis, considering not only the probability of rockfall events occurring, but more accurately capturing the consequences of the event to the public. Developments within the Dam Safety community have a perspective on the management of risks associated with large infrastructure that can assist in further development of rockfall management by: 1) assisting in a better understanding how to capture uncertainty, 2) give a robust and formalized methodology for capturing subjective probabilities and 3) assist in the development of consequences analysis. This paper suggests some paths forward that will better fit rockfall management into a more risk-based approach and identifies some areas where further developments are likely to result in additional improvements. A semi-quantitative screening level assessment is also proposed for rockfall assessments that can aid in this development.

Introduction – Dam and Levee Safety Risk Management Approach

Rockfall hazard assessments and Dam/Levee Safety Risk assessments have taken two separate approaches to risk management in the US in recent years. The rockfall hazard assessments focused on a relative ranking-based expert assessment with a weighted scoring methodology; the Dam/Levee Safety community focused on a risk-based approach that uses expert assessments of hazard and consequences to come up with a final comparative number comparing the risk posed by the infrastructure with what are termed the "tolerable risk." Tolerable risks are those that (USACE, 2014):

1) society is willing to accept to obtain the benefits of infrastructure or other activity,
2) society does not view as negligible,
3) society generally views as being properly managed by the owner and
4) the owner continues to reduce to as low as practicable

Assessments of potential failure modes of sites are compared with tolerable risk guidelines for a risk framework that prioritizes mitigation based on a judgement of the tolerability of the risk by the public (USACE, 2014). This approach allows a portfolio-based approach prioritizing limited infrastructure dollars to sites that exceed the tolerable risk guidelines. This is accomplished through explicit detailing of potential failure modes that could cause a dam or levee to fail as well as combining that information with the consequences of that failure. Risk in this context is a:

> "...measure of the probability and severity of an adverse effect to life, health, property, or the environment. Quantitatively, **Risk = Hazard x Potential Worth of Loss. This can be also expressed as 'Probability of an adverse event times the consequences if the event occurs'"** (Fell, Ho, Lacasse, & Leroi, 2005).

Figure 1 below shows components of risk as understood and applied on USACE Dam and Levee Safety Projects (USACE, 2014). In a quantitative risk assessment, expert elicitation is used to identify potential failure modes and the probability of the failure mode occurring. In a further for a step, event trees are constructed to detail how a failure might progress and capture subjective probability assessments of the team as to the likelihood of the event. The probability is a measure of the degree of belief in a prediction, which must be dictated by the evidence of events that can or may occur in the future. In USACE, life safety is paramount and guides the decisions for mitigation.

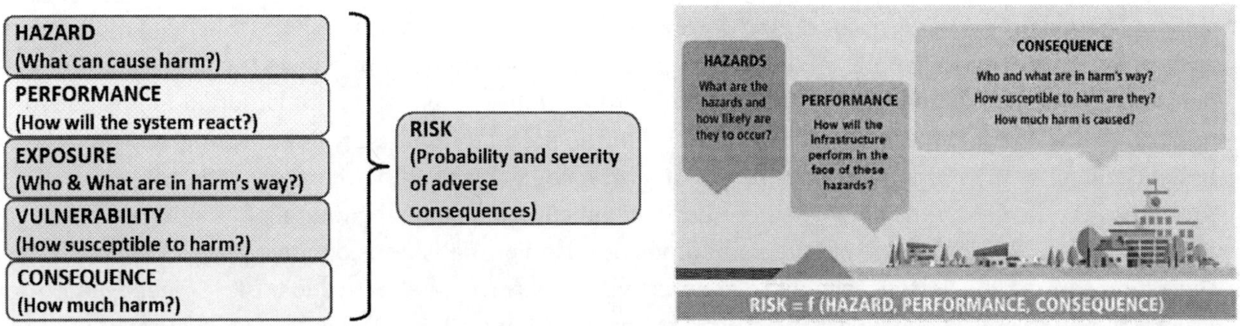

Figure 1. Risk Terminology– Risk is a Function of the Hazard, Performance of the Infrastructure and Consequences of a failure mode

For a fully quantitative risk assessment each potential failure mode (PFM) at a site has an event tree constructed and nodes within that tree are elicited for their probability of occurrence. An example for Internal Erosion failure of a dam or levee is shown in Figure 2 below. The events are laid out in sequence and probabilities of each are estimated at each node with values that must be between 0 and 1.

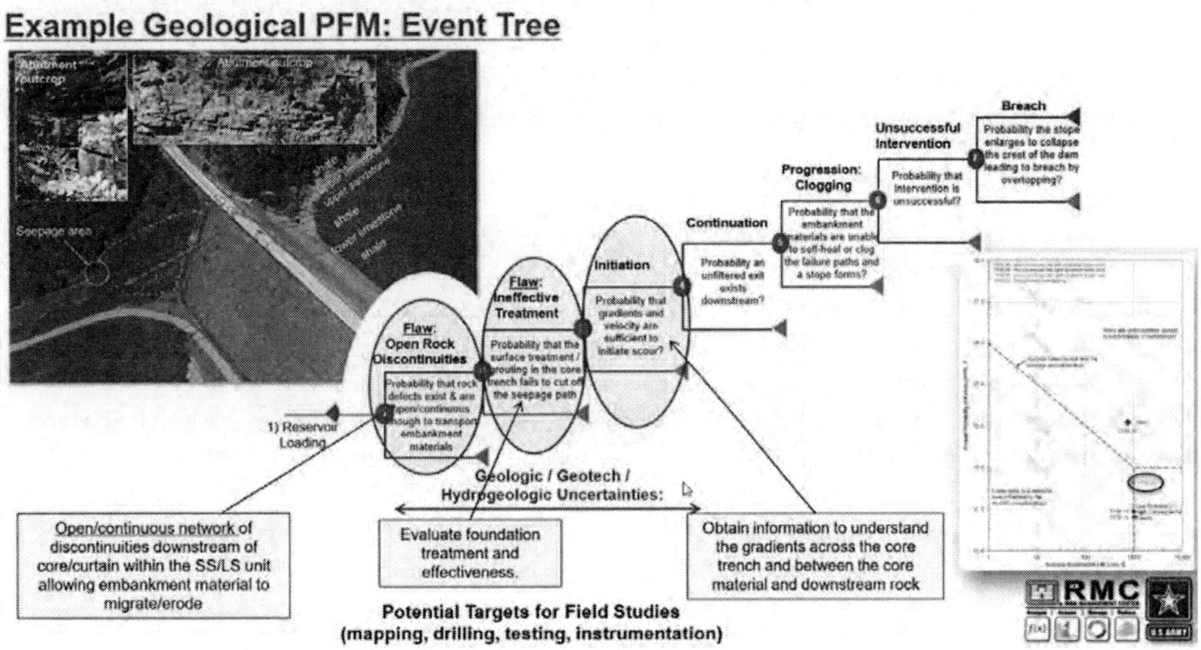

Figure 2. Internal Erosion event tree for a PFM (Potential Failure Mode) caused by an upstream to downstream path in the rock (Loar, 2017)

The pattern of the failure modes follows the sequence of Flaw → Initiation → Continuation → Progression → Unsuccessful Intervention → Breach occurs. All these nodes must have some probability greater than zero for breach to occur. The nodes are then multiplied to calculate the overall annual probability of failure. One advantage of this approach is that it forces the elicitors to think through the sequence of the problem and to estimate smaller "chunks" of the probabilities, supporting more accurate and consistent estimation. This provides not only a better estimate of the overall subjective probability of occurrence, but also makes it clear what actions might be taken to reduce the uncertainty. In Figure 2, for example, areas where field studies might reduce the uncertainty are identified.

One aspect of the system that must be clearly understood, is that the basis for this methodology is the subjective assessments that occur during an "expert elicitation" process. These must be evidence-based and are the collective judgments of a cadre of experts who are trained not only in the assessment of dams and levees, but who also must be trained in the cognitive traps of the human brain that can lead to poor estimations of risk. Successful application of the method requires both a thorough understanding of the potential failure modes of a rock slope and how to accurately estimate probabilities. The field of risk management both inside and outside of engineering has suffered from inconsistent application, tools that are sometimes not used properly and cognitive biases that can lead the estimators astray. Fortunately, there are methods to determine effectiveness of risk management programs and to improve estimator's assessment or risk (Hubbard, 2009). Thus, while a number is produced at the end, the annual probability of failure, we need to understand that it is an expert subjective assessment of this probability. The methodology is a consistent framework and a guide to good judgement. It provides a systematic method to both understand and communicate uncertainties in the problem. Thus, it does not require that we solve every problem deterministically, there can be unknowns and a variety of uncertainties, but we can still estimate probability.

> *"This perspective holds that the real promise of probabilistic methods is in their ability to lend a sense of what is important to uncertainty, whether this lies with the data or with mechanisms and processes. This view of probability of ultimately diagnostic, as opposed to prescriptive. Its purpose is to aid understanding first and thereby decisions second"* (Vick, 2002)

Weighted Scoring Systems and Rockfall Hazard Assessments

Rockfall rating systems have taken a much different approach and a considerable body of work exists with rockall hazard assessments, particularly since the work by Brawner and Wyllie (Brawner & Wyllie, 1975), (Wyllie, 1987) on rock slope stability assessments for the Canadian Railway and with the publication of the Rockfall Hazard Rating System (Pierson, Davis, & Van Vickle, 1990) and (Pierson & Van Vickle, 1993). These methods and the adaptations made by other highway agencies have been effective in bounding the problem, ranking sites and in assisting the users in qualitatively assessing the hazards posed. With perhaps one of the best effects of the implementation of these systems being an intelligent assessment by trained personnel as to the relative hazard of each site. Before these systems were implemented, there were few methods by which to have an "apples-to-apples" comparisons between sites, making an asset management or an overall risk management strategy for mitigation on a state, region or even highway basis nearly impossible.

A very quick screening methodology with a letter used to indicate severity is common to many of these systems (A,B,C etc.) and scoring with a weighted scoring system based on exponents of 3. This was pioneered by Wyllie (Wyllie, 1987) and implemented in Oregon and included in the FHWA manual for training personnel for State Departments of Transportation. Factors which are judged to contribute to the hazard are classified with a simplified scoring system from relatively low hazard for that factor (a score of 3^1=3) to a relatively high contribution of that factor (a score of 3^4 = 81). The scaling of the problem is thus expressed as a larger number where worse conditions are encountered. However, a score of 9 does not indicate that the condition is 3x worse that one with a score of 3. Plotting the scores of one factor along a line we can see the relative weights in judgement between a relatively low hazard and a relatively high hazard factor (Figure 1). This weighted score for each factor are then summed to receive a final score. While there have been many adaptations to this system most of these systems maintain the exponential scoring methodology from the progenitor system. Even the RHRS allowed ratings of 1 to 100 for each factor with the simplified charts only showing the 3, 9, 27, 81 criteria as an aid to judgement in applying the system.

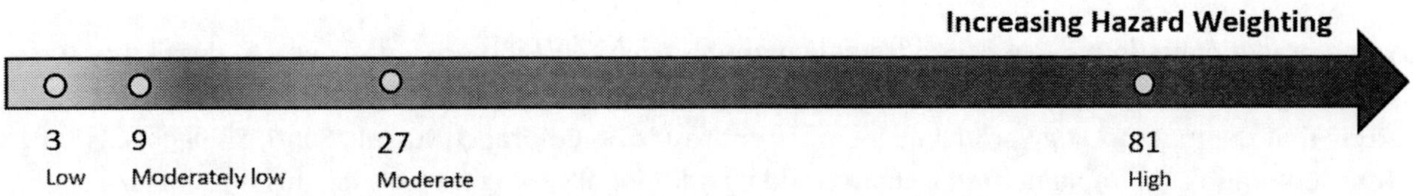

Figure 3. The "Hazard Continuum"- the relative weight of each score shown visually on a line

An example of this scoring criteria and how it is applied is show below from the Oregon Rockfall Hazard Rating System. For the RHRS, there were 10 factors (for Geologic Character only the worst scoring of Case 1 and Case 2 was used) giving a maximum score of up to 1000, or 810 depending if the full scale or the rating criteria from the charts was used. Table 1 below shows a specific example from the RHRS. Boxes highlighted in yellow are the scoring criteria for an example rock slope.

Table 1. Oregon Rockfall Hazard Rating System – Summary Sheet

Category			**Rating Criteria and Score**			
			3	**9**	**27**	**81**
Slope Height			25	50	75	100
Ditch Effectiveness			Good catchment	Moderate Catchment	Limited Catchment	No Catchment
Average Vehicle Risk			25% of the time	50% of the time	75% of the time	100% of the time
Percent of Decision Sight Distance			Adequate - 100% of low design value	Moderate – 80% of low design value	Limited – 60% of low design value	Very Limited – 40% of low design value
Roadway Width			44 feet	36 feet	28 feet	20 feet
Geologic Charachter (only worst case used for scoring)	Case 1	Structural Condition	Discontinuous Joints, favorable orientation	Discontinuous Joints, random orientation	Discontinuous Joints, adverse orientation	Continuous joints, adverse orientation
		Rock Friction	Rough irregular	Undulating	Planar	Clay Infilling or slickensides
	Case 2	Structural Condition	Few differential erosion features	Occasional differential erosion features	Many differential erosion features	Major differential erosion features
		Difference in Erosion Rates	Small difference	Moderate Difference	Large Difference	Extreme Difference
Block Size / Volume of Event			1 foot / 3 cubic yards	2 feet / 6 cubic yards	3 feet / 9 cubic yards	4 feet / 12 cu yards
Climate / Presence of Water on Slope			Low to moderate precipitation; no freezing; no water on slope	Moderate precipitation or short freezing periods; intermittent water on slope	High precipitation or long freezing periods or continual water on slope	High precipitation and long freezing periods or continual water on slope and long freezing periods
Rockfall History			Few Falls	Occasional	Many Falls	Constant Falls

Score = *9+81+27+27+9+27+9+81+27+27* **= 324**

All the systems that implement this type of scoring systems use the final weighted scoring number as a relative ranking system: a methodology to put sites in order. The function of the exponential scoring weighting is to simply make a rapid differentiation of the number. It does not indicate that a site is, for example, 3 times worse than another site, even if mathematically true of the numbers. A site with a scoring, for example of 972 would be worse than one scoring 324, but how much additional risk is not quantified. The final numbers only serve as a rank order, rather than relative risk or hazard.

Some practitioners found, even with training manuals from FHWA, there was a need to detail the geological factors in more detail than the original system. Russell, Santi and Higgins noted in a 2008 study that the methodology produced inconsistent results in Colorado (Russell, Santi, & Higgins, 2008). Thus, Colorado DOT implemented a change adding additional geological factors. This is a similar approach taken by the Tennessee Department of Transportation (Vandewater, Dunne, Mauldon, Drumm, & Bateman, 2005) where five specific geological failure modes (plane shear, wedge, toppling, differential erosion and raveling) were added. Both systems kept, however, the basic scoring mechanism, even if some details were changed.

Wyllie later pointed out that this formulation, along with most of the other systems that use this weighted scoring intermingle "hazard factors" and "risk factors" (Wyllie, 2015). He lists hazard factors as those criteria which correlate directly to the slope conditions (Slope Height, Geologic Character, Block Size, Climate/Presence of Water on Slope, Rockfall History) and risk factors as those that are related to the path of the vehicle and traffic conditions (Average Vehicle Risk, Decision Site Distance, Roadway Width and Ditch Effectiveness). Thus, we have two main groupings of factors in most of these systems:

1) those that reflect the characteristics and geometry of the rock slope where the event can initiate and
2) those that reflect characteristics of the roadway that may influence either the likelihood that a vehicle is in harm's way or influence the ability for the driver to react.

He suggested that these factors should be summed into the two categories and multiplied Hazard Factors x Risk Factors. For the example in Table 1 above, then the score would become 25,920 as shown below in Table 2. This takes the existing framework and recasts to a similar risk framework as used in the dam and levee safety community where Risk = Probability of Failure (Hazard & Performance) X Consequences while maintaining the base exponential scoring system.

However, as can be seen below while some of the "Risk Factors" are more explicitly factors which describe the geometry of the problem (roadway width), rather than the consequences of failure. For example, while a wider roadway would allow for additional maneuvering room, a large slide which covered the whole roadway would negate the effect of this factor. However, decision site distance is a mathematical equation that considers the time a driver will have to react to an obstruction and average vehicle risk is an equation which accounts for the percentage of time that a vehicle may be underneath a slope. The primary and significant advantage of this formulation is that builds on the considerable body of work already accomplished by State DOT's and can be accomplished by recalculation of existing data, rather than requiring additional assessments. Wyllie, however does state that the overall number here is still less important than the relative ranking. While this recasting does help, the methodology does not entirely bring the older hazard assessments into the same kind of risk framework used successfully for dam and levee safety management. This is primarily since most of the "Risk Factors" themselves are not an assessment of the actual risk presented to the public by rock slope failures. They are still only qualitative descriptions that capture when a condition is worse. Additionally, the consequences of failure are hidden within other risk factors, thus in many ways this is still a measure of the hazard, rather than the risk to the public (life safety) or to the transportation system (economic impact). In this case only "Average Vehicle Risk" is an actual calculation of likelihood of consequences occurring as it calculates how likely a vehicle is to be underneath a slope during an event.

Table 2. Hazard Factors and Risk Factors Weighted Scores (Wyllie, Rock Fall Engineering, 2015)

Hazard Factors	Score
Slope Height	9
Geologic Character (here Case 1)	9 + 27
Block Size	81
Climate/Presence of Water	27
Rockfall History	27
Sum	**180**
Risk Factor	Score
Ditch Effectiveness	81
Average Vehicle Risk	27
Percent Decision Site Distance	27
Roadway Width	9
Sum	**144**
	Total Score
Total Score = 180 * 144	**25,920**

The Unstable Slope Management system implemented by Federal Lands takes a similar approach for by multiplying the "Condition Index" in a simple formula to obtain a Priority number (Thompson, 2016) where more detailed costs are not available (Equation 1). While there is a discussion of a more quantitative risk-based approach, the preliminary scoring system also uses the exponential weighting system (Beckstand, et al., 2019). The quantitative risk-based approach (QRA) detailed in the manual uses a four factor probability estimate to calculate the annual risk to an individual of using a facility that can have an unstable slope event (Beckstand, et al., 2019) shown in Equation 2 below. This is a life safety formulation of risk, using subjective probability assessments of the component factors that is intended to capture the life safety risk to individuals by a slope.

Equation 1. Simple Prioritization Formula

$$Priority = Condition\ index \times \frac{ADT}{Size}$$

Condition index = Unstable Slope Score
ADT = Average Daily Traffic
Size = area of slope

Equation 2. QRA of Risk Estimate in the Unstable Slope Management System (Beckstand, et al., 2019)

$$R_{(air)} = P_{(occ)} * R_{(loc)} * R_{(pres)} * R_{(vul)}$$

Where:

$R_{(air)}$ = annual fatality risk or injury risk or risk of damage the facilities

$P_{(occ)}$ = annual probability of occurrence
$P_{(loc)}$ = probability of person being in the path of the hazard
$P_{(pres)}$ = occupancy rate or rate of presence – how long a person will be in the hazard zone
$P_{(vul)}$ = probability of person getting killed or injured by the event

This methodology breaks the risk down into hazard ($P_{(occ)}$, $P_{(loc)}$, $P_{(pres)}$) and consequence factors ($P_{(vul)}$) and in this is an improvement over earlier formulations that mixed risk and hazard. It is a truly risk-based approach. It is however, not as detailed as the methodologies used in dam/levee safety practice which quantify the failure modes more fully. It does not, for example, explicitly lay out the more and less likely factors that lead to these estimates. Nor does it explicitly lay out the steps of a potential failure mode. Instead, the system combines the probability of occurrence of an incident that can injure someone into three factors $P_{(occ)}$, $P_{(loc)}$, $P_{(pres)}$ with the remaining factor estimating the consequence of the event $P_{(vul)}$. This is a formulation using life safety risk, however, the first factor $P_{(occ)}$ can be very difficult to estimate as the failure modes and steps within that failure mode are not broken down further. For example, the event tree used in Figure 2 is used to estimate the equivalent here of $P_{(occ)}$. Thus, while it is a useful formulation that expresses the actual risk in terms of life safety, rather than a weighted hazard scoring approach, it is possible to improve on the risk estimates by further defining $P_{(occ)}$.

Another method to separate the hazard from the consequences was completed as part of the implementation of the Tennessee Rockfall Hazard Rating System (Bateman, 2005). In this formulation the exponential scoring system was also used for an additional rating criterion called the "Rockfall Closure Impact" which was an explicit attempt to capture in a number the consequences of a rockfall event (Table 3). However, again, this approach takes a rank scoring rather than an estimate of probabilities. In this method the TRHRS Score and the RCI score are plotted and compared with the cost of mitigation in order to decide which sites to mitigate and in which order (Figure 4). This methodology keeps the "hazard" factors somewhat separate from the "consequence factors." The RCI score is not multiplied by the TRHRS score, they are treated as separate entities and compared on a graph in order to make decisions on where to concentrate repairs. Thus, this is at least a little more of a risk-based formulation as understood by the dam and levee safety literature. Like Wyllie's recasting, it does not change the underlying assumptions of the system, that a description of the factors and weighting by an exponential method accurately captures the differences in hazard presented by the sites. It also does not really quantify the risk to the public, other than high hazard scores and high consequence scores are going to have more impact on the public. This also mingles life safety and economic costs. Here the RCI focuses on factors which will have an impact to the economics of the problem – how long a road facility will be blocked and on the significance of the facility. All these factors are components of the problem, but all criteria are listed as equally weighted for the final score.

One of the advantages of this formulation, despite the "ranking order" nature of the scoring system is that it fosters more risk informed decision making, by attempting to explicitly compare the hazards with the consequences and comparing that to a cost that can be used for prioritization. In Figure 4 with a \$1M budget, Site Number 1 would be the most obvious place to prioritize, as it combines the highest hazard with the highest consequences. With a \$1.2M Budget, Sites 1 and 3 would be chosen. Likewise, we see that the point with the highest hazard has a cost of \$6.8M, however, for less than that amount all the sites with an RCI above 250 could be repaired (all the sites contained in the blue box in Figure 4). With this method we can get a sense of the "funding gap" but it doesn't easily quantify risk buydown.

Table 3. Rockfall Closure Impact – State of Tennessee

Criteria	Score = 3	Score = 9	Score = 27	Score = 81
ADT (Average Daily Traffic)	Little Traffic ADT>300	Some Traffic ADT 300-1000	Moderate Traffic ADT 1000-3000	Major Traffic ADT >3000
Impedance	Shoulder	1 lane	2 lanes	>=3 lanes or Total
Impedance Duration	Hours	1 Day	Days	Weeks
Detour Length	Very Short <1 mile or lane still open	Short 1-2 miles	Medium 3-4 miles	Long or None >4 miles
Facility Degradation RF-DF= DOF	0	1	2	3

DOF=Degree of Facility (RF=Degree of Road Facility; DF= Degree of Detour Facility)

0 = Local Roads / 1 Lane Road
1 = 2 Lane, no shoulder
2 = 2 lane, adequate shoulder
3 = 3 lane
4 = 4 lane
5 = 4 lane, divided highway, 5 lane highway
6 = Interstate

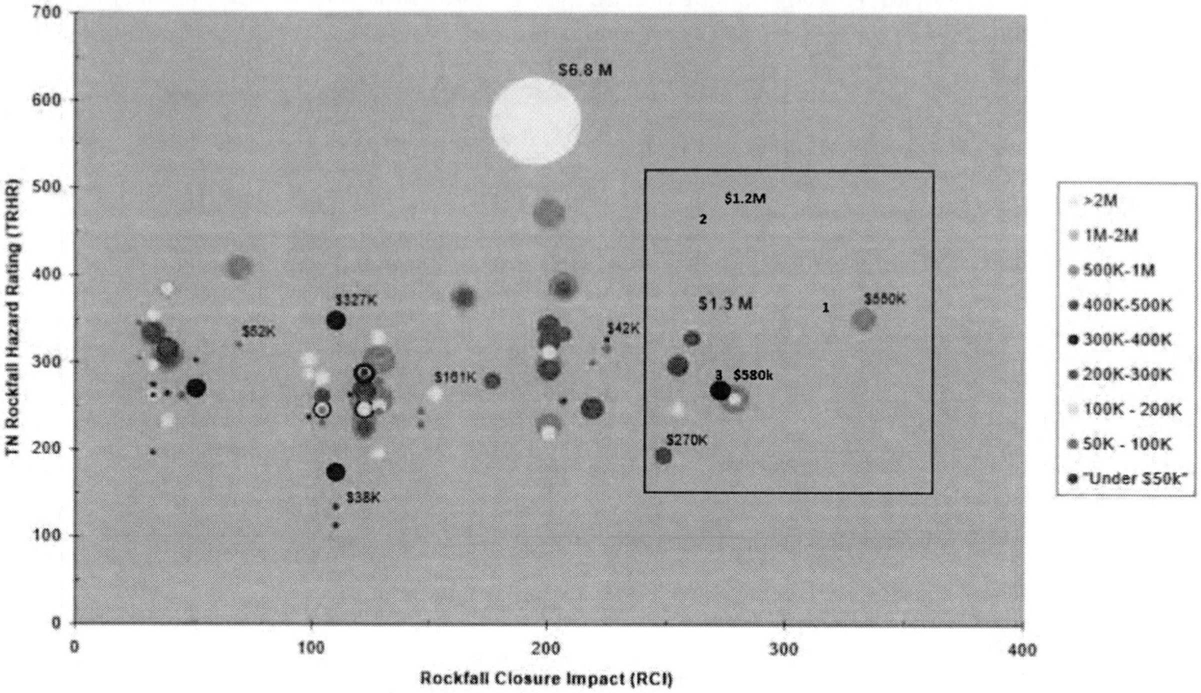

Figure 4. TRHRS, RCI and Cost Plot – all the sites contained in the blue box can be repaired for less than the $6.8M site

Bringing Rockfall Assessments Closer to Risk Assessments – Why change anything?

With nearly 30 years since the RHRS was published, and even longer since the original work of Wyllie, why make a change now? Systems have been implemented in multiple states and have been largely successful at identifying problems and giving DOT's a path forward for a more rational basis for management of these sites. The growth of these systems has forced practitioners to think about how to manage geotechnical infrastructure on a portfolio level and has given owners tools that were desperately needed in order to take repairs out of a "reaction" footing which was the only methods most owners had due to a lack of data.

Qualitative Systems have not fully "answered the questions"

There are a plethora of systems that have been implemented along the same qualitative lines since the progenitor systems were published. Further iterations of these systems have continued, but typically maintain the same basis for scoring and mixing hazard and risk together. An excellent overview of these in the US and worldwide can be found in a 2015 paper by Ferrari, Giacomini and Thoeni titled "Qualitative Rockfall Hazard Assessment: A Comprehensive Review of Current Practices" (Ferrari, Giacomini, & Thoeni, 2015). The continued work on these systems for applications by different owners and agencies indicates that the systems that were previously developed, while helpful, are not fully meeting the need.

Systems need to give a whole portfolio view of Risk and quantify "Risk-Buydown" of mitigation strategies

Part of the reason is continued improvements in assessments of sites to make better informed decisions on needed mitigations. None of the methodologies except the Federal lands Unstable Slope Management system above really capture numerically the "risk-buydown" of certain types of mitigations. Sites can be scored differently, but again, we cannot take these overall hazard numbers and say, "this mitigation reduces the risk to the public by 50%." Without a clear means to express what the money spent on mitigation is purchasing, it becomes difficult to compete for dollars for repair. Further, it becomes more difficult to have a supportable and explainable methodology for an agency to make decisions on spending that compare the risk reduction of many different activities. For example, adding a rockfall fence as opposed to adding traffic lights, improving grades/site distances on problematic roads instead of adding rockfall drapery at different road segments. None of these weighted scoring systems capture actual or subjective judgements of like safety risk in a way that can be easily compared with the other risks that must be managed by an agency. Risk-buydown for investment is a powerful communication tool to obtain and explain the need for funding as well as for assuring the public that risks are being managed appropriately and well. The newly published Unstable Slope Management System published by Federal lands recognizes these issues as it adds a quantitative risk assessment and takes a much more asset management-based approach (Beckstand, et al., 2019). Risk buydown can be calculated with that framework, however the "lumping" of the probability of occurrence into one probability estimate may make the methodology difficult to implement consistently.

Asset Management Frameworks rely on Risk

The other difficulty of these Hazard Rating Systems is that we can't really take the jump into Geotechnical Asset Management with only a hazard rating. This has been recognized by the Colorado DOT (Ortiz, 2016) and Federal Lands (Beckstand, et al., 2019) among others as state DOT's and Federal Agencies begin to implement geotechnical asset management. An inventory is a needed and necessary first step, but with no way to mathematically express risk-buydown it's hard to defend choices between mitigation strategies such as:

1) Recut the slope (eliminate the risk) or between options such as
2) add additional ditch width, rock bolts and high energy rockfall fencing (mitigate the risk option – high cost) or
3) conduct extensive scaling along a road segment every three years (mitigate the risk option – medium cost).

We can weigh costs at individual sites, we can make judgements that the risk will be significantly reduced, but outside of a true risk framework we can't easily quantify how much risk reduction each strategy brings. Interestingly Paul Thompson called out the cost gap needs for reconstruction of unstable slopes in Alaska as 3x that of the bridge inventory (Thompson, 2016). Yet, bridge programs across the country are generally far better funded than unstable slopes, including both landslides and rockfall. Under the framework used by most DOTs we can't easily compare the risk to the public of an agency's unstable slopes/rockfall as compared to the risks posed by structurally deficient bridges. The exception to this as discussed above is the Federal Lands Unstable Slope Management System which has a means to estimate life safety risks (Beckstand, et al., 2019).

We also don't have within the Rockfall Assessments a set of "Tolerable Risk Guidelines." This was one of the early criticisms of even starting rockfall assessments as "we don't have that many fatalities" - one of the statements that this author encountered frequently during the early days of implementation within the State of Tennessee. Yet, as the RCI scoring shows above in Figure 4, the impact to the roadway network and the economic impacts of road closures can be considerable. Consequences can be measured in life safety risk or in economic impacts. This was unfortunately aptly demonstrated in Tennessee when a slide along US 64 through the Ocoee River Gorge shut down the roadway for 5 months causing major socioeconomic impacts with a 1.5 hour detour each way for citizens driving to work in the neighboring county, and impacting emergency services by cutting off Ducktown, TN from its hospital located on the other side of the slide - necessitating helicopter evacuation for any critical patients (Bateman, Smerekanicz, & Sneyd, 2010). Fortunately, no one was injured during this rockslide event, but the impact to the public was considerable. Another slide in 2011 on I-75 in Tennessee, also with no fatalities, had and ADT of 28,000 affected the public for an estimated 5.5 months (Anderson, 2016) a considerable economic disruption.

Rockfall management has a vital part to play within the GAM (Geotechnical Asset Management) framework whose goals are (Vessely, et al., 2019):

- A process to measure and manage involuntary safety risk exposure across the entire asset class;
- Lessened traveler delay and closure times, resulting in improved network operational performance;
- Reduced adverse economic impacts to users, private enterprise, and communities;
- Fewer impacts and damages to other transportation assets;
- Optimized resources, improved sustainability, and well-maintained reputation;
- Enhancement of data-driven decisions that support agency and executive objectives;
- A greater understanding of risk exposure levels and distribution, and the ability to manage those risks; and
- The ability to start very simple and adapt the GAM process over time as the economic benefits are realized.

In addition, it increases the resiliency of the roadway network when risks are better managed. Because we cannot quantify actual risk reduction, it becomes exceedingly difficult to analyze the effect of differing levels of investment. For example, on a roadway segment such as US 64 in Tennessee it is very difficult to quantify the annual investments needed in order to "lessen traveler delay and closure times" or to compare it with other roadway segments. This puts the rockfall systems (and other landslide management systems) at a disadvantage for continued investment if these cannot be articulated numerically and clearly compared with the other risks/assets that an agency must manage. Preferred repair option costs for sites help but are not the entire answer.

The Path Forward – Risk-Informed Decision Making for Managing Rock slopes

We must recognize the need for change in order to better fit within both Asset Management and an overall Risk Management framework of our infrastructure projects and this will take some re-thinking of how sites are assessed in rockfall inventories. While this necessitates a fundamental change, as President Dwight Eisenhower observed in a speech to the National Defense Executive Reserve Conference in Washington, D.C. in 1957 "Plans are worthless, but planning is everything...to keep yourself steeped in the problem that you may be one day be called upon to solve – or to help solve" (Eisenhower, 1957). All the work done with these systems has been work that has been successful in putting the state of practice ahead of where we started and can be used to build this new framework.

Literature on implementing risks within Dam and Levee Safety management as well as with landslides exists and isn't new. The dam safety risk management programs are well established and have overcome some of the initial hurdles that have been perceived as barriers to adopting a true risk-based approach. Fell in 1994 suggested this approach for landslides and suggests using the same expert elicitation type of methodology described above used successfully with dams and levees (Fell, 1993). This work can be expanded to meet the need. A few guiding principles for further development of this endeavor may help:

1. **In any assessment, detail the potential failure modes for the site explicitly.** Use of event trees within dam and levee safety brings rigor to the process and ensures that evaluators have a clear understanding of both what and how a slope can fail. Both the CRHRS and TRHRS arrived at the need to evaluate details of slopes more than the original RHRS. These added a little time, though not very much, to evaluations of slopes and gave both states some additional management and informational tools. This kind of assessment and data gathering should continue. All failure mode probabilities should be estimated by qualified engineering geologists, geotechnical engineers or geological engineers experienced in rock slope mitigation.
2. **Scale evaluations to the need.** Both the rockfall and dam/levee assessments maintain a multi-level assessment tool with a differentiation between both preliminary and detailed screening. These have been largely successful in defining problems and give an agency flexibility to make rational decisions on limited labor resources. This paper will demonstrate a screening assessment that can be used on a fully quantitative or semi-quantitative basis depending on the severity of the site.
3. **Establish Tolerable Risk Guidelines and how to Calculate Consequences.** Two main frameworks for managing consequences have been successfully used: Life Safety and Economic Impacts. Unstable slopes present both consequences to the agencies that must react to them, but a prioritization needs to be made as to which controls, or to establish the how the combination of both consequences should be handled. Further, we need principals such as "Do No Harm" for new designs, ensuring that any mitigations to rockfall sites, reconstruction of old roadways or construction of new roadways to not add substantially to the risk to the traveling public. Life Safety alone may not be enough to capture the risk that these sites pose to the transportation network and the public.
4. **Establish that hazards probabilities are estimated on an average annual basis.** This will require a much different way of thinking about the problem than picking a weighted score. Each node in the potential failure modes event tree should be estimated based on the likelihood of that node or step occurring on an annual basis. The primary advantage of this is that it allows considerably better risk communication and comparison with other risks managed by an owner. This is still a subjective judgement, but a guided subjective judgement made by trained professional geologists and engineers who understand how rock and earth slopes can fail. With this and methods to account for consequences, the actual amount of risk buydown can be calculated. This will allow an agency to assess levels of spending for mitigation and compare it corporately to how much risk that spending will reduce.
5. **Establish a "recurrence" interval for evaluations.** One of the most successful aspects of risk management within USACE is the implementation of risk assessments in the Periodic Assessment (PA) of dam safety projects. These projects receive a risked based evaluation a minimum of once every 10 years. This acknowledges that sites can change over time, and that a single evaluation at one point in time are not enough to track and manage risks over time.

Detail the potential failure modes for the site explicitly

The detailing of specific failure modes present at the site is of considerable assistance when communicating risk to another trained practitioner, since knowledge of the failure types can constrain the range of possible solutions. Within dam / levee safety this also allows an overall assessment of the risk of a site to be compared, even where there are multiple significant failure modes. It also makes mitigation alternatives much easier to assess as all the potential failure modes at a site are evaluated. Within the area of rock slope stability, we can describe much of what can go wrong under 5 main failure modes:

1. **Raveling** – classical rockfall where there is no defined structural control that can lead to a larger failure. These occur in reaction to weathering of a slope by wind, freeze-thaw rain and other drivers such as vegetation growth and blasting damage.
2. **Plane Shear Failure** – A structurally controlled failure the rock slope where a failure can occur along a plane within the rock mass. This was the type of failure observed at the US 64 Ocoee Gorge Slide (Bateman, Smerekanicz, & Sneyd, 2010)
3. **Wedge Failure** – A structurally controlled failure of a rock slope where failure occurs along two intersecting planes within the rock mass.
4. **Toppling** – A structurally controlled failure of a rock slope where a block or column of rotates outward about a fixed base.
5. **Differential Weathering / Secondary Toppling** – Failure of rock on a slope due to undermining by the weathering of a less resistant layer. This can cause either rock to rotate outward, or for a cantilevered section to fail as underlying rock is removed.

These are the main failure mode classifications that were used in the TRHRS (Vandewater, Dunne, Mauldon, Drumm, & Bateman, 2005). It should be noted that most rockfall hazard rating systems explicitly or implicitly incorporate more rock slope stability failure modes than just what can be classically defined as rockfall from the very beginning. The RHRS calls out structural controls to failure in Case 1 and notes events that can and are larger than 4 feet/12 cubic yards. Sites with these kinds of failures have been routinely incorporated into rockfall hazard systems, which perhaps should be called rock stability hazard systems. The system implemented in Tennessee can and has been mined for the types of failures present in rock slopes statewide providing invaluable data to planners for assessing potential difficulties presented by new roadway construction. However, these are not the only failures that can occur on a site. This same framework can be expanded to all potentially unstable slopes. Wyllie and Mah point out that there can be circular failures in weak rocks, closely fractured rock and in rock fills (Wyllie & Mah, 2004). These and other slope failure modes can be easily added into this framework to expand it to all landslide failure modes as recommended by Fell et.al (Fell, Ho, Lacasse, & Leroi, 2005), (Fell, 1993).

A semi-quantitative approach for rockfall, however, should estimate probabilities of each failure mode type. As part of that process, more and less likely factors should be identified that inform these subjective probability estimates. For an example of this, see the framework with an event tree with more/less likely factors that can inform below in Figure 5. These factors will need to be customized for the site conditions and listed as the factors that were considered in making the probability assessment. Providing these factors aids future geologists and engineers in understanding both the information and the uncertainty that went into the risk estimate. Similarly, we can use the same pattern with structurally controlled failure modes such as Plane Shear Failure where more and less likely factors or shown (Figure 6). Figure 7 below shows only the failure modes and nodes: Wedge Failure, Toppling Failure and Differential Weathering/Secondary Toppling. These are all the probability that the failure mode will occur, separated from the consequences.

Failure Modes	Node 1 - viability (Flaw)		Node 2 - parameters (Flaw)		Node 3 - parameters (Initation)		Node 4 - parameters (Initation & Continuation)		Node 5 - intervention fails		Node 6 - failure occurs	
Rockfall (raveling)											=P(1)*P(2)*P(3)*P(4)*P(5)	
Descriptions	P(rockfall can occur on face this year)		P(rockfall large enough to present life safety risk)		P(root jacking, ice jacking, weath face can cause Rock to loosen from more stable position)		P(rock will have enough momentum to reach road/building/parking lot in case of failure)		P(measures taken to contain RF are insufficient)		P(rock has sufficient momentum to strike vehicle/person or to puncture building from above or side)	
Considerations	More likely	Less likely	More likely	Less likely	More likely	Less likely	More likely	Less likely	More likely	Less likely	More likely	Less likely
that make a particular event more or less likely	Blasting damage	Good presplit face	Rock blocks on slope at least 1' on a side evident	Small rocks	Trees and large shrubs evident on face	Little to no veg	Tall slope	Short slopes	Ditch is undersized for height of slope	Ditch is adequate size by design charts for 95% catchment	Rock strikes evident in pavement	Old pavement with few indications of RF reaching roadway
	Highly weathered face	Little weathering on face	Rock blocks of sufficient size show at least partial dilation from face	Smooth face, no displacement of blocks evident on face	Intermittent or constant water on face - when no precipitation	Dry face	Protrusions that can give falling rock horizontal momentum - inadequate "catchment" benches	Smooth face (for vertical slope)	Ditch is "hard" giving rock adequate opportunity to bounce out and into road	Ditch has energy reducing measures - loose gravel, grass etc.	Rocks located beside road/ parking area/ facility in center or opposite side (could have fallen or moved there during cleanup	No rocks located on or beside far side of facility
	Loose rocks evident on face	Few loose rocks evident, no large cracks, dilutions	Rocks of sufficient size evident in ditch		Significant ice formation noted during winter months	Little to no ice	Intermediate benches on face too full to catch rock	Clean intermediate benches	No catchment berm, or other structure present	Well sized rockfall fence present	CRSP shows rock falls onto road/parking lot or facility, or bounces into road, does not roll onto facility	CRSP shows rock rolls onto facility
	Rocks visible in ditch	No rocks in ditch, hasn't been recent cleanup			"Wet" veg on face - too small to root jack, but shows where water source occurs		1/4 or increased slope angle	Vertical face	No drapery to slow down or redirect rock movement	Drapery present and at least partially effective, rock too small to destroy drapery	Maintenance reports events or accident due to falling rock at this location.	No reports by maintenance in "well monitored" area
	Impact marks on roadway	Pavement age sufficient to show no impractical marks in roadway					.75/1 slope angle		Drapery or rockfall fence (etc.) too small, or damaged			

Figure 5. Rockfall (raveling) Potential Failure Mode Event Tree with more and less likely factors that can influence estimation

	Node 1 - viability (Flaw)		Node 2 - parameters (Flaw)		Node 3 - parameters (Flaw)		Node 4 - parameters (Initation and Continuation)		Node 5 - intervention		Node 6 - failure occurs	
Plane shear failure											=P(1)*P(2)*P(3)*P(4)*P(5)	
Descriptions	P(plane shear failure mode is kinematically viable)		P(failure plane angle is near the phi strength of the failure plane)		P(asperities on face do not provide sufficient strength to hold if phi equal angle)		P(water conditions will exceed strength of failure plane during the year)		P(measures taken to contain plane shear failure are insufficient)		P(failure reaches facility and covers at least one lane or part of facility likely to be occupied)	
Considerations	More likely	Less likely	More likely	Less likely	More likely	Less likely	More likely	Less likely	More likely	Less likely	More likely	Less likely
that make a particular event more or less likely	Exposed failure plane evident in the face from previous failure that extends under rock seen on face	No smaller exposures of potential failure plane evident on face	Exposed failure plane is weathered, fault gouge, weathered bedding plane etc.	Exposed failure plane is small at tight	Little waviness in failure plane	Significant waviness in failure plane	Wet face	Dry face	There is insufficient ditch width to contain the whole plane shear failure	There is a large distance between the plane shear failure and the roadway/building/parking lot	Stability analysis shows failure is viable under the right conditions.	Stability analysis shows failure is unlikely
	Inclined bedding plane, fault or joint intersects rock cut/ slope at an angle and daylights		Exposed failure plane is altered from rock to much weaker material	Little alteration at the rock failure plane	Asperities are expected to be small and potentially weathered	Asperities are expected to be larger and may be weathered	Water flow from face has significant constriction that could increase water pressure in failure plane in the event of a significant recharge	Water flow is unrestricted from the face even in large rain events	Berm or rockfall barrier insufficient to contain volume of failure.	Large barrier, or rock shed exists that protects traffic, people and facility in the event of failure	Records indicate plane shear failure has occurred on the site in adjacent areas	No records of this type of failure in this area
	Failure plane is within +/- 20 degrees of angle of slope	Failure plane further more than +/- 20 degrees of slope face orientation.	Water is evident leaking from the face from this failure plane	Dry face even during wet time of year	Existing plane shear failure (perhaps smaller) close to remaining rock shows few asperities	Existing planes similar to failure mode plane (joint sets, bedding etc) show significant asperities.	Vegetation growth along failure plane	No vegetation on face	Rock anchors of insufficient size to hold rock that can move - poorly grouted, poor material choice	Rock anchors are well designed by current standards and sufficient to restrain failure.	Evidence on the slope indcates this failure has occurred at some unknown point in the past.	No evidence on slope of previous potential failures.
			Dilation evident on sides, top of at exposed failure plane	No dilation present			Water likely to have near immediate access to plane during a large storm event, allowing water pressure to build up (eg in the case of a tension crack)	Good water control above the face and little interconnection within the ground make it unlikely water pressure could build up in the failure plane.	Rock anchors old and corrosion increasingly likely to make the anchorage insufficient. Corrosion evident at face or in similar locations where installation methods and age similar. Anchors past design life.	Little corrosion in anchors at slope nor at similar installations of similar age. Anchors recent enough that corruption protection is likely still adequate		
			Failure plane angle exceeds typical phi for material evident - e.g. phi = 27 for soils, phi = tilt test angle rock	Failure plane angle less than typical phi for material present								

Figure 6. Plane Shear Failure Potential Failure Mode Event Tree with More and Less Likely Factors

	Node 1 - viability (Flaw)	Node 2 - parameters (Flaw)	Node 3 - parameters (Flaw)	Node 4 - parameters (Initation and Continuation)	Node 5 - intervention	Node 6 - failure occurs
Wedge Failure						=P(1)*P(2)*P(3)*P(4)*P(5)
Descriptions	P(wedge failure mode is kinematically viable)	P(failure plane angle is near the phi strength of the failure plane)	P(asperities on face do not provide sufficient strength to hold if phi equal angle)	P(water conditions will exceed strength of failure plane during the year)	P(measures taken to contain plane shear failure are insufficient)	P(failure reaches facility and covers at least one lane or part of facility likely to be occupied)
Toppling Failure						=P(1)*P(2)*P(3)*P(4)*P(5)
Descriptions	P(toppling failure mode is kinematically viable)	P(toppling failure large enough to present life safety risk or close road/lane)	P(there is insufficient rock fabric connection or adhesion to prevent failure should movement initiate)	P(water conditions, ice jacking or root jacking will be sufficient to initiate movement)	P(measures taken to contain toppling shear failure are insufficient)	P(failure reaches facility and covers at least one lane)
Differential Weathering or Secondary Toppling						=P(1)*P(2)*P(3)*P(4)*P(5)
Descriptions	P(differential weathering or secondary toppling is kinematically viable)	P(rock failure large enough to present life safety risk or close road/lane)	P(there is insufficient rock fabric connection or adhesion to prevent failure should movement initiate)	P(water conditions, ice jacking or root jacking will be sufficient for movement movement)	P(measures taken to contain failure are insufficient)	P(failure reaches facility and covers at least one lane)

Figure 7. Wedge, Toppling and Differential Erosion/Secondary Toppling Event Trees.

One substantial difference in how these event trees are laid out with those for a dam/levee safety failure mode is that except for raveling, initiation and continuation are combined into one node. In many cases on a rock slope, once the rock failure starts to move, there is often little that will arrest the momentum of the movement. Thus, for most of these failure modes, these initiation and continuation have been combined. Rock bolts, meshing and rock fences would all come under the heading of intervention.

The data gathered already by previous rockfall hazard ratings has an important part to play in this assessment of risk. No estimate is worthwhile without supporting data, and ditch widths, height of slope effectiveness of the ditch, roadway widths, etc. are all input in the system, but used to support the judgement of the risk estimators. One factor that is not shown in these event trees is the size of the event. The reason for this is two-fold: 1) the size of the event is more properly accounted for in the consequences (i.e. it will close a lane, several lanes or the road) and 2) the event must be large enough, combined with other site factors such as the effectiveness of the ditch or rockfall mitigations already in place to reach the roadway. If not, the probability estimated in Node 5 – Intervention Fails will be very, very small and the site will have little risk. For a life safety footing the size of the event matters, but even a small rock in the wrong place through a windshield is enough to cause a fatality.

Establish how to Calculate Consequences.
This may be one of the most challenging aspects of implementation of a new system. However, there are economists and analysts within agencies that can be of assistance. Before a major rehabilitation of an interstate is conducted, for instance, full depth pavement replacement or multiple bridge replacements, economic evaluations are completed to study the impact of these changes and to calculate cost benefit ratios. The same can be done for segments of roadways when looking at economic consequences. Life safety considerations are harder, which is why parameters like decision site distance, average daily traffic (ADT), and average vehicle risk (AVR) have been calculated in the past. Of these, ADT and AVR appear to be the two best “stand in” parameters for economic impacts and life safety respectively, while a larger consequence model is developed. AVR has been used in the past and was part of the original RHRS and is an early attempt to capture the life safety risks of an event by calculating the amount of time a car is passing below a problematic slope.

Equation 3. Average Vehicle Risk

$$AVR = \frac{ADT \ x \ SL/24}{PSP} * 100$$

Where: ADT = Average Daily Traffic (cars/day)
SL = Slope Length (km)
PSP = Posted Speed Limit (km/hr)

The formula while useful has certain flaws, it does not account for vehicles traveling at a speed other than the posted speed limit and it also does not account for the effects of congestion where cars may be stopped beneath the slope in traffic for a sometimes extended period of time (Russell, Santi, & Higgins, 2008) and can therefore significantly underestimate the risk to a vehicle where speeds are slower or congestion occurs (Pantelidis, 2011). ADT (Average Daily Traffic) has similar flaws and neither of these parameters account for different conditions where visibility may be impacted such as fog, rain, or night time conditions. Nor do either account for different traffic counts and night or day. This is one of the strengths of the dam/levee consequence modeling as daytime and nighttime risks to people are explicitly included in the consequence estimates. Thus, these is a need to establish a better methodology to calculate life safety risk if that is going to be used as the criteria. This is the reason for the formulation of risk with the Federal Lands in Equation 2, which lays out three factors to estimate to calculate the life safety risk and consequences. However, on a life safety basis alone, the repair of rockfall and unstable slopes may not compare well with other improvements to the highway network. These events are very consequential and can cause long closures when looking at impact to the public on factors other than direct life safety. Accidents on secondary and tertiary roads not meant to accommodate heavy traffic diverted from a closure can also occur, but these are not accounted for on a direct life safety basis that focuses the probabilities that someone will be in the path of the rockfall. As second order effects are notoriously difficult to capture, in this case economic impacts may be a more rational means to assess and choose sites for mitigation.

Economic impacts are also more easily obtained and ADT can be a good initial screening level stand in until a better consequence estimates are calculated. Larger traffic volumes indicate a more important roadway. However, there are other factors such as length of detour, number of lanes closed etc. that also contribute to the impact of a roadway closure, which was the reason for the development of the RCI (Rockfall Closure Impact, Table 3 above) in Tennessee. There are a variety of methods in the literature for calculating the costs of road closures, with readily available models. One of the most promising for use with rockfall analysis is the QUADRO model developed in Scotland (Winter, et al., 2016). This model calculates in terms of economic costs:

1. Delays to road users – including time differences in travel priced to the value of the user's time based on the types of vehicles, occupants and trip purpose data
2. Fuel carbon emissions -accounting for additional fuel used where there is a lane or road closure.
3. Costs of accidents in terms of both additional delays and direct costs such as property damage, insurance etc. and
4. Allocates traffic to diversion routes representing the time needed for alternate routes in the event of a complete or partial lane closure.

The model is considerably more sophisticated than the RCI rating, using economic data that is available to a transportation agency to more accurately predict consequences. While no model will be exact, models such as this are used regularly to predict economic consequences in other arenas. One of the biggest advantages of using a model of this type is that the model is purely on the consequences part of the risk equation for economic impacts. The cause of the road/lane closure is not relevant to the model. Thus, we can estimate the probability of the failure occurring with and now combine it with economic risk. Best of all, this kind of model can be used for all our geotechnical asset management systems, as it can capture costs for a retaining wall failure, or from a landslide as easily as from a rockfall.
It's available for free download from: https://www.tamesoftware.co.uk/quadro/quadro.html

Another aspect that needs to be established are tolerable risk guidelines. Questions such as "How long a total road closure is acceptable?" need to be answered. While this depends on the location, infrastructure surrounding the problem, the cost of repairs and impacts to the road network, the tolerance for economic impacts needs to be assessed. The factors mentioned above for dam and levee safety can be adapted here as well:

1) TRG 1: society is willing to accept to obtain the benefits of infrastructure or other activity – we must make a judgment on what society will consider an acceptable risk in order to obtain the benefits of a highway system. There are not insignificant numbers of fatalities on roadways due

to accidents, but in general, roads are not closed until a level of accidents or fatalities reach over this background level.

2) TRG 2 - society does not view as negligible the risks presented, - Rockfall accidents receive wide press coverage as much because the fatality method is unexpected. Sites with more fatalities caused by the geometry of the roadway often receive less attention. Most members of the public do not consider the hazards presented by slopes of this type to be "normal" and thus are weighted by the public overall more heavily than the fatality rates might indicate. Further – as many of these events can result in unexpected and significant impacts to the road network, they are unplanned risks that receive more attention.
3) TRG 3 - society generally views as being properly managed by the owner (because the risks are being located, assessed and managed) – This is where management systems come into play as the owner can discuss how these risks are managed, tracked and mitigated. A good example of this is a recent newspaper article in Tennessee titled: "Tennessee is ready for Rockslides but Landslides are Another Story" https://www.wjhl.com/wjhl-weather/tdot-is-ready-for-rockfalls-but-landslides-are-another-story/ (Fuller, 2019).
4) TRG 4: the owner continues to reduce to as low as practicable (the owner has a program in place to repair and mitigate sites before the events occur). This is where smaller maintenance items can come into play, cleaning out ditches, scaling segments of roadway periodically that have smaller height cut slopes. These types of activities assist in the further reduction of risk and should be part of a properly managed program. This is where these rockfall systems fit into the overall asset management framework, because once risk-buydown criteria are establish, an agency can focus on those activities that reduce the overall risks from the portfolio, rather than just focusing on the "worst first."

Scale evaluations to the need

Data that have been gathered previously on slope heights, slope angles, ditch effectiveness, block size etc. are all useful in this framework, because they inform the estimated probabilities. Thus, this method can be used along with the information that has already been gathered at sites over the past 30 years that rock slope ratings have been implemented. A screening level assessment should be possible, for example with the data gathered in Tennessee looking at the old hazard rating system along with photographs of the site to assess these probabilities. However, the highest probability sites could then be assessed in detail, with site visits, supporting numerical analyses and additional data gathering. These activities within the framework will sharpen the pencil on the subjective risk assessment, reducing the uncertainties as much as practicable.

In order to begin without a proper consequence model, we need some stand-in for the "economic impact" or we must rely on a much more subjective judgement of consequences on a life safety basis. For a screening level assessment this paper proposes simply using ADT as the initial screening while the estimate of the annual probability of failure is used to capture the hazard. Note here the ADT is shown along with the average frequency of traffic based on that number. This does not consider congestion, or speed limit, so it's still a crude tool and is not enough for detailed assessments. However, this can be used to illustrate how we can categorize rockfall sites by an overall risk rating (RFR – Rockfall Risk) as shown in Figure 8. This graphic shows a comparison with the old RHRS categorization of A, B and C sites. In order to bring this to a semi-quantitative method the annual probabilities ranges should be assigned to these qualitative assessments (very high, high, moderate, low, very low). In this example, high traffic count roadways are weighted heavily because they have the highest impact on economic activity when closed and the highest impact on traffic when detours are required. Individual failure modes should be plotted as well as the combined probabilities of failure of the site. This gives both an overall view of the risk of a site, coupled with an understanding of the components that make up that risk. In the Unstable Slope Management formulation, $P_{(occ)}$ could be plotted on the y-axis with the ADT comparison on the X-axis. This could be done quickly for a screening tool, or the failure modes could be laid out and estimated, then compared to the ADT for a more accurate assessment.

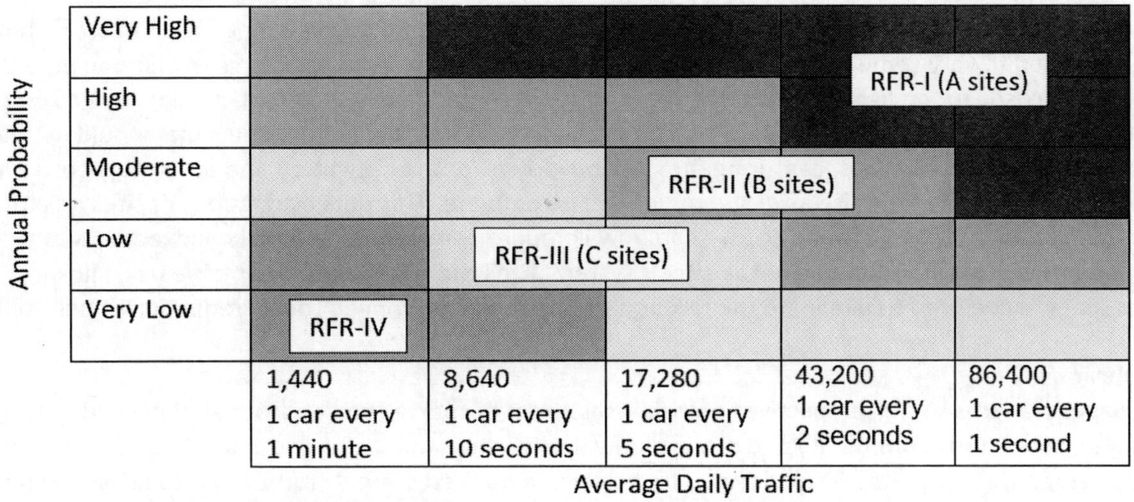

Figure 8. Rockfall Risk Screening Tool - Conceptual

Again, a model such as that produced by QUADRO will be more accurate than ADT in capturing economic impacts, and this would be included on the X axis rather than ADT. This would bring the rating into a fully quantitative stance. This might be done only for those sites that classify as RFR-I(A) or both RFR-I(A) and RFR-II(B). For a life safety evaluation, we would need to assign values based on the concept of incremental life loss. That is, the life loss that can be caused by failure at the site, over and above what would be expected for probability of life loss just by being in a vehicle and traveling through that segment of roadway. However, again, this formulation would leave out indirect fatalities and injuries due to heavy traffic diverted onto smaller roadways that may have inadequate capacity.

Establish that hazard probabilities are estimated on an average annual basis.
With estimated probabilities it can be easy to overestimate the importance of a factor where large timelines are involved. People are not great estimators of probabilities that contain several factors, which is why event trees (decision trees) break down the probabilities into more manageable chunks. This is the reason to use the kind of risk formulation in the Federal Lands system, but to expand that system and break down $P_{(occ)}$ into smaller more easily estimated parts. Estimating the probability of failure in a given year can help with a "reality check" as well. For example, where the raveling risk of rocks reaching the roadway and impacting traffic is estimated to be 0.0333 per year and we have 30 years of maintenance data with relatively accurate records, we can compare that probability estimate with the number of recorded events to see if our estimates make sense. This is where systems such as the Federal Lands one shines as it sets up a system by which incidents are tracked. This will give a needed reality check for estimators over time. Also, annual probabilities here would likely be considerably higher on the y axis than those used for dam and levee safety. This is to be expected because few rock slopes have the same life safety implications in the event of failure as a dam, failure of a rock slope is likely to produce considerably fewer fatalities and a lesser economic impact. Both a life

safety consequence and an economic model need to be developed and used for consequences. While life safety is extremely important, it may be that the economic impacts are a better choice for managing risks, unless indirect life safety implications (such as increased accidents on a detour route) can be effectively captured.

Establish a "recurrence" interval of evaluations.
Finally, there is a need to explicitly account for a re-evaluation of slopes after some period has elapsed. Every 10 years is the frequency used by USACE for routine re-evaluation with additional evaluations allowed where certain events occur or distress features that indicate a developing problem are noted. This may need to vary somewhat in high consequence areas, but a 5 year or less re-evaluation period may be too costly for an agency to absorb. After 10 years there is likely to be noticeable changes on sites if no further work is done. A site should also be evaluated after significant activities that would mitigate the risk have been completed, capturing the "risk buydown" of the mitigation and accurately portraying the risks that remain at the site after the mitigation is complete. This puts evaluations of sites on a more asset management friendly footing as we can now compare annual risks and make judgements as to how expensive a repair is needed at the site. It systematizes the concept of acceptable risk, allowing funds that may be used to eliminate the risk at a site to be better applied to risk reduction at another.

Conclusions
This paper presents some perspectives on risk assessment of rockfall gained from working within the dam and levee safety community and offers a new approach on managing rockfall risks and using rockfall assessment systems. There is need to modernize our assessment framework, to bring rockfall assessments truly into a risk management stance. This can build on earlier work with screening assessments scaled to the level of need and more detailed assessments of more significant sites. While work remains to be done, this paper offers some paths forward for future development to allow better risk-informed decision making. Once event trees and consequence calculations are set for rock slope evaluations, similar event trees can be constructed for other slope failures, extending the assessment system for landslides as well, meeting an additional need. Moving in this direction and implementing a truly risk based approach would allow Rockfall Inventories to truly become part of a Geotechnical Asset Management System and inform decision makers about the true costs of under-investing in this area. While still relying on subjective judgement, this methodology guides that judgement and captures it in a number that can be compared to other sites, rather than just producing a ranking. It also will allow agencies to quantify the risk buy down of mitigation alternatives both on a site, and on a portfolio basis.

Works Cited

Bateman, V. (2005). Rockfall Closure Impact and Economic Assessment for Rockfall Sites in Tennessee. *55th Highway Geology Symposium* (pp. 385-400). Kansas City, KS: Highway Geology Symposium.

Bateman, V., Smerekanicz, J., & Sneyd, D. (2010). Emergency Rock Slope Stabilization in the Ocoee Gorge. *Proceedings of the 61st Highway Geology Symposium* (pp. 331-348). Oaklahoma City, OK: Highway Geology Sumposium.

Brawner, C. O., & Wyllie, D. C. (1975). Rock Slope Stability of Railway Projects. *Proceedings of the American Railway Engineering Association* (p. 8). Vancouver, Canada: American Railway Engineering Association.

Eisenhower, D. (1957). Remarks at the National Defense Executive Reserve Conference. *Public Papers of the Presidents of the United States, Dwight D. Eisenhower*. Washington, DC.

Fell, R., Ho, K., Lacasse, S., & Leroi, E. (2005). A framework for landslide risk assessment and management. *Proceedings, International Conference on Landslide Risk Management,.* Vancouver, CA: Taylor & Francis.

Ferrari, F., Giacomini, A., & Thoeni, K. (2015). Qualitative Rockfall Hazard Assessment: A Comprehensive Review of Current Practices. *Rock Mechanics and Rock Engineering*.

Hubbard, D. (2009). *The Failure of Risk Management: Why it's Broken and How to Fix It.* Hoboken, NJ: John Wiley and Sons.

Loar, T. (2017). USACE Dam Risk Assessment Overview and how Engineering Geology Contributes to the Level of Confidence and Results. *Association of Environmental and Engineering Geologists Annual Meeting.* Colorado Springs, CO: AEG.

Pantelidis, L. (2011). A critical review of highway slope instability risk assessment systems. *Bulletin of Engineering Geology and the Environment*, 395-400.

Pierson, L. A., Davis, S. A., & Van Vickle, R. (1990). *Rockfall hazard rating system implementation manual. Report FHWA-OR-EG-90-01.* Washington, D.C.: Federal Highway Administration, US Department of Trans-portation.

Pierson, L., & Van Vickle, R. (1993). *Rockfall Hazard Rating System - Participants' Manual.* Washington, DC: Federal Highway Agency (FHWA) Office of Technology Applications.

Russell, C. P., Santi, P., & Higgins, J. D. (2008). *CDOT-2008-7: Modification and Statistical Analysis of the Colorado Rockfall Hazard Rating System.* Denver, CO: Colorado Department of Transportation .

USACE. (2014). *ER 1110-2-1156: Safety of Dams - Policy and Procedures.* Washington, DC: US Army Corps of Engineers.

Vandewater, C. J., Dunne, W., Mauldon, M., Drumm, E., & Bateman, V. (2005). Classifying and Assessing the Geologic Contribution to Rockfall Hazard. *Environmental & Engineering Geoscience*, 141-154.

Vessely, M., Robert, W., Richrath, S., Schaefer, V., Smadi, O., Loehr, E., & Boeckmann, A. (2019). *Geotechnical Asset Management for Transportation Agencies, Volume 2 Implementation Manual.* Washington, DC: National Cooperative Highway Research Progran, National Academy of Science. Retrieved from https://doi.org/10.17226/25364.

Vick, S. G. (2002). *Degrees of Belief: Subjective Probability and Engineering Judgement.* Washington, DC: ASCE Press.

Winter, M. G., Shearer, B., Palmer, D., Peeling, D., Harmer, C., & Sharpe, J. (2016). The Economic Impact of Landslides and Floods on the Road Network. *Procedia Engineering*, 1425–1434. Retrieved from https://www.tamesoftware.co.uk/

Wyllie, D. (1987). Rock Slope Inventory. *Proceedings, Federal Highway Administration Rockfall Mitigation Seminar, FHWA Region 10,* (p. 25). Portland, OR.

Wyllie, D. (2015). *Rock Fall Engineering.* Boca Raton, FL: CRC Press.

Wyllie, D., & Mah, C. (2004). *Rock Slope Engineering, 4th Ed.* New York, NY: Spon Press.

Slide Ridge Culvert Replacement

Robert E. Kimmerling, P.E., L.E.G.
Kimmerling Consulting LLC
13229 42nd Avenue W
Mukilteo, WA 98275
(206)-406-6245
rkimmerling@gmail.com

Prepared for the 70th Highway Geology Symposium, October, 2019

Acknowledgements

The author would like to thank the individuals/entities for their contributions in the work described:

Paula Cox, P.E. – Chelan County Public Works
Anne Streufert, P.E., S.E. – KPFF Consulting Engineers
Jason Pang, P.E. – KPFF Consulting Engineers
Pat Flanagan, P.E. – Indicator Engineering, Inc.
Joanna Curran, L.G. – Indicator Engineering, Inc.

Disclaimer

Statements and views presented in this paper are strictly those of the author, and do not necessarily reflect positions held by their affiliations, the Highway Geology Symposium (HGS), or others acknowledged above. The mention of trade names for commercial products does not imply the approval or endorsement by HGS.

Copyright Notice

Copyright © 2019 Highway Geology Symposium (HGS)

All Rights Reserved. Printed in the United States of America. No part of this publication may be reproduced or copied in any form or by any means – graphic, electronic, or mechanical, including photocopying, taping, or information storage and retrieval systems – without prior written permission of the HGS. This excludes the original author.

ABSTRACT

The South Lakeshore Road, along the west side of Lake Chelan, Washington, is prone to repeated closures and emergency repair work due to debris flows originating from the aptly named Slide Ridge above the highway. Previous attempts to control the debris flows included construction of a linear flow channel from the apex of the alluvial fan below Slide Ridge to a catchment basin upstream of the highway. However, recurring debris flow events over the last 15 years overwhelm the catchment basin, plug the existing culvert conveyances and over-top the highway resulting in closure and the need to remove the debris deposits and clear the culverts.

The goal of the culvert replacement project is to remove the under-sized culvert and catchment basin and provide conveyance that will promote unimpeded delivery of debris flow sediments to the bottom of the alluvial fan. Design challenges for the project include a mapped geologic feature named the "Granite Slide" which is identified as an "incipient blockslide" with the potential to "fail and descend to the lake either gradually or swiftly".

Seismic design considerations are also complicated by the proximity of the project site to the postulated epicenter of the largest historical earthquake in Washington State. The fault structure responsible for this event, which occurred in 1872, has not been identified and therefore remains unrepresented in probabilistic seismic hazard maps and codified design criteria. These design challenges are discussed, and the current culvert replacement configuration is outlined.

INTRODUCTION

South Lakeshore Road is the primary local access and only paved highway along the south and west side of Lake Chelan. Lake Chelan is located in Chelan County, Washington State as shown in Figure 1. Chelan County has been repeatedly burdened with emergency response to debris flows that cause road closures of South Lakeshore Road. With the exception of forest roads, which are only seasonally passable, detours are virtually non-existent in the event that South Lakeshore Road is closed. The debris flows originate from Slide Ridge, which rises over 3,200 feet above Lake Chelan as can be seen in the LIDAR imagery in Figure 2. Slide Ridge historically produces debris torrents during thunderstorm and rain-on-snow events on an almost annual basis.

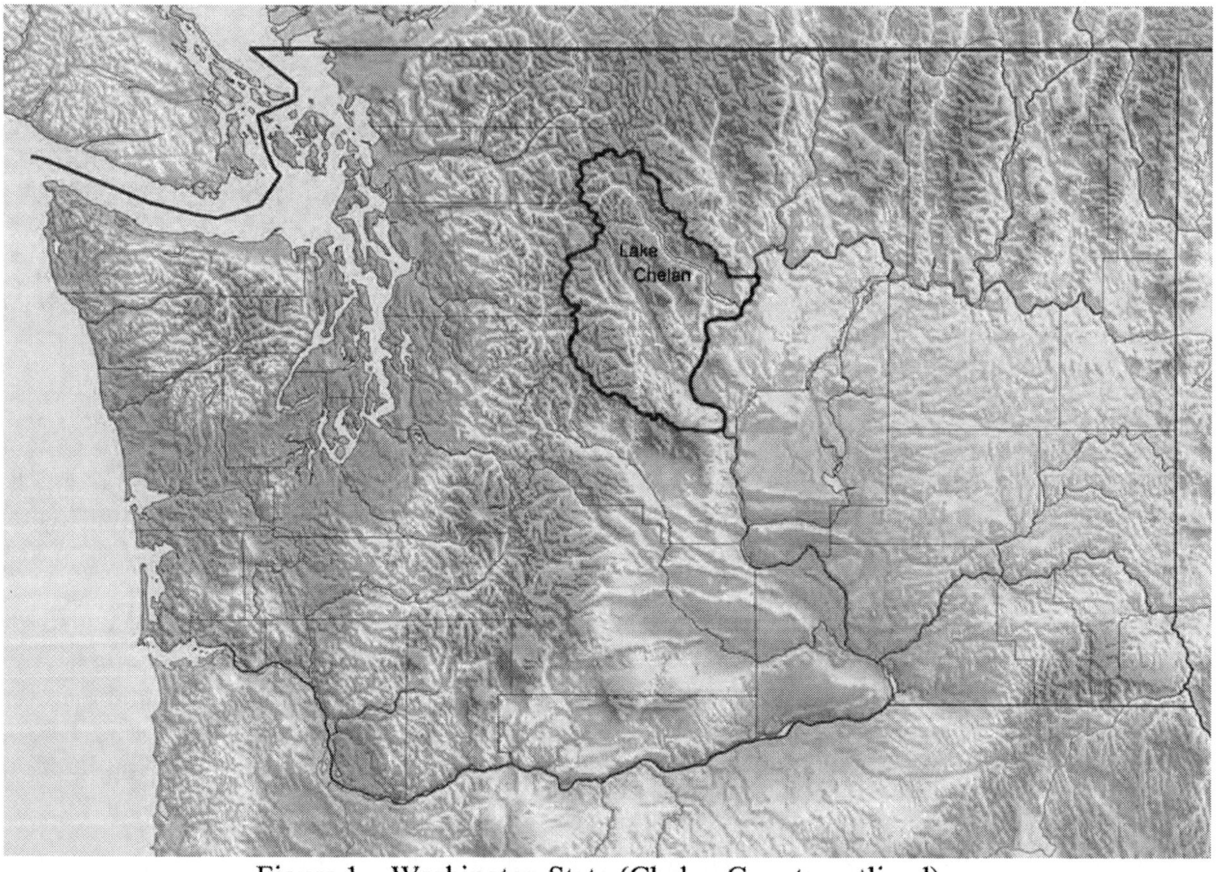

Figure 1 – Washington State (Chelan County outlined)

Debris flows are discharged down a combined natural and constructed drainage channel with levees (visible in Figure 2) that empties into a containment basin on the west (upstream) side of South Lakeshore Road. This combined channel and containment basin was constructed in 1994 from the apex of the alluvial fan below Slide Ridge to South Lakeshore Road in an attempt to protect the highway, the traveling public and private property downstream. The containment basin drains beneath South Lakeshore Road via a 96-inch corrugated metal pipe culvert, shown in Figure 3, intended to convey runoff and debris flow material downstream to Lake Chelan.

Figure 2 – Slide Ridge LIDAR imagery

Figure 3 – Existing Culvert and Catchment Basin

Over the last 15 years, these debris flow events have occurred every 1 to 4 years, with estimated volumes between 1,050 to 15,900 cubic yards of debris. Depending on the size of the event, the debris was deposited in the containment basin, and over-topping the roadway for the larger events. These events restrict access along South Lakeshore Road until emergency response establishes a temporary single-lane bypass on the downstream side of the paved roadway that is used until the main road is cleared of debris and restored to two-lane operation. Frequent removal of slide material and channel maintenance is costly for the County and the reoccurring overtopping is a major safety hazard to the public. The previous efforts to manage the hazard have therefore only been marginally successful.

The existing containment basin is capable of holding approximately 4,000 cubic yards of sediment, based on a 2017 survey by Chelan County. During significant debris flow events, the existing culvert is inadequate to convey the material and becomes blocked. The volume of debris has overwhelmed the basin and overtopped the road in the majority of the events over the last 15 years. The maximum height of debris deposited on the roadway was approximately 12 feet in the 2005 event. The relative locations of South Lakeshore Road, the flow channel and the catchment basin are shown in Figure 4. Figure 5 shows the catchment basin and mud caking on a utility pole that indicates the height of debris flow over the roadway from a prior debris flow event.

Figure 4 – Project Area

Figure 5 – Catchment Basin

Project Setting and History

Chelan County is the third largest county in the State of Washington. Lake Chelan lies entirely within the County and is not only the dominant water body in the County, but is also the largest natural water body in the State (*1*). Both the lake and the county derive their name from a Salish Indigenous word, "Tsi – Laan," meaning "Deep Water,"(*2*) which is certainly the case, as the lake is the deepest in Washington State and the third deepest in the United States (*1*). The lake level was raised 21 feet in 1927 by construction of a concrete dam at the outlet of the lake (*1*). Below the dam the Chelan River flows to its confluence with the Columbia River.

Lake Chelan occupies a 50+ mile long valley of alpine glaciation origin, although advance of the Okanogan lobe of the Cordilleran ice sheet did intrude on the eastern end of the valley which resulted in the formation of two morphometric basins separated at "The Narrows," shown in Figure 6 by a submerged feature called the "Sill," shown in Figure 7 (*3*). Coincident with the "the Narrows" is a coalesced alluvial fan locally referred to as Hollywood and Shrine Beaches. These alluvial fans, in turn, lie below Slide Ridge as shown in Figures 2 and 4.

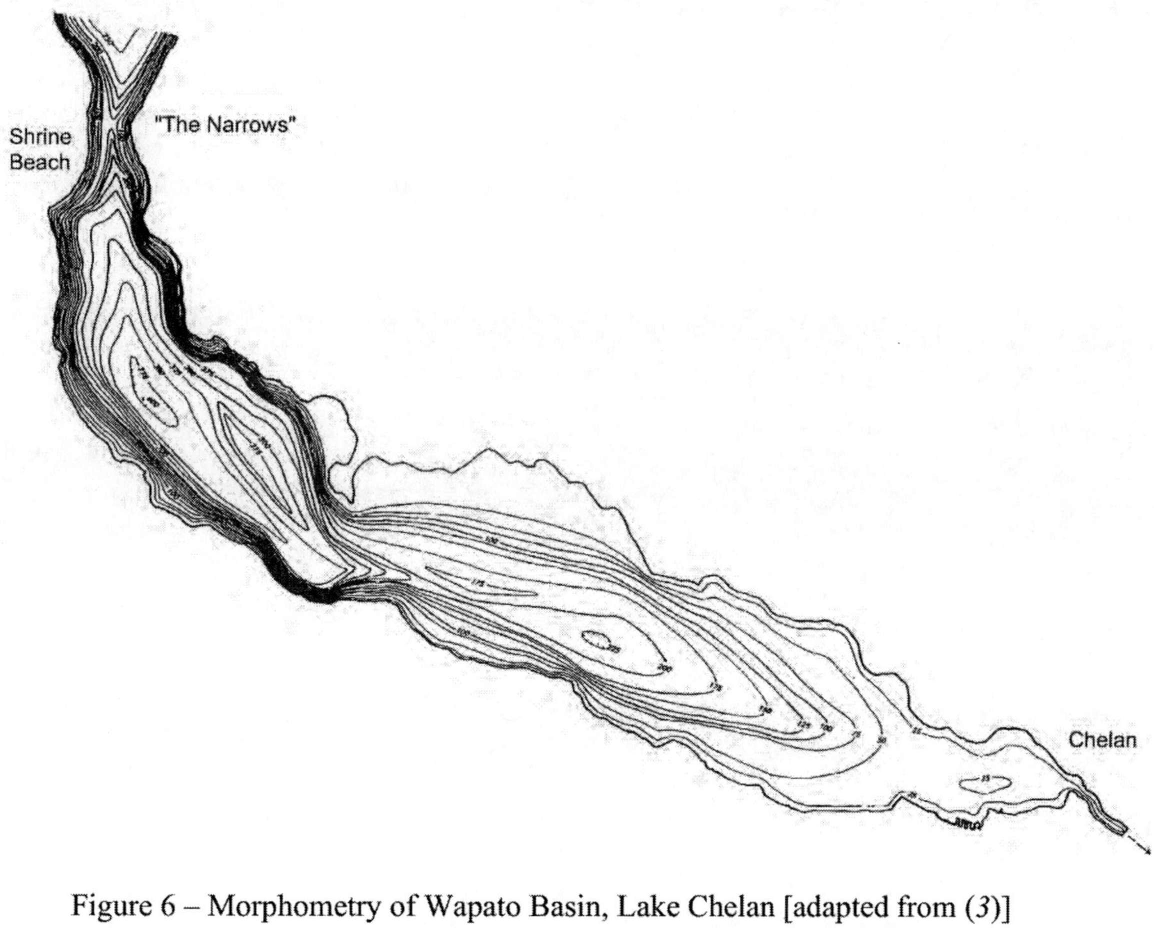

Figure 6 – Morphometry of Wapato Basin, Lake Chelan [adapted from (*3*)]

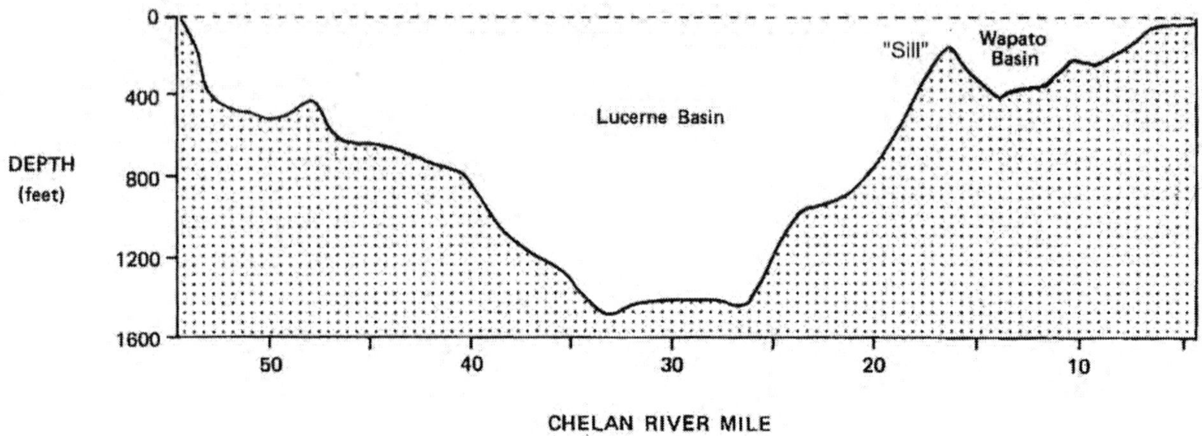

Figure 7 – Profile of Lake Chelan [adapted from (*3*)]

GEOLOGY AND GEOLOGIC HAZARDS

Mapping of the geology of the Slide Ridge area is available at the 1:100,000 scale of the Chelan 30′ by 60′ Quadrangle (*4*). A cropped and enlarged portion of this geologic map is provided in Figure 8. The alluvial fans of Shrine and Hollywood Beaches are visible at the location of "The Narrows", with Slide Ridge to the left of the fans. Although difficult to discern, a feature labeled "Granite Slide" is present to the left of Shrine Beach and includes the mapped outcrop of a rhyolite dike denoted "Tcrd".

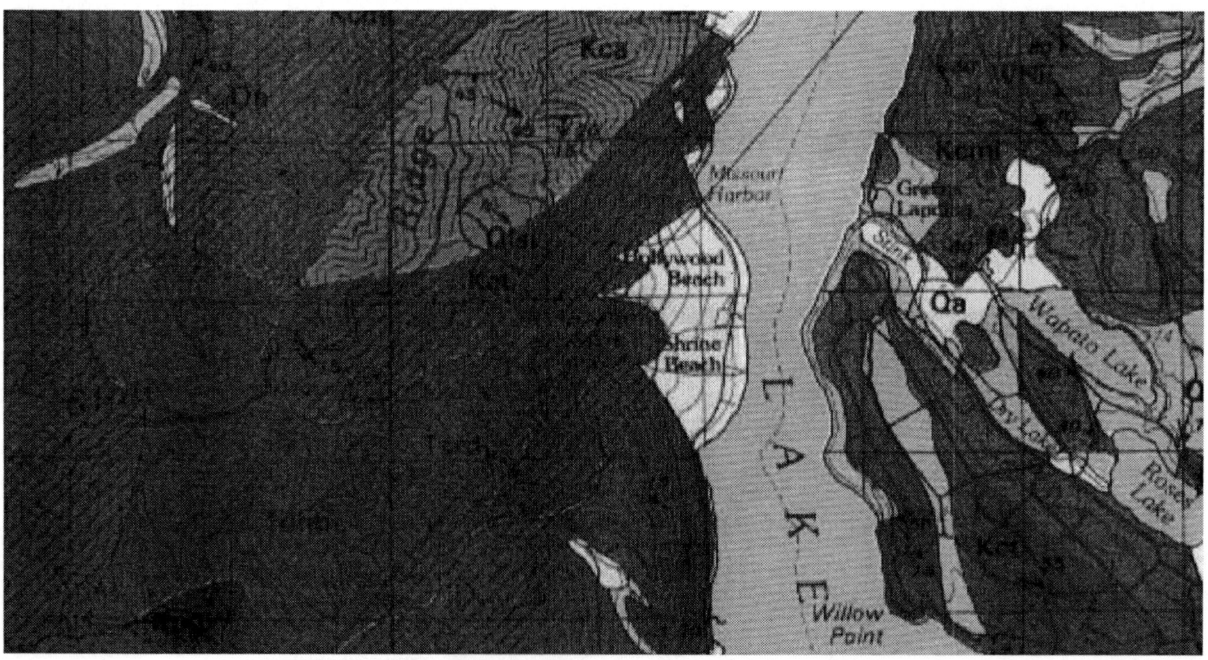

Figure 8 – Site Geology [excerpted from (*4*)]

The project area, including Slide Ridge, is part of the Chelan Mountains Terrain (*4*). Bedrock and surficial geologic units identified by this mapping in the Slide Ridge area include:

- Amphibolite and Hornblendite Migmatite (Kca) – Includes pods and lenses of hornblendite and dark amphibolite ranging from centimeters to several hundred meters across.
- Tonalite (Kct) – Hornblende-biotite and biotite tonalite. Rock is commonly strongly gneissic in outcrop. Locally the tonalite is cut by lighter colored tonalite dikes.
- Rhyolite dikes (Tcrd) – Predominantly white to yellow or brown rhyolite with small phenocrysts of plagioclase and/or quartz.
- Incipient blockslides (Qlsi) – Large nonrotated mass of bedrock extensively crevassed as a result of slight movement toward nearby free faces. Crevasse-arrow symbol shows direction of movement.
- Alluvium (Qa) – Alluvium includes poorly sorted gravelly sand or sandy gravel of alluvial fans. The fans of Shrine and Hollywood Beaches are of this material, and also constitute the lower slopes of the glacially eroded trough now occupied by Lake Chelan.

Several geologic hazards are present that could affect the project area, including mass wasting (debris flows and landslides) and strong ground motion associated with earthquakes.

Debris Flows

Debris flows are the most common mass wasting events originating from Slide Ridge. Triggering of the flows is closely associated with intense precipitation events. Rapid runoff from cloudburst storms or heavy rain-on-snow events mobilizes loose material on the raw and almost entirely denuded steep slopes in the drainage basin below Slide Ridge. The debris torrent is funneled to the apex of the alluvial fan below where the basin is closely and hardly flanked by bedrock outcrops on both sides of narrowest part of the flow path just above the apex of the alluvial fan.

The slope angles in the upper basin, above the apex, are around 45 degrees and steeper. The gradient of the flow path flattens to an average of about 20 percent between the apex and the catchment basin. Below the roadway, the channel flattens further to between 10 and 14 percent.

The frequency and magnitude of Slide Ridge debris flow events was analyzed by the project team by hand-fitting a log-normal distribution to the total debris flow volumes using the 15-year record of events. The resulting frequency analysis is shown in Figure 9. The analysis estimates a 10,000 cubic yard event to have an approximate 5-year return period, while a 100-year event would mobilize approximately 20,000 to 25,000 cubic yards of debris.

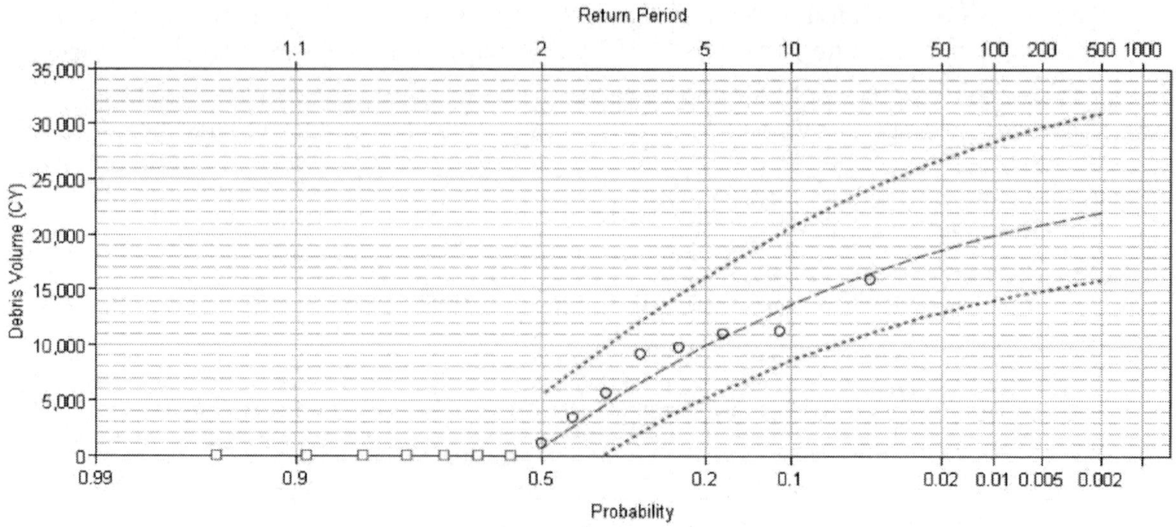

Figure 9 – Frequency Analysis of Debris Volume

These historic events mobilize up to two- to four-foot boulder-size material based on observations of the debris observed in the catchment basin (Figure 3). However, considerably larger material is visible in the channel side-slopes at locations mid-distance down the alluvial fan as can be seen in the photograph in Figure 10. Some of these blocks are as large as automobiles and are therefore unlikely to be mobilized by debris flows of the sizes recorded over the last 15 years. This suggests that the Slide Ridge basin is capable of producing much larger flows or mass wasting events.

Figure 10 – Large Block Material in Channel Side Slopes

Landslides

The significance of the "Granite Slide" feature is detailed in the pamphlet text that accompanies the Chelan 30′ by 60′ Quadrangle (*4*) as follows:

- "Two incipient blockslides perched on steep slopes 1,000 m above water bodies—one above Lake Chelan, another above Lake Wenatchee—could be severe hazards during future large earthquakes. Although both of these incipient slides may have been in their present form and positions during the largest historic earthquake of the region (in 1872), it is only a matter of time before they will fail and descend to the lakes either gradually or swiftly."
- "The Columbia is now a series of reservoirs, and any future slides will descend into lake water, where displacement could be locally devastating. Should one of the incipient blockslides above Lake Chelan or Lake Wenatchee suddenly detach, it would probably acquire great speed and momentum on its descent. When the slide enters a lake, water would be suddenly displaced to generate a wave that could devastate the shoreline area for many meters if not tens of meters above lake level."

The resulting wave, not unlike a large seiche, would also very likely over-top and fail the 21-foot concrete dam at the lake outlet, releasing something on the order of 125 to 150 acre-feet (more than 5.5 million cubic feet) of water into the Columbia River and the downstream pool of the Rocky Reach Dam. While the emptying of Lake Chelan would take some time, other studies of dam failure scenarios further up the Columbia River conclude that intermediate power generation dams like Rocky Reach would all over-top and fail if one of these larger upstream dams were to fail (*5*).

Site Seismicity

The project site is located within the uplifted bedrock complex of the Cascade Range. This area is not as seismically active as the area west of the Cascades but does experience seismic activity. The nearest mapped potentially active fault is the Class B Straight Creek/Evergreen Fault system. This is a north-south trending feature mapped about 50 miles west of the site (*6,7*).

However, the largest historical earthquake observed to date in Washington State, with an estimated magnitude of approximately 6.5 to 7.0, with modified Mercalli intensities up to VIII, occurred on December 14, 1872 in the northern Cascade Mountains (*8*). Some recent research and thinking suggests that this event may have taken place on a postulated Chelan Seismic Zone (*9*). Even more recent research by the USGS narrows the potential epicenter for the 1872 event to a zone that coincides with a prolific zone of micro-earthquakes referred to as the Entiat cluster located about 15 miles to the south of the Slide Ridge area (*8*). The Entiat cluster earthquakes are relatively small, with magnitudes of 3.5 or less, as shown in Figure 11 (*10*). Figure 11 also provides regional context and proximity to the Hollywood Beach and Slide Ridge area.

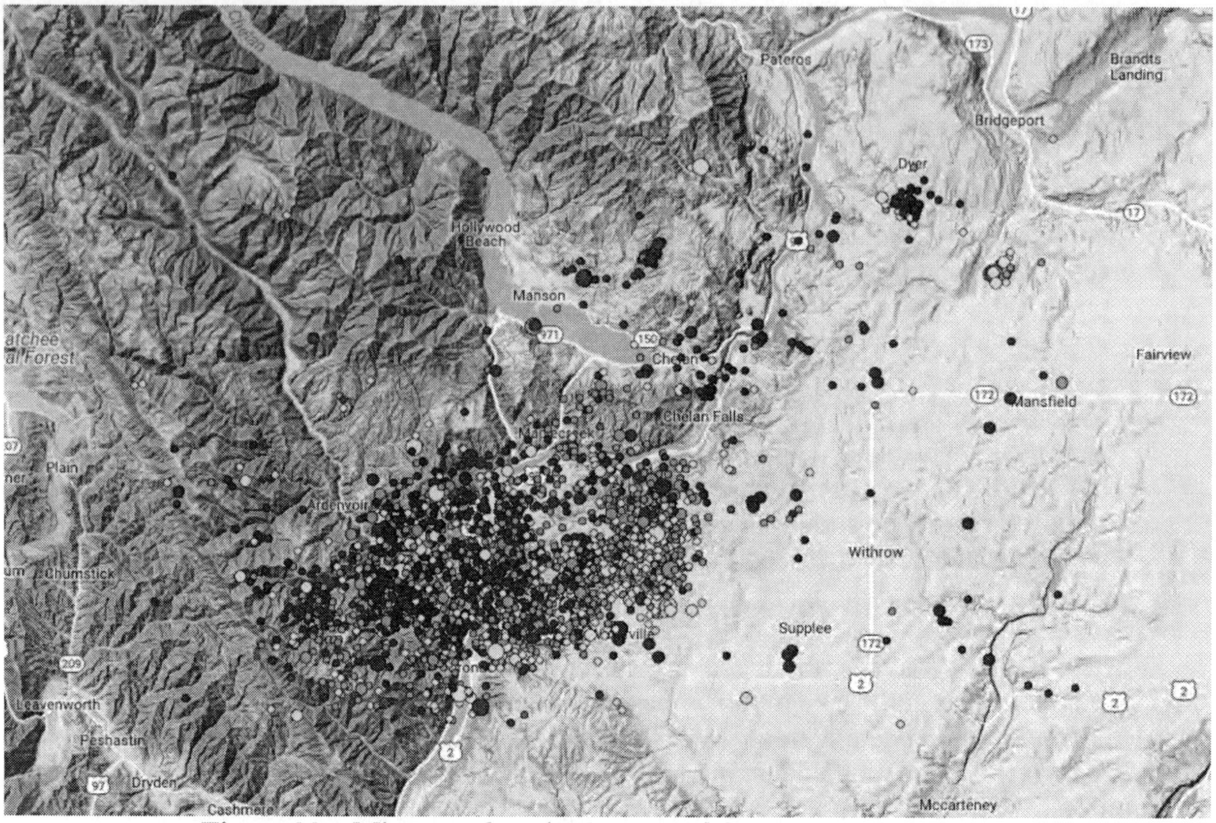

Figure 11 – Micro-earthquake Swarm of the Entiat Cluster (*10*)

There is more detail regarding current thinking about the 1872 event in a presentation by Mr. Tom Brocher[1] available here: https://earthquake.usgs.gov/contactus/menlo/seminars/1103 . Of particular interest to the Slide Ridge project is a landslide that was triggered by the 1872 event just north of the town of Entiat along the Columbia River. This landslide blocked the flow of the Columbia River for several hours and remnants of the slide blocks are still visible in the pool of Rocky Reach Dam. While the strong ground motion of the 1872 event was sufficient to mobilize a landslide in this location, the Granite Slide (described above) did not fail. Brocher et al. (*8*) also conclude that the aftershock sequence that is associated with the 1872 event is consistent with low slip-date faults that have relatively long recurrence intervals, on the order of a few to several thousand years.

Code-based seismic design criteria for this type of project is available in the AASHTO bridge design guidelines (*11*). AASHTO, similar to IBC guidelines, follows a probabilistic approach to seismic design. The earthquake hazard at a particular site is assigned according to the proximity of known seismic sources and the strong ground motion potential of each of those sources. The deaggregation of strong motion potential at the project site, following the AASHTO criteria, which is based on a 7 percent probability of exceedance of in 75 years, or an approximate return interval of 975 years, is shown in Figure 12 (*12*).

[1] USGS, Menlo Park, California

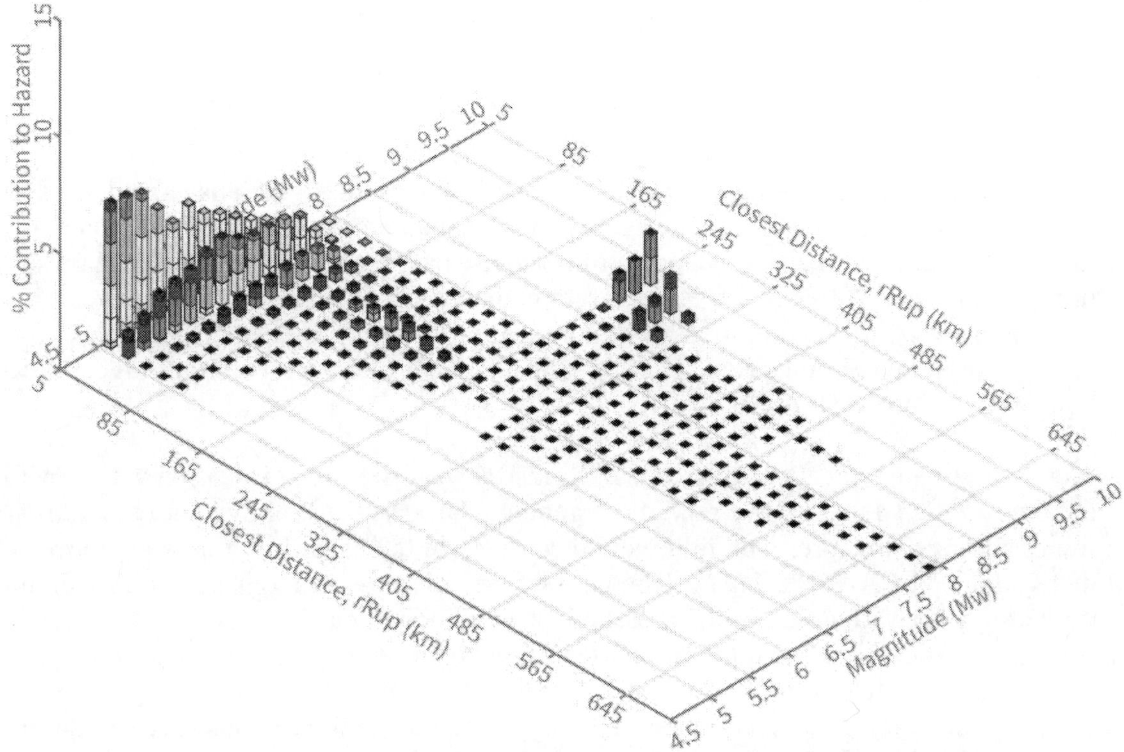

Figure 12 – Deaggregation of Probabilistic Seismic Hazard at Slide Ridge Site (*12*)

In Figure 12 there are two primary sources of earthquake hazard according to the probabilistic model. The first is near source random fault potential represented by the spikes in contribution hazard with relatively small Richter magnitudes at distances less than 50 kilometers from the project site. This proportion of the hazard is associated with the Entiat cluster of micro-earthquake swarms as well as a possibility of small events that could occur in the close vicinity of any particular site according to the general seismicity of the region. The second source is the known fault potential from the Cascadia Subduction Zone (contribution spikes in the Richter magnitude range of 8.5 to 9.5 at a distance of roughly 250 kilometers).

It is important to note that there is no significant probabilistic contribution in the range of Richter magnitude 7.0 at a distance of approximately 15 to 25 kilometers, which is the approximate energy potential and distance of the current thinking for the epicenter of the historic 1872 earthquake event. This is because the source fault, its length and slip rate has yet to be certainly identified. Until such time as this information has been determined, the strong ground motion potential associated with the largest known historic earthquake in the state of Washington remains absent in probabilistic hazard models.

RISK MANAGEMENT AND DESIGN DECISIONS

As outlined above, there is no lack of geologic hazard potential at the Slide Ridge culvert replacement site. In summary, the following hazards needed to be addressed for final design of the culvert replacement:

- Recurring debris flows,
- Large scale landslide potential, and
- Strong ground motion due to seismicity.

A consistent approach for managing the risk presented by these various hazards was provided through logical design criteria decisions. The fundamental purpose of the project is to replace the existing culvert that is not capable of routinely passing the frequent debris flows emanating from the Slide Ridge basin with a larger conveyance. Through an alternatives analysis, a single span bridge, combined with some realignment of the flow channel, was selected as the optimum configuration. The fundamental basis of design is for a bridge project which is provided by AASHTO (*11*).

New bridges are typically designed for 75- to 100-year life. An evaluation was therefore conducted that compared the relative risk of the geologic hazards listed above with the normal design life of a bridge structure. The following discussion of the hazards is framed in terms of reliability as adopted by AASHTO (*11*). The new structure is expected to have certain minimum levels of reliability for various conditions, termed "load combinations" in AASHTO. These load combinations are divided into the three categories (termed "limit states") of service, strength and extreme. Service limit state conditions or load combinations are those that the bridge is expected to withstand without damage or a reduction in the expected design life of the structure. Service conditions are typically defined in terms of deformations, such as settlement of foundations or expansion and contraction of the structure elements in response to thermal effects. Strength limit states ensure that the bridge will tolerate load conditions such as traffic live load with a suitable margin of safety. Extreme limit states include loads or conditions that are relatively rare and that the bridge is expected to survive without collapse or imposing the risk of loss of life to the traveling public. Extreme events include scour potential under flood conditions, strong ground motions due to earthquake and other transient, but rare conditions.

Debris Flows

While transient in nature, the frequency of the debris flows at the project site demands a reliability of the new bridge structure that is higher as compared to other extreme events such as the design seismic event which may only affect the structure once or twice (if at all) during its design life. While there is evidence of mass wasting events that are larger than the debris flows in the historic record (See Figure 10 and associated discussion, above), the frequency analysis shown in Figure 9 indicates that there is a low probability that debris flows exceeding about 20,000 to 25,000 cubic yards would occur at return intervals less than 500 years. This is consistent with an extreme event flood condition termed the "check flood" in AASHTO (*11*), which also has an expected return interval of 500 years. Therefore, provided the new bridge opening is sufficient for conveyance of the 500-year event, the design of the structure would be consistent with the reliability expectation for flood conditions.

Landslides

As described above, the primary landslide risk at the project site is associated with the mapped geologic feature named "Granite Slide". As described in (*4*), the main concern is that a seismic event could trigger catastrophic collapse of this feature. However, the 1872 seismic event discussed above did not trigger such movement, even though a slide large enough to temporarily dam the Columbia River was initiated by that event. When considered in conjunction with the recent thinking about the recurrence potential of the fault structure responsible for the 1872 event (*8*), it appears unlikely that such an event would recur within the design life for the new bridge.

It should be noted that the failure scenario of Granite Slide would certainly obliterate the bridge, the project area and much else as well. Such a scenario could not be accounted for as part of the bridge design criteria. The purpose of the evaluation described above was to determine whether the risk was being managed in a manner consistent with other risks present at the site.

Design Seismic Event

As discussed herein, the probabilistic seismic hazard available in the AASHTO (*11*) code-based design provisions does not account for the proximity of the project area to the likely source zone of the largest historic earthquake to occur in Washington State. The design concern was therefore whether the code-based seismic design criteria underestimates the actual hazard at the site. Recent research and work by the USGS (*8*) indicates that the recurrence expectation for whatever fault structure was responsible for the 1872 event is quite long, on the order of a couple to a few thousand years. This conclusion would result in a relatively low contribution by the fault to the probabilistic seismic hazard at the site and therefore use of the code-based seismic design criteria is reasonable and consistent for design of the new bridge.

Structure Type, Size and Location

The recommended alternative for the culvert replacement is a 108-foot single-span bridge structure founded on spread footing abutments. South Lakeshore Road would remain in its current alignment with traffic during construction routed along the downstream side of the existing roadway where by-pass traffic is currently detoured when a debris flow blocks the main roadway. A channel with 1H:1V sloped banks and a 30-foot bottom width would be constructed beneath the bridge and lined with concrete to maintain the integrity of the slopes and provide scour protection for the shallow spread footings supporting the abutments. Channel regrading and construction would occur upstream and downstream to tie in the existing channel, fill the catchment basin and reconstruct portions of the channel levees. The roadway and channel configurations are shown in Figure 13. The profile (elevation view) of the bridge is shown in Figure 14.

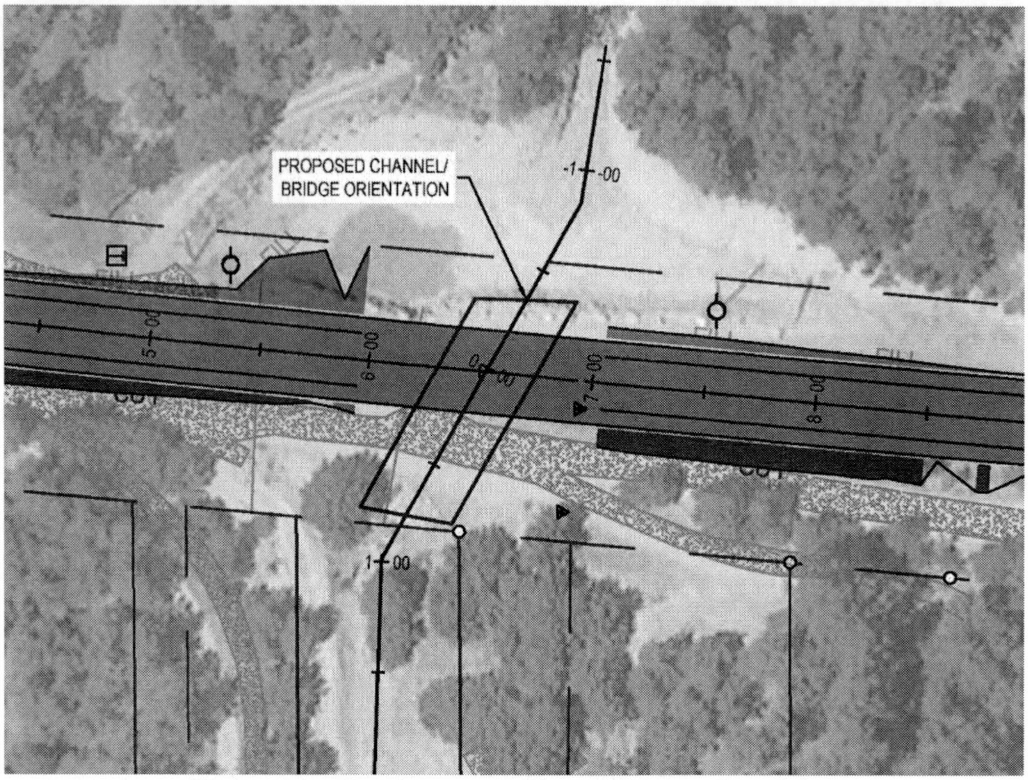

Figure 13 – Roadway and Channel Reconfiguration

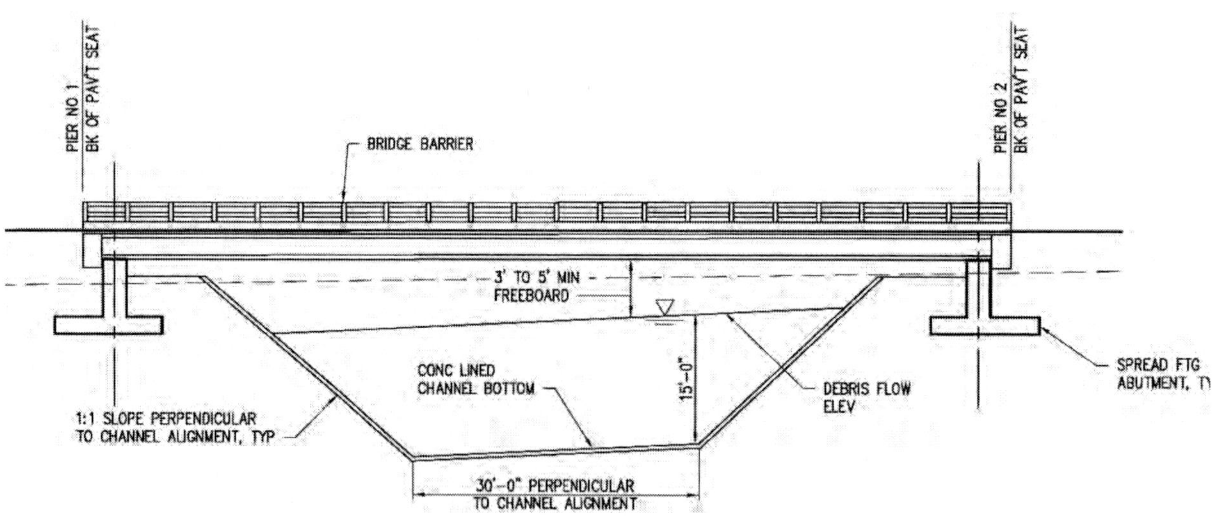

Figure 14 – Bridge Profile

CONCLUSION

The normal process of geologic data collection and review that was conducted for the Slide Ridge culvert replacement project disclosed several hazards that were concerning, even alarming. Through a consistent approach of evaluating the probabilistic risk of each of the hazards and aligning the design criteria for the new bridge accordingly, it was possible to target reliability and performance expectations that were compatible between the various extreme event scenarios associated with the identified geologic hazards.

REFERENCES:

1. Kendra, W. and Singleton, L. *Morphometry of Lake Chelan*, Ecology Report 87-1, Washington State Department of Ecology, Olympia, Washington, 1987.
2. USDA Forest Service, *Native American Legends* (PDF), Pacific Northwest Region.
3. Booth, D.B., Troost, K.G., Clague, J.J., Waitt, R.B. The Cordilleran Ice Sheet, *Development in Quaternary Science*, Volume 1, Elsevier Science B.V., 2003.
4. Tabor, R.W., Frizzell, Jr., V.A., Whetten, J.T., Waitt, R.B., Swanson, D.A., Byerly, G.R., Booth, D.B., Hetherington, M.J., and Zartman, R.E., *Geologic Map of the Chelan 30-Minute by 60-minute Quadrangle, Washington*, Miscellaneous Investigations Series, Map I-1661, U.S. Department of the Interior, U.S. Geologic Survey, 1997.
5. Structure Stability Branch, *Emergency Preparedness Brief with Inundation Map from Standard Operating Procedures, Grand Coulee Dam*, U.S. Dept. of Interior, Bureau of Reclamation, Boise, ID, May 1982.
6. Tabor, R.W.; Frizzel, V.R., Jr.; Booth, D.B., Waitt, R.B., Whetten, J.T. and Zartman, R.E., *Geologic Map of the Skykomish River 30- by 60-Minute Quadrangle, Washington*, Miscellaneous Investigations Series Map I-1963, U.S. Department of the Interior, U.S. Geological Survey, 1993,
7. Lidke, D.J., compiler, 2009, *Fault number 553, Straight Creek Fault, in Quaternary fault and fold database of the United States*, U.S. Geological Survey website, https://earthquakes.usgs.gov/hazards/qfaults. Accessed 12/02/2016 03:08 PM.
8. Thomas M. Brocher, Margaret G. Hopper, S. T. Ted Algermissen, David M. Perkins, Stanley R. Brockman, Edouard P. Arnold, *Aftershocks, Earthquake Effects, and the Location of the Large 14 December 1872 Earthquake near Entiat, Central Washington*, Bulletin of the Seismological Society of America ; 108 (1): 66–83, 2017.
9. Crider, J.G.; Crosson, R.S.; Brooks, J., *The Chelan Seismic Zone, The Great Terrace and the December 1872 Washington State Earthquake* (abst). Geological Society of America Abstracts with Programs, 2003.
10. UW Seismological Laboratory. *Pacific Northwest Seismic Network*. University of Washington, Seattle, WA. https://pnsn.org/. Accessed July 1, 2019.
11. AASHTO, *LRFD Bridge Design Specifications*, 8th edition, American Association of State Highway and Transportation Officials, Washington, D.C., 2017.
12. USGS, *Unified Hazard Tool*, U.S. Department of Interior, U.S. Geologic Survey, Reston, VA. https://earthquake.usgs.gov/hazards/interactive/. Accessed July 1, 2019.

Evidence for the value of risk-based life-cycle management for geohazards and geotechnical assets

Mark Vessely, P.E.
BGC Engineering
701 12th Street
Golden, CO 80401
(720)-598-5982
mvessely@bgcengineering.com

Sarah Newton, P.Eng
BGC Engineering
Suite 500 - 1000 Centre St. NE
Calgary, AB, Canada, T2E 7W6
snewton@bgcengineering.ca

Scott Anderson, Ph.D., P.Eng.
BGC Engineering
701 12th Street
Golden, CO 80401
(720)-598-5982
saanderson@bgcengineering.com

Prepared for the 70th Highway Geology Symposium, October, 2019

Acknowledgements

The authors would like to thank the individuals/entities for their contributions in the work described:

Mike Porter, P.Eng., LEG – BGC Engineering
Matthew Lato, Ph.D., P.Eng – BGC Engineering

Disclaimer

Statements and views presented in this paper are strictly those of the author(s), and do not necessarily reflect positions held by their affiliations, the Highway Geology Symposium (HGS), or others acknowledged above. The mention of trade names for commercial products does not imply the approval or endorsement by HGS.

Copyright Notice

Copyright © 2019 Highway Geology Symposium (HGS)

All Rights Reserved. Printed in the United States of America. No part of this publication may be reproduced or copied in any form or by any means – graphic, electronic, or mechanical, including photocopying, taping, or information storage and retrieval systems – without prior written permission of the HGS. This excludes the original author(s).

ABSTRACT

Geohazards are subgroup of natural hazards associated with geotechnical, hydrotechnical, tectonic, snow and ice, and geochemical processes that originate within or outside of the property boundary of the infrastructure owner. Separate from geohazards are geotechnical assets, which are constructed and owned earthen structures such as retaining walls, embankments, slopes, and constructed subgrades with deterioration vulnerabilities. Common owners of geotechnical assets include road and rail transportation systems, water storage and conveyance entities, cities and communities, and energy transmission and mineral resource investment interests. These same infrastructure owners and their assets, geotechnical or otherwise, can be impacted by geohazard events.

The relationship between geohazard events and deteriorating geotechnical assets can be either mutually independent or mutually dependent; however, the performance threats are similar in either relationship and include risks to worker and public safety, service interruption, poor asset conditions whether geotechnical or other critical assets, and unplanned total project or asset lifecycle cost escalation.

It is this similarity in adverse impacts that historically resulted in a merged management approach by North American transportation infrastructure owners for certain types of geotechnical assets and geohazards. In some cases, this historical management approach was initiated more than 40 years ago by motivated and capable geologists and engineers who recognized the opportunity to reduce future impacts.

More recently, the practice of asset management has become a means for geologists, engineers, and maintenance managers to take a more proactive role in reducing the impacts from geotechnical asset deterioration and geohazard events. Asset management is an internationally standardized practice that enables owners to shift away from reacting to failures as they occur and move to a proactive and systematic work process that preserves assets and finds cost-effective treatments to prolong the asset useful life, while also aligning with strategic goals.

The international standard for asset management defines the process as a coordinated activity by an organization that enables that owner to realize the value from assets. To participate in asset management and ultimately obtain new resources for proactive management, understanding and communicating the value of activities and decisions is an important step for those who are interested incorporating the asset management culture.

The definition of value differs depending on the objectives and stakeholders; yet, understanding the definition of value to the organization is essential for successful implementation of hazard and asset management. To enable productive discussions with decision makers and executives on the value of managing geohazards and geotechnical assets, this paper will present examples of measurable value creation that has resulted from existing geohazard and geotechnical asset management systems in highway, rail, and pipeline infrastructure. The data from these examples will be useful for geologists and engineers to present the benefits of risk-based management in quantifiable values that connect to objectives, customers, and stakeholders.

INTRODUCTION

It is well reported in the media that many domestic infrastructure owners – from road and rail systems, water and energy utilities, and states and cities – are challenged by aging infrastructure assets and insufficient funds for routine maintenance, renewal projects, and new construction. As evidence of the extent of the challenges, the grade point average for the American Society of Civil Engineers (ASCE) Infrastructure Report Card has fluctuated between D and D+ since 1998 with the cost to improve all infrastructure increasing every year of issuance and a current need of over $4 ½ trillion (*1*). For departments of transportation (DOTs), the operational consequences of such under-funding include deferral of maintenance and preservation work on existing assets, delaying new projects, and minimizing internal spending on new programs. These operational consequences translate into stakeholder consequences such as early asset failures and risks to safety and travel reliability.

In the U.S., bridges and pavements have been managed by DOTs as required through U.S. Federal authorization since 1970 and 1991 respectively, yet as recent as 2017, *Transportation Research Board (TRB) Report 859 – Consequences of Delayed Maintenance on Highway Assets* suggested the intrinsic assumptions in the current practice of transportation asset management often under emphasize the value of maintaining what exists now (*2*). As a result, a chronic under-investment in preservation work is occurring because new assets are not maintained following best practices and the deferred maintenance backlog is not addressed. To address this chronic problem, TRB Report 859 recommends that cost-benefit analyses need to "illuminate the chronic effects of under-investment in maintenance."

Also well reported and documented are the tragedies that result from the intersection of infrastructure systems and natural hazards. While the news about these disasters may seem to be about distant problems in countries with lesser income levels or for a few unfortunate individuals "in the wrong place at the wrong time," the reality is that having the advanced infrastructure systems that come with higher living standards does not automatically provide protection from harm. To suggest otherwise would require one to forget about the 43 people who lost their lives in the 2014 Oso landslide in Washington or the 23 people who lost their lives in the 2018 debris flows in Montecito California. Both of these events offer important lessons and considerations in reducing harm (3)(4).

For highways and railway systems, there are many asset types with deterioration vulnerabilities that impact performance of the owning organization and the safety of users. For instance, TRB Report 859 identifies over 25 asset groups that have quantifiable consequences from deferred maintenance. Within this group of 25 assets are slopes, embankments, and walls – assets that have been identified as geotechnical assets by others (*5, 6, 7, 8*). Further, the distributed nature of transportation systems results in variable exposure to threats from natural hazards that intersect with the highways and/or railways at discrete locations. As separate threats with similar consequences for a DOT, the deterioration of geotechnical assets and impacts from natural hazards can result in significant travel disruption, harm to users and adjacent property owners, and escalating ownership costs when measured over the long-term.

While the TRB Report names over 25 asset groups, only two asset groups – bridges and pavements – have the benefit of being identified with management requirements within Federal funding authorizations. Thus, reliance on only Federal authorization requirements for an organizational wide asset management strategy means a DOT is likely omitting and ultimately accepting unknown levels of risk from the many other assets on the system, now and into the future.

Given the growing funding needs for all infrastructure types and the potential for chronic under-investment in asset maintenance to exist within a DOT, there are significant challenges for those responsible for the management of geotechnical assets and hazards. However, the cumulative magnitude of consequences from adverse geotechnical asset and hazard performance presents a distinct opportunity for DOTs to make informed management decisions based on measurable value improvements.

Investment in geotechnical asset and hazard risk reduction can show improvements in public safety, operational costs, and delay metrics and these benefits can be compared to other asset groups in project selection processes. As documented later in this paper, evidence suggests there are substantial opportunities to reduce life-cycle costs and failure frequencies. As the practice of asset management continues to integrate into public agencies, demonstrating favorable return on investment and alignment with the objectives of the organization may become the required practice for all assets, and this is possible for geotechnical assets and geohazard sites.

When viewed internationally, the practice of asset management extends well beyond the individual legacy bridge and pavement asset classes named in Federal authorizations with many successful organization-wide (multi-asset) examples, international guidance standards, and certification programs. With respect to risk management, geotechnical engineers have a long history of recognizing life-cycle risks as demonstrated by Arthur Casagrande when he presented on calculated risk for highway embankment construction in his 1964 Terzaghi Lecture (*9*). Drawing on these management practices, case history and process examples now exist that are supported by evidence-based outcomes that can be applied to similar situations.

This paper introduces the background on the relationship between infrastructure systems, geohazards, and geotechnical assets and provides examples where owners have found measurable value by implementing risk-based life-cycle management processes and tools. Through these examples, those charged with the responsibility for the geotechnical aspects of an infrastructure system can better advocate for the benefits of new investment; both internally to executives and financial managers who allocate funds and externally to stakeholders whose favorable support is essential for the long-term reputation and viability of the organization.

BACKGROUND

Distributed Infrastructure

Examples of distributed infrastructure systems include highway and road networks, rail and transit systems, pipelines, canals and levees, and electrical transmission lines. The performance of these infrastructure systems is critical to the safety and economic viability of communities, businesses, and ultimately nations.

Distributed infrastructure systems are often complex and extend across varying physiographic and geologic terrains that are subject to risk from natural hazards such as unstable ground, flooding and scour, or seismicity. Further, the constructed and aging assets comprising these infrastructure systems are subject to hazards from deterioration leading to physical failure or a shortened service life. Together, the natural hazard and asset deterioration threats create risk to the performance of a distributed infrastructure system and ultimately the safety and economic well-being of users and owners (10, 11).

Geohazards

Across countries and infrastructure systems, geohazards are generally identified as hazards that originate from geologic or other natural processes. An example of a general geohazard definition used by FHWA is: *geological and climatic conditions that have the potential to cause damage to property, infrastructure and the environment; loss of life; and economic losses (12).*

The FHWA definition is similar to other general uses of the term, such as: *"Geohazards" are events caused by geological features and processes that present severe threats to humans, property and the natural and built environment. Earthquakes, floods, landslides, volcanoes, avalanches and tsunamis are typical examples of such events (13).*

While the authors are familiar with the general use of the term geohazards among individual DOTs and railways, there does not appear to be a uniform or authoritative definition used across the transportation infrastructure sector. Rather, the majority of prior work by DOTs directed at geohazards management has typically focused on programs for specific hazards such as rockfall sites, unstable slopes, and abandoned mines. (*14, 15*).

Pipeline infrastructure systems are similar to highway and rail infrastructure due to their distributed nature and continuous reliance on ground support. Within the pipeline industry, robust geohazard management programs began in the early 2000's and in about the last 5 to 10 years, geohazard management has progressed into a standard practice across operators in Canada and the U.S. (*16*).

When used in the pipeline industry, the term geohazard is used to define a broad subgroup of natural hazards with the following definition: *geotechnical, hydrotechnical, tectonic, snow and ice, and geochemical processes that can affect the safety of construction or operational personnel, impact construction schedules and costs, threaten the integrity of operating pipelines and associated infrastructure, and/or impact the environment (17).* Most geohazards affecting pipelines are natural processes triggered by storms or seismic activity, but others, such as cut and fill slope failures or mine subsidence along the pipeline rights of way (ROW), can be triggered by construction and site activities or by third party actions (*18*)

A partial list of geohazards used for pipeline systems is provided in (*18*) and presented in Table 1.

Table 1 - Geohazard Taxonomy from Pipeline Industry	
Geohazard Classification	**Type of Geohazard Process**
Geotechnical Hazards	Frost Heave Thaw Settlement Solifluction Rock Fall Rock Slide/Creep Earth Slide/Creep Earth Flow Debris Slide Ground Subsidence (Karst/Mines)
Hydrotechnical Hazards	Debris Flow Scour Channel Degradation Bank Erosion Encroachment Avulsion Shoreline Wave Erosion
Tectonic Hazards	Liquefaction Lateral Spreading Surface Fault Rupture Strong Ground Motion Volcanic Eruption
Snow and Ice Hazards	Snow Avalanche Ice Fall
Erosion Hazards	Surface Water Erosion Groundwater Erosion Wind Erosion and Dune Migration
Geochemical Hazards	Acid Rock Drainage and Metal Leaching

Geotechnical Assets

The AASHTO transportation asset management guide: a focus on implementation (*19*) defines an "asset" as: *An asset is the physical transportation infrastructure (e.g., travel way, structures, other features and appurtenances, operations systems, and major elements thereof); more generally, can include the full range of resources capable of producing value-added for an agency: e.g., human resources, financial capacity, real estate, corporate information, equipment and materials, etc.; an individual, separately-managed component of the infrastructure, e.g., bridge deck, road section, streetlight.*

Separately, the International Standards Organization (ISO) Standard 55000 (*20*) defines an asset as: *an item, thing or entity that has potential or actual value to an organization; value can be tangible or intangible, financial or non-financial, and includes consideration of risks and liabilities*

The term value is common to both the AASHTO and ISO asset definitions. Thus, a geotechnical asset needs to show value to an organization in terms of either tangible or intangible and financial or non-financial metrics.

For a DOT or railway, geotechnical assets are the retaining walls, embankments, cut slopes, and constructed subgrades within the organizations ROW or easement (*5, 6*). Figures 1 and 2 present schematic examples of embankment and cut slope assets. Like other asset groups within a transportation agency, geotechnical assets are designed, constructed, and maintained by a transportation agency and have an initial and operating value to the agency. In addition to supporting a roadway or railway, geotechnical assets also contribute to the performance of culverts, stormwater drainage systems, and utilities that are often contained within the asset. Geotechnical assets are not unique to transportation organizations and also exist in water storage and conveyance systems, cities and communities, and energy transmission and mineral resource entities.

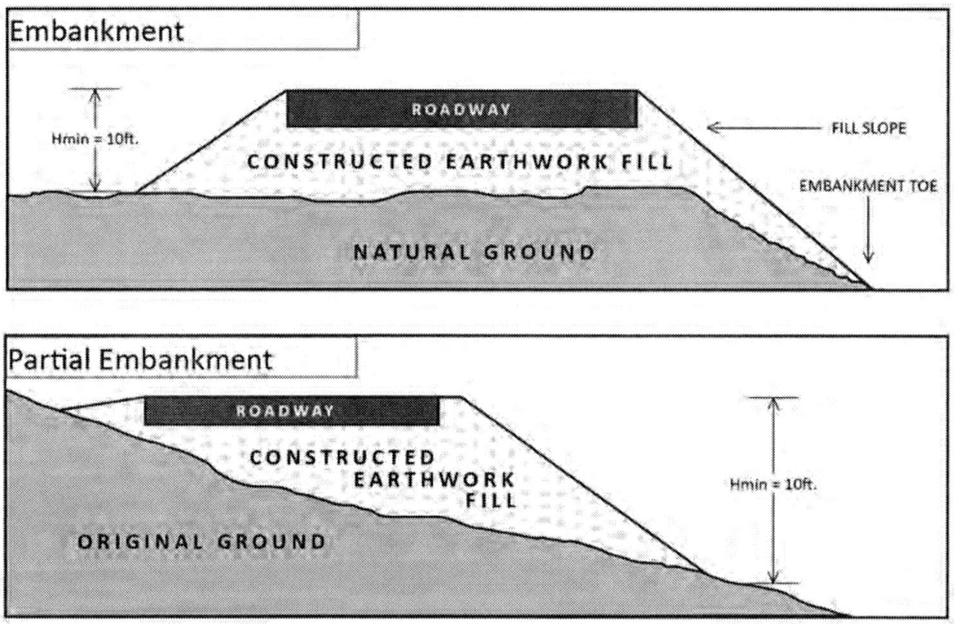

Figure 1 – Embankment Asset Schematics from *TRB Research Report 903* (5).

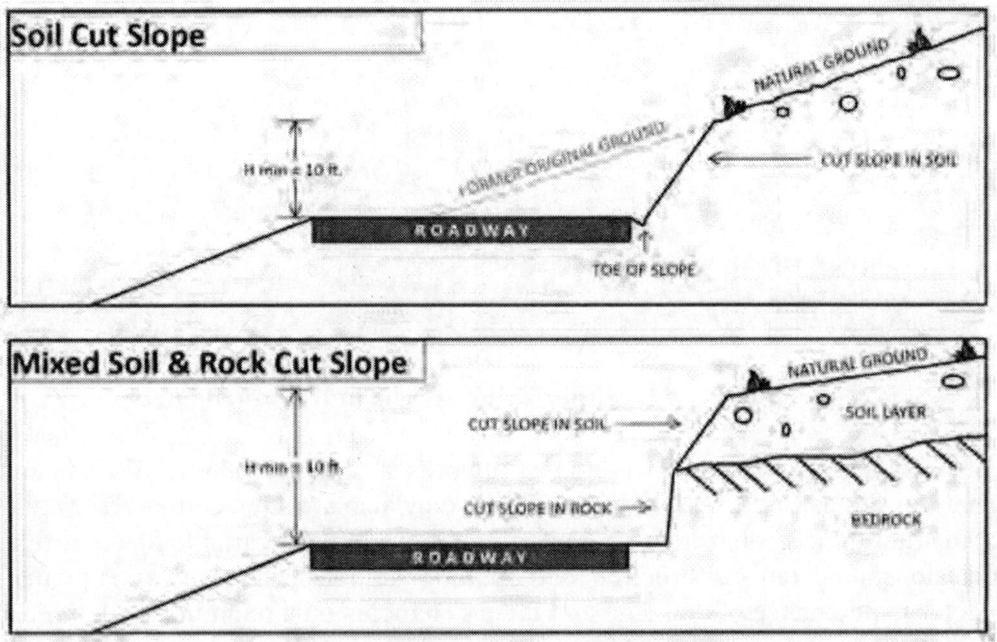

Figure 2 – Slope Asset Schematics from TRB Research Report 903 (5).

The good performance of geotechnical assets contributes to the safe and reliable operation of a transportation network, while poor performance will impact user and worker safety, damage other assets, increase delay and closure times, and threaten environmental resources. With respect to the capital and operational expenses of an agency, poor management of geotechnical assets can lead to escalating maintenance costs or premature replacement, resulting total ownership costs that are greater than the ownership cost if properly managed.

Relationship between Geohazards and Geotechnical Assets

In general, geohazards are an adverse event or process with a frequency and magnitude relationship. Geohazards can exist within the ROW or boundary of a DOT and they also frequently exist beyond the boundary, but with impacts that extend into the ROW. In the case of a geohazard extending beyond the ROW, a DOT does not generally have full access rights and/or open permission to modify the hazard location outside the boundary without prior approval (5, 6).

Conversely, a geotechnical asset is a constructed and owned earth-based structure. For geotechnical assets, a DOT has control over how they are built, maintained, and managed, in addition to full access rights. This important distinction between owned assets and beyond the ROW hazards is presented in Figure 3, which illustrates the difference between rockfall hazard from a highway cut slope asset and a natural hazard beyond the boundary.

Figure 3 – Distinguishing Assets and Hazards.

Because geotechnical assets are constructed earthworks such as retaining walls, embankments, or modified natural slopes (in the case of a cut slope), they also are subject to deterioration processes, in addition to geohazards. The rate of deterioration is a variable that is influenced through decisions in design, construction, and operations. For instance, an asset owner can decide to defer maintenance of drainage and facing elements on a retaining wall, which ultimately will shorten the useful life of the wall.

The poor performance of a geotechnical asset can result in events that manifest as geohazard processes. Examples include rockfall from a cut slope, a landslide in an embankment, or subgrade settlement resulting from thawing ground. As a result, the distinction between geotechnical assets and geohazards is often not made, particularly once the owner is forced to rapidly respond to a failure event and attempting to restore use.

There are many geotechnical assets located in areas with a nil to very low risk from geohazards. In these areas, deterioration threats are the critical process addressed through asset management. There are also many areas where geotechnical assets co-exist with geohazards. In these situations, both the risk from geohazards and asset deterioration should be managed to align with owner objectives. If not managed to the same performance objectives, there is the potential for a well-maintained and good performing asset to experience an unanticipated impact from a geohazard.

These aggregated impacts from event based geohazards and deterioration effects are conceptualized in Figure 4. This figure, commonly termed a 'bathtub curve,' illustrates the different types of failure modes that exist for a system or asset and how the sum of those failure modes results in a net or observed failure rate that changes with time. An infrastructure owner with assets at point 'A' or 'B' on the figure will have the perception that failure impacts are relatively constant or even decreasing because of the benefit of new assets. However, an owner at point 'C' will be observing increasing failure rates because of the contribution from an increasing deterioration failure rate.

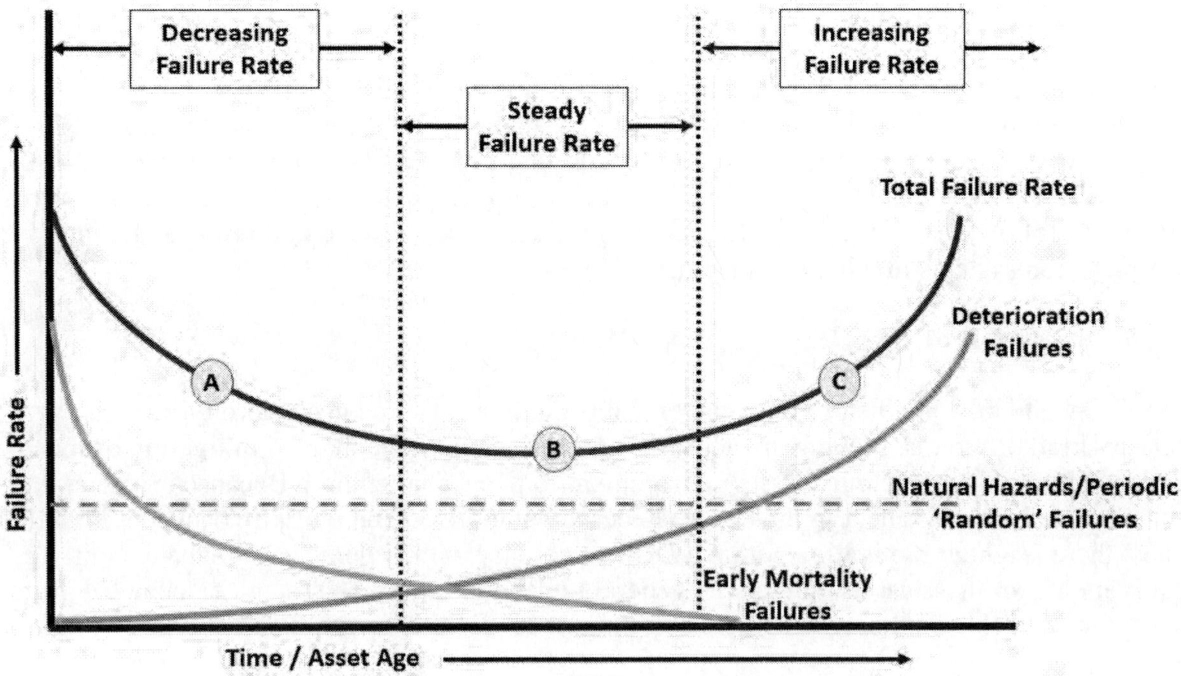

Figure 4 – Bathtub Curve concept applied to geohazards and assets.

Either individually or together, geohazards and geotechnical assets generate similar consequences and risk to the performance of a DOT or other infrastructure owner. These consequences are represented as the 'Total Failure Rate' curve on Figure 4. All DOTs manage the threats and the resulting performance risks, whether actively through informed processes and programs, or reactively through the default acceptance of the consequences after the event. For an active management program, DOTs have the opportunity to manage geohazards and assets to the lowest possible life-cycle cost to the desired level of risk tolerance, meaning the asset and/or asset contribution to overall value is optimized. Referring back to Figure 4, this means owners have an opportunity to lower the location of the natural hazard failure line and also change the slope and lower the deterioration failure rate. The resulting effect will be a lower total failure rate and the benefits that result from a lower failure rate.

Emergence of Asset Management in DOTs

The U.S. Federal authorization, MAP-21, was signed into law in July 2012. MAP-21 and the successor and current authorization, the FAST Act, required state DOTs to have asset management plans for bridge and pavement assets and encouraged states to add other assets within the ROW into these plans. As a result, at least one DOT has incorporated non-bridge and pavement assets into their transportation asset management plans (10, 21) and others have authored plans that could be incorporated in the future (22). To support the inclusion of geotechnical assets in a DOT transportation asset management plan, TRB Report 903 provides an implementation manual and tools for DOTs to begin geotechnical asset management (*5*).

While the requirements for asset management in U.S. state DOTs were initiated in 2012, the practice of asset management was well established in other infrastructure systems and countries prior to the Federal authorization requirements. Early practice standardization work began with British Standards Institute (BSI) standards, which were first published in 2004 and based on input from more than 50 public and private entities spanning 10 countries and 15 sectors (*23*). The BSI work was later adopted and superseded by the ISO, leading to the current ISO 55000 series on asset management. While relatively new for domestic DOTs, asset management across a wide portfolio of assets is a standard practice in many industries and countries, including transportation systems in other countries (24, 25).

EVIDENCE FOR MANAGEMENT VALUE

Despite the long history of management programs for some geohazard types, such as DOT rockfall programs, seeking greater levels of investment in geohazard mitigation or for implementation of geotechnical asset management remains challenging. Because bridge and pavement asset management in the U.S. is Federally authorized and partially backed with federal funds, there is a high motivation within a DOT to comply with Federal asset management requirements for these assets. Given the general under-funding of DOTs and extensive deferred maintenance backlogs, there are many competing needs for the remaining asset types within a DOT.

Separately, historical performance measures for rockfall hazard locations have often been related to technical measures such as factor of safety, condition, or a hazard index, and therefore do not directly connect to strategic and user measures involving value or system performance. As a result, geo-professionals can show improvements to technical measures over the life-cycle but not necessarily the strategic or user measures impacted by the program. The longevity of existing DOT rockfall programs would suggest agencies see value; however, the authors are not aware of research that has quantified the value of these programs in relation to strategic and user performance metrics such as reduced delay and closure, fewer accidents and injuries, and operational cost and maintenance improvements.

With the emergence of asset management practices within U.S. DOTs, geo-professionals have an additional means to gain support for geohazard and asset management initiatives through the demonstration of measurable value improvements that are possible over the life-cycle, in addition to improvement in technical measures. Using the ISO definition of asset management, ideally value should be communicated in terms that are tangible or intangible, financial or non-financial, and include the consideration of risk. In general, value is a measure that connects to system user measures (traveling public for a DOT) and the strategic measures of the organization, such as preserving the system, safety, and reducing delay.

The following discussion presents examples where management programs for geohazards and geotechnical assets has provided measurable evidence of value at the strategic levels of an organization.

Embankment and Slope Management in United Kingdom

Embankments and slopes have been included in risk-based asset management programs for railway and highway agencies in the United Kingdom since the 1990s. Two of the larger transportation agencies in the U.K, Highways England and Network Rail, manage nearly 250,000 geotechnical slope and embankment assets in total. In general, the age of embankments and slopes for U.K highway networks is similar to that of the same asset types in the U.S. For the rail networks, many of embankments in the United Kingdom are greater than a century old with the majority of construction occurring pre-1900s and with a peak of construction works in the 1840s (*24*).

As an agency, Highways England operates with a perspective that roadway construction is primarily complete, and the agency directs effort on system improvements, optimization, and maintenance (*25*). Further, the Highways England geotechnical asset management program has developed in stages, initiating from a program directed at developing the inventory and evolving to a program focused on business outcomes related to service levels for users. Similarly, Network Rail asset management systems have advanced with several process improvements, including recent changes to risk assessment processes that incorporate evidence-based performance data to calibrate life-cycle models (*7*).

The measurable value-based outcomes from approximately two-decades of asset management for embankments and slopes on transportation networks in the United Kingdom include the following.

- A significant reduction in the number expensive reactionary emergency stabilization projects, including several years with no slope failures recorded for the London Underground network (*26*)
- A measured cost savings of approximately 60% for asset rehabilitation projects constructed in a proactive program, versus those constructed in a reactive and unplanned manner (*26*).
- Removal of railway speed reductions due to earth structure conditions on the London Underground system, reductions in wear and tear for rolling stock, and improved ride quality for users (*26*).
- Annual asset ownership cost savings of approximately 80% when using a proactive maintenance strategy over a reactionary strategy on a motorway corridor (*27*).
- A declining 5-year moving average for the number derailments resulting from earthwork failures (*28*).

Geohazard Management for Pipeline Systems

Similar to transportation networks, the performance of pipeline systems can be adversely impacted by geohazards causing leaks or ruptures. In addition to recovery costs to restore the pipeline, other impacts include damage to third parties and property, environmental contamination, service and revenue disruption, and safety threats to employees and the general public. The common types of geohazards that impact pipelines are moving ground from geotechnical hazards such as earth and rock slides and hydrotechnical hazards at watercourse crossings where processes such as scour and channel degradation expose buried pipelines.

An average of 74 pipeline failures resulting from geohazards occur globally each year, with 16 failures per year occurring in the United States. In Canada, the United States, and western Europe, the industry average annual failure rate is reported to be 0.02 per 620 miles (1,000 km) per length of pipeline (*18*). The significant factors that contribute to the failure rate have been shown to be terrain type and planning and construction practices. Geohazard risk management practices have also been shown to measurably reduce operational pipeline failure rates when applied appropriately (*18*). These geohazard management programs follow a complete framework that includes hazard identification, levels of assessment, and risk management that align with strategic goals for the pipeline operator. Use of a complete framework is important as the partial step of inventory creation alone does not appear to have a marked impact on failure rates (*29*).

As an example of the value for a program of geohazard management for distributed infrastructure, evidence across several pipeline operators using the same geohazard management framework and tools indicates a measurable reduction in the frequency of failures. In aggregate, this subject geohazard management framework is being used by over 20 different operators in North America with a current field inspected inventory size of approximately 52,500 geohazard sites on over 77,500 miles of operational pipelines (29). The observed geohazard failure frequencies for pipelines that are actively managed under this subject framework is approximately 0.005 per 620 miles (1,000 km) per year. Conversely, the geohazard failure rate for pipelines prior incorporation into the full framework is 0.026 per (620 miles) 1,000 km per year. The results suggest that pipelines in a complete geohazard risk management program are realizing reductions in the rate of pipeline failure by a factor of 5.

DISCUSSION

When managing the value of geotechnical assets and all assets exposed to geohazards, evaluation of the risk over the life-cycle is essential. In the authors' experience, when delivering new highway projects, selection of alignments, structure types, and construction methods has typically been influenced by past experience of the agency and the need to align with available construction funds or budgets formulated during the initial planning stages. This approach will result in an unanticipated bias that considers the initial cost components of a project over the whole life costs and risk.

Once in the operational phase, the cost and risk of a program of geotechnical assets and/or the exposure to geohazards for a DOT is often managed in reactionary basis that relies on contingency or other types of discretionary funds. Further, research conducted as part of TRB Report 903 (*3*) indicated geo-professionals within a DOTs are not able to reliably track operations and maintenance expenses and other enterprise impacts associated with geotechnical assets and geohazard sites in their agency. As a result, measurement of risk exposure and the data-driven processes to better manage that risk exposure is challenging. Use of technology provides an opportunity to address these challenges as software used by pipeline and railroad owners is demonstrating improved risk measurement and management opportunities (29, 30).

Where managed in an enterprise program with performance measures that connect to high level strategic objectives, the outcomes from other infrastructure systems suggest there is a valuable opportunity for DOTs to reduce life-cycle risks from geotechnical assets and geohazards.

The two examples presented herein – roadway and railway systems in the U.K. and distributed pipeline networks in North America – indicate that risk-based geotechnical asset and geohazard management programs are generating measurable improvement in metrics that are important to users of the system and executives responsible for strategic goals. Further, the evidence suggests that it is the full cycle of asset and/or risk management that results in measurable improvement.

The documented improvements, as much as 80% life-cycle cost reductions and failure frequencies reduced by a factor of 5, are not trivial improvements. Taking advantage of these improvements is an important step for organizations that already are underfunded or suffer from chronic maintenance deferrals as these improvements are essentially self-funding when viewed in the context of the life-cycle. For instance, a study by the Colorado DOT estimated an annual statewide economic impact from geohazards of approximately $30M per year with higher impacts in years with significant events (31). Projecting even a portion of the benefits observed from the examples presented herein to this level of economic impact, is a basis for justifying the initial investment needed to begin or expand geotechnical asset and geohazard risk management programs.

Across industries and professions, communication of value is essential to the long-term viability of work activities and projects. While common geotechnical or geologic performance measures, such as hazard score, factor of safety, and quantification of ground movement are important to geo-professionals tasked with making engineering and science-based decisions, these technical measures do not easily communicate life-cycle value to the non-technical stakeholder. For example, the medical industry relies on external measures related to patient safety and efficacy of treatment as means to communicate value. Outward facing measures in commercial aviation relate to traveler safety and reliability of travel times, which are valuable to the consumer. Similarly, DOTs have many external measures, often set through political processes, that address safety, travel time, system reliability, and preservation.

Analysis, measurement and communication of the value of geohazard and geotechnical asset management in the context easily understood measures is a recommended step that is be beneficial to selecting projects and operational activities that preserve these assets and hazard sites to the lowest possible life-cycle cost and in a manner that aligns with the risk tolerance of the agency.

REFERENCES:

1. ASCE's 2017 Infrastructure Report Card. (2017). [online] Available at: https://www.infrastructurereportcard.org/ [Accessed June 1, 2019].

2. National Academies of Sciences, Engineering, and Medicine. 2017. Consequences of Delayed Maintenance of Highway Assets. Washington, DC: The National Academies Press. https://doi.org/10.17226/24933.

3. Lato, M. J., Anderson, S., & Porter, M. J. (2019). Reducing landslide risk using airborne lidar scanning data. Journal of Geotechnical and Geoenvironmental Engineering, 145(9), 1-8. https://doi.org/10.1061/(ASCE)GT.1943-5606.0002073

4. Keaton, J. R., Ortiz, R. M., Turner, B., Alessio, P., Gartner, J. E., Duffy, J. D., . . . & Watts, T. (2019). Overview of geotechnical effects of the January 9, 2018, debris-flow and flash-flood disaster in Montecito, California. In J. W. Kean, J. A. Coe, P. M. Santi, & B. K. Guillen (Eds.), Debris-flow hazards mitigation: Mechanics, monitoring, modeling, and assessment: Proceedings of the Seventh International Conference on Debris-flow Hazards Mitigation (AEG Special Publication 28, pp. 500-507).

5. National Academies of Sciences, Engineering, and Medicine 2019. Geotechnical Asset Management for Transportation Agencies, Volume 2: Implementation Manual. Washington, DC: The National Academies Press. https://doi.org/10.17226/25364

6. Anderson, S.A., V.R. Schaefer. and S.C. Nichols. 2016. "Taxonomy for Geotechnical Assets, Elements, and Features." In Transportation Research Board 95th Annual Meeting Compendium of Papers, Washington, D.C., 16 p.

7. Network Rail, 2017. Earthworks Asset Policy (March 2017). Network Rail Infrastructure Limited.

8. UK Department for Transport. 2003. "Maintenance of Highway Geotechnical Assets." In Design Manual for Roads and Bridges, Vol. 4, Sec. 1, Pt. 3. UK Department for Transport, London, England.

9. Casagrande, A., 1965, Role of the 'Calculated Risk' in Earthwork and Foundation Engineering, Journal of the Soil Mechanics and Foundation Division, ASCE, Vol. 91, No. SM4, July, pp. l-40.

10. Anderson, S., Vessely, M., and Ortiz, T. 2017. "Communication and Management of Geotechnical Risks to Transportation Performance Objectives." Geo-Risk 2017: Geotechnical Risk Assessment and Management. ASCE GSP 285, Pg 279-290.

11. Anderson, S.A. 2016. "Communicating Multiobjective Risk: A New Geotechnical Need for Transportation." Transportation Research Record, No. 2580. pp. 80–88.

12. Mohamed, K. *FHWA Geohazards, Extreme Events and Climate Change Program*, Geohazards Impacting Transportation in Appalachia, Blacksburg, VA. August 2017.

13. Nadim, F., Einstein, H.H., Pottler, R., Klapperich, H., Kramer, S. Proceedings of Geohazards 2006, Technical, Economical and Social Risk Evaluation, Lillehammer, Norway, 18-21 June 2006, Berkeley Electronic Press, in CD.

14. Pierson, L. A. 1991. The Rockfall Hazard Rating System. Oregon Department of Transportation, Salem, OR.

15. Western Transportation Institute. Unstable Slope Management Program for Federal Land Management Agencies, Phase 2. Montana State University. https://westerntransportationinstitute.org/research_projects/unstable-slope-management-program-for-federal-land-management-agencies-phase-2/ (accessed June 1, 2019).

16. Porter, M., Ferris, G., Leir, M., Leach, M. and Haderspock, M. "Updated estimates of frequencies of pipeline failures caused by geohazards." Proceedings of the International Pipeline Conference, Volume 2: Pipeline Safety Management Systems; Project Management, Design, Construction and Environmental Issues; Strain Based Design; Risk and Reliability; Northern Offshore and Production Pipelines. IPC2016-64085: pp. V002T07A003. Calgary, Alberta, Canada, September 26-30, 2016.

17. Porter, M., Leir, M., Baumgard, A., and Ferris, G. 2014. Integrating terrain and geohazard knowledge into the pipeline lifecycle. Proceedings of the 6th Canadian GeoHazards Conference - GeoHazards 6, Kingston, Canada. June 15-18, 2014.

18. Porter M, Ferris G, Leir M, Leach M, Haderspock M. 2016. Updated Estimates of Frequencies of Pipeline Failures Caused by Geohazards. ASME. International Pipeline Conference, Volume 2: Pipeline Safety Management Systems; Project Management, Design, Construction and Environmental Issues; Strain Based Design; Risk and Reliability; Northern Offshore and Production Pipelines.

19. AASHTO. 2011. AASHTO Transportation Asset Management Guide: A Focus on Implementation. American Association of Highway and Transportation Officials, Washington, D.C.

20. International Standards Organization (ISO). 2018. ISO 55000:2014(en). Asset Management—Overview, Principles, and Terminology. ISO website https://www.iso.org/obp/ui/#iso:std:iso:55000:ed-1:v2:en:term:3.2.1 (accessed January 15, 2018).

21. Vessely, M., Widmann, B., Walters, B., Collins, M., Funk, N., Ortiz, T., and Laipply, J. 2015. Wall and Geotechnical Asset Management Implementation at the Colorado Department of Transportation. Transportation Research Record, 2529(1), 27–36.

22. Thompson, Paul D., 2017. Geotechnical Asset Management Plan, Technical Report. Report STP000S(802)(B). Prepared for the Alaska Department of Transportation and Public Facilities, June 30, 2017.

23. British Standards Institution (BSI). (2008). PAS 55-1:2008: Asset Management, Part 1: Specification for the Optimized Management of Physical Assets. BSI: London, England.

24. Power, C., J. Mian, T. Spink, S. Abbott, and M. Edwards. 2016. "Development of an Evidence-Based Geotechnical Asset Management Policy for Network Rail, Great

Britain." Presented at Advances in Transportation Geotechnics 3: The 3rd International Conference on Transportation Geotechnics (ICTG 2016), Portugal. In: Procedia Engineering, Vol. 143, pp. 726–733.

25. Power, C., D. Patterson, D. Rudrum, D. Wright. 2012. "Geotechnical Asset Management for the UK Highways Agency." In: Radford, T. A. (ed.). Earthworks in Europe. Geological Society, Engineering Geology Special Publication 26, London, United Kingdom. DOI: 10.1002/gj.2667.

26. McGinnity, B.T., Fitch, T.R., and Rankin, W.J., 1998. A systematic and cost-effective approach to inspecting, prioritizing, and upgrading London Underground's earth structures. In proceeding of the seminar "Value of Geotechnics in Construction." Institution of Civil Engineers, November 4, 1998. Pp 309-332.

27. Patterson. D. and Perry, J. 1998. Geotechnical data and asset management systems for highways. Proceedings of the maintenance engineers conference. 1998.

28. Abbott, S. 2017. Railway Geotechnical Asset Management in Great Britain. ASCE Geo-Institute GeoStrata, January/February 2017.

29. Newton, S., Zahradka, A., Ferris, G., and Porter, M., 2019. Use of a geohazard management program to reduce pipeline failure rates. Proceedings of the Conference on Asset Integrity Management-Pipeline Integrity Management Under Geohazard Conditions AIM-PIMG2019 March 25 - 28, 2019, Houston, TX, USA

30. Lato, M, Quinn, P, Porter, M, Newton, S, Dixon, R, Wessels, SDN, Wessels, L, Sirois, D & Leveque, M 2019, 'Geohazard risk management for linear transportation', in J Wesseloo (ed.), Proceedings of the First International Conference on Mining Geomechanical Risk, Australian Centre for Geomechanics, Perth, pp. 323-336.

31. Vessely, M., S. Richrath, and E. Weldemicael. 2017. "Economic Impacts from Geologic Hazard Events on Colorado Department of Transportation Right-of-Way." Transportation Research Record: Journal of the Transportation Research Board, No. 2646. Transportation Research Board of the National Academies of Science, Engineering, and Medicine, Washington, DC, pp. 8–16.

Ground Anchor Testing
Matching Test Elements to Ensure Critical Anchor Attributes are Verified - or

"Why we don't push ropes"

Peter Ingraham,
Senior Geological Engineer and Principal, Golder Associates Inc,
670 North Commercial Street, Suite 103
Manchester, New Hampshire, 03101
(603) 668-0880
pingraham@golder.com

Martin Woodard
Chief Rockfall Engineer, GeoStabilization Intl.
Radford, Virginia
(540) 315-0270
marty@gsi.us

David Scarpato
President & Principal Geohazard Engineer, Scarptec, Inc.
P.O. Box 326
Monument Beach, Massachusetts, 02532
(603) 361-0397
Dave@scarptec.com

David Wood
President, David F. Wood Consulting Ltd.
55 Gloucester Court
Sudbury, Ontario
P3E 5M2, CANADA
(705) 698-0909
info@dfwood.com

Prepared for the 70th Highway Geology Symposium, October 2019

Acknowledgements

The authors wish to acknowledge the Association of Geohazard Professionals for their contribution to this paper. This paper was developed as part of the Association's Anchor Committee goals which were to provide industry guidance on rock anchor testing.

Disclaimer

Statements and views presented in this paper are strictly those of the author(s), and do not necessarily reflect positions held by their affiliations, the Highway Geology Symposium (HGS), or others acknowledged above. The mention of trade names for commercial products does not imply the approval or endorsement by HGS.

Copyright

Copyright © 2019 Highway Geology Symposium (HGS)

All Rights Reserved. Printed in the United States of America. No part of this publication may be reproduced or copied in any form or by any means – graphic, electronic, or mechanical, including photocopying, taping, or information storage and retrieval systems – without prior written permission of the HGS. This excludes the original authors(s)

ABSTRACT

Ground anchors come in many variations from simple guywire anchors used to support telephone poles to complex multi-strand tendons tensioned to millions of pounds to tie down large structures. Post-tensioned elements became popular in structures in the 1960s and '70s and in 1976 the Post-Tensioning Institute (PTI) was founded to establish guidelines and standards for post-tensioned structural elements. The complexity and at times difficult construction of such anchors led the PTI to develop testing methods to verify the construction, condition and capacity of all the technical components of the anchors as installed.

The PTI testing guidelines remain the standard for complex tieback and tiedown anchors that incorporate free stressing lengths and load transfer requirements, and they have been adopted by most Transportation Departments in the US. Lack of a similar standard for less complex anchors has let DOTs to specify PTI test protocols for all ground anchors, often testing the performance of anchor components/attributes not present or not important in simple anchors.

This paper discusses PTI testing elements, their purpose, and how to develop testing programs for simpler anchors applications to verify anchor construction QA, functionality, and serviceability as well as provide information to trouble-shoot anchor installation.

INTRODUCTION

Ground anchors come in many variations from simple guy wire anchors supporting telephone poles to complex multi-strand tendons tensioned to millions of pounds to tie a dam down during a probable maximum flood event. In 1961 Tom Lang published a treatise on tensioned rock bolting (Lang, 1961) presenting results of extensive testing conducted during the Snowy Mountains Hydroelectric Development construction in New Zealand. The primary application for tensioned rock bolts in that project was to develop a "reinforced rock unit", which was considered a zone of compression between the rock surface and rock bolt anchor end to hold fractured rock together and mobilize its inherent strength (Figure 1). Concurrently, post-

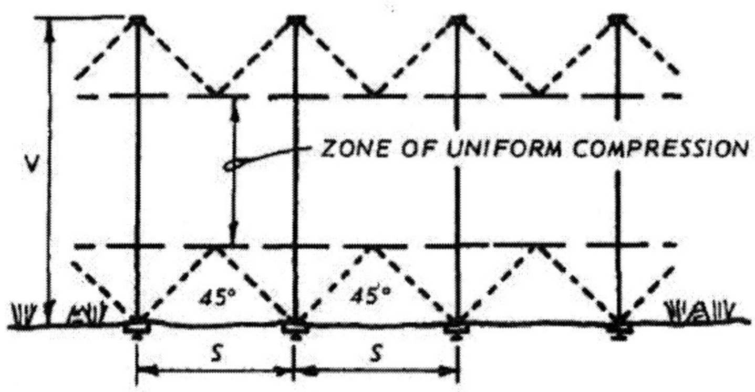

Figure 1 – Zone of Compression (after Lang, 1961)

tensioned elements started to become popular in civil design structures in the 1960s and '70s to ensure structural concrete members remained in compression when subjected to bending forces or to provide retaining forces for elements such as retaining/tieback walls.

Meeting an industry need, the Post-Tensioning Institute (PTI) was founded in 1976 and included materials suppliers, manufacturers, and construction industry representatives. Work by David Weatherby and others promoting the use of post-tensioned tieback walls for top-down construction and permanent structures led to preparation of an FHWA Research Document in 1982 geared toward the civil and structural aspects for transportation applications of post-tensioning (Weatherby, 1982). The adaptation of tensioned elements to support civil structures required development of complex anchors (Figure 2) serving as structural members to meet the requirements of tieback walls and post-tensioned beams including:

- Transferring externally applied tensile loads to an anchor zone some distance behind a wall (e.g., beyond the Rankine wedge in soil behind the wall);
- Establishing a free-stressing length in each anchor between a tieback wall and the anchor zone;
- Establishing an anchor zone where the load in the anchor is transferred to a soil/rock mass; and
- Constructing a load transfer head assembly that can bear on the wall/beam end and impart the anchor load as a pressure applied to the face of the wall or between beam ends.

In 1980 PTI developed testing methods to ensure the complex anchors and tendons being constructed met the requirements above and each component was functioning (Post-Tensioning Institute, 2004). As with all construction, Departments of Transportation were interested in having appropriate testing methods for tiebacks and tendons to verify compliance with plans and specifications and confirm the function of as-installed ground anchors. With limited options to choose from they adopted the FHWA document design guidelines as well as PTI

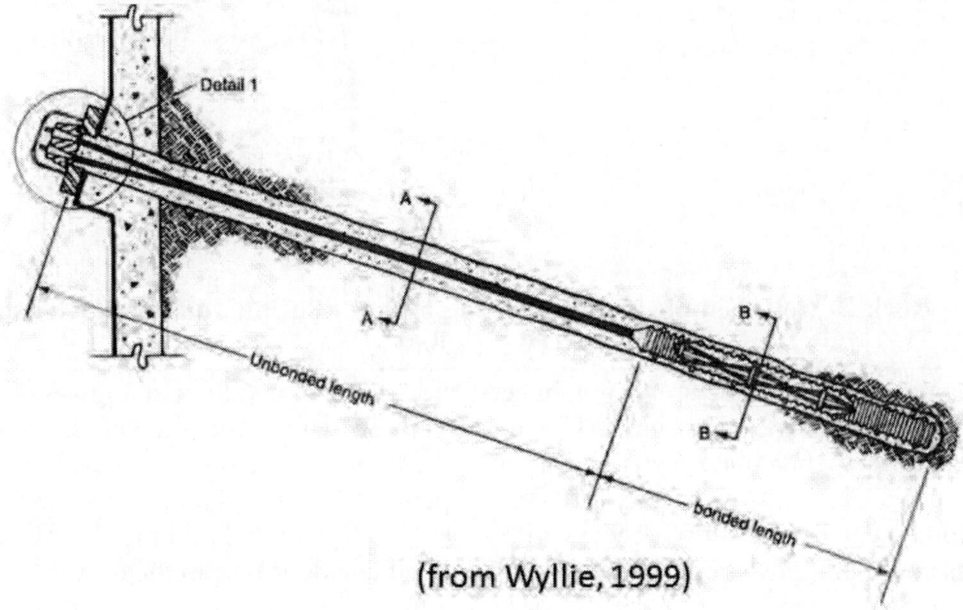

Figure 2 – Post-Tensioned Tendon (Wyllie, 1999)

Recommendations for Prestressed Rock and Soil Anchors (Post-Tensioning Institute, 2004) design guidelines and testing methods for post-tensioned tieback wall and tiedown anchors.

In some jurisdictions, the testing requirements have also been required for less complex anchors, implementing verification test protocols for components of anchors that are not present in simple anchors. This is particularly evident with passive (untensioned) rock bolts often called "rock dowels" that are widely used by the industry. Rock dowels are frequently used to "pin" down unstable rock blocks on slopes and to reinforce the rock mass around underground openings, usually at a substantially reduced installation cost and schedule in comparison to anchors. Other functions of a rock dowel commonly include anchor systems for draped and pinned mesh, post and anchors for flexible rockfall barriers, and many other cases. The function of a grouted rock dowel could be in pure tensile reinforcement to keep two blocks on either side of a planar joint from dilating so it can't slide over asperities. By holding the two sides together, the asperities interlock and sliding would require shearing intact rock through the asperities; however, a passive dowel element is also free to undergo bending and shear, underscoring the importance of proper dowel installation and orientation with respect to load application (Figure 3).

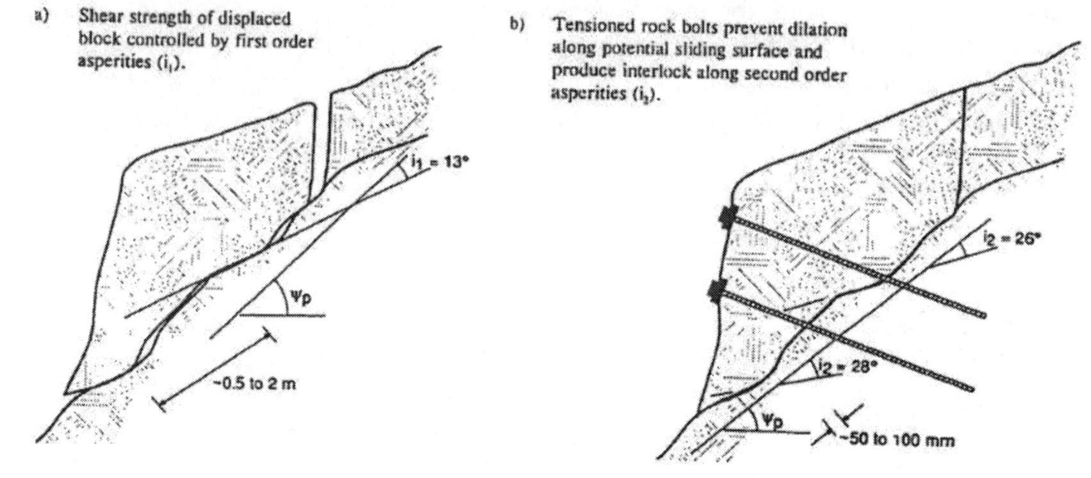

Figure 3 – Rock Dowel – Simple installation to prevent dilation and interlock asperities. (FHWA, 1998)

Although dowels can be used for enhanced shear resistance (albeit at a lower allowable shear capacity than for case of tension (ACI Committee 355, 1997), rock dowels are ideally oriented at a non-perpendicular downslope angle to the failure plane surface in order to maximize the available tensile capacity of the bar. Spang and Egger (Spang, 1990) looked at dowel reinforcement as a "nailing" process with rock dowels installed oblique to a planar joint. The analysis considers rock strength, shear, bending, and tensile forces in the dowel (Figure 4) as well as the strength of grout around the bar. In both applications, the principal consideration is

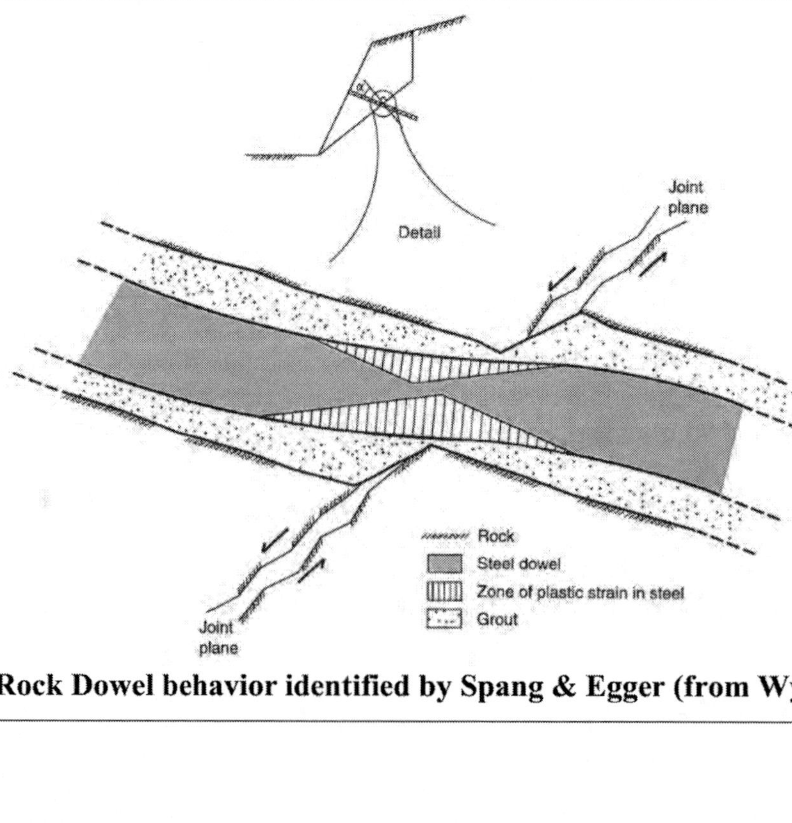

Figure 4 – Rock Dowel behavior identified by Spang & Egger (from Wyllie, 2017)

the rock and grout strength as well as bar strength and ductility, suggesting that testing should focus on those attributes. It further illustrates the need to understand how a passive anchor functions, it's complexity, and key attributes that need to be inspected during installation and testing and a means to select an appropriate testing program tailored to verify anchor suitability.

This paper will review common installations of tensioned rock bolts, rock dowels, soil nails, true uplift anchors and wire rope anchors to evaluate appropriate testing methods. The review continues with complex anchors and the elements of testing/verification for each component of PTI tests. From this review and a knowledge of the purpose and key components of anchors being installed, the authors hope the reader will be able to confidently develop an appropriate testing program to check the critical attributes of their ground anchors.

COMMON INSTALLATIONS

Tensioned Rock Bolts

Tensioned rock bolts are used to actively reinforce a rock mass by applying a force at the rock bolt head and bearing plate assembly and transferring that load to an anchor zone at the far end of the bolt. As noted above, the application of load creates a reinforced rock unit in a zone of compression between the bolt head and anchor (Figure 1). Many different types of anchor mechanisms have been used since rock bolts were developed (Hoek & Bray, 1981) including expansion shell anchors, quick set resin, and cement grout. Free stressing lengths within these systems include but are not limited to, bare bar, slow set resin (using quick set resin for the bonded anchor length), and a greased sheathed section cast into a fully tremmie-grouted hole (cementitious or tremmied resin). Once the anchor mechanisms have been set, the bolt can be tensioned. Testing of these anchors is done based on a design load or a higher test load. In either case the test load does not exceed 80 percent of the Guaranteed Ultimate Test Strength (GUTS) of the reinforcing element. In many cases excess capacity is not tested, rather, the desired lock off load is set with a jack and locked off with a nut, washer and bearing plate. Expansion shell anchors are sometimes specified with a torque value to achieve the correct anchorage for a tension value in the bar. With resin-grouted anchors, the hole is loaded with epoxy cartridges (fast set at the bottom, slow set above the anchor zone) and the bar is spun into the hole to break the epoxy cartridges, release hardener and mix the epoxy. There is a window of time, generally an hour, to complete tensioning after the fast setting resin has hardened and formed the anchor zone. A fully grouted anchor (grouted anchor zone) with a free stressing length can only be tested after the grout has cured to reach sufficient strength for testing.

Two stage grouting, where an anchor zone is constructed with resin or grout is possible but requires planning and care during execution. When the anchor zone has developed strength, the remainder of the hole can be tremmie grouted and the bar immediately tensioned – or, the bar can be tensioned, and the bar grouted after lockoff. The latter method can prove difficult because grouting through the bearing plate is necessary.

Tensioned rock bolts are usually less complicated than the complex post-tensioned tendons that the PTI Guidelines were developed for. Also, because their cost is much less than a large tendon, rigorous testing may not be as critical when an additional bolt can be added in less time and for less money than testing. These simple tension elements can be used for rock slope stabilization applications, for example, where an unstable rock slab may be present. The outcome of tensioning is to develop a zone of compression in the rock mass and interlock discrete blocks in the mass into a reinforced rock unit (Lang, 1961) As such, tensioning to the design load, locking off and liftoff testing are generally sufficient. The PTI guidelines for this simple type of application are therefore not necessarily needed and simplification of the tensioning/testing protocols should be considered to focus on verifying tension in the bolt and lock off efficiency. This is particularly applicable where rock bolts are installed for high-angle work on rappel. Testing after a rock bolt is fully grouted and the grout has cured is only feasible for single stage-grouted bolts with permanent free stressing lengths.

Rock Dowels

Rock dowels, or passive, untensioned rock bolts, are reinforcing elements that function by limiting joint dilation and mobilizing intact rock strength. In most cases they are used to hold together blocks forming a jointed rock mass, providing confinement and mobilizing friction and rock strength for stability. With limited confinement, rock mass cannot deform, loosen over time and fall out, and the rock mass/slope area that is reinforced remains integral and intact. Considering asperities on a planar joint surface, if the two sides of the plane are held together, the adjacent blocks cannot slide because the joint cannot dilate to allow the asperities to "ride over" one another and the joint must shear through intact asperities in order to slide (Figure 3). This simple model was further explored by Spang & Egger, to assess the function of fully grouted dowels as "nails" similar to those used in wood frame construction. They investigated the orientation of dowels installed oblique rather than normal to a joint plane and the complex interaction between the dowel and the rock mass. The net reinforcement identified by Spang and Egger is developed by a complex reaction putting the bar in tension, bending and shear, confinement of the bar by strong grout, and increasing shear resistance in the joint plane on the downslope side of the bar. Key parameters in their analyses include:

- Angle of friction along the shear plane (can contribute up to 50% improvement in stability with the fully grouted dowel);
- Bolt inclination (can contribute up to 20% improvement in stability – reducing shear and bending where steel is weakest);
- Elastic properties (composite stiffness) of the grout and rock;
- Angle of dilatancy;
- Working capacity and deformability of the steel;
- Bolt diameter; and,
- Thickness of the grout collar.

Note that the grout to rock bond is not included on the list - suggesting that tension testing is unnecessary. Furthermore, from review of this approach, setting up a jack and pulling on an inclined dowel could adversely stress and potentially damage the reinforcement.

Considering that passive rock dowels represent *reinforcement* to preclude rock mass degradation by tensile forces inducing dilation, the authors recommend treating rock dowels similarly to reinforcing steel in a concrete structure – understanding the rock, grout, and bar strength, and the bond between the dowel and the rock. To be comfortable with the simple dowel installation- (often designed using bond value and steel capacity) -and not adversely affect the condition of installed dowels, the authors recommend testing those important material properties (or verifying them through submitted test results) and performing sacrificial testing to confirm grout to ground bond. The test method outlined in ASTM D-4435-13 "Standard Test Method for Rock Bolt Anchor Pull Test recommends sacrificial testing to confirm ultimate grout to ground bond (ASTM, 2013) (Wood, 2005). Some practitioners also refer to this simple bond test as a "static bond test" or just a "pull test", without cyclical loads and creep testing. The test procedure recommends having a short grout bond length so the bond value can be verified well within the elastic load range of the selected dowel bar diameter and working strength, and having an ungrouted (bare bar) free stressing length in the upper test zone at least six feet long to allow for load transmission and to avoid issues where the jack chair rests on the installed grout column.

Consideration of rock lithology should be made and whether the bond can be affected by exposed rock degradation due to slaking or decomposition. In these cases, additional testing (e.g. lab tests of intact rock samples and/or sacrificial bond testing of holes allowed to age) should be considered to identify changes in the bond value with time between drilling and grouting of the dowels.

Final construction quality assurance should consist of verifying dowel installation parameters (length, subdrill length, orientation (azimuth, rake), stickup, etc., bar properties (mill sheets and/or testing) and adequate grout strength to compare to sacrificial tests. Applying the full-scale tension testing sequence to fully grouted dowels will only mobilize a fraction of the near surface bond, and de-bond/fracture some of the upper grout to bar bond as the load is transferred to the grout, neither of which provides useful information.

Soil Nails

Soil nails are passive dowels installed into soil or weak rock mass. The function of the soil nail is to stabilize the near surface soil mass, provide anchorage for a facing to retain soils between the nails, and create a monolithic soil mass to improve slope stability. In traditional soil nailing where a reinforced shotcrete facing is applied to the soil slope in intimate contact with the soil, often implementing top-down construction, the dowels are not post-tensioned but rather develop their mobilized strength through a small amount of surficial slope displacement.

For other applications where high strength mesh is applied to a slope surface in lieu of shotcrete, the mesh is drawn taught against the slope to achieve intimate contact against the soil face by the mesh so that the soil face between the nails is retained. The mesh is commonly drawn taught by tensioning a mesh bearing plate on the soil nails such that a nominal (e.g., 9-12-kip) load is put on the upper soil nail bar. The need for only a nominal load to draw the mesh taught suggests elaborate testing of the installed nails is not necessary. Rather, torque-tensioning of the mesh is recommended, and additional nails where necessary. For soil nail installation monitoring, sacrificial testing similar to that recommended for dowels and monitoring of grout quality/strength and installation is appropriate.

Wire Rope Anchors

Wire rope anchors are commonly used in our industry to absorb loads that are not aligned with the axis of the anchor itself. Bar anchors are very susceptible bending forces, particularly where high fiber stresses are concentrated at a grouted hole collar. Testing at an Ontario Power development (Wood, 2005) (Ingraham, 2014) saw a 1-inch-diameter 150 ksi bar fail at less than 5-kips load in bending (Figure 5). Wire rope anchors do not suffer the same capacity loss when loading is less than perfectly axial. Most wire mesh drapes, flexible rockfall barriers, and debris

Figure 5 – Bending Failure of 150 ksi Bar

flow/shallow landslide barriers call for upslope wire rope anchors and terminal anchors for top- and bottom support ropes by manufacturer design. Current practice leans toward double leg anchors to eliminate a weak point at the swaged loop end of single leg anchors. Wire rope anchors are generally aligned with the anchor loads they will receive but may need to be raked steeper downward than the load in order to have sufficient depth and confinement to achieve design loads. A small "active" bearing pressure wedge of soil or rock may develop along the radial transition length of wire rope as it bends; however, this length of grout column disturbance is typically very small in comparison to the embedment depth (e.g. 5% of embedment depth).

Testing of fully grouted wire rope anchors, particularly in the direction of their axial installation can be cumbersome. Muhunthan et. al. suggested different arrangements to facilitate pull-testing wire rope anchors in the direction they will be loaded (Figure 6); however, it is

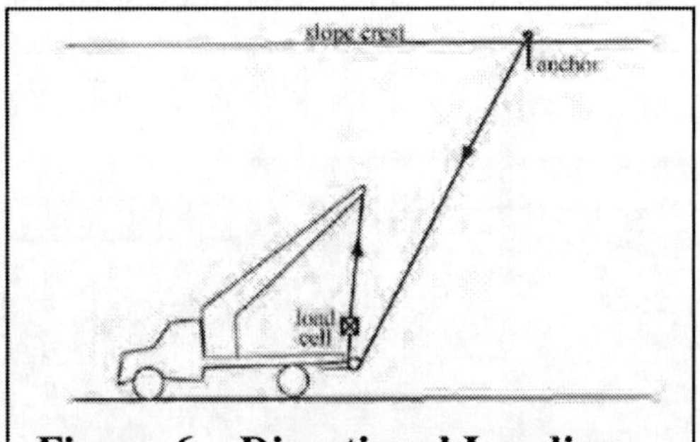

Figure 6 – Directional Loading

worthwhile to remember that wire ropes do not behave elastically. They will elongate a little and take a "set" as the wire rope wraps elongate, narrow, and bite into each other becoming tighter. So, at a maximum, testing to the design load to see if the anchor holds is all that is necessary; (i.e., static bond test), and consideration of only confirmatory bond testing in the anchor zone materials is appropriate together with construction observation, materials verification, and grout testing.

True Passive Uplift Anchors

Anchors used to tie down structures or literally resist pullout when a tensile load is applied to a bar or wire rope are often passive elements, though some uplift anchors are

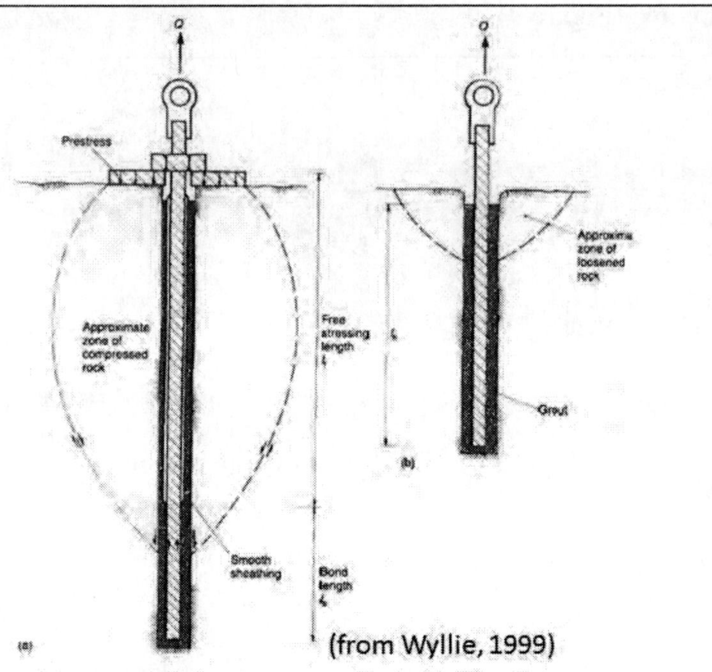

Figure 7 – Tensioned and Passive Uplift Anchors

tensioned (Figure 7, Wyllie, 1999), as was also outlined for complex post-tensioned anchors above for large structural loads. Post tensioned uplift anchors create a zone of compression similar to rock bolt installations to increase the effective diameter of the anchor. Uplift anchors mobilize a cone of rock, the shape of which is dependent upon the properties of the rock

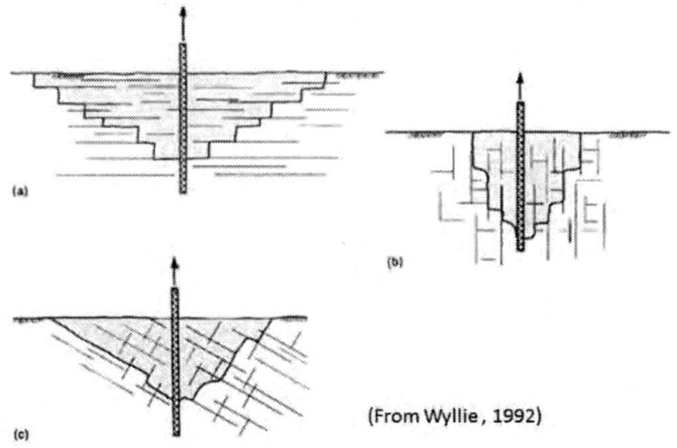

Figure 8 – Pullout Cone Geometry

including intact strength and joint orientations within the rock mass (Figure 8). Resistance to pullout is developed by the weight of the rock mass mobilized and the strength of the rock mass around the pullout cone (area of the cone surface). For multiple anchors where the cones overlap, or where the anchor loads come out of a rock slope face and are oblique, the surface area of the coalescing cones is developed using modelling techniques or AutoCAD, and Hoek-Brown strength criteria are applied (Figure 9).

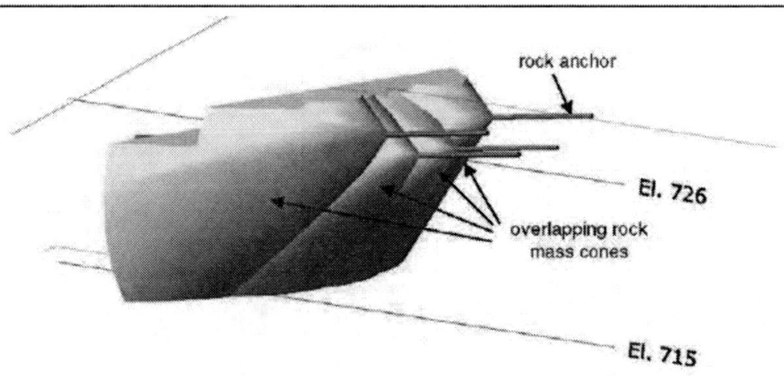

Figure 9 – Coalescing Cone Influence

Cone pullout testing is by nature sacrificial, complex, requires significant infrastructure, and at a project scale is generally not feasible. On a smaller scale it is invariably expensive and therefore, rare. Figure 10 shows the basic arrangement for a small- scale test, which is complicated by having to span the tested cone with reaction points outside the cone and applying

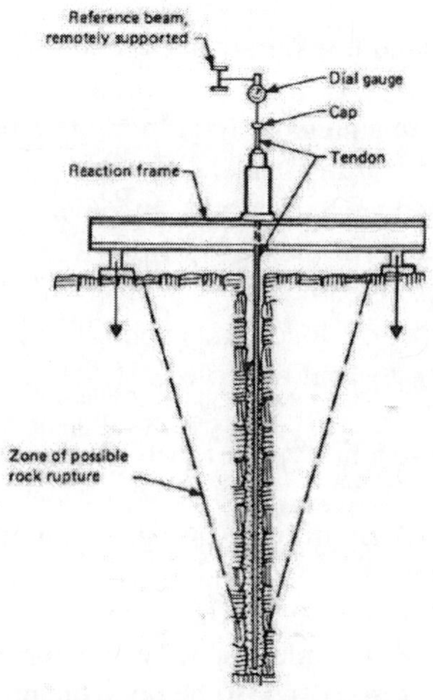

Figure 10 – Cone Pullout Test

loads with exceptionally robust jacking equipment. The authors have only heard of one full-scale test being completed over their respective careers, and certainly not on steep slopes. For tensioned anchors, a simple tensioning and lockoff Construction Quality Assurance (CQA) regimen is appropriate. In general, grout quality/strength and sacrificial bond value testing are appropriate for anchor installation CQA, together with drilling observation and review of bar properties.

COMPLEX POST-TENSIONED ANCHORS

PTI Testing

Post-tensioned anchors need to transfer loads behind a Rankine wedge (tieback wall), beyond a failure plane/back joint, or to a depth sufficient to mobilize a large pullout cone (e.g. dam tiedown). These types of anchors are used to resist large externally applied loads, typically imparted by structures like walls, uplift sections in dams and wind turbines. To carry the load to a designed point behind the anchor head assembly, an unbonded stressing length is needed where the anchor can elongate freely, store the tension needed in the anchor, and transfer the tensile load. The complexity of such anchors and their installation/construction led the PTI to develop testing methods to verify the condition and capacity of all the technical components of the anchors and ensure they have been installed as specified.

PTI testing guidelines (Post-Tensioning Institute, 2004) remain the "gold" standard for complex tieback and tiedown anchors and they have been adopted by most Transportation Departments in the United States for post-tensioned elements. Testing protocols have been designed to provide data to allow an engineer to:

- Establish whether the load transfer, capacity, and free stressing lengths have been adequately developed;
- Troubleshoot issues with an anchor (e.g., binding tendons in the free stressing length),
- Measure friction losses within an anchor (wobble), development length (depth of anchor elongation into a bonded anchor length), and anchor capacity;
- Check design values used for rock-grout bond strength and subsequent estimates for minimum bond length embedment;
- Measure lock-off and lift-off loads and anchor head effectiveness; and
- Assess load and anchor creep if observed.

PTI guidance calls for several types of tests. These include:

- Performance Testing – A rigorous testing program, conducted on the first few anchors tested to confirm design adequacy, and generally 2% of the total number of anchors, in order to thoroughly examine the performance of an anchor and examine free stressing and development lengths, wobble, capacity, creep and lockoff load. Performance tests, which are cyclical load tests, can also be done on any anchor that exhibits problems during Proof Testing to aid in troubleshooting anchor performance;
- Proof Testing – a focused test on all installed anchors to verify free stressing length development, capacity, creep, and lockoff load. The requirement for proof testing of all anchors is based on the fact that all the anchors already need to be brought up to the design lock-off load for performance during the service period and this test is just one additional step and
- Lift-Off Testing – a test to check and verify that lockoff load on a completed post-tensioned anchor is achieved and holding. Lift-off's are relatively simple and can also be completed post-installation over the service life of the anchor/structure to measure load loss over time.

Performance Testing

For complex anchors, performance tests are to be conducted on select production anchors and specifically on the first two or three anchors, as determined by the engineer and commonly a minimum of 2% of installed anchors thereafter. The test provides a check on design, installation accuracy, and a substantial amount of data on load transfer, load losses (wobble) due to friction in the free stressing length, anchor zone development, and capacity (Post-Tensioning Institute, 2004). The thoroughness of the Performance Test can also facilitate troubleshooting of an anchor if hysteresis loading indicates less than the required elastic reaction of the free stressing length as loads are increased and decreased incrementally up to the Test Load (TL) at 133% of the Design Load (DL). Loading stages for a typical Performance Test are presented in Table 1.

Table 1. Performance Test steps (PTI, 2004)

Load	Total Movement at load cycle Maximum δ_t	Residual Movement at AL after cycle Maximum δ_r	Elastic Movement at load cycle Maximum δ_e
AL, 0.25 DL, AL	δ_{t1}		$\delta_{t1} - \delta_{r1} = \delta_{e1}$
AL, 0.25 DL, 0.50 DL, AL	δ_{t2}	δ_{r1}	$\delta_{t2} - \delta_{r2} = \delta_{e2}$
AL, 0.25 DL, 0.50 DL, 0.75 DL, AL	δ_{t3}	δ_{r2}	$\delta_{t3} - \delta_{r3} = \delta_{e3}$
AL, 0.25 DL, 0.50 DL, 0.75 DL, 1.00 DL, AL	δ_{t4}	δ_{r3}	$\delta_{t4} - \delta_{r4} = \delta_{e4}$
AL, 0.25 DL, 0.50 DL, 0.75 DL, 1.00 DL, 1.20 DL, AL	δ_{t5}	δ_{r4}	$\delta_{t5} - \delta_{r5} = \delta_{e5}$
AL, 0.25 DL, 0.50 DL, 0.75 DL, 1.00 DL, 1.20 DL, 1.33 DL, AL	δ_{t6} Test Load (zero reading for Creep Test) δ_{tn} Final Load hold reading	δ_{r5}	$\delta_{t6} - \delta_{r6} = \delta_{e6}$
DL, Adjust to lock-off load		δ_{r6}	

The performance test is the "Full Monty" of post-tensioned anchor testing with each part of the test applied to verify whether the components of the complex anchors are properly constructed and are functioning as intended. The elastic behavior of the anchors is tested by measuring anchor elongation to ensure load transfer and capacity. Specifically, the test steps measure the following:

- Stepped loading – facilitates measurement of as-constructed free stressing strength, wobble (friction losses where the anchor sags between centralizers that are cast into the anchor during grouting), and development of the bonded anchor zone as the load is applied. All bonded anchor lengths experience some debonding of the bar/strand to grout at the top of the anchor zone as the load is transferred to the anchor length, and up to 50 percent of the anchor length is considered acceptable in many cases. Load steps also give a free stressing length time to allow an anchor to elongate as it should and overcome some wobble and redistribute stresses as tensioned.
- Cyclic loading – Largely for assessing how freely the bar/strands can elongate in the free stressing length, whether there is binding/sticking of the bar/strands leaving permanent elongation when unloaded and assessing wobble. If a free stressing length is not elongating at least 80 percent of the theoretical value based on design, cycling the load may also free up "sticking" spots and help troubleshoot an anchor not meeting the elongation criteria established by PTI.

- Test Load and Creep testing – allow an anchor to redistribute stresses through the anchor and verify the capacity of the bonded length. If the anchor cannot achieve the required test load, the anchor will not be able to be tensioned beyond the load at which the anchor begins to yield. The creep test is usually begun one minute after a test load is applied to allow the anchor components to redistribute stresses before timed measurements are begun. If the anchor yields more than 0.04 inches between 1 and 10 minutes (one log cycle), the creep test is extended to an hour to allow any binding in the system to be overcome, and the bond length to develop fully. Refer to PTI-14 for full details of performance testing.

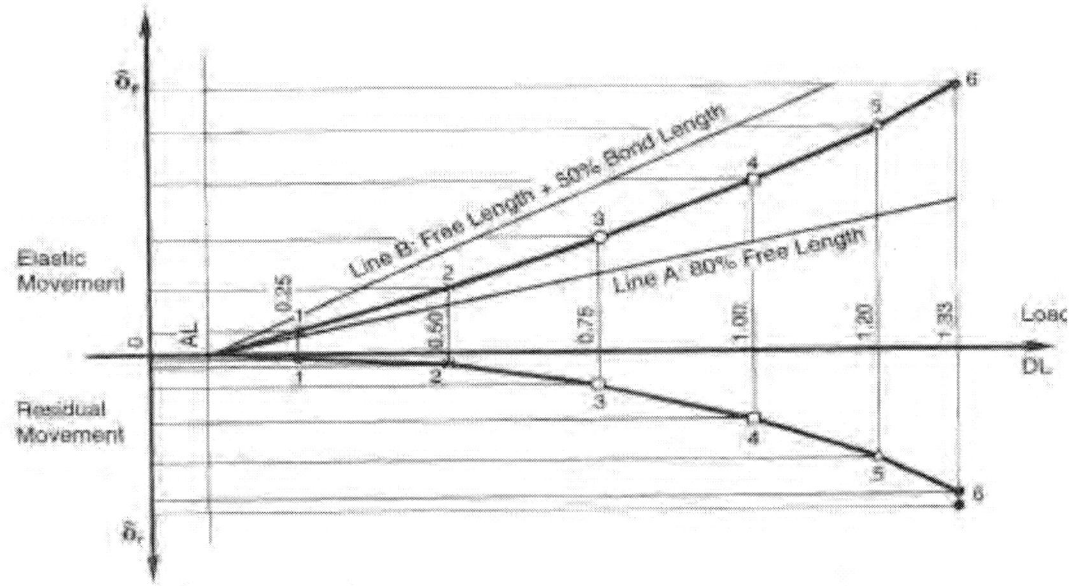

Figure 11 – Allowable Elongation Envelope (PTI-2004)

Proof Testing

Proof testing is typically conducted on the remaining production post-tensioned anchors as a function of their tensioning process. The proof test generally follows the performance test loading protocols to generally a maximum load of 133 percent of the design load; however, the hysteresis loading cycles are reduced to a single unloading curve after maximum proof load is achieved, or the cycled load is eliminated. This test provides Construction Quality Assurance (CQA) on each anchor as it is tensioned, and through plotting elongation with loading, can also verify that the anchor load is transferred to the bonded anchor length beyond a Rankine wedge or other critical length. Again, development of at least 80 percent of the free stressing length (minimum) or 100 percent of the free stressing length and up to 50 percent of the bonded anchor length (maximum) corresponding to confirmatory "A" and B" lines (Figure 8). The creep test is also incorporated in the proof test to ensure the bonded anchor length provides anchor capacity, and any "sticking" elements in the free stressing length are not slowly freeing up that could cause relaxation of a necessary design lockoff load.

Lift-Off Testing

Lift-off Tests are commonly done on a fraction of the anchors on a project to verify that the contractor's lock off procedures (tightening a nut or setting wedges on strands in an anchor head assembly) yield the correct applied load when the load is transferred to the anchor head assembly and the jacking mechanism is removed from an anchor. For this test, the anchor is slowly loaded until movement is registered in the anchor head or dowel, signifying the anchor is elongating and additional load is being added to the locked-off anchor. If the movement occurs below the design lockoff load, the lockoff procedures are checked and the anchor is re-tensioned, locked off and tested again.

Proof and liftoff tests include components of the performance test program and reflect a reduced testing program befitting their purpose of installation and capacity verification for quality assurance.

SUMMARY AND CONCLUSIONS

This paper presents a review of available anchor testing methods, what the test elements are intended to verify and what rock reinforcement types have anchor elements warranting testing. Complex anchors for tieback walls and dam tiedowns drove a need for thorough field testing and verification of their installation, stressing, and testing of their components and capacity. These complex anchors were intended to resist very large loads imparted by structures. Less complex anchors (frequently much less expensive too) have been subject to the complex testing methods as well, though they do not have many of the anchor components for which the complex tests were developed. Understanding the components of the rock bolts, whether they are tensioned or passive and their function as reinforcement or anchors subject to tensile loads will help the reader to select appropriate testing for their installation.

Consideration of testing rock/soil and grout properties, verification of ultimate grout to ground bond capacity and construction observation can simplify and streamline the QA/QC process and field installation. It can also help to avoid performing $1500 tests on $1000 rock dowels or wire rope anchors where it would be cheaper to add an additional dowel or anchor instead.

It is important to note that there is also a safety aspect to testing. A lot of energy is stored in a jack chair and anchor head assembly during testing. Tensioning complex anchors to hundreds of thousands of pounds or rock bolts to 50,000 pounds is inherently dangerous and many accidents have occurred when a bar extension fails, or a wire strand fails at the grips and launches out of the jack head assembly. Such dangers are multiplied when the contemplated work is on a rock slope in a high angle situation. Consider also the potential to fail or break the rock block being secured by imparting a large load on the block (Figure 11).

Figure 12 – Rock Broken by Anchor Tensioning

In the high angle arena, tension testing is acceptable for complex anchors but only to verify construction and lockoff load. Simple tensioned rock bolts should be tensioned and set. Stressing of tensioned rock bolts should not include spanning the bolts with large load frames and load capacity verification should be carefully weighed against simple tensioning and lockoff.

Passive rock dowels should be designed based on sacrificial bond tests and their installation should include grout, bar, and rock strength testing. In the authors' collective opinion, pulling on an a fully-grouted untensioned rock dowel makes as much sense as tensile testing reinforcing steel in structural concrete during construction.

A few simple questions can be the basis for a testing decision tree:

- If not a complex installation, is it post tensioned and locked off? – Consider less elaborate testing to tension and lock off simple tensioned anchors and verify bond capacities with sacrificial testing. An example would be a single stage grouted rock bolt with a bond breaker length.
- Is this an anchor that will be loaded after installation by an uplift force or pullout force, (e.g., guy wire anchor or wire rope anchor for a rockfall barrier fence) – conduct sacrificial bond (pullout) testing in accordance with ASTM D-4435-13 "Standard Test

Method for Rock Bolt Anchor Pull Test" and base design of the anchor on the testing. Tests will need to be conducted on each lithology the anchors are constructed in, and where rapid degradation of the rock may occur (e.g., slaking shales, close spaced joints parallel to anchor orientation) testing of sacrificial anchors in holes left open for various lengths of time may be warranted to gauge the effects of weathering/slaking on the installations.

- Is this a rock dowel or simple reinforcing element? – if yes, test the attributes used in the design – bond value with sacrificial testing by the ASTM method (or similar), but also test lab-determined grout unconfined compressive strength, steel and rock strength properties. No need to pull on a fully-grouted passive bar that will never be post-tensioned.
- Is this a complex, post-tensioned, anchor (free stressing length, load transfer to depth, intricate construction) intended to resist large external structural loads? - If yes, follow PTI guidance.

The PTI guidelines remain a reliable source for design and testing of complex post-tensioned anchorage elements; however, the PTI criteria are often overused for relatively simple ground anchorage elements. Keeping the testing program focused on rock reinforcement attributes and those ground anchor components present in the installation will allow simplification of the CQA testing program, increase cost effectiveness, reduce construction times and improve safety on high angle projects.

BIBLIOGRAPHY

ACI Committee 355. (1997). *State-of-the-Art report on anchorage to concrete.* American Concrete Institute.

Andrew, R. (2011). *Context Sensitive Rock Slope Design Solutions.* Lakewood, CO: Federal Highway Administration / Central Federal Lands Highway Division.

ASTM. (2013). *ASTM D4435-13e1. Standard test method for rock bolt anchor pull test.* West Conshohocken, PA: ASTM International .

Hoek, E., & Bray, J. (1981). *Rock Slope Engineering.* New York, NY: E & FN Spon.

Ingraham, P. T. (2014). Ground anchor selection and testing - matching ground anchors to their desired function and testing their serviceability. *65th Highway Geology Symposium Proceedings* (pp. 101-117). Laramie, Wyoming: Highway Geology Symposium.

Lang, T. (1961). Theory and practicce of rockbolting. *Transactions, Societry of Mining Engineers of A.I.M.E, 220*, 333-348.

Post-Tensioning Institute. (2004). *Recommendations for Prestressed Rock and Soil Anchors.* Phoenix, AZ: Post-Tensioning Institute.

Spang, K. a. (1990). Action of fully-grouted bolts in jointed rock and factors of influence. *Rock Mechanics and Rock Engineering Vol. 23, Issue 3*, 201-229.

Weatherby, D. a. (1982). *Tiebacks and their Application.* Washington DC: FHWA/RD-82/046 and 047.

Wood, D. J. (2005). Rock slope stabilization, Decew #2 Generating Station, St. Catharines, Ontario, Canada. *56th Highway Geology Symposium Proceedings* (pp. 73-88). Wilmington, North Carolina: Highway Geology Symposium.

Wyllie, D. (1999). *Foundations on Rock.* New York: E & FN Spon.

Laser Scanning of Highway Aggregates: Results of a Transportation Pooled Fund Study

Nancy J. McMillan, PhD, AOJN
Department of Geological Sciences
Box 30001, MSC 3AB
New Mexico State University
Las Cruces, NM 88003
(575)-646-5000
nmcmilla@nmsu.edu

Warren H. Chesner, PhD, PE
Chesner Engineering, P.C.
38 W. Park Avenue
Long Beach, NY 11561
(516)-431-4031
wchesner@chesnerengineering.com

Eileen O'Neil
Chesner Engineering, P.C.
2170 River Road
Coeymans, NY 12045
eolenterprises@yahoo.com

Prepared for the 70th Highway Geology Symposium, October, 2019

Acknowledgements

The authors would like to thank these individuals for their contributions of samples and discussion about the Pooled Fund SLT project:
Randy Billinger – KSDOT
Dan Sajedi – MDDOT
Amanuel Welderufael - MDDOT
Tom Festa – NYDOT
Bill Skerritt (deceased) – NYDOT
Mickey Cronin – OHDOT
The authors would like to thank for coding multivariate analysis software:
Chris Stein – Chesner Engineering PC

Disclaimer

Statements and views presented in this paper are strictly those of the authors and do not necessarily reflect positions held by their affiliations, the Highway Geology Symposium (HGS), or others acknowledged above. The mention of trade names for commercial products does not imply the approval or endorsement by HGS.

Copyright Notice

Copyright © 2019 Highway Geology Symposium (HGS)

All Rights Reserved. Printed in the United States of America. No part of this publication may be reproduced or copied in any form or by any means – graphic, electronic, or mechanical – including photocopying, taping or information storage and retrieval systems – without prior written permission of the HGS. This excludes the original authors.

ABSTRACT

The engineering properties of highway aggregates are used by State Departments of Transportation (DOT) for quality assurance/quality control of aggregates employed in highway construction. Some of the tests designed to measure these properties vary in complexity and can take several weeks or months to complete. An on-going Transportation Pooled Fund (TPF) Study uses emission spectra generated from targeting highway aggregates with a high-powered laser to classify and quantify the engineering properties of aggregates in minutes. Engineering properties being evaluated include specific gravity, Alkali Silica Reactivity (ASR), Acid Insoluble Residue (AIR), Dynamic Friction Value (DFV), D-cracking, presence of deleterious material, and non-approved aggregate identification in production samples. Participating TPF states include Kansas, Maryland, New York, and Ohio. The primary objective of the TPF effort is to determine whether laser scanning can be used to supplement and improve existing state DOT QA/QC programs.

The technology is based on a process referred to as Laser-Induced Breakdown Spectroscopy (LIBS). In this process, aggregate samples are targeted with a high-power laser pulse, which is used to excite atoms that make up the aggregate. This excitation results in the emission of light with unique spectra that can be used to identify the aggregate type and the engineering properties of the aggregates. The development of a spectra database of aggregate materials with known engineering properties provides the basis of employing multivariate modeling techniques to predict the properties of unknown aggregate materials.

An example of laser scanning is presented for each state; these case studies illustrate the methods and capabilities of the laser scanning system. Kansas aggregates are used to demonstrate the ability to identify non-approved aggregates that might be present in blended production samples. Maryland aggregates are used to construct and validate a model that predicts the amount of expansion due to ASR; New York aggregates are used to construct and validate a model that predicts the percent of AIR. Both models predict these values well. Ohio aggregates are used to illustrate how laser scanning can be used to detect deleterious materials such as chert. These case studies show that laser scanning is a reliable technique to supplement and enhance QA/QC analysis of highway aggregates.

INTRODUCTION

Many engineering properties of highway aggregates are used to monitor quality control and to determine the suitability of aggregates for various applications. These include a wide variety of properties, such as specific gravity, Acid Insoluble Residue (AIR), Alkali Silica Reaction (ASR), soundness, LA Wear, D-cracking susceptibility, and Dynamic Friction Value (DVF). The techniques used to acquire these data are determined by AASHTO Standard Methods and implemented in State DOT or contractor laboratories. Some tests are time-intensive, taking weeks to months to complete; such processes are expensive due to labor costs and can cause problems when data acquisition is required during a construction process. The Sample Laser Targeting (SLT) System instrument was developed to provide a rapid method for predicting these engineering properties.

Four state DOTs are involved in a Transportation Pooled Fund (TPF) study designed to test the SLT for use in improving aggregate QA/AC testing. The specific issues or problems identified by each state are presented in Table 1; parameters modeled in this paper are in italics. Kansas supplied aggregate samples used to develop a method to identify non-approved aggregates that might be present in blended production samples. Maryland supplied aggregates with measured ASR expansion percentages to investigate the ability to predict ASR expansion. Similarly, New York supplied aggregates with known AIR measurements. Finally, Ohio supplied aggregates used to build models that discriminate between chert and host limestone. This paper provides a brief overview of the SLT technology and presents examples of modeling results.

Table 1 – Aggregate Parameters Assessed for States in Pooled Fund Project. Items in Italics are Presented as Examples in this Paper.

State	Aggregate Parameter
All	Bulk specific gravity; AASHTO T84
All	Apparent specific gravity; AASHTO T84
All	Saturated surface dry specific gravity; AASHTO T84
Kansas	*Identification of non-approved aggregate in blended production samples*
Kansas	D-cracking susceptibility (durability and expansion during freeze-thaw)
Maryland	*Alkali-Silica Reaction (ASR); AASHTO 1260-2014*
Maryland	Dynamic Friction Value (DFV); Maryland Test Method MSMT-216
New York	*Acid Insoluble Residue (AIR); NYSDOT Materials Method No. NY 28-2017*
Ohio	*Identification of chert*

SAMPLE LASER TARGETING (SLT) SYSTEM TECHNOLOGY

Laser-Induced Breakdown Spectroscopy (LIBS)

LIBS is a laser ablation technique (Cremers and Radziemski, 2006) in which a high-power laser pulse in focused on the sample (Fig. 1). The pulse energy causes atoms to ablate off the surface and burn in a high-temperature plasma. In the plasma, electrons are excited to higher-energy orbitals and then decay back to lower-energy orbitals as the plasma cools. Photons with wavelengths proportional to the energy difference between the orbitals are emitted from atoms, captured by a lens, and transported through a fiber optic to a spectrometer. The light is diffracted and recorded on a charge-couple device, producing a spectrum with many peaks. The intensity of each peak is proportional to the number of photons generated from a specific orbital transition.

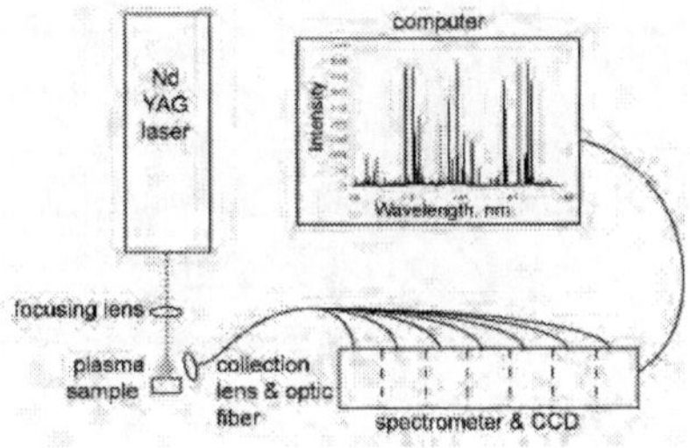

Figure 1 – Schematic Diagram of LIBS System (from McMillan et al., 2012).

LIBS spectra contain an enormous amount of information on the composition and structure of the analyzed material. Figure 2 shows representative spectra of four different aggregates: limestone, basalt, granite, and chert, with the major peaks labeled. Contained in the spectra are: 1) information on the concentrations of all naturally-occurring elements (Cremers and Radziemski, 2006), 2) information on some isotopic ratios (Russo et al., 2011), and 3) structural information (Serrano et al., 2015). Spectra can be modeled with multivariate statistical analysis to take advantage of this information. A single spectrum can be modeled to yield information on different engineering properties (specific gravity, ASR, AIR, etc.). Each model uses the parts of the spectrum that correlate with the engineering property and treats the rest of the spectrum as insignificant. Thus, LIBS represents an efficient method for highway QA/QC analysis.

Sample Laser Targeting (SLT) System

The Sample Laser Targeting (SLT) System is an automated sample delivery-LIBS system designed for analysis of highway aggregate samples. Figure 3A illustrates the design of the instrument. The laser, spectrometer and optical components are housed in separate cabinet boxes (Fig. 3B) that prevent dust contamination from the aggregate samples and heating of the spectrometer, which can cause the data to drift with changing temperature. A dust collection system removes the dust produced during sample ablation. Figure 3C is a photo of the SLT with the sample chamber open and a highway aggregate sample loaded on the circular sample tray. During analysis, the tray table rotates as the laser shoots directly down and light is collected by a lens near the sample. Rotation of the tray table continuously provides fresh sample to the laser, making it possible to acquire several thousand laser shots of a single sample without having to change samples.

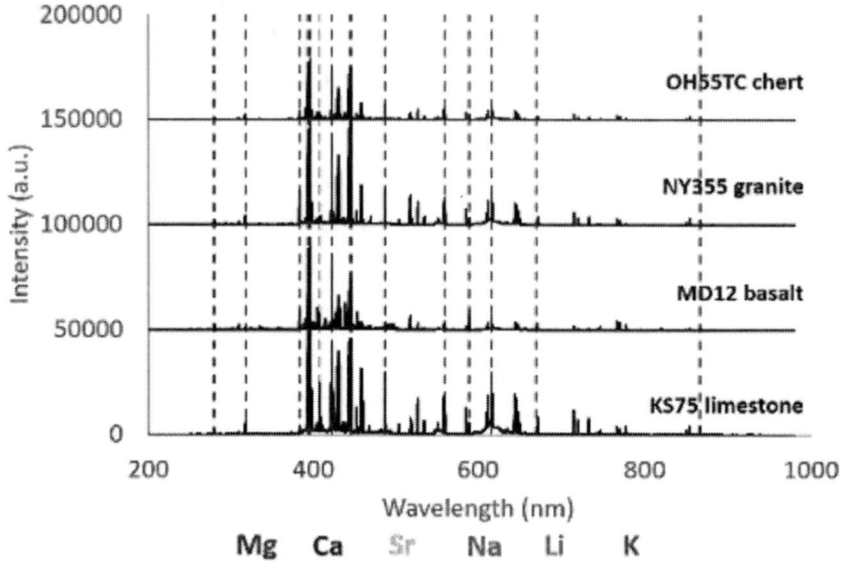

Figure 2 – Representative LIBS Spectra with Some Peaks Labeled.

Figure 3 – Sample Laser Targeting (SLT) System. A. Schematic diagram of the instrument's component parts. B. Schematic diagram of the laser cabinet with transparent faces. C. Photo of the SLT with an aggregate loaded in the sample tray.

MULTIVARIATE MODELING

Principal Component Analysis (PCA)

Multivariate analysis, or chemometric analysis, is an effective method for extracting information pertinent to a specific engineering property from LIBS spectra. This project uses two common chemometric techniques: Principal Component Analysis (PCA) and Partial Least Squares Regression (PLSR). PCA is an unsupervised technique that describes the spectral relationships between the samples; it is descriptive only (Esbensen, 2004). In PCA, the data are plotted in n-dimensional space, where n is the number of channels in the spectrum (14,336 channels for spectra generated by the SLT). A linear regression is then calculated through the data; this regression is called Principal Component 1, or PC 1. A second linear regression, PC 2, is calculated perpendicular to PC 1, and subsequent regressions (PC 3, PC 4, etc.) are calculated, all perpendicular to all other principal components. In general, most (>85%) of the variance in the data is contained in the first 4-6 principal components. Subsequent principal components model increasing amounts of noise in the data and are not as useful. Each sample has a position relative to each principal component. Thus, PCA results are often plotted as score plots, which are 2-dimensional plots showing the relationship of samples in PC 1 – PC 2 space. Score plots are interpreted in the same way as any two-dimensional plot: samples that plot close to each other are similar in composition and those that plot far from each other are different.

Figure 4 presents a PCA score plot calculated with samples from MDDOT. There are two types of aggregates in this example: carbonates (limestone, dolostone, and marble) and silicates (basalt, granite, serpentinite, gneiss, schist, hornfels, and quartzite). The carbonate and silicate samples plot in two discrete groups with separation along the PC 1 direction. This is an example of how PCA is used to observe groups of samples within a data set and discern which samples are similar in composition to each other.

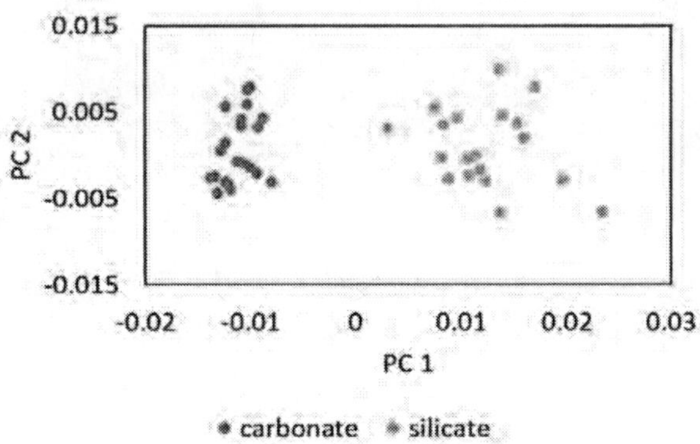

Figure 4 – Principal Component Analysis (PCA) Score Plot With Carbonate and Silicate Samples Provided by MDDOT.

Partial Least Squares Regression (PLSR)

PLSR (Wold et al., 2001) is a supervised multivariate analysis technique that is used in this study to predict the values of engineering properties of highway aggregates from LIBS spectra. In contrast to PCA, the regressions through the data set also use a matrix consisting of an independent variable for each sample. These independent variables change the direction of the linear regressions through the data set and thus are said to "supervise" the regression. The independent variable in this work is an engineering property, such as AIR % or ASR expansion %. The PSLR model locates the linear regression in a position that maximizes the correlation between the independent variable and the spectra. The independent variable can also be used to classify samples into groups, such as the chert group or the limestone group. In this case, the independent variable is set to "1" for one group and "0" for the other.

The PLSR model for AIR% calculated with NYSDOT aggregate samples in this study is shown in Figure 5 as an example. The model is calculated using 75% of the samples; the remainder are reserved for model validation. During calculation of the model, the value of the independent variable (AIR% in this case) is calculated for each of the calibration samples. Figure 5A plots these calculated values against the AIR% values measured at NYSDOT compared to the dashed 1:1 line, which indicates a perfect correlation. The model is fairly good, with a slope close to 1 (0.85) and a high correlation coefficient ($r^2 = 0.85$). This indicates a strong model.

In this study, PLSR models are validated using test sets. The 25% of the spectra not used in the calibration are input into the model, and the values of the independent variable are calculated for the test set samples. Because the measured AIR% values are known for these samples, it is possible to evaluate the strength of the model. The measured and predicted AIR% values for the test set samples are plotted in Figure 5B. The slope is high (0.90) and the correlation is fairly strong ($r^2 = 0.75$).

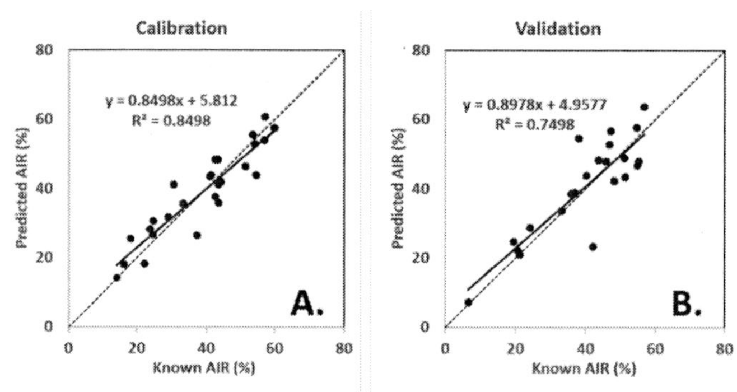

Figure 5 – Partial Least Squares Regression (PLSR) Model for AIR% Using Aggregate Samples Supplied by NYSDOT. A: Calibration. B: Test-set Validation.

RESULTS AND INTERPRETATIONS

KSDOT: Identification of Approved Aggregates and Non-Approved Aggregates

In Kansas limestone quarries, certain beds, or ledges, are tested and approved; other ledges did not pass testing and are non-approved. This study tested a method to ensure that suppliers are not using production samples that incorporate non-approved aggregates either in bulk or by blending them with an approved ledge aggregate. The problem is approached by calculating a PCA score plot with ledge and production samples.

Figure 6A shows a PCA score plot for ledge (black symbols) and production limestone samples (colored circles) from the Bethany Falls Quarry, KS. Each sample was scanned three or four times to capture the inherent heterogeneity of the samples. The approved ledge samples (KS95, KS96, and KS97) define a compositional range that runs along the lower half of the score plot. Production samples KS91, KS92, and KS93 fall within the range of the ledge samples. While one spectrum from KS93 (green) falls slightly to the right of the ledge range, the other three spectra plot well within the range. This one spectrum is interpreted to represent heterogeneity within the KS93 sample. All four samples of KS94 (yellow), however, fall above the range of approved ledge samples. KS94 is interpreted as either a non-approved aggregate or a mixture of an approved aggregate with a non-approved aggregate.

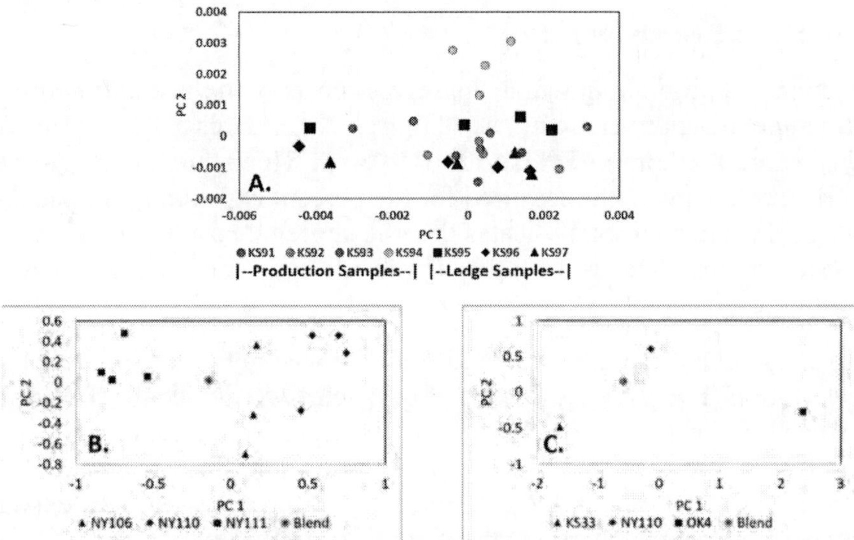

Figure 6 – PCA Score Plots that Compare Ledge and Production Aggregates. A: Ledge and Production Samples from Bethany Falls Quarry, KS. B: Physical Blend of Three New York Limestones. C: Physical Blend of Two Limestones and a Quartz Sandstone.

This method could be implemented real-time at a quarry or highway construction project. Approved ledge samples would be submitted and analyzed periodically to ensure that the PCA model represents current aggregates mined from the quarry. Production samples would be sampled and scanned frequently, perhaps daily or weekly, and a PCA model would be calculated with the addition of each set of samples to ensure that production compositions remain within the approved range. The time required for sample analysis and model calculation is approximately 30 minutes.

Two experiments were performed to test the idea that blends of aggregates will plot in the middle of the range defined by the aggregates in PCA score plots. The first experiment used NYSDOT limestones NY106, NY110, and NY111 as end-members. Each limestone was scanned three to four times. A physical blend of these limestones was created by weighing and mixing them in the proportions 33% NY106, 33% NY110, and 34% NY111. This blended aggregate was then scanned and a PCA score plot with the end-members and the blend (Fig. 6B). The blend falls in the center of the range defined by the end-members. A second, similar, experiment used limestones KS33 and NY110 and quartz-rich sandstone OK4 as end-members. Each sample was scanned once, and a 33:33:34 physical blend was made and scanned. In the PCA score plot (Fig. 6C), the blend falls within the range defined by the end-member aggregates, but falls close to the line defined by the two limestones. This is because the SLT laser light does not couple well with quartz and the spectrum from quartz sandstone OK4 has much lower intensities than the limestone spectra. Thus, the limestone spectra have greater leverage in the model and the model underestimates the amount of quart sandstone in the plot. The score plot does indicate, however, that spectra from blended aggregates do fall within the range defined by the end-members, despite issues of vastly different aggregate compositions and analytical issues such as laser coupling.

MDDOT: Alkali-Silica Reactivity (ASR)

Alkali-silica reaction is a problem in which aggregate reacts with the alkali hydroxides in concrete, resulting in expansion, cracking, and failure. Aggregates are tested for ASR in MDDOT using standard method AASHTO 1260-2014, in which cured concrete bars are soaked in 1 Molar NaOH solution and then measured for the percent expansion. Expansion of more than 0.20% after 14 days in solution indicates that the aggregate used in making the bar may be susceptible to ASR reaction. This is a good example of a time-intensive test that could be made more efficient with the SLT method.

Samples submitted by MDDOT with measured ASR expansion % were modeled using PLSR (Fig. 7). The calibration (Fig. 7A) has a slope of 0.87, close to the ideal 1:1 dashed line. The correlation coefficient indicates a strong correlation between the measured and predicted ASR values. Figure. 7B shows the results of test set validation. The slope is essentially equal to 1; however, the scatter is greater than in the calibration. Aggregates with ASR expansion % greater than 0.20 are considered to be deleterious; there are three such samples in the test set. Two of the three were correctly identified as having ASR expansion % greater than 0.20. Similarly one sample with measured ASR expansion % of 0.08 is predicted at greater than 0.20%. This is a high-quality sample that is classified as deleterious. With the exception of these two samples, all others are correctly classified relative to the 0.20% expansion limit.

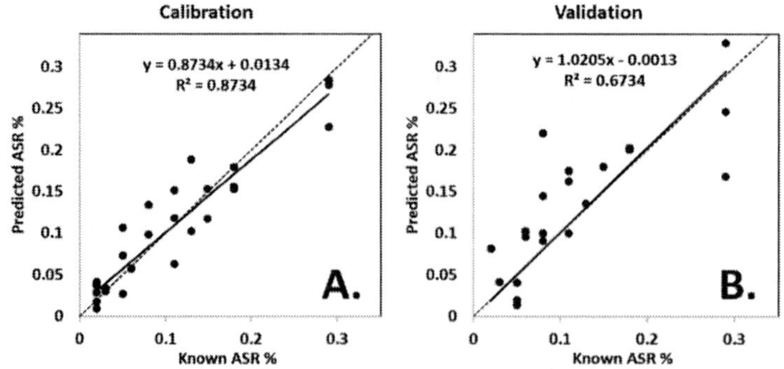

Figure 7 – Partial Least Squares Regression (PLSR) Model for ASR Expansion% Using Aggregate Samples Supplied by MDDOT. A: Calibration. B: Test-set Validation.

NYSDOT: Acid Insoluble Residue (AIR)

A major concern in New York is the friction properties of aggregates because limestone aggregates become polished with wear, resulting in slick roadways. NYSDOT uses AIR as measured by NYSDOT Materials Method No. NY 28-2017 to evaluate aggregates for friction properties. Aggregates are dissolved in HCl; the percent of mass remaining after complete dissolution is the AIR%. As dissolution continues, aggregate particles are coated with reactants which armor the limestone from the HCl solution, causing the reaction to cease. It is necessary to break the particles open periodically to expose fresh limestone to the HCl. The process can take days to weeks to go to completion.

Limestone samples submitted by NYSDOT were scanned and a PLSR model was calculated. The predicted and measured AIR% values of the calibration samples (Fig. 5A) are well correlated ($r^2 = 0.85$) and the slope close to 1 (0.85). Test set validation results are shown in Fig 5B. The slope is 0.9, with some scatter in the data ($r^2 = 0.75$). All samples with measured AIR% greater than 20% have predicted values greater than 20%; the model does an excellent job predicting high-AIR samples. Three samples have measured AIR% close to the 20% limit (Table 2). Samples NY121 and NY129 have predicted AIR% within 1% of the measured values. Sample NY131 has a predicted AIR% that is 5.7% higher than the measured value, placing it in the approved category. If SLT analysis of aggregates for AIR were to be implemented, samples with predicted AIR% values of 15-25% should be analyzed by the NYSDOT Materials Method.

Table 2 – Predicted AIR% of NYSDOT Samples Near the 20% limit

Sample	AIR % Measured, NYSDOT	AIR %, Predicted, SLT
NY121	20.6	22.1
NY129	21.1	20.9
NY131	19.4	24.7

OHDOT: Presence of Deleterious Materials: Chert

Chert can be a deleterious material, causing ASR cracking of aggregates. OHDOT submitted samples of chert and limestone that were separated from cherty limestones beds. Identification of chert in Ohio aggregates was developed using a PLSR model that classifies an aggregate as chert or as limestone. This PLSR model uses the independent variable "1" for chert samples and "0" for limestone samples. Figure 8A shows the results of model calibration. The independent variable calculated for each calibration sample is plotted against sample number. All limestone samples have low values, clustering around 0, and all chert sample have high values, clustering around 1. The dashed line indicates the VAD (Value of Apparent Distinction), 0.55 in this case. The VAD is used to classify samples as chert (independent variable ≥ 0.55) or limestone (independent variable < 0.55). Test set validation of this model is shown in Fig. 8B. All limestone samples in the test set have Chert Index values < VAD, and all chert samples have Chert Index values > VAD. This model is 100% successful.

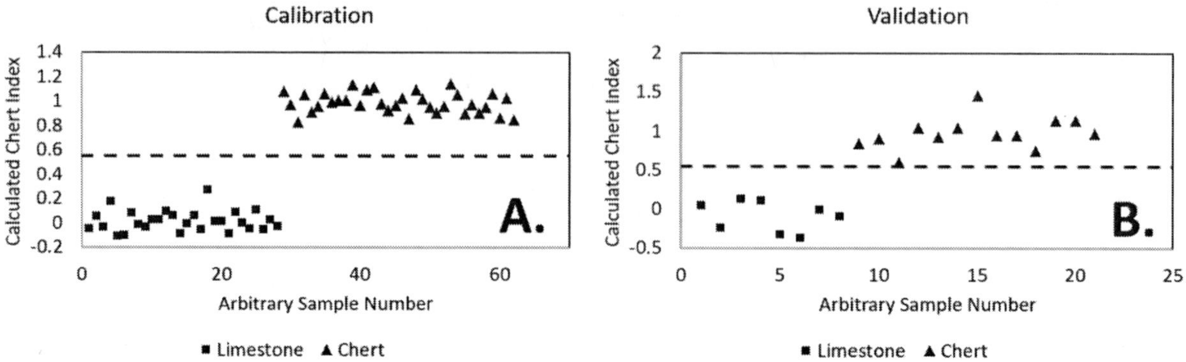

Figure 8 – Partial Least Squares Regression (PLSR) Models for the Presence of Chert and Type of Chert in OHDOT samples. A: Calibration of Chert Model. B. Test-set Validation of Chert Model.

CONCLUSIONS

This work demonstrates that multivariate analysis of SLT scans of highway aggregates has great potential to aid state DOTs in quality assurance/quality control of aggregates. Four examples of the type of analysis performed by the SLT were presented. Kansas aggregates were used to identify non-approved aggregates of blends incorporating non-approved aggregates by comparing their SLT spectra to those of approved aggregates using PCA. A PLSR model that predicts the ASR expansion % was calibrated and validated using Maryland samples. Models were able to predict which samples exceed the 0.20% expansion limit for use in construction. New York samples were used to calibrate and validate a model that predicts the AIR% as a measure of the friction properties of an aggregate. The model was successful, even for samples near the 20% AIR limit. Finally, PLSR models were calibrated and validated that can classify aggregates as chert or limestone (100% successful).

SLT scanning and chemometric modeling take approximately 30 minutes per sample. The multivariate analysis is almost instantaneous for most models because the calibrations are developed before analysis of unknown samples. This study shows that predicted values for some engineering properties or other questions such as the source of a given aggregate can be provided rapidly and prevent the use of non-approved or inferior aggregates in highway construction projects.

REFERENCES

Cremers, D. A. and L. J. Radziemski. *Handbook of Laser-Induced Breakdown Spectroscopy*. Wiley, 2006.

Esbensen, K.H. *Multivariate data analysis – in practice*. Camo Process (Oslo), 2004.

McMillan, N. J., C. Montoya, and W. H. Chesner. Correlation of limestone beds using laser-induced breakdown spectroscopy and chemometric analysis. *Applied Optics* vol. 51 (7), 2012, pp. B213-B222.

Russo, R. E. A. A. Bol'shakov, X. Mao, C. P., McKay, D. L., Perry, and O. Sorkhabi. Laser Ablation Molecular Isotopic Spectrometry. *Spectrochimica Acta Part B* vol. 66, 2011, pp. 99-104.

Serrano, J., J. Moros, and J. J. Laserna. Sensing signatures mediated by chemical structure of molecular solids in laser-induced plasmas. *Analytical Chemistry* vol. 87, 2015, pp. 2794-2801.

Wold, S. M. Sjöström, and L. Eriksson. PLS-regression: A basic tool of chemometrics. *Chemom. Intell. Lab. Syst.* Vol. 58, 2001, pp. 109-130.

BETTER HIGHWAYS THROUGH APPLIED GEOLOGY
2019 HIGHWAY GEOLOGY SYMPOSIUM
Oregon highways connecting coasts, mountains, and deserts
70th HGS - PORTLAND, OREGON

70th Highway Geology Symposium

Better highways through applied geology

70th HGS Field Trip Guide

"The Magnificent Columbia River Gorge"

Wednesday, October 23, 2019

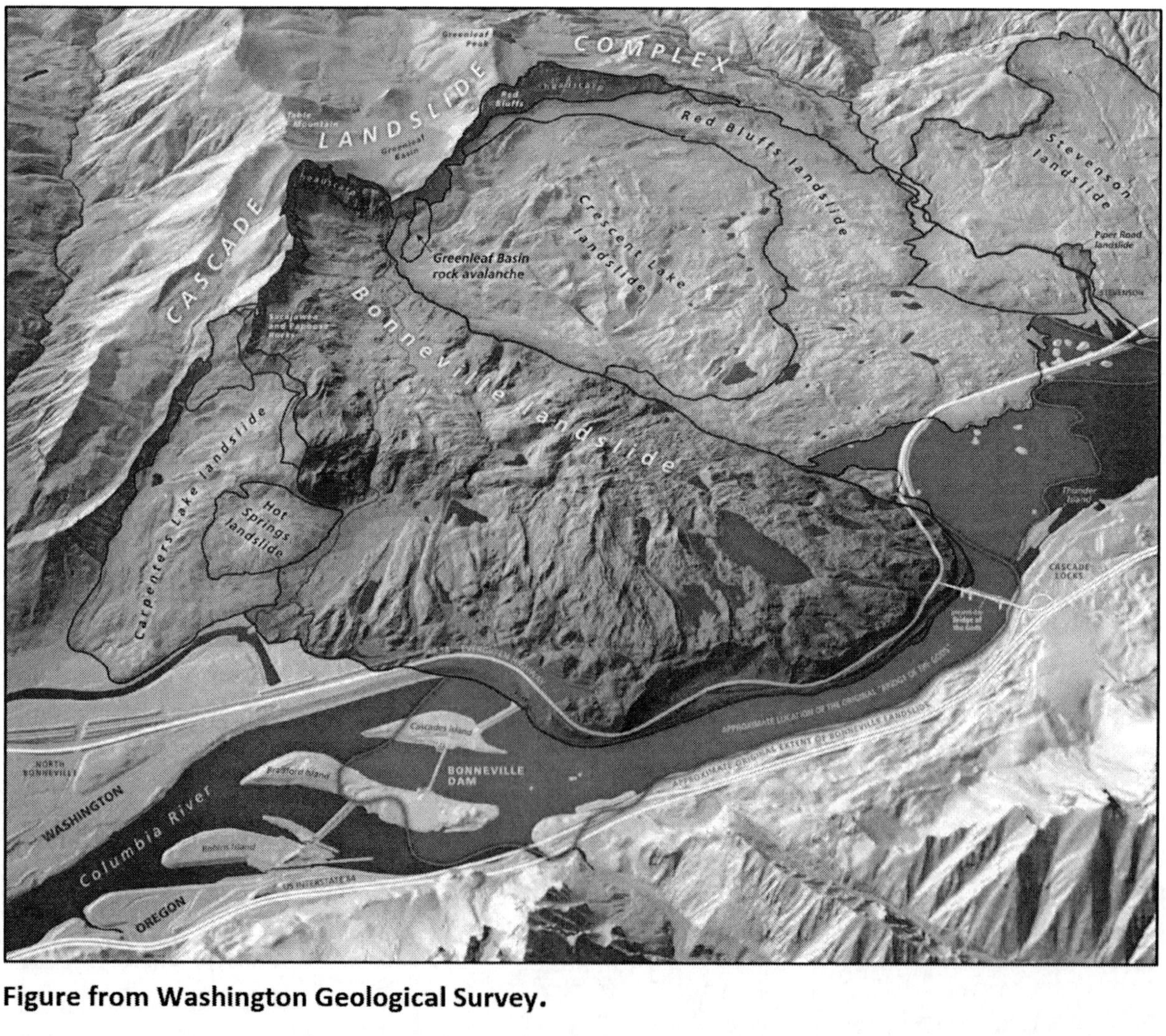

Figure from Washington Geological Survey.

HGS 2019 – Columbia River Gorge Field Trip Stops

Google Maps Link for Field Trip: https://goo.gl/maps/kc7KtTMnyKEjG7QF9

7:30 AM:	**Depart Portland Marriott Downtown Waterfront**
8:15 AM:	**Stop 1 (Buses 1 & 2): Portland Women's Forum State Scenic Viewpoint (25.3 miles)**
8:15 AM:	**Stop 1 (Buses 3 & 4): Vista House via HCRH (26.5 miles)**
9:00 AM:	**Buses 1 & 2 depart Portland Women's Forum State Scenic Viewpoint to Vista House**
9:00 AM:	**Buses 3 & 4 depart Vista House to Portland Women's Forum State Scenic Viewpoint**
9:05 AM:	**Stop 2: Buses 1 & 2 Vista House (1.2 miles)**
9:05 AM:	**Stop 2: Buses 3 & 4 Portland Women's Forum State Scenic Viewpoint (1.2 miles)**
9:50 AM:	**All buses depart from stop 2**
10:15 AM:	**Stop 3 (All buses): Multnomah Falls Lodge (13.5 miles)**
11:25 AM:	**Depart Multnomah Falls Lodge.**
11:40 AM:	**Stop 4 and 5 (All Buses): LUNCH. The Bonneville Landslide and Cascade Locks (13 miles)**
1:10 PM:	**Depart Cascade Locks – Marine Park Pavilion and Bonneville Landslide**
1:25 PM:	**Stop 6 (All Buses): HCRH State Trail, I-84 MP 52.8 (9.2 miles)**
2:25 PM:	**Depart HCRH State Trail, I-84 MP 52.8**
2:55 PM:	**Stop 7 (All Buses): Rowena Loops Viewpoint via Exit 76 (23.3 miles)**
3:40 PM:	**All buses depart Rowena Loops Viewpoint**
5:20 PM:	**Stop 8 (All Buses): Arrive at Portland Marriott Downtown Waterfront (78.2 miles)**

Figure from Washington Geological Survey

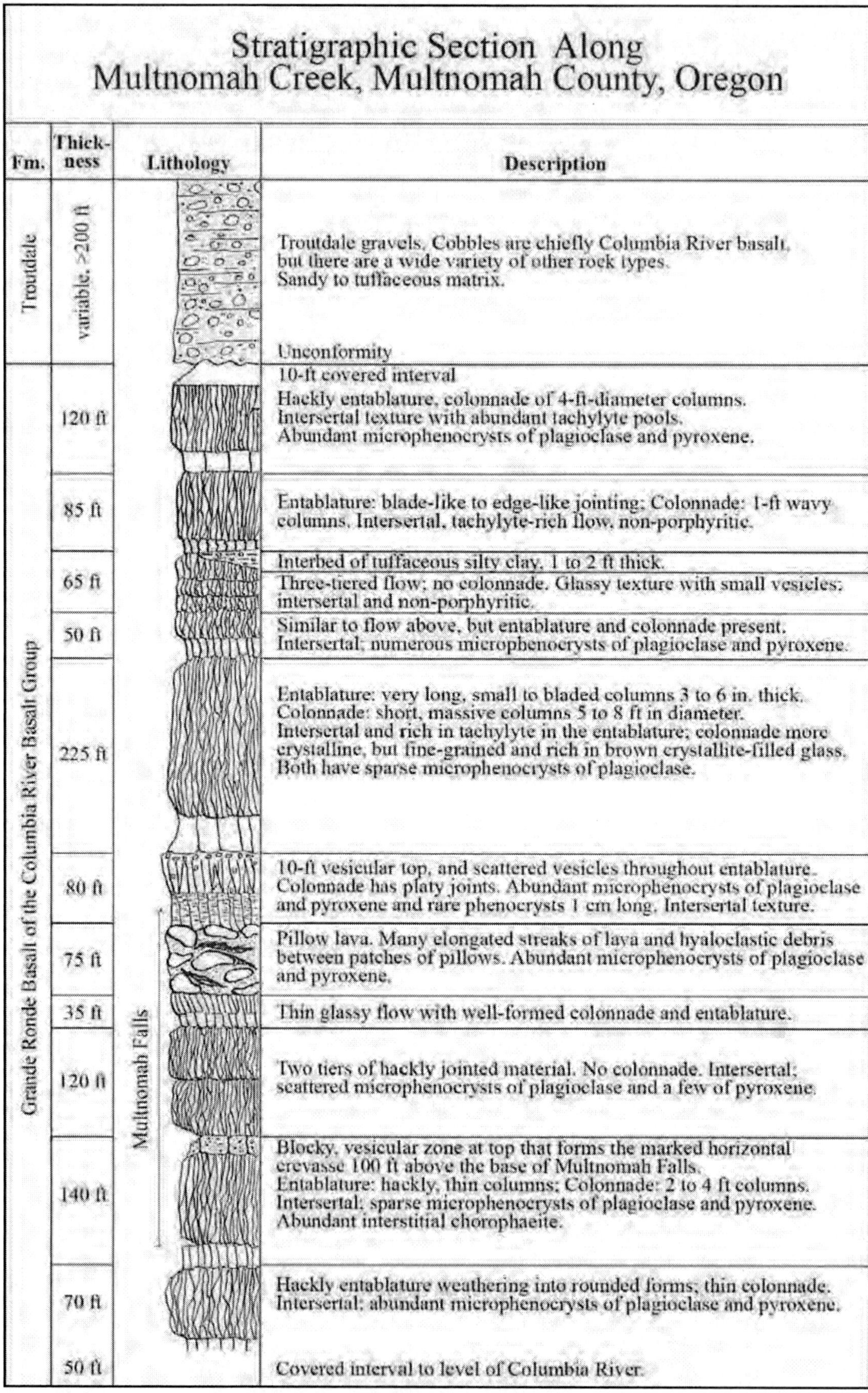

Figure from Washington Geological Survey

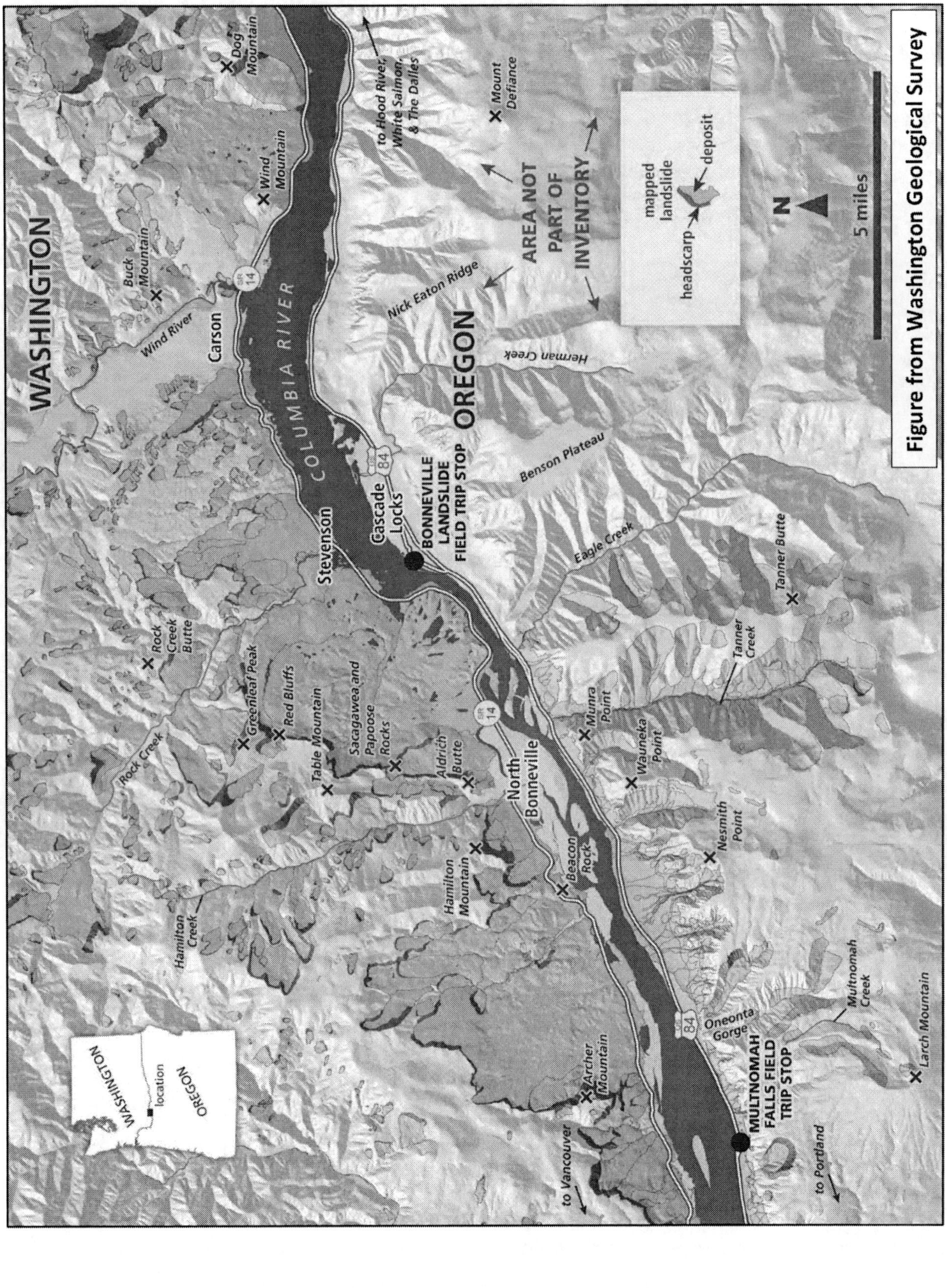
WASHINGTON
OREGON
COLUMBIA RIVER
Wind River
Carson
Buck Mountain
Wind Mountain
Dog Mountain
to Hood River, White Salmon, & The Dalles
Mount Defiance
AREA NOT PART OF INVENTORY
Nick Eaton Ridge
Herman Creek
Benson Plateau
Eagle Creek
Tanner Butte
Tanner Creek
Stevenson
Cascade Locks
BONNEVILLE LANDSLIDE FIELD TRIP STOP
Rock Creek Butte
Greenleaf Peak
Red Bluffs
Table Mountain
Sacagawea and Papoose Rocks
Aldrich Butte
Rock Creek
North Bonneville
Munra Point
Wauneka Point
Nesmith Point
Beacon Rock
Hamilton Mountain
Hamilton Creek
Oneonta Gorge
Multnomah Creek
Larch Mountain
Archer Mountain
MULTNOMAH FALLS FIELD TRIP STOP
to Vancouver
to Portland
WASHINGTON
location
OREGON
mapped landslide
headscarp
deposit
N
5 miles
14
84
Figure from Washington Geological Survey

Stop 1/2: Portland Women's Forum State Scenic Viewpoint (Chanticleer Point)

8:15 AM: **Stop 1 (Buses 1 and 2): Portland Women's Forum State Scenic Viewpoint (25.3 miles, 35 minutes, no restroom facilities). Total duration at stop 1 is 45 minutes.**

Note *the historic guardrail sections and mile markers and the creeping landslide headscarp that we cross at the top of the large Rooster Rock Landslide complex as we leave the Women's Forum State Scenic Viewpoint and head toward the Crown Point Vista House*

9:05 AM: **Stop 2: (Buses 1 and 2): Crown Point Vista House via HCRH – (1.2 miles, 5 minutes, restroom facilities available – elevator available for ADA if needed). Total duration at Stop 2 is 45 minutes.**

Aerial photograph courtesy of Oregon Department of Transportation (ODOT) of the Portland Women's Forum State Scenic Viewpoint parking area looking northward.

Presenter: Robert Hadlow (ODOT) – *Upper Lot at the placard*

At Portland Women's Forum State Scenic Viewpoint, participants will learn about the Historic Columbia River Highway (HCRH). The talk will focus, in part, on why this location was so important in 1913 for rallying support for the highway's construction. Portland Women's Forum State Scenic Viewpoint was formerly the location of Chanticleer Inn, a rural restaurant that served country dinners to those folks who had automobiles at the time and could venture out of Portland on county roads. Chanticleer Inn was also the location where Sam Hill, the highway's promoter, assembled a group of good roads advocates in August 1913 to persuade the Multnomah County commissioners to provide funding for its initial construction. Hill's strategy paid off. Shortly, the commissioners raised the money through bonds. Surveying for the new highway started in September 1913 and construction began in October 1913. Road advocates celebrated the HCRH's completion in Multnomah County in 1916.

<u>Presenter: Scott Burns</u> (Portland State University (PSU)) – *<u>Lower Lot at the placard</u>*

This site is one of the most spectacular sites in the whole Columbia River Gorge National Scenic Area. It is also one of the five most photographed sites in Oregon. Here we will talk about many things we can see. We compare the Oregon side, which is quite vertical (mainly stacked layers of Columbia River Basalt on top of one another) vs the Washington side, which is low angle (mainly landslides riding on the Ohanapecosh Formation).

We will discuss the story of the two floods that formed the Gorge. The first set of floods were from 15-16.7 million of years ago (Ma), and those were the flood basalts that formed the bedrock, Columbia River Basalt. They came from fissures where Oregon, Washington and Idaho come together, and they flowed to here. The second set of floods were the great Missoula Floods which came through here and deepened and widened the Gorge that was already here. There were 40 floods that reached here between 15,000-18,000 thousand years ago. The floods came as outburst floods from the breaking up of the ice dam that had formed Glacial Lake Missoula in Montana (Flathead Lake Area). The floods covered 16,000 square miles in the Pacific Northwest.

Other features we will talk about are Mt. Zion (cinder cone), Larch Mountain (shield volcano), Crown Point inter-canyon flow, Rooster Rock Landslide, Beacon Rock volcanic plug, sand dunes along the river, and the winds of winter in the Gorge.

Stop 1/2: Crown Point Vista House

8:15 AM: **Stop 1 (<u>Buses 3 and 4</u>): Vista House via HCRH - (26.5 miles, restroom facilities available – elevator available for ADA if needed). Total duration at stop 1 is 45 minutes.**

<u>Note</u> *the historic guardrail sections and mile markers and the creeping landslide headscarp that we cross at the top of the large Rooster Rock Landslide complex as we leave the Women's Forum State Scenic Viewpoint.*

9:05 AM: **Stop 2: (<u>Buses 3 and 4</u>): Crown Point Vista House to Women's Forum – (1.2 miles, 5 minutes, no restrooms available). Total duration at Stop 2 is 45 minutes.**

<u>Presenter: Bob Hadlow</u> (ODOT)

Overview of the Columbia River Highway and History of Vista House – *<u>Inside on main level</u>*.

At Vista House, participants will learn about the significance of this wonderful building that opened on Crown Point in 1918. Vista House is a monument to Oregon Pioneers and its rotunda contains artwork dedicated to their memory. The building also has an outdoor observation deck that offers wonderful views up and down the Columbia River. Finally, Vista House was equipped with the most modern of public comfort stations in its day, when traveling outside cities was very primitive.

<u>Above:</u> **Aerial photograph courtesy of ODOT of the Crown Point Vista House looking eastward.**
<u>Below:</u> **Historic photograph of Crown Point Vista House.**

Historic photograph of Crown Point Columbia Highway before Crown Point Vista House construction.

<u>Presenter: Nathan Jenks</u> (Bonneville Power Administration (BPA)); <u>*inside on the main level.*</u>

The Crown Point Viaduct Restoration Project was a collaborative effort between ODOT, the FHWA – WFLHD, Oregon State Parks, and a number of other stakeholders and interested parties. The project involved restoring the approximately 600-foot-long, 8-foot-wide, cast-in-place viaduct which was originally constructed in 1914. The viaduct provides an elevated pedestrian walkway with sweeping views of the Columbia River Gorge.

The overall goal of the project was to repair and strengthen the aging viaduct and associated retaining wall while maintaining the appearance and as much of the original structure as possible. The viaduct was in poor condition prior to restoration with cracked and damaged support columns, beams, and deck as well as inadequate foundation support. Portions of the dry-stacked stone retaining wall beneath the viaduct were also in poor condition with failed sections, bulging, settlement areas, and loss of backfill, which in-turn threatened the roadway immediately adjacent to the viaduct.

The viaduct is located on a sharp curve, and the ground surface slopes steeply away on the west, north, and east sides. A vertical cliff face dropping to the valley floor is located several tens of feet from the viaduct on the north and east sides. The site is mantled with gravel and cobbles deposited by the Pleistocene-age Missoula floods, underlain by the Miocene-age Grande Ronde Basalt member of the Columbia River Basalt Group. Stratigraphy at the site varies but generally consists of the following as measured from the roadway elevation immediately adjacent to the viaduct: 4 to 8 feet of roadway fill, 20 to 50 feet of Missoula Flood Deposits, 3 to 5 feet of weathered/decomposed basalt, over strong to very strong basalt.

The primary geotechnical elements of the project included reestablishing adequate foundation support for the 29 column pairs, as well as reconstruction of approximately 110 feet of retaining wall with wall heights up to about 18 feet. Structural repairs and rehabilitation to the columns, beams, and deck/sidewalk were also included. Numerous alternatives were considered and generally fell within three overall concepts, 1) remove and replace existing roadway, fill, and retaining wall and reconstruct or repair various elements, 2) provide supplemental foundation support as needed and construct a soil-nail wall to stabilize the failed wall areas, and 3) construct a reticulated micropile wall.

Approach #2 was the selected alternative based primarily on cost and constructability considerations, but also for its benefits in retaining the historic character of the structure. Both options 2 and 3 allowed for continued, one-lane traffic, during construction which was an important requirement for ODOT and other stakeholders.

The final design for the geotechnical elements included providing supplemental foundation support through the use of micropiles at all 29 of the interior column foundations and 15 of the 29 exterior column foundations as shown on the attached figure. The existing dry-stacked stone retaining wall was partially removed using a top down approach and a soil-nail wall was constructed for the full height to provide a durable retaining wall system. The dry-stacked stone that was removed for construction of the soil-nail wall was then replaced as facing to retain the historic appearance and materials as used in the original construction. While the construction was not without challenges, the overall project was considered a great success with positive feedback from the stakeholders and the public.

Photos of the viaduct prior to restoration project. Note the additional support at exterior columns in the left photograph on the left and the failed wall section in the photograph on right.

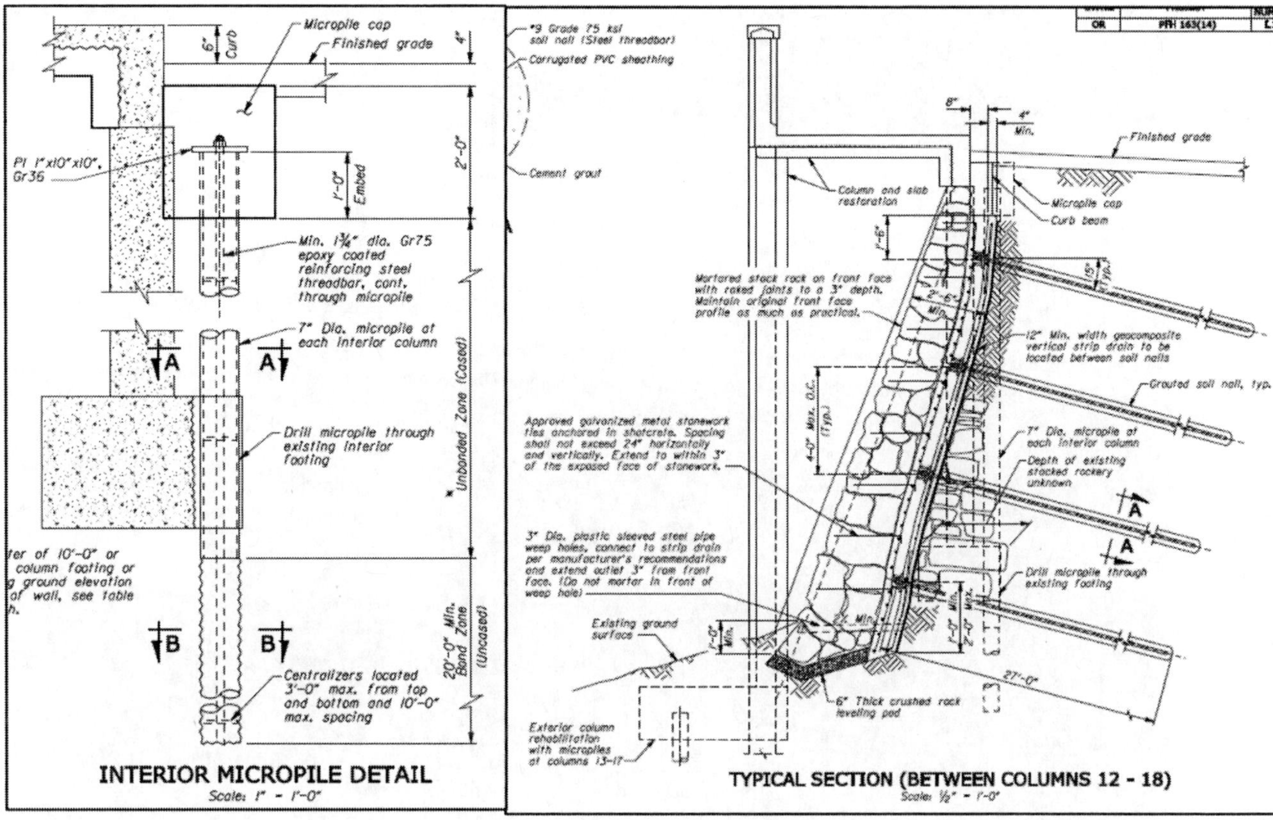

Plan details. <u>**Left detail**</u> **shows interior column micropile support.** <u>**Right detail**</u> **shows soil-nail wall stabilization and exterior micropile support.**

Construction photo showing completed micropiles and micropile cap on the interior side.

Construction photo showing soil-nail wall construction beneath viaduct on the exterior side.

9:50 AM: **<u>ALL BUSES</u> Depart Vista House and Women's Forum Scenic Viewpoint.**

<u>Note</u> *the beginning of cascades section on Interstate 84 at Latourell Falls and the first signs of Eagle Creek Wildfire begins at Shepperd's Dell.*

10:15 AM: **Stop 3 (All Buses): Multnomah Falls Lodge via Corbett on-ramp to I-84 – 13.6 miles, 21 minutes, restroom facilities available. Total duration at stop 3 is 60 minutes.**

Stop 3: Multnomah Falls Lodge

<u>Presenters: Ryan Cole</u> (United States Forest Service (USFS)) and **<u>Stephen Hay</u>** (ODOT) - *<u>Base of Falls discussion followed by a short hike to Benson Bridge to look at the different flexible rockfall fences and take pictures of the Falls from Benson Bridge if time allows.</u>*

Multnomah Falls is over 600 feet high and is fifth highest waterfall in the US. We will discuss the site geology, provide an overview of the Eagle Creek Fire - Response and Repairs and Future Preparedness.

The Eagle Creek Fire began September 2, 2017 near the town of Cascade Locks, and burned 48,861 acres of federal, state, county and private lands within the Columbia River Gorge National Scenic Area. Multiple federal, state, and local agencies responded to the fire which impacted Interstate 84 (I-84), the primary east-west transportation corridor from Portland, local communities, and significant cultural landmarks such as the Multnomah Falls Lodge. Mandatory evacuations were ordered throughout the affected area and I-84 was closed for weeks due to fire related damage. A United States Forest Service (USFS) Burned Area Emergency Response (BAER) team, with cooperation from the (ODOT), was established in September

<u>Left:</u> Aerial photograph courtesy of ODOT of Multnomah Falls. <u>Right:</u> Photograph looking up at the Falls and Benson Bridge.

2017 to assess the risk of post-fire threats to Critical Values at Risk. Significant geohazards, including debris flows, debris dam outburst flooding, and rockfall were identified throughout the burn area. Post-fire emergency treatments initiated by the USFS and ODOT included hazard tree falling and rockfall mitigation.

Fire impacts at the historic Multnomah Falls Lodge only consisted of smoke damage due to the courageous efforts by fire strike teams to save the historic structure. The Lodge was closed for three months and the HCRH at this location remained closed until November 2018. Geotechnical specialists from the USFS and the ODOT assessed impacted infrastructure throughout the burn area. Hazards identified at Multnomah Falls included hazard trees and rockfall. Hazard trees have been removed and rockfall fences placed along the south side of the HCRH and upslope of the Lodge. Approximately 3,000 lineal feet of 10 foot-high flexible rockfall fences have been installed and more frequent rockfall is estimated to continue for 5 to10 years post-fire. Significant progress has been made to reduce the risk of hazard trees and rockfall falling within the fire perimeter, but much work remains.

Looking down the rockfall chute where a 10 foot-high flexible rockfall fence was constructed above the historic Multnomah Falls Lodge.

Approximately 3000 linear feet of 10 foot-high flexible rockfall fence along the Historic Columbia River Highway near Multnomah Falls Lodge.

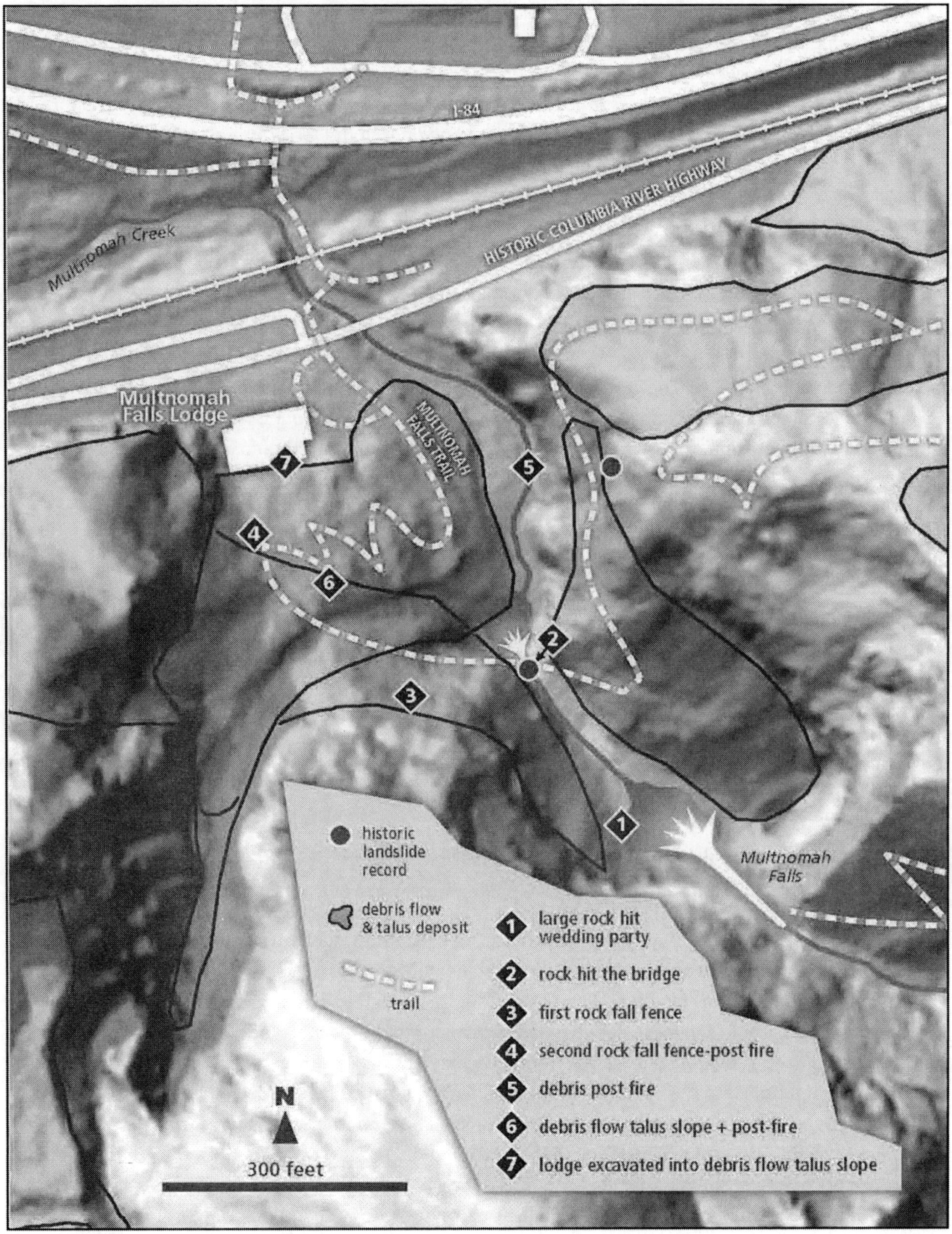

Figure from Washington Geological Survey.

11:25 AM: Depart Multnomah Falls Lodge

Note *the long sections of flexible rockfall fences and recent hazard tree removal as we head east of Multnomah Falls; the Oneonta Tunnel closure and rehab that is ongoing following the fire, the Dodson Debris Flow Fan and events from 1996 to present; and that we are approaching the nexus of the wildfire initiation as we look east as we pass John B. Yeon Trailhead on our way to Cascade Locks. Also note that people will begin to see a parallel trail that weaves in and out of the forest as we head east on I-84. This is referred to as the Historic Columbia River Highway State Trail and we will be hearing more about this in the afternoon.*

THE DODSON DEBRIS FLOWS – "A Story Along the Way..."

In the middle of the Gorge is a large debris flow/alluvial fan with the hamlet of Dodson on it. This large debris flow and alluvial fan most likely has formed from repeated debris flows over the past 15,000 years building it after the last Missoula Flood flowed through the Gorge. In February 1996, Northwest Oregon and Southwest Washington State received a huge amount of precipitation from a large "Pineapple Express," or Atmospheric River, from the central Pacific in four days. In the Gorge, we got over 35 inches of precipitation. It caused seven major debris flows on this fan.

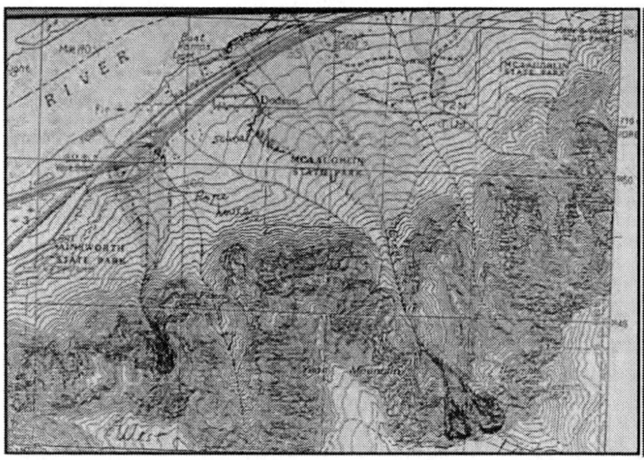

Left: Topographic map with alluvial fan and debris flow channels highlighted in yellow. Right: Oblique photograph taken from Washington looking southwest at Dodson alluvial fan complex.

The first one was the Royce Debris Flow which occurred just after noon on the second day of the storm. A house was inundated by at least 4 pulses over the next three days. The debris flow moved about 6.5 miles per hour (mph). This blocked I-84 and the train tracks. Later that same day at 10 PM, another debris flow came from a landslide dammed lake upstream that breached and sent a huge debris flow down the drainage at over 35 mph down Tumult Creek. This debris flow impacted several trucks on I-84 and six cars of a train into the Columbia River. Thankfully, no one was killed. In the intersection near the Royce Debris Flow, where the old highway intersects I-84, another debris flow occurred in 2001 and filled up the whole on-ramp intersection with debris, but did not impact I-84. With the great relief, the presence of colluvium hollows that collect debris during periods of acquiescence between large storm events in the headwaters of several creeks and streams in the Gorge, and the propensity for large tropical storms to impact the NW each year, debris flows continue to be a major geohazard that shapes and impacts the Gorge.

<u>Left:</u> Tumult Creek Debris Flow across I-84 and railroad. <u>Right:</u> Royce Debris Flow (west end of fan).

Figure from Washington Geological

11:40 AM: **Stop 4 and 5 (All Buses): LUNCH. The Bonneville Landslide and Cascade Locks – Marine Park Pavilion (Lunch) via HRCH, get on I-84 at Exit 35 East and get off at Exit 44, Cascade Locks 13.3 miles, 14 minutes, restroom facilities available, total duration at Stops 4 & 5 is 90 minutes.**

STOP 4/5: LUNCH AND BONNEVILLE LANDSLIDE

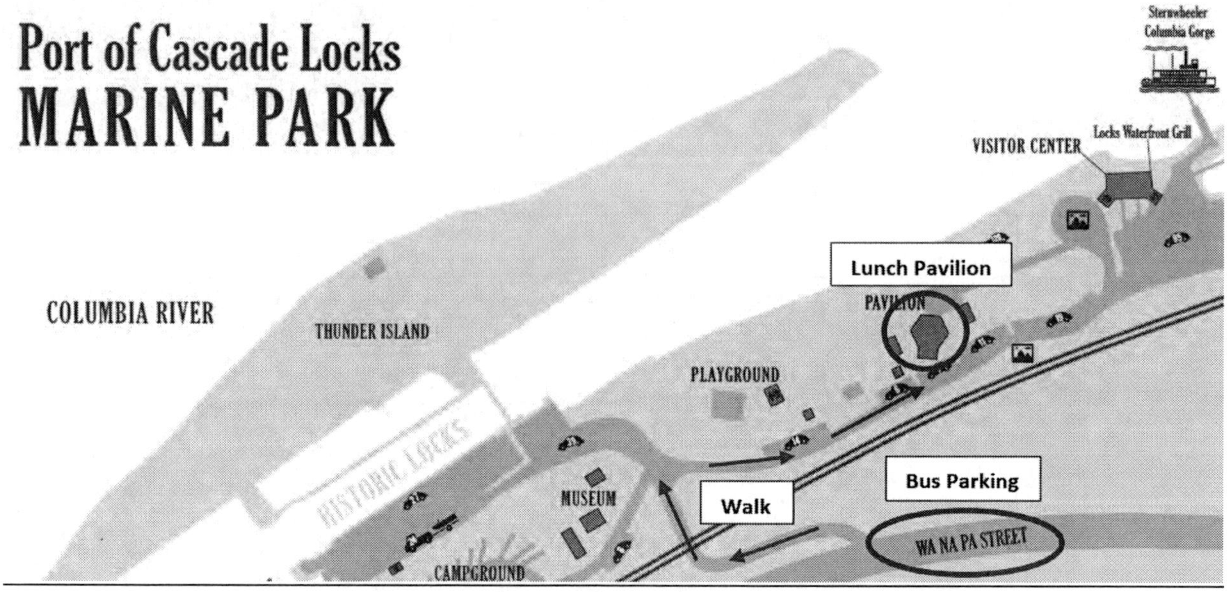

Map of Bus Parking and Pavilion Locations for Stop 4 (Lunch) and Stop 5.

<u>**Presenter: Scott Burns**</u> (PSU) – <u>*Meet at the Pavilion after lunch.*</u>

The Bonneville Landslide headscarp with geologic stratigraphy mapped on the right.

The Bonneville Landslide is the most studied landslide in the Pacific Northwest because we built our first dam on the Columbia River buttressed up against it in 1937 – Bonneville Dam. One did not want to have a landslide that might reactivate and cause the dam to fail. This landslide came down from the Washington side and crossed the Columbia River and totally stopped flow for a number of years. It killed the forests upstream (noted by Lewis and Clark when they came through here in 1805). It created a lake all the way to The Dalles, Oregon, over 55 miles upstream. Eventually, the dam broke and a wall of water 50 feet-high flowed down into what is now Portland. It created the Cascades of the Columbia, the most treacherous reach of the river to cross. Dating of logs from the landslide and tree trunks of dead trees in back of the dam give an approximate date of the landslide to be 1450 AD. It most likely was initiated by an earthquake that happened then up near Toppenish, Washington. **In**terferometric **S**ynthetic **A**perture **R**adar (**InSAR**) shows that parts of the landslide are still moving today.

Aerial photograph courtesy of ODOT looking to the northwest from Oregon at the Bonneville Landslide in the left half of the photograph.

Year of Observation	Observations and Interpretations	Geochronological Age (years BP)
1805	Lewis and Clark estimate the landslide occurred 20–30 years before their arrival, based on communication with local residents and observations of the exhumed sunken forest. Their interpretations were cited in Charles Lyell's seminal 1830–1833 Principles of Geology (Schuster and Pringle, 2002; Pringle, 2009)	175
1958	First radiocarbon date from sunken forest from wood salvaged before the flooding of the Columbia River by the 1930s construction of Bonneville Dam (Lawrence and Lawrence, 1958)	700
1984	Second radiocarbon dates published from sunken forest wood salvaged from the 1978 construction of the second powerhouse of the Bonneville Dam (Minor, 1984)	800
1998	Third radiocarbon dates published (Pringle and Schuster, 1998)	300-500
2001	Lichenometry dates interpreted to correlate the Bonneville landslide event with the 1700 Cascadia subduction earthquake (Reynolds, 2001)	287
2003	Minimum age based on interpretation of downstream deposits from Bonneville landslide dam breach that underlie dated pumice deposits from Mount St. Helens (Pierson and others, 2003).	370
2004	Re-dating of original Lawrence samples to mid-1400s (O'Connor, 2004).	495-535
2015	Dendrochronology and wiggle-matching radiocarbon dates (Reynolds and others, 2015)	513-529

Bonneville Landslide figure from Washington Geological Survey.

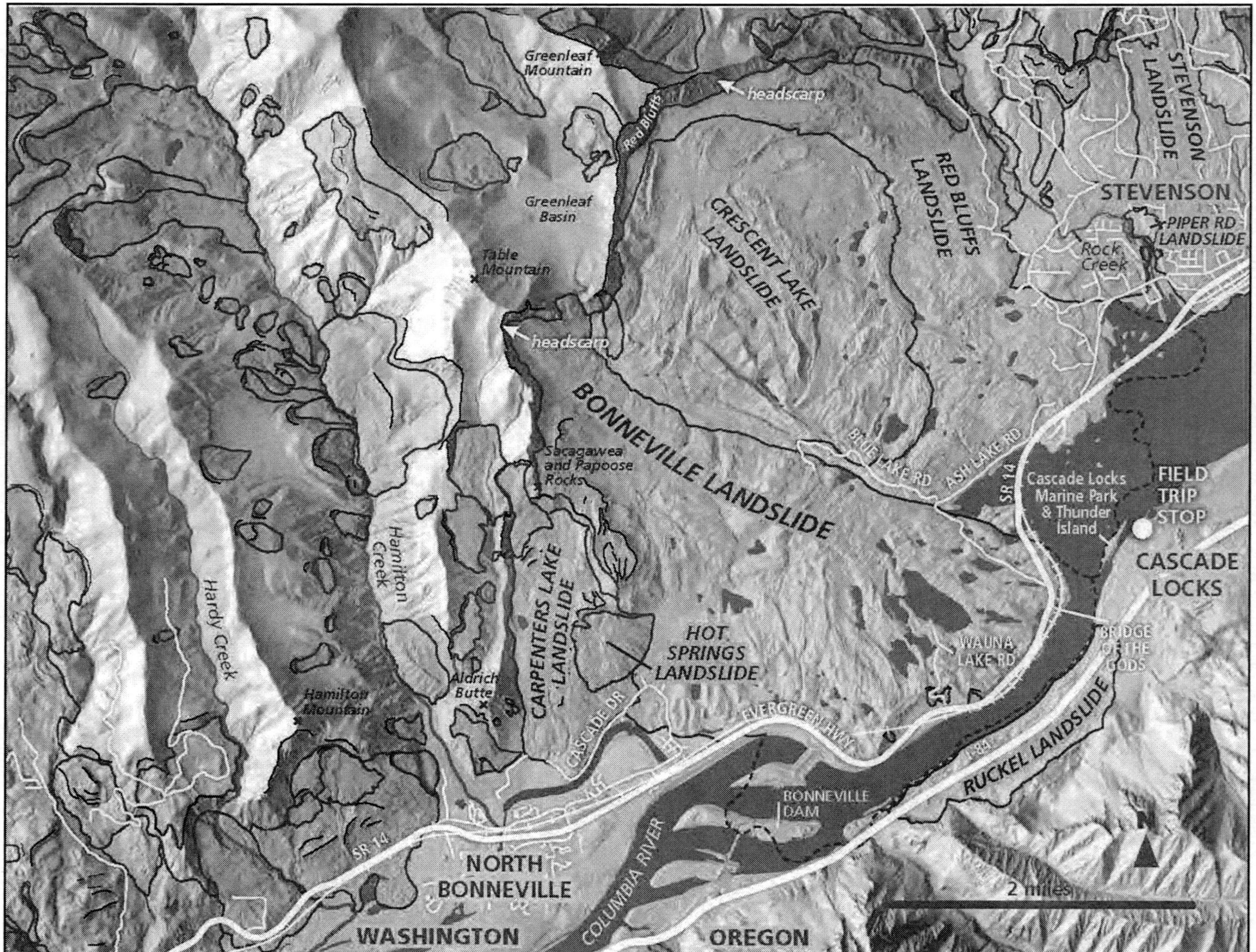

Figure from Washington Geological Survey

1:10 PM: **Depart Bonneville Landslide and Cascade Locks - Marine Park Pavilion**

Note *existing rockfall fences, bin retaining walls on the cut side and the trail weaving toward and away from the highway, mostly on the original HCRH alignment.*

1:25 PM: **Stop 6: HCRH State Trail, I-84 MP 52.8, 9.2 miles, 12 minutes, no restroom facilities. The total duration at Stop 6 is 60 minutes.**

STOP 6: HISTORIC COLUMBIA RIVER HIGHWAY STATE TRAIL

Presenters: Eric Lim (WFLHD) and **Brent Black** (Cornforth Consultants) – *Meet on the Trail.*

Eric Lim: Discussion on the upcoming sections in design and construction and the type of work being undertaken to capture the historic features of the original Historic Columbia River Highway (HCRH) design and construction for this regional trail system.

The HCRH State Trail Project is a collaborative effort between the FHWA –WFLHD, ODOT, Oregon State Parks and Recreation Department, U.S. Forest Service, Hood River County, The Historic Columbia River Highway Advisory Committee, and a number of other stakeholders and interested parties. The project involves restoring the alignment of the Historic Columbia River Highway, the nation's first scenic highway, into a pedestrian and bike trail connecting the cities of Troutdale and The Dalles.

The remaining segments of the trail in design between Wyeth and Hood River, include Segments E, F, G, and H. Numerous challenges faced include rockfall areas adjacent to the trail, bridges over Perham Creek and Mitchell Creek, numerous cut and fill retaining structures adjacent to I-84, and a tunnel through Mitchell Point. The Historic Columbia River Highway formerly was tunneled through the 900-

foot tall basalt promontory, but the tunnel was later blasted away during construction of I-84. The tunnel was famous for its adits or windows cut out of the rock to allow light to the driver and provide views of the gorge. We will discuss current design efforts are focusing on a new tunnel with adits similar to the historic tunnel, as well as balancing rockfall mitigation, and scenic area visual, environmental, and cultural concerns throughout these segments.

Historic photograph of the Mitchell Point Tunnel with adits. Sources for figures can be found at https://www.oregon.gov/ODOT/Regions/Documents/HCRH/Trail-Plan-Part-1-Wyeth-Hood-River-HCRH.pdf

Presenter: Brent Black - HCRH State Trail - Wyeth Section Design and Construction

The HCRH State Trail Plan centers on 11 miles of the trail between Exit 51 on I-84 (Wyeth, OR) and Ruthton County Park (Hood River, OR). This trail alignment is divided into eight segments (A through H) to facilitate a phased project approach and support civil, geotechnical, and structural design efforts and construction. These segments were further subdivided into reaches with similar geologic conditions, topographic features, and/or proposed improvements. Segments A through D consists of a 4-mile section of the trail that starts at Exit 51 on I-84 and ends at Starvation Creek State Park. This segment was subdivided into 14 reaches where rockfall from the adjacent slopes were a major concern. The longest reach is 1,350 feet long and the maximum slope height extends 1,150 feet above the trail alignment. A segment of the trail was constructed on an 850-foot long bench that was excavated within a near-vertical cliff using controlled blasting techniques.

Rockfall risk reduction designs capable of precluding 99% of impacting rockfalls and 90% of rolling rockfalls from reaching the trail and I-84 were desired (i.e. 90% retention of rolling rocks). This was accomplished using a combination of scaling, rock bolting, gabion basket walls, draped high-tensile strength mesh, mid-slope attenuator fences, and flexible rockfall barriers along the uphill side of the trail depending on site conditions. Context sensitive measures were incorporated into the design to reduce the visual impacts of the rockfall risk reduction elements. They included the staining of all metal components, rock-faced gabion baskets, and contouring draped mesh sections so it conforms to the slope shape. Over 3,000 linear feet of low-deflection, flexible rockfall fence barrier without upslope tiebacks was constructed. Other challenging geotechnical elements included: over 2,000 linear feet of MSE walls and reinforced soil slopes, a tied-back gravity wall, a 750-foot long viaduct supported on spread footings with micropile and tie-downs, a 400-foot long soldier pile tie-back wall, one micropile supported bridge, and one bridge on conventional spread footings.

Before and after construction photos from left to right along Wyeth Section from top to bottom.

Construction started in the summer of 2017 and is scheduled for completion in Fall 2019. The Eagle Creek Fire that occurred in September 2017 hampered construction efforts and increased the concerns associated with rockfall and hazard trees. An approximately 200-foot wide section of the extensive talus slope along Shellrock Mountain mobilized during the fire and travelled up to 15 feet beyond the original toe of the talus slope, which required the trail grade to be raised approximately five feet through this area.

2:25 PM: Depart HCRH State Trail, I-84 MP 52.8 for Rowena Loops viewpoint.

Note *that we will pass through Hood River, famous for its wind surfing and access to Mt. Hood and the Columbia Gorge for recreation. Notice the change in vegetation and climate as we head east toward Mosier and The Dalles. Between Hood River and Mosier, an early segment of the HCRH trail section, about 5.5 miles long, reopened the portion through the historic Mosier Twin Tunnels that can be accessed from Hood River and Mosier to the east. The twin tunnels can be seen up in the basaltic cliffs as we drive toward Mosier on I-84.*

Note *that as we drive up the Rowena Loops you will see a risk reduction rockfall mitigation that will be discussed at Stop 7.*

Stop 7: Rowena Crest Overlook

2:55 PM **Stop 7: Rowena Loops Viewpoint via Exit 76 - Rowena Interchange to WB HCRH to Rowena Crest Overlook (23.3 miles, 30 minutes, portable restrooms are available). Total duration of stop 7 is 30 minutes.**

Left: Aerial photograph of Rowena Loops courtesy of ODOT. Red polygon is the location of a risk reduction rockfall project by ODOT. Right: Photograph of the lower Rowena Loops taken from the upper viewpoint.

Presenters: Robert Hadlow (ODOT), **Jamie Schick** (McMillen Jacobs Associates), and **Scott Burns** (PSU)

Robert Hadlow: At Rowena Crest, participants will learn about how the Rowena Loops that take the highway from the crest to river level, maintain engineer Samuel Lancaster's design standards of maximum 5 percent grades and minimum 200-foot turning radiuses on the HCRH. Lancaster established grade and curvature standards years before the American Association of State Highway Officials (AASHO) adopted road design standards for application across the country.

Jamie Schick: The Historic Columbia River Highway (US 30) traverses the Columbia River Gorge along the south side of the Columbia River. Towards the eastern end of the Gorge it passes over the crest of Rowena Bluffs, a broad plateau punctuated with a near vertical 500-foot cliff adjacent to the river. This cliff consists of a series of Miocene-aged Columbia River basalt flows. East of this cliff the highway drops steeply back to the river, passing beneath several steep cliff faces that represent the headscarp of an ancient landslide. This includes a nearly 300-foot-high cliff near mile point 64.7 (red polygon in the picture above).

On May 7, 2017, the lower 75 feet of the cliff failed and fell onto the roadway, burying both lanes for nearly 200 feet with almost 20 feet of debris (figure below). This section of the cliff is dominated by columnar jointing that have developed columns exceeding eight feet in diameter. These columns terminate into a flow contact at approximately 75 feet above road grade. The flow contact includes a five to 15-foot-thick zone of highly weathered, closely fractured vesicular basalt. The failure undermined this flow contact, creating a nearly 10-foot overhang and was at risk of undermining the remainder of the cliff above.

Pre and post rock failure views looking north along the highway.

Initial response consisted of a site reconnaissance on May 8th to evaluate the site conditions and develop an approach to stabilizing the slope and reopening the road. The contractor, Triptych, mobilized on May 11 from southwest Oregon to begin safety scaling both above and within the failure area. Most of this work focused on the lower 150 feet of the cliff, accessed using a 145-foot boom lift. Scaling of the overhang was limited because of the safety risk to workers and the risk of exacerbating this condition by overscaling and destabilizing more of the cliff face. This scaling effort enabled an earthwork contractor to safely enter the failure area and remove the rockfall debris.

Initial mitigation response efforts focused on stabilizing the remaining lower columns as well as the columns above the interflow. Approximately 600 linear feet of rock bolts were designed to support this approach and were installed by Rock Supremacy from Bend, Oregon, between approximately May 17th and May 23rd (figure on next page). Rock bolt lengths varied between 10 and 25 feet and were tensioned to 30 kips. Rock bolt heads, nuts and plates were coated with Natina over the galvanization to reduce visual impacts. Final testing and tensioning were completed by May 26th, allowing the highway to reopen in time for the Memorial Day holiday.

Ongoing installation of rock bolts using 145-ft man-lift.

A second phase of investigation and design was completed to address the overhanging highly weathered interflow. This included additional site mapping and scanning of the project area using a drone-mounted light detection and ranging (Lidar) system operated by ODOT's Geometronics Group based in Salem, Oregon. The resultant surface created by the imaging was used to develop sections for design as well as to obtain accurate quantities for estimating purposes.

Final design included an anchored high-tensile strength wire mesh system across the weathered flow top, as well as additional scaling and spot rock bolting. The system was coated with Natina staining over the galvanization to minimize its visual impacts (figure on the next page). Wire mesh anchor nails were designed for a 6.5 foot spacing and extended to 15 feet from the slope face to anchor the system to competent rock. This phase of construction was initiated on January 8, 2018, proceeded without delays, and was completed by Rock Supremacy by early February 2018.

The completed installation of anchored, high tensile strength wire mesh with Natina stain over the galvanization, minimizing its visual impacts.

__Scott Burns:__ At the eastern end of the Gorge is Rowena Loops overlooking the Columbia River and Lyle, Washington, where the Klickitat River (which drains Mt. Adams) flows into the Gorge. At this site we will see a major fault (Chenoweth Fault) that turns the basalt flows on end, a major landslide created by the last Missoula Flood and has the old highway winding down through it, and mima mounds on the plateau. Formation of mima mounds is one of the incredible mysteries of geology. Note the vegetation we have been traveling through in the last 20 miles. We have decreased annual precipitation from 40 to 20 inches/year.

3:40 PM: **All buses depart Rowena Crest Viewpoint WB via HCRH to Mosier and get on I-84 at Exit 69 to head to Portland Marriott Downtown Waterfront (78.2 miles, 81 minutes with no traffic).**

STOP 8: MARRIOTT DOWNTOWN WATERFRONT

5:20 PM **Stop 8: Arrive at Portland Marriott Downtown Waterfront**

Speaker Bios

Scott Burns, RG, LG, CEG, PhD: Scott just completed his 49th year of teaching at the university level, with past positions in Switzerland, New Zealand, Washington, Colorado and Louisiana before coming to Portland State University 29 years ago. He has a BS and MS from Stanford University and his PhD from the University of Colorado. He is an engineering geologist and environmental geologist who also studies soils. His areas of expertise are landslides, radon gas, heavy metals in soils, Missoula Floods, and terroir of wine. He has over 100 publications including two books and has had 48 MS and PhD students complete degrees under him. He has been chair of three different geology departments and also has been an Associate Dean. He has been president of AEG, chair of the engineering geology division of GSA, and also president of IAEG (first American president in its 54 year history). He was chair of the HGS the last time there was a meeting in Portland in the early 1990's. burnss@pdx.edu

Robert W. Hadlow, PhD: ODOT – Historian: For nearly 30 years, Robert W. Hadlow has researched and written on historic road resources throughout the United States. In 2001, Oregon State University Press published his Ph.D. dissertation from Washington State University as *Elegant Arches, Soaring Spans: C. B. McCullough, Oregon's Master Bridge Builder.* Hadlow is the senior historian with the Oregon Department of Transportation, where he completes Section 106 and Section 4(f) compliance work. In 2000, he prepared a National Historic Landmark nomination for the Columbia River Highway Historic District. Hadlow is a member of the Transportation Research Board's Standing Committee on Historic and Archaeological Preservation in Transportation. When he is not pursuing transportation history, you might see him out on the backroads around Portland driving his 1939 Buick Roadmaster. Robert.W.Hadlow@ODOT.state.or.us

Nathan Jenks, LG, PE: Currently a Geotechnical Engineer for the Bonneville Power Administration in Vancouver, Washington since September 2019. Previously he was the Geotechnical Group Manager at the Western Federal Lands Highway Division of the FHWA. He has been with Western Federal Lands for about 13 years. Nathan holds undergraduate and graduate degrees in Geological Engineering from the Colorado School of Mines and his experience in geotechnical engineering has been focused on roadway and bridge projects in the Northwest US and Alaska. Nathan is a registered professional engineer and licensed geologist in the state of Washington. nkjenks@bpa.gov

Ryan Cole, CEG: USFS Engineering Geologist: Ryan Cole began his career in geology in 2008 with the US Geological Survey's Oregon Water Science Center, while working towards obtaining a BS in geology from Portland State University, which he completed in 2009. Ryan began working for the US Forest Service in 2011, and earned his MS in geology in 2013, also from Portland State University. Ryan has held a variety of geology-related positions within the agency, which includes work in the mining, energy, abandoned mine reclamation, CERCLA, and engineering fields. He is currently an engineering geologist on the Mount Hood National Forest and Columbia River Gorge National Scenic Area. racole@fs.fed.us

Stephen Hay, CEG: ODOT – Region 1 Geo/Hydro/Hazmat Unit Manager. Stephen Hay graduated from Portland State University in 1995 with a BS in Geology and BS in Geography. He spent three years as a geotechnical consultant working throughout the northwest before joining the Oregon Department of Transportation as an Engineering Geologist in Portland (Region 1). For the past 20 years Stephen has worked on ODOT transportation projects throughout the state and specializes in geotechnical hazard mitigation with an emphasis on rock slopes and landslides. Additionally, he is responsible for short and long-term aggregate resource identification and development. Stephen is a Certified Engineering Geologist (Oregon) and Licensed Engineering Geologist (Washington). Stephen.Hay@odot.state.or.us.

Eric Lim, PE: Senior Geotechnical Engineer at the Western Federal Lands Highway Division of the FHWA. Eric has been with Western Federal Lands for about 4 years after spending 15 years as a consulting geotechnical engineer in the Pacific Northwest. Eric graduated from Humboldt State University in 1999 and is a registered Professional Engineer in Washington and Oregon, and a registered Geotechnical Engineer in Oregon. Eric.Lim@dot.gov

Brent Black, CEG: Brent is a Senior Engineering Geologist for Cornforth Consultants, Inc. and its division Landslide Technology where he has worked for nearly 27 years. His areas of expertise include rockfall and rock slope evaluation and mitigation, landslide stabilization, site investigation and instrumentation, geologic hazards, and construction quality assurance. Brent serves on the board of the Engineering Geology committee for the TRB and is the chair for the subcommittee on Landslides. He is also serving on TRB subcommittees for Rockfall Management and GAM and was recently a panel member on the NCHRP committee for developing guidelines for certification and management of flexible rockfall protection systems. bblack@CornforthConsultants.com

Jamie Schick, CEG: James Schick is an engineering geologist for McMillen Jacobs Associates with over twenty years' experience in the practical application of the geological sciences to both large and small-scale engineering, permitting, and environmental projects for both the public and private sector. He has expertise in detailed site characterizations as well as broad general surveys for projects involving transportation, tunnels, dams, trenchless crossings, pipelines, industrial facilities, and power generation sites. Jamie has extensive experience with unstable rock and soil slope investigations and remediation and previously worked for the Oregon Department of Transportation as part of their unstable slopes GAM team and as the state rock slope geologist. Schick@mcmjac.com

Field Trip Planning Committee Bios

Douglas A. Anderson, LEG: FHWA – Western Federal Lands Highway Division – Engineering Geologist. Doug has been the engineering geologist for the Western Federal Lands Highway Division of the Federal Highway Administration since 2013, with 25 years of engineering geology and geotechnical experience with an emphasis in rock slope stability, rockfall hazards, blasting, landslide mitigation, geotechnical investigation and instrumentation technics, material source development and rehabilitation, large earthwork projects, and geotechnical asset and performance management. For 14 years, prior to joining Western Federal Lands, Doug worked his way up to the assistant chief engineering geologist for the Washington State Department of Transportation in Olympia, Washington after spending nearly four years as a geologist with the US Forest Service on the Mt. Hood National Forest on the heels of completing his bachelor's degree in geology at Portland State University in 1995. Mr. Anderson is a licensed geologist and engineering geologist in Washington State and has passed the Cold Region's Engineering Course required for licensing in the State of Alaska.

Stephen Hay, CEG: ODOT – Region 1 Geo/Hydro/Hazmat Unit Manager. *See speaker bio above.*

Evan Garich, PE: FHWA – Western Federal Lands Highway Division – Geotechnical Engineer. Evan Garich is a geotechnical engineer that has worked with Western Federal Lands Highway Division of the Federal Highway Administration since 2016. His areas of expertise include seismic hazard analysis and deep foundation design. He is responsible for the design and construction of transportation related assets in the northwest U.S and Alaska as a key member of the Western Federal Lands Geology and Geotechnical Team. He came to Western Federal Lands after working for WSP, formerly Parsons Brinkerhoff, for nine years working up to a lead geotechnical engineer after completing his degrees from Texas A&M and Portland State University in 2007. Mr. Garich is a professional engineer in Oregon and Washington and has passed the Cold Region's Engineering Course required for licensing in the State of Alaska. Furthermore, his favorite color is green!

Marc Fish, LEG: Washington State Department of Transportation – Assistant Chief Engineering Geologist. Marc works as an Engineering Geologist for the Washington State Department of Transportation and manages their unstable slopes program. He is a licensed Engineering Geologist in Washington State and has over 22 years of experience managing risk relating to unstable slopes and developing cost effective remediation designs for rock slopes, landslides, and debris flows along Washington State highways.

References

DeRoo, Thomas G., 2015, *Geology of the Columbia River Gorge and Mt Hood Volcano – A Field Trip Guide Book*, unpublished internal document prepared by the USFS - Mt. Hood National Forest for Geofest 2015.

Jacobacci, K.E., Gallin, W., Mickelson, K.A., Contreras, T.A. Burns., B., Calhoun, N., May 2019, *Landslides in the Columbia River Gorge, Field Trip Guide*, Washington Geological Survey & Oregon Department of Geology and Mineral Industries, 2019 GSA Cordilleran Section Meeting.

Acknowledgements

Special thanks to the Washington State Geologic Survey and Oregon Department of Geology and Mineral Industries for sharing their knowledge and previous Geologic Society of America Field Trip Guide and illustrations, as cited above. We would be remiss if we didn't also thank Thomas G. DeRoo, Retired Geologist from the Mt. Hood National Forest, for sharing his wealth of knowledge and previous field trip guide information that he developed for the Columbia River Gorge and Mt. Hood Volcano, as also cited above. We appreciate them letting us stand on their shoulders to ***hopefully*** make a more interesting and educational field trip for **YOU**, the HGS Participants!

Above all, the 2019 HGS Field Trip Planning Committee hopes you have a great field trip experience this year!

Highway Geology Symposium
C/O John Pilipchuk
1589 Mail Service Center
Raleigh, NC, 27699-1589

ISBN 978-1-7138-0875-6

CURRAN ASSOCIATES INC.
proceedings.com